Financial Econometrics, Mathematics and Statistics

Cheng-Few Lee · Hong-Yi Chen ·
John Lee

Financial Econometrics, Mathematics and Statistics

Theory, Method and Application

Springer

Cheng-Few Lee
Department of Finance and Economics
Rutgers Business School
Rutgers University
Piscataway, NJ, USA

Hong-Yi Chen
Department of Finance
National Chengchi University
Taipei, Taiwan

John Lee
Center for PBBEF Research
Morris Plains, NJ, USA

ISBN 978-1-4939-9427-4 ISBN 978-1-4939-9429-8 (eBook)
https://doi.org/10.1007/978-1-4939-9429-8

Library of Congress Control Number: 2019932616

This Springer imprint is published by the registered company Springer Science+Business Media, LLC part of Springer Nature.
The registered company address is: 233 Spring Street, New York, NY 10013, U.S.A.

Preface

We draw upon our years of teaching, research, and practice on the subjects of financial econometrics, mathematics and statistics for this textbook. Overall, our goal is to provide an advanced level book that reviews, discusses, and integrates financial econometrics, mathematics and statistics.

We focus on five principles to frame our presentation of this book:

(1) To discuss the basic methodology of financial econometrics, mathematics and statistics,
(2) To show how econometric methodologies can be used in finance and accounting-related research, which includes single equation, multiple regression, simultaneous regression, panel data analysis, time-series analysis, spectral analysis, nonparametric analysis, semiparametric analysis, GMM analysis, and other methods,
(3) To show how financial mathematics such as Itô's calculus is important to derive the intertemporal capital asset pricing model and option pricing model,
(4) To demonstrate how statistics distribution, such as normal distribution, stable distribution, and lognormal distribution, has been used in research related to portfolio theory and risk management,
(5) To show how binomial distribution, lognormal distribution, noncentral chi-square distribution, Poisson distribution, and others have been used in studies related to option and futures.

In order to comprehend this book, the reader needs two semesters of econometrics, two semesters of mathematical statistics, and one semester of multivariate statistics.

We divide this book into four parts: Regression and Financial Econometrics; Time-Series Analysis; Statistical Distributions, Option Pricing Model, and Risk Management; and Statistics, Itô's Calculus, and Option Pricing Model.

PART I: Regression and Financial Econometrics

There are seven chapters in this part. In Chap. 2, we discuss the assumptions of the multiple regression model, estimated parameters of the multiple regression model, the standard error of the residual estimate, and the coefficient of determination. We also investigate tests on sets and individual regression coefficients and the confidence interval for the mean response and prediction interval for the individual response.

In Chap. 3, we discuss various topics associated with the regression analysis, including multicollinearity, heteroscedasticity, autocorrelation, model specification and specification bias of the regression model, nonlinear regression models, lagged dependent variables in the regression model, dummy variables in the regression model, and regression model with interaction variables. We also apply the regression approach to investigate the effect of alternative business strategies and apply the logistic regression model to credit risk analysis.

In Chap. 4, we extend single-equation models to simultaneous equation models. Specifically, we discuss simultaneous equation system, two-stage least squares method, and three-stage least squares method. In Chap. 5, we discuss an econometric approach to financial analysis, planning, and forecasting. The issue of simultaneity and the dynamics of corporate-budgeting decisions will be explored by using finance theory. We also investigate the interrelationships among the programming, the simultaneous equations, and the econometrics approaches.

Chapter 6 addresses one of the important issues related to panel data analysis. We introduce the dummy variable technique and the error component model for analyzing pooled data. We investigate the possible impacts of firm effect and time effect on choosing the optimal functional form of a financial research study. In this chapter, we also discuss the criteria of using fixed effects or random effects approach.

Chapter 7 discusses how errors-in-variables estimation methods are used in finance research. We show how errors-in-variables problems can affect the estimators of the linear regression model, as well as discuss the effects they have on the empirical research cost of capital, asset pricing, capital structure, and investment decision. Chapter 8 provides three alternative errors-in-variables estimation models in testing the capital asset pricing model. Specifically, we present three alternative correction methods for the errors-in-variables problem. In Chap. 9, we discuss the issue of spurious regression and data mining in both conditional asset pricing models and simple predictive regression. We also discuss the impact of spurious regression and data mining on conditional asset pricing.

PART II: Time-Series Analysis and Its Applications

The purpose of Chap. 10 is to describe the components of time-series analyses and to discuss alternative methods of economic and business forecasting in terms of time-series data. Specifically, we discuss a classical description of three time-series components, the moving-average and seasonally adjusted time series, linear and log-linear time trend regressions, exponential smoothing and forecasting, autoregressive forecasting model, ARIMA model, and composite forecasting.

In Chap. 11, we attempt to achieve two goals. First, we present alternative theories for deriving optimal hedge ratios. We discuss various estimation methods and the relationship among lengths of hedging horizon, maturity of futures contract, data frequency, and hedging effectiveness. Second, we show how SAS program can be used to estimate hedge ratio in terms of ARCH method, GARCH method, EGARCH method, GJR-GARCH method, and TGARCH method.

PART III: Statistical Distributions, Option Pricing Model and Risk Management

Statistical distributions such as binomial distribution, multinomial distribution, normal distribution, lognormal distribution, Poisson distribution, central chi-square distribution, noncentral chi-square distribution, copula distribution, nonparametric distribution, and other distributions are important in finance research. In Chap. 12, we will discuss how binomial and multinomial distribution can be used to derive the option pricing model. In Chap. 13, we show how to use two alternative binomial option pricing model approaches to derive Black–Scholes option pricing model.

In Chap. 14, we will discuss how normal and lognormal distribution can be used to derive the option pricing model. In Chap. 15, we will show how copula distribution can be used to do credit risk analysis. In Chap. 16, we will show how multivariate analyses such as factor analysis and discriminant analysis can be used to do financial rating analysis. In Part IV, we will continue to discuss how statistics distribution can be used to derive option pricing model. In addition, we will also show how Itô's calculus can be used to derive option pricing model.

PART IV: Statistics, Itô's Calculus, and Option Pricing Model

In Chap. 17, we will show how characteristic function and noncentral chi-square can be used to analyze stochastic volatility option pricing model. In Chap. 18, we will discuss alternative methods to estimate implied variance. In Chap. 19, we will show the numerical valuation of Asian options with higher moments in the underlying distribution. Both European and American options will be discussed in this chapter. In Chap. 20, we will first review Itô Lemma and stochastic differential equation, and then we will show how this mathematical technique can be used to derive option pricing model. In Chap. 21, we will discuss the relationship between binomial option pricing model and Black–Scholes option pricing model. In addition, we also show how to use stochastic calculus to derive Black–Scholes model in detail. In Chap. 22, we will show how to use noncentral chi-square distribution to derive constant elasticity of variance option pricing model. In Chap. 23, we will discuss option pricing and hedging performance under stochastic volatility and stochastic interest rates. Finally, in Chap. 24, we will show how nonparametric distribution can be used to derive option bounds. Some empirical studies or option bounds are also provided.

This textbook can be used for the quantitative finance program and Ph.D. programs in economics, statistics, and finance. It demonstrates how to apply different econometrics and statistical methods in finance research. In addition, applications of Itô's calculus in deriving option pricing model are also discussed in some detail.

It is well known that financial econometrics, mathematics and statistics are three of the most important tools to solve theoretical and practical issues of quantitative finance. This textbook uses real-world data to show how these three quantitative tools can be used to solve quantitative finance issues in asset pricing, option pricing models, risk management, and other quantitative finance issues. This textbook uses a traditional approach by combining research papers from journals, handbooks, and textbooks to streamline these topics. We take advantage of using our own research papers, edited handbook, and textbook to formulate a meaningful, unique, comprehensive textbook. We heavily draw upon *Handbook of Quantitative Finance and Risk Management* (Springer 2009), *Statistics for Business and Financial Economics, 3rd ed.* (Springer 2013), and *Essentials of Excel, Excel VBA, SAS and Minitab for Statistical and Financial Analyses* (Springer 2016), as well as *Review of Quantitative Finance and Accounting* and other journals with which we have published relevant papers.

Note that this textbook is intended to be used in its entirety instead of chapter by chapter. Readers may find the aforementioned Springer volumes' useful references during the learning process. For example, there are several chapters wherein we draw from *Handbook of Quantitative Finance and Risk Management*—here, the reader can refer to the handbook. Similarly, Chaps. 2 and 3 are expanded versions of *Statistics for Business and Financial Economics, 3rd ed.*

There are undoubtedly some errors in the finished product such as conceptual, grammatical, or methodological. We would like to invite readers to send their suggestions, comments, criticisms, and corrections to the author, Professor Cheng F. Lee at the Department of Finance and Economics, Rutgers University at 100 Rockafeller Road, Room 5188, Piscataway, NJ 08854. Alternatively, readers can send this information by email to either Cheng Few Lee (cflee@business.rutgers.edu) or Hong-Yi Chen (fnhchen@nccu.edu.tw).

Piscataway, USA
Taipei, Taiwan
Morris Plains, USA
May 2019

Cheng-Few Lee
Hong-Yi Chen
John Lee

Contents

1 Introduction to Financial Econometrics, Mathematics, and Statistics

Contents

C.-F. Lee et al., *Financial Econometrics, Mathematics and Statistics*,
https://doi.org/10.1007/978-1-4939-9429-8_1

Abstract

In this introduction chapter, we give an overall view of financial econometrics and statistics as indicated in the chapter outline. We then discuss the material covered in this book. There are 24 chapters, which are divided into four sections. These four sections are: regression and financial econometrics, time-series analysis and its applications, statistical distributions, option pricing model and risk management, and statistics, Itô's calculus and option pricing model.

1.1 Introduction

Financial econometrics and statistics have been widely used in empirical research in both finance and accounting. Specifically, econometric methods are important tools for asset pricing, corporate finance, options and futures, and conducting financial accounting research. Econometric methods used in finance and accounting-related research include single equation multiple regression, simultaneous regression, panel data analysis, time-series analysis, spectral analysis, nonparametric analysis, semiparametric analysis, GMM analysis, and other methods.

Statistics distributions, such as normal distribution, stable distribution, and lognormal distribution, have been used in research related to portfolio theory and risk management. Binomial distribution, lognormal distribution, noncentral chi-square distribution, Poisson distribution, and others have been used in studies related to option and futures. Moreover, risk management research has used Copula distribution and other distributions.

This book presents principles of econometrics and statistics, with a focus on methods and applications in financial research. This book can be divided into four parts. First, we start from topics related to regression and financial econometrics. Second, we discuss time-series analysis. Third, we show how binomial, multinomial, and lognormal distributions can be applied in option pricing models and the risk management. Fourth, we discuss the application of statistics analyses and Itô's calculus to the risk management.

In this chapter, we will present topics related to financial econometrics and statistics. Theoretical models, empirical methods, and financial applications will be discussed in the following chapters. In Sect. 1.2, we will discuss the regression models and topics related to financial econometrics. In Sect. 1.3, we will discuss topics related to statistical distributions. In Sect. 1.4, we show methods of econometrics, mathematics, and statistics can be applied to finance and accounting research. In Sect. 1.5, we provide an overview of the rest of chapters in this book. Section 1.7 concludes this chapter. In addition, in Appendix, we provide keywords of this book.

1.2 Regression and Financial Econometrics

1.2.1 Single-Equation Regression Methods

There are important issues related to single-equation regression estimation method. They are (a) heteroscedasticity, specification error, measurement error, skewness and kurtosis effect, nonlinear regression and Box–Cox transformation, structural change, generalize fluctuation, probit and logit regression, Poisson regression, and fuzzy regression.

a. Heteroscedasticity
   White (1980) and Newey and West (1987) are two important papers discussing how the heteroscedasticity test can be performed. Specifically, Newey and West (1987) paper discusses heteroscedasticity when there are serial correlations.
b. Specification error
   Specification error occurs when there is missing variable in a regression analysis. We can refer to the papers by Thursby (1985), Fok et al. (1996), Cheng and Lee (1986), and Maddala and Rao (1996) for testing the existence of specification error.
c. Measurement error
   Measurement error problem involves the existence of imprecise independent variables in a regression analysis. Lee and Jen (1978), Kim (1995, 2010), Miller and Modigliani (1966), and Chen et al. (2015) have explored how measurement error methods can be applied to finance research. Chen et al. (2015) have discussed alternative errors-in-variable estimation methods and their application in finance research.
d. Skewness and kurtosis effect
   Both skewness and kurtosis are two important measurement variables to prepare stock variation analysis. Lee (1976a, b), Sears and Wei (1988), and Lee and Wu (1985) discuss the skewness and kurtosis issue in asset pricing.
e. Nonlinear regression and Box–Cox transformation
   Nonlinear regression and Box–Cox transformation are important tools for finance, accounting, and urban economic research. Lee (1976a, b, 1977a, b), Lee et al. (1990), Frecka and Lee (1983), and Liu (2006) have discussed how nonlinear regression and Box–Cox transformation techniques can be used to improve the specification of finance and accounting research. In addition, Kau and Lee (1976), and Kau et al. (1986) have explored how Box–Cox transformation can be used to conduct the empirical study of urban structure.
f. Structural change
   Yang (1989), Lee et al. (2011) have discussed how the structural change model can be used to improve the empirical study of dividend policy and the issuance of new equity. Chow (1960) have proposed a dummy variable approach to examine the existence of structure change for regression analysis. Zeileis et al. (2002) have developed software programs to perform the Chow test and other structural change models which have been frequently used in finance and economic research. In addition, Hansen (1996, 1997, 1999, 2000a, b) has explored the issue of threshold regressions and their applications in detecting structure change for regression.
g. Generalize fluctuation
   Kuan and Hornik (1995) have discussed how the generalize fluctuation test can be used to perform structural change to regression.
h. Probit and logit regression
   Probit and logit regressions are frequently used in credit risk analysis. Ohlson (1980) used the accounting ratio and macroeconomic data to do credit risk analysis. Shumway (2001) have used accounting ratios and stock rate returns for credit risk analysis in terms of probit and logit regression techniques. Most recently, Hwang et al. (2008, 2009) and Cheng et al. (2010) have discussed probit and logit regression for credit risk analysis by introducing nonparametric and semiparametric techniques into this kind of regression analysis.
i. Poisson regression
   Lee and Lee (2012) have discussed how the Poisson regression can be performed regardless of the relationship between multiple directorships, corporate ownership, and firm performance.
j. Fuzzy regression
   Shapiro (2005), Angrist and Lavy (1999), and Van Der Klaauw (2002) have discussed how Fuzzy regression can be performed. This

method has potential to be used in finance accounting and research.

1.2.2 Simultaneous Equation Models

Besides single-equation regression models, we can estimate simultaneous equation models. There are two-stage least squares estimation (2SLS) method, seemly unrelated regression (SUR) method, three-stage least square estimation (3SLS) method, disequilibrium estimation method, and generalized method of moments.

a. Two-stage least squares estimation (2SLS) method
   Miller and Modigliani (1966) have used 2SLS to study cost of capital for utility industry. Lee (1976a) has applied 2SLS to started market model. Moreover, Chen et al. (2007) have discussed 2SLS method for investigating corporate governance.
b. Seemly unrelated regression (SUR) method
   Seemly unrelated regression has frequently used in economic and financial research. Lee and Zumwalt (1981) have discussed how the seemly unrelated regression method can be applied in asset pricing determination.
c. Three-stage least squares estimation (3SLS) method
   Chen et al. (2007) have discussed how the three-stage least squares estimation method can be applied in corporate governance research.
d. Disequilibrium estimation method
   Fair and Jaffee (1972), Amemiya (1974), Quandt (1988), Mayer (1989), and Martin (1990) have discussed how alternative disequilibrium estimation method can be performed. Sealey (1979), Tsai (2005), and Lee et al. (2011) have discussed how the disequilibrium estimation method can be applied in asset pricing test and banking management analysis.
e. Generalized method of moments
   Hansen (1982) and Hamilton (1994, Chap. 14) have discussed how generalized method of moments method can be performed. Chen et al. (2007) have used the two-stage least squares estimation (2SLS), three-stage squares method and GMM method to investigate corporate governance.

1.2.3 Panel Data Analysis

There are several important issues related to panel data analysis, such as fixed effects model, random effects model, and clustering effect model. Three well-known textbooks by Wooldridge (2010), Baltagi (2008), and Hsiao (2014) have discussed the applications of panel data in finance, economics, and accounting research.

a. Fixed effects model
   Chang and Lee (1977) and Lee et al. (2011) have discussed the role of the fixed effects model in panel data analysis of dividend research.
b. Random effects model
   Arellano and Bover (1995) have explored the random effects model and its role in panel data analysis. Chang and Lee (1977) have applied both fixed effects and random effects model to investigating the relationship between price per share, dividend per share, and retained earnings per share.
c. Clustering effect model
   Petersen (2009), Cameron et al. (2011), and Thompson (2011) review the clustering effect model and its impact on panel data analysis.

1.2.4 Alternative Methods to Deal with Measurement Error

LISREL model, multifactor and multi-indicator (MIMIC) model, partial least square method, and grouping method can be used to deal with measurement error problem.

a. LISREL model
   Titman and Wessels (1988), Chang (1999), Chang et al. (2009), Yang et al. (2010) have

described the LISREL model and its way to resolve the measurement error problems of finance research.

b. Multifactor and multi-indicator (MIMIC) model
Wei (1984) and Chang et al. (2009) have applied in the multifactor and multi-indicator (MIMIC) model in capital structure and asset pricing research.

c. Partial least square method
Lambert and Lacker (1987), Ittner et al. (1997), and Core (2000) have applied the partial least square method to deal with measurement error problems in accounting research.

d. Grouping method
Black et al. (1972), Blume and Friend (1973), Fama and MacBeth (1973), Lee (1973, 1977b), Chen (2011), and Lee and Chen (2011) analyze grouping method and its way to deal with measurement error problem in capital asset pricing tests.

In addition, there are other errors-in-variables methods, such as classical method, instrumental variable method, mathematical programming method, maximum likelihood method, GMM method, and Bayesian statistic method. Chen et al. (2015) have discussed above-mentioned methods in detail.

1.2.5 Time-Series Analysis

There are various important models in time-series analysis, such as autoregressive integrated moving average (ARIMA) model, autoregressive conditional heteroscedasticity (ARCH) model, generalized autoregressive conditional heteroscedasticity (GARCH) model, fractional GARCH, and combined forecasting model. Anderson (1994) and Hamilton (1994) have discussed the issues related to time-series analysis.

Myers (1991) discloses ARIMA's role in time-series analysis: Lien and Shrestha (2007) discuss ARCH and its impact on time-series analysis. Lien (2010) discusses GARCH and its role in time-series analysis. Leon and Vaello-Sebastia (2009) further research into GARCH and its role in time series in a model called fractional GARCH.

Granger and Newbold (1973), Granger and Newbold (1974), and Granger and Ramanathan (1984) have theoretically developed combined forecasting methods. Lee et al. (1986) have applied combined forecasting methods to forecast market beta and accounting beta. Lee and Cummins (1998) have shown how to use the combined forecasting methods to perform cost-of-capital estimates.

1.3 Financial Statistics

1.3.1 Statistical Distributions

Cox et al. (1979) and Rendleman and Barter (1979) have used binomial, normal, and lognormal distributions to develop an option pricing model. Some researchers provide studies on these different statistical distributions. Black and Sholes (1973) have used lognormal distributions to derive the option pricing model. Finally, Aitchison and Brown (1973) is a well-known book to investigate lognormal distribution. Schroder (1989) has derived the option pricing model in terms of noncentral chi-square distribution.

In addition, Fama (1971) has used stable distributions to investigate the distribution of stock rate of returns. Chen and Lee (1981) have derived statistics distribution of Sharpe performance measure and found that Sharpe performance measure can be described by Wishart distribution. Kao and Lee (2018) review three methods to estimate Sharpe ratio, including (i) the exact non-central t distribution by Lee and Chen (1979), (2) the asymptotic normal distribution under *i.i.d* normally distributed excess returns by Jobson and Korkie (1981), (iii) the asymptotic normal distribution under *i.i.d* non-normally distributed excess returns, and (iv) the asymptotic normal distribution under strictly-stationary-ergodic excess returns using the generalized method of moment (Lo 2002;

Christie 2005; and Opdyke 2007). Kao and Lee (2018) find the asymptotic normal distribution method proposed by Christie (2005) and Opdyke (2007) can successfully reflect the skewed and leptokurtic characteristics of stock excess returns. Kao and Lee (2019) apply a strictly-stationary-ergodicity GARCH (1, 1) process to derive the asymptotic Sharpe ratio distribution under a more generalized process of stock returns. They also empirically show how this more generalized Sharpe ratio distribution is affected by the tail behaviors of stock returns.

1.3.2 Principle Components and Factor Analysis

Anderson's (2003) book entitled "*An Introduction to Multivariate Statistical Analysis*" has discussed principle components and factor analysis in details. Pinches and Mingo (1973), Chen and Shimerda (1981), and Kao and Lee (2012) have discussed how principle components and factor analysis can be used in finance and accounting research.

1.3.3 Nonparametric and Semiparametric Analyses

Hutchinson et al. (1994) and Ait-Sahalia and Lo (2000) have discussed how nonparametric can be used in risk management and the evaluation of financial derivatives. Hwang et al. (2007, 2010) and Chen et al. (2010) have used semiparametric to conduct credit risk analysis.

1.3.4 Cluster Analysis

The detailed procedures to discuss how cluster analysis can be used to find groups in data can be found in the textbook by Kaufman and Rousseeuw (1990). In addition, Brown and Goetzmann (1997) have applied cluster analysis in mutual fund research.

1.4 Applications of Financial Econometrics, Mathematics and Statistics

In this section, we will briefly discuss how different methodology of financial econometrics, mathematics, and statistics can be applied to the topics of finance and accounting research. Topics include asset pricing, corporate finance, investment and portfolio research, option pricing, futures and hedging, mutual fund, credit risk modeling, and other applications.

1.4.1 Asset Pricing

Methodologies used in asset pricing research include heteroscedasticity, specification error, measurement error, skewness and kurtosis effect, nonlinear regression and Box–Cox transformation, structural change, two-stage least squares estimation (2SLS) method, seemly unrelated regression (SUR) method, three-stage least squares estimation (3SLS) method, disequilibrium estimation method, fixed effects model, random effects model, clustering effect model of panel data analysis, grouping method, ARIMA, ARCH, GARCH, fractional GARCH, and Wishart distribution.

1.4.2 Corporate Finance

Methodologies used in corporate finance research include heteroscedasticity, specification error, measurement error, skewness and kurtosis effect, nonlinear regression and Box–Cox transformation, structural change, probit and logit regression for credit risk analysis, Poisson regression, fuzzy regression, two-stage least squares estimation (2SLS) method, seemly unrelated regression (SUR) method, three-stage least squares estimation (3SLS) method, fixed effects model, random effects model, clustering effect model of panel data analysis, and GMM analysis.

1.4.3 Financial Institution

Methodologies used in financial institution research include heteroscedasticity, specification error, measurement error, skewness and kurtosis effect, nonlinear regression and Box–Cox transformation, structural change, probit and logit regression for credit risk analysis, Poisson regression, fuzzy regression, two-stage least squares estimation (2SLS) method, seemly unrelated regression (SUR) method, three-stage least squares estimation (3SLS) method, disequilibrium estimation method, fixed effects model, random effects model, clustering effect model of panel data analysis, and semiparametric analysis.

1.4.4 Investment and Portfolio Management

Methodologies used in investment and portfolio management include heteroscedasticity, specification error, measurement error, skewness and kurtosis effect, nonlinear regression and Box–Cox transformation, structural change, probit and logit regression for credit risk analysis, Poisson regression, and fuzzy regression.

1.4.5 Option Pricing Model

Methodologies used in option pricing research include ARIMA, ARCH , GARCH, fractional GARCH, spectral analysis, binomial distribution, Poisson distribution, normal distribution, log-normal distribution, chi-square distribution, noncentral chi-square distribution, and nonparametric analysis.

1.4.6 Futures and Hedging

Methodologies used in future and hedging research include heteroscedasticity, specification error, measurement error, skewness and kurtosis effect, nonlinear regression and Box–Cox transformation, structural change, probit and logit regression for credit risk analysis, Poisson regression, and fuzzy regression.

1.4.7 Mutual Fund

Methodologies used in mutual fund research include heteroscedasticity, specification error, measurement error, skewness and kurtosis effect, nonlinear regression and Box–Cox transformation, structural change, probit and logit regression for credit risk analysis, Poisson regression, fuzzy regression, and cluster analysis.

1.4.8 Credit Risk Modeling

Methodologies used in credit risk modeling include heteroscedasticity, specification error, measurement error, skewness and kurtosis effect, nonlinear regression and Box–Cox transformation, structural change, two-stage least squares estimation (2SLS) method, seemly unrelated regression (SUR) method, three-stage least squares estimation (3SLS) method, disequilibrium estimation method, fixed effects model, random effects model, clustering effect model of panel data analysis, ARIMA, ARCH, GARCH, and Semiparametric analysis.

1.4.9 Other Applications

Financial econometrics is also important tool to conduct research in trading cost (transaction cost) modeling, hedge fund research, microstructure, earnings announcement, real option research, financial accounting, managerial accounting, auditing, and term structure modeling.

1.5 Overall Discussion of This Book

In this section, we classify Chaps. 2–24 into four parts. Part I is regression and financial econometrics. Part II is time-series analysis. Part III is statistical distributions, option pricing model and risk management. Part IV is statistics, Itô's calculus and option pricing model.

1.5.1 Regression and Financial Econometrics

In Chap. 2, we discuss multiple linear regressions. We discuss the assumptions of the multiple regression model, estimated parameters of the multiple regression model, the standard error of the residual estimate and the coefficient of determination. We also investigate tests on sets and individual regression coefficients and the confidence interval for the mean response and prediction interval for the individual response. We consider business and economic application and use computer programs to do multiple regression analyses. In addition, in the appendix, we show the derivation of the sampling variance of the least squares slope estimations and discuss the cross-sectional relationship among price per share, dividend per share, and retained earnings per share.

In Chap. 3, we discuss various topics associated with the regression analysis, including multicollinearity, heteroscedasticity, autocorrelation, model specification and specification bias of the regression model, nonlinear regression models, lagged dependent variables in the regression model, dummy variables in the regression model, and regression model with interaction variables. We also apply the regression approach to investigate the effect of alternative business strategies and apply the logistic regression model to credit risk analysis. In addition, in the appendix, we discuss dynamic ratio analysis, the term structure of interest rate, partial adjustment dividend behavior model, logistic model and probit model. Moreover, we present SAS code for hazard model in bankruptcy forecasting.

In Chap. 4, we extend single-equation models to simultaneous equation models. Specifically, we discuss simultaneous equation system, two-stage least squares method, and three-stage least squares method. Applications of simultaneous equation in finance research are also discussed.

In Chap. 5, we discuss econometric approach to financial analysis, planning, and forecasting. The issue of simultaneity and the dynamics of corporate-budgeting decisions will be explored by using the finance theory. We discuss how Zellner's (1962) seemingly uncorrelated regression (SUR) technique can be used in financial analysis and planning and explore applications of a structural simultaneous equation model for financial planning and forecasting. We also investigate the interrelationships among the programming, the simultaneous equations, and the econometrics approaches. Possible applications of financial-planning models to strategic business decisions are reviewed and discussed. Finally, we apply financial-planning method to Johnson & Johnson as a case study.

Chapter 6 addresses one of the important issues related to panel data analysis. We introduce the dummy variable technique and the error component model for analyzing pooled data. We investigate the possible impacts of firm effect and time effect on choosing the optimal functional form of a financial research study. We also discuss how Hausman's test can be used to determine whether fixed effects model or random effects model should be used, and how to obtain efficient firm fixed effects estimator and efficient correlated random effects estimator. Moreover, we refer Lee et al. (2011) to show how fixed effects model can be used and demonstrate cluster effect can be tested.

Chapter 7 discusses how errors-in-variables estimation methods are used in finance research. We show how errors-in-variables problem can affect estimators of the linear regression model. We provide seven alternative correction methods in dealing with errors-in-variables problems. We also discuss the effects of errors-in-variables problems on the empirical research of cost of capital, asset pricing, capital structure, and investment decision.

Chapter 8 provides three alternative errors-in-variables estimation models in testing the capital asset pricing model. Specifically, we present three alternative correction methods for the errors-in-variables problem. We conduct an empirical analysis and provide results of testing asset pricing models.

In Chap. 9, we discuss the issue of spurious regression and data mining in conditional asset pricing models. We discuss the issues of data mining and spurious regression in the simple predictive regression. We also discuss the impact of spurious regression and data mining on conditional asset pricing. Moreover, we conduct a simulation experiments for various form of conditional asset pricing models. We then discuss and evaluate solutions to the problems of spurious regression and data mining.

1.5.2 Time-Series Analysis and Its Application

The purpose of Chap. 10 is to describe components of time-series analyses and to discuss alternative methods of economic and business forecasting in terms of time-series data. Specifically, we discuss a classical description of three time-series components, the moving average and seasonally adjusted time series, linear and log linear time trend regressions, exponential smoothing and forecasting, autoregressive forecasting model, ARIMA model, and composite forecasting. In addition, we address the Holt–Winters forecasting model for seasonal series.

In Chap. 11, we present alternative theories for deriving the optimal hedge ratios. We discuss various estimation methods and the relationship among lengths of hedging horizon, maturity of futures contract, data frequency, and hedging effectiveness. In addition, we discuss theoretical models of the time-series analysis and summarize empirical models of the time-series analysis. Moreover, we apply SAS program to estimate hedge ratio by EGARCH, GJR-GARCH, and TGARCH.

1.5.3 Statistical Distributions and Option Pricing Model

In Chap. 12, we discuss binomial distribution and show simple binomial option pricing model. The generalized binomial option pricing model and multiple option pricing models are demonstrated. In addition, we introduce a lattice framework for option pricing.

In Chap. 13, we present two option pricing models, the two-state option pricing model by Rendleman and Barter (1979) and the binomial option pricing model by Cox et al. (1979). We also provide a detailed comparison between these two models.

In Chap. 14, we discuss normal, lognormal distributions, and the relationship between two distributions. We also discuss multivariate normal and lognormal distributions and demonstrate the normal distribution as an application to the binomial and Poisson distributions. Moreover, we present applications of the lognormal distribution in option pricing and introduce Microsoft Excel program for calculating the American call options.

Chapter 15 aims to (1) construct an efficient model to describe the dependence structure; (2) use this constructed model to analyze overall credit, marginal, and industrial risks; and (3) build up an automatic tool for banking system to analyze its internal credit risks. We present the methodology of the copula function to analyze credit risks and discuss experimental results.

In Chap. 16, we explore the theory and methodology of factor analysis and discriminant analysis. The linear algebra needed for factor analysis, discriminant analysis, and portfolio analysis are reviewed in accordance with the basic concepts of algebra. The theory and methodology of two-group discriminant analysis will be explored in accordance with both the dummy regression method and the analysis-of-variance (eigenvalue) method. We then extend two-group discriminant analysis to *k-group* discriminant analysis. In addition, the theory and the methodology of principal component and factor analysis are investigated.

1.5.4 Statistics, Itô's Calculus and Option Pricing Model

In Chap. 17, we discuss nonclosed-form type and closed-form type option pricing models. In addition, we derive a unique solution for the stock price function to evaluate the market price of the risk.

In Chap. 18, we review several alternative methods to estimate implied variance of option. We classify them into two different estimation routines: numerical search methods and closed-form derivation approaches. In addition, we show how the MATLAB computer program can be used to estimate implied variance. We also apply real data from American option markets to compare the performances of three typical alternative methods: regression method proposed by Lai et al. (1992), MATLAB computer program approach, and approximation method derived by Ang et al. (2013).

Chapter 19 introduces a numerical algorithm with higher moments in the underlying asset distribution to evaluate Asian options. We introduce the pricing process of a binomial tree for an Asian option and Edgeworth binomial model for pricing Asian options with higher moment consideration. In addition, we discuss the lower and upper bounds of the prices of European-Asian options and American-Asian options. The numerical examples for our approach are then provided.

Chapter 20 discusses the Itô process and financial modeling. We introduce Itô lemma and use stochastic differential equation to represent stock-price behavior. We then present the pricing of an option.

In Chap. 21, we discuss three different approaches for deriving option model. Specifically, we discuss the relationship between binomial option pricing model and Black–Scholes option pricing model. We compare Cox et al. method and Rendleman and Bartter method for deriving Black–Scholes option pricing model. We also discuss lognormal distribution method to derive Black–Scholes option pricing model. In addition, we use the de Moivre–Laplace theorem to prove that the best fit between the binomial and normal distributions occurs when binomial probability is 0.5.

In Chap. 22, we discuss the constant elasticity of variance (CEV) diffusion and its transition probability density function. We present the noncentral chi-square approach to option pricing model. In addition, we provide detailed proof of Feller's lemma.

In Chap. 23, we discuss the option pricing and hedging performance under stochastic volatility and stochastic interest rate (SVSI model). We develop the SVSI option pricing formula and discuss issues pertaining to the implementation of the formula, and derive the hedge ratios analytically. Moreover, we evaluate the static pricing and the dynamic hedging performance of the four models.

In Chap. 24, we discuss how nonparametric statistic can be used to do option pricing model analysis. Specifically, we compare the lower bound from nonparametric method to several lower bounds from previous studies. Finally, we apply the model to real data using histograms of the realized stock returns and investigate how many observations violate the lower bound.

1.6 Conclusion

This chapter has discussed important financial econometrics and statistics which have been used in finance and accounting research. We first discuss the regression models and topics related to financial econometrics, including single-equation regression models, simultaneous equation models, panel data analysis, alternative methods to deal with measurement error, and time-series analysis. We then introduce topics related to financial statistics, including statistical distributions, principle components and factor analysis, nonparametric and semiparametric analyses, and cluster analysis.

In addition, financial econometrics, mathematics, and statistics are the important tools to conduct research in finance and accounting areas. We briefly introduce applications of econometrics, mathematics, and statistics models in finance and accounting research. Research topics include asset pricing, corporate finance, financial institution, investment and portfolio management, option pricing model, futures and hedging, mutual fund, credit risk modeling, and others.

We also provide an overview of this book. The rest of 23 chapters can be divided into four parts: (1) regression and financial econometrics; (2) time-series analysis; (3) statistical distributions, option pricing model and risk management; and (4) statistics, Itô's calculus and option pricing model. Models, methodologies, and applications for each topic will be discussed in detail in the following 23 chapters. Chapters 2–9 discuss issues related to regression and financial econometrics. Chapters 10 and 11 introduce time-series analysis and hedge ratio estimation. Chapters 12–16 show how statistical distributions can be applied in option pricing models. Finally, Chaps. 17–24 discuss the application of statistics analyses and Itô's calculus to the risk management and topics related to the option pricing model.

Appendix: Keywords for Chaps. 2–24

This appendix provides keywords for Chaps. 2–24. The number behind each keyword is the chapter it is associated with.

American option (19), Approximation approach (18), Arbitrage (24), ARCH method (11), Asian options (19), Asset allocation (9), Autocorrelation (3), Autoregressive conditional heteroscedasticity (ARCH) (11), Autoregressive forecasting model (10), Bayesian approach (7), Binomial distribution (12, 14), Binomial option pricing model (12, 13, 21), Black and Scholes model (12, 24), Black–Scholes formula (13, 14), Black–Scholes option pricing model (20, 21), Capital asset pricing model (7, 8), Capital structure (7), CARA utility function (11), Characteristic function (17), Classical method (7), Closed-form option pricing model (17), Clustering effect (6), Coefficient of determination (2), Coincident indicators (10), Co-integration and error assertion method effectiveness (11), Confidence interval (2), Constant elasticity of variance model (22), Copula (15), Cost of capital (7), Credit risk (3, 15), Credit VaR (15), Cross-sectional data (6, 10), Cyclical component (10), Default correlation (15), Distribution of underlying asset (24), Dummy variables (3), Edgeworth binomial model (19), EGARCH method (11), Endogenous variables (5), Error component model (6), Errors-in-variables (7), Estimate implied variance (18), European option (24), Exogenous variables (5), Expected terminal option price (20), Exponential smoothing (10), Exponential smoothing constant (10), Factor analysis (16), Fixed effects (6), GARCH method (11), Generalized autoregressive conditional heteroscedasticity (GARCH) (11), GJR-GARCH method (11), Grouping method (7, 8), Hedge ratio (11), Hedge ratios (23), Hedging (23), Hedging performance (23), Heston model (17), Heteroscedasticity (3), Holt–Winters forecasting model (10), Hypothesis test (2), Instrumental variable method (7, 8), Interaction variables (3), Investment equation (7), Irregular component (10), Itô's lemma (17, 20), Kernel pricing (24), k-Group discriminant analysis (16), Lagging indicators (10), Lattice framework (12), Leading indicators (10), Linear algebra (16), LISREL method (7), Logistic regression (3), Lognormal distribution (14), Lognormal distribution method (21), Lower bound (24), Mathematical programming method (7), MATLAB approach (18), Maximum likelihood method (7, 8), Maximum mean extended-gini coefficient hedge ratio (11), Mean squared error (10), Measurement error (7), Minimum generalized semivariance hedge ratio (11), Minimum value at risk hedge ratio multivariable spew–normal distribution method (11), Minimum-variance Hedge ratio (11), Model identification (4), Moment generating function (17), Monte Carlo simulations (9), Multicollinearity (3), Multinomial distribution (12), Multinomial Option Pricing Model (12), Multiple regression (2), Multivariate lognormal distribution (14), Multivariate normal distribution (14), Noncentral chi-square distribution (22),

Bibliography

Aitchison, J., & Brown, J. A. C. (1973). *Lognormal distribution, with special reference to its uses in economics*. Cambridge University Press.

Ait-Sahalia, Y., & Lo, A. W. (2000). Nonparametric risk management and implied risk aversion. *Journal of Econometrics, 94,* 9–51.

Amemiya, T. (1974). A note on a fair and Jaffee model. *Econometrica, 42,* 759–762.

Anderson, T. W. (1994). *The statistical analysis of time series*. Wiley-Interscience.

Anderson, T. W. (2003). *An introduction to multivariate statistical analysis*. Wiley-Interscience.

Ang, J. S., Jou, G. D., & Lai, T. Y. (2013). A comparison of formulas to compute implied standard deviation. *Encyclopedia of Finance*, 765–776.

Angrist, J. D., & Lavy, V. (1999). Using Maimonides' rule to estimate the effect of class size on scholastic achievement. *Quarterly Journal of Economics, 14*(2), 533–575.

Arellano, M., & Bover, O. (1995). Another look at the instrumental variable estimation of error-components models. *Journal of Econometrics, 68*(1), 29–51.

Bakshi, G., Cao, C., & Chen, Z. (1997). Empirical performance of alternative option pricing models. *Journal of Finance, 52*(5), 2003–2049.

Bakshi, G., Cao, C., & Chen, Z. (2010). Option pricing and hedging performance under stochastic volatility and stochastic interest rate. In Cheng F. Lee, Alice C. Lee, & John Lee (Eds.), *Handbook of quantitative finance and risk management*. Singapore: Springer.

Baltagi, B. (2008). *Econometric Analysis of Panel Data* (4th ed.) Wiley.

Black, F., & Scholes, M. (1973). The pricing of options and corporate liabilities. *Journal of Political Economy, 81*(3), 637–654.

Black, F., Jensen, M. C., & Scholes., M. (1972). The capital asset pricing model: Some empirical tests. In M. C. Jensen (Ed.), *Studies in the theory of capital markets (Praeger)*.

Blume, M. E., & Friend, I. (1973). A new look at the capital asset pricing model. *Journal of Finance, 28,* 19–33.

Brick, I. E., Palmon, O., & Patro, D. K. (2015). The motivations for issuing putable debt: An empirical analysis. In Cheng F. Lee & John Lee (Eds.), *Handbook of financial econometrics and statistics*. Singapore: Springer.

Brown, S. J., & Goetzmann, W. N. (1997). Mutual fund styles. *Journal of Financial Economics, 43*(3), 373–399.

Cameron, A. C., Gelbach, J. B., & Miller, D. L. (2011). Robust inference with multiway clustering. *Journal of Business & Economic Statistics, 29,* 238–249.

Cederburg, S., & O'Doherty, M. S. (2015). Asset-pricing anomalies at the firm level. *Journal of Econometrics, 186,* 113–128.

Chacko, G., & Viceira, L. M. (2003). Spectral GMM estimation of continuous-time processes. *Journal of Econometrics, 116*(1), 259–292.

Chang, C. F. (1999). *Determinants of capital structure and management compensation: The partial least squares approach* (Ph.D. dissertation). Rutgers University.

Chang, H. S., & Lee, C. F. (1977). Using pooled time-series and cross section data to test the firm and time effects in financial analysis. *Journal of Financial and Quantitative Analysis, 12,* 457–471.

Chang, C., Lee, A. C., & Lee, C. F. (2009). Determinants of capital structure choice: A structural equation modeling approach. *Quarterly Review of Economic and Finance, 49*(2), 197–213.

Chen, H. Y. (2011). *Momentum strategies, dividend policy, and asset pricing test* (Ph.D. dissertation). Rutgers University.

Chen, S. N., & Lee, C. F. (1981). The sampling relationship between Sharpe's performance measure and its risk proxy: Sample size, investment horizon and market conditions. *Management Science, 27*(6), 607–618.

Chen, K. H., & Shimerda, T. A. (1981). An empirical analysis of useful finance ratios. *Financial Management, 10*(1), 51–60.

Chen, W. P., Chung, H., Lee, C. F., & Liao, W. L. (2007). Corporate governance and equity liquidity: Analysis of S&P transparency and disclosure rankings. *Corporate Governance: An International Review, 15*(4), 644–660.

Chen, Hong-Yi, Gupta, Manak C., Lee, Alice C., & Lee, Cheng-Few. (2013). Sustainable growth rate, optimal growth rate, and optimal payout ratio: A joint optimization approach. *Journal of Banking and Finance, 37,* 1205–1222.

Chen, H. Y., Lee, A. C., & Lee, C. F. (2015). Alternative errors-in-variables models and the applications in finance research. *The Quarterly Review of Economics and Finance, 58,* 213–227.

Cheng, D., & C. F. Lee, C. (1986). Power of alternative specification errors tests in identifying misspecified market models. *The Quarterly Review of Economics and Business, 16*(3), 6–24.

Cheng, K. F., Chu, C. K., & Hwang, R. C. (2010). Predicting bankruptcy using the discrete-time semi-parametric hazard model. *Quantitative Finance, 10,* 1055–1066.

Chow, G. C. (1960). Tests of equality between sets of coefficients in two linear regressions. *Econometrica, 28,* 591–605.

Christie, S. (2005). Is the Sharpe ratio useful in asset allocation? MAFC Research Paper No. 31, Applied Finance Center, Macquarie University.

Chu, C. C. (1984). *Alternative methods for determining the expected market risk premium: Theory and evidence* (Ph.D. dissertation). University of Illinois at Urbana-Champaign.

Core, J. E. (2000). The directors' and officers' insurance premium: An outside assessment of the quality of corporate governance. *Journal of Law, Economics, and Organization, 16*(2), 449–477.

Cox, J. C., Ross, S. A., & Rubinstein, M. (1979). Option pricing: A simplified approach. *Journal of Financial Economics, 7*(3), 229–263.

Fair, R. C., & Jaffee, D. M. (1972). Methods of estimation of market in disequilibrium. *Econometrica, 40,* 497–514.

Fama, E. F. (1971). Parameter estimates for symmetric stable distributions. *Journal of the American Statistical Association, 66,* 331–338.

Fama, E. F., & MacBeth, J. D. (1973). Risk, return, and equilibrium: Empirical tests. *Journal of Political Economy, 81,* 607–636.

Fok, R. C. W., Lee, C. F., & Cheng, D. C. (1996). Alternative specifications and estimation methods for determining random beta coefficients: Comparison and extensions. *Journal of Financial Studies, 4*(2), 61–88.

Frecka, T. J., & Lee, C. F. (1983). Generalized financial ratio adjustment processes and their implications. *Journal of Accounting Research, 21,* 308–316.

Granger, C. W. J., & Newbold, P. (1973). Some comments on the evaluation of economic forecasts. *Applied Economics, 5*(1), 35–47.

Granger, C. W. J., & Newbold, P. (1974). Spurious regressions in econometrics. *Journal of Econometrics, 2,* 111–120.

Granger, C. W. J., & Ramanathan, R. (1984). Improved methods of combining forecasts. *Journal of Forecasting, 3,* 197–204.

Hamilton, J. D. (1994). *Time series analysis*. Princeton University Press.

Hansen, L. P. (1982). Large sample properties of generalized method of moments estimators. *Econometrica, 50*(4), 1029–1054.

Hansen, B. E. (1996). Inference when a nuisance parameter is not identified under the null hypothesis. *Econometrica, 64*(2), 413–430.

Hansen, B. E. (1997). Approximate asymptotic p-values for structural change tests. *Journal of Business and Economic Statistics, 15,* 60–67.

Hansen, B. E. (1999). Threshold effects in non-dynamic panels: Estimation, testing, and inference. *Journal of Econometrics, 93,* 345–368.

Hansen, B. E. (2000a). Sample splitting and threshold estimation. *Econometrica, 68*(3), 575–603.

Hansen, B. E. (2000b). Testing for structural change in conditional models. *Journal of Econometrics, 97,* 93–115.

Heston, S. L. (1993). A closed-form solution for options with stochastic volatility with applications to bond and currency options. *Review of Financial Studies, 6*(2), 327–343.

Hsiao, C. (2014). *Analysis of panel data—Econometric society monographs* (3rd ed).

Hutchinson, J. M., Lo, A. W., & Poggio, T. (1994). A nonparametric approach to pricing and hedging derivative securities via learning networks. *Journal of Finance, 49*(3), 851–889.

Hwang, R. C., Cheng, K. F., & Lee, J. C. (2007). A semiparametric method for predicting bankruptcy. *Journal of Forecasting, 26,* 317–342.

Hwang, R. C., Wei, H. C., Lee, J. C., & Lee, C. F. (2008). On prediction of financial distress using the discrete-time survival model. *Journal of Financial Studies, 16,* 99–129.

Hwang, R. C., Cheng, K. F., & Lee, C. F. (2009). On multiple-class prediction of issuer crediting ratings. *Journal of Applied Stochastic Models in Business and Industry, 25,* 535–550.

Hwang, R. C., Chung, H., & Chu, C. K. (2010). Predicting issuer credit ratings using a semiparametric method. *Journal of Empirical Finance, 17*(1), 120–137.

Ittner, C. D., Larcker, D. F., & Rajan, M. V. (1997). The choice of performance measures in annual bonus contracts. *Accounting Review, 72*(2), 231–255.

Jobson, J. D., & Korbie, B. M. (1981). Performance hypothesis testing with sharpe and treynor measures. *Journal of Finance, 36*(4), 889–908.

Kao, L. J., & Lee, C. F. (2012). Alternative method to for determining industrial bond ratings: Theory and empirical evidence. *International Journal of Information Technology & Decision Making, 11,* 1215–1235.

Kao, L. J., & Lee, C. F. (2018). Alternative methods to derive statistical distribution of sharpe performance measure: Review, comparison, and extension. Working Paper, Rutgers University.

Kao, L. J., & Lee, C. F. (2019). Sharpe ratio under non-linear dependent and heavy-tailed asset returns obeying GARCH process: Theory and empirical evidence. Working Paper, Rutgers University.

Kau, J. B., & Lee, C. F. (1976). Functional form, density gradient and the price elasticity of demand for housing. *Urban Studies, 13*(2), 181–192.

Kau, J. B., Lee, C. F., & Sirmans, C. F. (1986). Urban econometrics: Model developments and empirical results. In *Research in urban economics* (Vol. 6). JAI Press.

Kaufman, L., & Rousseeuw, P. J. (1990). *Finding groups in data: An introduction to cluster analysis* (9th ed.). Wiley-Interscience.

Kim, D. (1995). The errors in the variables problem in the cross-section of expected stock returns. *Journal of Finance, 50*(5), 1605–1634.

Kim, D. (1997). A reexamination of firm size, book-to-market, and earnings price in the cross-section of expected stock returns. *Journal of Financial and Quantitative Analysis, 32*(4), 463–489.

Kim, D. (2010). Issues related to the errors-in-variables problems in asset pricing tests. In *Handbook of quantitative finance and risk management* (Part V, Chapter 70, pp. 1091–1108).

Kuan, C. M., & Hornik, K. (1995). The generalized fluctuation test: A unifying view. *Econometric Reviews, 14,* 135–161.

Lai, T.-Y., Lee, C. et al. (1992). An alternative method for obtaining the implied standard deviation. *Journal of Financial Engineering* 1: 369–375.

Lambert, R., & Larcker, D. (1987). An analysis of the use of accounting and market measures of performance in executive compensation contracts. *Journal of Accounting Research, 25,* 85–125.

Lee, C. F. (1973). *Errors-in-variables estimation procedures with applications to a capital asset pricing model* (Ph.D. dissertation). The State University of New York at Buffalo.

Lee, C. F. (1976a). A note on the interdependent structure of security returns. *Journal of Financial and Quantitative Analysis, 11,* 73–86.

Lee, C. F. (1976b). Functional form and the dividend effect of the electric utility industry. *Journal of Finance, 31*(5), 1481–1486.

Lee, C. F. (1977a). Functional form, skewness effect and the risk-return relationship. *Journal of Financial and Quantitative Analysis, 12,* 55–72.

Lee, C. F. (1977b). Performance measure, systematic risk and errors-in-variable estimation method. *Journal of Economics and Business*, 122–127.

Lee, A. (1996). *Cost of capital and equity offerings in the insurance industry* (Ph.D. dissertation). The University of Pennsylvania in Partial.

Lee, C. F., & Chen, S. N. (1979). Sampling properties of composite performance measures and their implications. Working Papers 541, The University of Illinois at Urbana Champaign.

Lee, A. C., & Cummins, J. D. (1998). Alternative models for estimating the cost of capital for property/casualty insurers. *Review of Quantitative Finance and Accounting, 10*(3), 235–267.

Lee, C. F., & Jen, F. C. (1978). Effects of measurement errors on systematic risk and performance measure of a portfolio. *Journal of Financial and Quantitative Analysis, 13*(2), 299–312.

Lee, K. W., & Lee, C. F. (2012) *Multiple directorships, corporate ownership, firm performance*. Working paper, Rutgers University.

Lee, C. F., & Lee, J. C. (2015). *Handbook of financial econometrics and statistics* (Vol. 1). New York: Springer Reference.

Lee, C. F., & Wu, C. C. (1985). The impacts of kurtosis on risk stationarity: some empirical evidence. *Financial Review, 20*(4), 263–269.

Lee, C. F., & Zumwalt, J. K. (1981). Associations between alternative accounting profitability measures and security returns. *Journal of Financial and Quantitative Analysis, 16,* 71–93.

Lee, C. F., Newbold, P., Finnerty, J. E., & Chu, C. C. (1986). On accounting-based, market-based and composite-based beta predictions: Method and implications. *Financial Review, 21,* 51–68.

Lee, C. F., Wei, K. C. J., & Bubnys, E. L. (1989). The APT versus the multi-factor CAPM: Empirical evidence. *Quarterly Review of Economics and Business, 29.*

Lee, C. F., Wu, C. C., & Wei, K. C. J. (1990). Heterogeneous investment horizon and capital asset pricing model: Theory and implications. *Journal of Financial and Quantitative Analysis, 25,* 361–376.

Lee, C. F., Lee, A. C., & Lee, J. (2010). *Handbook of quantitative finance and risk management.* New York: Springer.

Lee, C. F., Gupta, M. C., Chen, H. Y., & Lee, A. C. (2011). Optimal payout ratio under uncertainty and the flexibility hypothesis: Theory and empirical evidence. *Journal of Corporate Finance, 17*(3), 483–501.

Lee, C. F., Tsai, C. M., & Lee, A. C. (2013). Asset pricing with disequilibrium price adjustment: Theory and empirical evidence. *Quantitative Finance, 13,* 227–239.

Lee, C. F., Lee, A. C., & Lee, J. (2019). *Handbook of financial econometrics, mathematics, statistics, and technology.* Singapore: World Scientific (Forthcoming).

Leon, A., & Vaello-Sebastia, A. (2009). American GARCH employee stock option valuation. *Journal of Banking and Finance, 33*(6), 1129–1143.

Lien, D. (2010). An note on the relationship between the variability of the hedge ratio and hedging performance. *Journal of Futures Market, 30*(11), 1100–1104.

Lien, D., & Shrestha, K. (2007). An empirical analysis of the relationship between hedge ratio and hedging horizon using wavelet analysis. *Journal of Futures Market, 27*(2), 127–150.

Lin, F. C., Chien, C. C., Lee, C. F., Lin, H. C., & Lin, Y. C. (2019). Tradeoff between reputation concerns and economic dependence for auditors—Threshold regression approach. In *Handbook of financial econometrics, mathematics, statistics, and technology.* Singapore: World Scientific (Forthcoming).

Liu, B. (2006). *Two essays in financial econometrics: I: Functional forms and pricing of country funds. II: The term structure model of inflation risk premia* (Ph.D. dissertation). Rutgers University.

Lo, A. W. (2002). The statistics of sharpe ratios. *Financial Analysts Journal, 58*(4), 36–47.

Maddala, G. S., & Rao, C. R. (1996). *Handbook of statistics 14: Statistical methods in finance*. Elsevier Science & Technology.

Martin, C. (1990). Corporate borrowing and credit constrains: Structural disequilibrium estimates for the U.K. *Review of Economics and Statistics, 72*(1), 78–86.

Mayer, W. J. (1989). Estimating disequilibrium models with limited a priori price-adjustment information. *Journal of Econometrics, 41*(3), 303–320.

Miller, M., & Modigliani, F. (1966). Some estimates of the cost of capital to the utility industry, 1954–57. *American Economic Review, 56*(3), 333–391.

Myers, R. J. (1991). Estimating time-varying optimal hedge ratios on futures markets. *Journal of Futures Markets, 11*(1), 39–53.

Newey, W. K., & West, K. D. (1987). A simple, positive semi-definite, heteroskedasticity and autocorrelation consistent covariance matrix. *Econometrica, 55*(3), 703–708.

Ohlson, J. S. (1980). Financial ratios and the probabilistic prediction of bankruptcy. *Journal of Accounting Research, 19*(Spring), 109–131.

Opdyke, J. D. (2007). Comparing sharpe ratios: so where are the p-values? *Journal of Asset Management, 8*(5), 208–336.

Petersen, M. A. (2009). Estimating standard errors in finance panel data sets: Comparing approaches. *Review of Financial Studies, 22,* 435–480.

Pike, R., & Sharp, J. (1989). Trends in the use of management science techniques in capital budgeting. *Managerial and Decision Economics, 10,* 135–140.

Pinches, G. E., & Mingo, K. A. (1973). A multivariate analysis of industrial bond ratings. *Journal of Finance, 28*(1), 1–18.

Quandt, R. E. (1988). *The Econometrics of disequilibrium.* New York: Basil Blackwell Inc.

Rendleman, R. J., Jr., & Barter, B. J. (1979). Two-state option pricing. *Journal of Finance, 24,* 1093–1110.

Riahi-Belkaoui, A., & Pavlik, E. (1993). Effects of ownership structure, firm performance, size and diversification strategy on CEO compensation: A path analysis. *Managerial Finance, 19*(2), 33–54.

Rubinstein, M. (1994). Implied binomial trees. *Journal of Finance, 49,* 771–818.

Schroder, M. (1989). Computing the constant elasticity of variance option pricing formula. *Journal of Finance, 44*(1), 211–219.

Sealey, C. W., Jr. (1979). Credit rationing in the commercial loan market: Estimates of a structural model under conditions of disequilibrium. *Journal of Finance, 34,* 689–702.

Sears, R. S., & John Wei, K. C. (1988). The structure of skewness preferences in asset pricing model with higher moments: An empirical test. *Financial Review, 23*(1), 25–38.

Shapiro, A. F. (2005). *Fuzzy regression models.* Working paper, Penn State University.

Shumway, T. (2001). Forecasting bankruptcy more accurately: A simple hazard model. *The Journal of Business, 74,* 101–124.

Thompson, S. B. (2011). Simple formulas for standard errors that cluster by both firm and time. *Journal of Financial Economics, 99*(1), 1–10.

Thursby, J. G. (1985). The relationship among the specification error tests of Hausman, Ramsey and Chow. *Journal of the American Statistical Association, 80*(392), 926–928.

Titman, S., & Wessels, R. (1988). The determinants of capital structure choice. *Journal of Finance, 43*(1), 1–19.

Tsai, G. M. (2005). *Alternative dynamic capital asset pricing models: Theories and empirical results* (Ph.D. dissertation). Rutgers University.

Van Der Klaauw, W. (2002). Estimating the effect of financial aid offers on college enrollment: A regression-discontinuity approach. *International Economic Review, 43*(4), 1249–1287.

Wei, K. C. (1984). *The arbitrage pricing theory versus the generalized intertemporal capital asset pricing model: theory and empirical evidence* (Ph.D. dissertation). University of Illinois at Urbana-Champaign.

White, H. (1980). A heteroskedasticity-consistent covariance matrix estimator and a direct test for heteroskedasticity. *Econometrica, 48*(4), 817–838.

Wooldridge, J. M. (2010). *Econometric analysis of cross section and panel data* (2nd ed.). The MIT Press.

Yang, C. C. (1989). *The impact of new equity financing on firms' investment, dividend and debt-financing decisions* (Ph.D. dissertation). The University of Illinois at Urbana-Champaign.

Yang, C. C., Lee, C. F., Gu, Y. X., & Lee, Y. W. (2010). Co-determination of capital structure and stock returns —A LISREL approach: An empirical test of Taiwan stock markets. *Quarterly Review of Economic and Finance, 50*(2), 222–233.

Zeileis, A., Leisc, F., Hornik, K., & Kleiber, C. (2002). Strucchange: An R package for testing for structural change in linear regression models. *Journal of Statistical Software, 7,* 1–38.

Zellner, A. (1962). An efficient method of estimating seemingly unrelated regressions and tests for aggregation bias. *Journal of the American Statistical Association*, *57*(298), 348–368.

Part I
Regression and Financial Econometrics

There are eight chapters in this part. In Chap. 2, we discuss the assumptions of the multiple regression model, estimated parameters of the multiple regression model, the standard error of the residual estimate, and the coefficient of determination. We also investigate tests on sets and individual regression coefficients and the confidence interval for the mean response and prediction interval for the individual response.

In Chap. 3, we discuss various topics associated with the regression analysis, including multicollinearity, heteroscedasticity, autocorrelation, model specification and specification bias of the regression model, nonlinear regression models, lagged dependent variables in the regression model, dummy variables in the regression model, and regression model with interaction variables. We also apply the regression approach to investigate the effect of alternative business strategies and apply the logistic regression model to credit risk analysis.

In Chap. 4, we extend single equation models to simultaneous equation models. Specifically, we discuss simultaneous equation system, two-stage least squares method, and three-stage least squares method. In Chap. 5, we discuss an econometric approach to financial analysis, planning, and forecasting. The issue of simultaneity and the dynamics of corporate-budgeting decisions will be explored by using finance theory. We also investigate the interrelationships among the programming, the simultaneous equations, and the econometrics approaches.

Chapter 6 addresses one of the important issues related to panel data analysis. We introduce the dummy variable technique and the error component model for analyzing pooled data. We investigate the possible impacts of firm effect and time effect on choosing the optimal functional form of a financial research study. In this chapter, we also discuss the criteria of using fixed effect or random-effect approach.

Chapter 7 discusses how errors-in-variables estimation methods are used in finance research. We show how errors-in-variables problems can affect estimators of the linear regression model as well as discuss the effects they have on the empirical research cost of capital, asset pricing, capital structure, and investment decision. Chapter 8 provides three alternative errors-in-variables estimation models in testing the capital asset pricing model. Specifically, we present three alternative correction methods for the errors-in-variables problem. In Chap. 9, we discuss the issue of spurious regression and data mining in both conditional asset pricing models and simple predictive regression. We also discuss the impact of spurious regression and data mining on conditional asset pricing.

2 Multiple Linear Regression

Contents

Abstract

In this chapter, we first discuss the basic assumption of multiple regression. We then specify the model of regression and show how this regression coefficient can be estimated. Appendix 1 derives the sampling variance of the least squares slope estimations, and Appendix 2 shows how multiple regression can be used to investigate the cross-sectional relationship among price per share, dividend per share, and return earning per share.

C.-F. Lee et al., *Financial Econometrics, Mathematics and Statistics*,
https://doi.org/10.1007/978-1-4939-9429-8_2

2.1 Introduction

The simple regression model is examined with one independent variable (such as amount of fertilizer) and one dependent variable (such as yield of corn). In many cases, however, more than one factor can affect the outcome under study. In addition to fertilizer, rainfall and temperature certainly influence the yield of corn. In business, not only rates of return for the stock market at large affect the return on General Motors or Ford stock. Other variables, such as leverage ratio, payout ratio, and dividend yield, also contribute. Therefore, regression analysis with more than one independent variable is an important analytical tool.

The model that extends a simple regression to use with two or more independent variables is called a multiple linear regression. Simple linear regression analysis helps us determine the relationship between two variables or predict the value of one variable from our knowledge of another. Multiple regression analysis, in contrast, is a technique for determining the relationship between a dependent variable and more than one independent variable. In addition, it can be used to employ several independent variables to predict the value of a dependent variable.

In Sect. 2.2, we discuss the assumptions of the multiple regression model. In Sect. 2.3, we discuss estimated parameters of the multiple regression model. Section 2.4 discusses the standard error of the residual estimate and the coefficient of determination. Section 2.5 investigates tests on sets and individual regression coefficients. Section 2.6 discussed the confidence interval for the mean response and prediction interval for the individual response. In Sect. 2.7, we consider business and economic applications. In Sect. 2.8, we use computer programs to do multiple regression analyses. In Sect. 2.9, we conclude this chapter. Appendix 1 shows the derivation of the sampling variance of the least squares slope estimations, and Appendix 2 discusses the cross-sectional relationship among price per share, dividend per share, and retained earnings per share.

*For basic concepts and models of simple regression, readers can refer to Chap. 15 of Statistics for Business and Financial Economics by Lee et al. (2013).

2.2 The Model and Its Assumptions

In this section, we first review the simple regression model and extend it to a multiple regression model. Then we define and analyze the regression plane for two independent variables. Finally, the important assumptions we must make to use the multiple regression model are explored in some detail.

The Multiple Regression Model

In multiple regression, simple regression is extended by introducing more than one independent variable. It is well known that a simple linear regression model can be defined as $Y_i = \alpha + \beta x_i + e_i$ and its estimate as $y_i = a + bx_i + e_i$. The sample intercept a and the sample slope b are estimates for a and β, respectively.

The normal equations used to estimate unknown parameters a and β are

$$na + b\sum_{i=1}^{n} x_i = \sum_{i=1}^{n} y_i$$

and

$$a\sum_{i=1}^{n} x_i + b\sum_{i=1}^{n} x_i^2 = \sum_{i=1}^{n} x_i y_i$$

The foregoing equations from simple linear regression are the starting point for our exploration of multiple regression in this chapter.

Suppose an individual's annual salary (Y) depends on the number of years of education (X_1) and the number of years of work experience (X_1) the individual has had. The population regression model is

$$Y_i = \alpha + \beta_1 X_{1i} + \beta_2 X_{2i} + \epsilon_i \quad (2.1)$$

and its estimate is

$$y_i = a + b_1 x_{1i} + b_2 x_{2i} + e_i \quad (2.2)$$

where Eqs. (2.1) and (2.2) represent the multiple population regression line and the multiple sample regression line, respectively. In Eq. (2.1), α is the intercept of the regression; β_1 is the slope that represents the conditional relationship between Y and X_1, assuming X_2 is fixed; and β_2 is the slope that represents the conditional relationship between Y and X_2, assuming X_1 is fixed. If the model defined in Eq. (2.1) is linear, then the relationship between Y and each of the independent variables can be described by a straight line. In other words, the conditional mean of the dependent variable is given by the following population regression equation:

$$E(Y_i|X_1 = x_1, X_2 = x_2) = \alpha + \beta_1 x_1 + \beta_2 x_2$$

The coefficients β_1 and β_2 are called partial regression coefficients. They indicate only the partial influence of each independent variable when the influence of all other independent variables is held constant. Just as in simple regression, the multiple sample regression line of Eq. (2.2) can be used to estimate the multiple population regression line of Eq. (2.1).

The Regression Plane for Two Explanatory Variables

Let us say that the stock price per share (y) can be modeled as a function of both dividend per share (x_1) and retained earnings (x_2) per share.[1]

$$y_i = a + b_1 x_{1i} + b_2 x_{2i} + e_i$$

where

$y_i(P_i)$ = stock price per share for the ith firm, $X_{1i}(D_i)$ = dividend per share for the ith firm, and $x_{2i}(RE_i)$ = retained earnings per share for the ith firm. (Retained earnings per share equals earnings per share minus dividend per share.) The first goal of the analysis is to obtain the estimated multiple regression model

$$\hat{y}_i = a + b_1 x_{1i} + b_2 x_{2i} \quad (2.2a)$$

The value of b_1 indicates that after the influence of the retained earnings per share is taken into account, a \$1 increase in the dividend per share (D_i) will increase the mean value of the price per share (P_i) by b_1 other things being equal. Similarly, a \$1 increase in retained earnings per share will increase the mean price per share by b_2. If there is only one explanatory variable, the estimated regression equation generates a straight line. There are two explanatory variables in Eq. (2.2a), so it represents a regression plane (three-dimensional regression graph). On this three-variable regression plane, a combination of three observations (one for the value of y, one for x_1, and one for x_2) represents a single point. This point can be depicted on a three-dimensional scatter diagram. In Fig. 2.1, the best-fitted regression plane would pass near the actual sample observation points indicated by the symbol $\times$, some falling above the plane and some below in such a way as to minimize L in

$$L = \sum_{i=1}^{n} (y_i - \hat{y})^2 \quad (2.3)$$

where y_i and $\hat{y}_i$ are as defined in Eqs. (2.2) and (2.2a), respectively.[2]

If there are k independent variables, then Eq. (2.1) can be generalized to

$$Y_i = \alpha + \beta_1 X_{1i} + \beta_2 X_{2i} + \cdots + \beta_k X_{ki} + \epsilon_i \quad (2.4)$$

The following section explains how regression parameters are estimated via the least squares estimation method.

[1]Practical examples based on Eq. (2.2) will be explored in the Applications section of this chapter.

[2]Using Eq. (2.3) to estimate regression parameters will be discussed in Sect. 2.3.

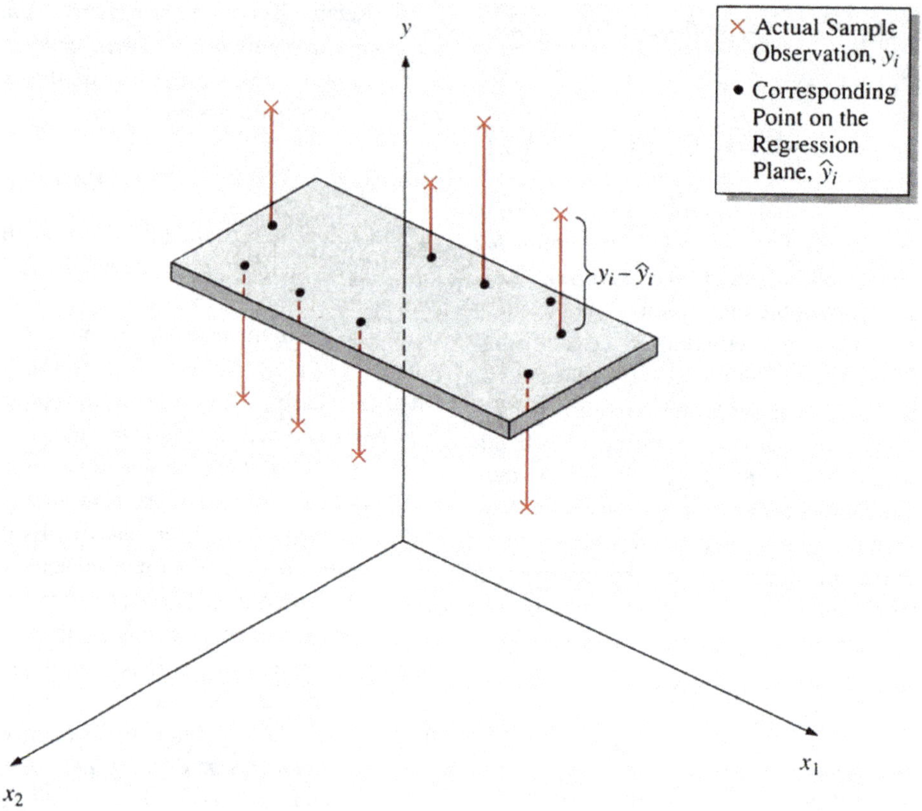

Fig. 2.1 Regression plane with $y_i(P_i)$ as dependent variable and with $X_{1i}(D_i)$ and $x_{2i}(\mathrm{RE}_i)$ as independent variables

Assumptions for the Multiple Regression Model As in simple regression analysis, we need five assumptions to perform a regression analysis of the model defined in Eq. (2.4).

1. The error term ε_i is distributed with conditional mean zero and variance σ_ϵ^2 for i = 1, 2, … , n.
2. The error term ε_i is independent of each of the k independent variables $X_1, X_2, \ldots, X_k$. In other words, there are no measurement errors associated with any independent variable. If there exists a measurement error of any independent variable, then the slope estimate will be biased. In Chap. 7 of this book, we will discuss alternative methods to deal this problem.
3. Any two errors ε_i and ε_i are not correlated with one another; that is, their covariance is zero: $\mathrm{Cov}(\varepsilon_i, \varepsilon_i) = 0$ for $i \neq j$. This assumption means that there is no autocorrelation (serial correlation) among residual terms.
4. The independent variables are not perfectly related to each other in a linear function. In other words, it is not possible to find a set of numbers $d_0, d_1, d_2, \ldots, d_k$ such that

$$d_0 + d_1X_{1i} + d_2X_{2i} + \cdots + d_kX_{ki} = 0,\quad i = 1, 2, \ldots, n$$

In practice, the linear relationship among independent variables is usually not perfect. When a perfect linear relationship occurs, a condition known as perfect collinearity exists. Multicollinearity is the condition in which two variables are highly correlated.

2.3 Estimating Multiple Regression Parameters

To estimate the best-fitted regression plane, we use the least squares method to estimate the regression parameters. The principle of using the least squares method for estimating the parameters of one population regression model is demonstrated in Eq. (2.3) and Fig. 2.1. Taking Eq. (2.2) as an example, we estimate the coefficients a, b_1, and b_2 by minimizing

$$L = \sum_{i=1}^{n} e_i^2 = \sum_{i=1}^{n} (y_i - a - b_1 x_{1i} - b_2 x_{2i})^2$$

The normal equations for estimating a, b_1, and b_2 are[3]

$$\begin{aligned} na + b_1 \sum_{i=1}^{n} x_{1i} + b_2 \sum_{i=1}^{n} x_{2i} &= \sum_{i=1}^{n} y_i \\ a \sum_{i=1}^{n} x_{1i} + b_1 \sum_{i=1}^{n} x_{1i}^2 + b_2 \sum_{i=1}^{n} x_{1i} x_{2i} &= \sum_{i=1}^{n} x_{1i} y_i \\ a \sum_{i=1}^{n} x_{2i} + b_i \sum_{i=1}^{n} x_{1i} x_{2i} + b_2 \sum_{i=1}^{n} x_{2i}^2 &= \sum_{i=1}^{n} x_{2i} y_i \end{aligned} \quad (2.5)$$

If we substitute $(x_{1i} - \bar{x})$, $(x_{2i} - \bar{x}_2)$, and $(y_1 - \bar{y})$ for x_{1i}, x_{2i}, and y_1, then the normal equations reduce to[4]

$$\begin{aligned} b_1 \sum_{i=1}^{n} x_{1i}'^2 + b_2 \sum_{i=1}^{n} x_{1i}' x_{2i}' &= \sum_{i=1}^{n} x_{1i}' y_i' \\ b_2 \sum_{i=1}^{n} x_{1i}' x_{2i}' + b_2 \sum_{i=1}^{n} x_{2i}'^2 &= \sum_{i=1}^{n} x_{2i}' y_i' \end{aligned} \quad (2.6)$$

There are two equations and two unknowns, b_1 and b_2, associated with this equation system. Hence, we can solve b_1 and b_2 uniquely by substitution.

$$b_1 = \frac{\left(\sum_{i=1}^{n} x_{1i}' y_i'\right)\left(\sum_{i=1}^{n} x_{2i}'^2\right) - \left(\sum_{i=1}^{n} x_{2i}' y_i'\right)\left(\sum_{i=1}^{n} x_{1i}' x_{2i}'\right)}{\left(\sum_{i=1}^{n} x_{1i}'^2\right)\left(\sum_{i=1}^{n} x_{2i}'^2\right) - \left(\sum_{i=1}^{n} x_{1i}' x_{2i}'\right)^2} \quad (2.7)$$

$$b_2 = \frac{\left(\sum_{i=1}^{n} x_{1i}'^2\right)\left(\sum_{i=1}^{n} x_{2i}' y_i'\right) - \left(\sum_{i=1}^{n} x_{1i}' x_{2i}'\right)\left(\sum_{i=1}^{n} x_{1i}' y_i'\right)}{\left(\sum_{i=1}^{n} x_{1i}'^2\right)\left(\sum_{i=1}^{n} x_{2i}'^2\right) - \left(\sum_{i=1}^{n} x_{1i}' x_{2i}'\right)^2} \quad (2.8)$$

From the estimated b_1 and b_2, we obtain the estimated regression line

$$\hat{y}_i' = b_1 x_{1i}' + b_2 x_{2i}' \quad (2.9)$$

It can be shown that the intercept of Eq. (2.2) is estimated as[5]

$$a = \bar{y} - b_1 \bar{x}_1 - b_2 \bar{x}_2 \quad (2.10)$$

Example 2.1 (*Annual Salary, Years of Education, and Years of Work Experience*)

Let us use the hypothetical data given in Table 2.1 to demonstrate the procedure for estimating a multiple regression. In Table 2.1, y represents an individual's annual salary (in thousands of dollars), x_1 represents that individual's years of education, and x_2 represents her or his years of work experience.

From the data of Table 2.1, we estimate the regression line

$$\hat{y}_i = a + b_1 x_{1i} + b_2 x_{2i} \quad (2.11)$$

The worksheet for estimating this regression line is given in Table 2.2. (This table is included

[3]Equation (2.5) is a three-equation simultaneous equation system with three unknowns. The values of these three unknowns can be obtained by solving this system of simultaneous equations, by using the formula derived in this section, or by using an appropriate computer package (see Sect. 2.8).

[4]In this new coordinate system, $\sum_{i=1}^{n} x_{1i}, \sum_{i=1}^{n} x_{2i}$, and $\sum_{i=1}^{n} y_i$ become $\sum_{i=i}^{n}(x_{1i} - \bar{x}_1) = 0$, $\sum_{i=1}^{n}(x_{2i} - \bar{x}_2) = 0$, and $\sum_{i=1}^{n}(y_i - \bar{y}) = 0$. If we set $x_{1i}' = x_{1i} - \bar{x}_1, x_{2i}' = x_{2i} - \bar{x}_2$, and $y_i' = y_i - \bar{y}$, then Eqs. (2.5) reduce to Eqs. (2.6).

[5]Using the definitions of $\hat{y}_i', x_{1i}'$, and x_{2i}', we can rewrite Eq. (2.9) as

$$(\hat{y}_i - \bar{y}) = b_1(x_{1i} - \bar{x}_1) + b_2(x_{2i} - \bar{x}_2)$$

which becomes

$$\hat{y}_i = (\bar{y} - b_1 \bar{x}_1 - b_2 \bar{x}_2) + b_1 x_{1i} + b_2 x_{2i} \quad (2.9')$$

Table 2.1 Data for Example 2.1

	x_{1i}	x_{2i}	y_i
	5	7	15
	10	5	17
	9	14	26
	13	8	24
	15	6	27
Total	52	40	109.0
Mean	10.4	8	21.8

to show how computers calculate mean, variance, and covariance. You do not need to remember the procedure.)

Substituting information from Table 2.2 into Eqs. (2.7), (2.8), and (2.10), we obtain

$$\hat{b}_1 = \frac{(62.4)(50) - (36)(-11)}{(59.2)(50) - (-11)^2} = \frac{3516}{2839} = 1.2385$$

$$\hat{b}_2 = \frac{(59.2)(36) - (-11)(62.4)}{(59.2)(50) - (-11)^2} = \frac{2817.6}{2839} = 0.99246$$

$$\hat{a} = 21.8 - (1.2385)(10.4) - (0.99246)(8) = 0.980$$

Hence, the regression line of Eq. (2.11) becomes

$$\hat{y}_i = 0.980 + 1.2385x_{1i} + 0.9925x_{2i} \quad (2.12)$$

The next section shows how to compute standard errors of estimates and the coefficients of determination.

2.4 The Residual Standard Error and the Coefficient of Determination

As in the case of simple regression, the standard error of estimate can be used as an absolute measure, and the coefficient of determination as a relative measure, of how well the multiple regression equation fits the observed data.

The Residual Standard Error

Just like simple regression, multiple regression can be used to break down the total variation of a dependent variable y_i into unexplained variation and explained variation.

Table 2.2 Worksheet for estimating a regression line (Example 2.1)

	x_{1i}	x_{2i}	y	a	b	c	aa	bb	cc
	5	7	15	−5.4	−1	−6.8	29.16	1	46.24
	10	5	17	−0.4	−3	−4.8	0.16	9	23.04
	9	14	26	−1.4	6	4.2	1.96	36	17.64
	13	8	24	2.6	0	2.2	6.76	0	4.84
	15	6	27	4.6	−2	5.2	21.16	4	27.04
Mean	10.4	8	21.8						
Total	52	40	109	0	0	0	59.2	50	118.8
		$(x_{1i} - \bar{x}_1)(y_i - \bar{y})$			$(x_{2i} - \bar{x}_2)(y_i - \bar{y})$			$(x_{1i} - \bar{x}_1)(y_i - \bar{y})$	
			ac			bc		ab	
			36.72			6.8		5.4	
			1.92			14.4		1.2	
			−5.88			25.2		−8.4	
			5.72			0		0	
			23.92			−10.4		−9.2	
Total			62.4			36		−11	

$$\underbrace{\sum_{i=1}^{n}(y_i-\bar{y})^2}_{\substack{\text{Sum of Squares}\\ \text{Total}\\ \text{(SST)}}} = \underbrace{\sum_{i=1}^{n}(y-\hat{y}_i)^2}_{\substack{\text{Sum of Squares}\\ \text{Error}\\ \text{(SSE)}}} + \underbrace{\sum_{i=1}^{n}(\hat{y}_i-\bar{y})^2}_{\substack{\text{Sum of Squares}\\ \text{due to Regresssion}\\ \text{(SSR)}}} \tag{2.13}$$

The estimated dependent variable ($\bar{y}_1$) of multiple regression is determined by two or more independent variables. SSR and SSE are the explained and unexplained sums of squares, respectively.

Using the definition of sum of squares error, we can define the estimate of the standard deviation of error terms, sometimes called the residual standard error, as

$$s_e = \sqrt{\frac{\sum_{i=1}^{n}(y_i-\hat{y}_i)^2}{n-3}} \tag{2.14}$$

Because there are three parameters—a, b_1, and b_2—for Eq. (2.2) that we must estimate before calculating the residual, the number of degree of freedom is $(n - 3)$. In other words, $(n - 3)$ sample values are "free" to vary. More generally, the number of degrees of freedom for estimating the residual standard error for Eq. (2.4) is $[n - (k + 1)]$.

Example 2.2 (Computing y_i, e_i, and e_i^2)

Using the data presented in Example 2.1, we can estimate y_i, e_i, and e_i^2 as shown in Table 2.3.

Here $\hat{y}_i$ is obtained by substituting x_{1i} and x_{2i} into Eq. (2.12). For example, 14.1198 = 0.980 + 1.2385(5) + 0.9925(7); $\hat{e}_i = y_i - \hat{y}_i$.

$$\sum_{i=1}^{5}(y_i-\hat{y}_i)^2 = (15 = 14.1198)^2 + (17 - 18.3272)^2$$
$$+ (26 - 26.0209)^2 + (24 - 25.0200)^2$$
$$+ (27 - 25.5120)^2 = 5.7912$$

Hence,

$$s_e = \sqrt{\frac{5.7912}{5-3}} = 1.7016$$

s_e is one of the important components in determining the distribution of estimated a, b_1, and b_2 and fitted dependent variable ($\hat{y}$).

The Coefficient of Determination

We can use Eq. (2.13) to calculate a relative measure of the goodness of fit for a multiple regression.

$$\begin{aligned} R^2 &= \frac{\sum_{i=1}^{n}(\hat{y}_i-\bar{y})^2}{\sum_{i=1}^{n}(y_i-\bar{y})^2} \\ &= \frac{\text{explained variation of } y \text{ (SSR)}}{\text{total variation of } y \text{ (SST)}} \\ &= 1 - \frac{\text{SSE}}{\text{SST}} \end{aligned} \tag{2.15}$$

The coefficient of determination R^2 is the proportion of total variation in y (SST) that is explained by the intercept and the independent variable x_1 and x_2. Note that both R^2 and s_e can be used to measure the goodness of fit for a regression. However, R_2 is a relative measure and s_e an absolute measure. Now we use the ANOVA table given in Table 2.4 to calculate the relationship between R^2 and s_e for the general multiple regression model in Eq. (2.4).

There are four columns in Table 2.4. Column (1) represents the sources of variation, column (2) alternative sums of squares that are identical to those discussed in Eq. (2.13), column (3) degrees of freedom associated with each source of

Table 2.3 Actual values, predicted values, and residuals for annual salary regression

	Actual value, y_i	Predicted value, $\hat{y}_1$	Residuals	
			e_i	e_i^2
	15	14.1198	0.8802	0.7748
	17	18.3272	−1.3272	1.7615
	26	26.0209	−0.0209	0.0004
	24	25.0200	−1.0200	1.0404
	27	25.5120	1.4880	2.2141
Total	109	–	–	5.7912

Table 2.4 Notation of analysis of variance table

(1)	(2)	(3)	(4)
Source of variation	Sum of squares	Degrees of freedom	Mean square
Due to regression	$SSR = \sum_{i=1}^{n} (\hat{y}_i - \bar{y})^2$	k	SSR/k
Residual	$SSE = \sum_{i=1}^{n} (y_i - \hat{y}_i)^2$	$n - k - 1$	SSR/($n - k -$ 1)
Total	$SST = \sum_{i=1}^{n} (y_i - \bar{y})^2$	$n - 1$	SST/(n − 1)

variation, and column (4) the mean squares. Note that alternative mean squares represent alternative variance estimates. Mean square due to the regression is also called explained variance; mean square due to the residuals is also called unexplained variance; and mean square due to the total variation can also be called variance of the dependent variable. Using those estimates, we can obtain an adjusted (or corrected) coefficient of determination $\overline{R}^2$.

$$\overline{R}^2 = 1 - \frac{\text{SSE}/(n-k-1)}{\text{SST}/(n-1)} = 1 - (1 - R^2)\frac{n-1}{n-k-1} \tag{2.16}$$

The difference between R^2 and $\overline{R}^2$ is that $\overline{R}^2$ is adjusted for degrees of freedom for both SSE and SST. $\overline{R}^2$ is always smaller than R^2. If the sample size becomes large, however, $\overline{R}^2$ approaches R^2. $\overline{R}^2$ can generally help us avoid overestimating the goodness of fit for a regression relationship by adding more independent variables (relevant or not) to a regression equation. Note that the standard error of estimate (Eq. 2.14) also has been adjured for the degrees of freedom $(n - k - 1)$.

If we divide components in Eq. (2.13) by their related degrees of freedom, then it can be shown that

$$\underset{\text{Total Variance}}{\frac{\sum_{i=1}^{n}(y_i-\bar{y})^2}{n-1}} \neq \underset{\text{Unexplained Variance}}{\frac{\sum_{i=1}^{n}(y_i-\hat{y}_i)^2}{n-k-1}} + \underset{\text{Explained Variance}}{\frac{\sum_{i=1}^{n}(\hat{y}_i-\bar{y})^2}{k}} \tag{2.17}$$

Table 2.5 Analysis of variance results

Source of variation	Sum of squares	Degrees of freedom	Mean square
Due to regression	113.0088	$k = 2$	56.5044
Residual	5.7912	5 − 2 − 1 = 2	2.8956
Total	118.8	5 − 1 = 4	29.7

so the adjusted coefficient of determination can be redefined as

$$\overline{R}^2 = 1 - \frac{\text{unexplained variance}}{\text{total variance}} \tag{2.16'}$$

Using the example of the last section, we can calculate the analysis of variance of Table 2.4 as shown in Table 2.5.

From Table 2.5, we can calculate R^2 and $\overline{R}^2$.

$$R^2 = 113.0088/118.8 = 0.95125$$

$$\overline{R}^2 = 1 - \frac{2.8956}{29.7} = 1 - 0.09749 = 0.90251$$

Both R^2 and $\overline{R}^2$ imply that more than 90% of the variation of annual salary can be explained by years of education and years of work experience. However, $\overline{R}^2$ is 4.874% smaller than that of R^2.

2.5 Tests on Sets and Individual Regression Coefficients

After having estimated the regression model, we would like to know whether the dependent variable is related to the independent variables. To find out, we can test whether an individual regression coefficient or a set of regression coefficients is significantly different from zero. The t-statistic is to test an individual coefficient, and the F-statistic to test linear restrictions on the parameters or regression coefficients. For this purpose, we need to assume that ε_i is normally distributed.

Logically, we perform the joint test first. If the joint test is not significant, then there is no need for the individual tests, and we normally abandon or modify the model. If the joint test is rejected, we must find out which regression coefficients are significant, so we perform individual tests.

Test on Sets of Regression Coefficients

Until now, our discussion has been limited to point estimation of multiple regression coefficients, the coefficient of determination, and the standard error of estimate. Now we will discuss how to use the F-statistic to test whether all true population regression (slope) coefficients equal zero. The F-test rather than the t-test is used. The null hypothesis for our case is

$$\begin{aligned} &H_0 : \beta_1 = \beta_2 = \cdots = \beta_k = 0 \\ &H_1 : \text{At least one } \beta \text{ is not zero.} \end{aligned} \tag{2.18}$$

If the null hypothesis is not true, then each $\hat{y}_i$ will differ from $\bar{y}$ substantially, and the explained variation $\sum_{i=1}^{n} (\hat{y}_i - \bar{y})^2$ will be large relative to the unexplained residual variation $\sum_{i=1}^{n} (y_i - \hat{y}_i)^2$. In other words, the R^2 indicated in Eq. (2.15) is relatively large. Thus, we can construct the F ratio as indicated in Eq. (2.19) to test whether the null hypothesis can be rejected.

$$F_{k,n-k-1} = \frac{\sum_{i=1}^{n} (\hat{y}_i - \bar{y})^2 / k}{\sum_{i=1}^{n} (y_i - \hat{y}_i)^2 / (n - k - 1)} \tag{2.19}$$

The F ratio we have constructed is the ratio of two mean square errors, as we noted in the last section, and they are two unbiased estimates of variances. Following the definition of the F distribution, we know that the F ratio has an F distribution with k and $(n - k - 1)$ degrees of freedom. This F ratio enables us to test whether at least one of the regression coefficients is significantly different from zero.

Consider the case $k = 2$. If there is no regression relationship (i.e., if $\beta_1 = \beta_2 = 0$) and because

$$\begin{aligned} \hat{y}_i &= a + b_1 x_{1i} + b_2 x_{2i} \\ &= \bar{y} + b_1 (x_{1i} - \bar{x}_1) + b_2 (x_{2i} - \bar{x}_2) \end{aligned}$$

the $\hat{y}_i$ will be close or equal to $\bar{y}$, so the F-value will be smaller or close to zero. Thus, we cannot reject the null hypothesis that ail regression coefficients are insignificantly different from zero.

Substituting related data from Table 2.5 into Eq. (2.19), we obtain

$$\begin{aligned} F &= \frac{113.0088/2}{5.7912/2} = \frac{56.5044}{2.8956} \\ &= 19.514 \end{aligned}$$

From F-distribution critical value table, we find that the critical value for a significance level of $\alpha = 0.05$ is $F_{0.05,2.2} = 19.0$, which is smaller than 19.514. Therefore, we can conclude that at least one of the regression coefficients is significantly different from zero. Thus, there is a regression relationship in the population, and the improvement of explanatory power achieved by fitting a regression plane is not due to chance. In other words, the null hypothesis that years of education and years of work experience contribute nothing to an individual's annual salary is rejected at a 5 percent level of significance.

Finally, the relationship between the R^2 indicated in Eq. (2.15) and the F-statistic in Eq. (2.19) can be shown to be[6]

$$F_{k,n-k-1} = \frac{n - k - 1}{k} \cdot \frac{R^2}{1 - R^2}$$

Hypothesis Tests for Individual Regression Coefficients

In the last section, we used the F-statistic to do a joint test about a regression relationship. Now we want to use the t-statistic to test whether multiple regression coefficients are significantly different from zero.

---

[6]Because $R^2 = 1 - \text{SSE}/\text{SST} = \text{SSR}/\text{SST}$,

$$\frac{R^2}{1 - R^2} = \frac{\text{SSR}/\text{SST}}{\text{SSE}/\text{SST}} = \frac{\text{SSR}}{\text{SSE}}$$

Hypothesis Testing Specification

We follow the procedure of the last chapter to define the null hypothesis and alternative hypothesis for testing individual multiple regression coefficients.

1. Two-tailed test

$$\begin{aligned} H_0 &: \beta_j = 0 \quad (j = 1, 2, \ldots, k) \\ \mathrm{H}_1 &: \beta_j \neq 0 \end{aligned} \tag{2.20}$$

2. One-tailed test

$$\begin{aligned} H_0 &: \beta_j = 0 \quad (j = 1, 2, \ldots, k) \\ \mathrm{H}_1 &: \beta_j > 0 \text{ or } \beta_j < 0 \end{aligned} \tag{2.21}$$

Let us look at Eq. (2.12) as an example. For convenience, the estimated regression line is repeated here.

$$\hat{y}_i = 0.980 + 1.2385x_1 + 0.9925x_2$$

In this equation, besides the estimated intercept (α) and slopes (β_1 and β_2), we have estimated the standard error of estimate for $\bar{y}_i$ as s_e = 1.7016. To perform the null hypothesis test, we need to know the sample distribution of β_j and the t-statistic as defined in the equation

$$t_{n-k-1} = (b_j - 0)/s_{bj} \tag{2.22}$$

where t_{n-k-1} represents a t-statistic with $(n - k - 1)$ degrees of freedom, k = the number of independent variables, and S_{bj} represents the standard error associated with bj. The concepts and procedure used to calculate S_{bj} are similar to those used for simple regression. However, S_{bj} is quite tedious to calculate by hand; fortunately, its value is readily available in the computer output of any standard regression analysis program. Thus in practice, we find t simply by finding the ratio of the coefficient to its estimated standard error. When the calculated value of t exceeds the critical value $t_{\alpha,n-k-1}$ indicated in the t-distribution table, the null hypothesis of no significance can be rejected. We conclude that the jth independent variable x_j does have an important influence on the dependent variable y_i after the influence of all other independent variables in the model is taken into account.

Performing the t-test for Multiple Regression Slopes

To perform the t-test for multiple regression coefficients b_1 and b_2, we estimate the sample variance of the coefficients b_1 and b_2 in accordance with Eqs. (2.23) and (2.24)[7]

$$\begin{aligned} \mathrm{Var}(b_1) = s_{b_1}^2 &= \frac{s_e^2}{(1 - r^2)\sum_{i=1}^{n}(x_{1i} - \bar{x}_1)^2} \\ &= \frac{s_e^2\left(\sum_{i=1}^{n} x_{2i}^{\prime 2}\right)}{\left(\sum_{i=1}^{n} x_{1i}^{\prime 2}\right)\left(\sum_{i=1}^{n} x_{2i}^{\prime 2}\right) - \left(\sum_{i=1}^{n} x_{1i}^{\prime} x_{2i}^{\prime}\right)^2} \end{aligned} \tag{2.23}$$

$$\begin{aligned} \mathrm{Var}(b_2) = s_{b_2}^2 &= \frac{s_e^2}{(1 - r^2)\sum_{i=1}^{n}(x_{2i} - \bar{x}_2)^2} \\ &= \frac{s_e^2\left(\sum_{i=1}^{n} x_{1i}^{\prime 2}\right)}{\left(\sum_{i=1}^{n} x_{1i}^{\prime 2}\right)\left(\sum_{i=1}^{n} x_{2i}^{\prime 2}\right) - \left(\sum_{i=1}^{n} x_{1i}^{\prime} x_{2i}^{\prime}\right)^2} \end{aligned} \tag{2.24}$$

where r represents the correlation coefficient between x_{1i} and x_{2i}. If the magnitude of r is great, a collinearity problem might exist. This issue will be discussed in detail in the next chapter.

Substituting the required numerical values obtained from Tables 2.2 and 2.3, we calculate sample variances of b_1 and b_2 for Eq. (2.16).

$$\begin{aligned} S_{b_1}^2 &= \frac{(2.8956)(50)}{(59.2)(50) - (-11)^2} \\ &= \frac{(2.8956)(50)}{2839} \\ &= 0.05100 \end{aligned}$$

and

$$\begin{aligned} s_{b_2}^2 &= \frac{(2.8956)(59.2)}{2839} \\ &= 0.06038 \end{aligned}$$

[7]Derivations of Eqs. (2.23) and (2.24) can be found in Appendix 1. Note that these two equations are generally estimated by computer packages (see Sect. 2.8). Manual approaches are presented here to show how sample variances of multiple regression slopes are actually calculated.

Then $S_{b_1} = 0.2258$ and $S_{b_2} = 0.2457$. Dividing b_1 and b_2 by S_{b_1} and S_{b_2}, we obtain t-values for b_1 and b_2.

$$t'_{b_1} = \frac{1.2385}{0.2258} = 5.4849$$
$$t_{b_2} = \frac{0.9925}{0.2457} = 4.0395$$

Because $n = 5$ and $k = 2$, from t-distribution critical value table, the critical value for a one-tailed test on either coefficient (at a significance level of $\alpha = 0.05$) is

$$t_{\alpha,n-k-1} = t_{0.05,2} = 2.920$$

We choose a one-tailed test because a priori theoretical propositions were that both x_1 and x_2 were positively related to y. Comparing 5.4849 and 4.0395 with 2.920, we conclude that both years of education and years of work experience are significantly related to an individual's annual salary.

Figure 2.2 presents all the estimates and hypothesis testing information we have discussed in the last three sections. This example certainly

```
MTB > READ C1-C3
DATA> 5 7 15
DATA> 10 5 17
DATA> 9 14 26
DATA> 13 8 24
DATA> 15 6 27
DATA> END
      5 rows read.
MTB > REGRESS C3 2 C1 C2;
SUBC> DW;
SUBC> PREDICT 6 5.
```

Regression Analysis

```
The regression equation is
C3 = 0.98 + 1.24 C1 + 0.992 C2

Predictor        Coef        StDev          T        P
Constant        0.980        3.439       0.29    0.802
C1             1.2385       0.2258       5.48    0.032
C2             0.9925       0.2457       4.04    0.056

S = 1.702       R-Sq = 95.1%      R-Sq(adj) = 90.3%

Analysis of Variance

Source        DF          SS          MS         F        P
Regression     2     113.009      56.504     19.51    0.049
Error          2       5.791       2.896
Total          4     118.800

Source        DF      Seq SS
C1             1      65.773
C2             1      47.236

Obs        C1         C3        Fit  StDev Fit   Residual    St Resid
  1       5.0     15.000     14.120      1.499      0.880        1.09
  2      10.0     17.000     18.327      1.076     -1.327       -1.01
  3       9.0     26.000     26.021      1.632     -0.021       -0.04
  4      13.0     24.000     25.020      0.961     -1.020       -0.73
  5      15.0     27.000     25.512      1.301      1.488        1.36

Durbin-Watson statistic = 2.39
```

Fig. 2.2 MINITAB output of multiple regression in terms of data given in Table 2.1

proves that multiple regression analysis can be more efficiently performed by using the MINITAB computer program.

2.6 Confidence Interval for the Mean Response and Prediction Interval for the Individual Response

Point Estimates of the Mean and the Individual Responses One of the important uses of the multiple regression line is to obtain predictions and forecasts for the dependent variable, given an assumed set of values of the independent variables. This kind of prediction is called the conditional prediction (forecast), just as in simple regression. Suppose the independent variables are equal to some specified values $x_{1,n+1}$ and $x_{2,n+1}$, and that the linear relationship among y_n, $x_{1,n}$, and $x_{2,n}$ continues to hold.[8] Then the corresponding value of the dependent variable $Y_{n+1,i}$ is

$$Y_{n+1,i} = \alpha + \beta_1 x_{1,n+1,i} + \beta_2 x_{2,n+1,i} + \varepsilon_{n+1,i} \tag{2.25}$$

which, given $x_{1,n+1}$ and $x_{2,n+1}$, has expectation

$$E(Y_{n+1}|x_{1,n+1}, x_{2,n+1}) = \alpha + \beta_1 x_{1,n+1} + \beta_2 x_{2,n+1} \tag{2.26}$$

Equation (2.26) yields the mean response $E(Y_{n+1}|x_{1,n+1}, x_{2,n+1})$ that we want to estimate when the independent variables are fixed at $x_{1,n+1}$ and $x_{2,n+1}$. Equation (2.25) yields the actual value (or individual response) that we want to predict.

To obtain the best point estimate, we first estimate the sample regression line as defined in Eq. (2.2). Then we substitute the given values $x_{1,n+1}$ and $x_{2,n+1}$ into the estimated Eq. (2.12), obtaining

$$\hat{y}_{n+1} = a + b_1 x_{1,n+1} + b_2 x_{2,n+1} \tag{2.27}$$

This is the best point estimate for both conditional expectation and actual-value forecasts. In other words, the forecast of conditional expectation value is equal to the forecast of actual value. However, the forecasts are interpreted differently. The importance of these different interpretations will emerge when we investigate the process of making interval estimates.

Interval Estimates of Forecasts
To construct a confidence interval for forecasts, it is necessary to know the distribution, mean, and variance of $\hat{y}_{n+1}$. The distribution of $\hat{y}_{n+1}$ is a t-distribution with $(n-3)$ degrees of freedom. The variance associated with $\hat{y}_{n+1}$ may be classified into three cases. First, we deal with a case in which the conditional mean $(\hat{y}_{n+1})$ is equal to the unconditional mean $(\bar{y})$. In the second and third cases, we deal with the conditional mean. However, case 2 involves the mean response and case 3 the individual response.

Case 2.1 [*Conditional Expectation (Mean Response) with* $\bar{x}_{1,n+1} = \bar{x}_1$ *and* $\bar{x}_{2,n+1} = \bar{x}_2$]

From the definitions of the intercept of a regression and the sample regression line, we have

$$\begin{aligned}\hat{y}_{n+1} &= (\bar{y} - b_1\bar{x}_1 - b_2\bar{x}_2) + b_1 x_{1,n+1} + b_2 x_{2,n+1} \\ &= \bar{y} + b_1(x_{1,n+1} - \bar{x}_1) + b_2(x_{2,n+1} - \bar{x}_2)\end{aligned}$$

If $x_{1,n} = \bar{x}_1$ and $x_{2,n} = \bar{x}_2$, then $\hat{y}_{n+1} = \bar{y}$. We can obtain the estimate of the variance for y_{n+1} as

$$s^2(\hat{y}_{n+1}) = s^2(\bar{y}) = s_e^2/n \tag{2.28}$$

Case 2.2 [*Conditional Expectation (Mean Response) with* $x_{1,n+1} \neq \bar{x}_1$ *and* $x_{2,n+1} \neq \bar{x}_2$]

In this case, the forecast value can be defined as

$$\hat{y}_{n+1} = \bar{y} + b_1(x_{1,n+1} - \bar{x}_1) + b_2(x_{2,n+1} - \bar{x}_2) \tag{2.29}$$

[8] $x_{1,n+1}$, and $x_{2,n+1}$, can be either given values or forecasted values. When a regression is used to describe a time-series relationship, they are forecasted values.

We therefore obtain the estimate of the variance for $\hat{y}_{n+1}$ in terms of sample standard variance of estimates S_e^2 as

$$s_1^2 = s^2(\hat{y}_{n+1}) = s_e^2\left[\frac{1}{n} + \frac{(x_{1,n+1}-\bar{x}_1)^2}{(1-r^2)C_1^2} + \frac{(x_{2,n+1}-\bar{x}_2)^2}{(1-r^2)C_2^2} - \frac{2(x_{1,n+1}-\bar{x}_1)(x_{2,n+1}-\bar{x}_2)r}{(1-r^2)C_1C_2}\right] \quad (2.30)$$

where $C_1 = \sqrt{\sum_{i=1}^{n}(x_{1,j}-\bar{x}_1)^2}$ $C_2 = \sqrt{\sum_{i=1}^{n}(x_{2,i}-\bar{x}_2)^2}$ and r = correlation coefficient between $x_{1,i}$ and $x_{2,i}$

Case 2.3 [*Actual Value (Individual Response) of* $\boldsymbol{y_{n+1}}$]

After we have derived the sample variance for $\hat{y}_{n+1}$, we derive the sample variance for individual response (observation), $y_{n+1,i}$ (which deviates from $\hat{y}_{n+1}$ by a random error e_i).

$$y_{n+1,j} = \hat{y}_{n+1} + e_i$$

The variance of an individual observation, $y_{n+1,i}$, includes the variance of the observation about the regression line (s_e^2) as well as $s^2(\hat{y}_{n+1,i})$. Because $\hat{y}_{n+1}$ and e_i are independent, $s^2(y_{n+1,j}) = s^2(\hat{y}_{n+1}) + s_e^2$. More explicitly,

$$s^2(\hat{y}_{n+1,i}) = s_1^2 + s_e^2 = s_2^2 \quad (2.31)$$

where s_1^2 is defined in Eq. (2.30).

Using Eqs. (2.28), (2.30), and (2.31), we can obtain a confidence interval for prediction as follows:

1. For prediction of the conditional expectation with $x_{1,n+1} = \bar{x}_1$ and $x_{2,n+1} = \bar{x}_2$, the confidence interval is

$$\hat{y}_{n+1} \pm t_{n-3,\alpha/2}\frac{s_e}{\sqrt{n}} \quad (2.32)$$

2. For prediction of the conditional expectation with $x_{1,n+1} \neq \bar{x}_2$ and $x_{2,n+1} = \bar{x}_2$, the confidence interval is

$$\hat{y}_{n+1} \pm (t_{n-3,\alpha/2})S_1 \quad (2.33)$$

where s_1 is defined in Eq. (2.30).

3. For prediction of the actual value y_{n+1}, the prediction interval is

$$\hat{y}_{n+1,i} \pm (t_{n-3,\alpha/2})S_2 \quad (2.34)$$

where S_2 is defined in Eq. (2.31).

To show how Eq. (2.34) is applied in constructing the confidence interval for forecasting the actual value of y_{n+1}, let us use the annual salary example (Table 2.2) to find the 95% prediction interval for annual salary, y_{n+1}, when a person has 6 years of education and 5 years of work experience. The predicted annual salary can be computed from Eq. (2.12).

$$\hat{y}_{n+1,i} = 0.980 + (1.2385)(6) + (0.9925)(5) = 13.3735 \text{ (in thousands of dollars)}$$

From Table 2.2, we have

$$\sum_{i=1}^{n}(x_{1i}-\bar{x}_1)^2 = 59.2,$$
$$\sum_{i=1}^{n}(x_{2i}-\bar{x}_2)^2 = 50, \quad n = 5$$
$$\sum_{i=1}^{n}(x_{1i}-\bar{x}_i)(x_{2i}-\bar{x}_2) = -11,$$
$$\bar{x}_1 = 10.4, \quad \bar{x}_2 = 8$$

Using this information, we calculate

$$C_1 = \sqrt{59.2} = 7.6942, \quad C_2 = \sqrt{50} = 7.0711$$

$$r = \frac{\sum_{i=1}^{n}(x_{1i}-\bar{x}_1)(x_{2i}-\bar{x}_2)}{\sqrt{\sum_{i=1}^{n}(x_{1i}-\bar{x}_1)^2\sum_{i=1}^{n}(x_{2i}-\bar{x}_2)^2}} = \frac{-11}{(7.6942)(7.0711)} = -0.2022, \quad r^2 = 0.0409$$

$$(x_{1,n+1}-\bar{x}_1)^2 = (6-10.4)^2 = 19.36$$
$$(x_{2,n+1}-\bar{x}_2)^2 = (5-8)^2 = 9$$

From Table 2.3, we have $S_e^2 = 5.7912/(5-3) = 2.8956$. Substituting this information into Eq. (2.31) yields

$$s_2^2 = (2.8956)\left[1 + \frac{1}{5} + \frac{19.36}{(1-0.0409)(59.2)} + \frac{9}{(1-0.0409)(50)} + \frac{9}{(1-0.0409)(50)} - \frac{2(-0.2022)(-4.4)(-3)}{(1-0.0409)(7.6942)(7.0711)}\right] = (2.8956)(1.83) = 5.2989$$

We will use $n = 5$, $s_2 = \sqrt{5.2989} = 2.3019$, and $\hat{y}_{n+1,i} = 13.3735$. From t-distribution critical value table, we have $t_{2,025} = 4.303$. Substituting this information into Eq. (2.34), we find that the annual salary is predicted with 95% confidence by the interval

$$13.3735 \pm (4.303)(2.3019) = 13.3735 \pm 9.9051$$
$$3.4684 \le y_{n+1,i} \le 23.2786$$

```
MTB > READ C1-C3
DATA> 5 7 15
DATA> 10 5 17
DATA> 9 14 26
DATA> 13 8 24
DATA> 15 6 27
DATA> END
      5 rows read.
MTB > BRIEF 3
MTB > REGRESS C3 2 C1 C2;
SUBC> DW.
```

Regression Analysis

```
The regression equation is
C3 = 0.98 + 1.24 C1 + 0.992 C2

Predictor        Coef        StDev          T        P
Constant        0.980        3.439       0.29    0.802
C1             1.2385       0.2258       5.48    0.032
C2             0.9925       0.2457       4.04    0.056

S = 1.702       R-Sq = 95.1%      R-Sq(adj) = 90.3%

Analysis of Variance

Source        DF        SS         MS        F        P
Regression     2   113.009     56.504    19.51    0.049
Error          2     5.791      2.896
Total          4   118.800

Source        DF    Seq SS
C1             1    65.773
C2             1    47.236

Obs        C1        C3        Fit   StDev Fit   Residual    St Resid
  1       5.0    15.000     14.120       1.499      0.880        1.09
  2      10.0    17.000     18.327       1.076     -1.327       -1.01
  3       9.0    26.000     26.021       1.632     -0.021       -0.04
  4      13.0    24.000     25.020       0.961     -1.020       -0.73
  5      15.0    27.000     25.512       1.301      1.488        1.36

Durbin-Watson statistic = 2.39

     Fit  StDev Fit        95.0% CI              95.0% PI
  13.373      1.551   (  6.699,  20.047)   (  3.466,  23.280)
```

Fig. 2.3 MINITAB output of $y_{n+1,i}$

When n is large, we can modify this expression by replacing t with the appropriate normal deviate z.

MINITAB output showing prediction results of $x_{1,n+1,i=6}$ and $x_{2,n+1,i=5}$ is presented in Fig. 2.3. The prediction interval shown in the last row of Fig. 2.3 is (3.466, 23.280), which is similar to what we calculated before.

In the next two sections, we will explore applications of multiple regression in business and economics. Section 2.8 explicitly treats the use of SAS and MINITAB computer programs to do multiple regression analyses.

2.7 Business and Economic Applications

Multiple regression analysis has been widely used in decision making in business and economics. Five examples are discussed in this section.

Application 2.1 Overall Job-Worth of Performance for Certain Army Jobs
Bobko and Donnelly (1988) employed multiple regression to estimate overall job-worth to the army of certain army jobs from attributes of those jobs.[9] Their final regression prediction model is

$$y_i = b_0 + b_1x_{1i} + b_2x_{2i} + b_3x_{3i} + b_4x_{4i} + b_5x_{5i} + b_6x_{6i} + b_7x_{7i} + e_i$$

where

y_i = job-value judgments of overall worth for the ith individual
x_{1i} = performance level for the ith job
x_{2i} = combat probability for the ith job
x_{3i} = enlistment bonus for the ith job
x_{4i} = reenlistment bonus for the ith job
x_{5i} = aptitude required for entry into the ith job
x_{6i} = cost of error for the ith job
x_{7i} = job variety for the ith job

[9] P. Bobko and L. Donnelly (1988), "Identifying Correlations of Job-Level, Overall Worth Estimates: Application in a Public Sector Organization," Human Performance 3, 187–204.

Bobko and Donnelly estimated this multiple regression model using data obtained from interviews. Their regression results are presented in Table 2.6. As would be expected, performance level was the single best predictor of 95 estimates of judgments of overall worth. The other job-level correlates were combat probability, enlistment bonus, reenlistment bonus, aptitude, cost of error, and task variety. The first six of these predictors had statistically significant regression weights (coefficients), p-value < 0.05, indicating their unique contribution to the prediction of overall worth estimates. However, task variety was not statistically significant.

Application 2.2 The Relationship Between Individual Stock Rates of Return, Payout Ratio, and Market Rates of Return
To demonstrate multiple regression analysis, a time-series regression for 1970–2009 is run, the dependent variable being the rate of return for the JNJ stock ($R_{j,t}$) and the independent variables being the payout ratio (dividend per share/earnings per share) for JNJ ($P_{j,j}$) and the rates of return on the S&P 500 index, $R_{m,t}$. The results are as follows:

$$R_{j,t} = \alpha_j + \gamma_j P_{j,t} + \beta_j R_{m,t} + \epsilon_{j,t}$$

Fortunately, the results do not have to be calculated by hand but can be obtained by using MINITAB. The MINITAB results are presented in Table 2.7. The parameter value for the market rates of return is 0.7329, which is called the beta coefficient. A 1% increase in the market rate of return will lead to a 0.7329% change in the rate of return of the JNJ stock, given the payout ratio, that is, the rate of return of JNJ stock is less volatile than that of the market. The payout ratio has a coefficient of −0.2133. This result implies that a 1% percent increase in the payout ratio will lead to a 0.2133 percent decrease in the mean rate of return on JNJ stock, given the market rate of return.

The independent variables are statistically significant at the 5% level. The t-value for the market is 2.0880, and the associated p-value is 0.0437, which means that the lowest level of significance at which the null hypothesis can be rejected is 4.37%. This suggests that the

Table 2.6 Best subset regression of overall worth on job-level predictors

Source	df	Sum of squares	F	P
Regression	7	16.007	274.12	0.0001
Error (residual)	87	0.726		

Variable	Regression weight	t-ratio	p
Performance level	0.013	1666.22	0.0001
Combat probability	0.039	21.19	0.0001
Enlistment bonus	0.034	18.52	0.0001
Re-enlistment bonus	0.016	15.73	0.0001
Aptitude	0.013	26.01	0.0001
Cost of error	0.029	5.61	0.0201
Task variety	0.016	2.51	0.1166

Source Bobko and Donnelly (1988), Human Performance
Note Adjusted $R^2 = 0.953$; $n = 95$ mean estimates of overall worth

Table 2.7 $R_{j,t} = \alpha_j + \gamma_j P_{j,t} + \beta_j R_{m,t} + \varepsilon_{j,t}$

Variable	Coefficient	Standard error	t-value	p-value
Constant	0.0777	0.2049	0.3793	0.7066
Payout ratio	−0.2133	0.5613	−0.3800	0.7061
Market rate of return	0.7329	0.3510	2.0880	0.0437
$R^2 = 0.106$				
$\overline{R}^2 = 0.057$				
F-value = 2.184				
Observations 40				

population coefficient for the market is not equal to zero. The t-statistic for the payout ratio, which is calculated by dividing the parameter value (−0.2133) by the standard error (0.5613), is −0.3800. Its p-value is 0.7061; thus, the null hypothesis cannot be rejected.

R^2 for the regression is 0.106. In other words, the independent variables explain about 10.6% of the variation in the rate of return on JNJ stock. The adjusted R-square, $\overline{R}^2$, which takes into account overfitting in the sample, is equal to 0.057.

The F-value, which tests the hypothesis that the population coefficients of the independent variables are both zero against the alternative that they are not, is equal to 2.184. The degrees of freedom associated with this F-value are $v_1 = 2$ and $v_2 = 37$. From F-distribution critical value table, we find that the critical value for the F-test is $F_{0.01,2,30} = 5.39$ and $F_{0.01,2,40} = 5.18$. Because the F-value for the regression is less than the critical value 5.39, the null hypothesis cannot be rejected.

Application 2.3 Analyzing the Determination of Price per Share To further demonstrate multiple regression techniques, let us say that a cross-sectional regression is run. In a cross-sectional regression, all data come from a single period. The dependent variable in this regression is the price per share (P_j) of the 30 firms used to compile the Dow Jones Industrial Average for the year 2009. The independent variables are the dividend per share (DPS_j) and the retained earnings per share (EPS_j) for the 30 firms. (Retained earnings per share is defined as earnings per share minus dividend per share.

Table 2.8 $P_j = a + b_1\text{DPS}_j + b_2\text{ EPS}_j + e_j$

Variable	Coefficient	Standard error	*t*-value	*p*-value
Constant	12.800	5.084	2.518	0.018
DPS	12.836	4.657	2.756	0.010
EPS	0.978	0.218	4.478	0.000
$R_2 = 0.724$				
$\bar{R}^2 = 0.703$				
F-value = 35.35				
Observations 30				

Price per share is the close price of the end of year 2009; dividend per share and retained earnings per share are based on 2009 annual balance sheet and income statement.) The sample regression relationship is

$$P_j = a + b_1 DPS_j + b_2\text{EPS}_j + e_j \quad (j = 1, 2, \ldots, 30)$$

Empirical results are presented in Table 2.8. The constant term is significant with a *t*-value of 2.518. This result means that the intercept term is statistically different from zero and the null hypothesis can be rejected at both a 10 and a 5% level. The retained earnings per share variable is highly significant with a *t*-value of 4.478 and a *p*-value of 0.000. Thus, we can reject the null hypothesis that the coefficient is equal to zero and accept the alternative hypothesis that it differs from zero and makes a contribution to price per share. The coefficient for this variable is 0.978; mean price per share increases \$0.978 when the retained earnings per share increases by \$1.00, given the dividend.

The coefficient for the dividend per share variable has a *t*-value of 2.756 and a *p*-value of 0.010. This is the lowest level of significance at which the null hypothesis can be rejected; thus, the null hypothesis is rejected at both a 10 and a 5% level. The coefficient for dividend per share is 12.836. When the dividend increases by \$1.00, the price per share tends to rise by \$12.836.

The value of R_2 is 0.724, which means that the model explains 72.4% of the observed fluctuations in the price per share. The adjusted *R*-square, $\bar{R}^2$, is 0.703. The *F*-value for the regression is 35.35. The numbers of degrees of freedom for the regression and residual are 2 and 27, respectively. The critical value for *F* at a 1% level of significance is 5.49. Because the regression *F*-value is greater than the critical value, the null hypothesis that the coefficients are equal to zero is rejected.

The relationship among price per share, dividend per share, and retained earning per share will be discussed in Appendix 2.

Application 2.4 Multiple Regression Approach to Evaluating Real Estate Property

To show how the multiple regression technique can be used by real estate appraisers, Andrews and Ferguson (1986) used the data in Table 2.9 to do the multiple regression analysis

$$y_i = b_0 + b_1 x_{1i} + b_2 x_{2i} + e_i$$

where

y_i = sale price for *i*th house
x_{1i} = home size for *i*th house
x_{2i} = condition rating for *i*th house

Table 2.9 Sale price, house size, and condition rating

Sale price, *y* (thousands of dollars)	Home size, x_1 (hundreds of sq. ft.)	Condition rating, x_2 (1–10)
60.0	23	5
32.7	11	2
57.7	20	9
45.5	17	3
47.0	15	8
55.3	21	4
64.5	24	7
42.6	13	6
54.5	19	7
57.5	25	2

Source R. L. Andrews and J. T. Ferguson, "Integrating Judgment with a Regression Appraisal." The Real Estate Appraiser and Analyst, Vol. 52, No. 2, Spring 1986 (Table 1)

```
MTB > READ C1-C3
DATA> 60.0 23 5
DATA> 32.7 11 2
DATA> 57.7 20 9
DATA> 45.5 17 3
DATA> 47.0 15 8
DATA> 55.3 21 4
DATA> 64.5 24 7
DATA> 42.6 13 6
DATA> 54.5 19 7
DATA> 57.5 25 2
DATA> END
      10 rows read.
MTB > BRIEF 1
MTB > REGRESS C1 2 C2 C3;
SUBC> DW.
```

Regression Analysis

```
The regression equation is
C1 = 9.78 + 1.87 C2 + 1.28 C3

Predictor        Coef          StDev          T        P
Constant        9.782          1.630        6.00    0.000
C2            1.87094        0.07617       24.56    0.000
C3             1.2781         0.1444        8.85    0.000

S = 1.081        R-Sq = 99.0%       R-Sq(adj) = 98.7%

Analysis of Variance

Source        DF          SS          MS          F        P
Regression     2      819.33      409.66     350.87    0.000
Error          7        8.17        1.17
Total          9      827.50


Durbin-Watson statistic = 1.56

MTB > CORRELATION C1-C3
```

Correlations (Pearson)

```
             C1        C2
C2        0.938
C3        0.373     0.043
```

Fig. 2.4 MINITAB output for Table 2.9

MINITAB regression outputs in terms of Table 2.9 are presented in Fig. 2.4. From this output, the estimated regression is

$$\begin{aligned}\hat{y}_i &= 9.782 + 1.87094x_{1i} + 1.2781x_{2i}\\ &\quad(6.00)\quad(24.56)\qquad(8.85)\end{aligned}$$

t-Values are in parenthesis.

From t-distribution critical value table, we find that $t_{0.025,7} = 2.365$. Because t-values for three regression parameters are larger than 2.365, all estimated parameters are significantly different from 0 at $\alpha = 0.05$. This estimated regression can be used to estimate the sale price for a house. For example, if $x_1 = 18$ and $x_2 = 5$, the predicted sale price is

$$\hat{y}_i = 9.781 + (1.87094)(18) + (1.2781)(5) = 49.8484$$

This implies that the estimated sale price is \$49,848.4 if the home size is 18,000 square feet and the condition rating is 5.

Application 2.5 Multiple Regression Approach to Doing Cost Analysis

To show how the multiple regression technique can be used to do cost analysis by accountants, we look at Benston's research. Benston (1966) used a set of sample data (as shown in Table 2.10) from a firm's accounting and production records to provide cost information about the firm's shipping department to do the multiple regression analysis

$$y_t = b_0 + b_1 x_{1t} + b_2 x_{2t} + b_3 x_{3t} + e_t$$

where

y_t = hours of labor in tth week
x_{1t} = thousands of pounds shipped in tth week

Table 2.10 Hours of labor and related factors cause costs to be incurred

Week	Hours of labor, y	Thousands of pounds shipped, x_1	Percentage of units shipped by truck, x_2	Average number of pounds per shipment, x_3
1	100	5.1	90	20
2	85	3.8	99	22
3	108	5.3	58	19
4	116	7.5	16	15
5	92	4.5	54	20
6	63	3.3	42	26
7	79	5.3	12	25
8	101	5.9	32	21
9	88	4.0	56	24
10	71	4.2	64	29
11	122	6.8	78	10
12	85	3.9	90	30
13	50	3.8	74	28
14	114	7.5	89	14
15	104	4.5	90	21
16	111	6.0	40	20
17	110	8.1	55	16
18	100	2.9	64	19
19	82	4.0	35	23
20	85	4.8	58	25

Source G. J. Benston (1966), "Multiple Regression Analysis of Cost Behavior," Accounting Review, Vol. 41, No. 4, 657–672. Reprinted by permission of the publisher

x_{2t} = percentage of units shipped by truck in tth week

x_{3t} = average number of pounds per shipment in tth week

MINITAB regression output is presented in Fig. 2.5. From p-values indicated in Fig. 2.5, we find that b_0 and b_3 are significantly different from 0 at $\alpha = 0.01$. Hence, we can conclude

```
MTB > READ C1-C4
DATA> 100 5.1 90 20
DATA> 85 3.8 99 22
DATA> 108 5.3 58 19
DATA> 116 7.5 16 15
DATA> 92 4.5 54 20
DATA> 63 3.3 42 26
DATA> 79 5.3 12 25
DATA> 101 5.9 32 21
DATA> 88 4.0 56 24
DATA> 71 4.2 64 29
DATA> 122 6.8 78 10
DATA> 85 3.9 90 30
DATA> 50 3.8 74  28
DATA> 114 7.5 89 14
DATA> 104 4.5 90 21
DATA> 111 6.0 40 20
DATA> 110 8.1 55 16
DATA> 100 2.9 64 19
DATA> 82 4.0 35 23
DATA> 85 4.8  58 25
DATA> END
     20 rows read.
MTB > REGRESS C1 3 C2 C3 C4;
SUBC> DW.
```

Regression Analysis

```
The regression equation is
C1 = 132 + 2.73 C2 + 0.0472 C3 - 2.59 C4

Predictor         Coef         StDev           T         P
Constant        131.92         25.69        5.13     0.000
C2               2.726         2.275        1.20     0.248
C3             0.04722       0.09335        0.51     0.620
C4             -2.5874        0.6428       -4.03     0.001

S = 9.810        R-Sq = 77.0%      R-Sq(adj) = 72.7%

Analysis of Variance

Source        DF          SS           MS          F         P
Regression     3      5158.3       1719.4      17.87     0.000
Error         16      1539.9         96.2
Total         19      6698.2


Durbin-Watson statistic = 2.43
```

Fig. 2.5 MINITAB output for Application 2.5

that the only important variable in determining the hours of labor required in the shipping department is the average number of pounds per shipment.

2.8 Using Computer Programs to Do Multiple Regression Analyses

2.8.1 SAS Program for Multiple Regression Analysis

In an example taken from Churchill's Marketing Research, data for the sales of click ballpoint pens (y), advertising (x_1, measured in TV spots per month), number of sales representatives (x_2), and a wholesaler efficiency index (x_3) were presented in Table 2.11.

We investigated not only the relationship between two variables (y and x_1, y and x_2, and y and x_3). Now we will expand that analysis by using the following three regression models:[10]

$$y_i = a + b_1 x_{1i} + e_i \tag{a}$$

$$y_i = a + b_1 x_{1i} + b_2 x_{2i} + e_i \tag{b}$$

$$y_i = a + b_1 x_{1i} + b_2 x_{2i} + b_3 x_{3i} + e_t \tag{c}$$

Equation (a) can be used to investigate the relationship between y and x_1.

Equation (b) can be used to analyze whether the second explanatory variable, x_2, improves the equation's power to explain the variation of sales. Equation (c) can be used to analyze whether the third explanatory variable, x_3, further improves that explanatory power. Part of the output of the SAS program for Eqs. (a–c) is presented in Fig. 2.6a–c. Figure 2.6a shows the regression results of Eq. (a), Fig. 2.6b the regression results of Eq. (b), and Fig. 2.6c the regression results of Eq. (c). Using these results, we will review and summarize simple regression and multiple regression results.

[10]In these regressions, we hold the price of a ballpoint pen and the income of a consumer constant, because this is a set of cross-sectional data.

Computer outputs of Fig. 2.6a–c present the following results of simple and multiple regression.

1. Estimated intercept and slopes;
2. F-values for the whole regression;
3. t-Values for individual regression coefficients;
4. ANOVA of regression;
5. R^2 and $\overline{R}^2$
6. p-Values;
7. Durbin–Watson D and first-order autocorrelation;
8. Standard error of residual estimate (mean square of error);
9. Root MSE $= \sqrt{\text{MSE error}}$. For example, for Eq. (b), Root MSE $= \sqrt{2039.85310} = 45.16473$. The root MSE estimate can be used to measure the performance of prediction.

These SAS regression outputs give us almost all the sample statistics we have examined so far. Now let us consider the practical implications of Eqs. (a–c).

Equation (b) specifies a regression model in which sales is the dependent variable and the independent variables are number of TV spots x_1 and number of sales representatives x_2. The fitted regression equation is

$$\hat{y} = \underset{(2.994)}{69.3} + \underset{(5.315)}{14.2x_1} + \underset{(5.393)}{37.5x_2} \quad F = 128.141$$

Here t-values are indicated in parentheses.

This regression indicates that when the number of TV spots increases by 1 unit, sales increase by \$14,200 on average. And when the number of sales representatives increases by 1 person, sales increase by \$37,500 on average.

The F-value for the regression of Eq. (b) is 128.141. There are 40 observations and 2 independent variables, so the number of degrees of freedom in the model is 40 − 2 − 1 = 37. By interpolation, it can be shown that the critical value of $F_{0.05,2,37}$ is 3.25 (F-distribution critical value table). Because the F-value for the regression is

Table 2.11 Territory data for click ballpoint pens

Territory	Sales, Y (thousands)	Advertising, X_1 (TV spots per month)	Number of sales representatives, X_2	Wholesaler efficiency index, X_3
005	260.3	5	3	4
019	286.1	7	5	2
033	279.4	6	3	3
039	410.8	9	4	4
061	438.2	12	6	1
082	315.3	8	3	4
091	565.1	11	7	3
101	570.0	16	8	2
115	426.1	13	4	3
118	315.0	7	3	4
133	403.6	10	6	1
149	220.5	4	4	1
162	343.6	9	4	3
164	644.6	17	8	4
178	520.4	19	7	2
187	329.5	9	3	2
189	426.0	11	6	4
205	343.2	8	3	3
222	450.4	13	5	4
237	421.8	14	5	2
242	245.6	7	4	4
251	503.3	16	6	3
260	375.7	9	5	3
266	265.5	5	3	3
279	620.6	18	6	4
298	450.5	18	5	3
306	270.1	5	3	2
332	368.0	7	6	2
347	556.1	12	7	1
358	570.0	13	6	4
362	318.5	8	4	3
370	260.2	6	3	2
391	667.0	16	8	2
408	618.3	19	8	2
412	525.3	17	7	4
430	332.2	10	4	3
442	393.2	12	5	3
467	283.5	8	3	3
471	376.2	10	5	4
488	481.8	12	5	2

Source C. A. Gilbert, in G. A. Churchill, Jr., Marketing *Research: Methodological Foundations,* 3rd ed., 1983, p. 563. Copyright © 1983 by the Dryden Press, reprinted by permission of the publisher

greater than the critical value, the hypothesis that the coefficients are equal to zero is rejected. From the t-values associated with estimated regression coefficients, we find that the estimated intercept and slopes are significant at $\alpha = 0\,0.01$.

Because the t-values of b_2 are significantly different from zero, we conclude that adding the number of sales representatives improves the equation's power to explain sales. This conclusion can also be drawn from the fact that $\overline{R}^2$ has increased from 0.7687 to 0.8670.

The fitted regression of Eq. (c) is

$$\begin{aligned} \hat{y} = {} & 31.1504 + 12.9682x_1 + 41.2456x_2 \\ & (0.911) \quad (4.738) \quad (5.666) \\ & + 11.5243x_3 \quad F = 89.051 \\ & (1.498) \end{aligned}$$

Again, t-values are indicated in parentheses.

Following Sect. 2.5, we first test the whole set of regression coefficients in terms of the F-statistic. From F-distribution critical value table, by interpolation, we find that $F_{0.05,3,36} = 2.88$. $F = 89.051$ is much larger than 2.88. This implies that we reject the following null hypothesis of our joint test $H_0 : \beta_1 = \beta_2 = \beta_3 = 0$

Now we can use t-statistics to test which individual coefficient is significantly different from zero. From t-distribution critical value table, by interpolation, we find that the critical value of t-statistic is $t_{0.005,36} = 2.72$. By comparing this critical value with 4.738, 5.666, and 1.498, we conclude that b_1 and b_2 are significantly different from zero and that b_3 is not significantly different from zero at $\alpha = 0.01$. In other words, the wholesaler efficiency index does not increase the explanatory power of Eq. (c).

MINITAB Program for Multiple Regression Prediction

MINITAB is used to run the regression defined in Fig. 2.6c and presented in Fig. 2.7. Besides regression parameters, we also predict y by assuming $x_1 = 13$, $x_2 = 9$, and $x_3 = 5$. The

(a)

```
Model: MODEL1
Dependent Variable: Y

                         Analysis of Variance

                          Sum of          Mean
Source          DF       Squares        Square      F Value      Prob>F

Model            1 463451.00888 463451.00888      130.644      0.0001
Error           38 134802.01487   3547.42144
C Total         39 598253.02375

    Root MSE       59.56023      R-square       0.7747
    Dep Mean      411.28750      Adj R-sq       0.7687
    C.V.           14.48141

                         Parameter Estimates

                   Parameter       Standard     T for H0:
Variable  DF        Estimate          Error   Parameter=0     Prob > |T|

INTERCEP   1      135.433596    25.90650568         5.228         0.0001
X1         1       25.307698     2.21415038        11.430         0.0001


Durbin-Watson D                  1.721
(For Number of Obs.)                40
1st Order Autocorrelation        0.133
```

Fig. 2.6 **a** SAS output for regression results of $y_i = a + b_1x_{1i} + e_i$, **b** SAS output for regression results of $y_i = a + b_1x_{1i} + b_2x_{2i} + e_i$, **c** SAS output for regression results of $y_i = a + b_1x_{1i} + b_2x_{2i} + b_3x_{3i} + e_i$

(b)

```
Dependent Variable: Y

                          Analysis of Variance

                              Sum of          Mean
       Source          DF    Squares        Square       F Value       Prob>F

       Model            2 522778.45899 261389.22949      128.141       0.0001
       Error           37  75474.56476   2039.85310
       C Total         39 598253.02375

           Root MSE      45.16473      R-square         0.8738
           Dep Mean     411.28750      Adj R-sq         0.8670
           C.V.          10.98130

                          Parameter Estimates

                        Parameter       Standard     T for H0:
     Variable  DF        Estimate          Error   Parameter=0     Prob > |T|

     INTERCEP   1       69.328469    23.15546229         2.994         0.0049
     X1         1       14.156185     2.66360071         5.315         0.0001
     X2         1       37.531322     6.95929855         5.393         0.0001


Durbin-Watson D                 2.125
(For Number of Obs.)               40
1st Order Autocorrelation      -0.083
```

(c)

```
Dependent Variable: Y

                          Analysis of Variance

                              Sum of          Mean
       Source          DF    Squares        Square       F Value       Prob>F

       Model            3 527209.08074 175736.36025       89.051       0.0001
       Error           36  71043.94301   1973.44286
       C Total         39 598253.02375

           Root MSE      44.42345      R-square         0.8812
           Dep Mean     411.28750      Adj R-sq         0.8714
           C.V.          10.80107
```

(d)

```
                          Parameter Estimates

                        Parameter       Standard     T for H0:
     Variable  DF        Estimate          Error   Parameter=0     Prob > |T|

     INTERCEP   1       31.150390    34.17504533         0.911         0.3681
     X1         1       12.968162     2.73723213         4.738         0.0001
     X2         1       41.245624     7.28010741         5.666         0.0001
     X3         1       11.524255     7.69117684         1.498         0.1428


Durbin-Watson D                 2.104
(For Number of Obs.)               40
1st Order Autocorrelation      -0.083
```

Fig. 2.6 (continued)

```
MTB > PRINT C1-C4
```

Data Display

Row	C1	C2	C3	C4
1	260.3	5	3	4
2	286.1	7	5	2
3	279.4	6	3	3
4	410.8	9	4	4
5	438.2	12	6	1
6	315.3	8	3	4
7	565.1	11	7	3
8	570.0	16	8	2
9	426.1	13	4	3
10	315.0	7	3	4
11	403.6	10	6	1
12	220.5	4	4	1
13	343.6	9	4	3
14	644.6	17	8	4
15	520.4	19	7	2
16	329.5	9	3	2
17	426.0	11	6	4
18	343.2	8	3	3
19	450.4	13	5	4
20	421.8	14	5	2
21	245.6	7	4	4
22	503.3	16	6	3
23	375.7	9	5	3
24	265.5	5	3	3
25	620.6	18	6	4
26	450.5	18	5	3
27	270.1	5	3	2
28	368.0	7	6	2
29	556.1	12	7	1
30	570.0	13	6	4
31	318.5	8	4	3
32	250.2	6	3	2
33	667.0	16	8	2
34	618.3	19	8	2
35	525.3	17	7	4
36	332.2	10	4	3
37	393.2	12	5	3
38	283.5	8	3	3
39	376.2	10	5	4
40	481.8	12	5	2

```
MTB > NAME C1'Y' C2'X1' C3'X2' C4'X3'
MTB > BRIEF 3
MTB > REGRESS C1 3 C2 C3 C4;
SUBC> DW;
SUBC> PREDICT 13 9 5.
```

Fig. 2.7 MINITAB output of $y_i = a + b_1x_{1i} + b_2x_{2i} + b_3x_{3i} + e_i$

The regression equation is
Y = 31.2 + 13.0 X1 + 41.2 X2 + 11.5 X3

Predictor	Coef	StDev	T	P
Constant	31.15	34.18	0.91	0.368
X1	12.968	2.737	4.74	0.000
X2	41.246	7.280	5.67	0.000
X3	11.524	7.691	1.50	0.143

S = 44.42 R-Sq = 88.1% R-Sq(adj) = 87.1%

Analysis of Variance

Source	DF	SS	MS	F	P
Regression	3	527209	175736	89.05	0.000
Error	36	71044	1973		
Total	39	598253			

Source	DF	Seq SS
X1	1	463451
X2	1	59327
X3	1	4431

Obs	X1	Y	Fit	StDev Fit	Residual	St Resid
1	5.0	260.30	265.83	14.96	-5.53	-0.13
2	7.0	286.10	351.20	12.82	-65.10	-1.53
3	6.0	279.40	267.27	11.35	12.13	0.28
4	9.0	410.80	358.94	11.54	51.86	1.21
5	12.0	438.20	445.77	15.11	-7.57	-0.18
6	8.0	315.30	304.73	13.18	10.57	0.25
7	11.0	565.10	497.09	16.43	68.01	1.65
8	16.0	570.00	591.65	15.24	-21.65	-0.52
9	13.0	426.10	399.29	13.88	26.81	0.64
10	7.0	315.00	291.76	13.24	23.24	0.55
11	10.0	403.60	419.83	15.63	-16.23	-0.39
12	4.0	220.50	259.53	18.78	-39.03	-0.97
13	9.0	343.60	347.42	8.26	-3.82	-0.09
14	17.0	644.60	627.67	18.77	16.93	0.42
15	19.0	520.40	589.31	17.23	-68.91	-1.68
16	9.0	329.50	294.65	15.87	34.85	0.84
17	11.0	426.00	467.37	14.98	-41.37	-0.99
18	8.0	343.20	293.21	11.61	49.99	1.17
19	13.0	450.40	452.06	11.57	-1.66	-0.04
20	14.0	421.80	441.98	13.88	-20.18	-0.48
21	7.0	245.60	333.01	13.61	-87.41	-2.07R
22	16.0	503.30	520.69	11.52	-17.39	-0.41

Fig. 2.7 (continued)

```
23      9.0    375.70   388.66    9.07   -12.96   -0.30
24      5.0    265.50   254.30   12.18    11.20    0.26
25     18.0    620.60   558.15   16.71    62.45    1.52
26     18.0    450.50   505.38   20.34   -54.88   -1.39
27      5.0    270.10   242.78   13.82    27.32    0.65
28      7.0    368.00   392.45   17.60   -24.45   -0.60
29     12.0    556.10   487.01   16.82    69.09    1.68
30     13.0    570.00   493.31   12.85    76.69    1.80
31      8.0    318.50   334.45    8.63   -15.95   -0.37
32      6.0    260.20   255.74   13.55     4.46    0.11
33     16.0    667.00   591.65   15.24    75.35    1.81
34     19.0    618.30   630.56   16.53   -12.26   -0.30
35     17.0    525.30   586.43   15.37   -61.13   -1.47
36     10.0    332.20   360.39    8.77   -28.19   -0.65
37     12.0    393.20   427.57    7.61   -34.37   -0.79
38      8.0    283.50   293.21   11.61    -9.71   -0.23
39     10.0    376.20   413.16   12.25   -36.96   -0.87
40     12.0    481.80   416.04   10.48    65.76    1.52

R denotes an observation with a large standardized residual

Durbin-Watson statistic = 2.10

     Fit  StDev Fit        95.0% CI              95.0% PI
  628.57      34.92   (  557.73,  699.40)  (  513.94,   743.19) XX
X  denotes a row with X values away from the center
XX denotes a row with very extreme X values
```

Fig. 2.7 (continued)

results are listed in the last row of Fig. 2.7. They are

1. $\hat{y}_{n+1,i} = 628.57$
2. $s(\hat{y}_{n+1}) = 34.92$
3. $s(\hat{y}_{n+1,i}) = \sqrt{s^2(\hat{y}_{n+1} + s_e^2)} = \sqrt{(34.92)^2 + 1973} = 56.50$
4. 95% confidence interval: (557.73, 699.40);
5. 95% prediction interval: (513.94, 743.19).

Stepwise Regression Analysis

In this example, we want to use stepwise regression to establish a statistical model to predict the sales of click ballpoint pens (y). We are considering three possible explanatory variables: advertising (x_1) measured in TV spots per month, the number of sales representatives (x_2), and a wholesaler efficiency index (x_3). The question is what variables should be included in the statistical model to explain the sales. The stepwise regression method suggests the following steps.

Step 1:
Run simple regression on each explanatory variable, and choose the model that explains the highest amount of variation in y. The R^2-value in each computer report is used to determine which variable enters the model first. Upon comparing R^2 values for the three models, we conclude that x_2, which has the highest R^2-value (0.7775), should enter the model first.

Independent variable	R^2	F-value
x_1	0.7747	130.644
x_2	0.7775	132.811
x_3	0.0000	0.000

Step 2:
The second variable to enter should be the variable that, in conjunction with the first variable, explains the greatest amount of variation in y.

Independent variables		R^2	F-value
x_2	x_1	0.8738	128.141
x_2	x_3	0.807	77.46

The R_2-values and F-values in the foregoing table are obtained from Fig. 2.6b and Fig. 2.8. The table shows the results when x_1 and x_3 are combined with x_2 to explain the variation in y. The combination of x_1 and x_2 clearly yields a higher R^2 (0.8738). This suggests that x_1 should be the second variable to enter.

Step 3:
In this step, we want to decide whether another variable should enter the model to explain y. Note that every time an additional variable is included in a model, R^2 increases. The question is whether the increase in R^2 justifies inclusion of the variable. We apply an F-test to answer this question.

$$F = \frac{(R_f^2 - R_R^2)(k_f - k_R)}{(1 - R_R^2)/(N - k_f - 1)}$$

where
$R_f^2 = R^2$ of the model with the new variable
$R_R^2 = R^2$ of the model without the new variable
k_f = number of the variables in the model with the new variable

```
MTB > BRIEF 2
MTB > REGRESS C1 2 C3 C4;
SUBC> DW.
```

Regression Analysis

```
The regression equation is
Y = 5.2 + 68.7 X2 + 22.1 X3

Predictor        Coef        StDev           T         P
Constant         5.19        42.40        0.12     0.903
X2             68.744        5.523       12.45     0.000
X3             22.079        9.252        2.39     0.022

S = 55.83          R-Sq = 80.7%      R-Sq(adj) = 79.7%

Analysis of Variance

Source       DF          SS          MS          F        P
Regression    2      482914      241457      77.46    0.000
Error        37      115339        3117
Total        39      598253

Source       DF      Seq SS
X2            1      465161
X3            1       17753

Unusual Observations
Obs         X2          Y        Fit   StDev Fit    Residual    St Resid
 21       4.00     245.60     368.49       14.28     -122.89      -2.28R
 25       6.00     620.60     505.97       15.79      114.63       2.14R

R denotes an observation with a large standardized residual

Durbin-Watson statistic = 2.22
```

Fig. 2.8 MINITAB output $y_i = a + b_1x_{1i} + b_2x_{2i} + b_3x_{3i} + e_i$

k_R = number of the variables in the model without the new variable

To determine whether x_3 should be included in the model, we need to compare the R^2 of the model with x_3 and R^2 of the model without x_3.

Independent variables	R^2
x_2 x_1	0.8738
x_2 x_1 x_3	0.8812

Using the foregoing formula, we compute

$$F = \frac{(0.8812 - 0.8738)/(3-2)}{(1-0.8812)/(40-3-1)} = 2.24 < F_{0.05,1,36} = 4.11$$

Because including x_3 does not increase R2 significantly, the null hypothesis that x_3 should not be included is not rejected in this case. Our conclusion from the stepwise regression analysis is that the best model should include only x_1 and x_2 as explanatory variables.

Some computer packages are programmed to perform the whole complicated stepwise regression in response to one simple command. Figure 2.9 shows the output of a stepwise regression analysis using MINITAB.

2.9 Conclusion

In this chapter, we examined multiple regression analysis, which describes the relationship between a dependent variable and two or more independent variables. Methods of estimating multiple regression (slope) coefficients and their standard errors were discussed in depth. The residual standard error and the coefficient of determination were also explored in some detail.

Both t-tests and F-tests for testing regression relationships were discussed in this chapter. We investigated the confidence interval for the mean response and the prediction interval for the individual response. And finally, we saw how multiple regression analyses can be used in business and economics decision making. In the next chapter, we will discuss other topics in applied regression analysis. Examples of applied multiple regression analysis in both business and finance will be discussed in some detail.

Fig. 2.9 Stepwise regression analysis

```
MTB > STEPWISE REGRESSION C1 3 C2 C3 C4

Stepwise Regression

 F-to-Enter:        4.00     F-to-Remove:       4.00

 Response is     Y      on  3 predictors, with N =    40

   Step          1          2
Constant     80.07      69.33

X2            66.2       37.5
T-Value      11.52       5.39
X1                       14.2
T-Value                  5.31

S             59.2       45.2
R-Sq         77.75      87.38
 More? (Yes, No, Subcommand, or Help)
SUBC> YES

 No variables entered or removed

 More? (Yes, No, Subcommand, or Help)
SUBC> NO
```

Appendix 1: Derivation of the Sampling Variance of the Least Squares Slope Estimations

We can obtain the correlation coefficient between x_1 and x_2 as

$$r = \frac{\sum_{i=1}^{n} (x_{1i} - \bar{x}_1)(x_{2i} - \bar{x}_2)}{C_1 C_2} \tag{2.35}$$

where

$$C_1 = \sqrt{\sum_{i=1}^{n} (x_{1i} - \bar{x}_1)^2} \tag{2.36a}$$

$$C_2 = \sqrt{\sum_{i=1}^{n} (x_{2i} - \bar{x}_2)^2} \tag{2.36b}$$

Substituting (2.35), (2.36a), and (2.36b) into Eq. (2.7) yields

$$\begin{aligned} b_1 &= \frac{C_2^2\left[\sum_{i=1}^{n} (x_{1i} - \bar{x}_1)(y_i')\right] - rC_1C_2\left[\sum_{i=1}^{n} (x_{2i} - \bar{x}_2)(y_i')\right]}{C_1^2C_2^2 - r^2C_1^2C_2^2} \\ &= \sum_{i=1}^{n}\left[\frac{(x_{1i} - \bar{x}_1)(y_i') - (rC_1/C_2)(x_{2i} - \bar{x}_2)(y_i')}{(1 - r^2)C_1^2}\right] \\ &= \sum_{i=1}^{n}\left[\frac{(x_{1i} - \bar{x}_1) - (rC_1/C_2)(x_{2i} - \bar{x}_2)}{(1 - r^2)C_1^2}\right]y_i' \end{aligned} \tag{2.37}$$

Substituting

$$\sum_{i=1}^{n} (x_{1i} - \bar{x}_1)y_i' = \sum_{i=1}^{n} (x_{1i} - \bar{x}_i)y_i$$

and

$$\sum_{i=1}^{n} (x_{2i} - \bar{x}_2)y_i' = \sum_{i=1}^{n} (x_{2i} - \bar{x}_2)y_i$$

into Eq. (2.37), and letting the coefficient of y_i equal B_{1i}, we obtain

$$b_1 = \sum_{i=1}^{n} B_{1i}y_i \tag{2.38}$$

Similarly,

$$\begin{aligned} b_2 &= \sum_{i=1}^{n}\left(\frac{(x_{2i} - \bar{x}_2) - \frac{rC_2}{C_1}(x_{1i} - \bar{x}_i)}{(1 - r^2)C_2^2}\right)y_i \\ &= \sum_{i=1}^{n} B_{2i}y_i \end{aligned} \tag{2.39}$$

Substituting Eq. (2.2) into Eqs. (2.38) and (2.39), we get

$$b_1 = a\sum_{i=1}^{n} B_{1i} + b_1\sum_{i=1}^{n} B_{1i}x_{1i} + b_2\sum_{i=1}^{n} B_{1i}x_{2i} + \sum_{i=1}^{n} B_{1i}e_i$$

and

$$\begin{aligned} b_2 = {} & a\sum_{i=1}^{n} B_{2i} + b_1\sum_{i=1}^{n} B_{2i}x_{1i} + b_2\sum_{i=1}^{n} B_{2i}x_{2i} \\ & + \sum_{i=1}^{n} B_{2i}e_i \end{aligned}$$

It can be easily seen that $\sum_{i=1}^{n} B_{1i} = 0$, $\sum_{i=1}^{n} B_{2i} = 0$, $\sum_{i=1}^{n} B_{1i}x_{1i} = 1$, $\sum_{i=1}^{n} B_{2i}x_{2i} = 1$, $\sum_{i=1}^{n} B_{2i}x_{1i} = 0$, and $\sum_{i=1}^{n} B_{1i}x_{2i} = 0$. Therefore, these two equations imply that

$$b_1 - E(b_1) = b_1 - \beta_1 = \sum_{i=1}^{n} B_{1i}e_i \tag{2.40}$$

and

$$b_2 - E(b_2) = b_2 - \beta_2 = \sum_{i=1}^{n} B_{2i}e_i \tag{2.41}$$

From Eq. (2.40), we obtain

$$\begin{aligned} \mathrm{Var}(b_1) &= E\left[\left(\sum_{i=1}^{n} B_{1i}e_i\right)^2\right] - \left[E\left(\sum_{i=1}^{n} B_{1i}e_i\right)\right]^2 \\ &= \sum_{j=1}^{n}\sum_{j=1}^{n} B_{1i}B_{1j}E(e_ie_j) - \left(\sum_{i=1}^{n} B_{1i}\right)^2 [E(e_i)]^2 \end{aligned} \tag{2.42}$$

In Eq. (2.8), the last equality holds because $E(e_i) = 0$ and $E(e_i e_j) = 0$ when $i \neq j$.

From the definition of B_{1i} in Eq. (2.38), we have

$$B_{1i}^2 = \frac{(x_{1i} - \bar{x}_1)^2 + r^2 C_1^2/C_2^2 (x_{2i} - \bar{x}_2)^2 - 2r(C_1/C_2)(x_{1i} - \bar{x}_1)(x_{2i} - \bar{x}_2)}{(1 - r^2)^2 C_1^4} \tag{2.43}$$

And from Eqs. (2.35), (2.36a), (2.36b), and (2.43)

$$\begin{aligned}\sum_{i=1}^{n} B_{1i}^2 &= \frac{C_i^2 + r^2(C_1^2/C_2^2)(C_2^2) - 2r(C^1/C_2)(r_1 C_1 C_2)}{(1 - r^2)^2 C_1^4} \\ &= \frac{C_1^2[1 - 2r + r^2]}{(1 - r^2)^2/C_1^4} \\ &= \frac{1}{(1 - r^2) C_1^2}\end{aligned} \tag{2.44}$$

Substituting Eq. (2.44) into Eq. (2.42) yields

$$\text{Var}(b_1) = \frac{s_e^2}{(1 - r^2)\sum_{i=1}^{n} (x_{1i} - \bar{x}_1)^2}. \tag{2.45}$$

Similarly, it can be proved that

$$\text{Var}(b_2) = \frac{s_e^2}{(1 - r^2)\sum_{i=1}^{n} (x_{2i} - \bar{x}_2)^2}. \tag{2.46}$$

Equations (2.45) and (2.46) are Eqs. (2.23) and (2.24), respectively.

If the correlation coefficient between x_1 and x_2—that is, r—is equal to zero, then $\text{Var}(b_1)$ and $\text{Var}(b_2)$ reduce to

$$\text{Var}(b_1) = \frac{s_e^2}{\sum_{i=1}^{n} (x_{1i} - \bar{x}_1)^2} \quad \text{and}$$

$$\text{Var}(b_2) = \frac{s_e^2}{\sum_{i=1}^{n} (x_{2i} - \bar{x}_2)^2}$$

This implies that the sample variance of multiple regression slopes reduces to a simple regression case.

Appendix 2: Cross-sectional Relationship Among Price Per Share, Dividend Per Share, and Return Earning Per Share

Based upon Lee and Lee (2017), we discuss the cross-sectional relationship among price per share, dividend per share, and return earning per share. First we discuss Gordon's empirical work and its extension; then we discuss M&M's empirical work.

Gordon (1959) was interested in the determinants of stock price per share. His method of testing for these determinants was one of cross-sectional regression, testing the relative importance of dividend per share and retained earnings per share by comparing the magnitude regression coefficient associated with dividend per share and retained earnings per share.

Gordon first sought to uncover what it was the market actually capitalized in arriving at stock values, both dividends and earnings—the dividends plus the growth opportunities, where the retained earnings figure was used as a proxy for growth potential, or simply the earnings by themselves. His sample was taken from four industries at two separate points in time, which allowed for analysis of the coefficient's stability and significance over time and across industries, as well as the analysis of the multiple correlations across time. He sought to make only some generalizations, although the results were to some extent inconclusive.

The dividends and earnings model as shown in Eq. (2.46) show the price of a share of stock to be linearly dependent upon dividends, earnings, and some constant term:

$$P = a_0 + a_1 D + a_2 \mathrm{Y}. \qquad (2.47)$$

The a_1 and a_2 terms are the point of interest; if a_1 is greater than a_2, then dividends have a relatively larger impact in the valuation procedure than do earnings, and vice versa. The multiple correlations for these eight regressions were quite high, averaging 0.905, with a range from 0.86 to 0.94. Of the dividend coefficients, seven were statistically significant at a 5% level (the only negative value being the insignificant value), while two of the earnings coefficients were not statistically significant. In each case, except that one where the dividend coefficient is insignificant, the dividend coefficient is greater in magnitude than the earnings coefficient. This evidence is shaky proof for dividend preference if one recognizes the multicollinearity problem and the apparent lack of consistency of the coefficients across industries or stability over time periods, even in any relative sense.

The dividend-plus-investment-opportunities approach, which, as we said earlier, uses the retained-earnings figure as a growth proxy, hinting again at an all internally financed firm, is symbolized by Eq. (2.47) below:

$$P = a_0 + a_1 D + a_2 (Y - D). \qquad (2.48)$$

The multiple correlation figures and the constant terms are the same as in the previous test; this might have been expected, since no new variables entered, and none left the formulation, but the relative dividend and retained-earnings coefficients do change in relative values. Actually, the retained-earnings coefficients were the same as the earnings coefficients in the earlier trial, but it is really the difference, or the ratio, of the two that we are interested in. The dividend coefficients also appear to be relatively more consistent from industry to industry and stable in relative terms through time. The new-found stability and consistency is appealing because we would like to believe stocks are priced in some logical and consistent manner, which, given this evidence, indicates that dividends are three to four times as important in the valuation procedure as are growth opportunities. Once again, multicollinearity is a problem that is left unquestioned.

Gordon dismisses the pure-earnings approach with the realization that investors would be indifferent between dividends and increases in market value of their holdings if they valued only earnings, dismissing market imperfections for now. If this were so, then the two relevant coefficients would tend to be equal, and since they were found to be drastically different in the second set of regressions, he discounts this as a viable valuation format.

In refining his model, it appears that Gordon was primarily interested in not allowing short-term aberrations to influence the parameter estimates, although he also adjusted all variables in accordance with a size factor, which, in some respects, could be interpreted as the inclusion of a previously omitted variable that could affect the stability of some of the coefficients. The new model appears as:

$$P = B_0 + B_1 \bar{d} + B_2 (d - \bar{d}) + B_3 (\bar{g}) + E_4 (g - \bar{g}), \qquad (2.49)$$

where

P = price per share/book value;
$\bar{d}$ = five-year average dividend/book value;
d = current year's dividend/book value;
$\bar{g}$ = five-year average retained earnings/book value;
g = current year's retained earnings/book value

The multiple correlations from the use of this model are slightly lower than those in the dividend-plus-growth case presented earlier, but most noticeably the constant terms are now all quite small, whereas, with the previous models, they were at times quite large, although statistical significance was not discussed. Only the five-year dividend is significant at a 5% level for all eight regressions with six of the current less the long-run average being significant and five of each of the growth coefficients being significant also, although not as much so as the dividend coefficients.

All things considered, the dividend factors appear to be the predominant influences on share values, although there are certain individual exceptions among the industries surveyed. The evidence presented here in the revised model and in the models presented earlier must be interpreted as supporting Gordon's contention that dividends do matter; in fact, given the methods used here, they are detected as being the most important variable that could be included in the valuation formula.

Friend and Puckett (1964) were concerned with unveiling the limitations to the type of analysis performed by Gordon and others. They pointed out a number of potential problems, the first being the accounting earnings and retained-earnings figures and the high degree of measurement error incurred when using these values as proxies for economic or true earnings. Again, it is assumed that investors value economic earnings and not accounting earnings per se. Risk measures are also missing from the analysis, as is any awareness of the dynamic nature of corporate dividend policy. In short, we must realize not only a multicollinearity problem and potential specification errors, but that omitted variables must also be accounted for, if the analysis is to be complete.

In a general sense, a composite firm-specific variable, E, could be included in the standard model, which contains dividends and retained earnings, yielding:

$$P = a_0 + a_1 D + a_2 R + F. \tag{2.50}$$

But to have any true economic meaning, this variable should be specified, so multiple statistical trials were run to see which economic variable would be best to include in the analysis. Attempts were also made to alter or adjust the retained-earnings figure so as to more accurately reflect its "true" value, as discussed in a preceding paragraph.

In running the statistical tests, five industry groups were analyzed at two points in time, along the same lines of reasoning as in Gordon's paper. The first trial was run using Eq. (2.48) as the model, without the F-coefficient. For each industry, the coefficients are relatively stable across time periods, and in two instances the retained-earnings coefficient is greater than the dividend coefficient. In particular, the electrical utility industry is seen to be such a case, and the work was then redone using logarithms, to see if the linearity assumption was responsible for this result. Unsurprisingly, the coefficients change in relative magnitudes; this is due to a combination of the utility industry's high dividend yield and the nature of the logarithmic transformation. Friend and Puckett concede there is no reason to prefer one method versus the other, and leave the functional form issue at that.

Returning to the omitted-variable issue, Friend and Puckett first included the previous period's price as the firm-specific variable. As a result, the retained-earnings coefficients are greater than the dividend coefficients, the latter being negative in six of ten instances. The multiple R-squared statistics are quite high but, by and large, the significance levels of the retained-earnings coefficients are not. Besides having very little economic rationale for the inclusion of a lagged price variable, the statistics leave sizable uncertainties as to the validity of this approach.

Recognizing that accounting numbers over the short run are subject to sizable measurement error, the authors normalized earnings over ten years and then ran the regressions again without a firm-specific variable. This rendered all coefficients statistically significant at a 5% level, and the dividend coefficients were consistently higher than the retained-earnings coefficients. This suggests the same conclusions and interpretation as were offered by Gordon. Adding the normalized-earnings price ratio lagged one period, it was found that the intercept or constant terms get very large, as do the dividend coefficients relative to the retained-earnings coefficients, and the R-squared term for each regression is exceptionally high. The earnings/price coefficient was found to be highly negative; all these findings yield questionable results. In a separate analysis of the chemical industry using dividends and normalized earnings, and then further adding the normalized earnings/price ratios, the retained-earnings coefficient was seen to dominate the dividend coefficient. Again, though, the other coefficients were given, and certainly appear to merit some attention.

In summation, Friend and Puckett argue for longer-term measures for variables of accounting nature and earnings-to-price ratios as a sort of risk or firm-specific variable, in that it shows the rate at which the market capitalizes the firm's earnings. In using this variable, it should be recognized that the dependent variable is being used to explain itself. This tends to show the retained-earnings figure as important as or more important in valuation than dividends, thus supposedly invalidating previous results to the contrary. The real question appears to be which approach possesses the best economic grounding. We would like to think theory precedes the empiricism; and to pursue this, one must have an appropriate theory, a question to which Friend and Puckett do not address themselves.

Lee (1976) cited the Friend and Puckett evidence in attacking the issue of functional form, concentrating on the electric utility industry where the risk differential between firms is often seen to be negligible, thereby eliminating a large portion of the firm-specific effects. Using the generalized functional form developed by Box and Cox (1964), the linear and log linear forms can be tested on a purely statistical basis, and the log linear form can be tested against a nonlinear form.

Quite interestingly, Lee found that the log linear form was statistically superior to the linear-form model in explaining the dividend effect. The results of this comparison were essentially the same as in Friend and Puckett's study. Using the linear form, nine of the ten years of data examined showed a stronger retained-earnings effect than that for dividends, while, in the log linear trial, all ten years showed stronger dividend effects. At this point, the only question remaining is whether either of these models accurately depicts the true functional form.

Using the true functional form from the generalized functional form method to compare the dividend and retained-earnings effects, it was found that in only four years was there any difference between the two effects; this leads to the conclusion that all models developed in linear or log linear form, and used to test this particular industry (and possibly other industries as well) are probably misspecified. This is a serious problem, in that the importance of dividends is muddied because the model is not correct. The true value of dividends can be inferred only if the model is correctly specified.

Part of the problem here is due to the nature of the industry, where high payouts are common and external financing is great and unaccounted for. These two factors serve to bias the dividend effect downward in the linear-form models.

The logarithmic-form models reduce the problem of weighting of regression coefficients due to size disparities among firms, a problem noticeably existent in the electric utility industry. However, it does have the disadvantage of being unable to cope with negative retained-earnings figures, a phenomenon that is a reality in the current environment.

These caveats to both methods of analysis should evoke more concern from empirical researchers in the future, since misspecification can have drastic effects on the conclusions reached.

Bibliography

Box, G. E., & Cox, D. R. (1964). An analysis of transformations. *Journal of the Royal Statistical Society: Series B (Methodological), 26*(2), 211–243.

Fogler, H. R., & Ganapathy, S. (1982). *Financial econometrics*. Englewood Cliffs, NJ: Prentice-Hall.

Friend, I., & Puckett, M. E. (1964). Dividends and stock prices. *American Economic Review*, 656–682.

Ghosh, S. K. (1991). *Econometrics: Theory and applications*. Englewood Cliffs, NJ: Prentice Hall.

Gordon, M. J. (1959). Dividends earnings and stock prices. *Review of Economics and Statistics*, 99–105.

Greene, W. H. (2017). *Econometric analysis* (8th ed.). New Jersey: Prentice Hall.

Gujarati, D., & Porter, D. (2011). *Basic econometrics* (5th ed.). MHE.

Johnston, J., & Dinardo, J. (1996). *Econometrics methods* (4th ed.). New York: McGraw-Hill.

Lee, C. F. (1976). Functional form and the dividend effect in the electric utility industry. *Journal of Finance,* 1481–1486.

Lee, A. C., Lee, J. C., & Lee, C. F. (2017). *Financial analysis, planning and forecasting: Theory and application* (3rd ed.). Singapore: World Scientific.

Lee, C. F., Finnerty, J., Lee, J., Lee, A. C., & Wort, D. (2013a). *Security analysis, portfolio management, and financial derivatives*. World Scientific.

Lee, C. F., & Lee, A. C. (2013). *Encyclopedia of finance* (2nd ed.). New York, NY: Springer.

Lee, C. F., & Lee, J. C. (2015). *Handbook of financial econometrics and statistics*. New York, NY: Springer.

Lee, C.-F., Lee, J., & Lee, A. C. (2013b). *Statistics for business and financial economics* (3rd ed). Berlin: Springer.

Theil, H. (1971). *Principles of econometrics*. Toronto, NY: Wiley.

Wooldrige, J. M. (2010). *Econometric analysis of cross section and panel data* (2nd ed.). The MIT Press.

3 Other Topics in Applied Regression Analysis

Contents

Abstract

Following the previous chapter, this chapter discusses other topics of multiple regression analysis. These topics include multicollinearity, heteroscedasticity, autocorrelation, model specification and specification bias, nonlinear

C.-F. Lee et al., *Financial Econometrics, Mathematics and Statistics*,
https://doi.org/10.1007/978-1-4939-9429-8_3

models, lagged dependent variables, dummy variables, and regression with interaction variables. In the appendices, we show how multiple regression can be used for finance applications such as dynamic ratio analysis, term structure of interest rate, partial adjustment dividend behavior model, logistic model, and probit model.

3.1 Introduction

Based upon the multiple regression concept and model discussed in the previous chapter, we discuss other related topics within this chapter. The main objectives in fitting a regression equation are (1) to estimate the regression coefficients and related parameters and (2) to predict the value of the dependent variable in terms of that of the independent variable (or variables). Several alternative specifications are possible in this kind of applied regression analysis, and a number of problems may occur.

In this chapter, we examine some of the problems associated with applying the multiple regression model. We also explore such related topics as lagged dependent variables and nonlinear regressions. Problems with the error term—specifically, violations of the assumptions of the regression model—can arise when we are running a regression. In this chapter, we discuss the detection of these problems, which include errors that are correlated and errors whose means and variance are not constant. Another problem we may encounter is a high correlation between independent variables. This problem can increase the value of standard errors and reduce the t-statistics of the parameters, leading to incorrect inferences in hypothesis testing.

Other topics we address in this chapter include specification bias and model building. We also show how a nonlinear functional form can be transformed into a linear regression analysis. In some cases, for example, both independent and dependent variables can be transformed by using logarithms, and the nonlinear relationship then becomes a linear relationship. Furthermore, a regression can have a lagged dependent variable as one of the independent variables when there is a relationship between previous observations in a time series and the value in the present period. In addition, regression with dummy and interaction variables is discussed in detail. The effect of alternative business strategies is investigated. Logistic regression is also discussed in this chapter.

In this chapter, we discussed various topics associated with the regression analysis. In Sect. 3.2, we discuss multicollinearity. In Sect. 3.3, we discuss heteroscedasticity. In Sect. 3.4, we discuss autocorrelation. In Sect. 3.5, we discuss model specification and specification bias of the regression model. Section 3.6 investigates nonlinear regression models. Section 3.7 considers lagged dependent variables in the regression model. Section 3.8 discusses dummy variables in the regression model. Section 3.9 discusses the regression model with interaction variables. Section 3.10 applies the regression approach to investigate the effect of alternative business strategies. Section 3.11 applies the logistic regression model to credit risk analysis. Finally, in Sect. 3.12, we conclude this chapter. There are five appendixes in this chapter. Appendix 1 discusses the dynamic ratio analysis. Appendix 2 presents the term structure of interest rate. Appendix 3 explores the partial adjustment dividend behavior model. Appendix 4 discusses the logistic model and probit model. Appendix 5 presents the SAS code for hazard model in bankruptcy forecasting.

3.2 Multicollinearity

Definition and Effect

The term **multicollinearity** refers to the effect, on the precision of regression parameter estimates, of two or more of the independent variables being highly correlated. For example, multicollinearity would be a problem if we were studying the cross-sectional relationship by regressing price per share against dividend per share and retained earnings per share. Because dividend per share and retained earnings per share are highly correlated, the precision of the least squares estimated regression coefficient might be affected.

If a set of independent variables is perfectly correlated, the least squares approach cannot be used to estimate the regression coefficients: The normal equations are not solvable. If independent variables move together, it is impossible to distinguish the separate effects of these variables on y. Perfect multicollinearity would occur, for example, if the following independent variables were specified to model the expenditures on food for a cross section of individuals.

$$x_1 = \text{average income in dollars}$$
$$x_2 = \text{average income in cents}$$

The variables x_1 and x_2 are perfectly correlated because x_2 = 100 times x_1 for each of the individuals in the data set. If both of these variables were included in a regression model, least squares results would not be obtainable because the two variables measure the same thing. Remember that the regression coefficient of x_2 is a slope term that measures the change in the dependent variable that is associated with a 1-unit change in x_2, other variables being held constant. But here it is impossible to keep the rest of the variables constant, because x_1 changes in the same direction and with the same magnitude as x_2. The solution? Simply delete one of the variables and run the regression again.

Unfortunately, most of the problems researchers face are not so easy to detect. Observations are more likely to be *highly* correlated than perfectly correlated. In such cases, least squares estimates can be obtained but are difficult to interpret.

For example, suppose national income in period t, y_t is modeled with independent variables x_{1t} = output of manufactured goods in period t and x_{2t} = output of durable goods in period t as defined in Eq. (3.1).

$$y_t = a + b_1 x_{1t} + b_2 x_{2t} + e_t \tag{3.1}$$

If the simple correlation r between x_{1t} and x_2t is 0.90, it can be concluded that the explanatory values of the two variables overlap considerably, probably because they are highly correlated and tend to measure the same thing.

In Eq. (3.1), the first coefficient of two highly correlated variables is the slope term b_1, which measures the change in the national income that is due to a 1-unit change in the output of manufactured goods, the output of durable goods being held constant. When one of the correlated variables changes, the other is likely to change in the same direction and with approximately the same magnitude. However, the standard error of the coefficient will tend to be great, leading to lower t-values for the coefficient.[1] This increase in the standard error results from the fact that estimates are sensitive to any changes in observations or model specification.

The sample variances of b_1 and b_2 ($S^2_{b_1}$ and $S^2_{b_2}$) of Eq. (3.1) can be defined as

$$S^2_{b_1} = \frac{S^2_e}{(1-r^2)\sum_{i=1}^{n}(x_{1t}-\bar{x}_1)^2} \tag{3.2}$$

$$S^2_{b_2} = \frac{S^2_e}{(1-r^2)\sum_{i=1}^{n}(x_{2t}-\bar{x}_2)^2} \tag{3.3}$$

where S^2_e = sample variance of e_t; $\bar{x}_1$ and $\bar{x}_2$ are means of x_{1t} and x_{2t}, respectively; and r is the correlation coefficient between x_{1t} and x_{2t}. In Eqs. (16.2) and (16.3), the factor $(1 - r^2)$ can be

[1] If x_1 and x_2 are highly correlated, then the regression can give weight to either x_1 or x_2 and it would not matter. Sampling idiosyncracies determine the choice. Hence, the large sampling error occurs.

used to measure the impact of collinearity on $S_{b_1}^2$ and $S_{b_2}^2$. If x_{1t} and x_{2t} are uncorrelated, then $(1 - r^2) = 1$. If x_{1t} and x_{2t} are perfectly correlated $(r^2 = 1)$, then $(1 - r^2) = 0$. In this case, the denominators of both $S_{b_1}^2$ and $S_{b_2}^2$ vanish, and both $S_{b_1}^2$ and $S_{b_2}^2$ equal infinity. In our national income example, $r^2 = 0.81$ and $(1 - r^2) = 0.19$. Therefore, the precision of estimated $S_{b_1}^2$ and $S_{b_2}^2$ is greatly affected by the collinearity between x_{1t} and x_{2t}.

Rules of Thumb in Determining the Degree of Collinearity

Several rules of thumb are helpful when we are testing for multicollinearity. These rules involve inspection of the correlation between the independent variables. First, multicollinearity is a problem if the correlation coefficient between any two independent variables is greater than 0.80 or 0.90. If there are more than two independent variables in a regression, as indicated in Eq. (3.4), then the simple correlation coefficient between any two independent variables is not sufficient to detect the existence of multicollinearity of a regression.

$$y_i = \alpha + \beta_1 x_{1i} + \beta_2 x_{2i} + \beta_3 x_{3i} + \epsilon_i \tag{3.4}$$

We must also consider that simple correlation coefficients generally fail to take into account the possible correlation between any one independent variable and all others taken as a group. Therefore, it is customary to regress each of the independent variables against all others and to note whether any of the resulting R^2-values are near 1. Using Eq. (3.4) as an example, we can define three multiple regressions in terms of x_{1i}, x_{2i}, and x_{3i}.

$$x_{1i} = a_0 + a_1 x_{2i} + a_2 x_{3i} \tag{3.5a}$$

$$x_{2i} = b_0 + b_1 x_{1i} + b_2 x_{3i} \tag{3.5b}$$

$$x_{3i} = c_0 + c_1 x_{1i} + c_2 x_{2i} \tag{3.5c}$$

$R_i^2 (i = 1, 2, 3)$ of these three regressions represents the **coefficient of determination** for the ith independent variable. It can be used to determine whether multicollinearity plagues Eq. (3.4).

In sum, to check for multicollinearity, we first calculate the three simple correlation coefficients between x_1 and x_2 (r_{12}), between x_1 and x_3 (r_{13}), and between x_2 and x_3 (r_{23}). Then, we find R^2 associated with Eqs. (3.5a), (3.5b), and (3.5c). In a sense, these two methods are similar, but the first is easier to understand. This estimated information can be used to determine the existence of multicollinearity.

One way to measure collinearity is to use the measurements of $(1 - R^2)$ indicated in Eq. (3.2) or (3.3) to construct a **variance inflationary factor** (VIF) for each explanatory variable in Eq. (3.4).

$$\text{VIF}_i = \frac{1}{1 - R_i^2} \tag{3.6}$$

where R_i^2 is as defined in Eqs. (3.5a), (3.5b), and (3.5c). If there are only two independent variables, R_i is merely the correlation coefficient. If a set of independent variables is uncorrelated, then VIF_i is equal to 1. If R_i^2 approaches 1, both VIF and the standard deviations of the slopes ($S_{b_1}^2$ and $S_{b_2}^2$) approach infinity. Researchers have used $\text{VIF}_i = 10$ as a critical-value rule of thumb to determine whether too much correlation exists between the ith independent variable and other independent variables.[2] The corresponding R_i values of VIF_i at least 10 are now illustrated. Thus, $1/(1 - R_i^2) \geq 10$ implies $1 - R_i^2 \leq 0.1$. This implies that $0.95 \leq R_i \leq 1$ or $-1 \leq R_i \leq -0.95$.

Example 3.1 Analyzing the Determination of Price per Share

Data on dividend per share (DPS), retained earnings (RE) per share, and price per share (PPS) from 2007 to 2009 for the 30 firms employed to compile the Dow Jones Industrial Average are used to estimate Eq. (3.7). Regression results are shown in Table 3.1.

[2] See D. W. Marquardt (1980), "You Should Standardize the Prediction Variables in Your Regression Models," discussion of "A Critique of Some Ridge Regression Methods," by G. Smith and F. Campbell, *Journal of the American Statistical Association* 75, 87–91.

Table 3.1 Regression results of Eq. (3.7)

(1)	(2)	(3)	(4)
	2007	2008	2009
Constant	19.413	14.536	10.465
	(3.46)	(3.26)	(2.23)
RE	8.14	6.389	7.605
	(6.20)	(7.79)	(5.57)
DPS	5.47	2.208	10.581
	(1.15)	(0.61)	(2.45)
R^2	0.70	0.77	0.77
F-statistic	31.33	44.46	45.53

The t-values are indicated in parentheses

$$\text{PPS}_{i,t} = a + b(\text{RE}_{i,t}) + c(\text{DPS}_{i,t}) + e_{i,t} \quad (3.7)$$

where $\text{PPS}_{i,t}$, $\text{DPS}_{i,t}$, and $\text{RE}_{i,t}$ represent price per share, dividends per share, and retained earnings per share for the ith firm in the tth year, respectively. This cross-sectional model states that the price per share is a function of dividends per share and retained earnings per share. The results of the regressions seem to be satisfactory with all of the independent variables and appear significant at the 5% level ($F_{0.05,2,27} = 3.35$, from Table A6). However, we must examine the correlations between $\text{DPS}_{i,t}$ and $\text{RE}_{i,t}$ to determine whether multicollinearity may be a problem.

The 2009 regression results indicated in column (4) of Table 3.1 are used to determine the degree of multicollinearity. To do this analysis, we assemble the correlation coefficients among PPS, DPS, and RE in Table 3.2. We note from the correlation matrix that each variable is perfectly correlated with itself; hence, we find three entries equal to 1.000 along the diagonal of the table. Substituting $r_{1,2} = 0.6351$ into Eq. (3.6), we obtain $\text{VIF}_1 = 1/[1 - (0.6351)^2] = 1.67603$. Because 1.67603 is much smaller than 10, we conclude that the degree of collinearity for this regression is relatively unimportant.

3.3 Heteroscedasticity

Definition and Concept

Heteroscedasticity arises when the variances of the error terms of a regression model are not constant over different sample observations. For example, heteroscedasticity would be a problem in a study of sales for a cross section of firms in an industry, because the error terms for large firms would be likely to have larger variances than those for small firms. In other words, the high volatility in sales for larger firms might pose problems for the researcher. The probable error terms are shown in Fig. 3.1. This figure indicates that the magnitude of error terms is a function of firm size. For example, it may be the case that $\sigma_i^2 < \sigma_j^2 < \sigma_k^2$.

Another commonly cited example of heteroscedasticity is the relationship between family expenditures and income. High-income families are likely to exhibit a higher variance in spending than lower-income families. Figure 3.2 shows a reasonable plot of level of income versus level of expenditures. This figure indicates that expenditures for consumers with lower income have smaller error terms.

Table 3.2 A simple correlation matrix

Variable	PPS, y	RE, x_1	DPS, x_2
PPS, y	$r_{y,y} = 1.000$	$r_{y,1} = .8488$	$r_{y,2} = 0.7132$
RE, x_1	–	$r_{1,1} = 1.000$	$r_{1,2} = 0.6351$
DPS, x_2	–	–	$r_{2,2} = 1.000$

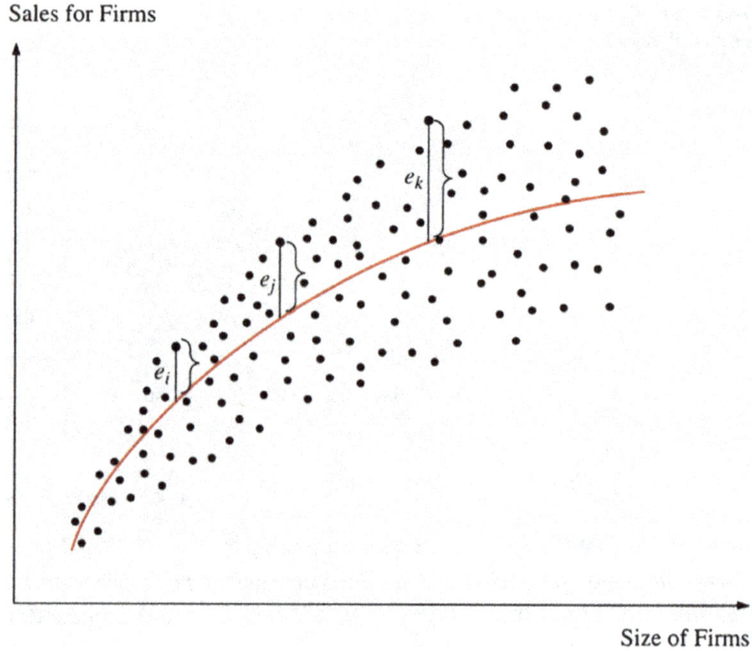

Fig. 3.1 A possible relationship between sales and firm size

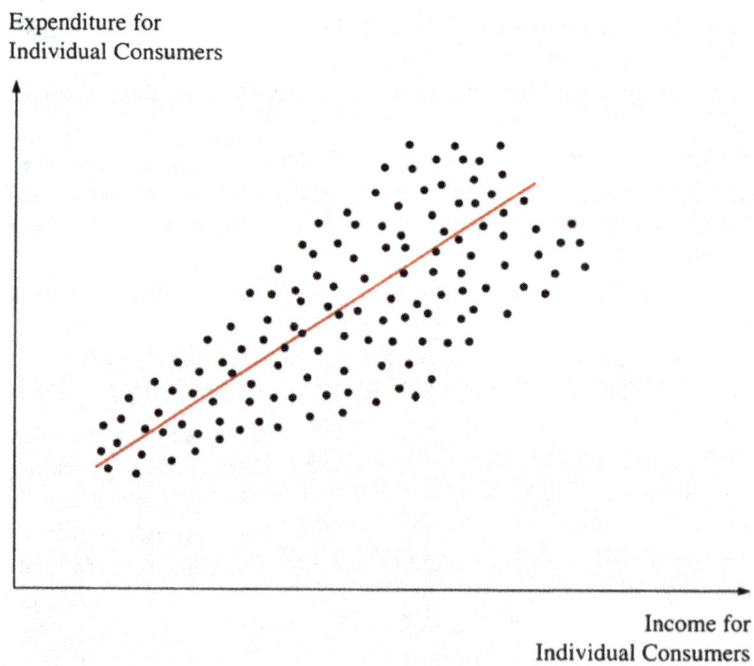

Fig. 3.2 Cross-sectional relationship between expenditure and income

Heteroscedasticity poses a problem when we are estimating parameters in the regression model, because the least squares estimation procedure places more weight on observations that have large errors and variances. Thus, the regression line is adjusted to give a good fit for the large-variance portion of the observations but largely ignores the small-variance part of the data. The result is that the variances of the estimates do not have a minimum variance.

Evaluating the Existence of Heteroscedasticity

The easiest way to check for heteroscedasticity is to look at a plot of the residuals against the independent variables or the expected values. To estimate the error term, we first compute the predicted dependent value by the regression model

$$\hat{y}_i = a + b_1 x_{1i} + b_2 x_{2i} + \cdots + b_k x_{ki} \quad (3.8)$$

Then, we calculate the error term by taking the actual value for y_i and subtracting the predicted value: $e_i = y_i - \hat{y}_i$. In practice, we generally plot the residuals' range e_i against independent variables on a series of graphs or examine the predicted values $\hat{y}_i$. If the residuals appear to be random and the width of the scatter diagram seems constant throughout the data—that is, if no pattern is apparent—then no heteroscedasticity is present.

A somewhat more involved evaluation for the existence of heteroscedasticity consists of the following steps:

1. Run a standard regression.
2. Calculate the residuals, $e_i = \hat{y}_i - y_i$.
3. Run a regression using the square of the residuals as the dependent variable and the estimated dependent variable $\hat{y}_i$ as the independent variable.
4. Estimate nR^2, where n is the sample size and R^2 is the coefficient of determination.
5. Use the χ^2 statistic with 1 degree of freedom to test whether nR^2 is significantly different from zero.

The use of this method to analyze the existence of heteroscedasticity is illustrated in the following example.

Example 3.2 Residual Heteroscedasticity Analysis for Price per Share

The regression results for Eq. (3.7) were run for the 30 Dow Jones industrials for 2007, 2008, and 2009. These results appear in Table 3.1.

To check for heteroscedasticity, we calculate the residuals of the regression and plot them against the predicted values $\hat{y}_{i,t}$. Residuals (e) and predicted PPS values ($\hat{y}$) for all three years are listed in Table 3.3. Figure 3.3 shows the plots of residuals from the least squares regression against $\hat{y}_{i,t}$ for 2007. Similar plots for 2008 and 2009 are given in Figs. 3.4 and 3.5. As can be seen from all three plots, there are patterns in the residuals. The pattern for 2007 is stronger than those for 2008 and 2009. We might conclude that the residuals plotted against predicted values are not random and do violate the standard assumptions of the regression model.

In addition to making this visual inspection of residuals, we must run a regression with the squared error terms as the dependent variable and $\hat{y}_i$ as the independent variable. Let us look at the results for 2009.

$$e^2_{i,t} = c + f\hat{y}_{i,t} + \text{residuals} \quad (3.9)$$

Using $e^2_{i,t}$ and $\hat{y}_{i,t}$ for 2009 as presented in Table 3.4, we estimate Eq. (3.9) and obtain

$$e^2_{i,2009} = -39.1519 + 4.0942\hat{y}_{i,2009} \quad R^2 = 0.1758$$
$$(t = 2.44)$$

Because there are n = 30 sets of observations, the test is based on nR^2 = (30) (0.1758) = 5.2740. From the table of critical values of χ^2, we find that for a test at the 5 percent level, $\chi^2_{1,0.05}$ = 3.84. Therefore, we can conclude that the residuals in the regression of price per share on dividends per share and retained earnings per share do not have the same variance, and the null hypothesis should be rejected. One way to deal with this problem is to use a two-stage procedure to estimate the parameters of regression models. In the first stage, we estimate the parameters of Eq. (3.7) and the predicted value $\hat{y}_{i,t}$ of the dependent variable. Predicted values of $y_{i,t}$ for 2009 are listed in Table 3.4. In the second stage, we estimate a transformed Eq. (3.7):

Table 3.3 Residuals (e) and predicted PPS values ($\hat{y}$) for 30 Dow Jones industrials

Year	2007		2008		2009	
Observations	$\hat{y}$	e	$\hat{y}$	e	$\hat{y}$	e
1	47.38	−10.83	47.38	−10.83	4.84	11.28
2	50.77	1.25	50.77	1.25	29.87	10.65
3	43.87	−0.18	43.87	−0.18	39.90	−6.77
4	70.69	16.77	70.69	16.77	42.62	11.51
5	71.79	0.77	71.79	0.77	39.27	17.72
6	63.99	−20.34	63.99	−20.34	33.19	8.48
7	103.62	−10.29	103.62	−10.29	78.61	−1.62
8	47.93	13.44	47.93	13.44	50.25	6.75
9	40.07	−5.68	40.07	−5.68	27.71	−0.25
10	54.17	−10.08	54.17	−10.08	42.50	−8.83
11	86.80	6.89	86.80	6.89	58.37	9.82
12	43.52	−6.45	43.52	−6.45	26.98	−11.85
13	43.62	8.06	43.62	8.06	38.26	9.20
14	45.89	−5.15	45.89	−5.15	22.40	−2.00
15	31.64	−4.98	31.64	−4.98	110.18	20.72
16	87.18	20.92	87.18	20.92	64.73	−0.32
17	58.14	8.56	58.14	8.56	63.87	−1.43
18	43.57	15.34	43.57	15.34	69.67	−33.13
19	37.93	20.18	37.93	20.18	66.73	15.94
20	76.30	8.02	76.30	8.02	8.68	6.38
21	59.80	−18.54	59.80	−18.54	28.28	−10.09
22	35.44	−12.71	35.44	−12.71	56.41	−5.31
23	52.62	8.57	52.62	8.57	43.94	−15.91
24	43.05	−1.49	43.05	−1.49	58.47	10.94
25	61.45	15.09	61.45	15.09	28.15	−4.38
26	45.92	1.77	45.92	1.77	18.45	3.56
27	33.26	−3.79	33.26	−3.79	72.00	−22.14
28	29.26	−0.35	29.26	−0.35	38.25	−11.07
29	82.88	−29.08	82.88	−29.08	46.07	1.05
30	38.34	−5.71	38.34	−5.71	30.41	−8.88

$$\frac{y_{i,t}}{\hat{y}_{i,t}} = a\frac{1}{\hat{y}_{i,t}} + b\frac{x_{1i,t}}{\hat{y}_{i,t}} + c\frac{x_{2i,t}}{\hat{y}_{i,t}} + e'_{i,t} \quad (3.7')$$

where $e'_{i,t}$ is an error term that approximates constant variance. Using 2009 data, we estimate Eq. (3.7′) and its results.

$$\frac{y_{i,2009}}{\hat{y}_{i,2009}} = \underset{(9.87)}{17.333}\left(\frac{1}{\hat{y}_{i,2009}}\right) + \underset{(4.15)}{4.553}\frac{x_{1i,2009}}{\hat{y}_{i,2009}} + \underset{(2.92)}{10.316}\frac{x_{2i,2009}}{\hat{y}_{i,2009}}$$

Fig. 3.3 Plots of residuals against the predicted PPS values, $\hat{y}$ (2007)

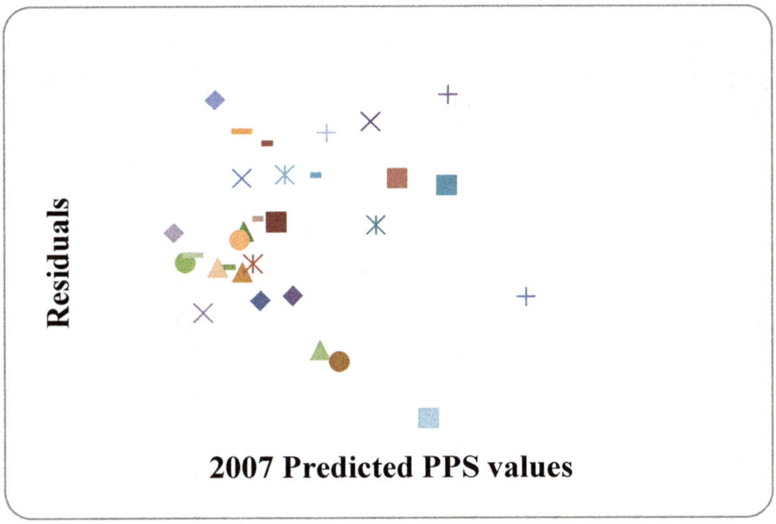

Fig. 3.4 Plots of residuals against the predicted PPS values, $\hat{y}$ (2008)

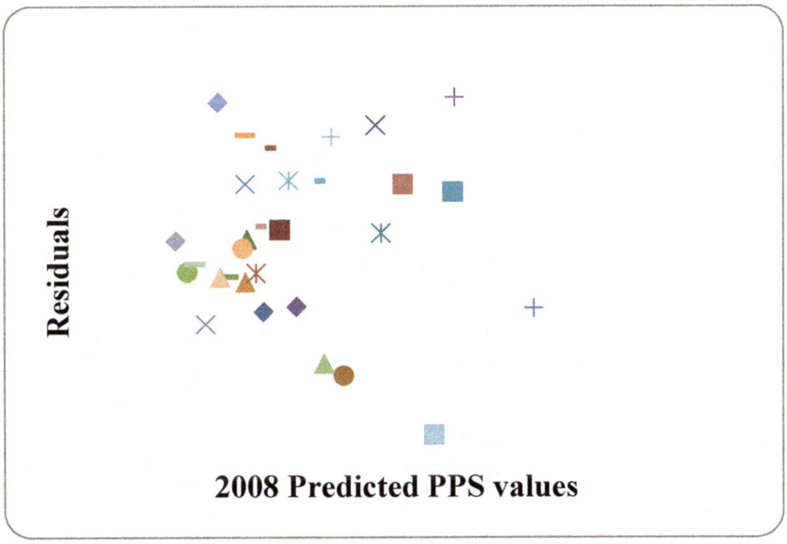

Fig. 3.5 Plots of residuals against the predicted PPS values, $\hat{y}$ (2009)

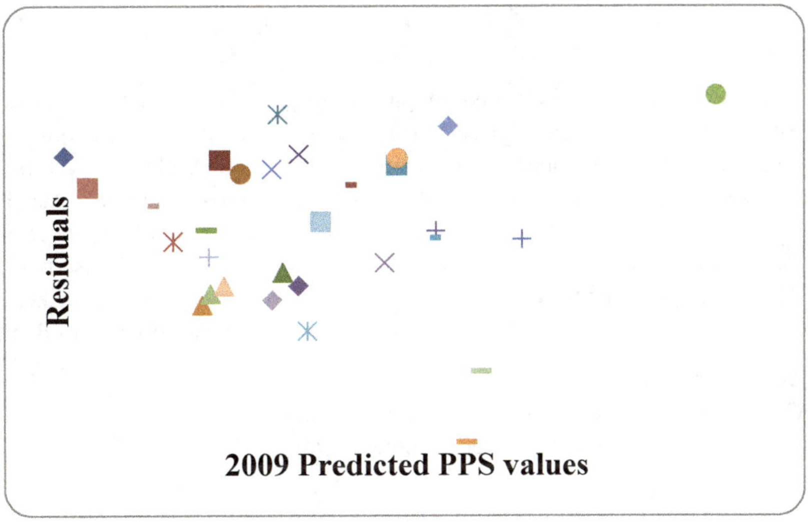

Table 3.4 $e_{i,t}^2$ and $\hat{y}_{i,t}$ for a test of heteroscedasticity for Eq. (3.7) in terms of 2009 data

$\hat{y}_{i,2009}$	$e_{i,2009}$	$e_{i,2009}^2$
4.84	11.28	127.3021
29.87	10.65	113.3852
39.90	−6.77	45.88082
42.62	11.51	132.5925
39.27	17.72	314.0423
33.19	8.48	71.99254
78.61	−1.62	2.638521
50.25	6.75	45.51427
27.71	−0.25	0.060564
42.50	−8.83	77.9001
58.37	9.82	96.33882
26.98	−11.85	140.305
38.26	9.20	84.5601
22.40	−2.00	3.99501
110.18	20.72	429.3093
64.73	−0.32	0.102427
63.87	−1.43	2.045569
69.67	−33.13	1097.628
66.73	15.94	254.0671
8.68	6.38	40.66668
28.28	−10.09	101.8989
56.41	−5.31	28.23771
43.94	−15.91	253.1638
58.47	10.94	119.5953
28.15	−4.38	19.20552
18.45	3.56	12.66711
72.00	−22.14	490.2615
38.25	−11.07	122.6317
46.07	1.05	1.100324
30.41	−8.88	78.80692

From the *t*-values indicated in parentheses, we find that the *t*-values for second-stage estimates are more efficient than those for the one-stage estimates indicated in Table 3.1.

3.4 Autocorrelation

Basic Concept

One of the assumptions of the regression model is that the errors are uncorrelated; in other words, the correlation between error terms is equal to zero. We are quite likely to encounter only uncorrelated errors when dealing with cross-sectional data. However, **autocorrelation**—the correlation of an error term and a lagged version of itself—is likely to occur with time-series data because errors made in a particular time period are readily carried over to future time periods. For example, an underestimate of the GDP in one year can generate more underestimates in future time periods.

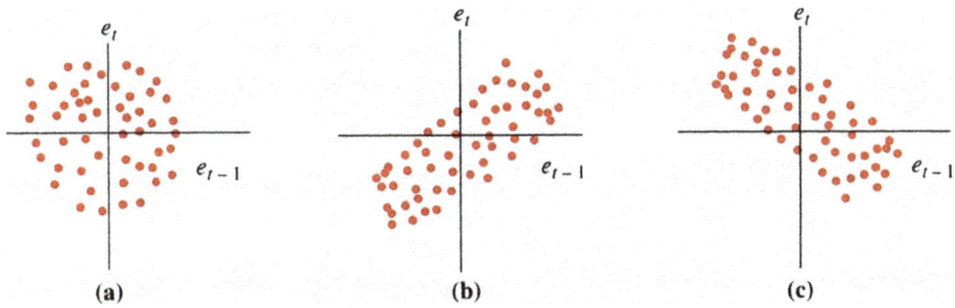

Fig. 3.6 **a** No autocorrelation, **b** positive autocorrelation, and **c** negative autocorrelation

Figure 3.6 shows examples of autocorrelation. The error in period t is graphed on the y-axis, the error in period $t - 1$ on the x-axis. If no autocorrelation exists, the plot is random, as shown in Fig. 3.6a. The upward slope of the diagram in Fig. 3.6b signals positive autocorrelation. This implies that a large error in period $t - 1$ will be associated with a large error in period t. Negative autocorrelation is apparent in Fig. 3.6c, where high errors in the previous period tend to result in low errors in the next period. Again, if plotting reveals a pattern, the regression assumption that the residuals are random and not correlated through time has been violated.

The Durbin–Watson Statistic

Detecting autocorrelation by inspecting errors is difficult; thus, the **Durbin–Watson statistic** (DW) is generally used to detect first-order autocorrelation. **First-order autocorrelation** occurs when correlation between errors is separated by one period. Here, we will discuss both first-order autocorrelation and the Durbin–Watson (DW) statistic in detail.

If e_t and e_{t-1} are residual terms in periods t and $t - 1$, respectively, then first-order correlation r_1 for these error terms is defined as follows:

$$r_1 = \frac{\sum_{t=2}^{n}(e_t - \bar{e}_t)(e_{t-1} - \bar{e}_{t-1})}{\sqrt{\sum_{t=2}^{n}(e_t - \bar{e}_t)^2 \sum_{t=2}^{n}(e_{t-1} - \bar{e}_{t-1})^2}} \quad (3.10)$$

where $\bar{e}_t$ and $\bar{e}_{t-1}$ are means of e_t and e_{t-1}, respectively.

The DW statistic can be used to test the null hypothesis that no first-order autocorrelation exists among the residuals of a regression. The statistic, calculated from the residuals, is

$$\mathrm{DW} = \frac{\sum_{t=2}^{n}(e_t - e_{t-1})^2}{\sum_{t=1}^{n} e_t^2} \quad (3.11)$$

where e_t and e_{t-1} are error terms in periods t and $t - 1$, respectively. This statistic is calculated by summing the difference in the error terms separated by one period and dividing by the squared error term. Figure 3.7 shows that the DW statistic falls between zero and 4. If no autocorrelation exists, DW equals 2, because the difference in the error terms is directly proportional to the error term in period t. Positive autocorrelation exists if DW is low, because the difference between the error term in period t and in period $t - 1$ tends to be very small. Negatively correlated errors are closer to 4 because the difference in the error terms tends to be large.

We can use the Durbin-Watson table to determine whether the null hypothesis is accepted or reject. These tables give two values, d_L and d_U, for different sample sizes n and number of independent variables k. If the DW statistic falls between d_U and $4 - d_U$, the null hypothesis that the correlation between lagged errors is equal to zero is accepted. If the statistic is less than d_L, that null hypothesis is rejected in favor of positive autocorrelation. Negative autocorrelation exists if the DW is greater than $4 - d_L$. Did you notice that there is an indeterminate zone in which no judgment can be made? This zone is between d_L and d_U, or $4 - d_U$ and $4 - d_L$. (The

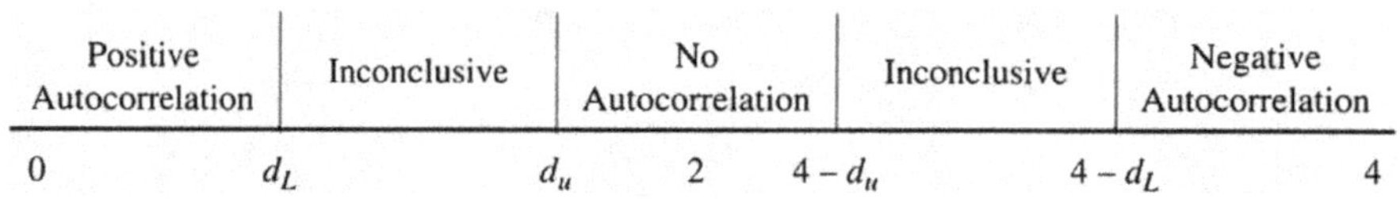

Fig. 3.7 Critical values of the Durbin–Watson statistic

probability level of d_U, d_L depends on whether we are performing a one- or a two-tailed test.)

Example 3.3 How to Detect First-Order Autocorrelation

Annual rates of return for both JNJ and MRK and market rates of return during 1990 to 2009 (Table 3.5) are used to estimate the market model of Eq. (3.12) for JNJ and MRK.

$$R_{i,t} = a_i + b_i R_{m,t} + e_{i,t} \tag{3.12}$$

where $R_{i,t}$ is the rate of return for the ith firm in period t, $R_{m,t}$ is the market rate of return, a_i and b_i are regression parameters, and $e_{i,t}$ is the error term.

MINITAB outputs of the estimated market models for JNJ and MRK are presented in Figs. 3.8 and 3.9. These two outputs indicate that the beta coefficients b_i for JNJ and MRK are 0.639 and 0.7897, respectively. In addition, we see that the DW statistics for JNJ and MRK are 2.5128 and 2.45845. Remember, this test determines whether there is evidence of autocorrelation among the residuals. In this example, there are one independent variable and 20 observations. By checking the Durbin-Watson table for a level of significance of 5 percent (under a two-tailed test, $\alpha = 10$ percent), we find that d_L is 1.20 and d_U is 1.41. Because both DW values are greater than d_U and less than $4 - d_U$, we conclude that there is no evidence of autocorrelation in either regression.

Dividend per share (DPS_t) and earnings per share (EPS_t) for both JNJ and MRK during 1990–2009 (Table 3.6) are used to estimate the regression specified in Eq. (3.13).

$$\mathrm{DPS}_{i,t} = a_i + b_i \mathrm{EPS}_{i,t} + e_{i,t} \tag{3.13}$$

where $\mathrm{DPS}_{i,t}$ and $\mathrm{EPS}_{i,t}$ are dividend per share and earnings per share for the ith firm in period t.

MINITAB output for estimated Eq. (3.13) is presented in Figs. 3.10 and 3.11. From Figs. 3.10 and 3.11, we know that DPS is highly correlated with EPS for both JNJ and MRK. And the DW statistics are 1.77395 and 1.06240, respectively.

In a two-tailed test, we look up in the Durbin–Watson table critical values for a 5 percent level of significance, the number of observations 20, and the number of independent variables 1. The critical values are 1.20 and 1.41. Remember that if the DW falls between the two values, the test is inconclusive. If it is less than 1.20, positive autocorrelation is a problem. The DW of JNJ is larger than 1.20, and the DW of MRK is below that value, so we conclude that positive autocorrelation exists among the residuals in the regression model of Eq. (3.13) for MRK.

When results of Eq. (3.13) for MRK, as indicated in Eq. (3.11), imply that the residuals of regression might be autocorrelated, least squares estimates and inferences based on them can be very unreliable. Under these circumstances, a modified model of Eq. (3.13) can be used to adjust for the autocorrelation.

$$\begin{aligned}\mathrm{DPS}_{i,t} &= \hat{\rho}\mathrm{DPS}_{i,t-1} \\ &= a_i(1-\hat{\rho}) + b_i(\mathrm{EPS}_{i,t} - \rho\mathrm{EPS}_{i,t-1}) \\ &\quad + e'_{i,t}\end{aligned} \tag{3.13a}$$

Table 3.5 Rates of return for JNJ, MRK, and S&P 500

Year	JNJ	MRK	S&P 500
1990	0.230	0.185	0.036
1991	0.617	0.879	0.124
1992	−0.551	−0.734	0.105
1993	−0.092	−0.183	0.086
1994	0.245	0.142	0.020
1995	0.585	0.754	0.177
1996	−0.410	0.235	0.238
1997	0.341	0.353	0.303
1998	0.288	0.410	0.243
1999	0.124	−0.537	0.223
2000	0.140	0.412	0.075
2001	−0.431	−0.357	−0.163
2002	−0.078	−0.013	−0.168
2003	−0.021	−0.158	−0.029
2004	0.249	−0.272	0.171
2005	−0.032	0.037	0.068
2006	0.122	0.418	0.086
2007	0.035	0.367	0.127
2008	−0.076	−0.451	−0.174
2009	0.108	0.254	−0.223
MEAN	0.070	0.087	0.066
MEDIAN	115	0.164	0.086
STD	0.303	0.425	0.151
CS.	−10.688	−1.743	−162.890
CV	4.353	4.879	2.284
Percentiles			
10th	−0.412	−0.459	−0.168
25th	−0.077	−0.205	0.008
50th	0.115	1.164	0.086
75th	0.246	0.378	0.173
90th	0.365	0.452	0.238

Regression Analysis: JNJ versus S&P

The regression equation is

JNJ = 0.0273 + 0.639 S&P

Predictor	Coef	SE Coef	T	P	VIF
Constant	0.02726	0.07227	0.38	0.71	
S&P	0.6386	0.4472	1.43	0.17	1

S = 0.294804 R-Sq = 10.2% R-Sq(adj) = 5.2%

Analysis of Variance

Source	DF	SS	MS	F	P
Regression	1	0.1772	0.1772	2.04	0.17
Residual Error	18	1.56437	0.08691		
Total	19	1.74158			

Obs	S&P	JNJ	Fit	SE Fit	Residual	St Resid
1	0.036	0.2301	0.0505	0.0673	0.1796	0.63
2	0.124	0.6168	0.1066	0.0709	0.5102	1.78
3	0.105	-0.5513	0.0944	0.0682	-0.6457	-2.25R
4	0.086	-0.0916	0.0821	0.0665	-0.1736	-0.6
5	0.02	0.2449	0.04	0.0691	0.2049	0.71
6	0.177	0.5846	0.14	0.0823	0.4445	1.57
7	0.238	-0.4098	0.1791	0.1011	-0.5888	-2.13R
8	0.303	0.3408	0.2205	0.1246	0.1203	0.45
9	0.243	0.2877	0.1823	0.1029	0.1054	0.38
10	0.223	0.1242	0.1695	0.0962	-0.0453	-0.16
11	0.075	0.1397	0.0753	0.066	0.0644	0.22
12	-0.163	-0.4312	-0.077	0.122	-0.3542	-1.32
13	-0.168	-0.078	-0.0798	0.1236	0.0018	0.01
14	-0.029	-0.0212	0.0088	0.0785	-0.03	-0.11
15	0.171	0.2486	0.1367	0.081	0.1119	0.39
16	0.068	-0.0325	0.0705	0.0659	-0.103	-0.36
17	0.086	0.1225	0.0819	0.0665	0.0406	0.14
18	0.127	0.0346	0.1085	0.0713	-0.0739	-0.26
19	-0.174	-0.0764	-0.0839	0.1261	0.0075	0.03
20	-0.223	0.1085	-0.1151	0.1452	0.2236	0.87

R denotes an observation with a large standardized residual.

Durbin-Watson statistic = 2.51280

Fig. 3.8 MINITAB output of market model for JNJ

Regression Analysis: MRK versus S&P

The regression equation is

MRK = 0.035 + 0.789 S&P

Predictor	Coef	SE Coef	T	P	VIF
Constant	0.0348	0.1027	0.34	0.738	
S&P	0.7886	0.6353	1.24	0.23	1

S = 0.418787 R-Sq = 7.9% R-Sq(adj) = 2.8%

Analysis of Variance

Source	DF	SS	MS	F	P
Regression	1	0.2702	0.2702	1.54	0.23
Residual Error	18	3.1569	0.1754		
Total	19	3.4271			

Obs	S&P	MRK	Fit	SE Fit	Residual	St Resid
1	0.036	0.1854	0.0635	0.0955	0.1219	0.3
2	0.124	0.8786	0.1328	0.1007	0.7458	1.83
3	0.105	-0.7338	0.1178	0.0969	-0.8516	-2.09R
4	0.086	-0.183	0.1025	0.0945	-0.2855	-0.7
5	0.02	0.1424	0.0506	0.0981	0.0919	0.23
6	0.177	0.7539	0.1741	0.117	0.5798	1.44
7	0.238	0.2353	0.2223	0.1437	0.013	0.03
8	0.303	0.3525	0.2735	0.177	0.0791	0.21
9	0.243	0.4097	0.2263	0.1461	0.1834	0.47
10	0.223	-0.5371	0.2105	0.1366	-0.7476	-1.89
11	0.075	0.4119	0.0942	0.0938	0.3177	0.78
12	-0.163	-0.3574	-0.0939	0.1733	-0.2635	-0.69
13	-0.168	-0.0133	-0.0974	0.1756	0.0841	0.22
14	-0.029	-0.1583	0.012	0.1114	-0.1703	-0.42
15	0.171	-0.272	0.17	0.115	-0.4419	-1.1
16	0.068	0.0369	0.0882	0.0936	-0.0513	-0.13
17	0.086	0.4183	0.1023	0.0944	0.3161	0.77
18	0.127	0.3674	0.1352	0.1013	0.2323	0.57
19	-0.174	-0.4508	-0.1025	0.1791	-0.3483	-0.92
20	-0.223	0.2541	-0.141	0.2062	0.395	1.08

R denotes an observation with a large standardized residual.

Durbin-Watson statistic = 2.45845

Fig. 3.9 MINITAB output market model for MRK

Table 3.6 EPS, DPS, and PPS for Johnson & Johnson, Merck, and the S&P 500

Year	Johnson & Johnson			Merck			S&500
	DPS	EPS	PPS	DPS	EPS	PPS	
1988	1.89	5.63	85.13	1.37	3.02	57.75	265.79
1989	1.10	3.19	59.38	1.70	3.74	77.50	322.84
1990	1.29	3.38	71.75	2.00	4.51	89.88	334.59
1991	1.51	4.30	114.50	2.34	5.39	166.50	376.18
1992	0.88	1.54	50.50	0.95	1.70	43.38	415.74
1993	1.00	2.71	44.88	1.06	1.86	34.38	451.41
1994	1.12	3.08	54.75	1.15	2.35	38.13	460.42
1995	1.25	3.65	85.50	1.24	2.63	65.63	541.72
1996	0.72	2.12	49.75	1.44	3.12	79.63	670.5
1997	0.83	2.41	65.88	1.70	3.74	106.00	873.43
1998	0.95	2.23	83.88	1.93	4.30	147.50	1085.5
1999	1.04	2.94	93.25	1.09	2.45	67.19	1327.33
2000	1.22	3.39	105.06	1.23	2.90	93.63	1427.22
2001	0.66	1.83	59.10	1.36	3.14	58.80	1194.18
2002	0.78	2.16	53.71	1.41	3.14	56.61	993.94
2003	0.91	2.39	51.66	1.45	3.03	46.20	965.23
2004	1.08	2.83	63.42	1.50	2.61	32.14	1130.65
2005	1.26	3.46	60.10	1.52	2.10	31.81	1207.23
2006	1.44	3.73	66.02	1.52	2.03	43.60	1310.46
2007	1.60	3.63	66.70	1.51	1.49	58.11	1477.19
2008	1.77	4.57	59.83	1.52	3.64	30.40	1220.04
2009	1.91	4.40	64.41	1.58	5.68	36.54	948.05

Source EPS, DPS, and PPS for Johnson & Johnson and Merck are from Standard & Poor's Compustat and Wharton Research Data Services (WRDS)

where $e'_{i,t} = e_{i,t} - \hat{\rho} e_{i,t-1}, \hat{\rho} = 1 - d/2 =$ estimated first-order autocorrelation. MINITAB output in terms of Eq. (3.13a) for MRK is presented in Fig. 3.12. From this figure, we find that

$$\text{DPS}_{i,t} - \hat{\rho}\text{DPS}_{i,t-1} = 0.387 + 0.233(\text{EPS}_{i,t} - \hat{\rho}\text{EPS}_{i,t-1})$$
$$(t = 5.26) \qquad \text{DW} = 1.39$$

This result implies that the DW statistic has improved substantially.

3.5 Model Specification and Specification Bias

A specification error is the error associated with either omitting a relevant variable from a regression model or including an irrelevant variable in it. When specifying a regression model (that is, when determining which variables should be included in the model), we must make two decisions: *which* variables to include and

Data Display

Row	EPS(JNJ)	lagEPS(JNJ)	DPS(JNJ)	lagDPS(JNJ)
1	3.38	3.19	1.29	1.1
2	4.3	3.38	1.51	1.29
3	1.54	4.3	0.88	1.51
4	2.71	1.54	1	0.88
5	3.08	2.71	1.12	1
6	3.65	3.08	1.25	1.12
7	2.12	3.65	0.72	1.25
8	2.41	2.12	0.83	0.72
9	2.23	2.41	0.95	0.83
10	2.94	2.23	1.04	0.95
11	3.39	2.94	1.22	1.04
12	1.83	3.39	0.66	1.22
13	2.16	1.83	0.78	0.66
14	2.39	2.16	0.91	0.78
15	2.83	2.39	1.08	0.91
16	3.46	2.83	1.26	1.08
17	3.73	3.46	1.44	1.26
18	3.63	3.73	1.6	1.44
19	4.57	3.63	1.77	1.6
20	4.4	4.57	1.91	1.77

Regression Analysis: DPS(JNJ) versus EPS(JNJ)

The regression equation is

DPS(JNJ) = 0.019 + 0.376 EPS(JNJ)

Predictor	Coef	SE Coef	T	P	VIF
Constant	0.0186	0.1025	0.18	0.858	
S&P	0.37612	0.03251	11.57	0.000	1.000

S = 0.123214 R-Sq = 88.1% R-Sq(adj) = 87.5%

Analysis of Variance

Source	DF	SS	MS	F	P
Regression	1	2.032	2.032	133.84	0
Residual Error	18	0.2733	0.0152		
Total	19	2.3052			

Fig. 3.10 MINITAB output of Eq. (16.13) for JNJ

Unusual Observations

Obs	EPS(JNJ)	DPS(JNJ)	Fit	SE Fit	Residual	St Resid
3	1.54	0.8769	0.5978	0.0559	0.2791	2.54R
20	4.4	1.9099	1.6735	0.0522	0.2364	2.12R

R denotes an observation with a large standardized residual.

Durbin-Watson statistic = 1.77395

Fig. 3.10 (continued)

what functional form—a log form, a squared term, or a lagged term—to use.

There is often a theoretical basis for selecting the independent variables for a regression model. For example, economic theory states that the demand for a product is a function of price, cost of substitute goods, income, and consumer tastes. In practice, of course, it may be impossible for the researcher to obtain information on all of these items, so **proxies** for the variables are used instead. Because it may be difficult to obtain data on the price of related goods, for example, some type of price index can be used as a proxy.

Model building is more art than science. The researcher tries to include all the variables that affect the outcome of the dependent variable, but no specification can perfectly determine the movements and attributes of the variable in question. The best the researcher can do is search for variables that seem consistent with underlying theory, practice, and common sense. Model specification is of great importance: If significant explanatory variables are left out, the model's worth is compromised even though least squares estimates of the parameters are obtained. Here again, good judgment and reliance on theory must guide the researcher.

Example 3.4 Impact of the Omission of Variables on Estimated Regression Coefficients

Suppose we omit retained earnings (RE) from Eq. (3.7) in Example 3.1 for the year 2009. The equation becomes

$$\text{PPS}_i = a' + b'\text{DPS}_i \qquad (3.7')$$

where

$$\begin{aligned} b' &= \frac{\text{Cov}(\text{DPS}_i, \text{DPS}_i)}{\text{Var}(\text{DPS}_i)} \\ &= \frac{12.3582}{0.47814} \\ &= 25.8462 \\ a' &= \overline{\text{PPS}} - \hat{b}'\overline{\text{DPS}} \\ &= 44.6360 - (25.8462)(1.2145) \\ &= 13.2458 \end{aligned}$$

By regressing RE_i on DPS_i, we obtain the auxiliary regression

$$\text{RE}_i = b_0 + b_1\text{DPS}_i \qquad (3.14)$$

where

$$\begin{aligned} b_1 &= \frac{\text{Cov}(\text{RE}_i, \text{DPS}_i)}{\text{Var}(\text{DPS}_i)} \\ &= \frac{0.9597}{0.47814} = 2.0072 \\ b_0 &= \overline{\text{RE}_i} - \hat{b}_1\overline{\text{DPS}_i} \\ &= 2.8033 - (2.0072)(1.2145) \\ &= 0.3656 \end{aligned}$$

From the specification analysis of Theil (1971),[3] the relationship among b, b', c, and b_1 can be defined as

$$b' = c + b_1 b \qquad (3.15)$$

where b' and b_1 are estimated in accordance with Eqs. (3.7′) and (3.14), respectively, and b and c are estimated by using Eq. (3.7). Substituting into the foregoing equation the estimated b and

[3] H. Theil (1971). *Principles of Econometrics* (New York: Wiley).

Data Display

Row	EPS(MRK)	lagEPS(MRK)	DPS(MRK)	lagDPS(MRK)
1	4.51	3.74	2	1.7
2	5.39	4.51	2.34	2
3	1.7	5.39	0.95	2.34
4	1.86	1.7	1.06	0.95
5	2.35	1.86	1.15	1.06
6	2.63	2.35	1.24	1.15
7	3.12	2.63	1.44	1.24
8	3.74	3.12	1.7	1.44
9	4.3	3.74	1.93	1.7
10	2.45	4.3	1.09	1.93
11	2.9	2.45	1.23	1.09
12	3.14	2.9	1.36	1.23
13	3.14	3.14	1.41	1.36
14	3.03	3.14	1.45	1.41
15	2.61	3.03	1.5	1.45
16	2.1	2.61	1.52	1.5
17	2.03	2.1	1.52	1.52
18	1.49	2.03	1.51	1.52
19	3.64	1.49	1.52	1.51
20	5.68	3.64	1.58	1.52

Regression Analysis: DPS(MRK) versus EPS(MRK)

The regression equation is

DPS(MRK) = 0.821 + 0.211 EPS(MRK)

Predictor	Coef	SE Coef	T	P	VIF
Constant	0.8209	0.1509	5.44	0	
EPS(MRK)	0.21133	0.04585	4.61	0	1

S = 0.232678 R-Sq = 54.1% R-Sq(adj) = 51.6%

Analysis of Variance

Source	DF	SS	MS	F	P
Regression	1	1.1503	1.1503	21.25	0
Residual Error	18	0.9745	0.0541		
Total	19	2.1248			

Unusual Observations

Fig. 3.11 MINITAB output of Eq. (3.13) for MRK

Obs	EPS(MRK)	DPS(MRK)	Fit	SE Fit	Residual	St Resid
20	5.68	1.5836	2.0213	0.1296	-0.4377	-2.27RX

R denotes an observation with a large standardized residual.

X denotes an observation whose X value gives it large leverage.

Durbin-Watson statistic = 1.06240

Fig. 3.11 (continued)

c (from Table 3.1), $b_1 = 2.0072$, and $b' = 25.8462$, we obtain

$$\begin{aligned} 25.8462 &\cong 10.581 + (2.0072)(7.605) \\ &= 25.8454 \end{aligned}$$

This implies that the misspecification error when RE_i is omitted from Eq. (3.7) is 15.2644 (25.8454 − 10.581) for b.

3.6 Nonlinear Models

Thus far we have assumed that there is a linear relationship between the dependent variable and a set of independent variables. This assumption yields a convenient approximation of the phenomena being modeled. However, there are times when other functional forms of the independent variable provide a better depiction of reality. In this section, we discuss nonlinear models, including quadratic and log linear models. We will continue to use the same regression concepts, such as hypothesis testing and confidence intervals, in our analysis.

The Quadratic Model

A quadratic model takes the form

$$y_i = a_0 + a_1 x_i + a_2 x_i^2 + e_i \tag{3.16}$$

The only difference between this model and the models previously specified is the squared term in this model. The square of the variable is calculated and used as another independent variable, and a regression on the data is run. The quadratic term traces a parabola, as shown in Fig. 3.13. If the parameter for the squared term has a positive value, the parabola opens upward. A negative parameter implies a downward-opening parabola. The linear model uses only the part of the data that is available to fit the curve. An example is shown in Fig. 3.14. Here, only the upward-sloping part is used by the linear model to fit the data.

Again, we must exercise judgment and common sense to determine whether a quadratic term is needed in the model. We may have some idea how the dependent variable will react to changes in the independent variable, and graphs of the data may give us more information to help us specify the model. A quadratic model might be of interest in a production function where output of a product is the dependent variable and an input is the independent variable. The **law of diminishing returns** states that after a certain point, the marginal product of the variable inputs declines when additional units of a variable input are added to fixed inputs. In agriculture for example, doubling the fertilizer doubles the output of corn at low levels of fertilizer use. However, further increases in fertilizer increase the output only marginally. If a regression were to be run, the sign of the quadratic term would be negative, reflecting the fact that the output increases at a decreasing rate. It should be noted that having x and x^2 in the regression introduces a certain degree of collinearity.

Example 3.5 A Nonlinear Market Model

Suppose the following regression model is run.

$$R_{i,t} = a + b_1 R_{m,t} + b_2 R_{m,t}^2 + e_{i,t} \tag{3.17}$$

where $R_{i,t}$ is the rate of return on Johnson & Johnson stock during 1970 to 2009, and $R_{m,t}$ is the rate of return on the market index (the S&P 500). The estimated results we get when we add a quadratic term for $R_{m,t}$ are

Data Display

Row	EPS(MRK)	DPS(MRK)	LolagEPS(MRK)	LolagDPS(MRK)	dif(EPS_MRK)	dif(DPS_MRK)
1	4.51	2	1.7513	0.79812	2.7587	1.19859
2	5.39	2.34	2.1156	0.93606	3.2744	1.40270
3	1.7	0.95	2.52771	1.09641	-0.82771	-0.14961
4	1.86	1.06	0.79565	0.44386	1.06435	0.61893
5	2.35	1.15	0.87109	0.49824	1.47891	0.64821
6	2.63	1.24	1.10092	0.53746	1.52908	0.70555
7	3.12	1.44	1.23162	0.58272	1.88838	0.85753
8	3.74	1.7	1.46125	0.67519	2.27875	1.02142
9	4.3	1.93	1.75192	0.79537	2.54808	1.13237
10	2.45	1.09	2.01569	0.90372	0.43431	0.18972
11	2.9	1.23	1.14841	0.51261	1.75159	0.72218
12	3.14	1.36	1.35901	0.57887	1.78099	0.78017
13	3.14	1.41	1.46997	0.63712	1.67003	0.77008
14	3.03	1.45	1.47197	0.6597	1.55803	0.78928
15	2.61	1.5	1.4213	0.67928	1.1887	0.81600
16	2.1	1.52	1.22409	0.70099	0.87591	0.81633
17	2.03	1.52	0.98671	0.71132	1.04329	0.80566
18	1.49	1.51	0.95011	0.71116	0.53989	0.79858
19	3.64	1.52	0.70022	0.70776	2.93978	0.80736
20	5.68	1.58	1.70632	0.71029	3.97368	0.87329

Regression Analysis: dif(DPS_MRK) versus dif(EPS_MRK)

The regression equation is

dif(DPS_MRK) = 0.387 + 0.233 dif(EPS_MRK)

Predictor	Coef	SE Coef	T	P	VIF
Constant	0.38673	0.08842	4.37	0	
dif(EPS_MRK)	0.23318	0.04433	5.26	0	1

S = 0.210802 R-Sq = 60.6% R-Sq(adj) = 58.4%

Fig. 3.12 MINITAB output of Eq. (3.13a) for MRK

Analysis of Variance

Source	DF	SS	MS	F	P
Regression	1	1.2296	1.2296	27.67	0
Residual Error	18	0.7999	0.0444		
Total	19	2.0295			

Unusual Observations

Obs	dif(EPS_MRK)	dif(DPS_MRK)	Fit	SE Fit	Residual	St Resid
3	-0.83	-0.1496	0.1937	0.121	-0.3433	-1.99 X
20	3.97	0.8733	1.3133	0.1118	-0.44	-2.46R

R denotes an observation with a large standardized residual.

X denotes an observation whose X value gives it large leverage.

Durbin-Watson statistic = 1.39124

Fig. 3.12 (continued)

$$R_{i,t} = \underset{(0.34)}{0.0233} + \underset{(2.29)}{0.865}R_{m,t} - \underset{(-0.93)}{1.928}R^2_{m,t}, \quad R^2 = 0.124$$

where t-values are in parentheses. From t table, by interpolation, we find that $t_{0.025,38} = 2.025$. Because only 2.29 is larger than 2.025, we conclude that estimated b_1 is significantly different from 0 at $\alpha = 0.05$, but estimated b_2 is insignificant. These results imply that $R^2_{m,t}$ should not be included in the regression because the t-value associated with this quadratic term is statistically insignificant at $\alpha = 0.05$.

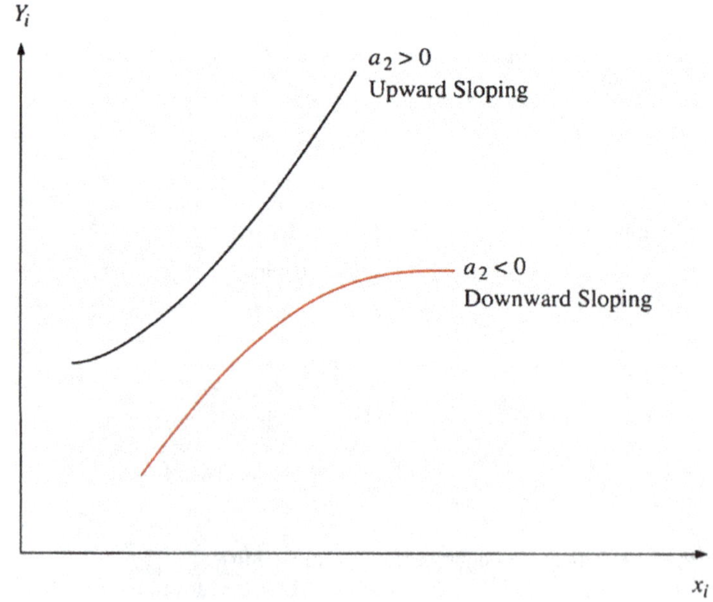

Fig. 3.13 Two different types of nonlinear curve

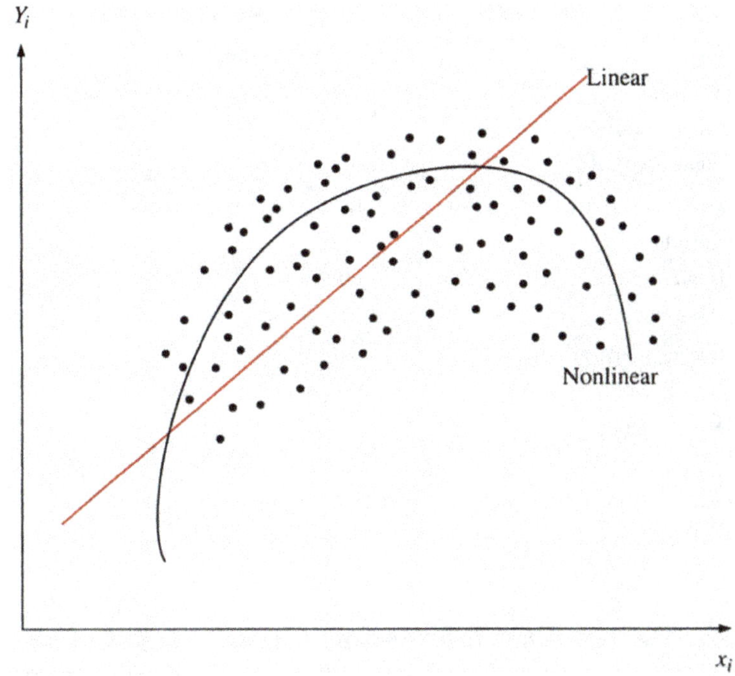

Fig. 3.14 Linear curve versus nonlinear curve

The Log Linear and the Log-Log Linear Model

A common transformed linear model involves the e-based logarithmic transformation of variables such as the one shown in Eq. (3.18).[4]

$$\log_e Y_i = \alpha + \beta_1(\log_e x_{1i}) + \beta_2(\log_e x_{2i}) + \cdots + \beta_n(\log_e x_{ni}) + \epsilon_i \tag{3.18}$$

The **log-log linear model** is a linear model with logarithmic transformation made on both dependent and independent variables. If only the dependent variable is being lognormally transformed, then we call this linear model a **log linear model**. As in the quadratic case, a visual inspection of the data may help determine whether a model should be specified in a log form.

The coefficients of a log-log linear model are elasticity coefficients, which give the percentage change in the dependent variable that is due to a 1% change in the independent variable. For example, suppose the demand relationship

$$Q = aP_1^b X^c P_2^d$$

has been specified, where Q is quantity purchased, P_1 is price, X is income, P_2 is the price of a competing good, and a, b, c, and d are parameters. Then, b, c, and d are elasticities of P_1, X, and P_2, respectively.[5]

[4]Equation (3.18) is obtained by taking the logarithmic transformation of a model of the equation

$$Y_i = \alpha_0 x_{1i}^{\beta_1} x_{2i}^{\beta_2} \cdots x_{ni}^{\beta_n}$$

and letting $\alpha = \log \alpha_0$.

[5]For example, the elasticity coefficient, e_p, is defined as $(dQ/dP_1)(P_1/Q)$. The first derivative is

$$dQ/dP_1 = abP_1^{b-1} X^c P_2^d$$

Substituting Q and this equation into the definition of e_p, we obtain

$$e_p = abP_1^{b-1} X^c P_2^d \left(\frac{P_1}{aP_1^b X^c P_2^d} \right) = b$$

Table 3.7 Cylinder volume and miles per gallon for 11 different kinds of cars

Car	Cylinder volume, x	Miles per gallon, y
VW Golf	97	37
Chevy Cavalier	173	19
Plymouth Horizon	97	31
Pontiac Firebird	151	23
Corvette	350	17
Honda Accord	119	27
Dodge Omni	97	31
Renault Alliance	85	35
Olds Firenza	173	19
Nissan Sentra	97	31
Ford Escort	114	32

Source 1986 Gas Mileage Guide, EPA Fuel Economy Estimates, US Dept. of Energy. Wards Automotive Yearbook, 1986

Example 3.6 The Relationship Between Cylinder Volume and Miles per Gallon

To study the relationship between cylinder volume and miles per gallon, we use the 1986 EPA mileage guide which gives the engine size and estimated city miles per gallon ratings for 11 gasoline-fueled subcompact and compact cars. Those data, as given in Table 3.7, were used to estimate the following regression relationships:

$$y_i = a + bx_i + e_i \tag{3.19a}$$

$$\log_e y_i = a' + b' \log_e x_i + e_i' \tag{3.19b}$$

where

$$y_i = \text{miles per gallon for } i\text{th kind of car}$$
$$X_i = \text{cylinder volume for } i\text{th kind of car}$$

Equation (3.19a) is a linear model and Eq. (3.19b) is a log-log linear model that is similar to Eq. (3.18). MINITAB regression outputs for Eqs. (3.19a) and (3.19b) are presented in Figs. 3.15 and 3.16, respectively. From these outputs, the estimates regression lines are

$$\hat{y}_i = \underset{(12.95)}{37.677} - \underset{(-3.95)}{0.07241}x_i \quad R^2 = 0.634$$

$$\log \hat{y}_i = \underset{(14.61)}{6.2020} - \underset{(-6.90)}{0.60133} \log x_i \quad R^2 = 0.841$$

and t-values are in parentheses.

From t table, we find $t_{.005,9} = 3.250$. Because all estimated t-values are larger than 3.250, all estimated parameters are significantly different from 0 at $\alpha = 0.01$. The scatter diagrams of residuals against the independent variable, which is not shown here, indicate that the logarithmic transformation will make the residuals become homoscedastic. Also note that the R^2 of the log-log linear model is 0.841, which is larger than that of the linear model (0.634). Finally, $\hat{b}' = -0.60133$ implies that with a 1% increase in cylinder volume, miles per gallon will decrease by 0.60133%.

Fig. 3.15 MINITAB output of Eq. (3.19a)

```
MTB > READ C1 C2
DATA> 97 37
DATA> 173 19
DATA> 97 31
DATA> 151 23
DATA> 350 17
DATA> 119 27
DATA> 97 31
DATA> 85 35
DATA> 173 19
DATA> 97 31
DATA> 114 32
DATA> END
     11 rows read.
MTB > BRIEF 2
MTB > REGRESS C2 1 C1;
SUBC> RESIDUAL C3;
SUBC> DW.

Regression Analysis

The regression equation is
C2 = 37.7 - 0.0724 C1

Predictor        Coef        StDev          T        P
Constant       37.677        2.909      12.95    0.000
C1           -0.07241      0.01833      -3.95    0.003

S = 4.410        R-Sq = 63.4%        R-Sq(adj) = 59.4%

Analysis of Variance

Source         DF         SS          MS         F        P
Regression      1     303.67      303.67     15.61    0.003
Error           9     175.06       19.45
Total          10     478.73

Unusual Observations
Obs         C1          C2          Fit   StDev Fit   Residual    St Resid
  5        350       17.00        12.33        4.05       4.67      2.68RX

R denotes an observation with a large standardized residual
X denotes an observation whose X value gives it large influence.

Durbin-Watson statistic = 2.78

MTB > GSTD
* NOTE  * Standard Graphics are enabled.
          Professional Graphics are disabled.
          Use the GPRO command to enable Professional Graphics.
```

3.7 Lagged Dependent Variables

In all the models we have discussed, the dependent variable was a function of independent variables in period t. However, for time-series data, we often want to lag the dependent variable by one period to estimate the effect on the variable from a previous period. The model is

$$Y_t = \alpha + \beta_1 X_{1t} + \beta_2 X_{2t} + \cdots + \beta_k X_{kt} + \gamma Y_{t-1} + \epsilon_t \tag{3.20}$$

Here, the dependent variable is a function of the X's and of the dependent variable lagged one period.

A regression can be run on the data to estimate the coefficients of Eq. (3.20). However, our interpretation of these estimated coefficients must be modified to take into account the long-run effect. The short-run (current) effect is that a 1-unit increase in X_k leads to a β_k-unit increase in Y. This is the usual interpretation of regression coefficients. The long-run effect of regression coefficients is

$$\beta_i^L = \frac{\beta_i}{1-\gamma}, \quad i = 1, 2, \ldots, k \tag{3.21}$$

where β_i^L represents the long-run coefficient that takes the lagged effect into account, and γ is the coefficient associated with the lagged dependent

Fig. 3.16 MINITAB output of Eq. (3.19b)

```
MTB > LET C4=LOGE(C1)
MTB > LET C5=LOGE(C2)
MTB > PRINT C1 C2 C4 C5

Data Display

 Row    C1    C2        C4        C5

   1    97    37   4.57471   3.61092
   2   173    19   5.15329   2.94444
   3    97    31   4.57471   3.43399
   4   151    23   5.01728   3.13549
   5   350    17   5.85793   2.83321
   6   119    27   4.77912   3.29584
   7    97    31   4.57471   3.43399
   8    85    35   4.44265   3.55535
   9   173    19   5.15329   2.94444
  10    97    31   4.57471   3.43399
  11   114    32   4.73620   3.46574

MTB > REGRESS C5 1 C4;
SUBC> RESIDUAL C6;
SUBC> DW.

Regression Analysis

The regression equation is
C5 = 6.20 - 0.601 C4

Predictor         Coef        StDev           T        P
Constant        6.2020       0.4246       14.61    0.000
C4            -0.60133      0.08711       -6.90    0.000

S = 0.1140       R-Sq = 84.1%      R-Sq(adj) = 82.4%

Analysis of Variance

Source         DF          SS          MS          F        P
Regression      1     0.61985     0.61985      47.66    0.000
Error           9     0.11706     0.01301
Total          10     0.73691

Unusual Observations
Obs         C4          C5          Fit   StDev Fit    Residual    St Resid
  5       5.86      2.8332       2.6794      0.0936      0.1538      2.36RX

R denotes an observation with a large standardized residual
X denotes an observation whose X value gives it large influence.

Durbin-Watson statistic = 2.31
```

variable as defined in Eq. (3.20). In a moment, we will offer two examples to show how Eq. (3.21) is calculated.

When a lagged dependent variable is used in the regression, the Durbin–Watson statistic is not a reliable indicator of autocorrelation. Another statistic—the **Durbin H**—is used instead. This statistic is

$$\mathrm{DH} = (1 - d/2)\sqrt{\frac{n}{1 - nV(\gamma)}} \tag{3.22}$$

where d is the Durbin–Watson statistic defined in Eq. (3.11), and $V(\gamma)$ is the least squares estimate of the variance of the coefficient of the lagged variable. Under the null hypothesis, H is normally distributed with mean zero and variance 1. Therefore, the Z statistic of normal distribution can be used to do the test.

Example 3.7 The Relationship Between Dividend per Share and Earnings per Share MINITAB outputs of two regressions of Eq. (3.23) for JNJ and MRK for period 1990–2009 are presented in Figs. 3.17 and 3.18, respectively.

$$\mathrm{DPS}_{i,t} = \alpha + \beta \mathrm{EPS}_{i,t} + \gamma \mathrm{DPS}_{i,t-1} + \epsilon_{i,t} \tag{3.23}$$

Data Display

Obs	EPS(JNJ)	DPS(JNJ)	Fit	SE Fit	Residual	St Resid
1	3.38	1.2877	1.2653	0.024	0.0223	0.24
2	4.3	1.5084	1.6199	0.0386	-0.1116	-1.28
3	1.54	0.8769	0.7948	0.0698	0.0821	1.26 X
4	2.71	1.0003	0.9812	0.0278	0.0191	0.21
5	3.08	1.1155	1.1383	0.0239	-0.0227	-0.25
6	3.65	1.2546	1.3575	0.028	-0.103	-1.13
7	2.12	0.7157	0.9052	0.0401	-0.1895	-2.19R
8	2.41	0.83	0.8352	0.0365	-0.0053	-0.06
9	2.23	0.9514	0.8118	0.0321	0.1397	1.55
10	2.94	1.0429	1.0782	0.025	-0.0353	-0.38
11	3.39	1.2163	1.2514	0.0259	-0.0351	-0.38
12	1.83	0.6605	0.7998	0.0453	-0.1394	-1.66
13	2.16	0.7796	0.7377	0.0402	0.0419	0.48
14	2.39	0.9129	0.8482	0.0331	0.0647	0.72
15	2.83	1.0824	1.0309	0.0263	0.0515	0.56
16	3.46	1.2591	1.2861	0.0257	-0.027	-0.29
17	3.73	1.4411	1.427	0.0276	0.0141	0.15
18	3.63	1.6044	1.45	0.0319	0.1545	1.72
19	4.57	1.7718	1.8034	0.0478	-0.0317	-0.38
20	4.4	1.9099	1.7993	0.0535	0.1106	1.4

Regression Analysis: DPS(JNJ) versus EPS(JNJ), lagDPS(JNJ)

The regression equation is

DPS(JNJ) = - 0.161 + 0.323 EPS(JNJ) + 0.304 lagDPS(JNJ)

Predictor	Coef	SE Coef	T	P	VIF
Constant	-0.16118	0.09389	-1.72	0.104	
EPS(JNJ)	0.32326	0.02918	11.08	0	1.34
lagDPS(JNJ)	0.30374	0.08443	3.6	0.002	1.34

S = 0.0955327 R-Sq = 93.3% R-Sq(adj) = 92.5%

Analysis of Variance

Source	DF	SS	MS	F	P
Regression	2	2.1501	1.075	117.79	0
Residual Error	17	0.1552	0.0091		
Total	19	2.3052			

Fig. 3.17 MINITAB output of Eq. (16.23) for JNJ

Source	DF	Seq SS
EPS(JNJ)	1	2.032
lagDPS(JNJ)	1	0.1181

R denotes an observation with a large standardized residual.
X denotes an observation whose X value gives it large leverage.
Durbin-Watson statistic = 1.85326

Fig. 3.17 (continued)

Equation (3.23) is obtained by adding a variable for lagged dividend per share (DPS_{t-1}) to the right-hand side of Eq. (3.13).

The results shown in Figs. 3.17 and 3.18 indicate that t-statistics of the γ coefficient for JNJ and MRK are 3.60 and 0.90, respectively. Therefore, lagged dividend per share is important in explaining dividend per share in period t for only JNJ.

Long-run coefficients (accumulated effect over all future periods) associated with EPS in terms of Eq. (3.21) are calculated as follows:

$$\frac{0.323}{1-0.304} = 0.04641 \text{ long-run coefficient for JNJ}$$
$$\frac{0.203}{1-0.146} = 0.2377 \text{ long-run coefficient for MRK}$$

These long-run coefficients imply that a $1.00 increase in EPS will spell a total increase of $0.4641 and $0.2377 in DPS for JNJ and MRK, respectively. Total dividend increases are much higher than short-run (current) increases of $0.323 and $0.203 for JNJ and MRK.

The partial dividend adjust model discussed in this example has been generalized to integrate dividend behavior model. A detailed discussion of these models can be found in Appendix 3.

Example 3.8 Time Aggregation and the Estimation of the Market Model

We can use market model to investigate the relationship between rates of return for individual securities and market rates of return. The coefficient of market rates of return in market model is called **Beta** coefficient, which is used to determine the degree of nondiversifiable risk of a firm. Cartwright and Lee (1987)[6] used data for heavily and lightly traded firms to evaluate the effects of temporal aggregation on beta estimates. They indicated the importance of price adjustment delays in the trading process and found that temporal aggregation has important effects on the market model. A regression of Eq. (3.24) is used to find the impact coefficient of temporal aggregation

$$R_{i,t} = \alpha_i + \beta_{i1} R_{i,t-1} + \beta_{i2} R_{m,t} + \epsilon_t \tag{3.24}$$

where $R_{i,t}$ is the rate of return for the ith firm in period t, $R_{i,t}$ is the lagged rate of return for the ith firm, $R_{m,t}$ is the market rate of return, and $e_{i,t}$ is the error term. The **Beta** coefficient β_{i2} is the traditional risk measure, and $\beta_{i2}/(1-\beta_{i1})$ is the long-run coefficient to represent the impact of temporal aggregation on systematic risk measure.

Here, we use the annual rates of return in 1970–2009 for Johnson & Johnson and Merck as the examples. MINITAB outputs of two regressions of Eq. (3.24) for JNJ and MRK are presented in Figs. 3.19 and 3.20, respectively.

The results shown in Figs. 3.19 and 3.20 indicate that t-statistics of β_{i1} coefficient for JNJ and MRK are −3.38 and −2.19, respectively. Therefore, lagged rate of return is important in explaining rate of return in period t for both JNJ and MRK.

Long-run coefficients associated with annual market rates of return in terms of Eq. (3.24) are calculated as follows:

[6]See Cartwright and Lee (1987), "Time Aggregation and the Estimation of the Market Model: Empirical Evidence." *Journal of Business and Economic Statistics*, vol. 5 (1), 131–143.

Data Display

Obs	EPS(MRK)	DPS(MRK)	Fit	SE Fit	Residual	St Resid
1	4.51	1.9967	1.795	0.0869	0.2018	0.93
2	5.39	2.3388	2.0168	0.1339	0.322	1.68
3	1.7	0.9468	1.3172	0.1728	-0.3704	-2.35RX
4	1.86	1.0628	1.1461	0.1078	-0.0833	-0.4
5	2.35	1.1465	1.2626	0.0872	-0.1161	-0.54
6	2.63	1.243	1.3317	0.0753	-0.0887	-0.4
7	3.12	1.4403	1.4454	0.065	-0.0051	-0.02
8	3.74	1.6966	1.6002	0.0615	0.0964	0.43
9	4.3	1.9277	1.7514	0.0802	0.1763	0.8
10	2.45	1.0934	1.4094	0.0987	-0.316	-1.49
11	2.9	1.2348	1.3788	0.0808	-0.144	-0.66
12	3.14	1.359	1.4482	0.066	-0.0892	-0.4
13	3.14	1.4072	1.4664	0.0561	-0.0592	-0.26
14	3.03	1.449	1.4511	0.0536	-0.0021	-0.01
15	2.61	1.4953	1.3719	0.0568	0.1234	0.54
16	2.1	1.5173	1.275	0.0704	0.2423	1.09
17	2.03	1.517	1.264	0.0733	0.2529	1.14
18	1.49	1.5097	1.1543	0.0927	0.3554	1.66
19	3.64	1.5151	1.59	0.0581	-0.0749	-0.33
20	5.68	1.5836	2.0052	0.1315	-0.4217	-2.18R

Regression Analysis: DPS(MRK) versus EPS(MRK), lagDPS(MRK)

The regression equation is

DPS(MRK) = 0.630 + 0.203 EPS(MRK) + 0.146 lagDPS(MRK)

Predictor	Coef	SE Coef	T	P	VIF
Constant	0.6297	0.2605	2.42	0.027	
EPS(MRK)	0.20315	0.04696	4.33	0	1.039
lagDPS(MRK)	0.1463	0.162	0.9	0.379	1.039

S = 0.233882 R-Sq = 56.2% R-Sq(adj) = 51.1%

Fig. 3.18 MINITAB output of Eq. (3.23) for MRK

Analysis of Variance

Source	DF	SS	MS	F	P
Regression	2	1.19487	0.59743	10.92	0.001
Residual Error	17	0.92991	0.0547		
Total	19	2.12478			

Source	DF	Seq SS
EPS(MRK)	1	1.15028
lagDPS(MRK)	1	0.4459

R denotes an observation with a large standardized residual.

X denotes an observation whose X value gives it large leverage.

Durbin-Watson statistic = 1.31920

Fig. 3.18 (continued)

$$\frac{0.9433}{1-(-0.4514)} = 0.6499 \text{ long-run coefficient for JNJ}$$

$$\frac{0.9542}{1-(-0.3228)} = 0.7213 \text{ long-run coefficient for MRK}$$

These long-run coefficients imply that a 1-unit change increase in annual market rates of return will spell a total change increase of 0.6499 and 0.7213 in annual rates of return for JNJ and MRK, respectively.

Data for heavily and lightly traded firms are used to evaluate the effects of temporal aggregation on beta estimates, *t*-values, and R^2 estimates. In addition to our analysis of the standard market model, dynamic and random coefficient models are estimated. This study evaluates differences in the short-term and long-term dynamic relationships between the market and each type of firm. It is found that temporal aggregation has important effects on both the specification of a market model and the stability of beta estimates.

The article provides evidence that the simple static market model is misspecified, and it is found that temporal aggregation does affect the specification of a market model and beta estimates. Both dynamic and random coefficient market models can be used to determine the appropriate horizon for specifying the market model and estimating beta coefficients in capital asset pricing.

Example Consumption Function Analysis

In this example, a consumption function is specified with a lagged dependent variable. A consumption function measures the change in consumption that is attributable to a 1-unit change in income. If the slope term in the regression for income is 0.75, for example, individuals tend to spend 75 cents out of every additional dollar earned (the slope term is called the *marginal propensity to consume*, or MPC). A regression of Eq. (3.25) is run with personal consumption (C_t) in the USA from 1962 to 2009 as the dependent variable, and with disposable income (DI_t) and personal consumption lagged one period as the independent variables.

$$C_t = \alpha + \beta_1 \mathrm{DI}_t + \beta_2 C_{t-1} + \epsilon_t \qquad (3.25)$$

The data used to run this regression are listed in Table 3.8, and the results are presented in Table 3.9. The critical *t*-value used to do the test is $t_{0.005,40} = 2.704$.

With a *t*-value of −2.72, the constant is statistically different from zero. Disposable income has a coefficient of 0.487. This implies that individuals will consume about 48.7 cents out of

Data Display

Row	Year	JNJ	MRK	S&P	JNJ(lag)	MRK(lag)
1	1970	-0.681455	-0.105661	-0.149428	0.698039	0.278422
2	1971	0.735557	0.274854	0.181086	-0.681455	-0.105661
3	1972	0.329385	-0.272215	0.110998	0.735557	0.274854
4	1973	-0.132041	-0.080191	-0.016209	0.329385	-0.272215
5	1974	-0.276278	-0.160629	-0.2288	-0.132041	-0.080191
6	1975	0.120214	0.064286	0.039952	-0.276278	-0.160629
7	1976	-0.119259	0.004346	0.18396	0.120214	0.064286
8	1977	0.001873	-0.162669	-0.037349	-0.119259	0.004346
9	1978	-0.017184	0.249828	-0.0222	0.001873	-0.162669
10	1979	0.101555	0.097778	0.072797	-0.017184	0.249828
11	1980	0.286313	0.20599	0.153092	0.101555	0.097778
12	1981	-0.619442	0.031377	0.078043	0.286313	0.20599
13	1982	0.362091	0.031582	-0.065131	-0.619442	0.031377
14	1983	-0.155059	0.101627	0.339988	0.362091	0.031582
15	1984	-0.087818	0.074488	0.000312	-0.155059	0.101627
16	1985	0.49125	0.493088	0.164402	-0.087818	0.074488
17	1986	0.272454	-0.080899	0.264933	0.49125	0.493088
18	1987	0.165022	0.300894	0.213633	0.272454	-0.080899
19	1988	0.162139	-0.627023	-0.073354	0.165022	0.300894
20	1989	-0.289582	0.371471	0.214643	0.162139	-0.627023
21	1990	0.230108	0.185441	0.036396	-0.289582	0.371471
22	1991	0.616842	0.878595	0.124301	0.230108	0.185441
23	1992	-0.551293	-0.733803	0.105162	0.616842	0.878595
24	1993	-0.091578	-0.18299	0.085799	-0.551293	-0.733803
25	1994	0.244915	0.142442	0.01996	-0.091578	-0.18299
26	1995	0.584558	0.753915	0.176578	0.244915	0.142442
27	1996	-0.409758	0.23528	0.237724	0.584558	0.753915
28	1997	0.340804	0.352548	0.302655	-0.409758	0.23528
29	1998	0.287688	0.409696	0.242801	0.340804	0.352548
30	1999	0.124207	-0.537078	0.222782	0.287688	0.409696
31	2000	0.139719	0.411867	0.075256	0.124207	-0.537078
32	2001	-0.431191	-0.357447	-0.163282	0.139719	0.411867
33	2002	-0.07801	-0.013313	-0.16768	-0.431191	-0.357447
34	2003	-0.021172	-0.158294	-0.028885	-0.07801	-0.013313
35	2004	0.248595	-0.271964	0.171379	-0.021172	-0.158294
36	2005	-0.032496	0.036942	0.067731	0.248595	-0.271964
37	2006	0.12248	0.418327	0.08551	-0.032496	0.036942
38	2007	0.034602	0.367425	0.12723	0.12248	0.418327
39	2008	-0.076435	-0.450781	-0.174081	0.034602	0.367425
40	2009	0.108473	0.254065	-0.222935	-0.076435	-0.450781

Fig. 3.19 MINITAB output of Eq. (3.24) for JNJ

Regression Analysis: JNJ versus JNJ(lag), S&P

The regression equation is

JNJ = 0.0159 - 0.451 JNJ(lag) + 0.943 S&P

Predictor	Coef	SE Coef	T	P
Constant	0.01585	0.04818	0.33	0.744
JNJ(lag)	-0.4514	0.1334	-3.38	0.002
S&P	0.9433	0.3078	3.06	0.004

S = 0.274393 R-Sq = 31.4% R-Sq(adj) = 27.7%

Analysis of Variance

Source	DF	SS	MS	F	P
Regression	2	1.27688	0.63844	8.48	0.001
Residual Error	37	2.78578	0.07529		
Total	39	4.06267			

Source	DF	Seq SS
JNJ(lag)	1	0.56965
S&P	1	0.70723

Obs	JNJ(lag)	JNJ	Fit	SE Fit	Residual	St Resid
1	0.698	-0.6815	-0.4402	0.127	-0.2413	-0.99
2	-0.681	0.7356	0.4943	0.1208	0.2413	0.98
3	0.736	0.3294	-0.2115	0.0975	0.5408	2.11R
4	0.329	-0.132	-0.1481	0.065	0.0161	0.06
5	-0.132	-0.2763	-0.1404	0.0992	-0.1359	-0.53
6	-0.276	0.1202	0.1782	0.0621	-0.058	-0.22
7	0.12	-0.1193	0.1351	0.0554	-0.2544	-0.95
8	-0.119	0.0019	0.0345	0.0564	-0.0326	-0.12
9	0.002	-0.0172	-0.0059	0.0513	-0.0113	-0.04
10	-0.017	0.1016	0.0923	0.0449	0.0093	0.03
11	0.102	0.2863	0.1144	0.0502	0.1719	0.64
12	0.286	-0.6194	-0.0398	0.0521	-0.5797	-2.15R
13	-0.619	0.3621	0.234	0.101	0.1281	0.5
14	0.362	-0.1551	0.1731	0.0943	-0.3282	-1.27
15	-0.155	-0.0878	0.0861	0.0539	-0.174	-0.65
16	-0.088	0.4912	0.2106	0.0587	0.2807	1.05
17	0.491	0.2725	0.044	0.0847	0.2284	0.88
18	0.272	0.165	0.0944	0.0637	0.0706	0.26
19	0.165	0.1621	-0.1278	0.0651	0.29	1.09
20	0.162	-0.2896	0.1451	0.0616	-0.4347	-1.63

Fig. 3.19 (continued)

21	-0.29	0.2301	0.1809	0.0633	0.0492	0.18
22	0.23	0.6168	0.0292	0.0498	0.5876	2.18R
23	0.617	-0.5513	-0.1634	0.0839	-0.3879	-1.48
24	-0.551	-0.0916	0.3456	0.0943	-0.4372	-1.7
25	-0.092	0.2449	0.076	0.049	0.1689	0.63
26	0.245	0.5846	0.0719	0.0565	0.5127	1.91
27	0.585	-0.4098	-0.0238	0.0878	-0.386	-1.48
28	-0.41	0.3408	0.4863	0.1149	-0.1455	-0.58
29	0.341	0.2877	0.0911	0.072	0.1966	0.74
30	0.288	0.1242	0.0962	0.066	0.0281	0.11
31	0.124	0.1397	0.0308	0.044	0.1089	0.4
32	0.14	-0.4312	-0.2012	0.0861	-0.23	-0.88
33	-0.431	-0.078	0.0523	0.0967	-0.1303	-0.51
34	-0.078	-0.0212	0.0238	0.0537	-0.045	-0.17
35	-0.021	0.2486	0.1871	0.0564	0.0615	0.23
36	0.249	-0.0325	-0.0325	0.0498	0	0
37	-0.032	0.1225	0.1112	0.046	0.0113	0.04
38	0.122	0.0346	0.0806	0.0469	-0.046	-0.17
39	0.035	-0.0764	-0.164	0.0857	0.0875	0.34
40	-0.076	0.1085	-0.1599	0.0975	0.2684	1.05

R denotes an observation with a large standardized residual.

Durbin-Watson statistic = 2.16451

Fig. 3.19 (continued)

every additional dollar in income. At 5.71, the t statistic is significant at every level of significance. The lagged consumption variable is also highly significant; it has a coefficient of 0.501. If disposable income increases by 1 unit in the current period, the expected increase in consumption is 0.487 in the current period, is $0.487 \times 0.501 = 0.244$ the next year, is $(0.501)^2 \times 0.487 = 0.122$ two years later, and so on. The total increase in all future consumption in terms of Eq. (3.21) is 0.487/(1–0.501) = 0.976. Note that the long-run coefficient, 0.976, is much larger than the short-run coefficient, 0.487.

Substituting $n = 48$, $d = 0.632$ and $V(\hat{\beta}_2) = (0.090)^2 = 0.0081$ into Eq. (3.22), we obtain

$$\mathrm{DH} = \left(1 - \frac{0.632}{2}\right)\sqrt{\frac{48}{1 - 48(0.0081)}} = 6.0616$$

Using the table of the standard normal cumulative distribution function, we find that DWH = 6.0616 is larger than $z = 3$ ($\alpha = 0.0013$). Hence, there is autocorrelation associated with this consumption function.

To adjust for the impact of autocorrelation, we can use a modified regression model to estimate the consumption function. It is

$$\begin{aligned} C_t - \hat{\rho}C_{t-1} = \alpha(1-\hat{\rho}) + \beta_1(\mathrm{DI}_t \\ - \hat{\rho}\mathrm{DI}_{t-1}) + \beta_2(C_{t-1} \\ - \hat{\rho}C_{t-2}) + \varepsilon_t' \end{aligned} \tag{3.25a}$$

where

$$\epsilon_t' = \epsilon_t - \hat{\rho}\epsilon_{t-1}$$

Plugging the data listed in Table 3.8 and $\hat{\rho} = 0.684$ into Eq. (3.25a) yields the results presented in Table 3.10. These results are more appropriate for null hypothesis testing than are those indicated in Table 3.9.

Regression Analysis: MRK versus MRK(lag), S&P

The regression equation is

MRK = 0.0190 - 0.323 MRK(lag) + 0.954 S&P

Predictor	Coef	SE Coef	T	P
Constant	0.01904	0.05497	0.35	0.731
MRK(lag)	-0.3228	0.1475	-2.19	0.035
S&P	0.9542	0.3473	2.75	0.009

S = 0.312749 R-Sq = 22.1% R-Sq(adj) = 17.8%

Analysis of Variance

Source	DF	SS	MS	F	P
Regression	2	1.02389	0.51195	5.23	0.01
Residual Error	37	3.61904	0.09781		
Total	39	4.64293			

Source	DF	SeqSS
MRK(lag)	1	0.28568
S&P	1	0.73822

Obs	MRK(lag)	MRK	Fit	SE Fit	Residual	St Resid
1	0.278	-0.1057	-0.2134	0.1004	0.1078	0.36
2	-0.106	0.2749	0.2259	0.0704	0.0489	0.16
3	0.275	-0.2722	0.0362	0.0588	-0.3084	-1
4	-0.272	-0.0802	0.0914	0.0724	-0.1716	-0.56
5	-0.08	-0.1606	-0.1734	0.113	0.0128	0.04
6	-0.161	0.0643	0.109	0.0594	-0.0447	-0.15
7	0.064	0.0043	0.1738	0.0636	-0.1695	-0.55
8	0.004	-0.1627	-0.018	0.0613	-0.1447	-0.47
9	-0.163	0.2498	0.0504	0.0646	0.1995	0.65
10	0.25	0.0978	0.0078	0.0564	0.0899	0.29
11	0.098	0.206	0.1336	0.0572	0.0724	0.24
12	0.206	0.0314	0.027	0.0535	0.0044	0.01
13	0.031	0.0316	-0.0532	0.0674	0.0848	0.28
14	0.032	0.1016	0.3333	0.1073	-0.2316	-0.79

Fig. 3.20 MINITAB output of Eq. (3.24) for MRK

15	0.102	0.0745	-0.0135	0.0556	0.088	0.29
16	0.074	0.4931	0.1519	0.0594	0.3412	1.11
17	0.493	-0.0809	0.1127	0.0974	-0.1936	-0.65
18	-0.081	0.3009	0.249	0.0764	0.0519	0.17
19	0.301	-0.627	-0.1481	0.0821	-0.4789	-1.59
20	-0.627	0.3715	0.4263	0.1317	-0.0548	-0.19
21	0.371	0.1854	-0.0662	0.0694	0.2516	0.83
22	0.185	0.8786	0.0778	0.0549	0.8008	2.60R
23	0.879	-0.7338	-0.1643	0.1283	-0.5695	-2
24	-0.734	-0.183	0.3378	0.1289	-0.5208	-1.83
25	-0.183	0.1424	0.0972	0.0619	0.0453	0.15
26	0.142	0.7539	0.1415	0.0618	0.6124	2
27	0.754	0.2353	0.0025	0.1184	0.2328	0.8
28	0.235	0.3525	0.2319	0.0945	0.1207	0.4
29	0.353	0.4097	0.1369	0.0834	0.2728	0.9
30	0.41	-0.5371	0.0994	0.083	-0.6364	-2.11R
31	-0.537	0.4119	0.2642	0.102	0.1476	0.5
32	0.412	-0.3574	-0.2697	0.1144	-0.0877	-0.3
33	-0.357	-0.0133	-0.0256	0.1057	0.0123	0.04
34	-0.013	-0.1583	-0.0042	0.0599	-0.1541	-0.5
35	-0.158	-0.272	0.2337	0.0723	-0.5056	-1.66
36	-0.272	0.0369	0.1715	0.07	-0.1345	-0.44
37	0.037	0.4183	0.0887	0.05	0.3296	1.07
38	0.418	0.3674	0.0054	0.0721	0.362	1.19
39	0.367	-0.4508	-0.2657	0.1138	-0.1851	-0.64
40	-0.451	0.2541	-0.0482	0.125	0.3022	1.05

R denotes an observation with a large standardized residual.

Durbin-Watson statistic = 1.95251

Fig. 3.20 (continued)

3.8 Dummy Variables

So far, we have used data that could take on any number of values. In this section, we will examine an independent variable that can take on either of just two values: 1 and 0. This binary variable is called a **dummy variable**, and it enables us to include information that is not quantitative. For example, a regression that models individuals' income might include a dummy variable for sex of the worker. The independent dummy variable for sex could take on the value 1 for a male worker and the value 0 for a female worker. (The assignment of dummy variables is arbitrary; we could—and in this day and age probably *should*—have reversed the assignment: 1 for female, 0 for male.) The regression is

$$Y_i = \alpha + \beta_1 X_{1i} + \beta_2 X_{2i} + \cdots + \beta_k X_{ki} + \gamma D_{1i} + \epsilon_i \tag{3.26}$$

where the betas (β_1, β_2, ..., β_k) are the coefficients for the quantitative variables, and γ is the parameter for the dummy variable:

Table 3.8 Personal consumption and disposable income (in billions of 2005 dollars)

Year	C_t	DI_t	C_{t-1}
1961	1821.2	2030.8	1784.4
1962	1911.2	2129.6	1821.2
1963	1989.9	2209.5	1911.2
1964	2108.4	2368.7	1989.9
1965	2241.8	2514.7	2108.4
1966	2369.0	2647.3	2241.8
1967	2440.0	2763.5	2369.0
1968	2580.7	2889.2	2440.0
1969	2677.4	2981.4	2580.7
1970	2740.2	3108.8	2677.4
1971	2844.6	3249.1	2740.2
1972	3019.5	3406.6	2844.6
1973	3169.1	3638.2	3019.5
1974	3142.8	3610.2	3169.1
1975	3214.1	3691.3	3142.8
1976	3393.1	3838.3	3214.1
1977	3535.9	3970.7	3393.1
1978	3691.8	4156.5	3535.9
1979	3779.5	4253.8	3691.8
1980	3766.2	4295.6	3779.5
1981	3823.3	4410.0	3766.2
1982	3876.7	4506.5	3823.3
1983	4098.3	4655.7	3876.7
1984	4315.6	4989.1	4098.3
1985	4540.4	5144.8	4315.6
1986	4724.5	5315.0	4540.4
1987	4870.3	5402.4	4724.5
1988	5066.6	5635.6	4870.3
1989	5209.9	5785.1	5066.6
1990	5316.2	5896.3	5209.9
1991	5324.2	5945.9	5316.2
1992	5505.7	6155.3	5324.2
1993	5701.2	6258.2	5505.7
1994	5918.9	6459.0	5701.2
1995	6079.0	6651.6	5918.9
1996	6291.2	6870.9	6079.0
1997	6523.4	7113.5	6291.2
1998	6865.5	7538.8	6523.4
1999	7240.9	7766.7	6865.5

(continued)

Table 3.8 (continued)

Year	C_t	DI_t	C_{t-1}
2000	7608.1	8161.5	7240.9
2001	7813.9	8360.1	7608.1
2002	8021.9	8637.1	7813.9
2003	8247.6	8853.9	8021.9
2004	8532.7	9155.1	8247.6
2005	8819.0	9277.3	8532.7
2006	9073.5	9650.7	8819.0
2007	9313.9	9860.6	9073.5
2008	9290.9	9911.3	9313.9
2009	9237.3	10,035.3	9290.9

Table 3.9 Results of regression of Eq. (3.24)

Variable	Coefficient	Standard error	t-value	p-value
Constant	−109.10	40.10	−2.72	0.009
DI_t	0.487	0.085	5.71	0.0000
C_{t-1}	0.501	0.090	5.54	0.0000
$R^2 = 0.9995$				
$\overline{R}^2 = 0.9991$				
Observations 48				
First-order autocorrelation $(\hat{\rho}) = 0.684$				
DW = 0.632				

Table 3.10 Results of regression of Eq. (3.25a)

Variable	Coefficient	Standard error	t-value	p-value
Constant	−48.87	22.18	−2.20	0.033
DI_t	0.638	0.088	7.28	0.358
C_{t-1}	0.337	0.092	3.67	0.001
$R^2 = 0.9956$				
$\overline{R}^2 = 0.9954$				
Observations 48				
DW = 1.627				

$$D_1 = 1 \quad \text{if the worker is a male}$$
$$D_1 = 0 \quad \text{if the worker is a female}$$

Dummy variables indicate whether a shift in the intercept term is attributable to the characteristic of the dummy. The dummy variable having a statistically significant coefficient of 1214, for example, would indicate that the intercept for males is \$1214 higher than the intercept for females. Figure 3.21 plots salary and years of experience. The female intercept is given by α. If the intercept for the dummy were

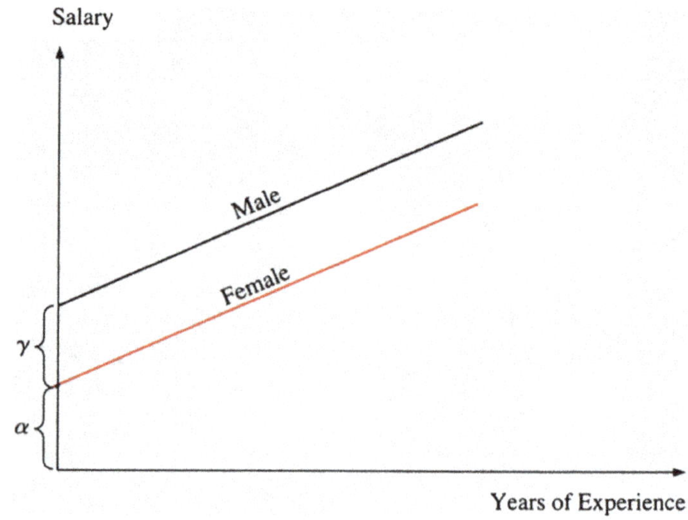

Fig. 3.21 Relationship between salary and years of experience

negative, the male intercept would be lower than the female. The dummy variable in Eq. (3.26) deals with the intercept term, not the slope term, so the dummy indicates only a shift in the intercept term. In other words, a positive and statistically significant coefficient for the dummy variable indicates that even though the salaries for males and females are affected by the same factors in the same way—that is, they have the same slope coefficients—males begin with a higher level of earnings and maintain that difference across all values for years of experience.

Example 3.10 Analysis of the Money Supply
In October 1979, the Federal Reserve Board of Governors switched from targeting interest rates to targeting the money supply. Before this period, the Fed adhered to a policy of increasing the money supply by an amount that would keep interest rates stable; hence, it "targeted" interest rates. After October 1979, the Fed focused on increasing the money supply at a fixed rate and let interest rates seek their own equilibrium level.

$$M_{3t} = \alpha + \beta_1 GNP_t + \beta_2 PRIME_t + \gamma DUM_t + \epsilon_t \tag{3.27}$$

In this model, we investigate whether the Fed's policy change had an effect on the money supply. M_3 is the money supply, GNP is the gross national product, PRIME is the prime interest rate, and DUM is a dummy variable in which 1 equals the years 1979–1990 and 0 the years 1959–1978. A significant positive sign would indicate that the money supply was greater after the change. A negative sign would indicate that the money supply was less.

The regression of Eq. (3.27) is run using the annual data for 1959–1990 presented in Table 3.11. The regression results appear in Table 3.12.

The relationship between GNP and the money supply is extremely strong. There is a negative relationship between the money supply and the prime interest rate. In addition, the dummy variable has a significant t-value at $\alpha = 1\%$, indicating that the money supply did increase after the Federal Reserve Board changed its policy.

3.9 Regression with Interaction Variables

The regression models specified thus far assume that there is no interaction between the independent variables. This assumption is not always realistic. In many situations, the relationship between one of the independent variables and the dependent variable is dependent on the value of

Table 3.11 GNP, prime rate, and M_3 (1959–1990), MTB > PRINT C1–C6

Data display

Row	Year	GNP	PRIMER$_t$	Dummy	GNP PRIME	M3
1	59	1629.1	4.48	1	7298.4	140.0
2	60	1665.3	4.82	1	8026.7	140.7
3	61	−1706.7	4.50	1	7689.2	145.2
4	62	1799.4	4.50	1	8097.3	147.9
5	63	1873.3	4.50	1	8429.9	153.4
6	64	1973.3	4.50	1	8879.9	160.4
7	65	2087.6	4.54	1	9477.7	167.9
8	66	2208.3	5.63	1	12,432.7	172.1
9	67	2271.4	5.61	1	12,742.6	183.3
10	68	2365.6	6.30	1	14,903.3	197.5
11	69	2423.3	7.96	1	19,289.5	204.0
12	70	2416.2	7.91	1	19,112.1	214.5
13	71	2484.8	5.72	1	14,213.1	228.4
14	72	2608.5	5.25	1	13,694.6	249.3
15	73	2744.1	8.03	1	22,035.1	262.9
16	74	2729.3	10.81	1	29,503.7	274.4
17	75	2695.0	7.86	1	21,182.7	287.6
18	76	2826.7	6.84	1	19,334.6	306.4
19	77	2958.6	6.83	1	20,207.2	331.3
20	78	3115.2	9.06	1	28,223.7	358.5
21	79	3192.4	12.67	0	40,447.7	382.9
22	80	3187.1	15.27	0	48,667.0	408.9
23	81	3248.8	18.87	0	61,304.9	436.5
24	82	3166.0	14.86	0	47,046.8	474.5
25	83	3279.1	10.79	0	35,381.5	521.2
26	84	3501.4	12.04	0	42,156.9	552.1
27	85	3618.7	9.93	0	35,933.7	620.1
28	86	3717.9	8.33	0	30,970.1	724.7
29	87	3845.3	8.22	0	31,608.4	750.4
30	88	4016.9	9.32	0	37,437.5	787.5
31	89	4117.7	10.87	0	44,759.4	794.8
32	90	4155.8	10.01	0	41,599.6	825.5

Table 3.12 Results of regression of Eq. (3.26)

	Coefficient	t-value	p-value
Constant	−212.74	−4.84	0.000
GNP	0.2356	13.57	0.000
PRIME	−19.066	−6.23	0.000
DUM	198.41	6.60	0.000
$R^2 = 0.924$	$F = 176.1$	DW = 0.17	

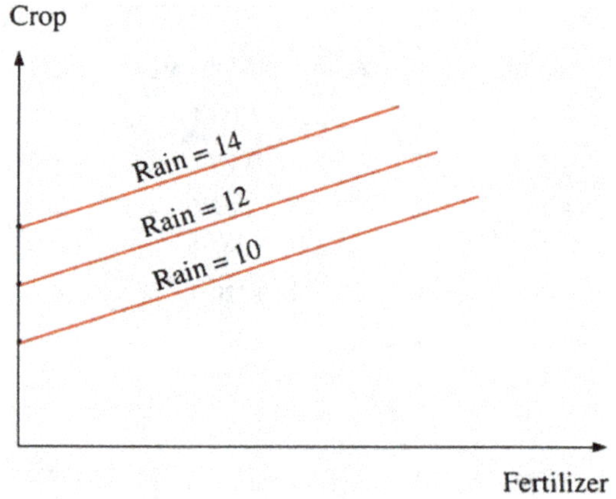

Fig. 3.22 Impact of fertilizer on output without interaction effect

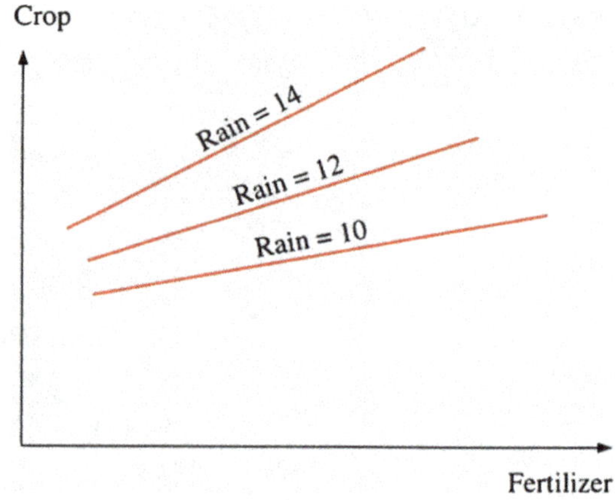

Fig. 3.23 Impact of fertilizer on output with interaction effect

another independent variable. This situation reflects **interaction**.

For example, suppose the following multiple regression model is specified.

$$\mathrm{CROP}_t = \alpha + \beta_1 \mathrm{RAIN}_t + \beta_2 \mathrm{FERT}_t + \epsilon_t \quad (3.28)$$

In this model, the amount of corn a farmer produces (CROP_t) is a function of the amount of rain received in a growing season (RAIN_t) and the amount of fertilizer used (FERT_t). Note that there is no interaction in this model; the fertilizer affects the output of corn, but this effect does not depend on how much rain fell (see Fig. 3.22). In Fig. 3.22, fertilizer is graphed on the x-axis, crop production on the y-axis. The rate of increase in crop production is constant for any change in the amount of rain.

However, interaction results if more rain makes the fertilizer more productive and increases corn production. We can model this interaction between the two variables by adding an interaction term.

$$\mathrm{CORN}_t = \alpha + \beta_1 \mathrm{RAIN}_t + \beta_2 \mathrm{FERT}_t + \beta_3 (\mathrm{FERT}_t \times \mathrm{RAIN}_t) + \epsilon_t \quad (3.29)$$

To create an interaction term, we multiply the two observations whose interaction we wish to investigate. This term measures whether additional rain makes fertilizer more productive. In the model shown in Eq. (3.28), the change in the corn production that results from a change in the amount of fertilizer is given by the slope term β_2. A 1-unit change in the amount of fertilizer leads to a β_2-unit change in crop production. In the interaction model of Eq. (3.29), the change in the corn production that is associated with a 1-unit change in the amount of fertilizer is equal to (β_2 + $\beta_3 \mathrm{RAIN}_t$). If the interaction term has a positive sign, then the rain makes the fertilizer more effective. This effectiveness is shown in Fig. 3.23, where the dependent variable is graphed on the y-axis and fertilizer on the x-axis. As the amount of rain increases, the slope of the line increases, indicating that fertilizer has a greater impact when more rain is present. In general, the interaction term tests whether the slope parameter for one variable changes as a function of the other variable. The t statistic is used to determine the statistical significance of the coefficient associated with the interaction term.

Example 3.11 Analysis of the Money Supply, with Interaction

Equation (3.27) with interaction can be written as

$$\mathrm{M}_{3t} = \alpha + \beta_1 \mathrm{GNP}_t + \beta_2 \mathrm{PRIME}_t + \beta_3 (\mathrm{GNP}_t)(\mathrm{PRIME}_t) + \gamma \mathrm{DUM}_t + \epsilon_t \quad (3.30)$$

Fig. 3.24 MINITAB output for Eq. (3.30)

```
MTB > BRIEF 3
MTB > REGRESS 'M3' 4 'GNP' 'PRIMERt' 'DUMMY' 'GNPPRIME';
SUBC> DW.
```

Regression Analysis

```
The regression equation is
M3 = 198 + 0.132 GNP - 71.1 PRIMERt - 127 DUMMY + 0.0186 GNPPRIME

Predictor        Coef        StDev          T        P
Constant       197.93        92.18       2.15    0.041
GNP           0.13247      0.03680       3.60    0.001
PRIMERt        -71.14        17.12      -4.15    0.000
DUMMY         -126.55        35.20      -3.60    0.001
GNPPRIME     0.018589     0.006038       3.08    0.005

S = 35.31        R-Sq = 97.8%      R-Sq(adj) = 97.5%

Analysis of Variance

Source       DF          SS          MS          F        P
Regression    4     1493207      373302     299.46    0.000
Error        27       33658        1247
Total        31     1526864

Source       DF      Seq SS
GNP           1     1391385
PRIMERt       1       19320
DUMMY         1       70686
GNPPRIME      1       11815

Obs       GNP         M3        Fit   StDev Fit   Residual    St Resid
  1      1629     140.00     104.17       15.00      35.83        1.12
  2      1665     140.70      98.32       13.56      42.38        1.30
  3      1709     145.20     120.55       13.94      24.65        0.76
  4      1799     147.90     140.15       12.87       7.75        0.24
  5      1873     153.40     156.13       12.06      -2.73       -0.08
  6      1973     160.40     177.74       11.09     -17.34       -0.52
  7      2088     167.90     201.15       10.10     -33.25       -0.98
  8      2208     172.10     194.53        8.46     -22.43       -0.65
  9      2271     183.30     210.07        8.32     -26.77       -0.78
 10      2366     197.50     213.63        8.67     -16.13       -0.47
 11      2423     204.00     184.72       11.55      19.28        0.58
 12      2416     214.50     184.04       11.48      30.46        0.91
 13      2485     228.40     257.85        8.63     -29.45       -0.86
 14      2608     249.30     298.03        9.58     -48.73       -1.43
 15      2744     262.90     273.28       11.15     -10.38       -0.31
 16      2729     274.40     212.39       15.73      62.01        1.96
 17      2695     287.60     263.02       10.77      24.58        0.73
 18      2827     306.40     318.67       11.02     -12.27       -0.37
 19      2959     331.30     353.08       12.39     -21.78       -0.66
 20      3115     358.50     364.21       16.42      -5.71       -0.18
 21      3192     382.90     471.41       14.63     -88.51       -2.75R
 22      3187     408.90     438.54       15.66     -29.64       -0.94
 23      3249     436.50     425.55       25.21      10.95        0.44 X
 24      3166     474.50     434.80       15.60      39.70        1.25
 25      3279     521.20     522.45       15.03      -1.25       -0.04
 26      3501     552.10     588.93       10.62     -36.83       -1.09
 27      3619     620.10     638.88       11.81     -18.78       -0.56
 28      3718     724.70     673.57       16.93      51.13        1.65
 29      3845     750.40     710.14       16.79      40.26        1.30
 30      4017     787.50     762.98       14.34      24.52        0.76
 31      4118     794.80     802.18       21.93      -7.38       -0.27
 32      4156     825.50     809.66       18.51      15.84        0.53

R denotes an observation with a large standardized residual
X denotes an observation whose X value gives it large influence.

Durbin-Watson statistic = 1.17
```

MINITAB results for this equation are presented in Fig. 3.24. This output indicates that the t statistic associated with the interaction terra is 3.08 and that the p-value associated with the interaction term is 0.005. Hence, the coefficient associated with the interaction term is significantly different from 0 at $\alpha = 1\%$.

3.10 Regression Approach to Investigating the Effect of Alternative Business Strategies[7]

Johnson et al. (1989) used multiple regression with dummy variables to investigate the relationship between business strategy and wages within the context of a significant environmental change, deregulation of the airline industry (1978–1984). Their regression results were

$$\begin{aligned}\log_e \text{Wages} = {} & \underset{(62.19)}{4.5848^{***}} + \underset{(1.18)}{0.3\ \text{PROFITS}} \\ & -\underset{(-0.19)}{0.000\ \text{DEBT}} + \underset{(2.38)}{0.1348\ \text{PERCENT UNION}^{*}} \\ & -\underset{(-0.78)}{0.1250\ \text{LOAD FACTOR}} + \underset{(1.24)}{0.0000\ \text{SALES}} \\ & -\underset{(-0.75)}{0.1650\ \text{FUEL COST}} - \underset{(-1.88)}{0.9040\ \text{COST}^{*}} \\ & -\underset{(-1.84)}{0.1271\ \text{FOCUS}^{*}} - \underset{(-2.28)}{0.0952\ \text{STUCK}^{*}}\end{aligned} \tag{3.31}$$

Equation (3.31) is a log linear model discussed in Sect. 3.10. In this estimated multiple regression, t-values appear in parentheses beneath the coefficients $R^2 = 0.18$, $n = 92$. *** means $p < 0.001$; ** means $p < 0.01$; * means $p < 0.05$, one-sided test.

In this equation, cost, focus, and stuck are business strategic variables as defined in Table 3.13. Table 3.13 presents four alternative business strategies. They are (1) the cost leadership strategy (cost), to maintain the lowest position in the industry; (2) the product differentiation strategy (Diff.), to create a unique product or industrywide service through brand image (Coca-Cola is a good example), customer service, technology (Polaroid cameras) or other distinguishing features; (3) the focus business strategy (focus), to cater to a narrow strategic target with the aim of being more effective or efficient than those that are competing on a national basis; and (4) the stuck-in-the-middle (stuck) strategy, where no clearly defined strategic position exists.

Results of Eq. (3.31) indicate that estimated coefficients associated with cost, focus, and stuck are all significant at $p = 0.05$; therefore, we can conclude that different business strategies did affect wages in the airline industry during the years 1978–1984.

3.11 Logistic Regression and Credit Risk Analysis: Ohlson's and Shumway's Methods for Estimating Default Probability

Based upon dummy variable regression discussed in Sect. 3.8, researcher has developed more complex estimation methods such as logit and probit to determine the likelihood of company bankruptcy. Logit model is a methodology that uses maximum likelihood estimation, or the so-called conditional logit model. Two well-known models using logit models to estimate the probability of bankruptcy are Ohlson (1980) and Shumway (2001). Ohlson's model is a static logit model and Shumway's model is a dynamic logit regression which we will discuss in Appendices 3 and 4.

The econometric advantages of logit model over multivariate discriminant analysis (MDA) is used for the development of Altman Z-score. MDA imposes certain statistical requirements on the distributional properties of the predictors, violation of which will result in invalid or poor approximations. For example, MDA assumes the same variance–covariance matrices of the predictors for both failed and nonfailed groups; it

[7]This section is essentially based on N. B. Johnson et al. (1989), "Deregulation, Business Strategy and Wages in the Airline Industry," *Industrial Relations* 28, 3, 419–430.

Table 3.13 Strategic classification and mean wages by regulating period

Regulation			Deregulation		
Airline	Deflated average wage	Strategy	Airline	Deflated average wage	Strategy
US Air	14,099	Focus	National	12,345	Cost
Delta	12,133	Cost	US Air	11,911	Diff.
Ozark	12,094	Focus	American	11,797	Diff.
Frontier	11,959	Focus	Delta	11,400	Cost[a]
Texas Air	11,827	Focus	TWA	11,397	Diff.
American	11,711	Diff.	United	11,240	Diff.
National	11,643	Cost	Western	11,199	Stuck
Eastern	11,374	Stuck	Northwest	11,085	Cost
Piedmont	11,332	Focus	Braniff	11,019	Stuck
TWA	11,201	Diff.	Pan Am	11,017	Stuck
Western	11,104	Stuck	Ozark	11,014	Focus
United	11,010	Diff.	Pacific SW	10,842	Focus
Continental	10,911	Focus	Frontier	10,808	Focus
Pan Am	10,831	Focus	Eastern	10,785	Stuck
Northwest	10,722	Cost	Texas Air	10,423	Cost
Braniff	10,696	Focus	Republic	10,098	Focus
			Continental	9799	Stuck/Cost[b]
			Piedmont	9706	Focus
			Southwest	8902	Focus
			People	4105	Cost

Source Based on "Deregulation, Business Strategy and Wages in the Airline Industry," *Industrial Relations,* Vol. 28, No. 3, pp. 419–430, reprinted with permission from Basil Blackwell Ltd

[a]Delta was coded as a differentiator in alternative regressions

[b]Continental was coded as stuck in 1978–1982 and as cost in 1983–1984

Table 3.14 Regression results for Eq. (16.41

Regression output: $\ln(1+R_t) = a_0 + a_1(t) + a_2(1/t) + a_3(x) + e_t$			
Constant (a_0)		0.673866	
Standard error of Y estimate		0.034919	
R^2		0.76818	
Number of observations	223		
Degrees of freedom	219		
	a_1	a_2	a_3
X coefficient(s)	0.042615	−0.17833	0.03791
Standard error of coefficients	0.002674	0.017143	0.007759

also requires normal distributions of failure predictors which bias against the use of indicator independent variables. Moreover, the output of MDA model is a score which has little intuitive

Table 3.15 Estimated yield of a two-year, 3.875-% coupon note

$\ln(1 + R_t) = 0.673688 + 0.042615\ (2) - 0.17833\ (1/2) + 0.03791(3.875) = 0.8168$
To transform $\ln(1 + R_t)$ into the yield to maturity, take the exponential of both sides of the equation:
$R_t = \exp[0.8168] - 1 = 1.2632$ or 126.32%

interpretation. The matching procedure which is typically used in MDA technique can result in losing useful information such as losing meaningful predictors. The use of logistic regression, on the other hand, essentially overcomes the weaknesses of MDA discussed above.

Ohlson (1980) used data available prior to the date of bankruptcy to ensure the strict forecasting relationships. His sample included 105 bankruptcies and 2058 nonbankruptcies from the seventies (1970–76). Among the 105 bankrupt firms, 8 firms are traded in New York Stock Exchange, 43 are traded in American Stock Exchange, and 54 are traded over-the-counter market or regional exchanges. Nine variables defined below are used to develop the logit model.

X_1 = Natural log of (total assets/GNP implicit price deflator index). The index assumes a base value of 100 for 1968;
X_2 = Total liabilities/total assets;
X_3 = (Current assets − current liabilities)/total assets;
X_4 = Current assets/current liabilities;
X_5 = One if total liabilities exceeds total assets, zero otherwise;
X_6 = Net income/total assets;
X_7 = Funds provided by operations/total liabilities;
X_8 = One if net income was negative for the last two years, zero otherwise; and
X_9 = (Net income in year t − net income in $t-1$)/(absolute net income in year t + absolute net income in year $t-1$).

Three sets of coefficients are estimated using data one year prior to bankruptcy and or two years prior to bankruptcy. Intuitively, the model with estimates computed using data one year prior to bankruptcy performs the best, which is expressed as follows:

$$\begin{aligned} Y = {} & -1.32 - 0.407X_1 + 6.03X_2 - 1.43X_3 \\ & + 0.0757X_4 - 2.37X_5 - 1.83X_6 + 0.285X_7 \\ & - 1.72X_8 - 0.521X_9 \end{aligned} \tag{3.32}$$

where $Y = \log[P/(1-P)]$, P = the probability of bankruptcy.

Thus, the probability of bankruptcy (P) is calculated as $\exp(Y)/[1 + \exp(Y)]$ exp, and the model becomes relatively easy to interpret.

Ohlson found that using a probability cutoff of 3.8% for classifying firms as bankrupt minimized type I and type II errors of the model presented in Eq. (3.9). At this probability cutoff point, the model correctly classified 87.6% of his bankrupt firm sample and 82.6% of the nonbankrupt firms. Begley et al. (1996) applied Ohlson's logit model (1980) to predict bankruptcy for a holdout sample of 65 bankrupt and 1300 nonbankrupt firms in the 1980s. They found substantially higher type I and type II error rates than those in the original studies. They re-estimated the coefficients for each model using data for a portion of their 1980s sample. However, they found no performance improvement for the re-estimated Ohlson Model.

The logit model used by Ohlson (1980) is single-period static logit model. Shumway (2001) employed a discrete hazard model or multiple-period dynamic logit model. The concept of discrete-time hazard model originates from the survival model and is widely used in biological medication field. It was not until recent years that social science researchers started using it for analyzing variables' effect upon survival (e.g., Lancaster 1992). Cox and Oakes

(1984) calculate hazard rate to estimate the likelihood of survival and survival time. Shumway (2001) elaborates the econometric advantages of a hazard model over a static binary logit model. First, hazard models control for each firm's period at risk, while static models do not. Secondly, hazard models exploit each firm's time-series data by including annual observations as time-varying covariates. Thirdly, hazard models produce more efficient out-of-sample forecasts by utilizing much more data. Shumway (2001) proves that a multiperiod logit model is equivalent to a discrete-time hazard model. Therefore, his model used multiple years of data for each sample firm and treated each firm as a single observation.

Moreover, Shumway (2001) also corrected the problem in the traditional approaches of bankruptcy forecasting; that is, previous studies used only most of the accounting ratios used in the previous bankruptcy studies are found not significant. Shumway (2001) incorporated not only financial ratios but also market variables such as market size, past stock returns, and idiosyncratic returns variability as bankruptcy predictors.

The dependent variable in the prediction models is each firm-year's bankruptcy status (0, 1) in a given sample year. In a hazard analysis, for a bankrupt firm, the dependent variable equals 1 for the year in which it files bankruptcy, and the dependent variable equals 0 for all sample years prior to the bankruptcy-filing year. The nonbankrupt firms are coded 0 every year they are in the sample.

The Shumway (2001) study employed all available firms in a broad range of industries, resulting in a sample of 300 bankrupt firms for the period of 1962–1992. The study found that a multiperiod logit model outperformed MDA and single-period logit models and that a combination of market-based and accounting-based independent variables outperformed models that were only accounting-based. The Shumway model with incorporation of market-based and accounting-based predictors is expressed as follows:

$$Y = -13.303 - 1.982X_1 + 3.593X_2 - 0.467X_3 - 1.809X_4 + 5.79X_5 \tag{3.33}$$

where

$Y = \log[P/(1-P)]$,
P = The probability of bankruptcy;
X_1 = Net income/total assets;
X_2 = Total liabilities/total assets);
X_3 = The logarithm of (each firm's market capitalization at the end of year prior to the observation year/total market capitalization of NYSE and AMEX market);
X_4 = Past excess return as the return of the firm in year $t-1$ minus the value-weighted CRSP NYSE/AMEX index return in year $t-1$; and
X_5 = Idiosyncratic standard deviation of each firm's stock returns. It is defined as the standard deviation of the residual of a regression which regresses each stock's monthly returns in year $t-1$ on the value-weighted NYSE/AMEX index return for the same year.

To evaluate the forecast accuracy of the hazard model presented in Eq. (3.33), Shumway (2001) divided the test sample into ten groups based upon their predicted probability of bankruptcies using the model. The hazard model classifies almost 70% of all bankruptcies in the highest bankruptcy probability decile and classifies 96.6 of bankrupt firms above the median probability. Following Shumway (2001), researchers of bankruptcy prediction have been using multiple-period logit regression. For instance, Sun (2007) developed a multiperiod logistic regression model and found this model outperforms auditors' going concern opinions in predicting bankruptcy. Saunders and Allen (2002) and Saunders and Cornett (2006) have discussed alternative methods for determining credit risk.

Hwang et al. (2008) use the discrete-time survival model by Cox and Oakes (1984) to predict the probability of financial distress for each firm under study. The maximum likelihood method is employed to estimate the values of our model parameters. The resulting estimates are analyzed by their asymptotic normal distributions and are used to estimate the in-sample probability of financial distress for each firm under study. Using such estimated probability, a strategy is developed to identify failing firms and is applied to study the probability of bankruptcy in the future for firms listed in Taiwan Stock Exchange. Empirical studies demonstrate that our strategy developed from the discrete-time survival model can yield more accurate forecasts than the alternative methods based on the logit model in Ohlson (1980) and the probit model in Zmijewski (1984).

3.12 Conclusion

In this chapter, we extended the concepts and issues of simple and multiple regression. Specifically, we investigated other topics in regression analysis, such as multicollinearity, heteroscedasticity, autocorrelation, and misspecification. We also examined nonlinear regression, regression with lagged dependent variables, dummy variables, and interaction variables. Related economics and business examples were used to demonstrate how the new models and techniques presented in this chapter can be used to analyze data. Logistic regression and its application in credit analysis were also discussed in this chapter.

Based upon both Chaps. 2 and 3 of this book, we will discuss simultaneous equation models in Chaps. 4 and 5. In Chap. 6, we will discuss fixed effects and random effects in finance research. Issues of errors-in-variables estimation method will be discussed in Chaps. 7 and 8 in detail. In Chap. 9, we will discuss spurious regression and data mining in conditional asset pricing models.

Appendix 1: Dynamic Ratio Analysis

In the appendices, we discuss how the industry average of financial ratio can be used to do dynamic financial ratio analysis. To do the dynamic financial ratio analysis, the individual financial ratio is related to the industry average over time by a regression such as[8]

$$y_{i,t} = a_0 + a_1 x_{t-1} + a_2 y_{i,t-1} + e_t \qquad (3.34)$$

where

$y_{i,t}$ = A financial ratio for *i*th firm in period *t;*
x_{t-1} = Industry average for a financial ration in period $t-1$;
$y_{i,t-1}$ = A financial ratio for *i*th firm in period $t-1$.

Use the current ratio data of both Johnson & Johnson and Merck and its industry average data during 1990–2009. Using the Microsoft Excel program, we obtain regression results in terms of Eq. (3.34) as presented in Figs. 3.25 and 3.26. From the values of R^2 and *t*-statistics associated with regression coefficients a_1 and a_2, we can conclude that the current ratio regression for both Johnson & Johnson and Merck are suitable to be used to forecast future current ratio.

Using the current ratio of 2009 for JNJ (1.82) and industry average (2.18401), we can forecast the current ratio for JNJ in 2010 as follows:

$$\begin{aligned} \mathrm{CR}_{2010}(\mathrm{JNJ}) &= 1.9421 - 0.1189(2.18401) + 0.037(1.82) \\ &= 1.7498 \end{aligned}$$

Appendix 2: Term Structure of Interest Rate

The structure of interest rates is typically described by the yield curve. Typical yield curve diagram used to describe the relationship between yield to maturity and time to maturity

[8]See Lee Cheng F. and Joe E. Finnerty (1990), *Corporate Finance* (Harcourt Brace Javanovich, San Diego).

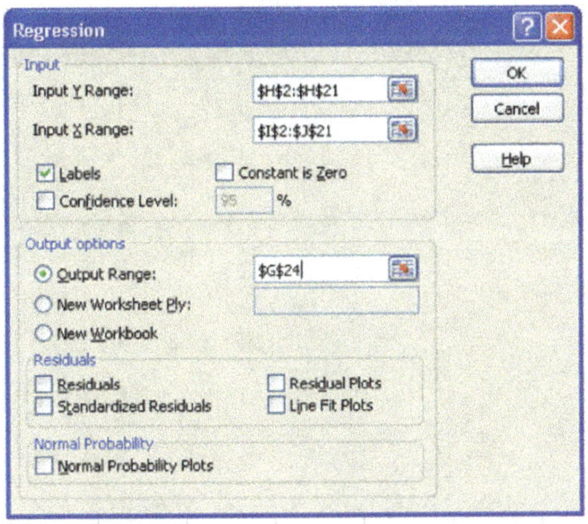

SUMMARY OUTPUT

Regression Statistics	
Multiple R	0.086185
R Square	0.007428
Adjusted R Square	-0.11664
Standard Error	0.326881
Observations	19

ANOVA

	df	*SS*	*MS*	*F*	*Significance F*
Regression	2	0.012794	0.006397	0.059868	0.942099
Residual	16	1.709615	0.106851		
Total	18	1.722409			

	Coefficients	*Standard Error*	*t Stat*	*P-value*	*Lower 95%*	*Upper 95%*	*Lower 95.0%*	*Upper 95.0%*
Intercept	1.942086	0.672174	2.88926	0.010676	0.517141	3.367031	0.517141	3.367031
X(t-1)	-0.11891	0.344571	-0.34511	0.734509	-0.84937	0.611543	-0.84937	0.611543
Y(t-1)	0.037048	0.264652	0.139988	0.890417	-0.52399	0.598084	-0.52399	0.598084

JNJ Current Ratio

t	Y(t)	X(t-1)	Y(t-1)
1991	1.835	1.65035	1.778
1992	1.582	1.58368	1.835
1993	1.624	1.8485	1.582
1994	1.566	1.89574	1.624
1995	1.809	1.45418	1.566
1996	1.807	1.52022	1.809
1997	1.999	2.023	1.807
1998	1.364	1.93359	1.999
1999	1.771	1.99812	1.364
2000	2.164	1.83149	1.771
2001	2.296	2.0827	2.164
2002	1.683	1.94572	2.296
2003	1.71	1.95464	1.683
2004	1.962	1.84799	1.71
2005	2.485	2.09756	1.962
2006	1.199	2.44232	2.485
2007	1.51	1.96882	1.199
2008	1.649	2.03753	1.51
2009	1.82	2.18401	1.649

Fig. 3.25 Current ratio regression for JNJ

term for Treasury securities. It can be shown that the following multiple regression model can be used to describe this relationship[9]

$$\ln(y_{it}) = a_0 + a_1 x_{1t} + a_2 x_{2t} + a_3 x_{3t} + e_t \quad (3.35)$$

where

y_{it} = Yield to maturity for *i*th bond to be mature in period *t;*
x_{1t} = Time to maturity for *i*th bond;
x_{2t} = $1/x_{it;}$
x_{3t} = Coupon rate for *i*th bond.

Using the data of Treasury bonds and notes as reported in *The Wall Street Journal* of February

[9]See Cheng F. Lee, Joseph E. Finnerty, and Donald H. Wort (1990), *Security Analysis and Portfolio Management* (Glenview. I11.: Scott, Foresman), Chap. 5.

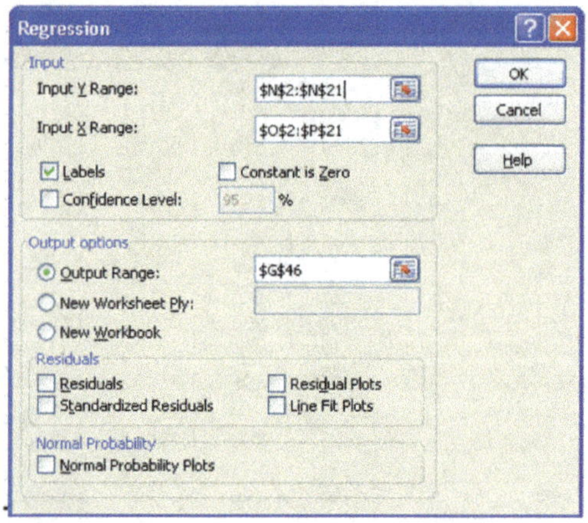

SUMMARY OUTPUT

Regression Statistics	
Multiple R	0.231794
R Square	0.053729
Adjusted R Square	-0.06456
Standard Error	0.22518
Observations	19

ANOVA

	df	*SS*	*MS*	*F*	*Significance F*
Regression	2	0.046065	0.023032	0.454234	0.642874
Residual	16	0.811299	0.050706		
Total	18	0.857364			

	Coefficients	*Standard Error*	*t Stat*	*P-value*	*Lower 95%*	*Upper 95%*	*Lower 95.0%*	*Upper 95.0%*
Intercept	1.244113	0.55122	2.257016	0.038344	0.075578	2.412648	0.075578	2.412648
X(t-1)	-0.10852	0.224029	-0.4844	0.634665	-0.58344	0.3664	-0.58344	0.3664
Y(t-1)	0.239664	0.2812	0.852291	0.40663	-0.35645	0.83578	-0.35645	0.83578

MRK Current Ratio			
t	Y(t)	X(t-1)	Y(t-1)
1991	1.532	1.65035	1.332
1992	1.216	1.58368	1.532
1993	0.973	1.8485	1.216
1994	1.27	1.89574	0.973
1995	1.515	1.45418	1.27
1996	1.6	1.52022	1.515
1997	1.475	2.023	1.6
1998	1.685	1.93359	1.475
1999	1.285	1.99812	1.685
2000	1.375	1.83149	1.285
2001	1.123	2.0827	1.375
2002	1.199	1.94572	1.123
2003	1.205	1.95464	1.199
2004	1.147	1.84799	1.205
2005	1.582	2.09756	1.147
2006	1.197	2.44232	1.582
2007	1.227	1.96882	1.197
2008	1.348	2.03753	1.227
2009	1.805	2.18401	1.348

Fig. 3.26 Current ratio regression for MRK

16, 2011, in Table 3.16, Lee et al. estimated Eq. (3.35) and the result is presented in Table 3.14.

Using the information in Table 3.14, the estimated yield of a two-year, 3.875% compound rate is 126.3% as presented in Table 3.15.

Appendix 3: Partial Adjustment Dividend Behavior Model

Based upon Lee et al. (1987) and Lee and Lee (2017), the partial adjustment dividend behavior model and its integration model will be discussed in this appendix.

Table 3.16 Yield, time to maturity, and coupon rates for Treasury bonds and notes as of February 16, 2011

(1) R_t	(2) t	(3) x	(1) R_t	(2) t	(3) x	(1) R_t	(2) t	(3) x
0.0618	0.117808	0.875	0.5159	1.369863	0.625	1.083	2.523288	4.25
0.043	0.117808	4.75	0.5034	1.369863	4.875	1.088	2.567123	3.125
0.1088	0.2	0.875	0.5273	1.410959	1.5	1.1674	2.608219	0.75
0.1194	0.2	4.875	0.5574	1.454795	0.625	1.157	2.649315	3.125
0.1154	0.284932	0.875	0.5595	1.454795	4.625	1.2163	2.690411	0.5
0.1096	0.284932	4.875	0.5931	1.49589	1.75	1.2179	2.734247	2.75
0.1355	0.367123	1.125	0.5662	1.49589	4.375	1.2563	2.731507	0.5
0.1353	0.367123	5.125	0.6087	1.539726	0.375	1.1699	2.731507	4.25
0.1458	0.452055	1	0.5916	1.539726	4.125	1.2661	2.772603	2
0.1381	0.452055	4.875	0.6444	1.580822	1.375	1.2953	2.813699	0.75
0.1461	0.493151	5	0.6574	1.621918	0.375	1.3104	2.857534	1.5
0.167	0.536986	1	0.6369	1.621918	4.25	1.3434	2.89863	1
0.1725	0.536986	4.625	0.7001	1.663014	1.375	1.3384	2.942466	1.75
0.1866	0.619178	1	0.711	1.706849	0.375	1.3997	2.983562	1.25
0.1914	0.619178	4.5	0.6779	1.706849	3.875	1.3096	2.983562	4
0.2108	0.70411	1	0.7456	1.747945	1.375	1.3889	3.019178	1.875
0.2038	0.70411	4.625	0.7099	1.747945	4	1.4282	3.10411	1.75
0.2229	0.745205	1.75	0.7496	1.789041	0.5	1.4757	3.186301	1.875
0.2453	0.786301	0.75	0.7088	1.789041	3.375	1.5079	3.271233	2.25
0.2423	0.786301	4.5	0.7942	1.830137	1.125	1.5515	3.353425	2.625
0.2656	0.827397	1.125	0.8023	1.873973	0.625	1.5977	3.438356	2.625
0.287	0.871233	1	0.7645	1.873973	3.625	1.5634	3.479452	4.25
0.2927	0.871233	4.625	0.8251	1.915068	1.375	1.6612	3.523288	2.375
0.3065	0.912329	1.125	0.8392	1.958904	0.625	1.6947	3.605479	2.375
0.3162	0.956164	0.875	0.7789	1.958904	2.875	1.7417	3.690411	2.375
0.3211	0.956164	4.75	0.7933	2	3.875	1.6727	3.731507	4.25
0.3354	0.99726	1.375	0.8481	2.063014	2.75	1.7794	3.772603	2.125
0.3423	0.99726	4.875	0.8981	2.10411	1.375	1.8158	3.857534	2.625
0.3504	1.035616	0.875	0.8785	2.147945	2.5	1.8648	3.942466	2.25
0.3486	1.035616	4.625	0.9392	2.189041	1.75	1.8246	3.983562	4
0.3626	1.076712	1.375	0.9126	2.230137	3.125	1.778	3.983562	11.25
0.3956	1.120548	1	0.9895	2.271233	1.375	1.9036	4.060274	2.375
0.409	1.161644	1.375	0.9264	2.271233	3.625	1.9303	4.145205	2.5
0.4309	1.20274	1	0.9576	2.315068	3.5	1.9814	4.227397	2.5
0.4157	1.20274	4.5	1.0293	2.356164	1.125	1.9563	4.268493	4.125
0.4516	1.243836	1.375	1.011	2.39726	3.375	2.0309	4.312329	2.125
0.4717	1.287671	0.75	1.0691	2.438356	1	2.0703	4.394521	1.875
0.469	1.287671	4.75	1.0369	2.482192	3.375	2.112	4.479452	1.75
0.4969	1.328767	1.875	1.113	2.523288	0.75	2.0764	4.520548	4.25

(continued)

Table 3.16 (continued)

(1) R_t	(2) t	(3) x	(1) R_t	(2) t	(3) x	(1) R_t	(2) t	(3) x
1.9998	4.520548	10.625	3.2892	8.273973	3.125	4.644	27.26849	4.5
2.1709	4.564384	1.25	3.3314	8.526027	3.625	4.6873	28.00548	3.5
2.2055	4.646575	1.25	3.1969	8.526027	8.125	4.6714	28.24932	4.25
2.2467	4.731507	1.25	3.3943	8.778082	3.375	4.6665	28.24932	4.5
2.3681	5.063014	2.625	3.4429	9.030137	3.625	4.6747	28.50137	4.375
2.409	5.147945	2.375	3.3005	9.030137	8.5	4.6685	29.00822	4.625
2.4363	5.230137	2.625	3.4965	9.276712	3.5	4.6793	29.26027	4.375
2.3992	5.271233	5.125	3.3523	9.276712	8.75	4.687	29.51233	3.875
2.3637	5.271233	7.25	3.5763	9.528767	2.625	4.6832	29.76438	4.25
2.4581	5.315068	3.25	3.4047	9.528767	8.75	4.6702	30.01644	4.75
2.4865	5.39726	3.25	3.616	9.780822	2.625			
2.5263	5.482192	3.25	3.6193	10.03288	3.625			
2.488	5.523288	4.875	3.4958	10.03288	7.875			
2.5655	5.567123	3	3.5484	10.27671	8.125			
2.5921	5.649315	3	3.5904	10.52603	8.125			
2.6241	5.734247	3.125	3.6425	10.77808	8			
2.5895	5.775342	4.625	3.7672	11.52603	7.25			
2.5587	5.775342	7.5	3.7995	12.27397	7.625			
2.666	5.816438	2.75	3.856	12.52603	7.125			
2.6863	5.90137	3.25	3.9425	13.02192	6.25			
2.7073	5.986301	3.125	4.035	14.27671	7.5			
2.6763	6.027397	4.625	4.0571	14.52877	7.625			
2.734	6.063014	3	4.1296	15.02466	6.875			
2.7579	6.147945	3.25	4.2027	15.52877	6			
2.793	6.230137	3.125	4.2129	15.52055	6.75			
2.7655	6.271233	4.5	4.2404	15.76986	6.5			
2.6806	6.271233	8.75	4.2513	16.02192	6.625			
2.8329	6.315068	2.75	4.2889	16.51781	6.375			
2.8672	6.39726	2.5	4.3119	16.76986	6.125			
3.0019	6.90137	2.75	4.3724	17.52055	5.5			
3.0288	6.986301	2.625	4.3924	17.7726	5.25			
3.0062	7.027397	3.5	4.3993	18.02466	5.25			
3.0511	7.271233	3.875	4.3882	18.52055	6.125			
2.9049	7.271233	9.125	4.4112	19.26575	6.25			
3.1047	7.523288	4	4.4605	20.02192	5.375			
3.1662	7.775342	3.75	4.6037	25.02466	4.5			
2.9979	7.775342	9	4.6089	26.0274	4.75			
3.2507	8.027397	2.75	4.6023	26.27397	5			
3.0691	8.027397	8.875	4.6436	27.02466	4.375			

Behavioral Considerations of Dividend Policy

Partial Adjustment and Information Content Models

Modigliani and Miller (1961) showed that dividends were not the only possible focus for valuation of a firm or its equity securities, but, as they found in their 1966 paper on cost of capital, there are other considerations in dividend policy, namely the information effects. To the extent that dividends convey information from the management of the firm to the actual owners, the dividend decision may be extremely important. This section attempts to describe the dividend policies of firms, seeking to uncover rational patterns. A number of long-standing theories as to how the dividend decision is made will be introduced, along with a recent synthesis of the two most elaborate theories. It should become readily apparent that a firm's dividend decision is not some haphazard or ad hoc decision. Firms pursue certain dividend policies for justifiably good reasons, such as the maximization of shareholder wealth subject to some (unspecified) uncertainty constraint.

Probably the most easily understood theory of dividend policy is simply termed the **residual theory of dividends**. As the name implies, dividends, when paid, more or less fall out of the system, since these are funds the firm has no immediate use for. Higgins (1972) supported the contention that dividends were, or should be, a residual payment, when he dispelled the notion that discount rates applicable to dividends would increase as their time to receipt lengthened, due to the added uncertainty induced by the widening of the probability distribution of expected returns. The contrary opinion had been formulated by Gordon (1963). Regardless of the intuitive appeal of the residual dividend theory, its validity or relevance is subject to question in light of the evidence that firms do manage their dividend payout. Primarily because this is so, we now leave the residual theory to consider two alternative explanations of the dividend behavior of firms, both of which have considerably more support on an empirical basis than does the residual theory.

A partial adjustment model of dividend behavior on the part of firms was investigated in some detail by Lintner (1956), as he studied the dividend patterns of twenty-eight well-known, established companies. From his analysis, he concluded that the major portion of the dividend of a firm could be modeled as follows:

$$D^* = rE_t, \tag{3.36}$$

and

$$D_t - D_{t-1} = a + b(D^* - D_{t-1}) + u_t \tag{3.37}$$

where

D^* = Firm's desired dividend payment;
F_t = Net income of the firm during period t;
r = Target payout ratio;
a = A constant relating to dividend growth;
b = Adjustment factor relating the previous period's dividend and the new desired level of dividends, where b is assumed to be less than one.

From this, we infer that firms set their dividends in accordance with the level of current earnings and that the changes in dividends over time do not correspond exactly with the change in earnings in the immediate time period. An alternative explanation to the b coefficient being the average-speed-of-adjustment factor, we can interpret the quantity $(1 - b)$ as a safety factor that management observes by not increasing the dividend payment to levels where it cannot be maintained. Together, the a and b coefficients can be used to test the hypothesis that management is more likely to increase dividends over time rather than cut them; this obviously contrasts with the major premise of the residual theory.

To eliminate the expectation terms and construct an empirically testable model, Eq. (3.36) is substituted into Eq. (3.37), yielding:

$$D_t - D_{t-1} = a + b(rE_t - D_{t-1}) + u_t, \quad (3.38)$$

which is expressed in difference form because the item of interest is changes in dividend levels and their cause, not the absolute levels themselves. Of the twenty-eight companies followed by Lintner, twenty-six appeared to have specific predetermined values for r, the target payout ratio, and the vast majority did update their dividend policies annually. While it could be argued that the desired level of dividends depends not only on current earnings but also on expected future levels of earnings, the empirical evidence has borne this theory out [see Brittain (1978) and Fama and Babiak (1968)]. A survey by Harkins and Walsh (1971) found that financial executives are well aware of this factor when they make dividend decisions; this casts doubt on the Lintner model. However, with the partial adjustment coefficient b being included in the model, we essentially have a proxy for expected earnings; this supports the theoretical development of the partial adjustment model.

Following the lines of criticism of Lintner's work, we can specify an adaptive expectations model, where dividends are predetermined, based on estimates of earnings as shown below, with all variables as defined before, the E^* being expected rather than actual earnings:

$$D_t = rE^* + u_t. \quad (3.39)$$

In this light, the dividend is intended to convey information pertaining to management's expectations of earnings and profitability, something akin to a specification of M&M's hypothesis on the informational content of dividends.

Laub (1976) and Pettit (1972, 1976) have shown that dividend payments do convey information about future earnings expected by management and that changes in dividends are a result of changes in longer-run expectations. Pettit's 1972 paper examined the returns to investors following dividend announcements, using the CAPM as the equilibrium return-generation process. While it was not clear that investors could obtain abnormal returns on a risk-adjusted basis by buying securities with large deviations in dividends from historical levels, the market was shown to impound the information of these announcements rapidly, particularly so in the case of initial payments. From this, we are forced to the conclusion that the market which appears to be quite efficient utilizes announcements of changes in dividend policy in the process of valuing a security. Though the dividend announcement may not give perfect information as to management's expectations, it is the best proxy available, and announcements of changes in dividend payments do appear to have informational content.

Returning to Eq. (3.39) we would like to find a way to deal with the expected earnings figure in the current time period, so we assume:

$$E_t^* = bE_t + (1 - b)E_{t-1}^*, \quad (3.40)$$

where b is again a coefficient of expectations, the proportion of the expectational error taken to be permanent. By specifying the transitory element, we now have an expression for expected or permanent level of earnings, partly based on current income, and partly based on the previous period's expected earnings, thus modifying previous expectations.

By substituting Eq. (3.39) into Eq. (3.40), multiplying both sides by $(1 - b)$, and lagging by one time period, we obtain:

$$D_t - D_{t-1} = rbE_t - bD_{t-1} + u_t + u_{t-1}(1 - b). \quad (3.41)$$

This closely resembles the partial adjustment model presented earlier, although the constant term denoting a reluctance to reduce dividends is missing, and the error term is specified differently. The b coefficient is a profit expectation coefficient rather than a speed-of-adjustment coefficient, which is the major distinguishing factor of this model. By assumption, D_t^* is subject only to random disturbances given by the u_t in the adaptive expectations model. Due to the conceptual oversights of each model, it is hypothesized that a more general model embodying the essential components from each model should be constructed.

Recognition of this shortfall in each model was initially discussed in Brittain (1978), though his discussion centered around the determinants of the payout ratio rather than the dividend itself, and he did not conduct tests relating to the present controversy. Ang (1975) did test to determine the difference between the two alternative hypotheses of dividend behavior. Using spectral analysis, he argued that his results, derived from quarterly FTC-SEC reports of manufacturing companies, proved that the informational content theory can be used to explain shorter-run dividend behavior and that the partial adjustment hypothesis can be used to explain longer-run dividend policy, both of which seem reasonable approximations. Oddly enough, though neither hypothesis was supported in an intermediate run, all of which reinforces the notion a more generalized model is needed.

An Integration Model

Djarraya and Lee (1981) followed the conceptual development of Waud (1966), in embodying the conceptual ingredients of both dividend behavior rationales, and constructed a more general specification of dividend determination. In a long-run framework, it is expected that the desired level of dividends can be expressed as a percentage of expected earnings, neither variable being observable for practical purposes. Including the major arguments of the partial adjustment and information content theories, we can write the following three equations:

$$D^* = rE_t^*, \tag{3.42}$$

$$D_t - D_{t-1} = a + b_1(D^* - D_{t-1}) + u_t, \tag{3.43}$$

$$E_t^* - E_{t-1}^* = b_2(E_t - E_{t-1}^*), \tag{3.44}$$

which, when combined and simplified, were shown to yield:

$$\begin{aligned} D_t - D_{t-1} &= ab_2 + (1 - b_1 - b_2)D_{t-1} \\ &\quad - (1 - b_2)(1 - b_1)D_{t-2} \\ &\quad + rb_1b_2E_t - (1 - b_2)u_{t-1} + u_t. \end{aligned} \tag{3.45}$$

In performing empirical tests, many varied conclusions can be reached depending on the relevant outcomes. Those most important for the analysis here include:

(i) If the b_2 coefficient of expectations is equal to one, then the generalized model reduces to the simpler partial adjustment model.

(ii) If the speed-of-adjustment coefficient, b_1, equals one and the intercept a equals zero, the model reduces to the alternative simplified model, an adaptive expectations model of earnings.

(iii) If the intercept term equals zero and the two b coefficients equal one, then dividend policy is actually a residual decision, much as Higgins suggested it should be.

(iv) If all coefficients are significantly different from zero and/or one, then the two previously discussed simplified models are insufficient to describe corporate dividend policy, and the generalized model offers a much-needed explanation.

As for the empirical tests Djarraya (1980) has shown that the ordinary least squares regression technique does not allow for the distinction between the b_1 and b_2 coefficients, Doran and Griffiths (1978) have shown that the OLS estimates of b_1 and b_2 are inconsistent. Therefore, the maximum likelihood estimator techniques required to perform the empirical tests were used in conjunction with Marquardt's (1963) nonlinear least squares regression technique for all eighty industrial firms included in the study.

Using unadjusted quarterly data, it was found that the estimated target payout ratio was between 0 and 1, as expected, with the mean value being 0.43. Breaking the results into deciles, it was found that the relationship between desired dividends and expected earnings per share varied from 0.19 for the lowest 10% of the price–earnings ratios to a high of 0.72, on average, for the highest 10% of the price–earnings ratios. In eighty percent of the firms examined, this relationship was found to be statistically significant, which, roughly translated, implies that the expected or permanent earnings, expressed as a weighted average of the current and previous period's expected earnings, have a significant impact on the desired dividend-per-share payout. The a coefficient incorporated from the Lintner model was, somewhat surprisingly, not significantly different from zero in the case of sixty percent of the firms, and in the lowest two deciles, it was, on average, less than zero. Such evidence supports the suggestion of Fama and Babiak that it should be excluded. As a caveat to taking this suggestion as valid for all tests, it is necessary to realize that firms with higher price–earnings multiples consistently showed positive and statistically significant coefficients for this parameter estimate, and in examining those firms individually, the constant term should probably not be suppressed.

The other factor from the Lintner model, the speed-of-adjustment coefficient, was found to be significantly different from one, with a mean of 0.612; this reveals that dividends, on average, tend to gravitate toward their desired levels in slightly greater than half of one year. In greater than seventy percent of the sample, this coefficient was statistically significant, which should lead to cautious use of the adapted expectation model by itself as a model of the dividend behavior of corporations.

Examining the coefficient of expectation, a result similar to that above was found, the mean of this estimated parameter being 0.16, where over sixty percent of the firms in the sample exhibited coefficients of expectation significantly different from zero. Again, as with the previous estimates, the higher the value of the *P/E* ratio, the higher the average level of significance of the coefficient. This systematic result may well come about because of the exclusion of outside influences not included in the model that have a natural dampening effect on P/E ratios; all these findings lend further credence to the generalized model.

In summary, it is reasonable to suggest that neither the partial adjustment model of dividend behavior nor the adaptive expectations model adequately explains the dividend behavior of firms. The results of Djarraya and Lee reveal the partial adjustment and expectations coefficients are significantly smaller than one and greater than zero. In effect, these results confirm the suspicion that a more generalizable model of dividend behavior on the part of firms is necessary, in order to understand the true nature of the dividend decision.

Appendix 4: Logistic Model and Probit Model

The likelihood function for binary sample space of bankruptcy and nonbankruptcy is

$$l = \prod_{i \in s_1} P(X_i, \beta) \prod_{i \in s2} (1 - P(X_i, \beta)) \quad (3.46)$$

where P is some probability function, $0 \leq P \leq 1$; and

$P(X_i, \beta)$ denotes probability of bankruptcy for any given X_i and β.

Since it is not easy to solve the selecting probability function P, for simplicity, one can solve the likelihood function in (3D1) by taking the natural logarithm. The logarithm of the likelihood function then is

$$L(l) = \sum_{i \in s_1} \log P(X_i, \beta) + \sum_{i \in s_2} \log(1 - P(X_i, \beta)) \quad (3.47)$$

where

S_1 is the set of bankrupt firms;
S_2 is the set of nonbankrupt firms.

The maximum likelihood estimators for β_s can be obtained by solving $\text{Max}_\beta L(l)$.

In logistic model, the probability of company i going bankrupt given independent variables X_i is defined as

$$P(X_i, \beta) = \frac{1}{1 + \exp(-\beta' X_i)} \quad (3.48)$$

The two implications here are (1) $P(.)$ is increasing in $\beta' X_i$ and (2) $\beta' X_i$ is equal to $\log\left[\frac{P}{(1-P)}\right]$.

We then classify bankrupt firms and non-bankrupt firms by setting a "cutoff" probability attempting to minimize type I and type II errors.

In probit models, the probability of company i going bankrupt given independent variables X_i is defined as

$$P(X_i, \beta) = \int_{-\infty}^{(X_i, \beta)} \frac{1}{\sqrt{2\pi}} e^{-\frac{z^2}{2}} dz \quad (3.49)$$

β, the cumulative standard normal distribution function.

Maximum likelihood estimators from probit model can be obtained similarly as in the logistic models. Although probit models and logistic models are similar, logistic models are preferred to probit models due to the nonlinear estimation in probit models (Gloubos and Grammatikos 1998).

Appendix 5: SAS Code for Hazard Model in Bankruptcy Forecasting

```
libname  jess 'D:\Documents and Settings\MAI\Desktop\sas code';
*test modified model's performance in the test sample;
*** Logistic Regression Analysis ***;
options pageno=1;
* Add prediction data set to original data;
data jess._prddata;
   _FREQ_ = 1;
   set jess.train_winsor1 jess.test_winsor1(in=inprd);
   _inprd_ = inprd;
   * set freq variable to 0 for additional data;
   if inprd then _FREQ_ = 0;
run;

proc logistic data=jess._prddata DESCEND;
   freq _FREQ_;
   model BPTSTATUS = NITA CASALES CACL CATA CASHTA LTDTA
LSALES CAR LNMCP/
    ctable pprob=0.001 to 0.99 by 0.01;
** Create output data set for predictions **;
   output out=jess.pred p=phat;
run;

proc means data=jess._prddata;
var NITA CASALES CACL CATA CASHTA LTDTA LSALES stress3 CAR
LNMCP;
run;

data jess.out_sample;
set jess.pred;
if _FREQ_ = 0;
run;

data jess.count1_1;
set jess.out_sample;
if phat >=0.021 and bptstatus = 1;
run;

data jess.count1_2;
set jess.out_sample;
if  bptstatus =1;
run;

data jess.count2_1;
set jess.out_sample;
if phat < 0.021 and bptstatus = 0;
run;

data jess.count2_2;
set jess.out_sample;
if  bptstatus =0;
run;
```

Bibliography

Altman, E. I. (1968). Financial ratios, discriminant analysis, and the prediction of corporate bankruptcy. *Journal of Finance, 23,* 589–609.

Altman, E. I. (1993). *Corporate financial distress and bankruptcy: A complete guide to predicting and avoiding distress and profiting from bankruptcy.* New York: Wiley.

Ang, J. S. (1975). Dividend policy: Informational content or partial adjustment? *Review of Economics and Statistics, 57,* 65–70.

Beaver, W. (1968a). Market prices, financial ratios and prediction of failure. *Journal of Accounting Research, 6*(2), 179–192.

Beaver, W. H. (1968b). Alternative accounting measures as predictors of failure. *The Accounting Review, 43,* 113–122.

Begley, J., Ming, J., & Watts, S. (1996). Bankruptcy classification errors in the 1980s: An empirical analysis of Altman's and Ohlson's models. *Review of Accounting Studies, 1,* 267–284.

Bharath, S. T., & Shumway, T. (2004). *Forecasting default with the KMV-Merton model.* Working paper, the University of Michigan.

Bharath, S. T., & Shumway, T. (2008). Forecasting default with the Merton Distance to Default model. *Review of Financial Studies, 21*(3), 1339–1369.

Bhattacharya, S. (1979). Imperfect information, dividend policy, and the bird-in-the-hand fallacy. *The Bell Journal of Economics*, 259–276.

Black, F., Jensen, M., & Scholes, M. (1972). The capital-asset pricing model: Some empirical tests. In M. Jensen (Ed.), *Studies in the theory of capital markets* (pp. 79–124). New York: Praeger.

Blume, M. P. (1974). The failing company doctrine. *Journal of Accounting Research, 43,* 1–25.

Brennan, M. (1970, December). Taxes, market valuation, and corporate financial policy. *National Tax Journal*, 417–427.

Brittain, J. A. (1978). *Corporate dividend policy.* Washington, D.C.: Brookings Institute.

Cartwright, P., & Lee, C.-F. (1987 January). Time aggregation and beta estimation: Some empirical evidence. *Journal of Business and Economics Statistics.*

Chen, K. H., & Shimerda, T. A. (1981). An empirical analysis of useful financial ratios. *Financial Management*, 51–60.

Cox, D. R., & Oakes, D. (1984). *Analysis of survival data.* New York: Chapman & Hall.

Daniels, K., Shin, T., & Lee, C. F. (1997, March). The information content of dividend hypothesis: A permanent income approach. *International Review of Economics and Finance*, 77–86.

DeAngelo, H., & DeAngelo, L. (2006). The irrelevance of the MM dividend irrelevance theorem. *Journal of Financial Economics, 79,* 293–315.

Djarraya, M. (1980). *Behavior models of dividend policy and implications to financial management.* Dissertation, The University of Illinois at Urbana-Champaign.

Djarraya, M., & Lee, C. F. (1981). Residual theory, partial adjustment, and information content on dividend-payment decisions: An integration and extension. BEBR, Faculty Working Paper 760, University of Illinois.

Doran, H. E., & Griffiths, W. E. (1978). Inconsistency of the OLS estimator of the partial-adjustment-adaptive-expectations model. *Journal of Econometrics, 7,* 133–146.

Fama, E. F., & Babiak, H. (1968, December). Dividend policy: An empirical model. *Journal of American Statistical Association, 63,* 1132–1161.

Fama, E. F., & Macbeth, I. D. (1973 May/June). Risk, return, and equilibrium: Empirical tests. *Journal of Political Economy*, 607–636.

Frecka, T., & Lee, C. F. (1983). Generalized ratio generation process and its implications. *Journal of Accounting Research*, 308–316.

Friend, I., & Puckett, M. E. (1964 September). Dividends and stock prices. *American Economic Review*, 656–682.

Gordon, M. J. (1959 May). Dividends, earnings, and stock prices. *Review of Economics and Statistics*, 99–105.

Gordon, M. J. (1963 May). Optimal investment and financing policy. *Journal of Finance*, 264–272.

Greene, W. H. (2017). *Econometric analysis* (8th ed.). New Jersey: Prentice Hall.

Gupta, M. C., Lee, A. C., & Lee, C. F. (2007). *Effectiveness of dividend policy under the capital asset pricing model: A dynamic analysis.* Working Paper, Rutgers University.

Han, D. (1985). *Dividend policy under the condition of capital market and signaling equilibria: Theory and evidence.* Unpublished Doctoral Dissertation, University of Illinois at Urbana-Champaign.

Harkins, E. P., & Walsh, F. J. (1971). *Dividend policies and practices.* The Conference Board, New York.

Higgins, R. C. (1972). The corporate dividend saving decisions. *Journal of Financial and Quantitative Analysis, 7,* 1531–1538.

Hillegeist, S. A., Keating, E. K., & Cram, D. P. (2004). Assessing the probability of bankruptcy. *Review of Accounting Studies, 9,* 5–34.

Honjo, Y. (2000). Business failure of new firms: An empirical analysis using a multiplicative hazards model. *International Journal of Industrial Organization, 18*(4), 557–574.

Horrigan, J. O. (1965). Some empirical bases of financial ratio analysis. *The Accounting Review, 40,* 558–586.

Hwang, R. C., Wei, H. C., Lee, J. C., & Lee, C. F. (2008). On prediction of financial distress using the discrete-time survival model. *Journal of Financial Studies, 16,* 99–129.

Jensen, M. C. (1972). Capital markets: Theory and evidence. *Bell Journal of Economics and Management Science*, 357–398.

Johnson, W. B. (1979). The cross-sectional stability of financial ratio patterns. *Journal of Financial and Quantitative Analysis, 14,* 1035–1048.

Joy, O. M., & Tollefson, J. O. (1975). On the financial applications of discriminant analysis. *Journal of Financial And Quantitative Analysis, 10,* 723–739.

Kao, C., Lee, C. F., & Wu, C. (1991). Rational expectations and corporate dividend policy. *Review of Quantitative Finance and Accounting, 1,* 331–348.

Kao, L-j, & Lee, C-f. (2012). alternative method for determining industrial bond ratings: Theory and empirical evidence. *International Journal of Information Technology & Decision Making, 11*(6), 1215–1235.

Kiefer, N. M. (1988). Economic duration data and hazard functions. *Journal of Economic Literature, 26,* 646–679.

King, B. F. (1966). Market and industry factors in stock price behavior. *Journal of Business,* (Suppl.), 139–190.

Lancaster, T. (1992). *The econometric analysis of transition data.* New York: Cambridge University Press.

Laub, P. M. (1976). On the informational content of dividends. *Journal of Business, 49*(1), 73–80.

Lee, A. C., Lee, J. C., & Lee, C. F. (2017). *Financial analysis, planning and forecasting: Theory and application* (3rd ed.). Singapore: World Scientific.

Lee, C.-F., Djarraya, M., & Wu, C. (July 1987). A further empirical investigation of the dividend adjustment process. *Journal of Econometrics.*

Lee, C. F., Wu, C., & Hang, D. (1993 March). Dividend policy under conditions of capital markets and signaling equilibrium. *Review of Quantitative Finance and Accounting*, 47–59.

Lee, C. F., Lee, J., & Lee, A. C. (2013). *Statistics for business and financial economics* (3rd ed.). Berlin: Springer.

Lee, C. F., Liaw, T., & Wu, C. (1992 September). Forecasting accuracy of alternative dividend models. *International Review of Economics and Finance*, 261–270.

Lee, C. F., & Kau, J. B. (1987). Dividend payment behavior and dividend policy of REITS. *The Quarterly Review of Economics and Business*, 6–21.

Lee, C. F., & Gilmore, R. H. (1986 May). Empirical tests of granger proposition on dividend controversy. *Review of Economics and Statistics*, 351–355.

Lee, C. F., & Primeaux, W. J., Jr. (1985). Relative importance of current vs. permanent income for dividend payment decision in the electric utility industry. *Journal of Behavior Economics*, 83–97.

Lee, C. F., & Forbes, S. (1982 June). income measures, ownership, capacity ratios and dividend decision of the non-life insurance industry: Some empirical evidence. *Journal of Risk and Insurance*, 269–289.

Lee, C. F., & Forbes, S. W. (1980 June). Dividend policy, equity value, and cost of capital estimates for the property and liability insurance industry. *Journal of Risk and Insurance*, 205–222.

Lee, C. F., & Gupta, M. C. (1977 September). An inter-temporal approach to the optimization of dividend with pre-determined investment: A further comment. *Journal of Finance*, 1348–1353.

Lee, C. F. (1976 December). Functional form and the dividend effect in the electric utility industry. *Journal of Finance*, 1481–1486.

Lee, C. F., & Chang, H. S. (1984). On dividend policy and capital market theory: A generalized error components model approach. *Journal of Business Research*, 505–516. (in press).

Lee, C. F., & Lee, A. C. (2013). *Encyclopedia of finance* (2nd ed.). New York: Springer.

Lee, C. F., & Primeaux, W. J., Jr. (1991). Current vs. permanent dividend payments behavioral model: Methods and application. *Advances in Quantitative Analysis of Finance and Accounting, I.*

Lee, C. F., Djarraya, M., & Wu, C. (1987 July). A further empirical investigation of the dividend adjustment process. *Journal of Econometrics*, 267–285.

Lintner, J. (1956). Distribution of income of corporations. *American Economic Review, 46,* 97–113.

Mai, J. S. (2010). Alternative *approaches to business failure prediction models.* Essay I of Dissertation, Rutgers University.

Marquardt, D. (1963). An algorithm for least-squares estimation of nonlinear parameters. *Journal of SIAM, 2,* 431–441.

Mehta, D. R. (1974). *Working capital management.* Englewood Cliffs, NJ: Prentice-Hall Inc.

Merton, R. C. (1974). On the pricing of corporate debt: The risk structure of interest rates. *Journal of Finance, 29,* 449–470.

Miller, M. H., & Modigliani, F. (1961). Dividend policy growth and the valuation of shares. *Journal of Business, 34,* 411–433.

Miller, M., & Modigliani, F. (1966 June). Some estimates of the cost of capital to the electric utility industry. *American Economic Review*, 334–391.

Miller, M. H., & Scholes, M. S. (1982). Dividends and taxes: Some empirical evidence. *Journal of Political Economy, 96,* 334–391.

Molina, C. A. (2005). Are firms underleveraged? An examination of the effect of leverage on default probabilities. *Journal of Finance, 3,* 1427–1459.

Moody (2000). RiskCalc for private companies.

Moody. (2003). Modeling default risk.

Ohlson, J. S. (1980). Financial ratios and the probabilistic prediction of bankruptcy. *Journal of Accounting Research, 19,* 109–131.

Orgler, Y. E. (1970). A credit-scoring model for commercial loans. *Journal of Money, Credit and Banking, 2,* 435–445.

Orgler, Y. E. (1975). *Analytical methods in loan evaluation.* Lexington, MA: Lexington Books.

Pettit, R. (1972 December). Dividend announcements, security performance, and capital-market efficiency. *The Journal of Finance*, 993–1007.

Pettit, R. R. (1976). The impact of dividend and earnings announcements: A reconciliation. *The Journal of Business, 49*(1), 86–96.

Pinches, G. E., & Mingo, K. A. (1973). A multivariate analysis of industrial bond ratings. *Journal of Finance, 28,* 1–18.

Pinches, G. E., Eubank, A. A., & Mingo, K. A. (1975). The hierarchical classification of financial ratios. *Journal of Business Research, 3,* 295–310.

Pinches, G. E., Mingo, K. A., & Caruthers, J. K. (1973). The stability of financial patterns in industrial organizations. *Journal of Finance, 28,* 389–396.

Pinches, G. E., Singleton, J. C., & Jahankhani, A. (1978). Fixed coverage as a determinant of electric utility bond ratings. *Financial Management, 8,* 45–55.

Pogue, T. F., & Soldofsky, R. M. (1969). What's in a bond rating? *Journal of Financial and Quantitative Analysis*, 201–228.

Roll, R., & Ross, S. A. (1980). An empirical investigation of the arbitrage pricing theory. *Journal of Finance, 35,* 1073–1103.

Saunders, A., & Allen, L. (2002). *Credit risk measurement: New approaches to value at risk and other paradigms* (2nd ed.). New York: Wiley.

Saunders, A., & Cornett, M. M. (2006). *Financial institutions management: A risk management approach*. New York, NY: McGraw-Hill.

Saunders, A., & Cornett, M. M. (2013). *Financial institutions management: A risk management approach* (8th ed.). New York:McGraw-Hill/Irwin.

Shumway, T. (2001). Forecasting bankruptcy more accurately: A simple hazard model. *The Journal of Business, 74,* 101–124.

Simkowitz, M. A., & Monroe, R. J. (1971 November). A discriminant analysis function for conglomerate targets. *Southern Journal of Business*, 1–16.

Sinkey, J. F. (1975). A multivariate statistical analysis of the characteristics of problem bank. *Journal of Finance, 30,* 21–36.

Stevens, D. L. (1970). Financial characteristics of merged firms: A multivariate analysis. *Journal of Financial and Quantitative Analysis, 5,* 36–62.

Sun, L. (2007). A re-evaluation of auditors' opinions versus statistical models in bankruptcy prediction. *Review of Quantitative Finance and Accounting, 28* (1), 55–78.

Trieschmann, J. S., & Pinches, G. E. (1973 September). A multivariate model for predicting financially distressed P-L insurers. *Journal of Risk and Insurance*, 327–338.

Van Horne, J. C. (2001). *Financial management and policy* (12th ed.). Englewood Cliffs, NJ: Prentice Hall Inc.

Waud, R. N. (1966). Small-sample bias due to misspecification in the 'Partial Adjustment' and 'Adaptive Expectations' models. *Journal of the American Statistical Association, 61,* 1130–1152.

Wilcox, J. W. (1971). A simple theory of financial ratios as predictors of failure. *Journal of Accounting Research*, 389–395.

Zavgren, C. V. (1983). The prediction of corporate failure: The state of the art. *Journal of Accounting Literature, 2,* 1–38.

Zmijewski, M. E. (1984). Methodological issues related to the estimation of financial distress prediction models. *Supplement to Journal of Accounting Research, 22,* 59–68.

Simultaneous Equation Models

4

Contents

Abstract

Based upon the discussion of single-equation models in the previous two chapters, in this chapter, we discuss simultaneous equation models, including the basic definition, their specification, and identification. The estimation methods for simultaneous equation models include two-stage least squares estimation method and three-stage least squares estimation method. Some examples of applying simultaneous equation models in finance research are also provided.

4.1 Introduction

Based upon Chen and Lee (2010) and Greene (2017), we extend single-equation models, which have been discussed in the previous two chapters, to simultaneous equation models. This chapter first introduces the concept of simultaneous equation system and the application of such system to finance research. Then, both two-stage and three-stage least square models have been extensively used in finance, accounting, and economic research. The pro and cons of these estimations

C.-F. Lee et al., *Financial Econometrics, Mathematics and Statistics*,
https://doi.org/10.1007/978-1-4939-9429-8_4

methods are summarized. Wang (2015) has discussed instrumental variable approach to correct for endogeneity in finance. In addition, Lee (1976) used two-stage least square to study interdependent structure of security returns. Lee et al. (2016) has discussed how 2LS and 3LS models in finance research in terms of investment, dividend, and financing policies. Lee et al. (2018) has applied simultaneous equation to perform the research related to inside debt, firm risk, and investment decision. Finally, the results of a study in executive compensation structure and risk-taking are used to illustrate the difference between single-equation and simultaneous equation method.

In this chapter, we extend single-equation models to simultaneous equation models. Section 4.2 discusses simultaneous equation system in detail. In Sect. 4.3, we discuss two-stage least squares method and three-stage least squares method. In Sect. 4.4, the applications of simultaneous equation in finance research are discussed. Finally, in Sect. 4.5, we conclude this chapter.

4.2 Discussion of Simultaneous Equation System

Empirical finance research often employs a single equation for estimation and testing. However, single equation rarely happens in the economic or financial theory. Using OLS method to estimate equation(s) which should otherwise be treated as a simultaneous equation system is likely to produce biased and inconsistent parameter estimators. Following Greene (2017), we start to illustrate the nature of simultaneous equation system in terms of simple Keynesian consumption function, specified as follows:

$$C_t = \alpha + \beta Y_t + \mu_t \tag{4.1}$$

$$Y_t = C_t + I_t \tag{4.2}$$

$$I_t = I_0 \tag{4.3}$$

where C_t is the consumption expenditure at time t, Y_t is the national income at time t, I_t is the investment expenditure at time t, which is assumed fixed at I_0, and μ_t is the stochastic disturbance term at time t. Equation (4.1) is the consumption function; Eq. (4.2) is the equilibrium condition (national income accounting identity); and Eq. (4.3) is the investment function. Some of the variables in the model are endogenous, and others are exogenous. For example, C_t and Y_t are endogenous, meaning they are determined within the model. On the other hand, I_t is the exogenous variable, which is not determined in the model, hence not correlated with μ_t.

A simultaneous equation bias arises when OLS is applied to estimate the consumption function because Y_t is correlated with the disturbance term μ_t, which violates the OLS assumption that independent variables are orthogonal to the disturbance term. To see this, through a series of substitutions, we can obtain the reduced form equations from the structural equation (4.1) through (4.3) as:

$$C_t = \frac{\alpha}{1-\beta} + \frac{\beta}{1-\beta} I_0 + \frac{\mu_t}{1-\beta} \tag{4.4}$$

$$Y_t = \frac{\alpha}{1-\beta} + \frac{1}{1-\beta} I_0 + \frac{\mu_t}{1-\beta} \tag{4.5}$$

Y_t is correlated with the disturbance term, μ_t, which violates the OLS assumption that independent variables and the disturbance term are uncorrelated. This bias is commonly referred to in the literature as the simultaneous equation bias. Furthermore, the OLS estimate of β in Eq. (4.1) is not only bias, but also inconsistent, meaning β estimate does not converge to the true β when the sample size increases to vary large.

4.3 Two-Stage and Three-Stage Least Squares Method

Greene (2017) has several chapters, which theoretically discuss both two-stage and three-stage least squares method. To resolve the simultaneous equation bias problem as illustrated in Sect. 4.1, in this section we discuss two popular simultaneous equation estimation methods. Different methods are available to handle the estimation of a simultaneous equation model: indirect least squares, instrumental variable procedure, two-stage least squares, three-stage least squares, limited information likelihood method, and full information maximum likelihood method, just to name a few. In this section, we will focus on two methods that popular statistical and/or econometric software are readily available.

4.3.1 Identification Problem

Before getting into the estimation methods, it is necessary to discuss the "identification problem." Identification problem arises when we cannot identify the difference between, say, two functions. Consider the demand and supply model of gold. The structure equations can be written as:

$$Q_d = \alpha_0 + \alpha_1 P + \varepsilon \tag{4.6}$$

$$Q_s = \beta_0 + \beta_1 P + e \tag{4.7}$$

$$Q_d = Q_s = Q \tag{4.8}$$

Equation (4.6) is the demand for gold function, where the demand Q_d is determined by the price of gold, P; Eq. (4.7) is the supply of gold, and it is a function of gold price; Eq. (4.8) is an identity stating the market equilibrium. Can we apply the OLS method to Eqs. (4.6) and (4.7) to obtain parameter estimates? To answer this question, we first obtain the "reduced form" equations for P and Q through substitutions.

$$P = \frac{\beta_0 - \alpha_0}{\alpha_1 - \beta_1} + \frac{\varepsilon - e}{\alpha_1 - \beta_1} \tag{4.9}$$

$$Q = \frac{\alpha_1\beta_0 - \alpha_0\beta_1}{\alpha_1 - \beta_1} + \frac{\alpha_1\varepsilon - \beta_1 e}{\alpha_1 - \beta_1} \tag{4.10}$$

Obviously, it is impossible to estimate Eqs. (4.9) and (4.10) using OLS method because there are four parameters $(\alpha_0, \alpha_1, \beta_0, \text{and}\, \beta_1)$ to be estimated, but there are only two equations. Therefore, we cannot estimate the parameters in the structure equations. This is the situation called "underidentification."

To differentiate demand equation from supply equation, now suppose we assume that demand curve for gold may shift due to the changes in economic uncertainty, which can be proxied by, say, stock market volatility, V, which is assumed to be exogenous. Hence, Eqs. (4.6) and (4.7) can be modified as:

$$Q_d = \alpha_0 + \alpha_1 P + \alpha_2 V + \varepsilon \tag{4.11}$$

$$Q_s = \beta_0 + \beta_1 P + e \tag{4.12}$$

The reduced form becomes

$$\begin{aligned} P &= \frac{\beta_0 - \alpha_0}{\alpha_1 - \beta_1} - \frac{\alpha_2}{\alpha_1 - \beta_1} V + \frac{\varepsilon - e}{\alpha_1 - \beta_1} \\ &= \gamma_0 + \gamma_1 V + \pi_1 \end{aligned} \tag{4.13}$$

$$\begin{aligned} Q &= (\beta_0 + \beta_1\gamma_0) + \beta_1\gamma_1 V + \frac{\alpha_1\varepsilon - \beta_1 e}{\alpha_1 - \beta_1} \\ &= \lambda_0 + \lambda_1 V + \pi_2 \end{aligned} \tag{4.14}$$

Because V is assumed exogenous and uncorrelated with residuals π_1 and π_2, OLS can be applied to the reduced form Eqs. (4.13) and (4.14) and obtain estimators of $\gamma_0, \gamma_1, \lambda_0$, and λ_1. Examine Eq. (4.14), we find that $\lambda_0 = \beta_0 + \beta_1\gamma_0$, and $\lambda_1 = \beta_1\gamma_1$. Since $\gamma_0, \gamma_1, \lambda_0$, and λ_1 are all obtained from the OLS estimates, β_0 and β_1 can be solved. Therefore, the supply function Eq. (4.14) is said to be identified. However, from the demand function, we find that

$$\gamma_0 = \frac{\beta_0 - \alpha_0}{\alpha_1 - \beta_1}, \text{ and } \gamma_1 = \frac{\alpha_2}{\alpha_1 - \beta_1}.$$

Since there are only two equations, we cannot possibly estimate three unknowns, α_0, α_1, and α_2,; hence, the demand function is not identified. Based upon the discussions of Eqs. (4.6) and (4.7), we thus know that in a two-equation model, if one variable is omitted from one equation, then this equation is identified. On the other hand, there is no omitted variable in Eq. (4.6), and hence, the demand function is not identified.

Now let us further modify Eqs. (4.6) and (4.7) as follows:

$$Q_d = \alpha_0 + \alpha_1 P + \alpha_2 V + \varepsilon \quad (4.15)$$

$$Q_s = \beta_0 + \beta_1 P + \beta_2 D + e \quad (4.16)$$

All variables are defined as before except now we have added a new variable D in the supply equation. Let D be the government deficit of Russia, which is assumed exogenous to the system. When Russia's budget deficit deteriorates, the government increases the gold production for cash. The reduced form of P and Q based upon Eqs. (4.11) and (4.12) is

$$\begin{aligned} P &= \frac{\beta_0 - \alpha_0}{\alpha_1 - \beta_1} - \frac{\alpha_2}{\alpha_1 - \beta_1} V + \frac{\beta_2}{\alpha_1 - \beta_1} D + \frac{\varepsilon - e}{\alpha_1 - \beta_1} \\ &= \gamma_0 + \gamma_1 V + \gamma_2 D + \pi_1 \end{aligned} \quad (4.17)$$

$$\begin{aligned} Q &= (\alpha_0 + \alpha_1 \gamma_0) + (\alpha_1 \gamma_1 + \gamma_2) V + \alpha_1 \gamma_2 D + \frac{\alpha_1 \varepsilon - \beta_1 e}{\alpha_1 - \beta_1} \\ &= \lambda_0 + \lambda_1 V + \lambda_2 D + \pi_2 \end{aligned} \quad (4.18)$$

Based upon the OLS estimates of Eqs. (4.17) and (4.18), we can obtain unique estimates for the structure parameters $\alpha_0, \alpha_1, \alpha_2, \beta_0, \beta_1$, and β_2, and hence, both the demand and supply functions are identified. In this case, we call the situation as "exactly identified."

In a scenario when there are multiple solutions to the structure parameters, the equation is said to be "overidentified." For example, in Eqs. (4.19) and (4.20), we modify the supply equation by adding another exogenous variable, q, representing lagged quantity of gold produced (i.e., supply of gold in the last period), which is predetermined.

$$Q_d = \alpha_0 + \alpha_1 P + \alpha_2 V + \varepsilon \quad (4.19)$$

$$Q_s = \beta_0 + \beta_1 P + \beta_2 D + \beta_3 q + e \quad (4.20)$$

The reduced form becomes

$$\begin{aligned} P &= \frac{\beta_0 - \alpha_0}{\alpha_1 - \beta_1} - \frac{\alpha_2}{\alpha_1 - \beta_1} V + \frac{\beta_2}{\alpha_1 - \beta_1} D + \frac{\beta_3 q}{\alpha_1 - \beta_1} + \frac{\varepsilon - e}{\alpha_1 - \beta_1} \\ &= \gamma_0 + \gamma_1 V + \gamma_2 D + \gamma_3 q + \pi_1 \end{aligned} \quad (4.21)$$

$$\begin{aligned} Q &= (\alpha_0 + \alpha_1 \gamma_0) + (\alpha_1 \gamma_1 + \gamma_2) V + \alpha_1 \gamma_2 D + \alpha_1 \gamma_3 + \frac{\alpha_1 \varepsilon - \beta_1 e}{\alpha_1 - \beta_1} \\ &= \lambda_0 + \lambda_1 V + \lambda_2 D + \lambda_3 q + \pi_2 \end{aligned} \quad (4.22)$$

Based upon Eqs. (4.21) and (4.22), we find $\alpha_1 \gamma_2 = \lambda_2$, and hence, structure equation parameter α_1 can be estimated as γ_2/λ_2. However, we also find $\alpha_1 \gamma_3 = \lambda_3$, and hence, α_1 can also take another value, γ_3/λ_3. Therefore, α_1 does not have a unique solution, and we say the model is "overidentified."

The condition we employ in the above discussions for model identification is the so-called order condition of identification. To summarize the order condition of model identification, a general rule is that the number of variables excluded from an equation must be $\geq$ the number of structural equations.[1]

Although "order condition" is a popular way of model identification, it provides only a necessary condition for model identification, not a sufficient condition. Alternatively, "rank condition" provides both necessary and sufficient conditions for model identification. An equation satisfies the rank condition if and only if at least one determinant of rank (M − 1) can be constructed from the column coefficients corresponding to the variables that have been excluded from the equation, where M is the number of equations in the system. However, "rank condition" is more complicate than "order condition," and it is difficult to determine in a large simultaneous equation model. The following example based upon Eq. (4.23) through (4.25) provides some basic ideas about "rank condition." Note Eq. (4.23) through (4.25) are

[1]The discussions of the order condition draw heavily from Ramanathan (1995).

similar to Eqs. (4.11), (4.12), and (4.8) with terms rearranged.

$$Q_d - \alpha_0 - \alpha_1 P - \alpha_2 V = \varepsilon \quad (4.23)$$

$$Q_s - \beta_0 - \beta_1 P = e \quad (4.24)$$

$$Q_d - Q_s = 0 \quad (4.25)$$

In the following table, all structural parameters are stripped from the equations and placed in a matrix.

	Variables				
Equations	Intercept	Q_d	Q_s	P	V
Equation (11.1)	$-\alpha_0$	1	0	$-\alpha_1$	$-\alpha_2$
Equation (11.2)	$-\beta_0$	0	1	β_1	0
Equation (11.3)	0	1	−1	0	0

Since variables Q_d and V are excluded from Eq. (4.24), the determinant of remaining parameters in columns Q_d and V is

$$\begin{vmatrix} 1 & -\alpha_2 \\ 1 & 0 \end{vmatrix} = \alpha_2$$

Because this has a rank of 2, which is equal to the number of equations subtracts 1, Eq. (4.2) is identified. On the other hand, a determinant of rank 2 cannot be constructed for Eq. (4.23) because it has only one zero coefficient, and hence Eq. (4.1) is "underidentified."[2]

4.3.2 Two-Stage Least Squares

Two-stage least squares (2SLS) method is easy to apply and can be applied to a model that is exactly or overidentified. To illustrate, let us use Eqs. (4.15) and (4.16) for demonstration, and rewrite them as follows:

$$Q_d = \alpha_0 + \alpha_1 P + \alpha_2 V + \varepsilon \quad (4.26)$$

$$Q_s = \beta_0 + \beta_1 P + \beta_2 D + e \quad (4.27)$$

Based upon the "order condition," Eqs. (4.26) and (4.27) each has one variable excluded from the other equation, which is equal to the number of equations minus one. Hence, the model is identified.

Since endogenous variable P is correlated with the disturbance term, the first stage for the 2SLS calls for the estimation of "predicted P" ($\hat{P}$) using a reduced form containing all exogenous variables. To do this, we can apply the OLS to the following equation:

$$P = \eta_0 + \eta_1 V + \eta_2 D + \tau \quad (4.28)$$

OLS will yield unbiased and consistent estimation because both V and D are exogenous, hence not correlated with the disturbance term τ. With the parameters in Eq. (4.28) estimated, the "predicted P" can be calculated as:

$$\hat{P} = \hat{\eta}_0 + \hat{\eta}_1 V + \hat{\eta}_2 D.$$

This $\hat{P}$ is the instrumental variable to be used in the second stage estimation and is not corrected with the structure equation disturbance term. Substituting $\hat{P}$ into Eqs. (4.26) and (4.27), we have

$$Q_d = \alpha_0 + \alpha_1 \hat{P} + \alpha_2 V + \varepsilon \quad (4.29)$$

$$Q_s = \beta_0 + \beta_1 \hat{P} + \beta_2 D + e \quad (4.30)$$

Since $\hat{P}$ is not correlated with the disturbance terms, OLS method can be applied to Eqs. (4.29) and (4.30).

4.3.3 Three-Stage Least Squares

The 2SLS method is a limited information method. On the other hand, the three-stage least squares (3SLS) method is a full information method.[3] A full information method takes into account the information from the complete

[2]For more detailed discussions of the rank condition, see econometric books such as Greene (2017), Judge et al. (1985), Fisher (1966), Blalock (1969), and Fogler and Ganapathy (1982).

[3]3SLS is a combination of 2SLS and the seemingly uncorrelated regression (SUR) method proposed by Zellner (1962). SUR method will be discussed in the next chapter in detailed.

system, and hence, it is more efficient than the limited information method. Simply put, 3SLS method incorporates information obtained from the variance–covariance matrix of the system disturbance terms to estimate structural equation parameters. On the other hand, 2SLS method assumes that ε and e in Eqs. (4.29) and (4.30) are independent and estimates structural equation parameters separately, and thus, it might lose some information when in fact the disturbance terms are not independent. This section briefly explains a 3SLS estimation method.

Let the structural equation, in matrix, be:

$$Y_i = Z_i\psi_i + \varepsilon_i \text{ where } i = 1, 2, \ldots m \quad (4.31)$$

In Eq. (4.31), Y_i is a vector of n observations on the left-hand side endogenous variables; Z_i is a matrix consisting of the right-hand side endogenous and exogenous variables, i.e., $Z_i = \left[y_i \vdots x_i\right]$; and ψ_i is a vector of structural equation parameters such that $\psi_i = \left[\alpha_i \vdots \beta_i\right]'$. Let

$$Y = \begin{pmatrix} Y_1 \\ \vdots \\ Y_m \end{pmatrix} \quad Z = \begin{pmatrix} Z_1 & \cdots & 0 \\ \vdots & \ddots & \vdots \\ 0 & \cdots & Z_m \end{pmatrix}$$

$$\psi = \begin{pmatrix} \psi_1 \\ \vdots \\ \psi_m \end{pmatrix} \quad \varepsilon = \begin{pmatrix} \varepsilon_1 \\ \vdots \\ \varepsilon_m \end{pmatrix}$$

Then,

$$Y = Z\psi + \varepsilon \quad (4.32)$$

If we multiply both sides of Eq. (4.32) by a matrix X', where

$$X' = \begin{pmatrix} x' & \cdots & 0 \\ \vdots & \ddots & \vdots \\ 0 & \cdots & x' \end{pmatrix}, \text{i.e.,}$$

$$X'Y = X'Z + X'\varepsilon \quad (4.33)$$

Then, the variance–covariance matrix of the disturbance term in Eq. (4.33), $X'\varepsilon$ will be

$$E(X'\varepsilon\varepsilon'X) = \Omega \otimes x'x = \begin{pmatrix} \sigma_{11}x'x & \cdots & \sigma_{1m}x'x \\ \vdots & \ddots & \vdots \\ \sigma_{m1}x'x & \cdots & \sigma_{mm}x'x \end{pmatrix} \quad (4.34)$$

The 3SLS structural equation parameters can thus be estimated as

$$\hat{\psi} = \{Z'X[\hat{\Omega}^{-1} \otimes (x'x)^{-1}]X'Z\}^{-1}Z'x[\hat{\Omega}^{-1} \otimes (x'x)^{-1}]X'Y \quad (4.35)$$

A question arises in the estimation process because the $\sigma's$ are unknown; hence, the matrix Ω^{-1} is also unknown. This problem can be resolved by using the residuals from the structural equations estimated by the 2SLS to form the mean sum of residual squares and use them to estimate Ω^{-1}. Standard econometric software such as SAS can be easily used to estimate Eq. (4.35). In sum, 3SLS takes three stages to estimate the structural parameters. The first stage is to estimate the reduce form system; the second stage uses 2SLS to estimate the $\hat{\Omega}$ matrix; and the third stage completes the estimation using Eq. (4.35). Since Ω contains information pertinent to the correlations between disturbance terms in the structural equations, 3SLS is called a full information method.[4]

Since 3SLS is a full information estimation method, the parameters estimated are asymptotically more efficient than the 2SLS estimates. However, this statement is correct only if the model is correctly specified. In effect, 3SLS is quite vulnerable to model misspecifications. This is because model misspecification in a single equation could easily propagate itself into the entire system.

[4]For more detailed discussions, see Ghosh (1991), Judge et al. (1985), and Greene (2017).

4.4 Application of Simultaneous Equation in Finance Research

In this section, we use an example employing the simultaneous equation model to illustrate how the system can be applied to finance research.

Corporate governance literature has long debated whether corporate executives' interest should be aligned with that of the shareholders. Agency theory argues that unless there is an incentive to align the managers' and shareholder's interests, facing the agency problem, managers are likely to exploit for personal interest at the expense of shareholders. One way to align the interest is to make the executive compensation incentive based. Chen et al. (2006) study the effect of bank executive incentive compensation on the firm risk-taking. A single-equation model of the effect of executive compensation on firm risk-taking would look like:

$$\text{Risk} = \alpha_0 + \alpha_1(\text{Comp}) + \alpha_2(\text{LTA}) + \alpha_3(\text{Capital}) + \alpha_4(\text{NI}) + \sum_{i=5}^{m} \alpha_i(\text{Dgeo}) + \sum_{i=m+1}^{n} \alpha_i(\text{Dyear}) + \mu \tag{4.36}$$

where Risk is measurements of firm risk; Comp is the executive incentive compensation (i.e., option-based compensation); LTA is the total assets in log form;

Capital is the bank's capital ratio; NI is the noninterest income, in percentage; Dgeo is a binary variable measuring bank's geographic diversification; and Dyear is a yearly dummy variable. Chen et al. (2006) argue that OLS estimates of Eq. (4.36) will produce simultaneity bias because executive compensation is endogenous to the model and is likely to be correlated with the disturbance term μ. Therefore, Chen et al. (2006) introduce another equation to measure executive compensation.

$$\text{Comp} = \beta_0 + \beta_1(\text{Risk}) + \beta_2(\text{LTA}) + \beta_3(\text{SP}) + \sum_{i=4}^{m} \beta_i(\text{Drate}) + \upsilon \tag{4.37}$$

where SP is the underlying stock price and Drate is a series of dummy variables measuring annual interest rates. For example, Rate92 is defined as the T-bill rate of 1992 if the data are from year 1992; otherwise, a value of 0 is assigned to Rate92. Interest rate dummies are used to control for the impact of interest rates on option value.

Taking Eqs. (4.36) and (4.37) together, we find that applying OLS to these two structural equations will not yield unbiased estimates because the right-hand-side variables include endogenous variables Comp and Risk. Therefore, Chen et al. (2006) apply 2SLS to these two equations, and Table 4.1 reports some of their results. Chen et al. first report OLS estimates of the risk equation and find that executive's compensation structure does not impact firm risk-taking. 2SLS results reported in Table 4.1, however, reveal that once the simultaneity of firm risk decision and executive compensation are taken into account, executive compensation does affect firm risk-taking. The incorrect inference derived from the OLS estimates is thus due to simultaneity bias.

Table 4.1 Simultaneous equation model showing the relation between total risk and option-based compensation estimated using two-stage least squares (2SLS)

Models	(1) σ_j and OPTION/TOTAL_COMP		(2) σ_j and ACCUMULATED_OPTION	
Equations	σ_j OPTION/TOTAL_COMP		σ_j ACCUMULATED_OPTION	
Variable	Equation 1	Equation 2	Equation 1	Equation 2
OPTION/TOTAL_COMP	0.00028	–	–	–
	(2.17)**			

(continued)

Table 4.1 (continued)

Models	(1) σ_j and OPTION/TOTAL_COMP		(2) σ_j and ACCUMULATED_OPTION	
Equations	σ_j	OPTION/TOTAL_COMP	σ_j	ACCUMULATED_OPTION
ACCUMULATED_OPTION		–	0.00021	–
			(4.86)***	
σ_j	–	1065.1	–	1877.65
		(3.13)***		(5.93)***
LN(TA)	−0.00123	2.191	−0.0015	2.5616
	(−3.73)***	(3.13)***	(−6.12)***	(3.93)***
CAPITAL_RATIO	−0.27	–	−0.08	–
	(−1.83)**		(−6.21)**	
NON_INT_INCOME%	0.0012	–	−0.0025	–
	(0.27)		(−0.83)	
GEO_DUMMY	−0.0009	–	0.0006	–
	(−0.99)		(0.85)	
STOCK_PRICE	–	0.0758	–	0.157
		(2.69)***		(5.99)***
D92/DRate92	0.0059	−1.842	0.0055	−0.3406
	(3.8)***	(−1.88)*	(5.54)***	(−0.37)
D93/DRate93	0.0056	−1.054	0.0052	0.6085
	(3.15)***	(−0.89)	(4.98)***	(0.55)
D94/DRate94	0.0022	0.7567	0.0024	0.4149
	(1.42)	(0.88)	(2.37)***	(0.52)
D95/DRate95	0.00006	−1.1032	0.0003	−0.3342
	(0.05)	(−1.77)*	(0.36)	(−0.57)
D97/DRate97	0.0033	0.404	0.0029	1.0367
	(2.66)***	(0.57)	(3.11)***	(1.58)
D98/DRate98	0.0095	1.592	0.0093	1.0771
	(7.68)***	(2.3)*	(10.15)***	(1.67)*
D99/DRate99	0.0071	3.591	0.0071	0.981
	(5.47)***	(4.7)***	(7.55)***	(1.38)
D00/DRate00	0.0136	1.259	0.0139	1.4515
	(10.53)***	(2.0)**	(14.65)***	(2.47)***
R^2 (%)	30	17	45	19

σ_j is a measure of total risk; OPTION/TOTAL_COMP is the percentage of total compensation in the form of stock options; ACCUMULATED_OPTION is the accumulated option value measuring the executive's wealth; LN(TA) is the natural log of total assets; CAPITAL_RATIO is the capital-to-assets ratio; NON_INT_INCOME% is the percentage of income that is from noninterest sources; GEO_DUMMY is a binary variable measuring geographic diversification; and Dum92–Dum00 are dummy variables coded as 1 or 0 for each year from 1992–2000. 1995 is the excluded year

***, **, * indicates significance at the 1, 5, and 10% levels, respectively

4.5 Conclusion

Rarely a single equation arises in economic theory. In a multi-equation system, OLS fails to yield unbiased and consistent estimators for the structural equations. Therefore, appropriate estimation methods must be applied to the estimation of structural equation parameters. This chapter first discusses situations where a simultaneous equation system may arise. We then explain why OLS estimation is not appropriate. Section B introduces two most frequently used methods to estimate structural parameters in a system of equations. Before 2SLS and 3SLS methods are synthesized, we explain the order condition and the rank condition of model identification. 2SLS and 3SLS are then introduced, and the differences between these two methods are discussed. Section C gives examples from the literature where applications of the simultaneous equation models in finance are shown. In the finance application, Chen et al. (2006) employ a two-equation model to examine the relationship between executives' incentive compensation and firm risk-taking. Because both executive compensation and firm risk are endogenous, 2SLS is more appropriate than OLS method.

When deciding on the appropriate method to estimate structural equations, one must be cautious that in the real world, the distinction between endogenous and exogenous variables are often not as clear-cut as one would like to have. Economic theory, therefore, must play an important role in the model construction. Furthermore, as explained in Section B, although full information methods produce more efficient estimation, they are not always better than the limited information method. This is because, for example, 3SLS is vulnerable to model specification errors. If an equation is misspecified, the error will propagate into the entire system of equations. In the next chapter, we will give more examples about applied simultaneous equation in finance research. In the next chapter, we will demonstrate how simultaneous equation models can be used to do financial analysis, planning, and forecasting.

In the next chapter, we will apply two-stage least squares (2SLS) and three-stage least squares (3SLS) to investigate how econometric methods and accounting data can be used for financial analysis, planning, and forecasting. In addition, the seemingly unrelated regression (SUR) estimation method will also be discussed in detailed.

Bibliography

Blalock, H. M. (1969). *Theory construction: From verbal to mathematical formulations*. Englewood Cliffs, New Jersey: Prentice Hall.

Chen, C. R., Steiner, T., & Whyte, A. (2006). Does stock option-based executive compensation induce risk-taking? An analysis of the banking industry. *Journal of Banking and Finance, 30,* 915–946.

Chen, C. R., & Lee, C. F. (2010). Application of simultaneous equation in finance research. In C. F. Lee et al. (Eds.), *Handbook of quantitative finance and risk management* (Vol. 3). New York: Springer.

Cochran, W. G. (1972). Some effects of errors of measurement on multiple correlation coefficients. *Journal of American Statistical Association, 65,* 22–34.

Fama, E. F. (1968). Risk, return and equilibrium: Some clarifying comments. *Journal of Finance, 23*(1), 29–40.

Fisher, F. M. (1966). *The identification problem in econometrics*. NY, New York: McGraw-Hill.

Fogler, H. R., & Ganapathy, S. (1982). *Financial econometrics*. Englewood Cliffs, New Jersey: Prentice Hall.

Ghosh, S. K. (1991). *Econometrics: Theory and applications*. Englewood Cliffs, New Jersey: Prentice Hall.

Greene, W. H. (2017). *Econometric analysis*. Upper Saddle River, New Jersey: Prentice Hall.

Haitovsky, Y. (1969). A note on the maximization of R. *The American Statistician, 23,* 20–21.

Judge, G. G., Griffiths, W. E., Hill, R. C., Lütkepohl, H., & Lee, T. (1985). *The theory and practice of econometrics*. NY, New York: Wiley.

Klein, L. R., & Nakamura, M. (1962). Singularity in the equation systems of econometrics: Some aspects of the problem of multicollinearity. *Internal Economic Review, 3,* 274–299.

Kmenta, J. (1971). *Element of econometric*. New York: The MacMillan Company.

Lee, C.-F., Lee, A. C., & Lee, J. (2010). *Handbook of quantitative finance and risk management*. New York: Springer.

Lee, C.-F. (1976). A note on the interdependent structure of security returns. *Journal of Financial and Quantitative Analysis*.

Lee, C.-F., Hu, C., & Foley, M. (2018). Inside debt, firm risk and investment decision. Working paper. Rutgers University.

Lee, C.-F., Liang, W. L., Lin, F. L., & Yang, Y. (2016). Applications of simultaneous equations in finance research: Methods and empirical results. *Review of Quantitative Finance and Accounting, 47*(4), 943–971.

Lee, C.-F., & Nieh, C.-C. (2001). Dynamics relationship between stock prices and exchange rates for G-7 countries. *The Quarterly Review of Economics and Finance, 41*(4), 477–490.

Lintner, J. (1965). Security prices, risk and maximal gains from diversification. *Journal of Finance, 20,* 587–616.

Mossin, J. (1966). Equilibrium in a capital asset market. *Econometrica, 34,* 768–783.

Ramanathan, R. (1995). *Introductory econometrics with application*. Fort Worth, Texas: The Dryden Press.

Sharpe, W. F. (1964). Capital asset prices: A theory of market equilibrium under conditions of risks. *Journal of Finance, 19,* 425–442.

Simkowitz, M. A., & Logue, D. E. (1973). The interdependent structure of security returns. *Journal of Financial and Quantitative Analysis, 8*(2), 259–272.

Wang, C. J. (2015). Instrumental variable approach to correct for endogeneity in finance. In C. F. Lee & J. Lee (Eds.), *Handbook of financial econometrics and statistics* (pp. 2577–2600). New York: Springer.

Zellner, A. (1962). An efficient method of estimating seemingly unrelated regressions and tests for aggregation bias. *Journal of the American Statistical Association*, 348–368.

Econometric Approach to Financial Analysis, Planning, and Forecasting

5

Contents

This chapter an update and extension of Chap. 26 of the book entitled, Financial Analysis, Planning and Forecasting by Lee et al. (2017).

C.-F. Lee et al., *Financial Econometrics, Mathematics and Statistics*,
https://doi.org/10.1007/978-1-4939-9429-8_5

Abstract

Following Chap. 4, we apply simultaneous equation models in discussing how econometrics methods and accounting data can be used for financial analysis, planning, and forecasting. The models used in this chapter include single-equation model, two-stage least squares model, three-stage least squares model, and SUR estimation method. In addition, we discuss the relationship among programming, simultaneous equation, and the econometric method.

5.1 Introduction

Based upon the concept and model of simultaneous equation discussed in the previous chapter, this chapter will show how econometric techniques can be used as an important alternative for financial analysis, planning, and forecasting. The econometric approach can be classified as both a single-equation and a simultaneous equation approach. The simultaneous equation approach can either explicitly or implicitly consider the interrelationships among financial variables. The implicit approach uses an estimation technique to take into account interrelationships for a set of equations; this set of equations does not explicitly specify their structural relationship. Then, this set of structural econometric relationships can be either separately or jointly estimated.

The simultaneous-equation models have been discussed in the last chapter. In addition, the last chapter also discussed some applications of simultaneous-equation models in finance research. The basic concepts needed to understand this chapter will be discussed in Sect. 5.2.

In Sect. 5.3, the issue of simultaneity and the dynamics of corporate-budgeting decisions will be explored by using the finance theory. Spies' empirical results will be used to show how the theoretical model can be empirically implemented. Section 5.4 discusses how Zellner's (1962) seemingly uncorrelated regression (SUR) technique can be used in financial analysis and planning. Notably, the role of firm-related variables in capital asset pricing and the role of capital structure in the financing decision are discussed in some detail. Section 5.5 explores applications of a structural simultaneous equation model for financial planning and forecasting. In Sect. 5.6, the interrelationships among the programming, the simultaneous equations, and the econometrics approaches are briefly discussed. In Sect. 5.7, possible applications of financial-planning models to strategic business decisions are reviewed and discussed. Finally, in Sect. 5.8, the results of this chapter are summarized. In addition, Appendix applies financial-planning method to Johnson & Johnson as a case study.

5.2 Simultaneous Nature of Financial Analysis, Planning, and Forecasting

5.2.1 Basic Concepts of Simultaneous Econometric Models

There are two alternative approaches, the implicit and the explicit, to deal with the interrelationships between accounting information and financial decision policies. The implicit approach uses the estimation technique of simultaneously estimating a set of regression equations; the explicit approach uses specification technique to derive a set of structural equations, which has been discussed in the previous chapter. The structural simultaneous equation system can be either separately estimated by two-stage least squares or simultaneously estimated by three-stage least squares.

In the structural simultaneous equation system, the variables can be classified into endogenous and exogenous variables. Following Chap. 4, exogenous variables are inputs to the system, and they are determined outside the equation system. The endogenous variables are derived inside the model and are to be estimated by using the information about the exogenous variables. Before the structural simultaneous equation system is set up, the identification condition should be examined. Following Theil (1971), Johnston (1972), and Chap. 4, an equation should be either just-identified or overidentified to be meaningfully used in the empirical analysis.

5.2.2 Interrelationship of Accounting Information

The major sources of accounting information have been discussed in Chap. 2 of Lee et al. (2017), where the simultaneous nature of financial ratio determination processes was also discussed. Accounting information provides the major inputs for both the programming approach and the simultaneous equation approach to financial planning and forecasting. The simultaneous nature of accounting data lies in the requirement that sources of funds should be equal to the uses of funds in the accounting system.

5.2.3 Interrelationship of Financial Policies

The interaction of investment policy, financing policy, and dividend policy has been theoretically investigated in the literature. The interrelationship of these three policies is the basic assumption for specifying the simultaneous equation financial-planning and forecasting model. Similarly, the interrelationship between financial policies is also the main theoretical justification for using either implicit or explicit simultaneous equation econometric techniques for financial planning and forecasting. In the econometric type of financial-planning and forecasting models, however, the interrelationships among alternative financial policies are assumed to be stochastic instead of deterministic.

5.3 The Simultaneity and Dynamics of Corporate-Budgeting Decisions

5.3.1 Definitions of Endogenous and Exogenous Variables

Greene (2017) has given the definition of both endogenous and exogenous variables. To investigate the interrelationship among investment, financing, and dividends decisions, Spies (1974) developed five multiple regressions to describe the behavior of five alternative financial-management decisions. The stacking technique, a variant of the ordinary least squares (OLS), is used by Spies to estimate all the equations at once.[1] Then, he used this model to demonstrate that investment, financing, and dividend policies generally are jointly determined within an individual industry or in the aggregated manufacture sector.

Through the partial adjustment model, the five endogenous variables (shown in Table 5.1) are determined simultaneously through the use of the sources-equals-uses accounting identity. This identity ensures that the adjustment of each component of the budgeting process (the endogenous variables) depends not only on the distance by which the component misses its target but also on the simultaneous adjustment of the other four components.

5.3.2 Model Specification and Applications[2]

(a) **An Adjustment Model of Capital Budgeting**

The problem of capital budgeting is one that affects the entire structure of the modern

[1]The stacking technique, which was first suggested by de Leeuw (1965), can be replaced by either the SUR or the constrained SUR technique. (See the next section and Appendix for detail.) It should be noted that these techniques themselves can be omitted from the lecture without affecting the substance of the econometric approach to financial analysis and planning.

[2]This section is essentially drawn from Spies' (1974) paper, reprinted with permission from the Journal of Finance and the author.

Table 5.1 Endogenous and exogenous variable

1. The endogenous variables are:
(a) $X_{1,t}$ = DIV_t = Cash dividends paid in period t;
(b) $X_{2,t}$ = IST_t = Net investment in short-term assets during period t;
(c) $X_{3,t}$ = ILT_t = Gross investment in long-term assets during period t;
(d) $X_{4,t}$ = $-\text{DF}_t$ = Minus the net proceeds from the new debt issues during period t;
(e) $X_{5,t}$ = $-\text{EQF}_t$ = Minus the net proceeds from new equity issues during period t

2. The exogenous variables are:
(a) $Y_t = \sum_{i=1}^{5} X_{i,t} = \sum X^*_{i,t}$, where Y = net profits + depreciation allowance; a reformulation of the sources = uses identity.
(b) RCB = Corporate bond rate (which corresponds to the weighted average cost of long-term debt in the FR Model and the parameter for average interest rate in the WS Model).
(c) RDP_t = Average dividend–price ratio (or dividend yield, related to the P/E ratio used by WS as well as the Gordon cost-of-capital model). The dividend–price ratio represents the yield expected by investors in a no-growth, no-dividend firm.
(d) DEL_t = Debt–equity ratio.
(e) R_t = The rates-of-return the corporation could expect to earn on its future long-term investment (or the internal rate-of-return).
(f) CU_t = Rates of capacity utilization (used by FR to lag capital requirements behind changes in percent sales; used here to define the R_t expected)

corporation; its solution determines the very nature of that corporation. Therefore, the capital budget has been treated at great length in the literature of finance. In most of this work, it has been explicitly recognized that the components of the capital budget are jointly determined. The investment, dividend, and financing decisions are tied together by the "uses-equals-sources" identity, a simple accounting identity which requires that all capital invested or distributed to stockholders be accounted for. Despite the obviousness of this relationship, however, very few attempts have been made to incorporate it into an econometric model. Instead, most of the empirical work in this area has concentrated on the components of the capital budget separately. It is the purpose of this section to describe Spies' (1974) econometric capital-budgeting model that explicitly recognizes the "uses-equals-sources" identity.

In his empirical work, Spies broke the capital budget into its five basic components: dividends, short-term investment, gross long-term investment, debt financing, and new equity financing. The first three are uses of funds, while the latter two are sources. The dividends component includes all cash payments to stockholders and must be nonnegative. Short-term investment is the net change in the corporation's holdings of short-term assets during the period. These assets include both inventories and short-term financial assets, such as cash, government securities, and trade credits. This component of the capital budget can be either positive or negative. Long-term investment is defined as the change in gross long-term assets during the period. Thus, the replacement of worn-out equipment is considered to be positive investment. Long-term investment can be negative but only if the sale of long-term assets exceeds this replacement investment.

The debt finance component is simply the net change in the corporation's liabilities. These liabilities include corporate bonds, bank loans, taxes owed, and other accounts payable. Since a corporation can either increase its liabilities or retire those already existing, this variable can be either positive or negative. Finally, new equity financing is the change in stockholder's equity minus the amount due to retained earnings. This should represent the capital raised by the sale of new shares of common stock. Although individual corporations frequently repurchase stock already sold, this variable is almost always positive when aggregated.

The first step is to develop a theoretical model that describes the optimal capital budget as a set of predetermined economic variables.[3] The first of these variables is a measure of cash flow, net profits plus depreciation allowances. This variable, denoted by Y, is exogenous as long as the policies determining production, pricing, advertising, taxes, and the like cannot be changed quickly enough to affect the present period's earnings. Since the data used in this work are quarterly, this does not seem to be an unreasonable assumption. It should also be noted that the "uses-equals-sources" identity ensures that:

$$\sum_{i=1}^{5} X_{i,t} = \sum_{i=1}^{5} X_{i,1}^{*} = Y_t \qquad (5.1)$$

where $X_{1,t}$, $X_{2,t}$, $X_{3,t}$, $X_{4,t}$, $X_{5,t}$, $X_{1,t}^{*}$ and Y_t are identical to those defined in Table 5.1.

Expanding Eq. (5.1), we obtain

$$\begin{aligned} & X_{1,t} + X_{2,t} + X_{3,t} + X_{4,t} + X_{5,t} \\ & \quad = X_{1,t}^{*} + X_{2,t}^{*} + X_{3,t}^{*} + X_{4,t}^{*} + X_{5,t}^{*} = Y_t. \qquad (5.1') \end{aligned}$$

The second exogenous variable in the model is the corporate bond rate, RCB_t. This was used as a measure of the borrowing rate faced by the corporation. In addition, the debt–equity ratio at the start of the period, DEL_t, was included to allow for the increase in the cost of financing due to leverage. The average dividend–price ratio for all stocks, RDP_t, was used as a measure of the rate-of-return demanded by investors in a no-growth, unlevered corporation for the average risk class. The last two exogenous variables, R_t and CU_t, describe the rate-of-return the corporation could expect to earn on its future long-term investment. The ratio of the change in earnings to investment in the previous quarter should provide a rough measure of the rate-of-return to that investment. A four-quarter average of that ratio, R_t, was used by Spies to smooth out the normal fluctuations in earnings. The rate of capacity utilization, CU_t, was also included in an attempt to improve this measure of expected rate of return. Finally, a constant and three seasonal dummies were included. The exogenous variables are summarized in the lower portion of Table 5.1. The matrix for this model is:

$$X_t^{*} = AZ_t, \qquad (5.2)$$

where

$$X^{*\prime} = (\mathrm{DIV}^{*}\ \mathrm{IST}^{*}\ \mathrm{ILT}^{*}\ -\mathrm{DF}^{*}\ -\mathrm{EQF}^{*}),$$
$$Z' = (1\ Q_1\ Q_2\ Q_3\ Y\ \mathrm{RCB}\ \mathrm{RDP}\ \mathrm{DEL}\ R\ \mathrm{CU})$$
$$A = \begin{bmatrix} a_{10} & a_{11} & \cdots & a_{19} \\ \vdots & & & \vdots \\ a_{50} & a_{51} & \cdots & a_{59} \end{bmatrix}$$

In order to better comprehend this model, let us look at the expanded version of Eq. (5.2):

$$\begin{aligned} \mathrm{DIV}_t^{*} &= a_{10} + a_{11}Q_1 + a_{12}Q_2 + a_{13}Q_3 + a_{14}Y_t \\ &\quad + a_5\mathrm{RCB}_t + a_{16}\mathrm{RDP}_t + a_{17}\mathrm{DEL}_t \\ &\quad + a_{18}R_t + a_{19}\mathrm{CU}_t, \\ \mathrm{IST}_t^{*} &= a_{20} + a_{21}Q_1 + a_{22}Q_2 + a_{23}Q_3 + a_{24}Y_t \\ &\quad + a_{25}\mathrm{RCB}_t + a_{26}\mathrm{RDP}_t + a_{27}\mathrm{DEL}_t \\ &\quad + a_{28}R_t + a_{29}\mathrm{CU}_t, \\ \mathrm{ILT}_t^{*} &= a_{30} + a_{31}Q_1 + a_{32}Q_2 + a_{33}Q_3 + a_{34}Y_t \\ &\quad + a_{35}\mathrm{RCB}_t + a_{36}\mathrm{RDP}_t + a_{37}\mathrm{DEL}_t \\ &\quad + a_{38}R_t + a_{39}\mathrm{CU}_t, \\ -\mathrm{DF}_t^{*} &= a_{40} + a_{41}Q_1 + a_{42}Q_2 + a_{43}Q_3 + a_{44}Y_t \\ &\quad + a_{45}\mathrm{RCB}_t + a_{46}\mathrm{RDP}_t + a_{47}\mathrm{DEL}_t \\ &\quad + a_{48}R_t + a_{49}\mathrm{CU}_t, \\ -\mathrm{EQF}_t^{*} &= a_{50} + a_{51}Q_1 + a_{52}Q_2 + a_{53}Q_3 + a_{54}Y_t \\ &\quad + a_{55}\mathrm{RCB}_t + a_{56}\mathrm{RDP}_t + a_{57}\mathrm{DEL}_t \\ &\quad + a_{58}R_t + a_{59}\mathrm{CU}_t. \end{aligned}$$

[3] Theoretical development of this optimal model can be found in Spies' (1971) dissertation.

Although this model itself is not a good description of actual capital-budgeting practices, it can be integrated with a partial adjustment model to derive a more realistic model.

The typical corporation is not flexible enough to achieve this optimal capital budget every quarter. Corporate planners are normally reluctant to deviate very much from their past levels of dividends, investment, and financing.[4] This kind of behavior can best be described by a partial adjustment model, as discussed in Chaps. 2 and 26 of Lee et al. (2017). In such a model, the capital budget depends on both the optimal budget and the actual budgets of the past. In its simplest form, the partial adjustment model makes the change in the level of X_i from period $(t - 1)$ to period t, a function of the difference between its desired level for period t and its actual level for period $(t - 1)$. Using a linear relationship, this becomes

$$X_{i,t} = X_{i,t-1} + \delta_i\left(X^*_{i,t} - X_{i,t-1}\right) \tag{5.3}$$

or

$$\begin{aligned}
&(\mathrm{a})\; X_{1,t} = X_{1,t-1} + \delta_1\left(X^*_{1,t} - X_{1,t-1}\right),\\
&(\mathrm{b})\; X_{2,t} = X_{2,t-1} + \delta_2\left(X^*_{2,t} - X_{2,t-1}\right),\\
&(\mathrm{c})\; X_{3,t} = X_{3,t-1} + \delta_3\left(X^*_{3,t} - X_{3,t-1}\right),\\
&(\mathrm{d})\; X_{4,t} = X_{4,t-1} + \delta_4\left(X^*_{4,t} - X_{4,t-1}\right),\\
&(\mathrm{e})\; X_{5,t} = X_{5,t-1} + \delta_5\left(X^*_{5,t} - X_{5,t-1}\right).
\end{aligned}$$

This assumes that the corporation adjusts the level of X_i, a flow variable, in the direction of its optimal level. The speed of this adjustment is measured by the parameter δ_1. However, Eq. (5.3) does not incorporate the "uses-equals-sources" identity. Suppose $X^*_{1,t} > X_{i,t-1}$ for all $i \neq 1$. If $\delta_1 < 1$, then X_1 will not completely adjust in period t and will remain below its optimal level. But the "uses-equals-sources" identity ensures that

$$\sum_{i=1}^{5} X^*_{i,t} = \sum_{i=1}^{5} X_{i,t} = Y_t.$$

Therefore, the fact that X_1 is less than desired implies that at least one other X_i is above its desired level. Suppose it is X_2 that alone compensates for the slow adjustment of X_1. Then, even though $X^*_{2,t} = X_{2,t-1}$, we will have $X_{2,t} > X_{2,t-1}$. In fact,

$$X_{2,t} = X_{2,t-1} + (1 - \delta_1)\left(X^*_{1,t} - X_{1,t-1}\right).$$

The result of the "uses-equals-sources" identity is that the adjustment of each X_i may depend on the distance of every X_i from its optimal level, not just its own level. Equation (5.3) should be rewritten as:

$$X_{i,t} = X_{i,t-1} + \sum_{j=1}^{5} \delta_{ij}(X^*_{j,t} - X_{j,t-1}) \quad (i = 1,2,3,4,5), \tag{5.4}$$

where[5]

$$\sum_{i=1}^{5} \delta_{ij} = 1$$

Putting Eq. (5.4) into matrix form and combining it with Eq. (5.2), we get

[4]Bower (1970) provides an interesting discussion of corporate decision making and its ability to adapt to a changing environment.

[5]The constraint on the values of $\pm ij$ is a result of the "uses-equals-sources" identity. Summing Eq. (5.4) over i gives

$$\sum_{i=1}^{5} X_{i,t} = \sum_{i=1}^{5} X_{i,t-1} + \sum_{i=1}^{5}\sum_{j=1}^{5} \delta_{ij}\left(X^*_{j,t-1}\right).$$

This can be rewritten as $\sum_i (X_{i,t} - X_{i,t-1}) = \sum_i (X^*_{j,t} - X_{j,t-1}) \sum_i \delta_{ij}$.The identity ensures that $\sum_j X_{j,t} = \sum_j X^*_{j,t}$, and therefore, $\sum_i (X_{i,t} - X_{i,t-1}) = \sum_j (X_{j,t} - X_{j,t-1}) \sum_i \delta_{ij}$.Changing the notation slightly, this becomes $\sum_j (X_{j,t} - X_{j,t-1}) = \sum_j (X_{j,t} - X_{j,t-1}) \sum_i \delta_{ij}$ or $1 = \sum_i \delta_{ij}$.

$$\begin{aligned} X_t &= X_{t-1} + D\left(X_t^* - X_{t-1}\right) \\ &= X_{t-1} + D(AZ_t - X_{t-1}) \\ &= X_{t-1} + DAZ_t - DX_{t-1} \\ &= (I - D)X_{t-1} + DAZ_t, \end{aligned} \quad (5.5)$$

where I is the 5×5 identity matrix and

$$D = \begin{bmatrix} \delta_{11} & \delta_{12} & \ldots & \delta_{15} \\ \vdots & & & \vdots \\ \delta_{51} & \delta_{52} & \ldots & \delta_{55} \end{bmatrix}$$

Equation (5.5) is an implicit, simultaneous, dynamic capital-budgeting model. The expansion is in Table 5.2.

It was argued above that the adjustment of each component of the capital budget depends not only on its own distance from the optimal level but also on the distances of the other components. This would imply that at least some of the off-diagonal elements of D are nonzero. The fact that the X_i are competing uses of the available funds would generally lead one to believe that they should be negative. If dividends were below their optimal level, for example, we might expect dividends to rise and investment to fall. At the same time, however, the normal assumption with models of this type is that $0 < \delta_{11} < 1$. But we have already seen that $\sum_{i=1}^{5} \delta_{ij} = 1$ for all j.

Therefore, it is not possible for all of the off-diagonal elements to be negative unless the δ_{ij} are greater than one. This means that the signs of the δ_{ij} coefficients cannot be predicted with absolute certainty. However, it should be noted that the primary reason most corporations hold liquid assets is to facilitate this type of adjustment process; the role of liquid assets is to give flexibility to the capital budget. Therefore, it is reasonable to expect short-term investment to take up the slack caused by the adjustments of the other components. Going back to the dividend example, we would expect short-term investment to rise to ensure that $\sum_{i=1}^{5} \delta_{i1} = 1$. It is then possible that $0 < \delta < 1$, and δ_{31}, δ_{41}, $\delta_{51} < 0$. Although this is not the only adjustment process a corporation can choose, it appears to be the most likely one.

(b) **Some Empirical Results**

For the purposes of estimation, Eq. (5.5) can be rewritten as

$$X = BX_{t-1} + CZ_t + U_t, \quad (5.6)$$

where B is a 5×5 matrix and C a 5×10 matrix of coefficients. The components of D and A can then be estimated by the following equations:

$$D = i - B, \quad (5.7)$$

$$A = D^{-1}C. \quad (5.8)$$

[An expanded form of Eq. (5.6) can be found in Table 5.3.] The estimation procedure used was the stacking technique that was first suggested by de Leeuw (1965). We have already seen that the "uses-equals-sources" identity implies that

$$\sum_{i=1}^{5} X_{it} = Y_t \text{ for every period } t.$$

In order to incorporate this into the estimation procedure, the estimators must be restricted in such a way that, across equations, the coefficients of Y add up to one, while the coefficients of the other predetermined variables add up to zero. In other words, we must ensure that

$$\sum_{i=1}^{5} b_{ij} = \sum \hat{C}_{ik} = 0 \text{ for all } j \text{ and all } k - 4,$$

and that

$$\sum_{i=1}^{5} \hat{C} = 1$$

Table 5.2 Expanded version of Eq. (5.5)

$$X_1 = \mathrm{DIV}_t = \mathrm{DIV}_{t-1} + \delta_{11}(\mathrm{DIV}_t^* - \mathrm{DIV}_{t-1}) + \delta_{12}(\mathrm{IST}_t^* - \mathrm{IST}_{t-1}) + \delta_{13}(\mathrm{ILT}_t^* - \mathrm{ILT}_{t-1}) + \delta_{14}(-\mathrm{DF}_t^* + \mathrm{DF}_{t-1}) + \delta_{15}(\mathrm{EQF}_t^* + \mathrm{EQF}_{t-1})$$

$$X_2 = \mathrm{IST}_t = \mathrm{IST}_{t-1} + \delta_{21}(\mathrm{DIV}_t^* - \mathrm{DIV}_{t-1}) + \delta_{22}(\mathrm{IST}_t^* - \mathrm{IST}_{t-1}) + \delta_{23}(\mathrm{ILT}_t^* - \mathrm{ILT}_{t-1}) + \delta_{24}(-\mathrm{DF}_t^* + \mathrm{DF}_{t-1}) + \delta_{25}(\mathrm{EQF}_t^* + \mathrm{EQF}_{t-1})$$

$$X_3 = \mathrm{ILT}_t = \mathrm{ILT}_{t-1} + \delta_{31}(\mathrm{DIV}_t^* - \mathrm{DIV}_{t-1}) + \delta_{32}(\mathrm{IST}_t^* - \mathrm{IST}_{t-1}) + \delta_{33}(\mathrm{ILT}_t^* - \mathrm{ILT}_{t-1}) + \delta_{34}(-\mathrm{DF}_t^* + \mathrm{DF}_{t-1}) + \delta_{35}(\mathrm{EQF}_t^* + \mathrm{EQF}_{t-1})$$

$$X_4 = -\mathrm{DF}_t = -\mathrm{DF}_{t-1} + \delta_{41}(\mathrm{DIV}_t^* - \mathrm{DIV}_{t-1}) + \delta_{42}(\mathrm{IST}_t^* - \mathrm{IST}_{t-1}) + \delta_{43}(\mathrm{ILT}_t^* - \mathrm{ILT}_{t-1}) + \delta_{44}(-\mathrm{DF}_t^* + \mathrm{DF}_{t-1}) + \delta_{45}(\mathrm{EQF}_t^* + \mathrm{EQF}_{t-1})$$

$$X_5 = -\mathrm{EQF}_t = -\mathrm{EQF}_{t-1} + \delta_{51}(\mathrm{DIV}_t^* - \mathrm{DIV}_{t-1}) + \delta_{52}(\mathrm{IST}_t^* - \mathrm{IST}_{t-1}) + \delta_{53}(\mathrm{ILT}_t^* - \mathrm{ILT}_{t-1}) + \delta_{54}(-\mathrm{DF}_t^* + \mathrm{DF}_{t-1}) + \delta_{55}(\mathrm{EQF}_t^* + \mathrm{EQF}_{t-1})$$

Table 5.3 Expanded form of Eq. (5.6)

$$\begin{bmatrix} \mathrm{DIV}_t \\ \mathrm{IST}_t \\ \mathrm{ILT}_t \\ -\mathrm{DF}_t \\ -\mathrm{EQF}_t \end{bmatrix} = \begin{bmatrix} b_{11} & b_{12} & b_{13} & b_{14} & b_{15} & c_{10} & c_{11} & c_{12} & c_{13} & c_{14} & c_{15} & c_{16} & c_{17} & c_{18} & c_{19} \\ b_{21} & b_{22} & b_{23} & b_{24} & b_{25} & c_{20} & c_{21} & c_{22} & c_{23} & c_{24} & c_{25} & c_{26} & c_{27} & c_{28} & c_{29} \\ b_{31} & b_{32} & b_{33} & b_{34} & b_{35} & c_{30} & c_{31} & c_{32} & c_{33} & c_{34} & c_{35} & c_{36} & c_{37} & c_{38} & c_{39} \\ b_{41} & b_{42} & b_{43} & b_{44} & b_{45} & c_{40} & c_{41} & c_{42} & c_{43} & c_{44} & c_{45} & c_{46} & c_{47} & c_{48} & c_{49} \\ b_{51} & b_{52} & b_{53} & b_{54} & b_{55} & c_{50} & c_{51} & c_{52} & c_{53} & c_{54} & c_{55} & c_{56} & c_{57} & c_{58} & c_{59} \end{bmatrix} \begin{bmatrix} \mathrm{DIV}_{t-1} \\ \mathrm{STI}_{t-1} \\ \mathrm{LTI}_{t-1} \\ \mathrm{DF}_{t-1} \\ \mathrm{EQF}_{t-1} \\ I \\ Q_1 \\ Q_2 \\ Q_3 \\ Y \\ \mathrm{RCB} \\ \mathrm{RDP} \\ \mathrm{DEL} \\ R \\ \mathrm{CU} \end{bmatrix}$$

This is a simple adding-up constraint.[6]

Now that we have completed our discussion of the nature of this model, we will consider some empirical results from using the model.

[6]This constraint ensures that the "uses-equals-sources" identity will hold for the estimated equations. First of all, we know that $\sum_i ij = 1$, since

$$ij = \begin{cases} 1 - b_{ij} & \text{for } i = j, \\ -b_{ij} & \text{for } i \neq j. \end{cases}$$

Therefore,

$$\sum_i ij = 1_i - \sum b_{ij} = 1 - 0 = 1.$$

In addition, it can be shown that $X_{i,t}^* = Y_t$. To show this, it is necessary only to show that

$$\sum_j jk = \begin{cases} 0 & \text{for all } k \neq 4, \\ 1 & \text{for all } k = 4. \end{cases}$$

Note that $ik = \sum_{jijajk}$. Since we have constrained

$$\sum_i C_{ik} = \begin{cases} 0 & \text{for all } k \neq 4, \\ 1 & \text{for all } k = 4. \end{cases}$$

We can see that

$$\sum_i C_{ik} = \sum_{ij} \sum_{ijajk} = \sum (\sum_{ij}) a_{jk} = \sum_j (1) a_{jk} = \sum_j a_{jk}.$$

Therefore,

$$\sum_j a_{jk} = \begin{cases} 0 & \text{for all } k \neq 4, \\ 1 & \text{for all } k = 4. \end{cases}$$

From all this it is clear that

$$\sum_i X_{i,t_i} = X_{i,t}^* = Y_t.$$

While the purpose of the model is to plan capital budgeting for a single firm, Spies' empirical results presented here used industry data instead of that of the individual firms. We will consider the application of the model to a single firm in Appendix.

The data used by Spies were taken mainly from the FTC-SEC Quarterly Financial Report for Manufacturing Corporations and covered the year 1969. Regression results are reported for the aggregated manufacturing sector and ten industry subsectors. Those industries and their SIC classification numbers are:

1. Food and Kindred Products (20),
2. Textile Mill Products (22),
3. Furniture and Fixtures (25),
4. Paper and Allied Products (26),
5. Chemicals and Allied Products (28),
6. Leather and Leather Products (31),
7. Stone, Clay, and Glass Products (32),
8. Other Machinery (35),
9. Electric Machinery, Equipment, and Supplies (36),
10. Motor Vehicles and Equipment (37).

This model is supposed to provide a realistic picture of the capital-budgeting practices of individual corporations, but Spies used aggregated data to test it. He indicated that there is no reason to expect the parameters of the model to be the same for all manufacturing corporations, or even for all corporations within a particular industry. There is no way to refute this argument on theoretical grounds. It is probably true, in theory at least, that individual firm data would have provided a better test of the model. However, such data might present serious problems for an empirical study of this kind. The capital-budgeting decisions of actual corporations are often dictated by outside influences and considerations that cannot be explained in a very general model of this kind. For example, a single large investment project could dominate a corporation's entire capital budget for several periods. Such things as labor negotiations and antitrust actions could also affect capital-budgeting decisions. It would be very difficult to incorporate all these factors into our general model. As a result, the capital budget of an individual corporation will often be very different from that predicted by our model. This could happen even if the model is an accurate description of the normal behavior of that corporation. Spies argued that aggregated data instead of individual firms' data have smoothed out the effects of these outside influences.

The estimates of the δ_{ij} coefficients are reported in Spies' (1974), Tables 1 through 5. In the estimation procedure employed here, we had to use the negative of the values of debt and equity financing in order to preserve the adding-up constraint. However, in reporting the results, we have adjusted the signs of these coefficients to eliminate this convention. In other words, the X_t and X_{t-1} vectors of Eq. (5.5) have been adjusted to make $X' = (\text{DIV IST ILT DF EQF})$. This transformed system of equations is equivalent to the old one and is much easier to interpret.

Spies used five tables to show the adjustment coefficients δ_{ij} for the aggregated manufacturing sector and the ten industry subsectors. Each table contains the estimates of δ_{ij} for $i = 1, 2, 3, 4, 5$, and a given value of j. For example, Table 5.4 shows that if the optimal level of dividends in the aggregated manufacturing sector exceeds last period's actual level by \$1.00 (that is, $\text{DIV}_t^* = \text{DIV}_{t-1} = 1$), then dividends will be raised \$0.83. In addition, short-term investment will rise by \$3.23, debt financing by \$2.09, and equity financing by \$0.97.

Table 5.4 contains the estimates of the coefficients of $\text{DIV}_t^* - \text{DIV}_{t-1}$. The own-adjustment coefficients of dividends are given in the first column.

They range from a low of 0.1894 for the chemical industry to a high of 1.2340 for "other machinery." Three of the industries have coefficients greater than one, but none of them is significantly different from one. Except for the chemical and paper industries, all the estimates are quite close to one. This would indicate that most corporations adjust dividends quite rapidly to their optimal level.

Table 5.4 Adjustment coefficients of $DIV_t^* - DIV_{t-1}$

	ΔDIV_t	ΔIST_t	ΔILT_t	ΔDF_t	ΔEQF_t
Food	1	2.8688 (4.14)[a]	−2.8688 (4.14)	0	0
Textile	1.1156 (8.97)	1.6469 (2.19)	1.9930 (2.29)	1.0763 (1.38)	−1.3041 (2.20)
Furniture	0.9194 (12.09)	2.9377 (2.34)	0	0	2.8531 (3.58)
Paper	0.2475 (0.37)	0	6.1983 (5.30)	1.8321 (1.76)	3.6137 (2.96)
Chemical	0.1894 (0.64)	0.8106 (2.74)	0	0	0
Leather	0.6412 (1.68)	1.6059 (0.90)	1.5603 (1.03)	5.4753 (3.05)	−2.6679 (2.12)
Stone, clay, and glass	0.7037 (3.32)	1.5594 (2.14)	−1.2531 (3.09)	0	0
Other machinery	1.2340 (5.22)	3.7843 (5.37)	0	4.0183 (5.23)	0
Electrical machinery	0.8355 (4.16)	0	0	−1.3451 (1.06)	1.1806 (0.87)
Motor vehicle	1.0158 (16.72)	0.9799 (4.15)	−0.6723 (2.63)	0	0.3234 (1.33)
Aggregated	0.8328 (6.67)	3.2308 (6.59)	0	2.0902 (7.64)	0.9734 (1.96)

[a]The numbers in parentheses are t-statistics. The 95% significance level is $t = 1.96$. The entries without t-statistics listed represent the coefficients of variables that were left out of the regression. These coefficients are equal to 1 for the own-adjustment coefficients and 0 for the others

Source Spies' (1974), Table 1. Reprinted by permission

The other components of the capital budget also adjust to a nonoptimality in dividends. The second column of Table 5.4 illustrates the reaction of short-term investment to such a nonoptimality. Except for the two zero values, all the estimates are positive and relatively large. This is strong support for the argument that corporations build up their supplies of liquid assets because of a planned increase in dividends. Long-term investment, on the other hand, seems to adjust downward. Eight of the ten industry coefficients are negative or zero, and one of the positive coefficients is insignificant. Thus, the evidence suggests that long-term investment might fall in the face of a planned increase in dividends. Since investment has to compete with dividends for funds, this result is not altogether surprising. Only the very large positive coefficient for the paper industry provides conflicting evidence.

By the same type of argument, it is clear that both debt and equity financing should rise in response to an expected increase in dividends. Nine of the ten industry adjustment coefficients for debt financing are positive or zero, and the other is not significant. In the equity finance equation, eight of the ten coefficients are positive or zero. In other words, both sources of funds are increased to meet the financing requirements of the new, liberalized dividend policy.

Results of the other four sets of partial adjustment coefficients can be interpreted in a similar manner. Spies' summarized empirical results are indicated in Table 5.5, which indicates

Table 5.5 Summary of results

	$DIV_t^* - DIV_{t-1}$	$IST_t^* - IST_{t-1}$	$ILT_t^* - ILT_{t-1}$	$DF_t^* - DF_{t-1}$	$EQF_t^* - EQF_{t-1}$
ΔDIV	Close to 1	Negative (but not significant)	Negative (but not significant)	Positive (but not significant)	Positive (but not significant)
ΔIST	Positive and large	Greater than 1	Positive and large	Negative and large	Negative and large
ΔILT	Negative	Negative[a]	Between 0 and 1	Positive	Positive
ΔDF	Positive[a]	Positive[a]	Positive[a]	Close to 1	Negative
ΔEQF	Positive	Positive	Positive[a]	Negative[a]	Close to 1

[a]The items marked with are those where at least five industries had estimated coefficients set equal to 0. Because of this large number of insignificant estimates, our conclusions about the signs of these coefficients must be viewed with caution

Source Spies' (1974), Table 6. Reprinted by permission

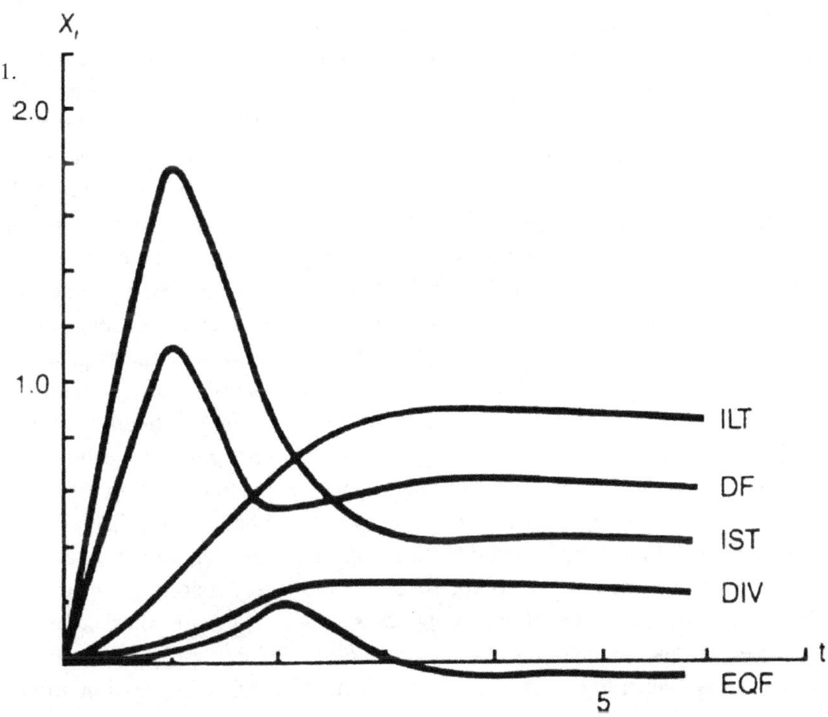

Fig. 5.1 Dynamics of corporate capital budging. *Source* Spies' (1974), Fig. 1. Reprinted by permission

that adjustment coefficients are remarkably consistent from industry to industry (at least in terms of signs). Of course, none of the industries or the aggregated manufacturing sector looks exactly like Table 5.5, but they all follow the general pattern to some degree. Hence, this general pattern has shed some light for financial managers on the various financial interrelationships.

Using Eq. (5.6), Spies performed a simple simulation to investigate the dynamic nature of simultaneous capital-budgeting decisions. Graphically, the simulation results are indicated in Fig. 5.1.

Figure 5.1 describes the changes in the capital budget that would result from a unit increase in gross earnings for the aggregated manufacturing sector. The difference between each component and its initial level is plotted against time. All the other exogenous variables are assumed to remain constant. The adjustment process seems quite clear. Dividends and long-term investment rise slowly but steadily to their new optimal levels. In the first few quarters, short-term investment takes up the slack caused by this slow adjustment. It rises sharply in the first quarter and remains well above its optimal level in the second. This means merely that the corporations are stockpiling liquid assets in preparation for the higher dividend and long-term investment levels. Debt financing rises sharply in the first quarter before falling to its new optimum. To a lesser degree, equity finance does the same thing. By the fourth quarter, all the components of the capital budget have just about reached their optimal levels. This is a much more rapid adjustment than those implied by most single-equation models. This is one of the most interesting results of the complete partial adjustment model.

(c) **Concluding Remarks**

In general, the model developed by Spies performs quite well. The evidence strongly supports a complete partial adjustment model. The individual adjustment coefficients are reasonable for most industries, and the systems are all stable. If the capital budget is out of equilibrium, dividends and long-term investment move steadily toward their equilibrium values. Debt and equity financing change quickly to help finance the other adjustments before moving to their own equilibrium levels. Finally, short-term investment takes up the slack. It adjusts rapidly to compensate for the slower adjustment of the other components and preserve the "uses-equals-sources" identity. Such behavior is generally consistent with actual capital-budgeting practices.

There are a number of interesting conclusions that can be drawn from Spies' empirical results. The first concerns the place of dividends in corporate planning. Apparently, dividends adjust to a new optimum almost immediately. It should be pointed out that the actual dividend policies of most corporations are asymmetric. Dividends are virtually rigid downward, while they are much more flexible upward. The preponderance of periods of rapid growth in earnings and dividends in our sample could very well affect the estimates of this speed of adjustment. In any event, dividends do not react significantly to a nonoptimality in any other component of the capital budget. It seems clear that the dividend level is of paramount importance to corporate planners. The evidence strongly supports the hypothesis that dividend policy is determined almost independently of the rest of the capital budget.

5.4 Applications of SUR Estimation Method in Financial Analysis and Planning

Zellner (1962) developed the seemingly uncorrelated regressions (SUR) technique. Telser (1964) has used iterative estimation procedure to show the process of SUR. More detail about SUR can be found in Greene (2017). SUR has been extensively used in financial analysis and planning (Peterson 1980). Frecka and Lee (1983) used the SUR method to improve the estimation efficiency of determining the dynamic financial ratio adjustment process. Gilmer and Lee (1986) use SUR to test dividend effect controversy. The main justification of this application is that the changes for the liquidity, leverage, activity, and profitability ratios are generally interrelated. Specifically, Gilmer and Lee (1986) follow Telser's iterative estimation of a set of linear regression equations to derive a multivariate regression model for testing the dividend effect. They apply 66 electric utility firms to the multivariate regression model and find the dividend effect is larger in magnitude than the retained earnings effect. The SUR results support Granger's proposition that the regression of prices on dividends and retained earnings produces larger dividend coefficients.

Lee and Vinso (1980) used the SUR to examine the role of firm-related variables in capital asset pricing. Taggart (1977) used the SUR approach to study the role of capital structure in financing decisions. In this section, the last two studies are used to show how the SUR method can be used to improve the empirical results of financial analysis and planning.

5.4.1 The Role of Firm-Related Variables in Capital Asset Pricing[7]

Lee and Vinso (1980) and Lee and Zumwalt (1981) used the SUR method to investigate the role of firm-related variables in capital asset pricing. Following Lee and Vinso, the simultaneous market model is defined as:

$$\begin{aligned} R_{1t} =& \alpha_1 + \beta_1 R_{mt} + \gamma_{11} X_{11} + \gamma_{12} X_{12} \\ &+ \gamma_{13} X_{13} + E_{1t}, \\ R_{2t} =& \alpha_2 + \beta_2 R_{mt} + \gamma_{21} X_{21} + \gamma_{22} X_{22} \\ &+ \gamma_{23} X_{23} + E_{2t}, \\ &\vdots \\ R_{nt} =& \alpha_n + \beta_n R_{mt} + \gamma_{n1} X_{n1} + \gamma_{n2} X_{n2} \\ &+ \gamma_{n3} X_{n3} + E_{nt}, \end{aligned} \tag{5.9}$$

where

R_{jt} = Return on the jth security over time interval t (j = 1, 2, …, n),
R_{mt} = Return on a market index over time interval t,
X_{j1t} = Profitability index of jth firm over time interval t (j = 1, 2, …, n),
X_{j2t} = Leverage index of jth firm over time period t (j = 1, 2, …, n),
X_{j3t} = Dividend policy index of jth firm over time period t (j = 1, 2, …, n),
γ_{jk} = Coefficient of the kth firm-related variable in the jth equation (k = 1, 2, 3),
β_j = Coefficient of market rate-of-return in the jth equation,
E_{jt} = Disturbance term for the jth equation, and a_j's are intercepts (j = 1, 2, …, n).

If the estimated coefficients of γ_{jk} are trivial and the variance–covariance matrix of Eq. (5.9) is a diagonal matrix, then we obtain Sharpe's (1963) diagonal model, which is:

$$R_{jt} = \alpha^{\alpha} + \beta' + E_{jt} \tag{5.10}$$

To investigate the SUR returns-generating process presented here, relative to the usual market model, annual data of stock price and firm-related variables (the profitability index, the leverage index, and the dividend policy index) from the period 1945–1973 for seven oil companies are used to calculate the related rates-of-return. The profitability index is defined as annual retained profit (retained earnings plus interest and preferred dividends) divided by total assets; the leverage index is defined as annual change of long-term debt plus annual change of outstanding preferred stock divided by total assets; and the dividend policy index is defined as annual change of total dividends divided by the book value of equity. Other firm-related variables can be added or substituted for those used here. The appropriate rates-of-return for each company are adjusted for dividends and stock splits. The annual Standard and Poor's index (SPI) with dividends is used to calculate the annual rate-of-return on the market.

To test the validity of the SUR market model specified in Eq. (5.9), OLS is first used to estimate the necessary parameters of seven oil companies (see Table 5.7). It can be seen that five of the seven R^2 associated with OLS estimates of the SUR market model are higher than those of the Sharpe model shown in Table 5.6. These results indicate that the firm-related variables increase the explanatory power of the market model. The residual correlation coefficient matrix for these seven companies (shown in Table 5.8) indicates these firms are highly interrelated in that 10 residual correlation coefficients involving all seven firms are significantly different from zero at the 0.05 level. The SUR estimation method can improve the efficiency of some estimators.

Parameter estimates utilizing the SUR method are also provided in Table 5.7. When the SUR estimation method is applied to the market model, in most cases the efficiency of the

[7]Major portion of this section was drawn from Lee and Vinso (1980). (Reprinted with permission from *Journal of Business Research*).

Table 5.6 OLS and SUR estimates of oil industry

		α_j	β_j	γ_{j1}	γ_{j2}	γ_{j3}	$\bar{R}^2$
R_1	OLS	−0.44 (1.64)	0.76 (2.47)[b]	6.04 (1.72)	0.45 (0.29)	−2.08 (−0.17)	0.22
	SUR	−0.28 (1.18)	0.76 (2.55)[b]	4.20 (1.36)	0.86 (0.62)	−6.97 (−0.66)	
R_2	OLS	−0.82 (−0.26)	0.54 (1.31)	1.06 (0.23)	−1.21 (−1.21)	5.29 (0.27)	0.02
	SUR	−0.16 (−0.80)	0.58 (1.43)	2.45 (0.85)	−0.81 (−1.27)	−8.50 (0.67)	
R_3	OLS	−0.22 (−1.72)	1.12 (3.98)[b]	2.05 (1.42)	2.51 (2.15)[b]	−0.51 (−0.06)	0.45
	SUR	−0.30 (2.63)[b]	1.12 (3.98)[b]	2.91 (2.39)[b]	2.66 (2.81)[b]	−2.53 (−0.41)	
R_4	OLS	−0.15 (−0.86)	0.72 (3.05)[b]	2.10 (0.78)	−0.23 (−0.27)	25.29 (1.81)[a]	0.30
	SUR	−0.05 (−0.35)	0.64 (2.75)[b]	−0.98 (−0.43)	0.42 (0.58)	26.35 (2.337)[a]	
R_5	OLS	−0.00 (−0.01)	1.16 (2.08)[b]	−0.75 (−0.22)	1.83 (1.34)	15.60 (0.89)	0.05
	SUR	−0.00 (−0.02)	1.16 (2.30)[b]	−0.86 (−0.41)	1.88 (2.49)[b]	0.91 (0.95)	
R_6	OLS	−0.14 (1.43)	0.70 (3.09)[b]	0.68 (0.55)	0.91 (1.30)	62.20 (1.71)	0.27
	SUR	−0.15 (1.99)[a]	0.70 (3.17)[b]	0.85 (1.01)	0.90 (2.02)	54.11 (2.40)[b]	
R_7	OLS	−0.11 (−0.91)	1.01 (4.75)[b]	1.34 (0.78)	0.64 (1.24)	4.53 (0.75)	0.46
	SUR	−0.06 (−0.58)	1.01 (4.68)[b]	0.65 (0.42)	0.96 (2.08)[b]	3.96 (0.74)	

t-Value appears in parentheses beneath the corresponding coefficients
[a]Denotes significance at 0.10 level of significance or better for two-tailed test
[b]Denotes significance at 0.05 level of significance or better for two-tailed test
From Lee and Vinso (1980), Table 3. Copyright 1980 by Elsevier Science Publishing Co., Inc., Reprinted by permission of the publisher

estimators appears to be increased. The gain associated with the SUR estimation method is measured using the t-statistic of the regression coefficient as the coefficient of determination for the SUR estimation method is not provided by the SUR computer program. The efficiency of SUR is greater than with OLS. Thus, the SUR market model developed here can result in more efficient estimators while also increasing the explanatory power of the market model.

Results of the SUR market model have a great deal of intuitive appeal. For example, Imperial Oil, which is a Canadian firm, shows the lowest correlation with other firms, as might be expected. In Table 5.7, the residuals of Phillips Petroleum are highly correlated with those of

Table 5.7 OLS parameter estimates of oil industry—Sharpe model[a]

		α'	β'	$\bar{R}^2$
R_1	Imperial Oil	0.04	0.74 (2.66)	0.18
R_2	Phillips Petroleum	−0.01	0.65 (1.67)	0.06
R_3	Shell Oil	−0.01	1.23 (4.24)	0.38
R_4	S.O. of IN	0.04	0.68 (2.86)	0.20
R_5	S.O. of OH	−0.02	0.87 (1.81)	0.07
R_6	Sun Oil	0.01	0.62 (2.71)	0.19
R_7	Union Oil of CA	0.01	1.02 (4.67)	0.43

[a]*t*-Value appears in parentheses beneath the corresponding coefficients
From Lee and Vinso (1980), Table 2. Copyright 1980 by Elsevier Science Publishing Co., Inc., Reprinted by permission of the publisher

Table 5.8 Residual correlation coefficient matrix after OLS estimate

	R_1	R_2	R_3	R_4	R_5	R_6	R_7
R_1	1.00	0.17	0.16	0.44[a]	0.55	0.11	0.14
R_2		1.00	0.20	0.23	0.74[a]	0.44[a]	−0.07
R_3			1.00	0.16	0.35[a]	0.57[a]	0.21
R_4				1.00	0.17	0.36[a]	0.33[a]
R_5					1.00	0.66[a]	0.62[a]
R_6						1.00	0.31[a]
R_7							1.00

[a]Denote significantly different from zero at 0.05 level of significance
From Lee and Vinso (1980), Table 4. Copyright 1980 by Elsevier Science Publishing Co., Inc., Reprinted by permission of the publisher

Standard Oil of Ohio and Sun Oil, but the SUR estimation does not improve the efficiency of estimators for Phillips Petroleum. One possible reason is that the financial-management policies of this company may be highly correlated with those of other companies in the oil industry. When the explanatory variables of a regression become more similar to those of other regressions in the same industry, the gain from the SUR estimation method will be smaller. While these results are interesting, they can be viewed with confidence only if the assumptions of the regression model are fulfilled. Tests of residuals show that the regression requirements are indeed met. Hence, the SUR market model can be used to determine the role of firm-related variables.

Now that the validity of the SUR market model has been shown, the three firm-related variables used by Simkowitz and Logue (1973) in capital asset pricing are of interest. The roles played by three firm-related variables are to identify the simultaneous equation system of the security market and to improve the explanatory power of the diagonal security market model. These same firm-related variables also are explicitly included in the SUR market model indicated in Eq. (5.9). Using the SUR estimates of the market model, the importance of these firm-related variables in the returns-generating process can be analyzed. The profitability index is significant in explaining the rate-of-return of Shell Oil; the dividend policy index is significant in explaining the rates-of-return of Shell Oil, Standard Oil of Ohio, Sun Oil, and Union Oil of California. These results imply that both leverage and dividend policies can be addition factors important in capital asset pricing. Both leverage and dividend policies are unique factors of an industry from a financial-management viewpoint. The market index itself cannot be used to accommodate the change of these two policies associated with a particular industry.

Thus, the SUR market model is formulated by introducing such accounting information as indices of profitability, leverage, and dividend policy into Sharpe's model. It explicitly takes into account the possible impact of accounting information on the behavior of security prices. This multi-index model differs from other multi-index models in several aspects. Firstly, the additional indices employed in the SUR market model are the accounting information of an individual firm rather than general economic-activity indicators. Secondly, the indices of accounting information are relatively orthogonal to the market rate-of-return, and the multicollinearity problem is much less essential relative to that of other multi-index models. Finally, the SUR estimation method can be used to take care of the interdependent relationship among securities of a particular industry. As quarterly data instead of annual data are employed to estimate the parameters for a particular industry, the gain associated with the SUR estimation method will become much more important (see Zellner 1962 for details). Since Lee and Vinso (1980) have shown that the SUR model is consistent with the multibeta interpretations of Sharpe (1977) and others, the results obtained here should be consistent with empirical tests of those models.

5.4.2 The Role of Capital Structure in Corporate-Financing Decisions

Taggart (1977) indicated that all empirical studies but Spies' (1974) model, in investigating the change of various balance-sheet items, take the size of the external financing deficit as exogenous. In addition, he argued that there are two possible weaknesses existing in Spies' model. Firstly, Spies' model does not make systematic use of the theory of optimal capital structure. Secondly, Spies did not allow for the possibility that balance-sheet interrelationships may enter through the error terms. To deal with these estimation weaknesses, Taggart showed that Zellner's SUR method can be used to improve the efficiency of estimation.

Taggart uses five equations to describe the behavior of (a) the change of long-term debt (ΔLDBT), (b) the change of gross stock issues (ΔGSTK), (c) stock retirements (SRET), (d) the change of liquid assets (ΔLIQ), and (e) the change in short-term debt (ΔSDBT). These equations can be defined as:

$$\begin{aligned}\Delta\text{LDBT} =& \alpha_1(\text{LDBT}^* - \text{LDBT}_{t-1}) + \alpha_2(\text{PCB}^* \\ & - \text{PCB}_{t-1} - \text{RE}) + \alpha_3\text{STOCKT} \\ & + \alpha_4\text{RT} + \varepsilon_1\end{aligned} \tag{5.11}$$

$$\begin{aligned}\Delta\text{GSTK} =& \beta_1(\text{LDBT}^* - \text{LDBT}_{t-1}) + \beta_2(\text{PCB}^* \\ & - \text{PCB}_{t-1} - \text{RE}) + \beta_3\text{STOCKT} \\ & + \beta_4\text{RT} + \varepsilon_2\end{aligned} \tag{5.12}$$

$$\begin{aligned}\text{STRET} =& \eta_1(\text{LDBT}^* - \text{LDBT}_{t-1}) + \eta_2(\text{PCB}^* \\ & - \text{PCB}_{t-1} - \text{RE}) + \eta_4\text{RT} + \varepsilon_3\end{aligned} \tag{5.13}$$

$$\Delta \text{LIQ} = \text{LIQ}^* + \gamma_2(\text{TC}^* - \text{TC}_{t-1}) + \gamma_3(\Delta A - \text{RE}) + \gamma_4 \text{RT} + \varepsilon_4 \quad (5.14)$$

$$\Delta \text{SDBT} = \Delta \text{LIQ}^* + \lambda_2(\text{TC}^* - \text{TC}_{t-1}) + \lambda_3(\Delta A - \text{RE}) + \lambda_4 \text{RT} + \varepsilon_5 \quad (5.15)$$

where

LDBT^* = bSTOCK (i/i) = A target for the book value of long-term debt,
STOCK = Market value of equity,
b = LDM/STOCK = Desired debt–equity ratio,
LDM = Market value of debt = (LDBT)(i/i),
i/i = Ratio between the average contractual interest rate on long-term debt outstanding and the current new-issue rate on long-term debt,
LDBT_{t-1} = Book value of long-term debt in previous period,
PCB^* = Permanent capital (book value) = net capital stock (NK) + the permanent portion of working assets (NWA),
PCB_{t-1} = Permanent capital in the previous period,
RE = Stock retirements,
STOCKT = Stock market timing variable = average short-term market value of equity divided by average long-term market value of equity,
RT = Interest timing variable, weighted average (with weight 0.67 and 0.33) of two most recent quarters' changes in the commercial paper rate,
TC^* = Target short-term capital = short-term asset–liquid assets,
TC_{t-1} = Short-term debt in the previous period,
ΔA = Changes in total assets,
ΔLIQ^* = Change of target liquidity assets.

Both OLS and Zellner's SUR method (1962) are used to estimate these behavioral equations in terms of quarterly aggregated data. To resolve the singular covariance problem caused by the balance-sheet constraint, Taggart omitted the stock-retirement equation when using the SUR method. To deal with the seasonal fluctuation problem associated with quarterly data, Taggart also introduced seasonal dummy variables into each equation.

Overall, Taggart's study contributes to the understanding of corporate-financing patterns by including a market value debt–equity ratio as a determinant of long-term debt capacity and using the SUR method, which explicitly accounts for balance-sheet interrelationships. Taggart infers from his estimates that firms base their stock- and bond-issue decisions on the need for permanent capital and on their long-term debt capacity. He has also found that retained earnings are the primary source of funds used by firms to increase the permanent capital. Both new bond and new stock issues are the supporting sources of funds in increasing the permanent capital. In addition, Taggart also found that liquid assets and short-term debt play an important role in absorbing short-run fluctuations in the external financing deficit.

5.5 Applications of Structural Econometric Models in Financial Analysis and Planning

5.5.1 A Brief Review

Miller and Modigliani (1966) and Lee and Wu (1985) used the simultaneous equation approach to improve the precision of the cost-of-capital estimate. Simkowitz and Logue (1973) and Lee (1976) used simultaneous equation specifications to investigate the interdependent structure of security returns. Oudet (1973) used a two-equation, simultaneous equation system to show that inflation had a negative effect on stock returns. Elliott (1972) used a simultaneous econometric model to evaluate the corporate financial performance of a firm. Davis et al. (1973) developed an econometric planning model for American Telephone and Telegraph (AT&T) Company.

5.5.2 AT&T's Econometric Planning Model

Following Davis et al. (1973), the econometric model of the Bell System has been developed as a planning tool to assist in evaluating the impact of general changes in the economy and other environmental changes for alternative Bell System policy. These policies are concerned with long-range planning. The model is called FORECYT, an acronym for Econometric forecasting Model for Corporate Policy Analysis for T (the ticker symbol for AT&T). The overall modeling approach is based upon the premise that the state of the economy determines an individual firm's demand, making it externally derived rather than created by the firm's supply capacity. With this model, Bell System demand is assumed dependent upon economic factors external to the corporation, and supply is a reaction, via corporate policy actions, to the demand.

FORECYT has a tripartite structure consisting of three submodels (Fig. 5.2), which are further divided into modules to facilitate disaggregation as interest dictates and data allow. The flowchart of three submodels and their related modules are indicated in Fig. 5.3.

From the flowchart indicated in Fig. 5.3, the three submodels can be discussed in some detail as follows:

(i) The environment model is a construct of the "world" or "environment" with which the corporation interacts. This model contains (a) the national economy module, (b) the regulatory module, and (c) the price–wage review module.

(ii) The corporate model is an econometric construct of the Bell System. This model contains (a) the demand module, (b) the price module, (c) the capital market module, (d) the financial module, (e) the revenue module, and (f) the production (or supply) module.

(iii) The management model is a logical construct providing policy variables for control (in the input module) and displaying corporate indicators of performance (in the output module). Policy variables include finance mix and factor mix; corporate indicators of performance include earnings per share and realized rates-of-return.

To specify this FORECYT, finance theories are directly needed to determine the capital market module and input module. Needless to say, economic indicators and AT&T's market and accounting information are definitely needed to use this kind of corporate-decision model.

Using Data Resources, Incorporated's (1972) economic forecasts and other related data as inputs, DCC obtained some interesting and useful financial-planning and forecasting results for AT&T.[8] Other updated concepts and corporate planning, using aggregated economic forecasting results, can be found in Eckstein's (1981) speech on decision support systems for corporate planning.

5.6 Programming Versus Simultaneous Versus Econometric Financial Models

Earlier in this chapter, we have discussed some econometric financial-planning models. In this section, we will attempt to compare these models, looking at their strengths and weaknesses.

Two types of programming were examined: linear- and goal programming. One of the major differences between these two methods is that goal programming allows for more than one objective function, while linear programming does not. Either method can be used for working-capital management and long-term planning.

These two programming methods also have a similar basic structure. Starting from a database, the basic financial statements are used to categorize the data, and then, finance theory is utilized to set up the constraints and goals that the

[8]The economic forecasts from other econometrics models (e.g., Chase Econometric and Wharton Econometrics can also be used as inputs for corporate-analysis planning and forecasting).

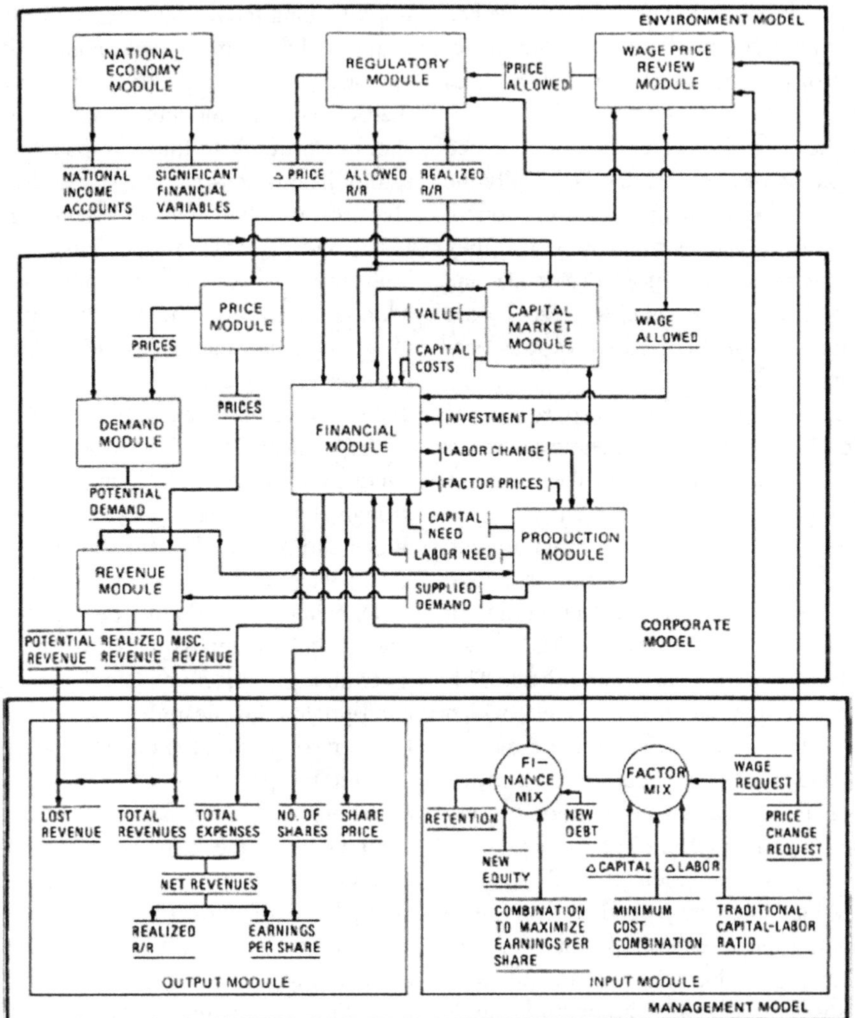

Fig. 5.2 Econometric Forecasting Model for Corporate Policy Analysis

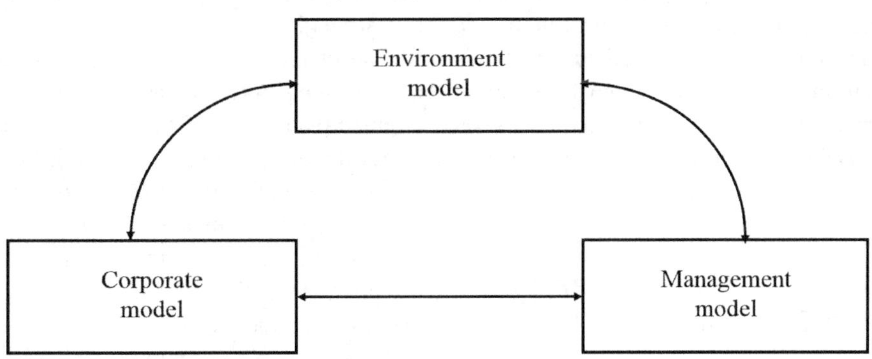

Fig. 5.3 Tripartite structure of FORECYT

firm would like to achieve. These constraints and goals come from two major sources—the objective function of the firm and the overall policies of the firm.

Usually the objective function is concerned with the valuation of the firm; that is, usually the overall goal for a firm is the maximization of shareholder wealth. As for major policies, they are the investment, financing, dividend, and production policies. These policies are set by top management and are influenced by various economic factors.

While under a programming approach the objective function is explicitly identified, under the simultaneous equation method the objective function is implicit from the set of equations.

One thing that both the programming approaches and the simultaneous equation approaches ignore is the statistical properties of the variables used in the model. These two methods take the values as deterministic. This leads one to look at econometric models, which take variables as stochastic. Econometric models attempt to integrate single-equation regression models and simultaneous equation models into a simultaneous equation system in two ways.

The first way is an implicit method, where a set of equations are put together without a specific structural form such as Zellner's seemingly unrelated regression. The second way is an explicit method, in which a structural form is used to improve the specification to formulate the structural type of simultaneous equation system. In sum, the programming approach, the simultaneous equation approach, and the econometric approach are three alternative modeling methods for financial analysis, planning, and forecasting. The relative empirical performance between these models in financial planning and forecasting is still to be investigated.

5.7 Financial Analysis and Business Policy Decisions

Traditionally, business policy has been approached with a case method that uses qualitative analysis. However, in the last decade or so, the use of quantitative analysis in business policy analysis has gained acceptance through the increased use of industry and national econometric planning models. The question is, how should a firm utilize financial-planning models in setting its business policy? Hopefully, we will be able to show that financial training is important for a solid understanding of business policy.

In their paper, Duhaime and Thomas (1983) address the issue of the role of financial analysis in strategic management. The authors evaluate financial management for a corporate viewpoint that is not just in financial terms, but rather, looks at the contribution of financial management to overall corporate goals such as innovation and growth, business efficiency, maintenance, operations and profitability control, and flexibility to changes in strategy.

Duhaime and Thomas define strategic management as encompassing "the firm's activities of determining, enacting, and maintaining an appropriate and desirable match between the firm's resources and skills and the opportunities and threats posed by its continually changing environment." The process of strategic management involves the steps of: (i) identification of problems and resources, (ii) formulation of alternative strategies, (iii) evaluating and choosing from these alternatives, (iv) implementing the chosen strategy, and (v) monitoring and evaluating the performance of the adopted strategy.

There are three areas of strategic management in which financial planning can be utilized: planning phases, organization level, and corporate strategic type.

The planning phase encompasses basic financial planning, forecast-based planning, externally oriented planning, and corporate planning.

As for the organizational level, the increased use of diversification and divisional organizational structures makes the use of financial-planning models more important when examining alternative diversification strategies or organizational structures.

Finally, the corporate strategic type is involved with the question of what businesses the firm should be involved in and how the firm should compete in those businesses.

Having set the picture of strategic management, let us look at the application of financial approaches to strategic management. In the context of strategic management, financial managers should match strategic objectives with corporate long-term resources since financial resources are often vital for achieving these objectives.

One problem with the use of financial approaches is that they may become too rigid in their criteria and limit proper options that a firm might take. Normative finance theory needs to include behavioral and political variables as well.

Many financial models take a deterministic approach, which is "bottom-up" in nature, in that they use accounting data and are decision support rather than optimizing models. These models are very complex.

Another set of models that are very helpful to the manager of a firm comprises sensitivity analysis and risk-analysis models that allow the manager to look at the different effects of different strategies and different inputs. Risk analysis can be used in the corporate development process of analyzing existing activities of the firm, choosing new activities, portfolios of new activities, and combined portfolios of new and existing activities.

What strategic risk analysis should do for the managers of a firm is highlight what alternatives should be considered, and look at the effects of uncertainty on contingency planning. Managers hope that the initial analysis will stimulate further investigation into the assumptions, values, and uncertainties, in order to come up with a better problem solution.

Another use of financial approaches is in strategic portfolio planning. A popular model in this area is the growth-per-share matrix developed by the Boston Consulting Group (see Zellner 1968; Zakon 1976; Hedley 1977). In this model, the markets where the firm operates are defined by two parameters: market growth rate and relative market share. Market growth rate is a proxy for cash use, such that the investment needed to maintain market share is a function of that market's growth rate. Relative market share is a proxy for cash generation, such that profitability of competing products is a function of their market share.

Planning models can also be used to analyze major decisions such as acquisitions, divestments, and expansions. These potential changes can be analyzed in terms of their expected financial and synergistic effects.

Duhaime and Thomas come to several conclusions in their paper of note. First, they indicated that there is a need for integrating various models of investment and financing in order to come up with an overall financial strategy. Financial management should be judged not only in terms of profitability, efficiency, and control, but also in terms of organizational adaptability and flexibility criteria.

Lastly, Duhaime and Thomas point out that strategic management includes more than strategic financial management; it also includes structural uncertainty and behavioral and political dimensions.

5.8 Conclusion

Based upon the information, theory, and methods discussed in previous chapters, we discussed how the econometrics approach can be used as an alternative to both the programming approach and simultaneous equation approach to financial planning and forecasting.[9] Both the SUR method and the structural simultaneous equation method were used to show how the interrelationships among different financial-policy variables can be more effectively taken into account. In addition, it is also shown that financial-planning and forecasting models can also be incorporated with the environment model and the management model to perform business policy decisions. In the next chapter, we will discuss fixed effects and random effects in finance research.

[9]Both the programming approach and simultaneous equation approach can be found in Chaps. 23 and 24 of Lee et al. (2017).

Appendix: Johnson & Johnson as a Case Study

Introduction

The purpose of this appendix is to use Johnson & Johnson's annual data as an example to show how Spies' model can be use to analyze an individual firm's dynamic capital budget decisions. Firstly, Johnson & Johnson's (J&J) operations are briefly reviewed. Secondly, both the balance sheet and the income statement for J&J during the period of 1997–2006 are used to evaluate its financial performance. Thirdly, both the endogenous and the exogenous variables needed to estimate the equation system. Implications of these regression results are also briefly analyzed.

Study of the Company's Operations

Johnson & Johnson was incorporated in New Jersey on November 10, 1887. The company is engaged in the manufacture and sale of a broad range of products in the healthcare and other fields in many countries of the world. J&J's worldwide operations are divided into three industry segments: consumer, pharmaceutical, and medical devices and diagnostics.

Consumer

Consumer products encompass baby and childcare items, skincare products, oralcare products, woundcare products, and women's healthcare products.

Pharmaceuticals

The pharmaceutical sector includes products in the following areas: antifungal, anti-infective, cardiovascular, contraceptive, dermatology, gastrointestinal, hematology, immunology, neurology, oncology, virology, pain management, psychotropic, and urology fields.

Medical Devices and Diagnostics

The medical devices and diagnostics segment includes suture and mechanical wound closure products, surgical equipment and devices, wound management and infection prevention products, interventional and diagnostic cardiology products, diagnostic equipment and supplies, joint replacements, and disposable contact lenses.

Table 5.9 shows that all three divisions of the company are quite profitable and its product lines well-diversified. One should also note that the international division contributes as much as the domestic division, making the company less susceptible to the ups and downs of the US economy.

Research activities are important to J&J's business and account for about 14% of sales. The company employs about 122,000 persons worldwide engaged in the research and development, manufacture, and sale of a broad range of products in the healthcare field.

Analysis of the Company's Financial Performance

Tables 5.10, 5.11A and B show J&J's balance sheet and common-size income statements for the period 1997–2006. Tables 5.12 and 5.13 analyze the company's performance as reflected in key ratios, trends, and the DuPont analysis.

As can be seen from Table 5.12, the company has a fairly comfortable current ratio and quick ratio. However, since 2006 the ratios have been declining slightly. On examining the receivables and inventory turnover ratios, one notes that the receivables turnover ratio has been increasing over the same period, implying that J&J is shortening the periods of credit to generate sales. Inventory turnover has fluctuated between 8.3 times and 12.7 times over the ten-year period. In other words, the average raw materials and finished goods inventory ranged between one to one-half months' sales, which show that the firm does not carry a high inventory of slow-moving stocks; it also holds sufficient inventory to avoid stock-outs. On the overall, J&J seems to have a fairly comfortable working-capital and short-term liquidity position.

The long-term leverage position can be analyzed by examining the debt–equity ratio. From

Table 5.9 Sales in different segment

Division	2003		2004		2005		2006	
	Sales	Profits	Sales	Profits	Sales	Profits	Sales	Profits
	%		%		%		%	
Consumer	18	13	18	11	18	12	18	10
Pharmaceuticals	47	56	47	58	44	48	44	48
Medical devices and diagnostics	36	31	36	31	38	40	38	43
Total	100	100	100	100	100	100	100	100
Domestic	60		59		56		56	
International	40		41		44		44	
Total	100		100		100		100	

Table 5.10 Balance sheet

		1997	1998	1999	2000	2001	2002	2003	2004	2005	2006
	Asset (Figures in millions $)										
1	Account receivables	3329	3661	4233	4464	4630	5399	6574	6831	7010	8712
2	Inventories	2516	2853	3095	2842	2992	3303	3588	3744	3959	4889
3	Other current assets	1819	2040	1993	2400	2879	3089	3310	3861	4287	5290
4	Total current assets (1 + 2 + 3)	10,563	11,132	13,200	15,450	18,473	19,266	22,995	27,320	31,394	22,975
5	Gross fixed assets	9444	10,024	11,046	11,248	12,458	14,314	17,052	18,664	19,716	24,028
6	Less: depreciation	3634	3784	4327	4277	4739	5604	7206	8228	8886	10,984
7	Net fixed assets (5 − 6)	5810	6240	6719	6971	7719	8710	9846	10,436	10,830	13,044
8	Other assets	5080	8839	9244	8900	12,296	12,580	15,422	15,561	15,801	34,537
9	Total assets (4 + 7 + 8)	21,453	26,211	29,163	31,321	38,488	40,556	48,263	53,317	58,025	70,556
	Liabilities										
10	Common equity	12,359	13,590	16,213	18,808	24,233	22,697	26,869	31,813	37,871	39,318
11	Preferred stock	0	0	0	0	0	0	0	0	0	0
12	Shareholder's equity (10 + 11)	12,359	13,590	16,213	18,808	24,233	22,697	26,869	31,813	37,871	39,318
13	Current liabilities	5283	8162	7454	7140	8044	11,449	13,448	13,927	12,635	19,161
14	Long-term liabilities	1126	1269	2450	2037	2217	2022	2955	2565	2017	2014
15	Other liabilities	2685	3190	3046	3336	3994	4388	4991	5012	5502	10,063
16	Total liabilities	9094	12,621	12,950	12,513	14,255	17,859	21,394	21,504	20,154	31,238
17	Working capital (4 −13)	5280	2970	5746	8310	10,429	7817	9547	13,393	18,759	3814

Table 5.11 Common-size income statement ($ in millions)

(A)

		1997		1998		1999		2000		2001	
		$	%	$	%	$	%	$	%	$	%
1	Total sales	22,629	100	23,657	100	27,471	100	29,139	100	33,004	100
2	Cost of goods sold	6085	26.89	6190	26.17	6998	25.47	7346	25.21	7931	24.03
3	Gross profit (1 − 2)	16,544	73.11	17,467	73.83	20,473	74.53	21,793	74.79	25,073	75.97
	Less: expenses										
4	Operating expenses	10,855	47.97	11,176	47.24	13,103	47.70	13,801	47.36	15,583	47.22
5	Interest expenses	160	0.71	181	0.77	278	1.01	242	0.83	248	0.75
6	Total expenses (4 + 5)	11,015	48.68	11,357	48.01	13,381	48.71	14,043	48.19	15,831	47.97
7	Operating profit (3 − 6)	5529	24.43	6110	25.83	7092	25.82	7750	26.60	9242	28.00
8	Depreciation	1067	4.72	1246	5.27	1444	5.26	1515	5.20	1605	4.86
9	Other income	114	0.50	−595	−2.52	105	0.38	387	1.33	261	0.79
10	Profit before tax (7 − 8 + 9)	4576	20.22	4269	18.05	5753	20.94	6622	22.73	7898	23.93
11	Tax	1273	5.63	1210	5.11	1586	5.77	1822	6.25	2230	6.76
12	Net profit (10 − 11)	3303	14.60	3059	12.93	4167	15.17	4800	16.47	5668	17.17
13	Tax rate (11/10)	28%		28%		28%		28%		28%	
14	Dividend	1137	5.02	1305	5.52	1479	5.38	1724	5.92	2047	6.20
15	Retained earnings (12 − 14)	2166	9.57	1754	7.41	2688	9.78	3076	10.56	3621	10.97
16	Payout ratio (14/12)	34%		43%		35%		36%		36%	

(B)

		2002		2003		2004		2005		2006	
		$	%	$	%	$	%	$	%	$	%
1	Total sales	36,298	160	41,862	177	47,348	172	50,434	173	53,194	161
2	Cost of goods sold	8785	38.82	10,307	43.57	11,298	41.13	11,861	40.70	12,880	39.03
3	Gross profit (1 − 2)	27,513	121.58	31,555	133.39	36,050	131.23	38,573	132.38	40,314	122.15
	Less: expenses										
4	Operating expenses	16,173	71.47	18,815	79.53	21,063	76.67	23,189	79.58	24,558	74.41
5	Interest expenses	258	1.14	315	1.33	323	1.18	165	0.57	181	0.55
6	Total expenses (4 + 5)	16,431	72.61	19,130	80.86	21,386	77.85	23,354	80.15	24,739	74.96
7	Operating profit (3 − 6)	11,082	48.97	12,425	52.52	14,664	53.38	15,219	52.23	15,575	47.19
8	Depreciation	1662	7.34	1869	7.90	2124	7.73	2093	7.18	2177	6.60
9	Other income	−129	−0.57	−248	−1.05	298	1.08	530	1.82	1189	3.60

(continued)

Table 5.11 (continued)

(B)

		2002		2003		2004		2005		2006	
		$	%	$	%	$	%	$	%	$	%
10	Profit before tax (7 − 8 + 9)	9291	41.06	10,308	43.57	12,838	46.73	13,656	46.87	14,587	44.20
11	Tax	2694	11.91	3111	13.15	4329	15.76	3245	11.14	3534	10.71
12	Net profit (10 − 11)	6597	29.15	7197	30.42	8509	30.97	10,411	35.73	11,053	33.49
13	Tax rate (11/10)	29%		30%		34%		24%		24%	
14	Dividend	2381	10.52	2746	11.61	3251	11.83	3793	13.02	4267	12.93
15	Retained earnings (12 − 14)	4216	18.63	4451	18.81	5258	19.14	6618	22.71	6786	20.56
16	Payout ratio (14/12)	36%		38%		38%		36%		39%	

Table 5.12 Ratio analysis

	1997	1998	1999	2000	2001	2002	2003	2004	2005	2006
Current ratio	2.00	1.36	1.77	2.16	2.30	1.68	1.71	1.96	2.48	1.20
Quick ratio	1.52	1.01	1.36	1.77	1.92	1.39	1.44	1.69	2.17	0.94
Sales/receivables	6.80	6.46	6.49	6.53	7.13	6.72	6.37	6.93	7.19	6.11
Sales/inventories	8.99	8.29	8.88	10.25	11.03	10.99	11.67	12.65	12.74	10.88
Debt/equity %	9.11	9.34	15.11	10.83	9.15	8.91	11.00	8.06	5.33	5.12
Trend analysis (1997 = 100)										
Sales trend	100.0	104.5	121.4	128.8	145.8	160.4	185.0	209.2	222.9	235.1
Net profit trend	100.0	92.6	126.2	145.3	171.6	199.7	217.9	257.6	315.2	334.6
Working-capital trend	100.0	56.3	108.8	157.4	197.5	148.0	180.8	253.7	355.3	72.2
Gross fixed assets trend	100.0	106.1	117.0	119.1	131.9	151.6	180.6	197.6	208.8	254.4
Net worth trend	100.0	110.0	131.2	152.2	196.1	183.6	217.4	257.4	306.4	318.1
Total assets trend	100.0	122.2	135.9	146.0	179.4	189.0	225.0	248.5	270.5	328.9
Dividends trend	100.0	114.8	130.1	151.6	180.0	209.4	241.5	285.9	333.6	375.3
Price/earnings multiple										
Average for the year	24.15	27.48	31.03	25.65	37.63	24.95	19.48	18.03	18.38	16.72
At year-end	26.67	30.72	30.78	30.19	30.31	23.25	19.13	20.46	17.07	17.75
Price range										
High	66.94	89.00	106.13	105.06	103.10	65.49	58.67	63.76	69.40	69.10
Low	49.75	64.00	78.06	67.38	50.00	41.85	48.73	49.50	60.04	56.80

Table 5.12, one can see that, while this ratio has been rising since 1999, it is trying to lower its financial leverage which is around 5% in 2006. While this shows very low financial risk, it could also mean that the company is not taking advantage of financial leverage, especially since the company does not have much business risk, as will be seen in the DuPont analysis.

Table 5.13 DuPont analysis

	1997	1998	1999	2000	2001	2002	2003	2004	2005	2006
Sales/total assets	1.05	0.90	0.94	0.93	0.86	0.90	0.87	0.89	0.87	0.75
Gross profit/sales %	0.73	0.74	0.75	0.75	0.76	0.76	0.75	0.76	0.76	0.76
EBIT/gross profit %	0.28	0.29	0.29	0.30	0.31	0.35	0.34	0.36	0.34	0.34
EBIT/total assets %	0.22	0.19	0.20	0.21	0.20	0.24	0.23	0.24	0.23	0.19
Net profit/EBIT %	0.71	0.61	0.70	0.74	0.72	0.68	0.66	0.66	0.78	0.81
Total assets/shareholders' equity	1.74	1.93	1.80	1.67	1.59	1.79	1.80	1.68	1.53	1.79
Net profit/shareholders' equity	0.27	0.23	0.26	0.26	0.23	0.29	0.27	0.27	0.27	0.28
Retained earnings/net profit %	0.66	0.57	0.65	0.64	0.64	0.64	0.62	0.62	0.64	0.61
Retained earnings/shareholders' equity %	0.18	0.13	0.17	0.16	0.15	0.19	0.17	0.17	0.17	0.17

Formulas used

(Sales/total assets) × (gross profit/sales) % × (EBIT/gross profit) = (EBIT/total assets) %

(EBIT/total assets) % × (net profit/EBIT) × (total assets/shareholders' equity) = (net profit/shareholders' equity)

(Net profit/shareholders' equity) % × (retained earnings/net profit) = (retained earnings/shareholders' equity) %

An analysis of the various trends show that sales have grown over twice and the growth has been very steady over the period. In fact, sales have been growing ever since 1950. Profits have kept pace with the sales growth, which is a healthy sign. Sales have grown faster than gross investment and working capital, implying that J&J is utilizing its capacity and working-capital funds better. A cursory glance of the balance sheet shows that the lower working-capital growth is due to the increase in current liabilities, which grew four times since 1997. Since the company has the funds to repay its account payables if necessary, this shows that J&J's creditors are granting it longer periods of credit, showing increasing faith in the company's stability.

Net worth (shareholders' equity), sales, and profits have grown similarly, and this is because the company has been paying stable rate of dividends as reflected in the payout ratio in Table 5.11 and the dividend trend in Table 5.12. The formulas at the end of Table 5.13 show that the DuPont analysis is divided into three sections:

1. Operating performance as reflected in the asset turnover, gross profit margin, and operating leverage,
2. Financial efficiency as reflected in the financial leverage and assets/equity ratios,
3. Growth as shown in the retention ratio.

Asset turnover has been decreasing during 1997–2006. Gross profit margin has been fairly steady. However, the operating leverage as reflected in the EBIT/gross profit ratio has been increasing since 1973. This has been offset by the decrease in asset turnover, resulting in a steady return on investment (EBIT/total assets).

The company's net profit/EBIT ratio has been improving, and this is because of decreasing debt leverage as described earlier. The assets/equity ratio has fluctuated slightly over the period. The combined impact of both these ratios resulted in J&J's return on equity (net profit/shareholders' equity) does not change too much in the last ten years. The company's retention ratio has also been fairly stable, as was also seen earlier. The overall effect is that the ratio of retained earnings to equity has been fairly steady over the last ten years, with a slight fall in 1998 and 2001.

The combined impact of all three aspects shows that the company has a very steady operating performance with improving financial efficiency.

The final aspect of the analysis of the company's financial performance is a comparison with the industry and the stock market as a whole. Value Line's comparison of J&J's rankings with the market as a whole is given in the following table:

Category	Rank
Timeliness	3
Safety	1
Financial strength	A++
Stock-price stability	95
Earnings predictability	100
Beta	0.60

Note Based upon value line report on November 30, 2007

From the above study of J&J's operations and performance, one can see that the company has an excellent track record, especially as evidenced by the industry and overall rankings shown above. Further, the hospital supplies industry to which J&J belongs is a very steady industry and not affected by cyclical influences and other factors like changes in consumer's tastes and preferences. While research and development is an important area in this industry, the technological developments are not so rapid to make this industry as volatile as some others, like the electronics industry.

This historical study has been restricted mainly to the last ten years because the observations made in this section will be used to supplement the analysis made using Spies' model. One should also note that many of the comments made on J&J's performance in the 2000s were also applicable to earlier time periods.

Variables and Time Horizon

The variables used in the empirical testing of the model were exactly the same as those used in Spies' model. The following section explains the methodology followed in computing the various variables used in the model.

Dividends (DIV). Only equity dividend was considered in the computation of this variable. Preferred dividend was paid only up to 1955, and therefore, for consistency's sake, preferred dividend was deducted from cash flow for the period and not added to equity dividends. Also, stock dividends were not considered in the computation.

Short-Term Investment (IST). Short-term investment is the net change in the corporation's holdings of current and other short-term assets.

Long-Term Investments (ILT). Long-term investments is defined as the change in gross long-term fixed assets and noncurrent marketable securities.

Debt Financing (DF). This component is simply the net change in the corporation's liabilities, both long-term and current.

Equity Financing (EQF). The change in stockholders' equity, minus the amount due to retained earnings, was used as the definition of equity financing. Though no new shares were issued to the public over the period studied, adjustments were made for stock splits, changes in common stock in treasury, and stock issued to employees under options exercised and stock compensation agreements.

Cash Flow (*Y*). This variable was calculated by adding depreciation to net profit for the period and adjusting for other noncash entries in the retained-earnings statement as well as for preferred dividends paid.

Corporate Bond Rate (RCB). This variable was not included in the model, since the corporate bond rate is not available in Compustat.

Debt–Equity Leverage (DEL). The debt–equity ratio at the beginning of each period was computed for this variable.

Dividend–Price Ratio (RDP). For this variable, the average dividend–price ratio for the period was used.

Rate-of-Return (*R*). The ratio of the change in earnings to long-term investment in the previous period was taken as a measure of the rate-of-return on investments.

Capacity Utilized (CU). Since J&J is a multi-product company, the ratio of sales to gross fixed assets was taken as a proxy for the capacity utilized.

Time Horizon. Annual data were used in the empirical study and the time period covered was 1950–1979.

Tables 5.14A and B list the data collected for the variables above. The data for 1969 have been omitted from the study because J&J consolidated its accounts that year, to include its foreign subsidiaries and this distorted the sources and uses of funds for that year.

Model and Empirical Results[10,11]

The model used in the empirical study was the regression model developed by Spies (1974). Following Eq. (5.16) of the text, the model is defined as

$$X_t = BX_{t-1} + CZ + U_t, \qquad (5.16)$$

where

X_t is a 5 × 1 matrix with the following variables:

$$X'_t = (\mathrm{DIV}_t \quad \mathrm{IST}_t \quad \mathrm{ILT}_t \quad -\mathrm{DF}_t \quad -\mathrm{EQF}_t)';$$

Z_t is a 6 × 1 matrix, which includes a constant (I) where

$$Z'_t = (1 \quad Y_t \quad \mathrm{RDP}_t \quad \mathrm{DEL}_t \quad R_t \quad \mathrm{CU}_t)';$$

B is a 5 × 5 matrix of coefficients and C is a 5 × 6 matrix of coefficients;

U_t is a 5 × 1 matrix which represents error terms.

Since annual data have been used, no dummy variables are used to remove seasonality.

The "sources-equals-uses" identity implies that $\Sigma_i X_{i,t} = Y_t$ must be true for every period t; here, $X_{i,t}$ is defined as the ith variable in the X-matrix at period t. In order to incorporate this into the estimation procedure, the estimators must be restricted in such a way that, across equations, the coefficients of Y add up to 1, while the coefficients of the other exogenous variables add up to zero. Algebraically, these constraints can be stated as:

$$\sum_{i=1}^{5} \hat{c}_{i2} = 1 \quad \text{and}$$

$$\sum_{i=1}^{5} \hat{b}_{ij} = \sum_{i=1}^{5} \hat{c}_{ik} = 0 \text{ for all } j \text{ and all } k \neq 2,$$

where b_{ij} is a coefficient in the B-matrix corresponding to the ith row and jth column and c_{ik} corresponds to the ith row and kth column in the C-matrix, and c_{i2} refers to the column for the variable Y_t as discussed in Sect. 5.3 of the text.

Equation (5.16) defines five multiple regressions, which are used to describe the behavior of five endogenous variables, *DIV*, *IST*, *ILT*, *DF*, and *EQF*. Exogenous variables for these multiple regressions are *Y*, *DEL*, *RDP*, *R*, and *CU*.

Based upon data listed in Tables 5.14A and B, both OLS and constrained SUR methods are used to estimate the coefficients of five multiple regressions. OLS regression results are listed in Table 5.15A, and the SUR results are listed in Table 5.15B. There are 10 OLS regression coefficient estimates that are significantly different from zero under 10% significance. However, there are 14 constrained SUR coefficient

[10] All formulas and definitions of variables used in this section are identical to those defined in Sect. 5.3 of the text.

[11] Taggart (1977) argued that the stacking technique cannot allow for the possibility that balance-sheet interrelationship may enter through the error terms. Hsieh and Lee (1984) show that the constrained SUR method is the generalized case of the stacking technique. Conceptually, the constrained SUR can be obtained by imposing constraints on regression coefficients of the SUR model. Hence, the constrained SUR replaces the stacking technique used by Spies' (1974). Note that the estimation procedure of both the SUR and the constrained SUR is not required for understanding Spies' dynamic capital-budgeting model.

Table 5.14 Endogenous and Exogenous Variables of Johnson & Johnson during 1949–1978

(A)

Year	DIV	IST	ILT	DF	EQF	Year	DIV	IST	ILT	DF	EQF
1949	2.52	−8.96	2.24	−7.97	−5.49	1978	100.72	221.70	140.88	140.82	23.41
1950	3.98	7360.00	13.40	12.96	−4.79	1979	122.20	216.69	274.90	205.77	55.96
1951	2.31	3.20	1.50	−0.84	−0.28	1980	136.90	252.83	215.72	186.09	18.66
1952	2.09	5.70	−0.80	1.78	−2.96	1981	158.60	231.20	246.70	219.10	−50.20
1953	3.45	2.90	−0.50	−3.20	0.14	1982	182.40	50.90	338.27	117.60	−19.43
1954	3.46	2.40	2.60	0.10	−1.10	1983	204.60	204.00	47.93	24.90	−57.37
1955	3.47	6.50	1.08	1.60	−1.74	1984	219.90	56.60	23.28	174.40	−389.13
1956	3.47	−2.00	14.48	1.90	1.12	1985	233.20	382.80	98.66	134.80	38.42
1957	3.81	7.80	0.61	−1.62	0.66	1986	244.70	305.10	160.27	1308.30	−611.53
1958	3.61	6.40	2.84	−0.32	0.62	1987	278.00	70.40	370.04	8.50	105.84
1959	4.74	9.00	8.71	2.08	4.92	1988	327.00	231.00	355.00	555.00	−629.00
1960	5.93	4.70	8.68	2.89	0.64	1989	373.00	273.00	432.00	155.00	−64.00
1961	5.95	4.90	12.62	2.30	4.86	1990	436.00	888.00	697.00	835.00	45.00
1962	5.98	4.60	6.30	−0.88	−0.22	1991	513.00	269.00	706.00	281.00	−222.00
1963	6.63	11.00	5.78	2.54	0.20	1992	587.00	490.00	903.00	1826.00	−898.00
1964	7.19	8.40	8.00	3.92	−5.11	1993	659.00	−206.00	355.00	−39.00	−731.00
1965	8.71	27.30	22.30	23.80	3.45	1994	727.00	1463.00	485.00	1872.00	275.00
1966	9.85	3.60	19.90	5.29	−8.16	1995	827.00	1258.00	400.00	282.00	347.00
1967	11.11	28.70	14.60	6.53	6.71	1996	974.00	1432.00	548.00	346.00	−122.00
1968	11.79	43.30	10.60	10.79	5.25	1997	1137.00	1193.00	96.00	−80.00	−643.00
1969	15.52	26.78	21.68	1.67	11.00	1998	1305.00	569.00	241.00	3527.00	−523.00
1970	18.89	174.78	50.98	82.80	77.38	1999	1479.00	2068.00	522.00	329.00	−65.00
1971	24.02	88.73	35.47	34.94	10.70	2000	1724.00	2250.00	223.00	−437.00	−481.00
1972	25.14	114.70	37.28	40.89	14.98	2001	2047.00	3023.00	2323.00	1742.00	1804.00
1973	29.88	77.46	131.12	72.13	16.99	2002	2381.00	793.00	1106.00	3604.00	−5752.0
1974	41.79	110.86	105.40	85.59	11.17	2003	2746.00	3729.00	1685.00	3535.00	−279.00
1975	49.18	111.48	34.35	−5.65	16.84	2004	3251.00	4325.00	426.00	110.00	−314.00
1976	61.01	120.36	58.86	28.11	6.73	2005	3793.00	4074.00	301.00	−1350.0	−560.00
1977	81.77	212.45	76.63	111.94	11.59	2006	4267.00	−8419.0	4437.00	11,084.0	−5339.0

(B)

Year	Y	DEL	RDP	R	CU	Year	Y	DEL	RDP	R	CU
1950	16,3562	8.05	1.63	30.87	2.778	1979	230.843	44.6637	2.52366	40.1784	4.4436
1951	3.23	34.2649	1.61798	18.3521	4.7892	1980	262	47.3051	2.23058	34.8091	4.16334
1952	2.94	35.0809	1.70213	17.8384	5.04737	1981	67.9	51.1294	2.29414	34.2099	4.04238
1953	3.7	29.5074	2.249	19.8	5.33244	1982	−61.5	50.3702	1.95466	29.339	3.65097
1954	3.99	28.1263	2.12903	21.2584	5.13165	1983	61	47.4146	2.62997	25.0535	3.58043
1955	5.34	28.0099	2.10191	24.0127	5.55309	1984	16	54.8914	3.2526	25.7282	3.55951
1956	5.92	27.0749	2.20736	26.8784	4.59888	1985	293.098	52.0519	2.4228	30.3358	3.49003
1957	4.88	23.5878	2.0059	21.0175	4.74863	1986	216.9	108.085	2.09524	15.5301	3.65439
1958	3.52	21.6554	1.22371	19.9298	4.7102	1987	−58	87.8336	2.15025	36.5037	3.56089
1959	5.75	20.7863	1.32231	23.3557	4.9703	1988	−93	103.226	2.25551	36.727	3.61011

(continued)

Table 5.14 (continued)

(B)

Year	*Y*	DEL	RDP	*R*	CU	Year	*Y*	DEL	RDP	*R*	CU
1960	5.11	21.2534	1.33779	20.8498	4.55423	1989	−77	90.9113	1.88632	35.9827	3.42832
1961	5.19	20.6873	1.03627	19.8683	4.59637	1990	374	94	1.82578	33.2364	3.45919
1962	5.77	18.9179	1.36519	18.7054	4.63494	1991	−237	86.8646	1.34498	35.324	3.39433
1963	7.08	18.8653	1.07843	19.9074	4.38462	1992	156	129.82	1.76238	21.2722	3.34216
1964	10.99	19.5993	1.07383	22.902	4.459	1993	−373	119.864	2.2507	31.1053	3.20881
1965	16.34	27.4276	0.8418	26.7298	4.47358	1994	264	119.994	2.06393	32.8852	3.20448
1966	20.84	27.5443	0.96491	26.1516	4.42311	1995	565	97.6009	1.49708	36.492	3.62625
1967	23.793	26.3234	0.67215	25.9867	4.13365	1996	810	84.6622	1.47739	41.3314	3.82587
1968	30.63	26.1589	0.61033	28.6994	4.33723	1997	742	73.582	1.29032	43.8471	3.89484
1969	37.023	22.9665	0.47222	31.9695	4.25344	1998	−826	92.8698	1.15648	40.097	3.79119
1970	50.964	32.3613	0.59649	40.7517	4.18981	1999	369	79.8742	1.1689	52.9479	4.08855
1971	65.113	33.3743	0.43655	39.7339	4.23275	2000	1048	66.5302	1.18025	57.1973	4.18003
1972	78.222	33.9194	0.34268	41.3766	4.37299	2001	−520	58.8247	1.18443	65.7922	4.27568
1973	97.229	36.9335	0.46563	45.0989	4.41428	2002	−864	78.6844	1.48017	60.3127	4.16739
1974	100.586	40.657	0.89645	35.1249	4.14812	2003	2483	79.6234	1.79055	59.7559	4.25168
1975	115.213	34.8153	0.94708	32.5037	4.2114	2004	3826	67.595	1.72658	61.9783	4.53699
1976	127.834	32.934	1.34615	34.2361	4.43747	2005	6852	53.2175	2.12146	73.55	4.65688
1977	160.822	36.5578	1.8241	37.5442	4.46691	2006	−11972	79.4496	2.20388	76.4596	4.07804
1978	195.918	40.0712	2.30508	40.6681	4.43689						

estimates that are significantly different from zero under the same significance level. This implies that the efficiency of the constrained SUR method is higher than that of the OLS method. In addition, the results in Table 5.15B indicate that the empirical results for DIV_t, DF_t, and EQF_t are more significant than those for IST_t and ILT_t. If quarterly instead of annual data are used to estimate these five equations, we would expect that the performance of empirical results would improve substantially. Considering that OLS and SUR methods neglect information contained in the other equations, we use two-stage least squares method to estimate the coefficients of five multiple regressions. Similar to SUR method, Table 5.15C shows that the 2SLS results for DIV_t, DF_t, and EQF_t are more significant than the 2SLS results for IST_t and ILT_t.

Finally, Spies (1974) indicated that Eq. (5.16) can be used to predict X_t given Z_t, and Z_{t-1}. The model used to do one-period forecasting can be easily derived from Eq. (5.16) as

Table 5.15 **A** OLS estimates for the period 1950–2006. **B** Constrained SUR estimates for the period 1950–2006. **C** Two-stage least square estimates for the period 1951–2006

(A)

X	Intercept	$X_{1,t-1}$	$X_{2,t-1}$	$X_{3,t-1}$	$X_{4,t-1}$	$X_{5,t-1}$	Y	DEL	RDP	R	CU
$X_{1,t}$	−84.810	1.134*	0.002	0.027*	0.003	0.001	0.000	−0.033	−1.897	0.805*	14.757
	(−1.77)	(131.34)	(0.83)	(2.35)	(0.65)	(0.31)	(1.17)	(−0.18)	(−0.41)	(2.00)	(1.67)
$X_{2,t}$	−4229.844	−0.883	0.081	0.416	0.780*	−0.175	0.057*	10.515	−131.906	33.828	726.736
	(−1.35)	(−1.57)	(0.45)	(0.55)	(2.55)	(−0.63)	(5.22)	(0.90)	(−0.43)	(1.29)	(1.26)
$X_{3,t}$	313.367	0.520*	−0.021	0.132	−0.325*	−0.182	−0.002	2.256	−5.631	5.398	−110.324
	(0.30)	(2.74)	(−0.34)	(0.52)	(−3.15)	(−1.94)	(−0.61)	(0.57)	(−0.05)	(0.61)	(−0.57)
$X_{4,t}$	1646.386	1.663*	−0.140	0.691	−0.992*	−0.439	−0.004	7.099	−193.664	−27.856	−191.113
	(0.56)	(3.13)	(−0.82)	(0.97)	(−3.44)	(−1.67)	(−0.36)	(0.65)	(−0.67)	(−1.13)	(−0.35)
$X_{5,t}$	−1273.649	−0.908*	0.081	−1.610*	0.468*	−0.251	0.003	5.010	−28.205	21.725	147.348
	(−0.69)	(−2.74)	(0.76)	(−3.62)	(2.61)	(−1.54)	(0.47)	(0.73)	(−0.16)	(1.41)	(0.43)
Total	−602.846	0.382	−0.040	−0.074	−0.043	−0.193	0.040	4.821	−74.769	6.325	94.281

(B)

X	Intercept	$X_{1,t-1}$	$X_{2,t-1}$	$X_{3,t-1}$	$X_{4,t-1}$	$X_{5,t-1}$	Y	DEL	RDP	R	CU
$X_{1,t}$	−84.810	1.134*	0.002	0.027*	0.003	0.001	0.000	−0.033	−1.897	0.805*	14.757
	(−1.97)	(146.20)	(0.93)	(2.62)	(0.73)	(0.34)	(1.30)	(−0.20)	(−0.45)	(2.23)	(1.86)
$X_{2,t}$	−4229.844	−0.883	0.081	0.416	0.780*	−0.175	0.057*	10.515	−131.906	33.828	726.736
	(−1.50)	(−1.74)	(0.50)	(0.61)	(2.84)	(−0.70)	(5.81)	(1.00)	(−0.48)	(1.44)	(1.40)
$X_{3,t}$	313.367	0.520*	−0.021	0.132	−0.325*	−0.182*	−0.002	2.256	−5.631	5.398	−110.324
	(0.33)	(3.05)	(−0.38)	(0.57)	(−3.51)	(−2.16)	(−0.68)	(0.64)	(−0.06)	(0.68)	(−0.63)
$X_{4,t}$	1646.386	1.663*	−0.140	0.691	−0.992*	−0.439	−0.004	7.099	−193.664	−27.856	−191.113
	(0.62)	(3.48)	(−0.91)	(1.08)	(−3.83)	(−1.86)	(−0.40)	(0.72)	(−0.75)	(−1.25)	(−0.39)
$X_{5,t}$	−1273.649	−0.908*	0.081	−1.610*	0.468*	−0.251	0.003	5.010	−28.205	21.725	147.348
	(−0.77)	(−3.05)	(0.84)	(−4.03)	(2.90)	(−1.71)	(0.52)	(0.81)	(−0.17)	(1.57)	(0.48)
Total	−725.710	0.305	0.001	−0.069	−0.013	−0.209	0.011	4.969	−72.261	6.780	117.481

(C)

X	Intercept	$X_{1,t-1}$	$X_{2,t-1}$	$X_{3,t-1}$	$X_{4,t-1}$	$X_{5,t-1}$	Y	DEL	RDP
$X_{1,t}$	−50.178	1.115	0.025	0.022	−0.004	−0.007	0.007	−0.010	−2.519
	(−1.84)	−105.54	(3.16)	(3.35)	(−1.46)	(−1.79)	(7.62)	(−0.10)	(−0.96)
$X_{2,t}$	−1423.4	0.207	−0.279	0.233	0.119	−0.011	0.705	8.318	−189.484
	(−1.09)	(0.41)	(−0.75)	(0.74)	(0.88)	(−0.06)	(15.11)	(1.71)	(−1.50)
$X_{3,t}$	−553.258	0.227	0.056	0.191	−0.123	−0.220	−0.200	2.89	11.709
	(−0.93)	(0.98)	(0.33)	(1.33)	(−1.99)	(−2.68)	(−9.38)	(1.30)	(0.20)
$X_{4,t}$	−902.203	1.838	−0.794	0.971	−0.424	−0.221	−0.568	7.221	−169.025
	(−0.80)	(4.18)	(−2.45)	(3.55)	(−3.62)	(−1.42)	(−14.04)	(1.71)	(−1.31)
$X_{5,t}$	−85.188	−1.476	0.79	−1.786	0.219	−0.505	0.253	5.688	−49.962
	(−0.07)	(−2.99)	(2.17)	(−5.83)	(1.67)	(−2.89)	(5.58)	(1.20)	(−0.41)
Total	−602.845	0.382	−0.040	−0.074	−0.043	−0.193	0.04	4.821	−79.856

*Significant at the 10% level (cutoff point = 1.96)

$$X_{t-1} = BX_t + CZ_{t+1} + U_t$$
$$= B^2X_{t-1} + BCZ_t + CZ_{t+1} + BU_t + U_{t-1}. \quad (5.17)$$

The applications of Eq. (5.17) to forecast are left to students themselves by using the data in Tables 5.14A and B.

Bibliography

Anderson, W. H. L. (1964). *Corporate finance and fixed investment: An econometric study*. Boston: Harvard Business School.

Annual report of Johnson & Johnson, 1950–2006.

Baumol, W. J. (1959). *Economic dynamics: An introduction* (2nd ed.). New York: Macmillan Company.

Bower, J. L. (1970). Planning within the firm. *American Economic Review, 60,* 186–194.

Brainard, W. C., & Tobin, J. (1968). Pitfall in financial-model building. *American Economic Review, 58,* 99–122.

Davis, B. E., Caccappolo, G. C., & Chaudry, M. A. (1973). An econometric planning model for American Telephone and Telegraph Company. *The Bell Journal of Economics and Management Science 4*(Spring 1973), 29–56.

de Leeuw, F. (1965). A model of financial behavior. In J. S. Duesenberry, G. Fromm, L. R. Klein, & E. Kuh (Eds.), *The Brookings S.S.R.C. quarterly econometric model of the United States*. Chicago: Rand-McNally Company.

Dhrymes, P. J., & Kurz, M. (1967). Investment, dividend, and external finance behavior of firms. In R. Ferber (Ed.), *Determinants of investment behavior: A conference of the universities, national bureau committee for economic research*. New York: Columbia University Press.

Duhaime, I. M., & Thomas, H. (1983). Financial analysis and strategic management. *Journal of Economics and Business, 35,* 413–440.

Eckstein, O. (1981). Decision-support systems for corporate planning. *Data Resources, U.S. Review*, 1.9–1.23. Also in C. F. Lee (ed.), *Financial analysis and planning, theory and application—A book of readings*. Reading, MA: Addison-Wesley Publishing Company.

Elliott, J. W. (1972). Forecasting and analysis of corporate financial performance with an econometric model of the firm. *Journal of Financial and Quantitative Analysis, 7,* 1499–1526.

Frecka, T., & Lee, C. F. (1983). A SUR approach to analyzing and forecasting financial ratios. *Journal of Economics and Business*.

Gilmer, R. H., & Lee, C.-F. (1986). Empirical tests of Granger's propositions on the dividend effect controversy. *The Review of Economics and Statistics, 68*(2), 351–355.

Gordon, M. J. (1962). *The investment, financing, and valuation of the corporation*. Homewood, IL: Richard D. Irwin Inc.

Greene, W. H. (2017). *Econometric analysis* (8th ed.). New York: Pearson.

Hedley, B. (1977). Strategy and business portfolio. *Long Range Planning*, 9–15.

Hendenshatt, P. H. (1979). *Understand capital markets*. Lexington, MA: Lexington Books.

Hsieh, C. C., & Lee, C. F. (1984). *Constrained SUR approach to dynamic capital-budgeting decision*. Mimeo.

Johnston, J. (1972). *Econometric methods* (2nd ed.). New York: McGraw-Hill.

Lee, C. F. (1976). A note on the interdependent structure of security returns. *Journal of Financial and Quantitative Analysis, 9,* 73–86.

Lee, C. F., & Vinso, J. D. (1980). Single vs. simultaneous-equation models in capital-asset pricing: The role of firm-related variables. *Journal of Business Research,* 65–80.

Lee, C. F., & Wu, C. (1985). The inpacts of kurtosis on risk stationarity: Some empirical evidence. *The Financial Review, 20*(4), 263–269.

Lee, C. F., & Zumwalt, J. K. (1981). Associations between alternative accounting profitability measures and security returns. *Journal of Financial and Quantitative Analysis, 16*(1), 71–93.

Lee, C.-F., Lee, A. C., & Lee, J. (2010). *Handbook of quantitative finance and risk management*. New York: Springer.

Lee, C.-F., Liang, W.-L., Lin, F.-L., & Yang, Y. (2016). Applications of simultaneous equations in finance research: Methods and empirical results. *Review of Quantitative Finance and Accounting, 47*(4), 943–971.

Lee, A. C., Lee, J. C., & Lee, C. F. (2017). *Financial analysis, planning and forecasting: Theory and application* (3rd ed.). Singapore: World Scientific.

Lee, C.-F., Hu, C., & Foley, M. (2018). *Inside debt, firm risk and investment decision* (Working paper). Rutgers University.

Miller, M. H., & Modigliani, F. (1966). Some estimates of the cost of capital for the electric utility industry, 1954–57. *American Economic Review,* 333–391.

Naylor, T. H. (Ed.). (1979). *Simulation models in corporate planning*. New York: Praeger Publishing Company.

Oudet, B. A. (1973). The variation of the return on stocks in a period of inflation. *Journal of Financial and Quantitative Analysis, 8,* 247–258.

Peterson, P. P. (1980). A re-examination of seemingly unrelated regressions methodology applied to estimation of financial relationship. *Journal of Financial Research 3*(Fall), 297–308.

Sharpe, W. F. (1963). A simplified model for portfolio analysis. *Management science, 9*(2), 277–293.

Sharpe, W. F. (1977). The capital asset pricing model: a "multi-beta" interpretation. In Financial Dec Making Under Uncertainty (pp. 127–135). Academic Press, New York.

Simkowitz, M. A., & Logue, D. E. (1973). The interdependent structure of security returns. *Journal of Financial and Quantitative Analysis, 8,* 259–272.

Spies, R. R. (1971). *Corporate investment, dividends, and finance: A simultaneous approach* (unpublished Ph.D. dissertation). Princeton University.

Spies, R. R. (1974). The dynamics of corporate capital budgeting. *Journal of Finance, 29,* 29–45.

Standard & Poor's Compustat, a division of The McGraw-Hill Companies, Inc.

Taggart, R. A., Jr. (1977). A model of corporate financing decisions. *Journal of Finance, 32,* 1467–1484.

Telser, L. G. (1964). Iterative estimation of a set of linear regression equations. *Journal of the American Statistical Association, 59,* 845–862.

The Center for Research in Security Prices (CRSP). (2007, November 16). *A research center at Chicago GSB. Value line investment survey.*

Theil, H. (1971). *Principles of econometrics.* New York: Wiley.

Wang, C. J. (2015). Instrumental variable approach to correct for endogeneity in finance. In C. F. Lee & J. Lee (Eds.), *Handbook of financial econometrics and statistics* (pp. 2577–2600). New York: Springer.

Zakon, A. J. (1976). *Capital-structure optimization.* Boston, MA: Boston Consulting Group.

Zellner, A. (1962). An efficient method of estimating seemingly unrelated regression and tests for aggregation bias. *Journal of American Statistical Association, 57,* 348–368.

Zellner, A. (1968). *Growth and financial strategies.* Boston: Boston Consulting Group.

Fixed Effects Versus Random Effects in Finance Research

6

Contents

Abstract

In this chapter, we discuss two alternative methods of panel data analysis. These two methods include both the fixed effects and random effects models. In addition, we discuss the dummy variable technique and the error component model. Finally, we discuss how these methods can be used to investigate alternative dividend policy hypotheses.

This chapter is drawn upon the papers by Chang and Lee (1977) and Lee et al. (2011).

C.-F. Lee et al., *Financial Econometrics, Mathematics and Statistics*,
https://doi.org/10.1007/978-1-4939-9429-8_6

6.1 Introduction

In financial analyses, both firm effect and time effect are of interest to the researcher[1]. Bower and Bower (1969) and Chung (1974) have used the residual technique to deal with the firm effect, but the statistical property of the technique is ambiguous. Chang and Lee (1977) have generalized the study by Bower and Bower (1969) and Chung (1974) to demonstrate how the pooled time-series and cross-sectional data can be used to test the importance of both the firm effect and the time effect in financial studies. They use data on electric utility industry as examples. They show that both firm effect and time effect are statistically significant in this set of data, and therefore, should be taken into account empirically when the effect of alternative corporate policies is evaluated. Further, they use the transformation technique developed by Box and Cox (1964) to integrate with the pooled time-series and cross-sectional data to draw additional methodological implications. Lee et al. (2011) have used fixed effects model to study dividend policy. The main points of this chapter are to discuss panel data model classification and different estimation methods in terms of the papers by Chang and Lee (1977) and Lee et al. (2011) and the book by Hsiao (2014).

In Sect. 6.2, the dummy variable technique and the error component model for analyzing pooled data will be introduced. In addition, the classification of panel data model will be discussed. In Sect. 6.3, financial data on the electric industry from 1963 to 1973 are used to investigate the impacts of firm effect and time effect on the variation of stock price per share and on the change of regression coefficients associated with dividends and retained earnings. In Sect. 6.4, methods of analyzing pooled time-series and cross-sectional data are combined with the transformation technique developed by Box and Cox (1964). Data from the previous section are used to investigate the possible impacts of firm effect and time effect on choosing the optimal functional form of a financial research study. In Sect. 6.5, we will discuss clustering effect and how standard errors can be adjusted when regression residuals are not independent. In Sect. 6.6, we will discuss how Hausman's test can be used to determine whether fixed effects model or random effects model should be used. In Sect. 6.7, efficient firm fixed effects estimator and efficient correlated random effects estimator are discussed. In Sect. 6.8, Lee et al.'s (2011) dividend policy model will be used to show how fixed effects model can be used and demonstrated clustering effect can be tested. Finally, in Sect. 6.9, we will present the summary and make some concluding remarks about how fixed effects versus random effects model should be used for finance and accounting research. In addition, we present hypothesis development and sample for optimal payout ratio under uncertainty and the flexibility hypothesis.

6.2 The Dummy Variable Technique and the Error Component Model

Suppose we have observations on N firms over T periods of time. The model for analyzing both firm and time effects of an industry can be written as:

$$P_{it} = \sum_{k=1}^{K} \beta_k \chi_{kit} + u_{it} \quad (i = 1, 2, \ldots, N;\ t = 1, 2, \ldots, T) \tag{6.1}$$

where P_{it} represents the stock price per share of the ith firm of the industry in period t; χ's are the factors affecting the stock price; and u_{it} is the disturbance term. In actuality, the factors affecting the stock price per share are often numerous and complex and may not be readily observable or measurable. Consequently, usually only a subset of these factors is included in regression analysis in empirical studies. In addition, when cross-sectional and time-series data are combined in the estimation of a regression equation, certain unobservable "other effects" may be present in

[1]The firm effect refers to the effect of factors affecting the behavior of an individual firm; it is constant over time. The time effect refers to the economic condition of particular time point; it varies over time.

the data.[2] Without considering those other factors, the ordinary least squares (OLS) estimates of the β's in (6.1), as indicated by Nerlove (1971) and Wallace and Hussain (1969), may be biased and inefficient. To consider other causal variables. Equation (6.1) is written as:

$$P_{it} = \sum_{k=1}^{K} \beta_k \chi_{kit} + w_i + v_t + u_{it} \quad (i = 1, 2, \ldots, N;\ t = 1, 2, \ldots, T) \tag{6.2}$$

where w_i represents more or less time invariant, unobserved firm effects; v_t represents more or less cross-sectional invariant, unobserved time effects on the stock price per share of the industry; and u_{it} represents the remaining effects which are assumed to vary in both cross-sectional and time dimensions. Other notations remain the same as in Eq. (6.1).

One way to estimate the parameters in Eq. (6.2) is through the treatment of w_i and v_t as constants. Under the assumption that u_{it} are independent with zero means and constant variances, least squares regression of P on χ's firm and time dummies can be used. This approach is known as the least squares with dummy variable technique (LSDV). As indicated by Maddala (1971), the use of this dummy variable technique eliminates a major portion of the variation among the dependent and explanatory variables if the between-firm and between-time period variation is large. In addition, in some cases, the loss of a substantial number of degrees of freedom occurs. Hence, LSDV is not an efficient method for estimating Eq. (6.2). In a Monte Carlo study, Nerlove (1971) also found that LSDV produces estimates with serious bias in finite samples.

Another approach to dealing with Eq. (6.2) is to treat w_i and v_t as random.[3] In this case, instead of N w's and T v's, we estimate only the means and the variances of the distributions of w's and v's. This is known as the error component model, in which the regression error is assumed to be composed of three components—one associated with time, another with cross-section, and the third variable both with the time and cross-sectional dimensions. Hence, in the error component model, Eq. (6.2) becomes:

$$P_{it} = \sum_{k=1}^{K} \beta_k \chi_{kit} + \varepsilon_{it} \tag{6.3}$$

$$\varepsilon_{it} = w_i + v_t + u_{it} \quad (i = 1, 2, \ldots, N;\ t = 1, 2, \ldots, T) \tag{6.4}$$

The assumptions on the components of the error term are that they are independent random variables with constant variances. Without loss of generality, it is also assumed that they have zero means. To estimate the parameters in (6.3), Aitken's generalized least squares (GLS) can be used. In matrix notation, Eq. (6.3) can be written as:

$$Y = \chi\beta + \varepsilon \tag{6.5}$$

where Y is an $NT \times 1$ vector, the elements of which are the observations on price per share of firm i in period t; χ is an $NT \times K$ matrix with the observations on the K explanatory variables; ε is an $NT \times 1$ vector containing the error terms. Under the assumptions on the error components, the variance–covariance matrix of the disturbance terms ε_{it} is the following $NT \times NT$ matrix:

$$E(\varepsilon\varepsilon') = \Omega = \begin{bmatrix} \sigma_w^2 A_T & \sigma_v^2 I_T & \cdots & \cdots & \sigma_v^2 I_T \\ \sigma_v^2 I_T & \sigma_w^2 A_T & & & \sigma_v^2 I_T \\ \vdots & \vdots & \ddots & & \vdots \\ \vdots & \vdots & & \ddots & \vdots \\ \sigma_v^2 I_T & \sigma_v^2 I_T & \cdots & \cdots & \sigma_w^2 I_T \end{bmatrix} \tag{6.6}$$

where I_T is a $(T \times T)$ identity matrix and A_T is a $(T \times T)$ matrix defined as:

$$A_T = \begin{bmatrix} \frac{\sigma^2}{\sigma_w^2} & 1 & \cdots & \cdots & 1 \\ 1 & \frac{\sigma^2}{\sigma_w^2} & \cdots & \cdots & 1 \\ \vdots & \vdots & \ddots & & \vdots \\ \vdots & \vdots & & \ddots & \vdots \\ 1 & 1 & \cdots & \cdots & \frac{\sigma^2}{\sigma_w^2} \end{bmatrix}$$

---

[2] For a discussion of the existence of unobservable effects, see Friend and Puckett (1964).

[3] For a discussion of this sort, see, for example, Balestra and Nerlove (1968).

in which $\sigma^2 = \sigma_w^2 + \sigma_v^2 + \sigma_u^2$. Given Eq. (6.6), it is well known that the generalized least squares estimate of β, if σ_w^2, σ_v^2, and σ_u^2 are known, is

$$\hat{\beta} = (\chi'\Omega^{-1}\chi)^{-1}(\chi'\Omega^{-1}Y) \tag{6.7}$$

with variance–covariance matrix

$$\mathrm{Var}(\hat{\beta}) = (\chi'\Omega^{-1}\chi)^{-1} \tag{6.8}$$

GLS estimates are more efficient than LSDV or OLS estimates because they enable us to extract some information about the regression parameters from the between-firm and between-time-period variation. In finite samples, Nerlove (1971) has also found that it produces little bias.

In actuality σ_w^2, σ_v^2, and σ_u^2 are usually unknown, but they can be estimated by the analysis of covariance techniques as follows (see, e.g., Amemiya 1972):

$$\hat{\sigma}_u^2 = \frac{1}{(N-1)(T-1)}\left[\sum_{i=1}^{N}\sum_{t=1}^{T} e_{it} - \frac{1}{T}\sum_{t=1}^{T} e_{it} - \frac{1}{N}\sum_{i=1}^{N} e_{it}\right]^2 \tag{6.9}$$

$$\hat{\sigma}_w^2 = \frac{1}{T}\left[\frac{1}{(N-1)T}\sum_{i=1}^{N}\sum_{t=1}^{T} e_{it}^2 - \hat{\sigma}_u^2\right] \tag{6.10}$$

$$\hat{\sigma}_v^2 = \frac{1}{N}\left[\frac{1}{N(T-1)}\sum_{t=1}^{T}\sum_{i=1}^{N} e_{it}^2 - \hat{\sigma}_u^2\right] \tag{6.11}$$

where e_{it} represents residuals obtained by applying the least squares method to the pooled data, assuming that w_i and v_t are constants to be estimated rather than random variables.

If $\hat{\sigma}_w^2$ and $\hat{\sigma}_v^2$ are estimated to equal zero, then Ω in (6.6) is a $NT \times NT$ identity matrix and hence Eqs. (6.7) and (6.8) are the same as the OLS estimators. On the other hand, if the estimate of σ_w^2/σ^2 approaches one and $\hat{\sigma}_v^2$ approaches zero, they are equivalent to LSDV with firm dummies; if $\hat{\sigma}_v^2/\sigma^2$ approaches one and σ_w^2 approaches zero, they are equivalent to LSDV with time dummies. Hence, in applying GLS rather than OLS or LSDV, the existence of other time or firm effects can be determined by the sample rather than assumed. The relative weights given to between and within firm and time period variations for the estimation of the parameters are determined by the data. In OLS, it is assumed that the between and within variations are just added up; in LSDV, the between variation is ignored completely (see Maddala 1971, pp. 341–344).

Equation (6.2) can be classified into seven different models as follows:

(a) w_i and v_t are identical to zero for all i and t (OLS).
(b) v_t are identical to zero and w_i are constant (LSDV with firm dummies).
(c) v_t are identical to zero, and w_i is a random variable with zero mean and constant variance (GLS with firm effects).
(d) w_i are identical to zero and v_t are constants (LSDV with year dummies).
(e) w_i are identical to zero and v_t is a random variable (GLS with year effects).
(f) Both w_i and v_t are constants (LSDV with both firm and year dummies).
(g) Both w_i and v_t are random variables (GLS with both firm and year effects).

6.3 Impacts of Firm Effect and Time Effect on Stock Price Variation

Either time-series or cross-sectional data are used generally to investigate variation of stock price per share within an industry. However, the firm effect and time effect have never been formally investigated by the technique described above. Gordon (1959), Durand (1959), and the others have claimed that stock price per share can be explained by dividends and retained earnings per share. Friend and Puckett (1964) have argued that some unobservable variables are of importance in explaining the behavior of stock price per share. They have further demonstrated that both firm and time effects are the most important unobservable effects to be considered. However, they could not find a satisfactory method to handle these effects.

Table 6.1 Estimated coefficients from the pooled data (Linear form)

	Based on	Constant	Coefficient for dividend	Coefficient for earning-dividend	$\bar{R}^2$
(a)	OLS	9.64 (0.99)	8.65 (0.78)	12.01 (0.93)	0.31
(b)	LSDV with firm effects	19.59 (1.49)	5.94 (1.23)	3.12 (1.30)	0.48
(c)	GLS with firm effects	16.25 (1.42)	7.05 (1.06)	5.73 (1.17)	0.48
(d)	LSDV with time effects	2.87 (0.64)	12.67 (0.49)	13.97 (0.58)	0.74
(e)	GLS with time effects	2.91 (2.54)	12.64 (0.49)	13.96 (0.58)	0.74
(f)	LSDV with both effects	−2.44 (0.70)	20.72 (0.56)	6.18 (0.55)	0.91
(g)	GLS with both effects	−1.70 (2.71)	19.83 (0.54)	6.82 (0.54)	0.91

Note Figures in parentheses are standard errors; LSDV is OLS with dummy variables

To allow both firm effect and time effect to be included in the Gordon-type model, following Eqs. (6.3) and (6.4), a generalized Gordon-type model is defined as:

$$P_{it} = \beta_1 + \beta_2 D_{it} + \beta_3 R_{it} + w_i + v_t + u_{it} \quad (6.12)$$

where P, D, and R represent per share price, dividend, and retained earnings, respectively; w, v, and u are the same as before. To test the importance of firm effect and time effect in explaining stock price variation, annual data of utility industry associated with P, D, and R are collected from Compustat tape. There are 110 firms used in this study.[4] The sample period is from 1963 to 1973 which allows different economic conditions to be reflected in the empirical study. Both linear and log linear forms are used for the estimation, based upon OLS, LSDV, and GLS methods. Estimates on Eq. (6.12) were obtained by assuming that (a) w_i and v_t are identical to zero for all i and t (OLS); (b) v_t are identical to zero, and w_i are constant (LSDV with firm dummies); (c) v_t are identical to zero, and w_i is a random variable with zero mean and constant variance (GLS with firm effects); (d) w_i are identical to zero, and v_t are constants (LSDV with year dummies); (e) w_i are identical to zero, and v_t is a random variable (GLS with year effects); (f) both w_i and v_t are constants (LSDV with both firm and year dummies); and (g) both w_i and v_t are random variables (GLS with both firm and year effects). The estimated results of the seven assumptions are summarized in Tables 6.1 and 6.2. Table 6.1 is based upon the linear form, and Table 6.2 is based upon the log linear form.

The figures of adjusted coefficients of determination ($\bar{R}^2$) in the tables demonstrate the importance of both firm effect and time effect in explaining the variation of stock price per share. Unless either firm effect or time effect is considered, the two explanatory variables can account for only 30% of the dependent variable. As the firm effect and/or the time effect are included, the explanatory power is improved significantly. The tables also demonstrate that if either of the two effects is not taken into account, the coefficient for dividend will be seriously underestimated in both linear and log linear forms. The coefficient for retained earnings, on the other hand, will be overestimated in both functional forms if the firm effect is not considered, and it will be slightly underestimated in the

[4]Sample lists of these firms are available from the authors.

Table 6.2 Estimated coefficients from the pooled data (log linear form)

	Based on	Constant	Coefficient for dividend	Coefficient for earning-dividend	$\bar{R}^2$
(a)	OLS	3.37 (0.02)	0.35 (0.03)	0.29 (0.02)	0.33
(b)	LSDV with firm effects	3.38 (0.02)	0.05 (0.05)	0.14 (0.03)	0.49
(c)	GLS with firm effects	3.37 (0.03)	0.16 (0.04)	0.1 (0.02)	0.49
(d)	LSDV with time effects	3.32 (0.01)	0.57 (0.02)	0.29 (0.01)	0.78
(e)	GLS with time effects	3.32 (0.08)	0.57 (0.02)	0.29 (0.01)	0.78
(f)	LSDV with both effects	3.18 (0.01)	0.83 (0.02)	0.14 (0.01)	0.94
(g)	GLS with both effects	3.19 (0.08)	0.81 (0.02)	0.14 (0.01)	0.94

Note Figures in parentheses are standard errors; LSDV is OLS with dummy variables

log linear form if the time effect is not taken into account. In sum, both firm and time effects are statistically significant in explaining the stock price variation and the omission of either effect will cause the estimated coefficients to be biased. Hence, both firm effect and time effect should be handled carefully in this kind of financial analysis. In the following section, the transformation technique developed by Box and Cox (1964) is integrated with the model discussed in this section to find an appropriate functional form for doing the pooled time-series and cross-sectional analysis.

6.4 Functional Form and Pooled Time-Series and Cross-Sectional Data

The choice of a linear or log linear functional form for financial analyses has often been arbitrary, usually based on the ease of estimation. Since the choice of one or the other forms might have serious implications on the effect of explanatory variables on the dependent variable, the choice of a proper functional form should be based on the sample and determined on statistical grounds. In order to do so, the deterministic portion of Eq. (6.12) is written in terms of the following general form according to the suggestion of Box and Cox (1964):

$$P_{it}^{\lambda} = \beta_0 + \beta_1 D_{it}^{\lambda} + \beta_2 R_{it}^{\lambda} \tag{6.13}$$

where λ is the functional form parameter to be estimated. Equation (6.13) includes both the linear and the logarithmic forms as a special case and provides a generalized functional form (GFF) for testing the dividend effect. For allowing the generalized functional form to be continuous at $\lambda = 0$, Box and Cox (1964) and Zarembka (1968) have shown that Eq. (6.13) can be rewritten as[5]

$$P_{it}^{(\lambda)} = \beta_0' + \beta_1 D_{it}^{(\lambda)} + \beta_2 R_{it}^{(\lambda)}$$
$$\text{where } P_{it}^{(\lambda)} = \frac{P_{it}^{\lambda}-1}{\lambda}, \quad D_{it}^{(\lambda)} = \frac{D_{it}^{\lambda}-1}{\lambda}, \tag{6.14}$$
$$R_{it}^{(\lambda)} = \frac{R_{it}^{\lambda}-1}{\lambda} \quad \text{and} \quad \beta_0' = \frac{(\beta_0+\beta_1+\beta_2)-1}{\lambda}$$

Following Box and Cox (1964) and Zarembka (1968), for the true functional form (i.e., the true "λ"), it is assumed that additive disturbance terms, w_i', v_t', and u_{it}', exist to allow Eq. (6.14) to be rewritten into a stochastic relationship:

[5]Zarembka (1968) has employed the generalized functional form technique to determine the true functional form for money demand. The proof of this statement can also be found in his paper.

Table 6.3 Estimated coefficients from the pooled data (generalized functional form)

λ	Constant							Coefficient for dividend						
		Cross-sectional effects		Time effects		Both effects			Cross-sectional effects		Time effects		Both effects	
	OLS	LSDV	GLS	LSDV	GLS	LSDV	GLS	OLS	LSDV	GLS	LSDV	GLS	LSDV	GLS
1.5	111.83 (3.03)	96.03 (4.02)	100.51 (5.05)	109.03 (2.06)	109.05 (14.21)	77.91 (2.05)	81.56 (15.25)	44.73 (3.81)	45.69 (6.03)	45.69 (5.18)	60.78 (2.63)	60.66 (2.62)	105.5 (3.22)	99.64 (3.06)
1.4	84.77 (2.14)	74.5 (2.86)	77.37 (3.54)	82.59 (1.43)	82.60 (10.00)	60.99 (1.41)	63.45 (10.72)	32.16 (2.77)	30.72 (4.40)	31.52 (32.72)	44.35 (1.88)	44.27 (1.87)	76.07 (2.28)	72.02 (2.16)
1.3	64.53 (1.51)	57.91 (2.04)	59.71 (2.49)	62.82 (0.99)	62.88 (7.05)	47.83 (0.97)	49.49 (7.55)	23.14 (2.02)	20.55 (3.20)	21.72 (2.75)	32.39 (1.34)	32.33 (1.34)	54.89 (1.61)	52.11 (1.53)
1.2	49.34 (1.07)	45.13 (1.45)	46.23 (1.75)	48.01 (0.69)	48.02 (4.97)	37.6 (0.67)	38.71 (5.32)	16.66 (1.47)	13.67 (2.33)	14.95 (2.00)	23.66 (0.96)	23.62 (0.96)	39.64 (1.13)	37.73 (1.08)
1. 1	37.92 (0.76)	35.26 (1.03)	35.92 (1.23)	36.88 (0.48)	36.89 (3.51)	29.65 (0.46)	30.39 (3.75)	12.0 (1.07)	9.04 (1.70)	10.27 (1.46)	17.3 (0.69)	17.27 (0.69)	28.65 (0.80)	27.34 (0.76)
1	29.3 (0.54)	27.65 (0.73)	28.03 (0.87)	28.49 (0.33)	28.5 (2.48)	23.46 (0.31)	23.96 (2.65)	8.65 (0.78)	5.94 (1.23)	7.05 (1.06)	12.66 (0.49)	12.64 (0.49)	20.72 (0.56)	19.83 (0.54)
0.9	22.78 (0.38)	21.77 (0.52)	21.98 (0.62)	22.15 (0.23)	22.15 (1.75)	18.65 (0.21)	18.98 (1.87)	6.25 (0.57)	3.88 (0.90)	4.83 (0.77)	9.26 (0.35)	9.25 (0.35)	15 (0.40)	14.39 (0.38)
0.8	17.82 (0.27)	17.22 (0.37)	17.32 (0.44)	17.34 (0.16)	17.34 (1.24)	14.9 (0.15)	15.11 (1.32)	4.51 (0.41)	2.51 (0.65)	3.31 (0.56)	6.79 (0.25)	6.77 (0.25)	10.87 (0.28)	10.45 (0.27)
0.7	17.04 (0.19)	13.7 (0.26)	13.74 (0.31)	13.67 (0.11)	13.67 (0.88)	11.97 (0.10)	12.11 (0.94)	3.27 (0.30)	1.62 (0.47)	2.26 (0.40)	4.97 (0.18)	4.96 (0.18)	7.88 (0.20)	7.60 (0.19)
0.6	11.15 (0.14)	10.96 (0.18)	10.97 (0.22)	10.86 (0.08)	10.87 (0.62)	9.68 (0.07)	9.77 (0.66)	2.36 (0.22)	1.03 (0.34)	1.55 (0.30)	3.65 (0.13)	3.64 (0.13)	5.71 (0.14)	5.52 (0.13)
0.5	8.92 (0.10)	8.82 (0.13)	8.82 (0.16)	8.71 (0.06)	8.71 (0.44)	7.88 (0.05)	7.94 (0.47)	1.72 (0.16)	0.65 (0.25)	1.06 (0.22)	2.67 (0.10)	2.67 (0.10)	4.15 (0.10)	4.01 (0.1)
0.4	7.21 (0.07)	7.16 (0.10)	7.15 (0.12)	7.04 (0.04)	7.04 (0.03)	6.46 (0.03)	6.51 (0.34)	1.25 (0.12)	0.41 (0.18)	0.72 (0.16)	1.96 (0.07)	1.96 (0.07)	3.01 (0.07)	2.92 (0.05)
0.3	5.87 (0.05)	5.86 (0.06)	5.84 (0.08)	5.75 (0.03)	5.75 (0.22)	5.34 (0.02)	5.37 (0.24)	0.91 (0.08)	0.25 (0.13)	0.5 (0.12)	1.44 (0.05)	1.44 (0.05)	2.18 (0.05)	2.12 (0.05)

(continued)

Table 6.3 (continued)

λ	Constant							Coefficient for dividend						
		Cross-sectional effects		Time effects		Both effects			Cross-sectional effects		Time effects		Both effects	
	OLS	LSDV	GLS	LSDV	GLS	LSDV	GLS	OLS	LSDV	GLS	LSDV	GLS	LSDV	GLS
0.2	4.83 (0.03)	4.83 (0.04)	4.82 (0.06)	4.74 (0.02)	4.74 (0.16)	4.46 (0.02)	4.48 (0.17)	0.66 (0.06)	0.15 (0.01)	0.34 (0.08)	1.06 (0.04)	1.06 (0.04)	1.59 (0.04)	1.54 (0.04)
0.1	4.01 (0.02)	4.02 (0.03)	4.01 (0.04)	3.94 (0.01)	3.94 (0.11)	3.74 (0.01)	3.76 (0.12)	0.48 (0.05)	0.09 (0.07)	0.23 (0.06)	0.78 (0.03)	0.78 (0.03)	1.15 (0.03)	1.12 (0.03)
0	3.37 (0.02)	3.38 (0.02)	3.37 (0.03)	3.32 (0.01)	3.32 (0.08)	3.18 (0.01)	3.19 (0. 08)	0.35 (0.03)	0.05 (0.05)	0.16 (0.04)	0.57 (0.02)	0.57 (0.02)	0.83 (0.02)	0.81 (0.02)
−0.1	2.85 (0.01)	2.86 (0.02)	2.86 (0.02)	2.82 (0.01)	2.82 (0.06)	2.72 (0.01)	2. 73 (0.06)	0.26 (0.02)	0.03 (0.04)	0.11 (0.03)	0.42 (0.01)	0.42 (0.01)	0.6 (0.01)	0.59 (0.01)
−0.2	2.44 (0.01)	2.45 (0.01)	2.44 (0.01)	2.41 (0.01)	2.41 (0.04)	2.35 0.00	2.35 (0.04)	0.19 (0.02)	0.02 (0.03)	0.08 (0.02)	0.31 (0.01)	0.3 (0.01)	0.44 (0.01)	0.43 (0.01)
−0.3	2.11 (0.01)	2.12 (0.01)	2.11 (0.01)	2.09 0.00	2.09 (0.02)	2.04 0.00	2.05 (0.03)	0.14 (0.01)	0.01 (0.02)	0.05 (0.02)	0.23 (0.01)	0.23 (0.01)	0.32 (0.01)	0.31 (0.01)
−0.4	1.84 (0.00)	1.85 (0.01)	1.84 (0.01)	1.83 0.00	1.83 (0.02)	1.79 0.00	1.8 (0.02)	0.1 (0.01)	0.003 (0.01)	0.04 (0.01)	0.17 (0.01)	0.17 (0.01)	0.23 (0.01)	0.22 (0.01)
−0.5	1.62 (0.00)	1.63 (0.00)	1.62 (0.01)	1.51 (0.00)	1.61 (0.01)	1.59 (0.00)	1.59 (0.01)	0.08 (0.01)	0.00 (0.01)	0.03 (0.01)	0.12 (0.00)	0.12 (0.00)	0.17 (0.00)	0.02 (0.00)
1.5	70.01 (5.91)	4.64 (8.02)	23.63 (7.24)	85.8 (4.05)	85.68 (4.05)	30.88 (4.08)	36.23 (3.96)	0.2980	0.4754	0.4795	0.6793	0.6798	0.8721	0.8726
1.4	49.41 (4.10)	5.29 (5.61)	18.24 (5.05)	60.01 (2.76)	59.93 (2.75)	22.79 (2.75)	26.34 (2.67)	0.3013	0.4750	0.4802	0.6959	0.6927	0.8808	0.8812
1.3	34.81 (2.84)	5.12 (3.91)	13.88 (3.52)	41.85 (1.87)	41.81 (1.87)	16.67 (1.85)	19.01 (1.80)	0.3046	0.4750	0.4794	0.7055	0.7055	0.8891	0.8897
1.2	24.47 (1.96)	4.56 (2.72)	10.45 (2.44)	29.11 (1.27)	29.08 (1.27)	12.08 (1.24)	13.61 (1.20)	0.3078	0.4752	0.4798	0.7175	0.7175	0.8971	0.8975
1. 1	17.16 (1.36)	3.84 (1.88)	7.78 (1.69)	20.19 (0.86)	20.17 (0.86)	8.68 (0.83)	9.67 (0.80)	0.3108	0.4757	0.4804	0.7287	0.7287	0.9047	0.9050

(continued)

Table 6.3 (continued)

λ	Constant							Coefficient for dividend						
		Cross-sectional effects		Time effects		Both effects			Cross-sectional effects		Time effects		Both effects	
	OLS	LSDV	GLS	LSDV	GLS	LSDV	GLS	OLS	LSDV	GLS	LSDV	GLS	LSDV	GLS
1	12.01 (0.93)	3.12 (1.30)	5.73 (1.17)	13.97 (0.58)	13.96 (0.58)	6.18 (0.55)	6.82 (0.54)	0.3136	0.4764	0.4813	0.7391	0.7391	0.9117	0.9119
0.9	8.38 (0.64)	2.46 (0.90)	4.19 (0.81)	9.63 (0.39)	9.63 (0.39)	4.37 (0.36)	4.78 (0.36)	0.3162	0.4773	0.4824	0.7484	0.7484	0.9181	0.9183
0.8	5.84 (0.44)	1.9 (0.61)	3.04 (0.55)	6.62 (0.27)	6.62 (0.27)	3.06 (0.24)	3.33 (0.24)	0.3186	0.4784	0.4837	0.7568	0.7568	0.9239	0.9241
0.7	4.05 (0.30)	1.44 (0.42)	2.18 (0.38)	4.54 (0.18)	4.53 (0.18)	2.13 (0.16)	2.3 (0.16)	0.3207	0.4796	0.4850	0.7641	0.7641	0.9290	0.9291
0.6	2.81 (0.21)	1.07 (0.29)	1.56 (0.26)	3.1 (0.12)	3.1 (0.12)	1.47 (0.11)	1.58 (0.10)	0.3225	0.4809	0.4864	0.7703	0.7703	0.9333	0.9334
0.5	1.94 (0.14)	0.79 (0.20)	1.1 (0.18)	2.11 (0.08)	2.11 (0.10)	1 (0.07)	1.08 (0.07)	0.3240	0.4822	0.4878	0.7754	0.7754	0.9368	0.9369
0.4	1.33 (0.10)	0.57 (0.13)	0.78 (0.12)	1.43 (0.06)	1.43 (0.06)	0.68 (0.05)	0.73 (0.05)	0.3252	0.4835	0.4892	0.7793	0.7793	0.9395	0.9396
0.3	0.92 (0.07)	0.41 (0.01)	0.54 (0.08)	0.97 (0.04)	0.97 (0.04)	0.46 (0.03)	0.49 (0.03)	0.3259	0.4847	0.4905	0.7819	0.7819	0.9413	0.9413
0.2	0.62 (0.04)	0.29 (0.06)	0.38 (0.06)	0.65 (0.03)	0.65 (0.03)	0.31 (0.02)	0.33 (0.02)	0.3273	0.4858	0.4916	0.7833	0.7837	0.9423	0.9424
0.1	0.43 (0.03)	0.2 (0.04)	0.26 (0.04)	0.44 (0.02)	0.44 (0.02)	0.21 (0.01)	0.22 (0.01)	0.3259	0.4867	0.4926	0.7835	0.7835	0.9423	0.9424
0	0.29 (0.02)	0.14 (0.03)	0.18 (0.02)	0.29 (0.01)	0.29 (0.01)	0.14 (0.01)	0.14 (0.01)	0.33	0.49	0.49	0.78	0.78	0.94	0.94
−0.1	0.19 (0.01)	0.1 (0.02)	0.12 (0.02)	0.19 (0.01)	0.19 (0.01)	0.09 (0.01)	0.1 (0.01)	0.3238	0.4880	0.4940	0.7798	0.7798	0.9396	0.9397
−0.2	0.13 (0.01)	0.07 (0.01)	0.08 (0.01)	0.13 (0.01)	0.13 (0.01)	0.06 (0.00)	0.06 (0.00)	0.3218	0.4883	0.4943	0.7761	0.7761	0.9369	0.9370

(continued)

Table 6.3 (continued)

λ	Constant							Coefficient for dividend						
		Cross-sectional effects		Time effects		Both effects			Cross-sectional effects		Time effects		Both effects	
	OLS	LSDV	GLS	LSDV	GLS	LSDV	GLS	OLS	LSDV	GLS	LSDV	GLS	LSDV	GLS
−0.3	0.08 (0.01)	0.04 (0.01)	0.05 (0.01)	0.08 (0.00)	0.08 (0.00)	0.04 (0.00)	0.04 (0.00)	0.3191	0.5353	0.4951	0.7710	0.7710	0.9333	0.9334
−0.4	0.06 (0.00)	0.03 (0.01)	0.04 (0.00)	0.06 (0.00)	0.06 (0.00)	0.02 (0.00)	0.03 (0.00)	0.3157	0.4879	0.4839	0.7646	0.7646	0.9288	0.9289
−0.5	0.04 (0.00)	0.02 (0.00)	0.02 (0.00)	0.04 (0.00)	0.03 (0.00)	0.02 (0.00)	0.16 (0.00)	0.3116	0.4872	0.4931	0.7570	0.7570	0.9234	0.9235

Notes Figures in parentheses are standard errors, and LSDV: OLS with dummy variables

Zarembka (1968) has employed the generalized functional form technique to determine the true functional form for money demand. The proof of this statement can also be found in his paper.

$$P_{it}^{(\lambda)} = \beta_0' + \beta_1 D_{it}^{(\lambda)} + \beta_2 R_{it}^{(\lambda)} + w_i' + v_t' + u_{it}' \tag{6.15}$$

where w_i', v_t', and u_{it}' are normally and independently distributed with zero means and variances σ'^2_w, σ'^2_v and σ'^2_u.

Using the maximum likelihood method, Box and Cox (1964) derived a maximum logarithmic likelihood for determining the functional form parameter:

$$\begin{aligned} L\max\left(\hat{\lambda}\right) &= \text{Constant} - n \log \hat{\sigma}_\tau(\lambda) \\ &\quad + (\lambda - 1) \sum_{i=1}^{n} \log P_{it} \end{aligned} \tag{6.16}$$

where n is the sample size, and $\hat{\sigma}_\tau(\lambda)$ is the estimated regression residual standard error of Eq. (6.15).[6] For calculating $\hat{\sigma}_\tau(\lambda)$, P_{it}, D_{it}, and R_{it} should be transformed in terms of Eq. (6.14). After $\hat{\sigma}_\tau(\lambda)$ is estimated, Eq. (6.16) will be employed to determine the optimum value of the functional form parameter, $\hat{\lambda}$. The optimum value of λ is obtained by plotting Eq. (6.16) for different values of λ to arrive at the maximized logarithmic likelihood over the whole parameter space. Using the likelihood ratio method, an approximate 95% confidence region for λ can be obtained from:

$$L\max(\hat{\lambda}) - L\max(\lambda) < 1/2\chi_1^2(0.05) = 1.92 \tag{6.17}$$

The 95% confidence region for λ will be used to determine the true functional form for the pooled time-series and cross-sectional models.

For determining the true functional parameters, P_{it}, D_{it}, and R_{it} are transformed in accordance with Eq. (6.14) by λ between −0.5 and 1.5 at intervals of length 0.1. These transformed data

[6] $\hat{\sigma}_\tau(\lambda)$ is obtained either from OLS, LSDV, or GLS.

are then used to estimate the relationship among P_{it}, D_{it}, and R_{it} in accordance with OLS, LSDV with cross-sectional effect, GLS with cross-sectional effect, LSDV with time effect, GLS with time effect, LSDV with both effects, and GLS with both effects. Twenty-one regressions are estimated for each case, and the results are listed in Table 6.3. To estimate the optimal functional form parameters for every case above mentioned, the logarithmic likelihoods are estimated in accordance with Eq. (6.16) and listed in Table 6.4. Using the χ^2 test indicated in Eq. (6.17) and Table 6.4, it is found that the linear form ($\lambda = 1$) has been rejected for all seven cases under 95% confidence interval. Under the same confidence interval, it also is found that the log linear form has been rejected for LSDV with cross-sectional effect, GLS with cross-sectional effect, LSDV with both effects, and GLS with both effects. These results have demonstrated that the functional parameter estimation method can be integrated with the pooled time-series and cross-sectional data to improve the specification of a financial relationship.

Table 6.4 Estimates of log likelihood

λ	(a)	(b)	(c)	(d)	(e)	(f)	(g)
1.5	−2751.31	−2584.67	−2580.22	−2303.24	−2303.23	−177.5	−1775.34
1.4	−2712.17	−2548.69	−2544.07	−2242.36	−2242.35	−1700.88	−1698.81
1.3	−2675.43	−2514.7	−2509.89	−2184.01	−2184	−1625.14	−1623.18
1.2	−2641.1	−2482.72	−2477.71	−2128.42	−2128.41	−1550.61	−1548.75
1.1	−2609.22	−2452.77	−2447.58	−2075.86	−2074.84	−1477.68	−1475.93
1	−2579.79	−2424.91	−2419.5	−2026.59	−2026.57	−1406.88	−1405.22
0.9	−2552.87	−2399.17	−2393.57	−1980.9	−1980.89	−1338.91	−1337.41
0.8	−2528.45	−2375.56	−2369.78	−1939.11	−1939.11	−1274.65	−1273.25
0.7	−2506.58	−2354.15	−2348.22	−1901.57	−1901.57	−1215.17	−1213.66
0.6	−2487.24	−2334.94	−2328.83	−1868.6	−1868.6	−1161.31	−1160.23
0.5	−2470.6	−2318.21	−2311.81	−1840.07	−1840.07	−1114.48	−1114.48
0.4	−2456.39	−2303.69	−2297.03	−1817.75	−1817.75	−1075.42	−1075.42
0.3	−2445.12	−2291.09	−2284.46	−1800.55	−1800.55	−1045.15	−1045.15
0.2	−2435.88	−2280.65	−2274.09	−1787.23	−1787.23	−1038.86	−1038.86
0.1	−2431.47	−2275.52	−2266.92	−1780.53	−1780.53	−1022.96	−1022.96
0	−2426.51	−2268.93	−2262.17	−1778.64	−1778.64	−1028.76	−1027.28
−0.1	−2426.33	−2267.13	−2260.53	−1784.11	−1784.11	−1043.58	−1043.58
−0.2	−2428.74	−2265.79	−2259.36	−1796.64	−1796.64	−1066.68	−1066.68
−0.3	−2431.75	−2271.82	−2265.64	−1810.28	−1810.28	−1105.29	−1105.29
−0.4	−2440.42	−2275.28	−2268.09	−1830.43	−1830.48	−1146.91	−1146.05
−0.5	−2450.64	−2281.53	−2276.93	−1856.91	−1856.91	−1194.85	−1194.85

Note
(a) = OLS
(b) = LSDV with cross-sectional effects
(c) = GLS with cross-sectional effects
(d) = LSDV with time effects
(e) = GLS with time effects
(f) = LSDV with both effects
(g) = GLS with both effects

6.5 Clustering Effect and Clustered Standard Errors

In this section, we will discuss clustering effect and how standard errors can be adjusted when regression residuals are not independent. Under an ordinary least square (OLS) model, standard errors will be consistent when residuals are independent and identically distributed. That is, in a panel regression, standard errors will not be biased only if residuals are uncorrelated across firms and months. However, empirically, the regression analysis in most of finance and accounting research contains unobservable variables that may clustered by firms and/or time. Even understanding the existence of dependent residuals in the regressions, Petersen (2009) shows that around 42% of finance researches do not adjust standard errors for the consideration of dependent residuals. This may result in biased standard errors and incorrect inferences.

Petersen (2009) provides theoretical model and conducts simulations to show that standard errors of Fama–MacBeth procedure are unbiased if there is only a time effect in the panel data set. However, Fama–MacBeth produces biased standard errors in the presence of a firm effect or both time and firm effects. Therefore, clustered standard errors will be good candidate when the panel data set has firm effect or both time and firm effects. Thompson (2011) provides standard errors that are robust to simultaneous correlated across time and firms. Specifically, the adjusted covariance matrix can be obtained as the covariance estimator clustered by time plus covariance estimator clustered by firms, and minus the heteroscedasticity robust covariance matrix (White covariance) in OLS model. That is, the estimated variance for an OLS estimator $\hat{\beta}$ is

$$\hat{V}\left(\hat{\beta}\right) = \hat{V}_{\text{firm}} + \hat{V}_{\text{time}} - \hat{V}_{\text{white}} \tag{6.18}$$

where $\hat{V}_{\text{firm}}$ and $\hat{V}_{\text{time}}$ are the estimated variances clustered by firm and time, and $\hat{V}_{\text{white}}$ is the heteroscedasticity robust covariance matrix in OLS model. In addition, Cameron et al. (2011) propose a variance estimator adjusting for two-way or multiway clustering effects.

Although Petersen (2009), Thompson (2011), and Cameron et al. (2011) point that standard error of two-way fixed effects estimates can be biased if two-dimensional clustering is not controlled for, cluster standard errors are still biased if there has not sufficient number of clusters. Petersen (2009) proposes an alternative way, using fixed effects dummies in one dimension and then estimate clustered standard errors in the other dimension, to deal with insufficient number of clusters. In addition, Thompson (2011) and Boehmer et al. (2013) empirically find that these clustering effects are not important for the large samples when fixed effects model is applied.

6.6 Hausman Test for Determining Either Fixed Effects Model or Random Effects Model

We may face a problem of decision between fixed effects model and random effects model in a panel data regression. As discussed in previous sections, random effects strictly assumes that individual (or time-series) effects are not correlated with the regressors. However, in practice, individual (or time-series) effects may not uncorrelated with the other regressors. The random effects model may lead an inconsistency. On the other hand, the fixed effects model can obtain consistent estimators no matter individual (or time-series) effects are correlated with regressors or not, while the dummy variable approach in the fixed effects model may lose degrees of freedom and be inefficient.

Considering consistency and efficiency of the model, a Hausman (1978) test can help us to decide using a fixed effects model (LSDV) or a random effects model (GLS). Specifically, Hausman test is under a null hypothesis that the unique errors are not correlated with the regressors. That is, under the null hypothesis, both LSDV model and GLS model are consistent, but LSDV is inefficient. In contrast, under the alternative hypothesis, LSDV model is consistent, but GLS is not.

$$W = \left(\hat{\beta}_{\mathrm{RE}} - \hat{\beta}_{\mathrm{FE}}\right)' \hat{\Sigma}^{-1} \left(\hat{\beta}_{\mathrm{RE}} - \hat{\beta}_{\mathrm{FE}}\right) \sim \chi^2(k) \quad (6.19)$$

where $\hat{\beta}_{\mathrm{RE}}$ and $\hat{\beta}_{\mathrm{FE}}$ are estimators for random effects model and fixed effects model, $\hat{\Sigma}$ is the covariance matrix for the difference vector between random effects model and fixed effects model, and k is number of regressors.

Therefore, if the Hausman test cannot reject the null hypothesis, W is insignificant, and a random effects model can be used. Otherwise, we should use fixed effects model. A more detailed discussion about the Hausman test can be found in Patrick (2018).

6.7 Efficient Firm Fixed Effects Estimator and Efficient Correlated Random Effects Estimator

Osterrieder et al. (2018) introduce two additional estimators, efficient firm fixed effects (EFE) estimator and efficient correlated random effects (ECRE) estimator in the panel data regression. Comparing to estimators of three regression models discussed in the previous sections, OLS, LSDV and GLS, Osterrieder et al. (2018) find that LSDV estimator is consistent but has higher uncertainty in simulations, while the OLS and GLS estimators are inconsistent. In contrast, the EFE and ECRE estimators are consistent with relatively lower uncertainty.

In addition, Osterrieder et al. (2018) introduce a *J*-test to decide using the EFE estimator or the ECRE estimator. Specifically, J-test is under a null hypothesis that the unobserved firm effect is a linear function of the time-series averages of the covariates and exogenous error term. Therefore, if we cannot reject the null hypothesis, we should use the EFE estimator. If we can reject the null hypothesis, we can use either EFE or ECRE estimators.

Osterrieder et al. (2018) subsequently apply EFE and ECRE to examine the relationship between CEO compensation and firm value. They find that the EFE estimator shows a much higher economic impact of CEO pay on firm value and suggest EFE estimator is the most appropriate in testing the relationship between CEO pay and firm value.

6.8 Empirical Evidence of Optimal Payout Ratio Under Uncertainty and the Flexibility Hypothesis

To examine the relationship between the payout ratio and other financial variables, Lee et al. (2011) propose fixed effects models of the payout ratio as follows[7]:

$$\begin{aligned}\ln&\left(\frac{\text{payout ratio}_{i,t}}{1-(\text{payout ratio}_{i,t})}\right)\\ &= \alpha + \beta_1 \mathrm{Risk}_{i,t} + \beta_2 D_{i,t}\left(g_{i,t} < c \cdot ROA_{i,t}\right) \cdot \mathrm{Risk}_i\\ &\quad + \beta_3 \mathrm{Growth}_{i,t} + \beta_4 \mathrm{Risk}_{i,t} \times \mathrm{Growth}_{i,t}\\ &\quad + \beta_5 \ln(\mathrm{Size})_{i,t} + \beta_6 ROA_{i,t} + e_{i,t}.\end{aligned} \quad (6.20)$$

In the regression, the dependent variable is the logistic transformation of the payout ratio. Independent variables include risk measure (beta coefficient or total risk), the interaction of dummy variable and risk measure, growth rate, the interaction of risk measure and growth rate, log of size, and rate of return on total assets.[8] Based upon the theoretical model and its implications discussed in Lee et al. (2011), we assume that c is equal to 1. The dummy variable (D_i) is

[7]The dummy variable $D_{i,t}\left(g_{i,t} < c \cdot ROA_{i,t}\right)$ used in Eq. (6.20) implies that the relationship between the payout ratio and risks is nonlinear (piecewise regression). In other words, the breakpoint of the structural change is at $g_{i,t} = c \cdot ROA_{i,t}$. Based upon our theoretical model, we assume that c is equal to 1 in our empirical work.

[8]Besides merely adding an interaction dummy as indicated in Eq. (6.20), we include an intercept dummy to take care of the individual effect of two groups. We also run regressions for high-growth firms and low-growth firms separately. Results from both models are qualitatively the same as those from Eq. (6.20) and also support Hypotheses 1–3.

equal to 1 if a firm's five-year average growth rate is less than its five-year average rate of return on assets and 0 otherwise[9]. Such structure allows us to analyze the relationship between payout ratio and growth rate and the relationship between payout ratio and risk under different growth rate levels.

The robust standard errors are very similar to standard errors for ordinary least square (OLS), suggesting that the fixed effects and control variables are removing most of the correlation that is present across observations. In addition, our data set cannot be meaningfully applied to the clustering effect model[10]; therefore, statistical inferences in this study are conducted using fixed effects standard errors.

Table 6.5 provides the results of fixed effects regressions for 2645 firms during the period 1969–2009. *Model* (1) and *Model* (2) show that the estimated coefficients of the growth rate are −0.03 with a *t*-statistics of −4.85 and −0.03 with a *t*-statistics of −4.85, respectively. Such significantly negative coefficients confirm Hypothesis 1, which states that high-growth firms will pay less in dividends for the consideration of flexibility. We also include an interaction term of risk and growth rate into *Model* (3) and *Model* (4). The results in *Model* (3) and *Model* (4) also support Hypothesis 1.

Models (1)–(4) show that the relationship between the payout ratio and the risk is significantly negative. The results are similar to the findings of Rozeff (1982), Jagannathan et al. (2000), and Grullon et al. (2002), indicating that dividend payouts are negatively correlated to firm risks; but our theoretical model shows that if firms follow their optimal dividend payout policy, the relationship between dividend payouts and firm risks depends on their growth rates relative to their rate of return on total assets as our theoretical analysis presented in Sect. 6.6. In Table 6.7, we find the number of firms with a higher growth rate with respect to their rate of return on assets is greater than the number of firms with a lower growth rate with respect to their rates of return on assets. When pooling high-growth firms and low-growth firms together, the negative risk effect of high-growth firms will dominate the positive risk effect of low-growth firms due to the larger proportion of high-growth firms in the observations. The results of the negative relationship between the payout ratio and the risk shown in *Models* (1)–(4) may therefore result from the greater proportion of high-growth firms. Based on our subsequent analysis, the effect of growth rates on dividend payout policies can be more accurately found when firms are separated into high-growth firms and low-growth firms relative to their rates of return on total assets of return on assets.

To test Hypotheses 2 and 3, we introduce an interaction term of the dummy variable and the risk. In *Model* (5) and *Model* (6), the estimated coefficients of risk are −0.23 with a *t*-statistics of −13.88 and −0.22 with a *t*-statistics of −12.69, respectively. The significantly negative coefficients support the hypothesis 2 that, because of the consideration of flexibility, the payout ratio and the risk are negatively correlated for firms with a higher growth rate relative to their rate of return on assets. In addition, significant and positive coefficients of the interaction term of the dummy variable and the risk indicate that, when the risk changes, the dividend policy for low-growth firms is different from that of high-growth firms. By summing the coefficient of risk and the coefficient of interaction term, we can obtain coefficients of 0.48 and 35.54 for beta and total risk, respectively, indicating the relationship between the payout ratio and the risk for low-growth firms is positive. That is, when the risk increases, low-growth firms will follow their optimal payout policies to increase their dividend payouts. Hypothesis 3 is thus confirmed in our empirical work. Blau and Fuller (2008) find that the flexibility hypothesis is more suitable than the free cash flow hypothesis to explain the

[9]For the discussion of hypotheses and data, please refer Appendix.

[10]Because our sample is an unbalanced panel data, the clustering computer program cannot meaningfully estimate the variance components, variance of firm $\left(\hat{V}_{\text{firm}}\right)$, variance of time $\left(\hat{V}_{\text{time}}\right)$, and heteroscedasticity robust OLS variance $\left(\hat{V}_{\text{white}}\right)$.

Table 6.5 Fixed effects regressions using theoretical structural change point

Model	Intercept	Beta	D * Beta	Growth * Beta	Total risk	D * total risk	Growth * total risk	Growth	ln(size)	ROA	Adj−R^2	F-test (p-value) p-value
(1)	0.32*** (2.70)	−0.18*** (−9.82)						−0.03*** (−4.85)	0.04*** (3.76)	−13.24*** (−63.14)	0.1945	
(2)	0.45*** (3.76)				−17.47*** (−14.64)			−0.03*** (4.68)	0.02** (2.10)	−13.07*** (−62.42)	0.1982	
(3)	0.31*** (2.64)	−0.14*** (−7.76)		−0.29*** (−8.84)				0.09*** (6.38)	0.04*** (3.47)	−12.96*** (−61.17)	0.1970	
(4)	0.45*** (3.77)				−17.14*** (−13.13)		−2.90 (−0.63)	<0.01 (0.01)	0.02** (2.06)	−13.06*** (−61.97)	0.1982	
(5) $c = 1$	0.33*** (2.94)	−0.23*** (−13.88)	0.71*** (57.57)					−0.02*** (−3.70)	0.002 (0.21)	−11.54*** (−57.83)	0.2867	<0.01
(6) $c = 1$	0.52*** (4.58)				−27.81*** (−24.15)	63.35*** (52.23)		−0.02*** (−3.53)	−0.002 (−0.27)	−11.40*** (−56.55)	0.2753	<0.01
(7) $c = 1$	0.32*** (2.91)	−0.22*** (−12.69)	0.70*** (56.96)	−0.13*** (−4.02)				0.03*** (2.35)	−0.001 (−0.10)	−11.43*** (−56.76)	0.2871	<0.01
(8) $c = 1$	0.51*** (4.53)				−29.61*** (−23.44)	63.68*** (52.35)	15.30*** (3.46)	−0.16*** (−3.89)	−0.001 (−0.06)	−11.47*** (−56.63)	0.2756	<0.01

This table presents results from fixed effects regressions of dividend payout ratios on firm characteristics for 2645 firms during the period from 1969 to 2009. The regressions are as follows

$$\ln\left(\frac{\text{payout ratio}_{i,t}}{1-(\text{payout ratio}_{i,t})}\right) = \alpha + \beta_1 \text{Risk}_{i,t} + \beta_2 D\left(g_{i,t} < c \cdot ROA_{i,t}\right) \cdot \text{Risk}_i + \beta_3 \text{Growth}_{i,t} + \beta_4 \text{Risk}_{i,t} \times \text{Growth}_{i,t} + \beta_5 \ln(\text{Size})_{i,t} + \beta_6 ROA_{i,t} + e_{i,t}$$

The dependent variable is the payout ratio with a logistic transformation. Following the theoretical model, we assume c is equal to 1. The dummy variable is equal to 1 if a firm's 5-year average growth rate is less than its 5-year average ROA, and 0 otherwise. The independent variables are beta risk (total risk), dummy times beta risk (total risk), interaction between growth rate and beta risk (total risk), growth rate, log of size, and the rate of return on assets. t-statistics are presented in parentheses, and the last column presents p-values of the F-test on the null hypothesis that the restricted *Models* (1)–(4) are not different from the restricted—*Models* (5)–(8)

*, **, and *** indicate significance at the 10, 5, and 1% levels, respectively

dividend policy, but their method cannot separate the dividend policy decisions between high-growth firms and low-growth firms. Consequently, the model used here can be regarded as a generalization of their results.

In *Model* (7) and *Model* (8) in Table 6.5, we further include the interaction term of the risk and the growth rate into the regressions. The results still support our hypotheses that the relationship between the payout ratio (Hypothesis 1) and the growth rate is negative and the relationship between the payout ratio and the risk depends on firm's growth rate with respect to its rate of return on total assets (Hypotheses 2 and 3). In addition, we find that the interaction terms of the risk and the growth rate are significantly different from zero. We also find that the adjusted R-squares for *Models* (5)–(8) are higher than those for models without dummies. *F*-tests also reject the null hypothesis that the regression with the interaction of the dummy variable and the risk is not different from the regression without the interaction of the dummy and the risk. We can thus conclude that the payout ratio is not linearly related to the growth rate or to the risk, and previous empirical studies on dividend policy using linear model may suffer from model misspecification.

Moving Estimate Process for Structural Change Model

We theoretically show that the structural change breakpoint for the relationship between the payout ratio and risks is at $g_{i,t} = ROA_{i,t}$. In empirical works, we use dummy variable approach to separate the sample into a high-growth $(g_{i,t} > c \cdot ROA_{i,t})$ group and a low-growth $(g_{i,t} < c \cdot ROA_{i,t})$ group, assuming c is equal to 1, and empirically test the relationship between the payout ratio and risks for high-growth firms and low-growth firms. Based on the moving estimate process developed by Chow (1960), Hansen (1996, 1999, 2000), and Zeileis et al. (2002), we here try to estimate the empirical breakpoint of the structural change and examine whether the empirical breakpoint is different from our theoretical breakpoint or not.

Table 6.6 Fixed effects regressions using empirical structural change point

Model	Intercept	Beta	*D* * Beta	Growth * Beta	Total risk	*D* * total risk	Growth * total risk	Growth	ln(size)	ROA	Adj-R^2	*F-test* (*p*-value)
(1) $c = 0.93$	0.29***	−0.21***	0.77***					−0.02***	0.005	−11.63***	0.2891	<0.01
	(2.58)	(−12.57)	(58.41)					(−3.46)	(0.55)	(−58.46)		
(2) $c = 0.97$	0.47***				−27.03***	66.22***		−0.02***	0.002	−11.50***	0.2758	<0.01
	(4.15)				(−23.53)	(52.43)		(−3.50)	(0.25)	(−57.13)		
(3) $c = 0.93$	0.28***	−0.20***	0.77***	−0.13***				0.04***	0.004	−11.51***	0.2896	<0.01
	(2.56)	(−11.39)	(57.82)	(−4.19)				(2.60)	(0.44)	(−57.33)		
(4) $c = 0.97$	0.46***				−28.67***	66.52***	13.00***	−0.15***	0.004	−11.56***	0.2761	<0.01
	(4.10)				(−22.76)	(52.53)	(3.17)	(−3.60)	(0.45)	(−57.17)		

This table presents results from fixed effects regressions of dividend payout ratios on firm characteristics for 2645 firms during the period from 1969 to 2009. The regressions are as follows

$$\ln\left(\frac{\text{payout ratio}_{i,t}}{1-(\text{payout ratio}_{i,t})}\right) = \alpha + \beta_1 \text{Risk}_{i,t} + \beta_2 D(g_{i,t} < c \cdot ROA_{i,t}) \cdot \text{Risk}_i + \beta_3 \text{Growth}_{i,t} + \beta_4 \text{Risk}_{i,t} \times \text{Growth}_{i,t} + \beta_5 \ln(\text{Size})_{i,t} + \beta_6 ROA_{i,t} + e_{i,t}$$

The dependent variable is the payout ratio with a logistic transformation. Based upon results from the moving estimate process, c is equal to 0.93 when using beta risk and 0.97 when using total risk. The dummy variable is equal to 1 if a firm's 5-year average growth rate is less than its 5-year average ROA, and 0 otherwise. The independent variables are beta risk (total risk), dummy times beta risk (total risk), interaction between growth rate and beta risk (total risk), growth rate, log of size, and the rate of return on assets. *t*-statistics are presented in parentheses, and the last column presents *p*-values of the *F*-test on the null hypothesis that the restricted *Models* (1)–(4) in Table 6A.3 are not different from the restricted *Models* (1)–(4) in Table 6A.4

*, **, and *** indicate significance at the 10, 5, and 1% levels, respectively

By using the moving estimate process, we can obtain an empirical estimate of $c = 0.93$ when using beta risk and $c = 0.97$ when using total risk.[11] That is, the breakpoint of the structural change is at $g_{i,t} = 0.93 \times ROA_{i,t}$ or $g_{i,t} = 0.97 \times ROA_{i,t}$. Then, we redo *Model* (5)–*Model* (8) of Table 6.5 by using the empirical structural change point instead of the theoretical structural change point. In Table 6.6, we compare to the theoretical model and find that the estimates and significances using the empirical breakpoint of the structural change are almost same as those using the theoretical breakpoint indicated in Table 6.5. Results obtained from the moving estimate process show the existence of an empirical breakpoint for the relationship between the payout ratio and risks and also confirm that the dummy variable used in Eq. (6.20) is both theoretically and empirically acceptable.

6.9 Conclusion

Based upon Chang and Lee (1977), Hsiao (2014), and Lee et al. (2011), we have discussed fixed effects versus random effects in finance research. First, we discussed two alternative techniques to analyze pooled time-series and cross-sectional data which are used to test the importance of firm effect and time effect in the financial analysis. These techniques are also integrated with the functional form parameter estimation method to show the importance of appropriate functional form in handling a pooled time-series- and cross-section-type econometric model. The data on the electric industry show that both the time effect and cross-sectional effect are of importance in explaining stock price variation. It is also found that linear form (and/or) log linear form is not always appropriate in testing the importance of both time effect and firm effect in financial analyses.

Lee et al. (2011) developed a dynamic stochastic model deriving relationships between corporate dividend payout and several important financial variables. The relationships derived from the model are then tested by extensive empirical research. In the empirical test, Lee et al. (2011) have used fixed effects to test the model. In addition, they also took care of the clustering effect of the fixed effects model. Overall, this chapter has discussed both fixed effects and random effects in some detail. It is well known that fixed effects model is generally less efficient than the random effects model; however, random effects model frequently faces the problem of endogeneity. Most recently, Osterrieder et al. (2018) have developed both efficient fixed effects and random effects models. Their random effects model has taken care of the problem of endogeneity. Therefore, we believe the efficient random effects model developed by Osterrieder et al. (2018) will be the best way to deal with the time effect and firm effect issue in corporate finance research using panel data.

Appendix: Optimal Payout Ratio Under Uncertainty and the Flexibility Hypothesis: Theory and Empirical Evidence

By Cheng-Few Lee, Manak C. Gupta, Hong-Yi Chen and Alice C. Lee

Hypothesis Development

A growing body of literature focuses on the determinants of optimal dividend payout policy. Rozeff (1982), Jagannathan et al. (2000), Grullon et al. (2002), Aivazian et al. (2003), Blau and Fuller (2008), and others empirically investigate the determination of dividend policy, but none of them has a solid theoretical model to support their findings. Based upon the theoretical model and its implications Lee et al. (2011), three testable hypotheses are developed as follows.

Hypothesis 1: Firms generally reduce their dividend payouts when their growth rates increase.

The negative relationship between the payout ratio and the growth ratio in our theoretical model implies that high-growth firms need to

[11]Gujarati (2009) shows that this kind of problem can be regarded as piecewise regression by using moving estimate processes.

reduce the payout ratio and retain more earnings to build up "precautionary reserves," but low-growth firms are likely to be more mature and already build up their reserves for flexibility considerations. Rozeff (1982), Fama and French (2001), Blau and Fuller (2008), and others argue that high-growth firms will have higher investment opportunities and tend to pay out less in dividends. Based upon flexibility concerns, we predict that high-growth firms pay higher dividends. This result is obtained when risk factor is not explicitly considered.

Lee et al. (2011) theoretically find that the relationship between the payout ratio and the risk can be either negative or positive, depending upon whether the growth rate is higher or lower than the rate of return on total assets. Based upon this finding, Lee et al. (2011) develop two other hypotheses.

Hypothesis 2: The relationship between the firms' dividend payouts and their risks is negative when their growth rates are higher than their rates of return on asset.

High-growth firms need to reduce the payout ratio and retain more earnings to build up "precautionary reserves," which become more important for a firm with volatile earnings over time. For flexibility considerations, high-growth firms tend to retain more earnings when they face higher risk. This theoretical result is consistent with the flexibility hypothesis.

Hypothesis 3: The relationship between the firms' dividend payouts and their risks is positive when their growth rates are lower than their rates of return on asset.

Low-growth firms are likely to be more mature and have most likely already built such reserves over time, and they probably do not need more earnings to maintain their low-growth perspective and can afford to increase the payout (see Grullon et al. 2002). Because the higher risk may involve higher cost of capital and make the free cash flow problem worse, for free cash flow considerations, low-growth firms tend to pay more dividends when they face higher risk. This theoretical result is consistent with the free cash flow hypothesis.

Sample Description

We collect the firm information, including total asset, sales, net income, and dividends payout, from Compustat. Stock price, stock returns, share codes, and exchange codes are retrieved from the Center for Research in Security Prices (CRSP) files. The sample period is from 1969 to 2009. Only common stocks (SHRCD = 10, 11) and firms listed on NYSE, AMEX, or NASDAQ (EXCE = 1, 2, 3, 31, 32, 33) are included in our sample. We exclude utility services (SICH = 4900–4999) and financial institutions (SICH = 6000–6999).[12] The sample includes those firm-years with at least five years of data available to compute average payout ratios, growth rate, return on assets, beta, total risk, size, and book-to-market ratios. The payout ratio is measured as the ratio of the dividend payout to the net income. The growth rate is the sustainable growth rate proposed by Higgins (1977). The beta coefficient and total risk are estimated by the market model over the previous 60 months. For the purpose of estimating their betas, firm-years in our sample should have at least 60 consecutive previous monthly returns. To examine the optimal payout policy, only firm-years with five consecutive dividend payouts are included in our sample.[13] Considering the fact that firm-years

[12]We filter out those financial institutions and utility firms based on the historical Standard Industrial Code (SIC) available from Compustat. When a firm's historical SIC is unavailable for a particular year, the next available historical SIC is applied instead. When a firm's historical SIC is unavailable for a particular year and all the years after, we use the current SIC from Compustat as a substitute.

[13]To avoid creating a large difference in dividend policy, on the one hand managers partially adjust firms' payout by several years to reduce the sudden impacts of the changes in dividend policy. On the other hand, they use not only one-year firm conditions but also multiyear firm conditions to decide how much they will pay out. In examining the optimal payout policy, we use the five-year rolling averages for all variables.

with no dividend payout one year before (or after) might not start (or stop) their dividend payouts in the first (fourth) quarter of the year, we exclude from our sample firm-years with no dividend payouts one year before or after to ensure the dividend payout policy reflects the firm's full-year condition.

Table 6.7 shows the summary statistics for 2645 sample firms during the period from 1969 to 2009. Panel A of Table 6.7 lists the number of firm-year observations for all sample high-growth firms and low-growth firms, respectively. High-growth firm-years are those firm-years that have five-year average sustainable

Table 6.7 Summary statistics of sample firm characteristics

Panel A. Sample size

Year	Number of firm-years			Year	Number of firm-years		
	All	Growth > *ROA*	Growth < *ROA*		All	Growth > *ROA*	Growth < *ROA*
1969	345	161	184	1990	690	522	168
1970	360	175	185	1991	668	511	157
1971	404	201	203	1992	653	494	159
1972	513	269	244	1993	642	460	182
1973	535	308	227	1994	655	479	176
1974	572	371	201	1995	651	483	168
1975	609	432	177	1996	693	530	163
1976	650	486	164	1997	725	582	143
1977	678	530	148	1998	743	620	123
1978	711	553	158	1999	725	612	113
1979	779	620	159	2000	709	607	102
1980	764	636	128	2001	659	569	90
1981	929	785	144	2002	599	503	96
1982	1203	1003	200	2003	571	475	96
1983	1151	933	218	2004	525	433	92
1984	1067	832	235	2005	481	391	90
1985	1010	744	266	2006	510	430	80
1986	958	669	289	2007	542	451	91
1987	897	645	252	2008	579	470	109
1988	847	615	232	2009	610	484	126
1989	721	531	190	All years	28,333	21,065	6728

Panel B. Descriptive statistics of characteristics of sample

	Payout ratio	Growth rate	ROA	Beta	Total risk	Size ($MM)	*M*/*B*
All sample (*N* = 28,333)							
Mean	0.3793	0.1039	0.0723	1.0301	0.0106	3072	1.7940
Median	0.3540	0.0886	0.0648	1.0251	0.0089	291	1.3539
Stdev	0.1995	0.7444	0.0389	0.4272	0.0078	14,855	1.9479

(continued)

Table 6.7 (continued)

Panel B. Descriptive statistics of characteristics of sample							
High-growth firms (*N* = 21,065)							
Mean	0.3180	0.1233	0.0698	1.0624	0.0112	3267	1.7951
Median	0.2996	0.1002	0.0638	1.0581	0.0095	314	1.3757
Stdev	0.1658	0.8060	0.0355	0.4352	0.0070	15,806	1.6496
Low-growth firms (*N* = 6728)							
Mean	0.5762	0.0413	0.0800	0.9265	0.0087	2447	1.7904
Median	0.5542	0.0524	0.0692	0.9375	0.0071	229	1.3007
Stdev	0.1690	0.4918	0.0476	0.3822	0.0099	11,250	2.6909

This table presents the descriptive statistics for those major characteristics of our sample firms. Sample includes those firms listed on NYSE, AMEX, and NASDAQ with at least five years of data available to compute average payout ratios, growth rate, return on assets, beta, total risk, size, and book-to-market ratios. All financial service operations and utility companies are excluded. Panel A lists the numbers of firm-year observations for all sample firms, high-growth firms, and low-growth firms, respectively, between year 1969 and year 2009. High-growth firm-years are defined as firm-years with sustainable growth rates higher than their rates of return on assets. Low-growth firm-years are defined as firm-years with sustainable growth rates lower than their rate of return on assets. Panel B lists the mean, median, and standard deviation values of the five-year average of the payout ratio, growth rate, rate of return on assets, beta risk, total risk, size, and book-to-market ratio. The payout ratio is measured as the ratio of the dividend payout to the earnings. Growth rate is the sustainable growth rate proposed by Higgins (1977). The beta coefficient and total risk are estimated by the market model over the previous 60 months. Size is defined as market capitalization calculated by the closing price of the last trading day of June of that year times the outstanding shares at the end of June of that year

growth rates higher than their five-year average rate of return on assets. Low-growth firm-years are those firms with five-year average sustainable growth lower than their five-year average rate of return on assets. The sample size increases from 345 firms in 1969 to 1203 firms in 1982, while declining to 610 firms by 2009. A total of 28,333 dividend paying firm-years are included in the sample. When classifying high-growth firms and low-growth firms relative to their return on assets, the proportion of high-growth firms increases over time. The proportion of firm-years with a growth rate higher than return on assets increases from less than 50% during the late 1960s and early 1970s to 80% in 2008. Panel B of Table 6.7 shows the five-year moving averages of mean, median, and standard deviation values for the measures of payout ratio, growth rate, rate of return, beta coefficient, total risk, market capitalization, and market-to-book ratio across all firm-years in the sample. Among high-growth firms, the average growth rate is 12.33%, and the average payout ratio is 31.80%, but for low-growth firms, the average growth rate is 4.13%, and the average payout ratio is 57.62%. High-growth firms undertake more beta risk and total risk, indicating that high-growth firms undertake both more systematic risk and unsystematic risk to pursue a higher rate of return.

Bibliography

Aivazian, V., Booth, L., & Cleary, S. (2003). Do emerging market firms follow different dividend policies from US firms? *Journal of Financial Research, 26*(3), 371–387.

Amemiya, T. (1972). The estimation of the variances in a variance-components model. *International Economic Review, 12,* 1–13.

Balestra, P., & Nerlove, M. (1968). Pooling cross-section and time-series data in the estimation of a dynamic model: The demand for natural gas. *Econometrica, 34,* 585–612.

Box, G. E. P., & Cox, D. R. (1964). An analysis of transformations. *Journal of the Royal Statistical Society: Series B, 26,* 211–243.

Boehmer, E., Jones, C. M., & Zhang, X. (2013). Shackling short sellers: The 2008 shorting ban. *Review of Financial Studies, 26,* 1363–1400.

Bower, R. S., & Bower, D. H. (1969). Risk and valuation of common stock. *The Journal of Political Economy, 77,* 349–362.

Blau, B. M., & Fuller, K. P. (2008). Flexibility and dividends. *Journal of Corporate Finance, 14,* 133–152.

Cameron, A. C., Gelbach, J. B., & Miller, D. L. (2011). Robust inference with multiway clustering. *Journal of Business & Economic Statistics, 29,* 238–249.

Chang, H. S., & Cheng-Few Lee. (September 1977). Using pooled time-series and cross section data to test the firm and time effects in financial analysis. *Journal of Financial and Quantitative Analysis.*

Chow, G. C. (1960). Tests of equality between sets of coefficients in two linear regressions. *Econometrica, 28,* 591–605.

Chung, P. S. (1974). An investigation of the firm effects influence in the analysis of earnings to price ratios of industrial common stocks. *The Journal of Financial and Quantitative Analysis, 9,* 1009–1029.

Durand, D. (1959). The cost of capital, corporation finance, and the theory of investment: Comment. *American Economic Review, 49,* 639–654.

Fama, E. F., & French, K. R. (2001). Disappearing dividends: Changing firm characteristics or lower propensity to pay? *Journal of Financial Economics, 60,* 3–43.

Friend, I., & Puckett, M. (1964). Dividends and stock prices. *American Economic Review, 54,* 656–681.

Gordon, J. J. (1959). Dividends, earnings and stock prices. *Review of Economic and Statistics, 41,* 99–105.

Grullon, G., Michaely, R., & Swaminathan, B. (2002). Are dividend changes a sign of firm maturity? *Journal of Business, 75,* 387–424.

Gujarati, D. N. (2009). *Basic econometrics.* Tata McGraw-Hill Education.

Hansen, B. E. (1996). Inference when a nuisance parameter is not identified under the null hypothesis. *Econometrica, 64,* 413–430.

Hansen, B. E. (1999). Threshold effects in nondynamic panels: Estimation, testing, and inference. *Journal of Econometrics, 93,* 345–368.

Hansen, B. E. (2000). Sample splitting and threshold estimation. *Econometrica, 68,* 575–603.

Hausman, J. A. (1978). Specification tests in econometrics. *Econometrica, 46,* 1251–1276.

Higgins, R. C. (1977). How much growth can a firm afford? *Financial Management, 6,* 7–16.

Hsiao, C. (2014). *Analysis of panel data—Econometric society monographs* (3rd ed.).

Jagannathan, M., Stephens, C. P., & Weisbach, M. S. (2000). Financial flexibility and the choice between dividends and stock repurchases. *Journal of financial Economics, 57*(3), 355–384.

Lee, C. F., Gupta, M. C., Chen, H. Y. & Lee, A. C. (June 2011) Optimal payout ratio under uncertainty and the flexibility hypothesis: Theory and empirical evidence. *Journal of Corporate Finance, 17*(3), 483–501.

Maddala, G. S. (1971). The use of variance components models in pooling cross-section and time-series data. *Econometrica, 39,* 341–358.

Nerlove, M. (1971). Further evidence on the estimation of dynamic economic relations from a time-series of cross-section. *Econometrica, 39,* 359–382.

Osterrieder, D., Palia, D., & Wu, G. (2018). *Evaluating panel regression models in corporate finance: Evidence from CEO pay.* Working paper.

Patrick, R. H. (2018). Durbin-Wu-Hausman specification tests. In C. F. Lee & J. Lee (Eds.) *Handbook of financial econometerics, mathematics, statistics, and technology.* World Scientific (Forthcoming).

Petersen, M. A. (2009). Estimating standard errors in finance panel data sets: Comparing approaches. *Review of Financial Studies, 22,* 435–480.

Rozeff, M. S. (1982). Growth, beta and agency costs as determinants of dividend payout ratios. *Journal of Financial Research, 5,* 249–259.

Thompson, S. B. (2011). Simple formulas for standard errors that cluster by both firm and time. *Journal of Financial Economics, 99,* 1–10.

Wallace, T. D., & Hussain, A. (1969). The use of error components models in combining cross-section with time-series data. *Econometrica, 37,* 55–72.

Zarembka, P. (1968). Functional form in the demand for money. *Journal of American Statistical Association, 63,* 502–511.

Zeileis, A., Leisc, F., Hornik, K., & Kleiber, C. (2002). strucchange: An R package for testing for structural change in linear regression models. *Journal of Statistical Software, 7,* 1–38.

7 Alternative Methods to Deal with Measurement Error

Contents

Abstract

This chapter discusses how errors-in-variables problems affect estimators in the regression model. In addition, we show alternative methods to deal with measurement error of estimation regression coefficient. These alternative methods include the classical estimation method, the constrained classical method, the grouping method, the instrumental variable method, the maximum likelihood method, the LISREL and MIMIC methods, and the Bayesian approach. Examples using these alternative methods in finance research are also discussed.

This chapter is an update and extension of the paper by Chen et al. (2015).

C.-F. Lee et al., *Financial Econometrics, Mathematics and Statistics*,
https://doi.org/10.1007/978-1-4939-9429-8_7

7.1 Introduction

Specification error and measurement error are two major issues in applying the econometric model to economic and finance research. Studies by Miller and Modigliani (1966) and Roll (1969) are two of the earliest finance-related research studies to apply errors-in-variables (EIV) model in their empirical works. Miller and Modigliani (1966) show that, in determining the cost of capital, anticipated average earnings are unobservable and using accounting estimates of earnings as the proxy may result measurement error problems. Roll (1969, 1977) and Lee and Jen (1978) show that the observed market rate returns in terms of stock market index are measured with errors since the stock market index does not include all assets which can be invested by investors. Roll (1969, 1977) argues that testing capital asset pricing model suffers from an EIV problem and concludes that no correct and unambiguous test of the theory can be accomplished. Lee and Jen (1978) have theoretically shown how beta estimates and Jensen performance measures can be affected by both constant and random measurement errors of the market rate of return and risk-free rate. Other issues such as the determination of the capital structure and investment functions also suffer EIV problems.[1]

Understanding the existence of measurement error problems on finance-related studies, a large extent of literature subsequently tries to mitigate biased results from measurement errors. For the issue of the estimation of the cost of capital, Miller and Modigliani (1966) use the instrumental variable approach to resolve the measurement error problem and get consistent estimators in determining the cost of capital. Zellner (1970) and Lee and Wu (1989) also use various estimation methods to deal with potential EIV problems on estimating the cost of capital. For the issue of the capital asset pricing test, Lee and Jen (1978) argue that both market return and beta coefficient are subjected to measurement error and show how the beta coefficient can be estimated. Lee (1984) shows that the most generalized beta estimate can be decomposed into three components: bias due to specification error, bias due to measurement error, and interaction bias. Therefore, the evidence of failure in capital asset pricing model or the finding of new risk factors might result from model misspecification error or EIV problem. Gibbons and Ferson (1985), Green (1986), Roll and Ross (1994), and Diacogiannis and Feldman (2013) have argued that market portfolio measure with errors is an inefficient portfolio and show how the inefficient benchmark can affect the theoretical CAPM derivation. For the issue of the determinants of the capital structure, Titman and Wessels (1988), Chang et al. (2009), and Yang et al. (2010) apply structure equation models to investigate determinants of the capital structure. For the measurement error problems related to Tobin's Q in investment function, Erickson and Whited (2000) use generalized method of moments (GMM) to obtain consistent estimators in testing *q* theory. Most recently, Almeida et al. (2010) propose an alternative instrumental method to deal with measurement error problems in Tobin's Q and support the *q* theory.

The main purpose of this paper is to study the existing EIV estimation methods and to discuss how these estimation methods have been used in finance research. We first show how EIV problems affect estimators in the regression model. We further provide seven alternative estimation methods dealing with EIV problems. Classical method, grouping method, instrumental variable method, mathematical programming method, maxima likelihood method, LISREL method, and Bayesian approach will be discussed and extended from the original simple regression case to multiple regression case. Finally, we conduct a survey on various studies and investigate the effect that resulted from EIV problems associated with cost of capital, capital asset pricing model, capital structure, and investment equation. We also investigate the correction models used in such studies to mitigate the problem raised from measurement errors.

[1]For the measurement problems related to the determinants of the capital structure, please see Titman and Wessels (1988), Chang et al. (2009), and Yang et al. (2010). For the measurement problems related to the investment function, please see Erickson and Whited (2000, 2002) and Almeida et al. (2010).

This chapter studies how EIV estimation methods are used in finance research. Section 7.2 shows the classical EIV problems and how they affect estimators of the linear regression model. Section 7.3 provides seven alternative correction methods in dealing with EIV problems. Section 7.4 presents the effects of EIV problems on the empirical research of cost of capital, asset pricing, capital structure, and investment decision. Finally, Sect. 7.5 presents the conclusion.

7.2 Effects of Errors-in-Variables in Different Cases

7.2.1 Bivariate Normal Case

Suppose we have a two-variate structural relationship

$$V_i = \alpha + \beta U_i \tag{7.1}$$

Both V_i and U_i are unobserved, while we can observe $Y_i = V_i + \eta_i$ and $X_i = U_i + \varepsilon_i$. We assume that

(a) $\varepsilon_i \sim N(0, \sigma_1^2)$ and $\eta_i \sim N(0, \sigma_2^2)$.
(b) $E(\varepsilon_i U_i) = 0,\ E(\varepsilon_i V_i) = 0,$
$E(\varepsilon_i \eta_i) = 0,\ E(\eta_i U_i) = 0$ and $E(\eta_i V_i) = 0$.
(c) $U_i \sim N(E(X), \sigma_U^2)$ and $V_i \sim N(\alpha + \beta E(X), \beta^2 \sigma_U^2)$.

The effects of measurement error on the estimates of α and β can be seen as follows:

(a) $$\text{plim}\hat{\beta} = \frac{\beta \sigma_U^2}{\sigma_U^2 + \sigma_1^2} \tag{7.2a}$$

(b) $$\text{plim}\hat{\alpha} = E(Y) - \frac{\beta \sigma_U^2}{\sigma_U^2 + \sigma_1^2} E(X). \tag{7.2b}$$

This implies that the asymptotic biases for β and α are

(a) $$\text{plim}\hat{\beta} - \beta = \frac{-\beta \sigma_1^2}{\sigma_U^2 + \sigma_1^2} \tag{7.3a}$$

(b) $$\text{plim}\hat{\alpha} - \alpha = \frac{\beta \sigma_1^2}{\sigma_U^2 + \sigma_1^2} E(X). \tag{7.3b}$$

Equations (7.3a) implies that $\hat{\beta}$ is downward-biased and $\hat{\alpha}$ is upward-biased.

7.2.2 Multivariate Case

Suppose we have a trivariate structural relationship

$$W_i = \alpha + \beta U_i + \gamma V_i \tag{7.4}$$

W_i, U_i, and V_i are unobserved, but we can observe $Z_i = W_i + \tau_i$, $X_i = U_i + \varepsilon_i$, and $Y_i = V_i + \eta_i$. U_i and V_i have a joint normal distribution with variances σ_U^2 and σ_V^2 and correlation coefficient ρ_{UV}. In the observed variables X, Y, and Z, the observed errors ε, η, and τ are independent normal variables with zero means and variance σ_1^2, σ_2^2, σ_3^2. X, Y, and Z have a multivariate normal distribution.

7.2.2.1 The Classical Case

The effects of measurement errors on the estimates of β and γ in the classical case can be seen from the following:

$$\text{plim}\ \hat{\beta} = \frac{(\sigma_V^2 + \sigma_2^2)\sigma_{WV} - (\sigma_{WV}\sigma_{UV})}{(\sigma_U^2 + \sigma_1^2)(\sigma_V^2 + \sigma_2^2) - (\sigma_{WV})^2} \tag{7.5a}$$

$$\text{plim}\ \hat{\gamma} = \frac{(\sigma_V^2 + \sigma_1^2)\sigma_{WV} - (\sigma_{WV}\sigma_{UV})}{(\sigma_U^2 + \sigma_1^2)(\sigma_V^2 + \sigma_2^2) - (\sigma_{UV})^2} \tag{7.5b}$$

From Eqs. (7.5a) and (7.5b), the asymptotic biases of $\hat{\beta}$ and $\hat{\gamma}$ can be defined as

$$\text{plim}\hat{\beta} - \beta = \frac{\sigma_{VW}\sigma_2^2 - \beta(\sigma_U^2\sigma_2^2 + \sigma_V^2\sigma_1^2 + \sigma_1^2\sigma_2^2)}{(\sigma_U^2\sigma_V^2 - \sigma_{VW}^2) + \sigma_U^2\sigma_2^2 + \sigma_V^2\sigma_1^2 + \sigma_1^2\sigma_2^2} \tag{7.6a}$$

$$\text{plim}\hat{\gamma} - \gamma = \frac{\sigma_{WV}\sigma_1^2 - \gamma(\sigma_U^2\sigma_2^2 + \sigma_V^2\sigma_1^2 + \sigma_1^2\sigma_2^2)}{(\sigma_U^2\sigma_V^2 - \sigma_{UV}^2) + \sigma_U^2\sigma_2^2 + \sigma_V^2\sigma_1^2 + \sigma_1^2\sigma_2^2} \quad (7.6b)$$

The direction of the biases of $\hat{\beta}$ and $\hat{\gamma}$ can be treated according to the following:

- Under the assumption that $\text{Cov}(UV) = 0$

(i) If $\sigma_1^2 = 0, \quad \sigma_2^2 > 0,$

$$\text{(a)} \quad \text{plim}\hat{\beta} - \beta = \frac{\sigma_2^2(\sigma_{WU} - \beta\sigma_U^2)}{\sigma_U^2(\sigma_V^2 + \sigma_2^2)} = 0, \quad (7.7a)$$

$$\text{(b)} \quad \text{plim}\,\hat{\gamma} - \gamma = \frac{-\gamma\sigma_U^2\sigma_2^2}{\sigma_U^2(\sigma_V^2 + \sigma_2^2)} = \frac{-\gamma\sigma_2^2}{(\sigma_V^2 + \sigma_2^2)}. \quad (7.7b)$$

Equation (7.7a) implies that $\hat{\beta}$ is an asymptotic unbiased estimator of β, while Eq. (7.7b) implies that $\hat{\gamma}$ is downward-biased estimator of γ.

(ii) If $\sigma_1^2 > 0, \sigma_2^2 = 0,$

$$\text{(a)} \quad \text{plim}\,\hat{\beta} - \beta = \frac{-\beta\sigma_1^2}{\sigma_V^2(\sigma_U^2 + \sigma_1^2)}, \quad (7.8a)$$

$$\text{(b)} \quad \text{plim}\,\hat{\gamma} - \gamma = \frac{\sigma_1^2(\sigma_{WV} - \gamma\sigma_V^2)}{\sigma_V^2(\sigma_U^2 + \sigma_1^2)} = 0. \quad (7.8b)$$

In accordance with Eqs. (7.8a, 7.8b) can be used to draw some meaningful conclusions about the biases of both $\hat{\beta}$ and $\hat{\gamma}$.

(iii) Finally, if $\sigma_1^2 > 0, \sigma_2^2 > 0,$

$$\text{(a)} \quad \text{plim}\,\hat{\beta} - \beta = -\frac{\beta\sigma_1^2}{\sigma_U^2 + \sigma_1^2}, \quad (7.9a)$$

$$\text{(b)} \quad \text{plim}\,\hat{\gamma} - \gamma = -\frac{\gamma\sigma_2^2}{\sigma_V^2 + \sigma_2^2}. \quad (7.9b)$$

In this case, both $\hat{\beta}$ and $\hat{\gamma}$ are downward-biased estimators of β and γ.

- Suppose that $\text{Cov}(UV) \neq 0$

(i) If $\sigma_1^2 = 0, \sigma_2^2 > 0,$

$$\text{(a)} \quad \text{plim}\,\hat{\beta} - \beta = \gamma b_{VU}\left(\frac{\sigma_2^2}{\sigma_2^2 + \sigma_V^2(1 - R_{UV}^2)}\right), \quad (7.10a)$$

$$\text{(b)} \quad \text{plim}\,\hat{\gamma} - \gamma = \frac{-\gamma\sigma_2^2}{\sigma_V^2 - b_{VU} + \sigma_2^2}, \quad (7.10b)$$

where b_{VU} is the auxiliary regression coefficient of a regressing V on U, and R_{UV}^2 is the correlation coefficient between U and V.

Equation (7.10a) implies that the direction of the bias of $\hat{\beta}$ depends upon the sign of both γ and b_{VU}; Eq. (7.10b) implies that γ is a downward-biased estimator of γ unless $(\sigma_V^2 - b_{VU} + \sigma_2^2)$ is greater than zero.

(ii) If $\sigma_1^2 > 0, \sigma_2^2 = 0,$

$$\text{(a)} \quad \text{plim}\,\hat{\beta} - \beta = \frac{-\beta\sigma_1^2}{\sigma_U^2 - b_{UV} + \sigma_1^2}, \quad (7.11a)$$

$$\text{(b)} \quad \text{plim}\,\hat{\gamma} - \gamma = \beta b_{UV}\left(\frac{\sigma_1^2}{\sigma_1^2 + \sigma_U^2(1 - R_{UV}^2)}\right) = 0, \quad (7.11b)$$

where b_{UV} = the auxiliary regression coefficient of regressing U on V.

(iii) If $\sigma_1^2 > 0$, $\sigma_2^2 > 0$,

(a)
$$\text{plim}\,\hat{\beta} - \beta = \frac{\gamma b_{VU} - \frac{\beta}{\sigma_U^2}(\sigma_V^2\sigma_1^2 + \sigma_1^2\sigma_2^2)}{\sigma_V^2 - b_{UV} + \sigma_2^2 + \frac{(\sigma_V^2\sigma_1^2 + \sigma_1^2\sigma_2^2)}{\sigma_U^2}}, \tag{7.12a}$$

(b)
$$\text{plim}\,\hat{\gamma} - \gamma = \frac{\beta b_{UV} - \frac{\gamma}{\sigma_V^2}(\sigma_U^2\sigma_2^2 + \sigma_1^2\sigma_2^2)}{\sigma_U^2 - b_{UV} + \sigma_1^2 + \frac{\sigma_V^2\sigma_1^2 + \sigma_1^2\sigma_2^2}{\sigma_V^2}}. \tag{7.12b}$$

From Eqs. (7.12a) and (7.12b), we can see that the directions of the biases of both $\hat{\beta}$ and $\hat{\gamma}$ are ambiguous.

7.2.2.2 The Constrained Classical Case

If we impose $\beta + \gamma = 1$ and assume that only V is measured with error, then (7.10a) should be rewritten as

$$\text{plim}\,\hat{\beta} - \beta = \frac{\sigma_\eta^2(1-\beta)}{\sigma_U^2 + \sigma_V^2 - 2\sigma_{UV} + \sigma_\eta^2} \tag{7.13}$$

The asymptotic bias of this case can be analyzed as follows:

(i) If $\beta > 0$, then $\hat{\beta}$ is unbiased when $\beta = 1$, $\hat{\beta}$ is upward-biased when $\beta < 1$ and $\hat{\beta}$ is downward-biased when $\beta > 1$.
(ii) If $\beta < 0$, then $\hat{\beta}$ is always upward-biased.

Concerning the coefficient of the reliability, Cochran (1970) has shown that measurement errors of both explained and explanatory variables will reduce the multiple correlations and increase the residual variance. In addition, he also points out that the good prediction formula is more sensitive to measurement errors than the poor one. Moreover, from the analysis of the effects of measurement error on both the simple regression coefficient and residual variance, in general, we can conclude that the *t*-statistic of the simple regression coefficient will be downward-biased if variables are measured with errors.

7.3 Estimation Methods When Variables Are Subject to Error

In this section, we will discuss alternative EIV estimation methods, classical method, grouping method, instrumental variable method, mathematical method, maxima likelihood method, LISREL method, and the Bayesian approach. In addition, we will extend from simple regression to multiple regression.

7.3.1 Classical Estimation Method

7.3.1.1 The Classical Method to a Simple Regression Analysis

In general, the classical method considers three cases: (i) either σ_1^2 or σ_2^2 is known; (ii) $\lambda = \frac{\sigma_2^2}{\sigma_1^2}$ is known; and (iii) σ_1^2 and σ_2^2 are known. We can obtain the estimate for β from Eqs. (7.2a) under every possible situation as:

(i)
$$\hat{\beta} = \frac{S_{XY}}{S_{XX} - \sigma_1^2}, \text{ when } \sigma_1^2 \text{ is known.} \tag{7.14}$$

$$\hat{\beta} = \frac{S_{YY} - \sigma_2^2}{S_{XY}}, \text{ when } \sigma_2^2 \text{ is known.} \tag{7.15}$$

(ii)
$$\hat{\beta} = \frac{(S_{YY} - \lambda S_{XX}) + \left\{(S_{YY} - \lambda S_{XX})^2 + 4\lambda S_{XY}\right\}^{1/2}}{2S_{XY}},$$
$$\text{when } \lambda = \frac{\sigma_2^2}{\sigma_1^2} \text{ is known.} \tag{7.16}$$

(iii) When both σ_1^2 and σ_2^2 are known, Kendall and Stuart (1961) regarded it as an overidentified situation unless a nonzero covariance between U_i and V_i is introduced. Bartnett (1967) followed Kiefer's

(1964) suggestion and derived a consistent estimator of $\hat{\beta}$ as one of the real roots of Eq. (7.17).

$$\hat{\beta}^4 - \left(\frac{1}{b_2} - \frac{\lambda}{b_1} - 2\lambda b_2\right)\hat{\beta}^3 - 3\lambda\left(1 - \lambda\frac{b_2}{b_1}\right)\hat{\beta}^2 + \lambda^2\left(\lambda\frac{b_2}{b_1^2} - \frac{1}{b_1} - 2b_2\right)\hat{\beta} - \lambda^3\frac{b_2}{b_1} = 0, \tag{7.17}$$

where $S_{XX} = \frac{\sum_{i=1}^n (X_i - \bar{X})^2}{n}$, $S_{YY} = \frac{\sum_{i=1}^n (Y_i - \bar{Y})^2}{n}$, $S_{XY} = \frac{\sum_{i=1}^n (X_i - \bar{X})(Y_i - \bar{Y})}{n}$, $m_{XX} = \mathrm{Var}(X) = \sigma_U^2 + \sigma_1^2$, $m_{YY} = \mathrm{Var}(Y) = \beta^2\sigma_U^2 + \sigma_2^2$, $m_{XY} = \mathrm{Var}(Y) = \beta\sigma_U^2$, $b_1 = \frac{m_{XY}}{m_{XX}}$, $b_2 = \frac{m_{XY}}{m_{YY}}$, and $\lambda = \frac{\sigma_2^2}{\sigma_1^2}$.

It has been shown that the meaningful solution of Eq. (7.17) is identical to Eq. (7.12a). The advantage of the knowledge of both σ_1^2 and σ_2^2 is that one obtains a more efficient estimator of $\hat{\sigma}_2^2$.

The analysis of Eqs. (7.14)–(7.17) furnishes us two important implications:

(i) If only U or V is subject to measurement error, then we know that the maxima likelihood estimator of β is equivalent to fitting a least squares line when using the error-free variable as a regressor; (ii) if both U and V are measured with errors, then the estimate of β lies between the values estimated from Eqs. (7.14) and (7.15). This situation can be further analyzed. The quadratic equation for Eq. (7.16) is

$$\hat{\beta}^2 S_{XY} + \hat{\beta}(\lambda S_{XX} - S_{YY}) - \lambda S_{XY} = 0 \tag{7.18}$$

(a) When either $\sigma_2^2 \to 0$ or $\sigma_1^2 \to \infty$, Eq. (7.15) shows $\hat{\beta} = \frac{S_{YY}}{S_{XY}}$.
(b) When $\sigma_1^2 \to 0$ or $\sigma_2^2 \to \infty$, Eq. (7.14) shows $\hat{\beta} = \frac{S_{XY}}{S_{XX}}$.

7.3.1.2 The Classical Method to a Multiple Regression Analysis

It is clear that U is orthogonal to V if $\rho = 0$. It is well known that this multiple regression reduces to two simple regression relationships. If $\rho \neq 0$, to identify β and γ, we need to know either the actual values of σ_1^2, σ_2^2, and σ_3^2 or the relative ratios among σ_1^2, σ_2^2, and σ_3^2. We will investigate the following cases:

(i) $\sigma_2^2 = 0,\ \sigma_1^2 > 0,\ \sigma_3^2 < 0$, and $\sigma_1^2 = \lambda\sigma_3^2$

(ii) $\sigma_1^2 > 0,\ \sigma_2^2 > 0$, and $\sigma_3^2 > 0$

(a) σ_1^2 and σ_2^2 are known
(b) $\sigma_2^2 = \lambda_1\sigma_1^2$ and $\sigma_3^2 = \lambda_2\sigma_1^2$

(Case i) Only Z and X are measured with errors.

The estimator of $\hat{\beta}$ can be one of the real roots of Eq. (7.18)

$$k_2\hat{\beta}^2 + k_1\hat{\beta} + k_0 = 0, \tag{7.19}$$

where

$$k_0 = \left(S_{XZ} - \frac{S_{XY}S_{YZ}}{S_{YY}}\right),$$

$$k_1 = -\left(S_{XX} - \frac{S_{XY}^2}{S_{YY}}\right) + \lambda\left(S_{ZZ} - \frac{S_{YZ}^2}{S_{YY}}\right), \text{ and}$$

$$k_2 = -\lambda k_0 = -\lambda\left(S_{XZ} - \frac{S_{XY}S_{YZ}}{S_{YY}}\right).$$

There are three cases to consider:

(a) $\lambda \to 0$ when $\sigma_1^2 \to 0$, then in this case from Eq. (7.19), we know that

$$\hat{\beta} = \frac{-S_{XY}S_{YZ} + S_{XZ}S_{YY}}{S_{XX}S_{YY} - S_{XY}^2} \tag{7.20}$$

(b) $\lambda \to \infty$ when $\sigma_3^2 \to 0$, in this case from (7.20), we know that

$$\hat{\beta} = \frac{-S_{ZZ}S_{YY} - S_{YZ}^2}{S_{XZ}S_{YY} + S_{XY}S_{YZ}} \quad (7.21)$$

(c) When both $\sigma_1^2 > 0$ and $\sigma_3^2 > 0$, then

$$\hat{\beta} = \frac{k_1 \pm \sqrt{k_1^2 + 4\lambda k_O^2}}{2\lambda k_O} \quad (7.22)$$

When β is determined, γ can also be estimated. After both β and γ are estimated, then α can be estimated by

$$\hat{\alpha} = \bar{Z} - \hat{\beta}\bar{X} - \hat{\gamma}\bar{Y}. \quad (7.23)$$

(Case ii) When Z, X, and Y are all observed with errors

(a) Both σ_1^2 and σ_2^2 are known.

We can obtain the two normal equations as follows:

$$S_{XZ} = \beta(S_{XX} - \hat{\sigma}_1^2) + \gamma S_{XY} \quad (7.24)$$

$$S_{YZ} = \beta S_{XZ} + \gamma(S_{YY} - \hat{\sigma}_2^2) \quad (7.25)$$

Solving Eqs. (7.24) and (7.25) by Cramer's rule, we have

$$\hat{\beta} = \frac{S_{XZ}S_{YY} - S_{XY}S_{YZ} - S_{XZ}\hat{\sigma}_2^2}{S_{XX}S_{YY} - \hat{\sigma}_1^2 S_{YY} - \hat{\sigma}_2^2 S_{XX} + \hat{\sigma}_1^2\hat{\sigma}_2^2 - (S_{XY})^2}$$
$$\hat{\gamma} = \frac{S_{YZ}S_{XX} - S_{XZ}S_{YY} - S_{YZ}\hat{\sigma}_1^2}{S_{XX}S_{YY} - \hat{\sigma}_1^2 S_{YY} - \hat{\sigma}_2^2 S_{XX} + \hat{\sigma}_1^2\hat{\sigma}_2^2 - (S_{XY})^2} \quad (7.26)$$

(b) Both $\hat{\sigma}_2^2 = \lambda_1\hat{\sigma}_1^2$ and $\hat{\sigma}_3^2 = \lambda_2\hat{\sigma}_1^2$ are known.

We can obtain β estimator as one of the real roots of the following cubic equation

$$\hat{\beta}^3 H_3 + \hat{\beta}^2 H_2 + \hat{\beta} H_1 + H_O = 0, \quad (7.27)$$

where

$$\begin{aligned}
H_3 &= S_{XY}S_{YZ}^2 - MS_{XZ}S_{YZ} - \lambda_1 S_{XY}S_{XZ}^2 \\
H_2 &= -S_{YZ}^2 + 2\lambda_2 S_{XY}^2 S_{YZ} + MTS_{YZ} - M\lambda_2 S_{XY}S_{YZ} \\
&\quad - \lambda_1 S_{XZ}^2 S_{YZ} + 2\lambda_1 TS_{XZ}S_{XY} \\
H_1 &= \lambda_2^2 S_{XY}^3 - 2\lambda_2 S_{XY}S_{YZ}^2 + MT\lambda_2 S_{XY} + \lambda_2 S_{XY}S_{YZ} \\
&\quad + \lambda_2 MS_{YZ}S_{XZ} - \lambda_1\lambda_2 S_{XZ}^2 S_{XY} \\
&\quad + 2\lambda_1 S_{XY}S_{XZ}^2 - \lambda_1 T^2 S_{XY} \\
H_O &= \lambda_2^2 S_{YZ}S_{XY}^2 + \lambda_2^2 S_{XZ}S_{XY} + \lambda_1\lambda_2 S_{XZ}^2 S_{YZ} \\
&\quad + T\lambda_1\lambda_2 S_{XZ}S_{YZ} - 2\lambda_1\lambda_2 TS_{XZ}S_{XY}
\end{aligned}$$

When $\lambda_2 = 0$, Eq. (7.27) will reduce to a quadratic equation in $\hat{\beta}$.

7.3.1.3 The Constrained Classical Method

Under the classical case (Case ii), if we only know $\hat{\sigma}_2 = \hat{\sigma}_1\lambda$, then we can identify β and γ by imposing $\beta + \gamma = 1$.

We can obtain a quadratic equation in $\hat{\beta}$

$$\begin{aligned}
&\hat{\beta}^2(1-\lambda)S_{XY} + \hat{\beta}(S_{YZ} - \lambda S_{XZ} + 2\lambda S_{XY} + S_{YY} \\
&\quad - \lambda S_{XX}) + \lambda(S_{XZ} - S_{XY}) \\
&= 0.
\end{aligned} \quad (7.28)$$

When $\lambda = 0$, Eq. (7.28) will reduce to

$$\begin{aligned}
\hat{\beta}S_{XY} &= -[S_{YZ} + S_{YY}] \\
\text{i.e. } \hat{\beta} &= -\frac{S_{YZ} + S_{YY}}{S_{XY}}
\end{aligned} \quad (7.29)$$

Imposing $\beta + \gamma = 1$, upon a multiple regression will help to identify the regression coefficients, but it should also be realized that the constrained regression technique will bias the estimates of the regression coefficients if the unrestricted estimator fails to satisfy the restriction $\beta + \gamma = 1$. The advantages and disadvantages of the constrained regression technique have been discussed by Theil (1971) in some detail.

7.3.2 Grouping Method

Following the structural relationship described in Eq. (7.29)

$$V_i = a + bU_i. \tag{7.30}$$

Both V_i and U_i are unobserved, and only $Y_i = V_i + \eta_i$ and $X_i = U_i + \varepsilon_i$ can be observed. There exists EIV bias when using Y_i and X_i, to investigate the relationship between V_i and U_i.

Wald (1940) proposes a two-portfolio grouping method in dealing with the EIV problem when both dependent and independent variables are subject to measurement errors. He suggests that the measurement error can be reduced by grouping observations into portfolios. In Wald's two-portfolio grouping method, he groups the independent variable either in descending or ascending order and divides the observations into two equal groups for both dependent and independent variables; therefore, the first-step estimator of the market model, estimated beta risk, can be written as:

$$\hat{b} = \frac{(\bar{Y}_1 - \bar{Y}_2)}{(\bar{X}_1 - \bar{X}_2)}, \tag{7.31}$$

where $\bar{X}_1$ and $\bar{X}_2$ are the arithmetic means of independent variables for the first and the second groups, respectively, and $\bar{Y}_1$ and $\bar{Y}_2$ are the arithmetic means of independent variables for the first and the second groups, respectively.

Grouping method is widely used in finance-related research. For example, to minimize the EIV problem in testing the asset pricing model, Black et al. (1972), Blume and Friend (1973), Fama and MacBeth (1973), and Litzenberger and Ramaswamy (1979) use two-pass procedure and the k-portfolio grouping method to examine the capital asset pricing model. By combining securities into portfolios, most of the firm-specific component of the returns can be diversified away and the precision of the beta estimates will be enhanced. The grouping method can, therefore, mitigate the problem raised from measurement errors in estimated beta.

However, some limitations affect the grouping method. First, the grouping method shrinks the range of estimators in the first step and reduces statistical power. To mitigate this problem, in two-pass procedure, the grouping method suggests sorting securities on the first-pass estimator first. Then, portfolios are formed by grouping securities with same level of first-pass estimators. This sorting procedure is now standard in empirical tests. Second, a trade-off exists between the bias and the variance of the first-pass estimator according to the number of portfolios. Shanken (1992) argues that the grouping method may cause a larger variation in the portfolio beta. As the number of portfolios (N) increases, the magnitude of the bias becomes greater, while the variance of the estimator becomes smaller, and vice versa; therefore, an optimal number of portfolios might exist in which a minimum mean squared error can be obtained. More specifically, when risk premium is estimated by the time-series mean of the cross-sectional regression estimates in testing capital asset pricing model, the mean squared error of the risk premium estimate would be dominated by its bias because its variance would monotonically decrease as the testing period becomes longer. Third, the formation of portfolios for the second-pass estimation might cause a loss of valuable information about cross-sectional behavior among individual securities, because the cross-sectional variations would be smoothed out. Fourth, Ahn et al. (2009) argue that the grouping method, although mitigating measurement error, may yield different results by using different portfolio grouping methods.

Although the grouping method suffers from the limitations discussed above, it still has some clear advantages. With the cross-sectional regression in the second pass, interpreting the results in economic terms is straightforward. Examining model misspecification by testing whether firm characteristics, such as firm size and book-to-market ratio, can explain returns across firms is also convenient. Moreover, the grouping method is intuitive and easy to implement with real data. The grouping method is therefore still preferred in many empirical studies.

7.3.3 Instrumental Variable Method

Durbin (1954) proposes an instrumental variable method to deal with the EIV problem in a regression model. In the instrumental variable method, the instrumental variable, T_i, is an observable variable known to correlate with V_i and U_i, but is independent of η_i and ε_i. Then, b can be estimated by

$$\hat{b} = \frac{\sum_{i=1}^{n}(T_i - \bar{T})(Y_i - \bar{Y})}{\sum_{i=1}^{n}(T_i - \bar{T})(X_i - \bar{X})} = \frac{\sum_{i=1}^{n}(T_i - \bar{T})(U_i - \bar{U}) + \sum_{i=1}^{n}(T_i - \bar{T})(\eta_i - \bar{\eta})}{\sum_{i=1}^{n}(T_i - \bar{T})(V_i - \bar{V}) + \sum_{i=1}^{n}(T_i - \bar{T})(\varepsilon_i - \bar{\varepsilon})}. \tag{7.32}$$

If $p\lim \sum_{i=1}^{n}(T_i - \bar{T})(U_i - \bar{U})$ exists, then $\hat{b}$ is a consistent estimator of b because both ε_i and η_i are independent of T_i. Equation (7.32) can be written in matrix form as follows:

$$\begin{bmatrix} \hat{\alpha} \\ \hat{\beta} \end{bmatrix} = (\mathbf{T}'\mathbf{X})^{-1}\mathbf{T}'\mathbf{Y}, \tag{7.33}$$

where

$$\mathbf{T}' = \begin{bmatrix} 1 & 1 & 1 & \cdots & 1 \\ T_1 & T_2 & T_3 & \cdots & T_n \end{bmatrix},$$
$$\mathbf{X}' = \begin{bmatrix} 1 & 1 & 1 & \cdots & 1 \\ X_1 & X_2 & X_3 & \cdots & X_n \end{bmatrix}, \text{ and}$$
$$\mathbf{Y}' = [Y_1 \quad Y_2 \quad Y_3 \quad \cdots \quad Y_n].$$

However, finding an instrumental variable uncorrelated with η_i and ε_i while highly correlated with V_i and U_i is difficult. Durbin (1954) suggests that if the order of U_i is the same as the order of X_i, then a better instrumental variable would be $T_i = i$, where X_i are ordered by magnitude. That is, $\mathbf{T}' = \begin{bmatrix} 1 & 1 & 1 & \cdots & 1 \\ 1 & 2 & 3 & \cdots & n \end{bmatrix}$ and $\mathbf{X}' = \begin{bmatrix} 1 & 1 & 1 & \cdots & 1 \\ X_1 & X_2 & X_3 & \cdots & X_n \end{bmatrix}$. This variable will lead to a more efficient estimate than that of the method of grouping. If we let $T_i = 1$ for X_i greater than its median, $T_i = 0$ for X_i equal to its median, and $T_i = -1$ for X_i smaller than its median, then the estimator of the instrumental variable method will be the same as the estimator of Wald's two-group grouping method. Therefore, Wald's two-group grouping method is a special case of the instrumental variable method. In other words, the instrumental variable method is more generalized than the grouping method.

Griliches and Hausman (1986) propose an instrumental variable approach to reduce the bias resulted from measurement error. In a penal data framework, they show that instrumental variable estimator is consistent if the measurement error $\varepsilon_{i,t}$ is *i.i.d* across i and t and unobserved independent variable is serially correlated. An instrumental variable using the lags or the difference of lags of the unobserved independent variable can result in a consistent estimator when T is finite and N approaches to infinite.

However, Griliches and Hausman's *i.i.d* assumption is too strong. Biorn (2000) further relaxes Griliches and Hausman's *i.i.d* assumption and, instead, assumes that the regressor has τ period moving-average process. Biorn (2000) shows that using the lags of the variables at least $\tau - 2$ periods as instruments can clear the memory of the moving-average process and obtain the consistent estimator.

Lewbel (1997) show that simple functions of the model data can be used as instruments for two-stage least squares (TSLS) estimation. Such instruments can be used for identification and estimation when no other instruments are available or improve efficiency.

Given the standard linear regression model with measurement error:

$$Y_i = a + b'W_i + cX_i + e_i, \text{ and} \tag{7.34}$$

$$Z_i = d + X_i + v_i, \tag{7.35}$$

in which Y_i, W_i, and Z_i are observable for $i = 1, \ldots, n$, while X_i, e_i, and v_i are unobservable. Equations (7.34) and (7.35) imply that

$$Y_i = \alpha + b'W_i + cZ_i + \varepsilon_i. \quad (7.36)$$

However, since both Z_i and ε_i depend on v_i, estimators of b and c from OLS regression are inconsistent. Lewbel (1997) shows that the consistent estimators can be obtained by using TSLS with instruments 1, W_i, and q, where q is some vector of instruments that are correlated with X_i but not correlated with e_i and v_i. q can be chosen as

$$\begin{aligned} q_{1i} &= (G_i - \bar{G}), \\ q_{2i} &= (G_i - \bar{G})(Z_i - \bar{Z}), \\ q_{3i} &= (G_i - \bar{G})(Y_i - \bar{Y}), \\ q_{4i} &= (Y_i - \bar{Y})(Z_i - \bar{Z}), \\ q_{5i} &= (Z_i - \bar{Z})^2, \text{ and} \\ q_{6i} &= (Y_i - \bar{Y})^2, \end{aligned}$$

where $G_i = G(W_i)$ for any given function G and $\bar{S}$ denote the sample mean of variables S_i (S_i can be Y_i, Z_i, or G_i).

Lewbel (1997) further empirically applies the instrumental variable method to testing elasticity of patent applications with respect to research and development (R&D) expenditures. He finds, using the TSLS instrumental variable model, the estimated elasticity yields very close to one. Therefore, the TSLS instrumental variable model can mitigate the effects of measurement error and confirm the relationship between patent and R&D.

In addition, Erickson and Whited (2000, 2002) propose a two-step generalized method of moment (GMM) estimators that exploit overidentifying information contained in the high-order moments of residuals obtained from perfectly measured regressors. Basing GMM estimation on residual moments of more than second order requires that the GMM covariance matrix be explicitly adjusted to account for the fact that estimated residuals are used instead of true residuals defined by population regressions. Erickson and Whited (2000) show that estimators obtained by using moments up to seventh order perform well in Monte Carlo simulations.

Almeida et al. (2010) use Monte Carlo simulations and empirically test investment models to compare the performance of the instrumental variables approach suggested by Biorn (2000) and generalized method of moments. They find that the instrumental variable method can obtain more consistent and efficient estimators than generalized method of moments when independent variables subject to measurement error.

However, it is difficult to obtain appropriate instrument variables, resulting in weak evidence in empirical research. Lewbel (2012) proposes a new method to deal with measurement error problems in regression model when instrumental variables are not available. Under the assumption of heteroscedastic errors, Lewbel (2012) shows that the regression model with measurement regressors can be identified and estimated by TSLS or GMM.

7.3.4 Mathematical Method

7.3.4.1 Bivariate Case

Deming (1943), York (1966), and Clutton-Brock (1967) have developed a weighted-regression-method-under-iteration approach. Deming (1943) proposed that the best straight line of Eq. (7.1) can be obtained by minimizing the sum in the following equation:

$$S = \sum_i \left\{ w(X_i)\left(\hat{U}_i - X_i\right)^2 + w(Y_i)\left(\hat{V}_i - Y_i\right)^2 \right\} \quad (7.37)$$

$\hat{U}_i$ and $\hat{V}_i$ are the adjusted value of X_i and Y_i which make the sum in Eq. (7.37) a minimum. Since we require $\hat{U}_i$ and $\hat{V}_i$ to lie on the best straight line, we must have

$$\hat{V}_i = \alpha + \beta \hat{U}_i, \quad (i = 1, \cdots, n) \quad (7.38)$$

Both $w(X_i)$ and $w(Y_i)$ are the weights of various observations. They are reciprocally proportional to the variance of their measurement error, respectively.

If these values of $\hat{U}_i$, $\hat{V}_i$, α, and β make S a minimum, we have

$$\beta^3 \sum_i \frac{k_i^2 x_i^2}{w(X_i)} - 2\beta^2 \sum_i \frac{k_i^2 x_i y_i}{w(X_i)} - \beta\left\{\sum_i k_i x_i^2 - \sum_i \frac{k_i^2 y_i^2}{w(X_i)}\right\} + \sum_i k_i x_i y_i = 0 \quad (7.39)$$

where $x_i = X_i - \bar{X}$, $y_i = Y_i - \bar{Y}$, $\bar{X} = \frac{\sum_i k_i X_i}{\sum_i k_i}$, $\bar{Y} = \frac{\sum_i k_i Y_i}{\sum_i k_i}$, and $k_i = \frac{w(X_i)w(Y_i)}{\beta^2 w(Y_i) + w(X_i)}$.

Equation (7.39) is the least squares cubic derived by York (1966). To solve Eq. (7.39), an initial value is assigned to β to estimate k_i. After obtaining the roots of Eq. (7.39), one of the legitimate solutions is assigned to estimate k_i and obtain new solutions for β again. A similar procedure is employed iteratively until a convergent solution is obtained.

The mathematical approach involves the estimation of the parameters of a function conditional on the maximum likelihood function adjusted for the true values. This method is different from the classical method in three ways. First, variances of measurement errors for every observation are different. Second, a weighted regression method is applied. Third, the iteration procedure is used to obtain a consistent estimator.

It can be proved that the mathematical programming method reduces to the classical method under certain conditions.

Three special cases for the observations X_i and Y_i are discussed in the following.

(i) Only Y_i has an EIV problem

In this case, we put more weight on X_i which has no EIV problem, and then $w(X_i) = \infty$, $k_i = w(Y_i)$. We can solve the least squares cubic

$$\beta = \frac{\sum_i w(Y_i) x_i y_i}{\sum_i w(Y_i) x_i^2}, \quad (7.40)$$

which is the estimated coefficient of weighted regression of Y_i on X_i.

(ii) Only X_i has an EIV problem

In this case, we put more weight on Y_i which has no EIV problem, and then $w(Y_i) = \infty$, $k_i = \frac{w(X_i)}{\beta^2}$. We can solve the least squares cubic

$$\beta = \frac{\sum_i w(X_i) y_i^2}{\sum_i w(X_i) x_i y_i}, \quad (7.41)$$

which is the inverse estimated coefficient of weighted regression of Y_i on X_i.

(iii) Both X_i and Y_i have EIV problem, and $w(X_i)/w(Y_i) = c$.

The least squares cubic becomes

$$\beta^2 + \beta \frac{\{c \sum_i k_i x_i^2 - \sum_i k_i y_i^2\}}{\sum_i k_i x_i y_i} - c = 0 \quad (7.42)$$

7.3.4.2 Multivariate Case

Lee (1973) extends the bivariate mathematical programming method (which was developed by Deming (1943), York (1966), and Clutton-Brock (1967), to a trivariate case. We define $w(Z_i)$, $w(X_i)$, and $w(Y_i)$ which are the weights of the various observations of Z_i, X_i, and Y_i. It is assumed W, U, and V are functionally rather than structurally related. The mathematical programming procedure begins by minimizing

$$S = \sum_i \left\{ w(X_i)(x_i - X_i)^2 + w(Y_i)(y_i - Y_i)^2 + w(Z_i)(z_i - Z_i)^2 \right\}$$
$$\text{s.t. } z_i = \alpha + \beta x_i + \gamma y_i \quad (7.43)$$

Solving Eq. (7.43), we can obtain

$$\beta^3 C_3 + 2\beta^2 C_2 + \beta C_1 + C_O = 0, \quad (7.44)$$

where

$$
\begin{aligned}
C_3 &= \sum_i \frac{K_i^2}{w(X_i)} \left(Q_i - \frac{R_i \sum K_i Q_i P_i}{\sum K_i R_i^2} \right)^2 \\
C_2 &= \sum_i \frac{K_i^2}{w(X_i)} \left(Q_i - R_i \frac{\sum K_i Q_i R_i}{\sum K_i R_i^2} \right) \\
&\quad \left(P_i - R_i \frac{\sum K_i R_i P_i}{\sum K_i R_i^2} \right) \\
C_1 &= \sum K_i P_i^2 - \frac{(\sum K_i Q_i R_i)^2}{\sum K_i R_i^2} - \sum \left(\frac{K_i^2}{w(X_i)} \right) \\
&\quad \left(\frac{R_i \sum K_i R_i P_i}{\sum K_i^2} - P_i \right)^2 \\
C_O &= \frac{\sum K_i Q_i R_i \sum K_i P_i Q_i}{\sum K_i R_i^2} - \sum K_i P_i Q_i \\
K_i &= \frac{w(X_i) w(Y_i) w(Z_i)}{w(X_i) w(Y_i) + \beta^2 w(Z_i) w(Y_i) + \gamma^2 w(Z_i) w(X_i)} \\
\bar{X} &= \frac{\sum_i X_i K_i}{\sum_i K_i},\ \bar{Y} = \frac{\sum_i K_i Y_i}{\sum_i K_i},\ \bar{Z} = \frac{\sum_i K_i Z_i}{\sum K_i} \\
P_i &= Z_i - \bar{Z},\ Q_i = X_i - \bar{X}, \\
&\text{and } R_i = Y_i - \bar{Y}
\end{aligned}
$$

This extension will reduce to Deming's (1943) weighted regression results when the quadratic term of equations is omitted, while Lee's (1973) result is more general than Deming's weighted multiple regression analysis.

7.3.5 Maximum Likelihood Method

In testing capital asset pricing model with dividend and tax, Litzenberger and Ramaswamy (1979), instead of using the grouping method or instrumental variables method, use maximum likelihood method to reduce the effect of errors-in-variables. Litzenberger and Rmaswamy (1979) show that, assuming that the variance of the measurement error in beta is known, the cross-sectional variance of true betas can be replaced by the difference in the variation of the observed betas and the variance of the measurement error. Then, the estimator in capital asset pricing model test, under such condition, is consistent by maximum likelihood method.

Kim (1995, 1997, 2010) further provides a maximum likelihood method to correct the EIV problem in testing the asset pricing model. Based upon two-pass capital asset pricing model, Kim (1995) shows that in a multifactor asset pricing model test the EIV leads to an underestimation of the independent variable with a measurement error and an overestimation of the independent variable without measurement error. To correct EIV biases, Kim (1995) extracts additional information about the relation between idiosyncratic error variance which can be obtained from the first step and the measurement error variance and incorporates such additional information into the second step of the capital asset pricing model test. Assuming the homoscedasticity of the disturbance term of the market model, Kim (1995) shows that the corrected factors for the traditional least squares estimators of the cross-sectional regression coefficients can be obtained by the maximum likelihood method. The closed-form estimators of the multifactor asset pricing model test can therefore be obtained. Assuming the first and second steps of the multifactor asset pricing model are

$$R_{i,t} = \alpha_i + \beta_i R_{m,t} + e_{i,t}, \tag{7.45}$$

and

$$R_{i,t} = \gamma_{0,t} + \gamma_{1,t} \hat{\beta}_{i,t-1} + \gamma_{2,t} V_{i,t-1} + e_{i,t}, \tag{7.46}$$

where $\beta_{i,t-1}$ is the market risk factor with measurement error, and $V_{i,t-1}$ is a risk factor with no measurement error for security i at time $t-1$. The adjusted estimators in the second step can be written as follows:

$$
\begin{aligned}
\hat{\gamma}_{1t} &= \frac{M + \left[M^2 + 4\delta_t m_{R\hat{\beta}}^2 \left(1 - \left(\hat{\rho}_{RV} \hat{\rho}_{\hat{\beta}V} / \hat{\rho}_{R\hat{\beta}} \right) \right)^2 \right]^{1/2}}{2 m_{R\hat{\beta}} \left(1 - \left(\hat{\rho}_{RV} \hat{\rho}_{\hat{\beta}V} / \hat{\rho}_{R\hat{\beta}} \right) \right)} \\
\hat{\gamma}_{2t} &= \left(m_{RV} - \hat{\gamma}_{1t} m_{\hat{\beta}V} \right) / m_{VV} \\
\hat{\gamma}_{0t} &= \tilde{R}_t - \hat{\gamma}_{1t} \tilde{\hat{\beta}}_{t-1} - \hat{\gamma}_{2t} \tilde{V}_{t-1}
\end{aligned}
\tag{7.47}
$$

where $M = m_{RR}(1-\hat{\rho}_{RV}^2) - \delta_t m_{\hat{\beta}\hat{\beta}}\left(1-\hat{\rho}_{\hat{\beta}V}^2\right)$,

$$m_{xy} = (1/N)\sum_{i=1}^{N}$$

$$\sum_{j=1}^{N} w_{ij}(x_i-\bar{x})(y_i-\bar{y})/\sum_{i=1}^{N}\sum_{j=1}^{N} w_{ij},$$

$$\bar{x} = \sum_{i=1}^{N}\sum_{j=1}^{N} w_{ij}x_i/\sum_{i=1}^{N}\sum_{j=1}^{N} w_{ij},$$

$$\bar{y} = \sum_{i=1}^{N}\sum_{j=1}^{N} w_{ij}\bar{x}_i/\sum_{i=1}^{N}\sum_{j=1}^{N} w_{ij},$$

w_{ij} is the (*i*, *j*) element of inverse matrix of residual variance in the first step, $\hat{\Sigma}_{\varepsilon}^{-1}$, and

$\hat{\rho}_{xy}^2 = m_{xy}/(m_{xx}m_{yy})^{1/2}$.

As a result, the maxima likelihood method can correct the problem on exaggerating the estimated coefficient associated with the variable without measurement error. Moreover, the absolute value of estimated intercept by maxima likelihood method is generally smaller than the absolute value of estimated intercept by traditional least squares.

7.3.6 LISREL and MIMIC Methods

Goldberger (1972) conceptually described the LISREL model as a combination of factor analysis and econometrics model. In addition, Anderson (1963) and Mandansky (1964) have shown that factor analysis is a generalized version of errors-in-variables (EIV) methods. In this section, we will review and discuss how LISREL and MIMIC methods can be used to deal with EIV in finance research.

The linear simultaneous equation system is widely used in finance- and accounting-related research. However, a serious limitation of the simultaneous equation approach is an EIV problem. For example, the theoretical determinants of capital structure in corporate finance can be attributed to unobservable constructs that are usually measured in empirical studies by a variety of observable indicators or proxies. These observable indicators or proxies can then be viewed as measures of latent variables with measurement errors. Maddala and Nimalendran (1996) show that the use of these indicators as theoretical explanatory variables may cause EIV problems. Bentler (1983) also emphasizes the estimated results of the traditional simultaneous equation model have no meaning when variables have measurement errors. Therefore, the latent variable covariance structure model is provided and applied in corporate finance. Titman and Wessels (1988), Chang et al. (2009), and Yang et al. (2010) mitigate the measurement problems of proxy variables and apply structure equation models (e.g., LISREL model and MIMIC model) to determine capital structure decision. Maddala and Nimalendran (1996) use the structure equation model to examine the effect of earnings surprises on stock prices, trading volumes, and bid–ask spreads.

Goldberger (1972) and Jöreskog and Goldberger (1975) developed a structure equation model with multiple indicators and multiple causes of a single latent variable, MIMIC mode, and obtained maximum likelihood estimates of parameters. Figure 7.1 shows the path diagram that depicts a simplified MIMIC model in which variables in a rectangular box denote observable variables, while variables in an oval box are latent constructs. In this diagram, observable variables X_1, X_2, and X_3 are causes of the latent variable η, while Y_1, Y_2, and Y_3 are indicators of η. In our study, *X*'s are determinants of capital structure (η), which are then measured by *Y*'s.

7.3.6.1 Structural Model (Lisrel Model)

A structural equation model is composed of two submodels–structural submodel and measurement submodel. The structural model can be defined as

$$\eta = \Gamma X + \zeta, \tag{7.48}$$

$$Y = \Lambda_y \eta + \varepsilon, \tag{7.49}$$

where *Y* is a vector of indicators of the latent variable η, and *X* is a vector of causes of η.

The latent variable η is linearly determined by a set of observable exogenous causes, $X = (x_1, x_2, \ldots, x_q)'$, and a disturbance ζ. The latent variable η, in turn, linearly determines a set of

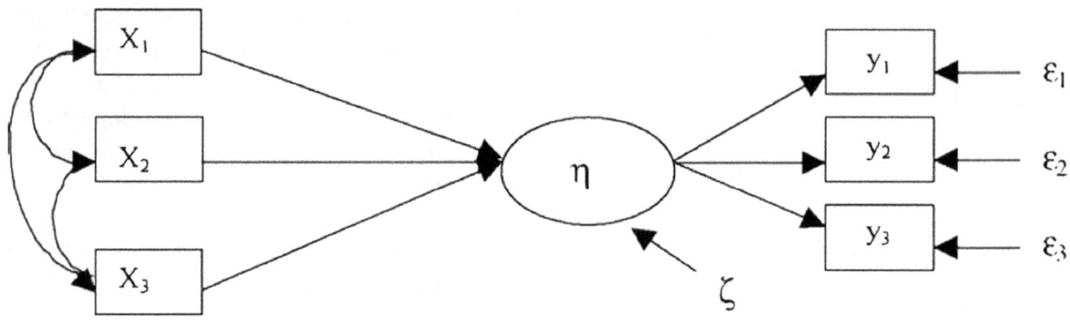

Fig. 7.1 Jöreskog and Sörbom (1978) show that the full structural equation (LIEREL model) can be restricted to a MIMIC model. We here discuss the structural model and show how structural model can be restricted to a MIMIC model

observable endogenous indicators, $Y = (y_1, y_2, \ldots, y_p)'$ and a corresponding set of disturbance, $\varepsilon = (\varepsilon_1, \varepsilon_2, \ldots, \varepsilon_p)'$.[2]

7.3.6.2 Mimic Model

Substituting Eq. (7.48) into Eq. (7.49), we obtain a reduced form:

$$Y = \Lambda_y \eta + \varepsilon = \Lambda_y(\gamma' X + \zeta) + \varepsilon = \Pi' X + z \tag{7.50}$$

In structural equation modeling, the total effect of a cause variable on an indicator can be measured as the sum of the direct effect and the indirect effect. Since a MIMIC model is a reduced form of a structural equation model, the total effect of MIMIC model, denoted as Π' in Eq. (7.50), comes merely from the indirect effect.

Since the scale of the latent variable is unknown, the factor indeterminacy is a common problem in the MIMIC model, as in other structure equation models. We can obtain infinite parameter estimates from the reduced form by arbitrarily changing the scale of the latent variables. However, by fixing the scales of latent variables, one can solve the indeterminacy problem. Two methods are usually adopted to fix the scale of latent variables. One method is normalization in which a unit variance is assigned to each latent variable, while another method is to fix a nonzero coefficient at unity for each latent variable.

In terms of estimation of the parameters, Jöreskog and Goldberger (1975) adopt the normalization method to deal with the factor indeterminacy problem and use maximum likelihood estimation method in structural equation modeling to estimate parameters. The maximum likelihood estimates for the parameters of the model are obtained at the minimization of the fit function as follows:

$$F = \log\|\Sigma\| + \mathrm{tr}\left(S\Sigma^{-1}\right) - \log\|S\| - (p - q), \tag{7.51}$$

where Σ is the population covariance matrix; S is the model-implied covariance matrix; p is the number of exogenous observable variables; and q is the number of endogenous observable variables.

Minimization of the fit function can be done by the LISREL program provided by Jöreskog and Sörbom (1981).

7.3.7 Bayesian Approach

Zellner (1970) uses the Bayesian approach to deal with measurement problems in the estimation of regression relationships containing unobservable independent variables. Zellner (1970) shows that the Bayesian approach can obtain optimal estimates under a finite sample.

[2]Stapleton (1978) further develops MIMIC with more latent variables.

Several studies use Bayesian approaches to examine cost of capital (e.g., Lee and Wu 1989) and asset pricing models (e.g., McCulloch and Rossi 1991; Geweke and Zhou 1996; Ang and Chen 2007; Cederburg and O'Doherty 2015).

Cederburg and O'Doherty (2015) introduce a hierarchical Bayesian approach and uses US firm-level data to reexamine financial anomalies under capital asset pricing model. The Bayesian approach can estimate all parameters simultaneously in one step and effectively avoid the errors-in-variables problem on the estimators induced from two-pass capital asset pricing test. Their empirical finding shows that the economic importance of CAPM anomalies is overstated.

Cederburg and O'Doherty (2015) uses a Bayesian approach to simultaneously estimate coefficients of the following three equation systems.

$$r_{i,t,y} = \alpha_{i,y} + \beta_{i,y} r_{m,t,y} + \varepsilon_{i,t,y}, \text{ where } \varepsilon_{i,t,y} \sim N\left(0, \sigma^2_{r_{i,y}}\right); \quad (7.52)$$

$$\alpha_{i,y} = X_{i,y} + \eta_{i,y}, \text{ where } \eta_{i,y} \sim N\left(0, \sigma^2_{\alpha,y}\right) \text{ and}; \quad (7.53)$$

$$\delta_y = \bar{\delta} + v_y, \text{ where } v_y \sim MVN(0, \mathbf{V}); \quad (7.54)$$

where $r_{i,t,y}$ is firm i's excess return in month t during the time period y, and $r_{m,t,y}$ denotes the monthly excess return for firm i during the time period y. $\alpha_{i,y}$ and δ_y are latent variables. $X_{i,y}$ is a vector including a constant and firm characteristics observable during the time period y. The model allows firm-level βs to vary over each time period, y.

In addition to deal with errors-in-variables problem, there are several advantages using the Bayesian approach to test the capital asset pricing model. First, the Bayesian approach allows βs to vary over time periods and firms and controls the inherent uncertainty associated with firm-level βs. Second, the Bayesian approach can modify the distribution assumptions in stock returns and market returns. Third, the Bayesian inference is free from the use of asymptotic approximations and therefore can be used under finite sample. Fourth, the Bayesian approach takes parameter uncertainty associated with all the model parameters into account.

7.4 Applications of Errors-in-Variables Models in Finance Research

For the last four and a half decades, alternative EIV methods have been used to correct estimation bias associated with empirical results in cost of capital, asset pricing model, and capital structure research. Here, we review these four kinds of research, cost of capital, asset pricing models, capital structure, and investment equation. EIV models used in finance research will be discussed in detail. Table 7.1 provides the literature summary of the application of EIV models.

7.4.1 Cost of Capital

Miller and Modigliani (1966) developed a theoretical expression for the value of a firm from which the firm's cost of capital could be derived. They assume a perpetual stream of earnings from real assets, and a constant capitalization rate (ρ), at which the market discounts the uncertain pure (unlevered) equity stream of earnings for some risk classes and perfect markets. It is thus possible to estimate the market capitalization rate (and thus the cost of capital) of a group of firms by performing a cross-sectional regression of the market value of the firm's equity on the expected average earnings of the firm, the market value of debt, and the growth resulting from the above-average investment opportunities. The above analysis suggests a cross-sectional regression:

$$(V - \tau_c D) = a_0 + a_1 \overline{X}(1 - \tau_c) + a_2(\text{growth potential}) + \varepsilon \quad (7.55)$$

where V is sum of the market value of all securities issued by the firm, τ_c is the corporate tax

Table 7.1 Applications of errors-in-variables models in finance research

Study	Issue	Method	Results
Miller and Modigliani (1966)	Determinants of cost of capital	Instrumental variable method	- The extrapolation of historical population trends is superior to the conventional use of change of capital, and share prices are not a positive function of dividends as often suggested
Black et al. (1972)	CAPM test	Grouping (10 groups)	- Reject both the CAPM and the zero-beta CAPM
Blume and Friend (1973)	CAPM test	Grouping (12 groups)	- Linear model is better than quadratic model in explaining expected return. - Reject both the CAPM and the zero-beta CAPM
Fama and MacBeth (1973)	CAPM test	Grouping (20 groups), period by period	- Find a linear relationship between the expected return and beta risk, beta is the only risk measure in explaining expected return, and risk premium is greater than zero. - CAMP and efficient capital market hold
Lee (1977)	CAPM test	Wald's Grouping/Instrumental Variable	- Adjust for measurement error of market return (first step). - Estimated risk premium is larger than realized risk premium. - Reject CAPM
Litzenberger and Ramaswamy (1979)	CAPM test	MLE, OLS, GLS (individual stock)	- Before-tax expected rates of return are linearly related to systematic risk and dividend yield. - MLE can obtain consistent estimators without losing efficiency. - CAPM is rejected because of nonzero $\hat{\gamma}_0$
Cheng and Grauer (1980)	CAPM test	Grouping (20 groups), Price-level testing (Invariance Law)	- Neither framework of Invariance Law or security market line can accommodate the possibility that the CAPM may hold for each period. - Reject CAPM
Gibbons (1982)	CAPM test	One-step Gauss–Norman procedure (40 groups)	- Gauss–Norman procedure can increase the precision of estimated risk premium. - Reject CAPM
Titman and Wessels (1988)	Determinants of capital structure	LISREL model	- Do not support for four of eight propositions on the determinants of capital structure - A firm's capital structure is not significantly related to its nondebt tax shields, volatility of earnings, collateral value of assets, and future growth
Lee and Wu (1989)	Determinants of cost of capital	MLE	- Obtain better cost-of-capital estimates
MacKinlay and Richardson (1991)	CAPM test	GMM	- Conclusions of mean-variance efficiency vary by settings
Shanken (1992)	CAPM test	MLE (individual stock)	- To deal with small-sample bias in the second-step cross-sectional regression estimates due to measurement error in the betas. - The adjustment does not have much effect on Fama and MacBeth's (1973) conclusion. - Support CAPM

(continued)

Table 7.1 (continued)

Study	Issue	Method	Results
Fama and French (1992)	CAPM test	Two-way grouping (10 × 10 groups)	- The market capitalization and the book-to-market ratio can replace beta altogether. - Reject CAPM
Jagannathan and Wang (1993)	CAPM test	Multifactor asset pricing model	- Including human capital and business cycle can increase explanatory power of expected return. - Support CAPM
Kim (1995, 2010)	CAPM test	MLE (individual stock —20 × 20 groups)	- MLE method can effectively adjust the errors-in-variables bias and CAPM holds. - Support CAPM
Kim (1997)	CAPM test	Multifactor, MLE	- Linear relationship between beta and expected return. - Book-to-market ratio has significant explanatory power for expected return, but size has not
Lewbel (1997)	Elasticity of patent applications to R&D expenses	Instrumental variable method	- TSLS estimator can mitigate the effects of measurement error. - The estimated elasticity of patent applications with respect to R&D expenditures yields very close to one
Erickson and Whited (2000)	Test *q* theory	GMM	- Cash flow does not affect firms' financial decision, even for financially constrained firms. - Support the *q* theory if measurement error is taken into account
Chang et al. (2009)	Determinants of capital structure	MIMIC model	- Seven constructs, growth, profitability, collateral value, volatility, nondebt tax shields, uniqueness, and industry, as determinants of capital structure have significant effects on capital structure decision
Yang et al. (2010)	Determinants of capital structure	LISREL model	- Stock returns, expected growth, uniqueness, asset structure, profitability, and industry classification are main determinants of capital structure. - Leverage, expected growth, profitability, firm value, and liquidity can explain stock returns. - The capital structure and stock return, in addition, are mutually determined by each other
Almeida et al. (2010)	Test *q* theory	GMM and instrumental variables method	- Estimators from GMM are unstable across different specifications and not economically meaningful. - Estimators from a simple instrumental method are robust and conform to *q* theory
Cederburg and O'Doherty (2015)	CAPM test	Bayesian approach	- Positive relationship between excess return and market risk - Support CAPM

rate, D is the market value of a firm's debt, and $\bar{X}$ is the expected level of average annual earnings generated by current assets.

The problems involved and the solutions provided by Miller and Modigliani (1966) illustrate several important issues involved in regression estimation. First, some quantifiable estimate must be made for the growth-component variable of the firm's value. The variable used by Miller and Modigliani (1966) for the investment opportunity level was a linear five-year average growth rate of assets, multiplied by the current total assets. It should be noted that this proxy was suggested by the stable pattern of investment present in the utility industry at the time in question; it should not be expected to be generally applicable to all types of industries.

Secondly, empirical research (e.g., Gordon 1962) suggests that market capitalization rates vary systematically with the size of the firm within the same industry. Thus, *a* constant coefficient a_0 was added to embody the scale effect, and the coefficient of the earning variable can be considered the marginal capitalization rate of the industry. The size and sign of the constant coefficient therefore contain information concerning the average capitalization rate. If the sign is negative, the marginal rate would be less than the average, leading to the conclusion that the capitalization rate tends to rise with increasing size of the firm.

Thirdly, efficient and unbiased estimates of the ordinary least squares (OLS) coefficients can be obtained only if the variance of the error term is constant, and the covariance between the independent variables and the error term is zero. In the regression, the standard deviation of the error term is not a constant, but varies proportionately to the size of the firm. To avoid heteroscedasticity of regression residuals, the equation must be adjusted to compensate for the dominance of the large companies. Miller and Modigliani (1966) use weighted least square to adjust the standard deviation of the error term to firm size (deflating each variable by the book value of total assets). Therefore, Eq. (7.55) can be adjusted to:

$$\frac{(V - \tau_c D)}{A} = \frac{a_0}{A} + a_1 \overline{X}\frac{(1 - \tau_c)}{A} + a_2 \frac{\Delta \overline{A}}{A} + u \tag{7.56}$$

where $u = \varepsilon / A$. With this reformulation, the regression equation is expected to be homogeneous, that is, to have no constant term, and the term A, total assets, is used to avoid heteroscedasticity.

An additional problem beyond that of heteroscedasticity is the possible errors of measurement associated with the earnings term. Since anticipated average earnings are essentially unobservable, accounting-statement estimates of earnings must be used instead. Therefore, the true relation between value and anticipated earnings, when replaced by the observable estimates, implies a simultaneous system of relationships:

$$V_i^* = \alpha X_i^* + \sum_j \beta_j Z_{ij} + u_i, \tag{7.57}$$

$$X_i = X_i^* + v_i, \tag{7.58}$$

$$X_i^* = \sum_j \delta_j Z_{ij} + w_i, \tag{7.59}$$

where $V_i^* = \frac{V_i - \tau_c D_i}{A_i}$, $X_i^* = \frac{\overline{X}(1-\tau_c)}{A_i}$ (the true anticipated earnings);

v_i = measurement errors associated with current earnings;

X_i = observable estimate of earnings derived from the accounting statements; and

Z_{ij} = other relevant variables determining earnings.

The firm's values as represented by the system of equations. Equations (7.57)–(7.59) are related to anticipated earnings and a set of explanatory variables which may also be correlated with the firm's anticipated earnings.

In addition, the earning variable used in the regression only approximates the true value of anticipated earnings, varying by the error of measurement, v_i. The system represents the simultaneous determination of two endogenous variables, V^* and X, by the Z_j exogenous variables. In regressing:

$$V_i^* = \alpha X_i + \sum \beta_j Z_{ij} + U' \tag{7.60}$$

the coefficients will be biased. The coefficient for earnings, α, will have a downward bias.

In an attempt to remedy the simultaneous equation bias, Miller and Modigliani (1966) use an instrumental variable approach. In this approach, the endogenous variable X is first regressed against all the instrumental variables, Z_j, to obtain estimates of the various coefficients. These estimates are then used to develop a new variable, X, which is

$$\hat{X}_i^* = \sum \hat{\delta}_j Z_{ij} \tag{7.61}$$

Depending on the choice of Z_j, the new estimate of earnings, $\hat{X}_i^*$, should be relatively free of the error measurement. It can then be used in the second-stage regression as the earnings variable. The resulting estimates of α and β can be shown to be consistent.

Miller and Modigliani (1966) hypothesized that the constant term was really zero. Below we show the values of estimated constant term for the direct least squares and two-stage estimates. The reduction of bias on the estimates through the use of the two-stage process also seems to support the hypothesis that the constant term is zero. Miller and Modigliani (1966) state that the reason the constant term was significantly different from zero for the direct least squares cases was that the error of measurement for earnings was large. This error is reduced by the two-stage process.

	Direct	Two-stage
1957	0.164	0.004
1956	0.057	0.054
1954	0.274	0.072

Higgins (1974) derives and tests a finite-growth model for the estimation of the cost of capital and share price of electric utility industry between 1960 and 1968. He suggests that the market value of equity is related to book value of equity, earnings, dividends, and changes in net assets. There is a linear relationship that can be shown as

$$\frac{S}{A} = a_0 + a_1\frac{1}{A} + a_2\frac{Y}{A} + a_3\frac{r\pi G_p}{A} + a_4\frac{\pi G_p}{A} + e, \tag{7.62}$$

where

S market value of equity;
A book value of total assets;
Y trend value of flow through earnings;
r trend value of flow through earnings divided by book value of equity;
G_p annualized growth in population in utility's service area during 1950–1960; and
π book value of equity/total assets.

A trend value means that it is equal to the estimated value for the current period derived from a regression of the most recent 5 years of the variable against time. That is $Y = a + bt + e$, where t is time in years. Higgins (1974) assumes that observations of a variable consist of a true component and a random element. If such random elements have zero mean and are serially uncorrelated, the smoothing procedure can reduce potential errors in measurement. Empirical results show that the extrapolation of historical population trends is superior to the conventional use of change of capital, and share prices are not a positive function of dividends as often suggested.

Zellner (1970) proposes the maximum likelihood method, instrumental variables method, and Bayesian approach to deal with potential errors-in-variables problems. Zellner (1970) shows that his methods utilize more information than traditional instrumental variables methods do in dealing with an errors-in-variables problem. Lee and Wu (1989) further apply Zellner's method to reexamine Miller and Modigliani's (1966) cost-of-capital estimation for utility industry and obtain better cost-of-capital estimates.

7.4.2 Capital Asset Pricing Model

The capital asset pricing model (CAPM) developed by Sharpe (1964), Lintner (1965), and Mossin (1966) implies that the expected returns on securities and their market risks (β) are positively and linearly correlated and that market risks have sufficient power to explain expected returns of securities. Black et al. (1972), Fama and MacBeth's (1973), and others use the two-step method to test CAPM. In the first step, estimated betas are obtained by time-series market model for each security. In the second step, the estimated betas are used in testing the linear relationship between betas and expected returns on securities. Because estimated betas are subjected to a measurement error (estimation error) problem, there exists an EIV problem in the second step. The EIV problem will result in estimating the explanatory power of beta and the estimated rate of return on beta risk. More specifically, the EIV problem leads to an underestimation of the coefficient associated with beta risk. Although the EIV problem exists in the two-step method of the asset pricing test, most researchers do not carefully use sophisticated econometric and statistical methods to deal with this kind of problem. Roll (1969, 1977) shows that the testing asset pricing model suffers an EIV problem, concluding that (1) no correct and unambiguous test of the theory has appeared in the literature, and (2) practically no possibility exists that such a test can be accomplished in the future. Lee (1984) shows that the most generalized beta estimate can be decomposed into the following: (1) true component, (2) bias resulting from measurement error, (3) bias resulting from specification error, and (4) interaction bias. Roll and Ross (1994) show that the measurement error problem of market rate of return can bias the empirical test of CAPM.

Several studies focus on beta estimation in the first step to solve the EIV problem in testing CAPM. Brennan (1970), Lee and Jen (1978), and Roll (1977) show that the possible measurement error in market beta risk is the unobserved market rate of return and risk-free rate of return.[3] To improve the beta estimator, Fabozzi and Francis (1978) and Lee and Chen (1979) use the random coefficient procedure to estimate random coefficient betas. Brennan and Schwartz (1977), Brennan (1979), and Brown and Warner (1980) also provide different types of market models which can produce different results in predicting rates of return, testing efficient market hypotheses, and measuring security price performance.

To test the linear relationship between cross-sectional rate of returns of securities and beta risks of CAPM, Black et al. (1972) provide a cross-sectional model as follows:

$$R_i - R_f = \gamma_0 + \gamma_1 \beta_i + \varepsilon_i, \tag{7.63}$$

where R_i is the rate of return for i security, R_m is the rate of return of risk-free security, β_i is beta for i security, and ε_i is the idiosyncratic error for i security. As implied by the theory, Black et al. (1972) try to test if

$$\gamma_0 = 0, \text{ and } \gamma_1 = R_m - R_f,$$

where R_m is the rate of return of market portfolio, and R_f is the rate of return of risk-free security.

To test CAPM, we first use market model to estimate beta for each security.

$$R_{i,t} = \alpha_i + \beta_i R_{m,t} + e_{i,t}. \tag{7.64}$$

To reduce the impact of the measurement error problem, Black et al. (1972) use the grouping method in the capital asset pricing model test. Black et al. (1972) use the previous five-year monthly returns for each security to obtain preranking beta from the market model in the January of each year. Securities are then sorted on their preranking beta and divided into ten portfolios. The subsequent twelve monthly returns of

[3]Roll (1969, 1977), and Lee and Jen (1978) show that the observed market rate returns in terms of stock market index are measured with errors since the stock market index does not include all assets which investors can invest. Lee and Jen (1978) have theoretically shown how beta estimate and Jensen performance measures can be affected by both constant and random measurement errors of R_m and R_f. Diacogiannis and Feldman (2013), Green (1986), Roll and Ross (1994), and Gibbons and Ferson (1985) have argued that market portfolio measure with errors is an inefficient portfolio and show how the inefficient benchmark can affect theoretical CAPM derivation. Diacogiannis and Feldman (2013) provide a pricing model that uses inefficient benchmarks, a two beta model, one induced by the benchmark, and one adjusting for its inefficiency.

each portfolio are equal-weighted. The next twelve monthly portfolio returns are again calculated by the preranking beta in the next year. This process is then repeated from January 1931 to January 1965. The post-ranking beta for each of the ten portfolios is estimated by the market model using its corresponding 420 monthly portfolio returns.

In the second step, the estimated post-ranking betas are used in testing the linear relationship between betas and expected returns. Then, the second-step regression in testing CAPM can be written as:

$$R_i - R_f = \gamma_0 + \gamma_1 \hat{\beta}_i + \varepsilon_i. \quad (7.65)$$

Black et al. (1972) regress the average thirty-five-year monthly returns of ten portfolios on the post-ranking betas of ten portfolios. Their cross-sectional regression results show that the estimated intercept term ($\hat{\gamma}_0$) is significantly different from zero and the estimated slope term ($\hat{\gamma}_1$) is significantly different from the average market premium during the estimation period. Although the results support the linear relationship between the systematic risk and expected return, CAPM cannot hold because of the nonzero intercept term and the lower market premium in the cross-sectional regression.

To cope with the measurement error problem, Blume and Friend (1973) also use the grouping method in testing the capital asset pricing model. The advantages of the grouping method are that (1) the estimated betas for portfolios have substantially smaller measurement errors, (2) the variance of the measurement errors for portfolios will be much smaller than the variance of the measurement errors for individual securities, and (3) the use of grouping method can adjust for survival bias problems. Different from the grouping method used by Black et al. (1972), Blume and Friend's (1973) grouping method in testing CAPM uses only a five-year period in the second step. The preranking beta of each security is obtained by the market model using previous five-year monthly data. Securities are then assigned into twelve portfolios based upon their preranking betas. The post-ranking beta for each portfolio is obtained by the market model using subsequent five-year monthly returns of each portfolio and market index. In the second step, Blume and Friend (1973) use the subsequent five-year period in the cross-sectional regression analysis. They take the five-year average monthly return for each security and run the average monthly returns on the corresponding post-ranking betas. Blume and Friend's (1973) grouping method shows that the relationship between average returns and betas for NYSE-listed common stocks are approximately linear, indicating that CAPM and its associated short-sales assumption may be held.

In addition to testing the linearity between systematic risk and expected return, Fama and MacBeth (1973) provide a cross-sectional regression in the second step, Eq. (7.66), to test the nonsystematic effects and the linearity between systematic risk and expected return.

$$R_{i,t} = \gamma_{0,t} + \gamma_{1,t}\beta_{i,t} + \gamma_{2,t}\beta_{i,t}^2 + \gamma_{3,t}s_{i,t} + \varepsilon_{i,t}, \quad (7.66)$$

where $R_{i,t}$ is the rate of return for i security at time t, $\beta_{i,t}$ is beta for i security at time t, and $s_{i,t}$ is the idiosyncratic risk for i security at time t. If the capital asset pricing holds, $\gamma_{0,t} = r_{f,t}$, $\gamma_{1,t} = R_{m,t} - r_{f,t}$, $\gamma_{2,t} = 0$, and $\gamma_{3,t} = 0$, where $r_{f,t}$ is the risk-free rate of return at time t and $R_{m,t}$ is the rate of return of the market at time t. In Fama and MacBeth's (1973) grouping method, the preranking beta of each security is also obtained by the market model using monthly data from 1926 to 1929, and securities are then assigned into 20 portfolios based upon their preranking betas. Using the subsequent monthly data from 1930 to year $k - 1$, the beta of each security used in year k is recalculated. The post-ranking beta of each portfolio is the simple average of individual betas of each portfolio. In each month, a cross-sectional regression is performed as follows:

$$R_{p,t} = \gamma_{0,t} + \gamma_{1,t}\beta_{p,t} + \gamma_{2,t}\beta_{p,t}^2 + \gamma_{3,t}s_{p,t} + \varepsilon_{p,t}, \quad (7.67)$$

where $R_{p,t}$ is the rate of return for portfolio p at time t, $\beta_{p,t}$ is beta for portfolio p at time t, and $s_{i,t}$ is the average idiosyncratic risk for securities within portfolio p from 1930 to year $t-1$. Instead of using average returns portfolios and systematic risks of portfolios in a period of time, Fama and MacBeth's grouping method uses monthly second-step regression to allow estimated coefficients, such as estimated risk-free rate and estimated market premium, to change period by period. Therefore, they obtain $\bar{\hat{\gamma}}_0$ by taking the time-series average of the estimators $\hat{\gamma}_{0,t}$ as the final estimate of γ_0, obtain $\bar{\hat{\gamma}}_1$ by taking the time-series average of the estimators $\hat{\gamma}_{1,t}$ as the final estimate of γ_1, and so on. $\bar{\hat{\gamma}}_1$ is then used to test whether the price of the beta risk is significantly positive, and $\bar{\hat{\gamma}}_0$ is used to test whether the intercept is zero. If the capital asset pricing model is valid, we expect results of a significantly positive price premium $\left(\bar{\hat{\gamma}}_1\right)$ and an insignificant intercept term $\left(\bar{\hat{\gamma}}_0\right)$. Jagannathan et al. (2009) show that the time series average of the cross-sectional estimators converges in probability to the true value of the estimator. Although their results support CAPM indicating a linear relationship between the systematic risk and the expected return, the lower value of time-series average of the market premium shows that the measurement error problem still exists after Fama–MacBeth's (1973) grouping method.

Considering the measurement errors of the market rate of return and risk-free rate of return, Lee (1977) uses two EIV estimation methods, Wald's two-group grouping method and Durbin's instrumental variable method, to adjust the estimated beta risk in the first step of the capital asset pricing test. Although correcting the measurement errors induced from the unobservable market rate of return, Lee (1977) finds that the predictive ability of the capital asset pricing model is still poor.

Litzenberger and Ramaswamy (1979) derive an after-tax version of CAPM and show that, in the equilibrium, the before-tax expected return on a security is linearly related to its systematic risk and its dividend yield. Litzenberger and Ramaswamy (1979) further empirically test both the before-tax and the after-tax versions of CAPM as indicated in the following.

Before tax:

$$R_{i,t} - r_{f,t} = \gamma_{0,t} + \gamma_{1,t}\beta_{i,t} + \varepsilon_{i,t}, \text{ and} \quad (7.68)$$

After tax:

$$R_{i,t} - r_{f,t} = \gamma_{0,t} + \gamma_{1,t}\beta_{i,t} + \gamma_{2,t}\left(d_{i,t} - r_{f,t}\right) + \varepsilon_{i,t}, \quad (7.69)$$

where $R_{i,t}$ is the rate of return of security i at time t, are the beta risk and the dividend yield of security i at time t, respectively. If the after-tax version of the capital asset pricing model holds, we will expect $\gamma_{0,t} > 0$, $\gamma_{1,t} > 0$, and $\gamma_{2,t} > 0$.

Litzenberger and Ramaswamy (1979) point out that, while the grouping method can eliminate measurement errors problems of unobserved beta, grouping method results in reduction of the efficiency because of losing information pertaining to individual security. Instead of grouping method, Litzenberger and Ramaswamy (1979) use maximum likelihood estimation in the second-step regression to test the before-tax and the after-tax versions of capital asset pricing model. Although maximum likelihood estimators are consistent, the average risk premium is small and not significantly different from zero.

Given the EIV bias in the two-step CAPM test, Gibbons (1982) introduces a one-step Gauss-Newton procedure and uses the maximum likelihood method to obtain the estimated price of systematic risk. Because the one-step Gauss-Newton procedure does not use estimated beta as an explanatory variable in a regression model, the measurement errors problem of estimated beta can be avoided. Shanken (1992) shows that using generalized least square (GLS) on the second step of CAPM test can yield an estimator identical to the Gauss-Newton estimator obtained by Gibbons' (1982) maximum likelihood method. Gibbons (1982) shows that the Gauss-Newton procedure increases the precision of estimated risk premium, but rejects the mean-variance efficiency of the market portfolio.

Shanken (1992), who provides a modified version of the two-step estimator by using maximum likelihood estimation, finds that Fama and MacBeth's two-step procedure overstates the precision of the estimator in the second-step and therefore provides an adjusted standard error for the estimator in the second-step regression. Jagannathan and Wang (1998) further release Shanken's assumption that asset returns are conditional homoscedasticity to derive a more general standard error for the second-step estimator by generalized least squares.

Although Black et al. (1972) and Fama and MacBeth (1973) found a significant positive relationship between beta risk and stock return during the early years from 1926 to 1968, Fama and French (1992) found a positive relationship between beta risk and stock return disappears during the year from 1963 to 1990. In addition, Fama and French (1992) used a two-way sort grouping method to control for size effect and find a weak relationship between beta risk and expected return. Before reaching the conclusion that the capital asset pricing has not been valid in the recent years, one possible reason that may be considered is that the measurement error problem cannot be fully eliminated by the grouping method, and results of CAPM test may vary depending on the portfolio formation technique (e.g., Ahn et al. 2009).

To deal with the problem of EIV in testing CAPM, Kim (1995) provides a maximum likelihood method, extracting information associated with the relationship between the measurement error variance and idiosyncratic error variance and incorporating such information into the maximum likelihood estimation in the second step of the capital asset pricing model test. Given the assumption that the disturbance term of the market model is homoscedasticity, the corrected factors for the traditional least squares estimators of the cross-sectional regression coefficients can be obtained. Although Kim's (1995) maximum likelihood method can only deal with the EIV problem of the estimated beta in the first pass, the maximum likelihood method can test, besides the capital asset pricing model, the multifactor asset pricing models. Kim (1995) uses maxima likelihood method to reexamine CAPM and multifactor asset pricing model and finds more support for the role of market beta risk and less support for the role of firm size. His results show that the prominent risk factors (e.g., size, book-to-market ratio, and momentum factors) might result a different explaining power for cross-sectional stock returns after correcting the EIV problem.

MacKinlay and Richardson (1991) use generalized method of moment (GMM) to test the mean-variance efficiency. They theoretically show that the estimator from GMM and the estimator from maximum likelihood method are equivalent when stock returns are conditionally homoscedasticity, but GMM can avoid the EIV problem by estimating coefficients in one step. Empirical and simulation results show that the conclusion mean-variance efficiency of market indexes is sensitive to the model settings.

Chen (2011) offers an empirical examination of various EIV estimation methods in the testing of CAPM, including the grouping method, the instrumental variable method, and the maximum likelihood method. Both potential measurement error problems of market return in the first pass and estimated beta in the second pass are corrected by either the grouping method or the instrumental variable method. Chen (2011) shows that empirical results support the role of market beta in the capital asset pricing model after correcting the EIV problem.

To deal with the measurement error problem associated with testing both CAPM and APT, Lee et al. (2015) and Wei (1984, Chap. 7) use the MIMC model discussed in previous section to test whether APT outperformed CAPM. The CAPM can be rewritten, in terms of the MIMIC model, as a simultaneous equation model:

$$\begin{aligned} \tilde{r}_i &= \beta_i \tilde{r}_m^* + \tilde{u}_i \\ \tilde{r}_m^* &= \tilde{r}_m + \tilde{e}_m, \quad (i = 1, \ldots, N) \end{aligned} \tag{7.70}$$

where $\tilde{r}_i$ is the realized excess return (total return less risk-free rate) on security i in a deviation form, $\tilde{r}_m$ is the realized excess return of the

NYSE composite index, and $\tilde{r}_m^*$ is the unobservable excess return on the market portfolio. There are N equations linking the individual security (or portfolio) return to the unobservable true market return, and one equation linking the unobservable true market portfolio return to the realized return of the NYSE composite index. After obtaining the estimated beta from simultaneous equation system of Eq. (7.70), a cross-sectional regression of the security return against its β will be used to test the CAPM as follows:

$$r_{it} = \hat{a}_{0t} + \hat{a}_{1t}\beta_i \qquad (7.71)$$

where r_{it} is the excess return on security i at time t, $\hat{a}_{0t}$ is the estimate of the intercept which is supposed to be zero, and $\hat{a}_{1t}$ is the estimate of the market risk premium. Four different estimation procedures, stationary OLS, nonstationary OLS, GLS, and MLE, are used to estimate Eq. (7.71).

The testing model in terms of the MIMIC model can be written as

$$\begin{aligned} \tilde{r}_i &= b_{i1}\tilde{f}_1 + \ldots + b_{ik}\tilde{f}_k + \tilde{u}_i, \quad (i = 1, \ldots, N) \\ \tilde{f}_j &= a_{j1}\tilde{I}_1 + \ldots + a_{jp}\tilde{I}_p + \tilde{e}_j, \quad (j = 1, \ldots, k) \text{ and } (h = 1, \ldots, p) \end{aligned} \qquad (7.72)$$

where $\tilde{f}_j$ is the jth unobservable factor, and $\tilde{I}_h$ is the hth macroeconomic indicator. There are N return equations plus k factor equations in the simultaneous equation system. The LISREL computer program of Jöreskog and Sörbom (1981) is used to estimate the parameters, a and b, in Eq. (7.72). A cross-sectional regression is also used to test the APT and to estimate the riskless rate and the factor risk premium by regressing the security return against its risks, b's. The a's coefficients in Eq. (7.72) can be used to explain the relationship between factors and indicators.

The major conclusions of Lee et al. (2015) and Wei (1984) are as follows:

(i) The APT performs better than the CAPM.
(ii) The beta estimated from the MIMC model by allowing measurement error on the market portfolio does not significantly improve the OLS beta estimate.
(iii) The MLE estimator does a better job than the OLS and GLS estimators in the cross-sectional regressions because the MLE estimator takes care of the measurement error in beta.
(iv) The empirical results support Stambaugh's (1982) argument that the inference about the tests of the CAPM is insensitive to alternative market indexes.
(v) The market variables play a major role in testing capital asset pricing relationship.

7.4.3 Capital Structure

Titman and Wessels (1988), Chang et al. (2009), and Yang et al. (2010) use structure equation models (e.g., LISREL model and MIMIC model) to mitigate the measurement problems of proxy variables when working on capital structure theory. Titman and Wessels (1988) use LISREL method to investigate determinates of capital structure. In the structure equation model, they use 15 indicators associated with eight latent variables and set 105 restrictions on the coefficient matrix. Empirical results, however, do not support four of eight propositions on the determinants of capital structure. Specifically, their results show that a firm's capital structure is not significantly related to its nondebt tax shields, volatility of earnings, collateral value of assets, and future growth. One possible reason for the poor results is that the indicators used in the empirical study do not adequately reflect the nature of the attributes suggested by financial theory.

Therefore, Chang et al. (2009) apply a MIMIC model with refined indicators to reexamine Titman and Wessel's (1988) work on determinants of capital structure. Chang et al. (2009) examine the seven indicator factors as follows: growth, profitability, collateral value, volatility, nondebt tax shields, uniqueness, and industry. Their empirical results show that the growth is the most influential determinant on capital structure, followed by profitability, and then collateral value. Under a simultaneous cause–effect framework, their seven constructs as determinants of capital structure have significant effects on capital structure decision.

Yang et al. (2010) apply a LISREL model to find determinants of capital structure and stock returns, and estimate the impact of unobservable attributes on capital structure decisions and stock returns. Using leverage ratios and stock returns as two endogenous variables and 11 latent factors as exogenous variables, Yang et al. (2010) find that stock returns, expected growth, uniqueness, asset structure, profitability, and industry classification are main determinants of capital structure, while leverage, expected growth, profitability, firm value, and liquidity can explain stock returns. In addition, the capital structure and stock return are mutually determined by each other.

7.4.4 Measurement Error in Investment Equation

Modern q theory, developed by Lucas and Prescott (1971) and Mussa (1977), shows that the shadow value of capital, marginal q, is the firm manager's expectation of the marginal contribution of new capital goods to future profits. Marginal q, therefore, should summarize the effects of all factors relevant to the investment decision. Lucas and Prescott (1971) and Hayashi (1982) show that the equality of marginal q with average q is under the assumptions of constant returns to scale and perfect competition. Because the marginal q is unobservable in the real world, most of empirical studies adopt Lucas and Prescott (1971) and Hayashi's (1982) assumption and use average q instead of marginal q to test q theory. In addition, if financial markets' valuation of the capital will be equal to the manager's valuation, average q, should equal an observable value, Tobin's q, defined as the ratio of the market value to the replacement value. Most empirical studies use Tobin's q as a proxy for marginal q to test the q theory of investment. However, their empirical results are inconsistent to the q theory (e.g., Fazzari et al. 1988; Schaller 1990; Blundell et al. 1992; Gilchrist and Himmelberg 1995).[4]

The model introduced by Fazzari et al. (1988) is

$$I_{it}/K_{it} = \eta_i + \beta q_{it}^* + \alpha CF_{it}/K_{it} + u_{it}, \quad (7.73)$$

where I_{it} represents the investments of firm i at time t, K_{it} is capital stock of firm i at time t, q_{it}^* is the marginal q_{it}, CF_{it} is cash flow of firm i at time t, η_i is the firm-specific effect, and u_{it} is the innovation term.

Almeida et al. (2010) show that OLS estimated coefficient of independent variable with measurement error, q_{it}^*, will be biased downward, and OLS estimated coefficient of the independent variable without measurement error, CF_{it}/K_{it}, will be biased upward. Following Eq. (7.73), if one of the independent variables has measurement error and the other independent variables has no measurement error; the asymptotic biases of estimated coefficients can be defined as:

$$\text{plim } \hat{\beta} - \beta = \frac{-\beta\sigma_\varepsilon^2}{\sigma_{CF/K}^2 - b_{CF/K,q^*} + \sigma_\varepsilon^2}, \text{ and} \quad (7.74)$$

$$\text{plim } \hat{\alpha} - \alpha = \beta b_{CF/K,q^*}\left(\frac{\sigma_\varepsilon^2}{\sigma_\varepsilon^2 + \sigma_{CF/K}^2(1 - R_{CF/K,q^*}^2)}\right), \quad (7.75)$$

in which $\sigma_{CF/K}^2$ is the variance of CF/K, σ_ε^2 is the variance of error term between unobserved marginal q and observable average q^*, $b_{CF/K,q^*}$ is the auxiliary regression coefficient of a regressing q^* on CF/K, and $R_{CF/K,q^*}^2$ is the correlation coefficient between q^* and CF/K. We know that $\sigma_{CF/K}^2 - b_{CF/K,q^*} + \sigma_\varepsilon^2$ is generally positive, so

[4]Empirical work in testing association between the investment decision and cash flow shows that cash flow has poor explanation in determining investment decision. In addition to cash flow, output, sales, and internal funds have significant explanation in determining investment decision.

the estimated coefficient of q^* is downward-biased. In addition, the direction of bias of estimated coefficient of CF/K will depend on the signs of β and $b_{CF/K,q^*}$. Given that q and cash flow are positively correlated, we can get the conclusion of Almeida et al. (2010) that $\hat{\beta}$ is downward-biased and $\hat{\alpha}$ is upward-biased.

Erickson and Whited (2000) argue that the measurement error of marginal q can result in different implications in empirical q models. They incorporate an EIV model to reexamine the empirical work done by Fazzari et al. (1988). By using generalized method of moment (GMM), Erickson and Whited (2000) obtain consistent estimators that the information contained in the third- and higher-order moments of the joint distribution of the observed regression variables. The estimator precision and consistency can be increased by exploiting the information afforded by an excess of moment equations over parameters. Results show that cash flow does not affect a firms' financial decision, even for financially constrained firms, and the q theory is held if measurement error is taken into account.

Almeida et al. (2010) use Monte Carlo simulations and real data to compare the performance of generalized method of moments and instrumental variables approach dealing with measurement error problems in investment equations. In Monte Carlo simulations, they find estimators of GMM proposed by Erickson and Whited (2000) are biased for both mismeasured and well-measured regressors when the data have individual-fixed effects, heteroscedasticity, or no high degree of skewness. In contrast, the instrumental variable method results fairly unbiased estimators under those same conditions. Almeida et al. (2010) further empirically examine the investment equation introduced by Fazzari et al. (1988) by using GMM and instrumental variable method. Almeida et al. (2010) adopt Biorn's (2000) method using the lags of the variable as instruments in testing the investment equation. Empirical results show that estimators from generalized method of moments are unstable across different specifications and not economically meaningful, while estimators from a simple instrumental method are robust and conform to q theory. Almeida et al. (2010) conclude that instrumental method yields more consistent estimators and support the q theory in the investment equation.

Peters and Taylor (2017) found that Tobin's q also helps explain intangible investment. In addition, they show that Tobin's q explains physical and intangible investment roughly equally well, and it explains total investment even better in terms of firm-level data. Compared with physical capital, intangible capital adjusts more slowly to changes in investment opportunities. The classic q theory performs better in firms and years with more intangible capital: Total and even physical investment is better explained by Tobin's q and is less sensitive to cash flow. At the macro level, Tobin's q explains intangible investment many times better than physical investment. They propose a simple, new Tobin's q proxy that accounts for intangible capital, and they show that it is a superior proxy for both physical and intangible investment opportunities. Their research implies that dependent variable uses only physical assets is measured with error; therefore, intangible assets should also be included in dependent variables. In their research, they show that R-square will increase about 30% if both tangible and intangible assets are included in the dependent variable relative to only use tangible assets as a dependent variable, which has been done by traditional research. In other words, their research shows measurement error of dependent variable can affect R-square estimates, which has been shown by Cochran (1970).

7.5 Conclusion

In this chapter, we investigated theoretical issues related to errors-in-variables (EIV) problem. We also reviewed empirical applications of using errors-in-variables (EIV) methods in finance research. In other words, we reviewed and discussed the existing EIV estimation methods and how the empirical research in finance has applied these estimation methods. We first showed how EIV problems affect the coefficients of independent variables in the regression model. Theoretically, we reviewed and extended classical method and mathematical programming method to deal with EIV problems. Then, we discussed the grouping method and instrumental variable methods to deal with EIV problems. Furthermore, we discuss maximum likelihood method, and LISREL method.

In addition, we also reviewed and investigated how alternative EIV models have been used in empirical finance research. We found that the empirical research of cost of capital, asset pricing, capital structure, and investment equation have used alternative EIV methods to improve the empirical results of these issues. Not only can the reader of this chapter understand the important research topics in finance, but the reader can also realize how measurement error problems affect the results of empirical work in such research topics. Finally, we suggest that future empirical studies on finance-related issues should pay more efforts to deal with EIV problems and obtain more robust empirical results. In the next chapter, we will discuss how to use EIV model discussed in this chapter to test capital asset pricing model.

Bibliography

Ahn, D., Conrad, J., & Dittmar, R. F. (2009). Basis asset. *Review of Financial Studies, 22,* 5122–5174.

Aigner, D. J. (1974). MES dominance of least squares with errors-of-observation. *Journal of Econometrics, 2,* 365–372.

Almeida, H., Campello, M., & Galvao, A. F., Jr. (2010). Measurement errors in investment equations. *Review of Financial Studies, 23,* 3279–3328.

Anderson, T. W. (1963). The use of factor analysis in the statistical analysis of multiple time series. *Psychometrika, 28,* 1–25.

Ang, A., & Chen, J. (2007). CAPM over the long run: 1926–2001. *Journal of Empirical Finance, 14*(1), 1–40.

Banz, R. W. (1981). The relationship between return and market value of common stocks. *Journal of Financial Economics, 9,* 3–18.

Barnett, V. D. (1967). A note on linear structural relationship when both residual variances are known. *Biomtrika, 54,* 670–672.

Bentler, P. M. (1983). Some contributions to efficient statistics in structural models: Specification and estimation of moment structures. *Psychometricka, 48,* 493–517.

Biorn, E. (2000). Panel data with measurement errors: Instrumental variables and GMM procedures combining levels and differences. *Econometric Reviews, 19,* 391–424.

Black, F. (1972). Capital market equilibrium with restricted borrowing. *Journal of Business, 50,* 80–83.

Black, F., Jensen, M. C., & Scholes, M. (1972). Chapter the capital asset pricing model: Some empirical tests. In *Studies in the theory of capital markets* (pp. 79–121). New York: Praegar.

Blume, M. E., & Friend, I. (1973). A new look at the capital asset pricing model. *Journal of Finance, 28,* 19–33.

Blundell, R., Bond, S., Devereux, M., & Schiantarelli, F. (1992). Investment and Tobin's Q: Evidence from company panel data. *Journal of Econometrics, 51,* 233–257.

Brennan, M. J. (1970). Taxes, market valuation and corporate financial policy. *National Tax Journal, 23,* 417–427.

Brennan, M. J. (1979). The pricing of contingent claims in discrete time models. *The Journal of Finance, 34,* 53–68.

Brennan, M. J., & Schwartz, E. S. (1977). Convertible bonds: Valuation and optimal strategies for call and conversion. *The Journal of Finance, 32,* 1699–1715.

Brown, S. J., & Warner, J. B. (1980). Measuring security price performance. *Journal of Financial Economics, 8,* 205–258.

Carhart, M. M. (1997). On persistence in mutual fund performance. *Journal of Finance, 52,* 57–82.

Cederburg, S., & O'Doherty, M. S. (2015). Asset-pricing anomalies at the firm level. *Journal of Econometrics, 186,* 113–128.

Chang, C., Lee, A. C., & Lee, C. F. (2009). Determinants of capital structure choice: A structural equation modeling approach. *Quarterly Review of Economics and Finance, 49,* 197–213.

Chen, H. Y. (2011). Momentum strategies, dividend policy, and asset pricing test. *Ph.D. Dissertation State University of New Jersey*, Rutgers.

Cheng, P. L., & Grauer, R. R. (1980). An alternative test of the capital asset pricing model. *The American Economic Review, 70*(4), 660–671.

Chen, H.-Y., Lee, A. C., & Lee, C.-F. (2015). Alternative errors-in-variables models and their applications in finance research. *The Quarterly Review of Economics and Finance, 58,* 213–227.

Chordia, T., & Shivakumar, L. (2006). Earnings and price momentum. *Journal of Financial Economics, 80,* 627–656.

Clutton-Brock, M. (1967). Likelihood distributions for estimating functions when both variables are subject to error. *Technometrics, 9,* 261–269.

Cochran, W. G. (1970). Some effects of errors of measurement on multiple correlation. *Journal of the American Statistical Association, 65,* 22–34.

Deming, W. E. (1943). *Statistical adjustment of data*. New York: Wiley.

Diacogiannis, G., & Feldman, D. (2013). Linear beta pricing with inefficient benchmarks. *Quarterly Journal of Finance, 3*(1), 1350004.

Durbin, J. (1954). Errors in variables. *Review of the International Statistical Institute, 22,* 23–32.

Erickson, T., & Whited, T. (2000). Measurement error and the relationship between investment and *q*. *Journal of Political Economy, 108,* 1027–1057.

Erickson, T., & Whited, T. (2002). Two-step GMM Estimation of the errors-in-variables model using higher-order moments. *Econometric Theory, 18,* 776–799.

Fama, E. F., & French, K. R. (1992). The cross-section of expected stock returns. *Journal of Finance, 47,* 427–466.

Fama, E. F., & MacBeth, J. D. (1973). Risk, return and equilibrium: Empirical tests. *Journal of Political Economy, 81,* 607–636.

Fabozzi, F. J., & Francis, J. C. (1978). Beta as a random coefficient. *Journal of Financial and Quantitative Analysis, 13,* 101–116.

Fazzari, S., Hubbard, R. G., & Petersen, B. (1988). Financing constrains and corporate investment. *Brookings Papers on Economic Activity, 1,* 141–195.

Geweke, J., & Zhou, G. (1996). Measuring the pricing error of the arbitrage pricing theory. *The Review of Financial Studies, 9*(2), 557–587.

Gibbons, M. R. (1982). Multivariate tests of financial models: A new approach. *Journal of Financial Economics, 10,* 3–27.

Gibbons, M. R., & Ferson, W. (1985). Testing asset pricing models with changing expectations and an unobservable market portfolio. *Journal of Financial Economics, 14,* 217–236.

Gilchrist, S., & Himmelberg, C. P. (1995). Evidence on the role of cash flow for investment. *Journal of Monetary Economics, 36,* 541–572.

Goldberger, A. S. (1972). Structural equation methods in the social sciences. *Econometrica, 40,* 979–1001.

Gordon, M. J. (1962). *The investment financing and valuation of the corporation*. Homewood, IL: Richard D.

Green, R. (1986). Benchmark portfolio inefficiency and deviations from the security market line. *Journal of Finance, 41,* 295–312.

Griliches, Z., & Hausman, J. A. (1986). Errors in variables in panel data. *Journal of Econometrics, 31,* 93–118.

Hayashi, F. (1982). Tobin's marginal *q* and average *q*: A neoclassical interpretation. *Econometrica, 50,* 213–224.

Higgins, R. C. (1974). Growth, dividend policy and capital cost in the electric utility industry. *Journal of Finance, 29,* 1189–1201.

Jagannathan, R., Skoulakis, G., & Wang, Z. (2009). The analysis of the cross section of security returns. In Y. Aït-Sahalia & L. Hansen (Eds.), *Handbook of financial econometrics* (North-Holland) (Vol. 2, pp. 73–134).

Jagannathan, R., & Wang, Z. (1998). An asymptotic theory for estimating beta-pricing models using cross-sectional regression. *Journal of Finance, 53,* 1285–1309.

Jöreskog, K. G., & Goldberger, A. S. (1975). Estimation of a model with multiple indicators and multiple causes of a single latent variable. *Journal of the American Statistical Association, 70,* 631–639.

Jöreskog, K. G., & Sörbom, D. (1978). *LISREL: Analysis of linear structural relationships by the method of maximum likelihood (Version IV)*. Chicago: National Educational Resources Inc.

Jöreskog, K. G., & Sörbom, D. (1981). *LISREL V. analysis of Linear structural relationships by maimum likelihood and least squares methods*. Mimeographed by Department of Statistics, Sweden: University of Ippsala

Kendall, M. G., & Stuart, A. (1958). *The advance theory of statistics, I*. London: Griffin.

Kendall, M. G., & Stuart, A. (1961). *The advance theory of statistics, II*. London: Griffin.

Kiefer, J. (1964). Review of Kendall and Stuart's advanced theory of statistics, II. *Annals of Mathematical Statistics, 35,* 1371–1380.

Kim, D. (1995). The errors in the variables problem in the cross-section of expected stock returns. *Journal of Finance, 50,* 1605–1634.

Kim, D. (1997). A reexamination of firm size, book-to-market, and earnings price in the cross-section of expected stock returns. *Journal of Financial and Quantitative Analysis, 32,* 463–489.

Kim, D. (2010). Issues related to the errors-in-variables problems in asset pricing tests. In C. F. Lee, A. C. Lee, & J. Lee (Ed.), *Handbook of quantitative finance and risk management*. Berlin: Springer.

Lee, C. F. (1973). Errors-in-variables estimation procedures with applications to a capital asset pricing model. *Ph.D. Dissertation State University of New York at Buffalo*.

Lee, C. F. (1975). *Three Different Estimation Methods when Variables are Subject to Errors*. Working paper.

Lee, C. F. (1977). Performance measure, systematic risk and errors-in-variable estimation method. *Journal of Economics and Business*, 122–127.

Lee, C. F. (1984). Random coefficient and errors-in-variables models for beta estimates: methods and application. *Journal of Business Research, 12,* 505–516.

Lee, Cheng F., & Chen, C. R. (1982). Beta stability and tendency: an application of a variable mean response regression model. *Journal of Economics and Business, 34,* 201–206.

Lee, C. F., & Chen, S. N. (1979). A random coefficient model for reexaming risk decomposition method and risk-return relationship test. *Financial Review, 14,* 65.

Lee, C. F., & Chen, S. N. (1980). A random coefficient model for reexaming risk decomposition method and risk-return relationship test. *Quarterly Journal of Economics and Business, 34,* 201–206.

Lee, C. F., & Chen, S. N. (1984). On the measurement errors and ranking of composite performance measures. *Quarterly Review of Economics and Business, 24,* 6–17.

Lee, C. F., & Jen, F. C. (1978). Effects of measurement errors on systematic risk and performance measure of a portfolio. *Journal of Financial and Quantitative Analysis, 13,* 299–312.

Lee, C. F., & Cheng, D. C. (1986). Ramsey's specification error test and alternative specifications of the market model: methods and application. *Quarterly Review of Economics and Business, 26,* 6–24.

Lee, C. F., Wei, K. C., & Chen, H. Y. (2015). Multi-factor, multi-indicator approach to asset pricing: Methods and empirical evidence. In C. F. Lee & J. Lee (Ed.), *Handbook of financial econometrics and statistics.* Singapore: Springer.

Lee, C. F., & Wu, C. C. (1989). Usinge Zellner's errors-in-variables model to reexamine MM's valuation model for the electric utility industry. *Advances in Financial Planning and Forecasting, 3,* 63–73.

Lewbel, A. (1997). Constructing instruments for regressions with measurement error when no additional data are available, with an application to patents and R&D. *Econometrica, 65,* 1201–1213.

Lewbel, A. (2012). Using heteroscedasticity to identify and estimate mismeasured and endogenous regressor models. *Journal of Business and Economic Statistics, 30,* 67–80.

Lewellen, J., & Nagel, S. (2006). The conditional CAPM does not explain asset-pricing anomalies. *Journal of Financial Economics, 82,* 289–314.

Lewellen, J., Nagel, S., & Shanken, J. (2010). A skeptical appraisal of asset-pricing tests. *Journal of Financial Economics, 96,* 175–194.

Lintner, J. (1965). The valuation of risky assets and the selection of risky investments in stock portfolio and capital budgets. *Review of Economics and Statistics, 47,* 13–37.

Litzenberger, R., & Ramaswamy, K. (1979). The effects of personal taxes and dividends on capital asset prices: theory and empirical evidence. *Journal of Financial Economics, 7,* 163–195.

Lo, A. W., & MacKinlay, C. A. (1990). Data-snooping biases in tests of financial asset pricing models. *Review of Financial Studies, 3,* 431–467.

Lucas, R. E., Jr., & Prescott, E. C. (1971). Investment under uncertainty. *Econometrica, 39,* 659–681.

MacKinlay, A. C., & Richardson, M. P. (1991). Using generalized method of moments to test mean-variance efficiency. *Journal of Finance, 46,* 511–527.

Maddala, G. S., & Nimalendran, M. (1996). Error-in-variables problems in financial models. In G. S. Maddala & C. R. Rao (Eds.), *Handbook of statistics.* New York: Elsevier.

Madansky, A. (1964). Instrumental variables in factor analysis. *Psychometrika, 29,* 105–113.

McCallum, B. T. (1972). Relative asymptotic bias from errors of omission and measurement. *Econometrica, 40,* 757–758.

McCulloch, R., & Rossi, P. E. (1991). A Bayesian approach to testing the arbitrage pricing theory. *Journal of Econometrics, 49*(1–2), 141–168.

Miller, M. H., & Modigliani, F. (1966). Some estimates of the cost of capital to the electric utility industry, 1954–57. *American Economic Review, 56,* 333–391.

Mossin, J. (1966). Equilibrium in a capital asset market. *Econometrica, 34,* 768–783.

Mussa, M. (1977). External and internal adjustment costs and the theory of aggregate and firm investment. *Economica, 44,* 163–178.

Peters, R. H., & Taylor, L. A. (2017). Intangible capital and the investment- q relation. *Journal of Financial Economics, 123,* 251–272.

Reinganum, M. R. (1981). Misspecification of capital asset pricing: empirical anomalies based on earning yield and market value. *Journal of Financial Economics, 9,* 19–46.

Richardson, D. H., & Wu, D. M. (1970). Least squares and grouping method estimation in the errors in variable model. *Journal of American Statistical Association, 65,* 724–784.

Roll, R. (1969). Bias in fitting the sharpe model to time-series data. *Journal of Financial and Quantitative Analysis, 4,* 271–289.

Roll, R. (1977). A critique of the asset pricing theory's tests; part 1: On past and potential testability of the theory. *Journal of Financial Economics, 4,* 129–176.

Roll, R., & Ross, S. A. (1994). On the cross-sectional relation between expected returns and betas. *Journal of Finance, 49,* 101–121.

Schaller, H. (1990). A re-examination of the q theory of investment using US firm data. *Journal of Applied Econometrics, 5,* 309–325.

Shanken, J. (1992). On the estimation of beta pricing models. *Review of Financial Studies, 5,* 1–33.

Shanken, J., & Zhou, G. (2007). Estimating and testing beta pricing models: Alternative methods and their performance in simulations. *Journal of Financial Economics, 84,* 40–86.

Sharpe, W. F. (1964). Capital asset prices: A theory of market equilibrium under conditions of risk. *Journal of Finance, 19,* 425–442.

Stapleton, D. C. (1978). Analyzing political participation data with a MIMIC model. In K. F. Schuessler (Ed.), *Sociological methodology.* San Francisco, CA: Jossey-Bass.

Stambaugh, R. (1982). On the exclusion of assets from tests of the two-parameter model: a sensitivity analysis. *Journal of Financial Economics, 10,* 237–268.

Titman, S., & Wessels, R. (1988). The determinants of capital structure choice. *Journal of Finance, 43,* 1–19.

Theil, H. (1971). *Principles of econometrics.* Wiley, Toronto, NY.

Wald, A. (1940). The fitting of straight lines if both variables are subject to error. *Annals Mathematical Statistics, 11,* 284–300.

Wei, K. C. (1984). The arbitrage pricing theory versus the generalized intertemporal capital asset pricing model:

Theory and empirical evidence. *Ph.D Dissertation University of Illinois at Urbana-Champaign.*

Yang, C. C., Lee, C. F., Gu, Y. X., & Lee, Y. W. (2010). Co-determination of capital structure and stock returns —A LISREL approach: An empirical test of Taiwan stock markets. *The Quarterly Review of Economics and Finance, 50,* 222–233.

York, D. (1966). Least-squares fitting of a straight line. *Canadian Journal of Physics, 44,* 1079–1086.

Zellner, A. (1970). Estimation of regression relationships containing unobservable variables. *International Economic Review, 11,* 441–454.

Three Alternative Methods in Testing Capital Asset Pricing Model

8

Contents

Abstract

Following the previous chapter, we show how three alternative errors-in-variable models can be used to test the capital asset pricing model. These three methods include the grouping method, the instrumental variable method, and the maximum likelihood method. In addition, we discuss how the errors-in-variable model can improve the capital asset pricing model tests at the individual stock level.

8.1 Introduction

The capital asset pricing model developed by Sharpe (1964), Lintner (1965), and Mossin (1966) implies that the expected returns on securities and their market risks are positively and linearly correlated and that market risks have sufficient power to explain expected returns of securities. Numerous empirical studies show that market risks fail to explain the average returns of their corresponding securities. For example, Fama and French (1992) propose size and book-to-market ratio as new risk factors in pricing

This chapter is an update and expansion of Chap. 3 of Chen's Ph.D. dissertation (2011).

C.-F. Lee et al., *Financial Econometrics, Mathematics and Statistics*,
https://doi.org/10.1007/978-1-4939-9429-8_8

expected returns; and Carhart (1997) and Chordia and Shivakumar (2006) suggest that momentum factors should be priced in expected returns.

Because the market risks of securities are not observed, most empirical studies use Fama and MacBeth's (1973) two-step method to test the capital asset pricing model. In the first step, estimated betas are obtained by time-series market model for each security. In the second step, the estimated betas are used in testing the linear relationship between betas and expected returns on securities. If estimated betas with measurement errors are used in the second step, however, an errors-in-variables problem will result in estimating the explanatory power of beta and the estimated rate of return on beta risk. More specifically, the errors-in-variables problem leads to an underestimation of the rate of return on beta risk and an overestimation of the rate of return on the risk factors without error.[1] Although the errors-in-variables problem exists in the two-step method of the asset pricing test, most researchers do not carefully use sophisticated econometric and statistical methods to deal with this kind of problem. Roll (1969, 1977) shows that the testing asset pricing model suffers an errors-in-variables problem, concluding that (1) no correct and unambiguous test of the theory has appeared in the literature, and (2) practically no possibility exists that such a test can be accomplished in the future. In addition, Lee (1984) shows that the most generalized beta estimate can be decomposed into four parts, true component, bias resulting from measurement error, bias resulting from specification error, and interaction bias. Therefore, besides model misspecification, the evidence of failure in the capital asset pricing model or the finding of new risk factors might result from the errors-in-variables problem in the capital asset pricing test.

To solve the errors-in-variables problem in testing the asset pricing model, several studies focus on beta estimation in the first step. Lee and Jen (1978) and Brennan (1970) show that the possible measurement error in market beta risk is the unobserved market rate of return and risk-free rate of return. To improve the beta estimator, Fabozzi and Francis (1978) and Lee and Chen (1979) use the random coefficient procedure to estimate random coefficient betas. Brennan (1979) and Brown and Warner (1980) also provide different types of market models which can produce different results in predicting rates of return, testing efficient market hypotheses, and measuring security price performance.

Besides improving the beta estimation in the first step, several estimation methods have been developed to solve the problem of errors-in-variables in the asset pricing test. Lee (1973) provides the classical method and mathematical method in correcting errors-in-variables problem. Black et al. (1972), Blume and Friend (1973), Fama and MacBeth (1973), and Litzenberger and Ramaswamy (1979) use the portfolio grouping procedure to minimize the measurement error of the beta coefficient. Shanken (1992) and Jagannathan and Wang (1998) derive asymptotic standard errors of estimated parameters in the second-step regression by maximum likelihood estimation and generalized least squares. In addition, Kim (1995, 1997, 2010) provides a maximum likelihood method to obtain consistent estimations of the test on the capital asset pricing model. Different from other studies, which focus on eliminating the measurement error from the first step of regression or adjusting the standard errors of estimators in the second step, Gibbons (1982) uses a one-step Gauss–Newton procedure. Because the market premium and market risk are estimated simultaneously, the one-step Gauss–Newton procedure can avoid the errors-in-variables problem inherent in the two-step regression approach.

In this chapter, we try to investigate how the errors-in-variables problem can affect the result of the test of the capital asset pricing model. Specifically, we apply three different errors-in-variables estimation methods: the grouping method, the instrumental variable method, and the maximum likelihood method. Using individual US stock price and market index data from 1931 to 2009, we offer an empirical examination of various errors-in-variables

[1]Chapter 7 provides a detailed explanation of the underestimation of beta risk and the overestimation of other risk factors without measured error.

estimation methods in the testing of the capital asset pricing model. Empirical results support the role of market beta in the capital asset pricing model by using maximum likelihood method.

This chapter provides three alternative errors-in-variables estimation models in testing the capital asset pricing model. Section 8.2 presents three alternative correction methods for the errors-in-variables problem. Section 8.3 includes an empirical analysis and results of testing asset pricing models. Section 8.4 discusses the issue of the investment horizon of the beta estimation. Finally, the summary and conclusion appear in Sect. 8.5.

8.2 Empirical Test on Capital Asset Pricing Model

From Sect. 7.4.2, we have discussed how the errors-in-variables problem affects the estimators of second-step regression in testing the asset pricing model. We find that if the estimated beta subjects to errors-in-variables, the results of the asset pricing test may underestimate the price of beta risk and overestimate the other risk factor without measurement error. Various correction methods for errors-in-variables problem are also discussed in Chap. 7, such as grouping method (Fama and MacBeth 1973), Wald's method (Lee 1977), maximum likelihood estimation (MLE) method (Litzenberger and Rmaswamy 1979; Shanken 1992; Kim 1995), price-level testing (Cheng and Grauer 1980), One-step Guass-Norman procedure (Gibbons 1982), generalized method of moments (GMM) (Mackinlay and Richardson 1991), multi-factor, multi-indicator approach (Lee et al. 2015), Bayesian approach (Cederburg and O'Doherty 2015), etc.

More recently, Gu et al. (2018) apply the machine learning technique to measure asset risk premium. Gu et al. (2018) conduct a comparative analysis of methods in the machine learning repertoire, including generalize linear models, dimension reduction, boosted regression trees, random forests, and neural networks. Empirical results show that machine learning has great potential for improving risk premium measurement and return prediction in terms of individual stocks.

In this section, based on two-step capital asset pricing test, we will empirically implement the grouping method, the instrumental variable method, and the maximum likelihood method to deal with measurement errors problem in traditional least square method.

8.2.1 Data

We collect stock prices, stock monthly returns, share codes, exchange codes, number of shares outstanding, and market monthly returns from the Center of Research in Security Prices (CRSP) files. The monthly risk-free rates are taken from Kenneth French's website.[2] The sample period is from 1931 to 2009 (NASDAQ stocks are included during the period 1973–2009). Only common stocks (SHRCD = 10, 11) and firms listed on NYSE, AMEX, or NASDAQ (EXCE = 1, 2, 3, 31, 32, 33) are included in the sample. We also exclude utility services (SICH = 4900–4999) and financial institutions (SICH = 6000–6999).[3] For the purpose of estimating beta in the cross-sectional regression estimation, stocks with at least 60 consecutive monthly returns are used in the cross-sectional regression estimation with individual stock. For the cross-sectional regression estimation for individual stocks, each firm's beta for month t is estimated by using the previous five-year return observations up to month $t - 1$.

[2]For the Kenneth French's website, please see http://mba.tuck.dartmouth.edu/pages/faculty/ken.french/data_library.html.

[3]We filter out those financial institutions and utility firms based on the historical Standard Industrial Code (SIC) available from COMPUSTAT. When a firm's historical SIC is unavailable for a particular year, the next available historical SIC is applied instead. When a firm's historical SIC is unavailable for a particular year and all the years after, we use the current SIC from COMPUSTAT as a substitute.

8.2.2 Grouping Method for Testing Capital Asset Pricing Model

Based upon Wald's (1940) grouping method, Black et al. (1972), Blume and Friend (1973), Fama and MacBeth (1973), and Litzenberger and Ramaswamy (1979) extend two groups to k groups to correct the measurement errors of the beta coefficient in the capital asset pricing test. They group the independent variable, estimated beta, either in descending or ascending order, and divide the observation into k equal groups for both dependent and independent variables; then OLS applies to the subgroup means. The price of beta risk can be written as

$$\hat{\gamma}_{1,t} = \frac{\sum_{p=1}^{k}\left(\bar{\hat{\beta}}_{p,t} - \bar{\hat{\beta}}_t\right)\left(\bar{R}_{p,t} - \bar{R}_t\right)}{\sum_{i=1}^{n}\left(\bar{\hat{\beta}}_{p,t} - \bar{\hat{\beta}}_t\right)^2}, \quad (8.1)$$

where $\bar{\hat{\beta}}_{p,t} = \frac{\sum_{j=1+(p-1)\frac{n}{k}}^{\frac{pn}{k}} \hat{\beta}_{j,t}}{\frac{n}{k}}$, $\bar{R}_{p,t} = \frac{\sum_{j=1+(p-1)\frac{n}{k}}^{\frac{pn}{k}} R_{p,t}}{\frac{n}{k}}$, $\bar{\hat{\beta}}_t = \frac{\sum_{j=1}^{n} \hat{\beta}_{j,t}}{n}$, and $\bar{R}_t = \frac{\sum_{j=1}^{n} R_{j,t}}{n}$.

If $k = 2$, then Eq. (8.1) will reduce to

$$\hat{\gamma}_{1,t} = \frac{\left(\bar{R}_{p,t} - \bar{R}_t\right)}{\left(\bar{\hat{\beta}}_{p,t} - \bar{\hat{\beta}}_t\right)}, \quad (8.2)$$

which is the estimator of Wald's 2-portfolios grouping method.

The grouping method can be used to deal with the measurement error problems in both market return and estimated beta. That is, the grouping method can be used in both the first step and the second step of the two-step asset pricing test. Because the true market portfolio is unobserved in the real world, we can use a proxy of the market return only in testing the capital asset pricing model. The rate of return of S&P 500 index, equal-weighted return on all NYSE, AMEX, and NASDAQ stocks, and value-weighted return on all NYSE, AMEX, and NASDAQ stocks are popular proxies of market return in finance studies. However, without taking the bond market, real estate, and other financial derivatives into account, such proxies of market return may yield measurement errors. To deal with the measurement error of the market return by grouping method, in the first step, we group 60 time-series observations of each stock into k (k = 2, 10, 15, 20, and 30) groups based on their corresponding market returns. The market-adjusted beta for each stock is estimated by the market model using k time-series observations.

Table 8.1 provides uncorrected and five different market-adjusted results of the capital asset pricing test by grouping method. The estimated parameters, $\bar{\hat{\gamma}}_0$ and $\bar{\hat{\gamma}}_1$, are time-series averages of the month-by-month cross-sectional regression coefficients during the period from 1931 to 2009. From the uncorrected ordinary least square model, similar to the previous studies which do not correct the errors-in-variables problem, we can find a significant intercept and a relatively insignificant estimated price of market beta risk. Compared with the average estimators without correction, the average estimators with correction by grouping methods, $\bar{\hat{\gamma}}_0$ and $\bar{\hat{\gamma}}_1$, show no significant improvement. The lack of difference between uncorrected and corrected results from Table 8.1 indicates that the failure of the capital asset pricing test might not occur because of the measurement errors of market returns. Instead, the measurement error of the estimated beta from market model may affect the result of the capital asset price test.

To deal with the errors-in-variables problem of estimated beta from the market model, we can use the grouping method in the second step of asset pricing test. In the first step, each firm's preranking beta for month t is estimated by the market model for previous five-year period. Then firms are formed into N (N = 2, 10, 20, 50, 100, and 200) beta-sorted portfolios in month t. the post-ranking beta of each beta-sorted portfolio in month t is the average of preranking betas of each portfolio. In the second step, the monthly cross-sectional regression coefficients for each portfolio, $\hat{\gamma}_{0t}$ and $\hat{\gamma}_{1t}$, are estimated through the ordinary least square.

Panel A of Table 8.2 presents six different beta-adjusted results of the capital asset pricing

Table 8.1 Grouping method in the first step for testing capital asset pricing model

Time period	Portfolio size (first step)	$R_{it} = \gamma_{0t} + \gamma_{1t}\hat{\beta}_{it-1} + e_{it}$				
		$\bar{\hat{\gamma}}_0$		$\bar{\hat{\gamma}}_1$		$\overline{R^2}$
1931–2009	60 (uncorrected)	0.7715	(5.74)	0.3513	(1.90)	0.0329
	2	0.8174	(6.42)	0.3049	(1.70)	0.0305
	10	0.7682	(6.00)	0.3542	(1.86)	0.0335
	15	0.7667	(5.90)	0.3562	(1.89)	0.0334
	20	0.7645	(5.83)	0.3582	(1.91)	0.0332
	30	0.7693	(5.82)	0.3533	(1.90)	0.0331
1931–1950	60 (uncorrected)	0.7655	(1.96)	1.0590	(1.82)	0.0587
	2	0.7457	(2.22)	1.0765	(1.79)	0.0574
	10	0.6951	(1.95)	1.1274	(1.84)	0.0601
	15	0.7188	(1.96)	1.1044	(1.83)	0.0597
	20	0.7084	(1.90)	1.1151	(1.86)	0.0593
	30	0.7490	(1.98)	1.0758	(1.83)	0.0594
1951–1970	60 (uncorrected)	0.8329	(5.09)	0.0466	(0.21)	0.0316
	2	0.8757	(5.03)	0.0025	(0.01)	0.0272
	10	0.8618	(5.22)	0.0195	(0.09)	0.0314
	15	0.8529	(5.21)	0.0287	(0.13)	0.0317
	20	0.8464	(5.18)	0.0343	(0.16)	0.0316
	30	0.8413	(5.16)	0.0387	(0.18)	0.0315
1971–1990	60 (uncorrected)	0.7575	(3.27)	−0.1441	(−0.52)	0.0192
	2	0.7250	(3.04)	−0.1122	(−0.46)	0.0172
	10	0.7365	(3.20)	−0.1242	(−0.45)	0.0197
	15	0.7308	(3.16)	−0.1174	(−0.42)	0.0194
	20	0.7477	(3.22)	−0.1344	(−0.49)	0.0194
	30	0.7431	(3.22)	−0.1300	(−0.47)	0.0193
1991–2009	60 (uncorrected)	0.7279	(1.07)	0.4486	(0.56)	0.0213
	2	0.9285	(1.33)	0.2502	(0.38)	0.0197
	10	0.7799	(1.18)	0.3961	(0.51)	0.0221
	15	0.7643	(1.15)	0.4119	(0.53)	0.0220
	20	0.7551	(1.13)	0.4210	(0.54)	0.0218
	30	0.7425	(1.10)	0.4328	(0.55)	0.0217

This table provides corrected estimators of the capital asset pricing test by using Wald's grouping method in the first step. 60 time-series observations of each stock are divided into k groups based on their corresponding market returns (k = 2, 10, 15, 20, and 30). The market-adjusted beta for each stock is estimated by the market model using k time-series observations. In the second step, the monthly cross-sectional regression coefficients for each stock, $\hat{\gamma}_{0t}$ and $\hat{\gamma}_{1t}$, are estimated through the ordinary least square. Parameters, $\bar{\hat{\gamma}}_0$ and $\bar{\hat{\gamma}}_1$, are time-series averages of the month-by-month cross-sectional regression coefficient estimates. The sample period runs from January 1931 to December 2009. Coefficients are presented in percentages, and t-statistics are presented in the parentheses. Results of subperiods are also provided

Table 8.2 Grouping method in the second step for testing capital asset pricing model

Panel A: Post-ranking beta is average beta

Time period	Portfolio size (second step)	$R_{it} = \gamma_{0t} + \gamma_{1t}\hat{\beta}_{it-1} + e_{it}$				
		$\bar{\hat{\gamma}}_0$		$\bar{\hat{\gamma}}_1$		$\overline{R^2}$
1931–2009	N (uncorrected)	0.7715	(5.74)	0.3513	(1.90)	0.0329
	2	0.6952	(5.55)	0.4253	(2.11)	–
	10	0.7645	(5.78)	0.3580	(1.89)	0.5633
	20	0.7676	(5.77)	0.3547	(1.89)	0.4502
	50	0.7653	(5.73)	0.3575	(1.91)	0.3117
	100	0.7708	(5.77)	0.3541	(1.90)	0.2184
	200	0.7642	(5.70)	0.3611	(1.95)	0.1428
1931–1950	N (uncorrected)	0.7655	(1.96)	1.0590	(1.82)	0.0587
	2	0.5934	(1.68)	1.2249	(1.91)	–
	10	0.7601	(1.99)	1.0643	(1.79)	0.5812
	20	0.7802	(2.03)	1.0434	(1.77)	0.4605
	50	0.7662	(1.98)	1.0583	(1.80)	0.3100
	100	0.7698	(1.99)	1.0617	(1.81)	0.2115
	200	0.7526	(1.94)	1.0800	(1.86)	0.1310
1951–1970	N (uncorrected)	0.8329	(5.09)	0.0466	(0.21)	0.0316
	2	0.8012	(5.01)	0.0795	(0.35)	–
	10	0.8204	(5.03)	0.0593	(0.27)	0.5117
	20	0.8178	(5.01)	0.0616	(0.28)	0.3947
	50	0.8317	(5.11)	0.0481	(0.22)	0.2624
	100	0.8323	(5.10)	0.0483	(0.22)	0.1778
	200	0.8327	(5.05)	0.0565	(0.26)	0.1127
1971–1990	N (uncorrected)	0.7575	(3.27)	−0.1441	(−0.52)	0.0192
	2	0.7235	(3.19)	−0.1130	(−0.39)	–
	10	0.7524	(3.30)	−0.1397	(−0.49)	0.5632
	20	0.7549	(3.29)	−0.1417	(−0.51)	0.4548
	50	0.7841	(3.24)	−0.1351	(−0.49)	0.3146
	100	0.7506	(3.25)	−0.1373	(−0.50)	0.2186
	200	0.7578	(3.28)	−0.1437	(−0.52)	0.1414
1991–2009	N (uncorrected)	0.7279	(1.07)	0.4486	(0.56)	0.0213
	2	0.6612	(0.47)	0.5142	(0.60)	–
	10	0.7231	(0.74)	0.4530	(0.54)	0.5989
	20	0.7150	(0.70)	0.4611	(0.56)	0.4929
	50	0.7125	(0.70)	0.4638	(0.56)	0.3625
	100	0.7285	(0.76)	0.4485	(0.55)	0.2683
	200	0.7203	(0.72)	0.4565	(0.56)	0.1885

Panel B: Post-ranking beta is market model beta

Time period	Portfolio size (second step)	$R_{it} = \gamma_{0t} + \gamma_{1t}\hat{\beta}_{it-1} + e_{it}$				
		$\bar{\hat{\gamma}}_0$		$\bar{\hat{\gamma}}_1$		$\overline{R^2}$
1931–2009	N (uncorrected)	0.7715	(5.74)	0.3513	(1.90)	0.0329
	2	0.0656	(0.50)	0.4237	(2.10)	–
	10	0.1360	(1.11)	0.3548	(1.88)	0.5634
	20	0.1382	(1.11)	0.3523	(1.88)	0.4508
	50	0.1320	(3.19)	0.3591	(0.72)	0.2436
	100	0.1444	(1.21)	0.3473	(1.87)	0.2184
	200	0.1401	(1.17)	0.3539	(1.91)	0.1426

(continued)

Table 8.2 (continued)

Panel B: Post-ranking beta is market model beta

Time period	Portfolio size (second step)	$R_{it} = \gamma_{0t} + \gamma_{1t}\hat{\beta}_{it-1} + e_{it}$				
		$\bar{\hat{\gamma}}_0$		$\bar{\hat{\gamma}}_1$		$\overline{R^2}$
1931–1950	N (uncorrected)	0.7655	(1.96)	1.0590	(1.82)	0.0587
	2	−0.2856	(−0.74)	1.2178	(1.89)	–
	10	−0.1168	(−0.34)	1.0533	(1.78)	0.5806
	20	−0.1032	(−0.30)	1.0383	(1.77)	0.4628
	50	−0.1315	(1.47)	1.0688	(1.01)	0.2134
	100	−0.1019	(−0.30)	1.0400	(1.80)	0.2096
	200	−0.1117	(−0.33)	1.0595	(1.83)	0.1301
1951–1970	N (uncorrected)	0.8329	(5.09)	0.0466	(0.21)	0.0316
	2	0.0702	(0.42)	0.0787	(0.34)	–
	10	0.0882	(0.55)	0.0598	(0.27)	0.5132
	20	0.0891	(0.56)	0.0587	(0.27)	0.3950
	50	0.1020	(1.16)	0.0460	(−0.21)	0.2076
	100	0.1065	(0.67)	0.0426	(0.20)	0.1788
	200	0.1038	(0.65)	0.0448	(0.21)	0.1127
1971–1990	N (uncorrected)	0.7575	(3.27)	−0.1441	(−0.52)	0.0192
	2	0.3665	(1.89)	−0.1137	(−0.39)	–
	10	0.3961	(2.11)	−0.1409	(−0.50)	0.5629
	20	0.4003	(2.16)	−0.1447	(−0.52)	0.4543
	50	0.3931	(3.86)	−0.1373	(−1.60)	0.2339
	100	0.3943	(2.15)	−0.1384	(−0.50)	0.2186
	200	0.4000	(2.18)	−0.1438	(−0.52)	0.1411
1991–2009	N (uncorrected)	0.7279	(1.07)	0.4486	(0.56)	0.0213
	2	0.1137	(0.16)	0.5164	(1.71)	–
	10	0.1784	(0.26)	0.4518	(1.54)	0.5986
	20	0.1680	(0.25)	0.4623	(1.59)	0.4932
	50	0.1660	(0.25)	0.4624	(1.61)	0.3626
	100	0.1807	(0.76)	0.4501	(1.57)	0.2689
	200	0.1699	(0.72)	0.4602	(1.61)	0.1887

This table provides corrected estimators of the capital asset pricing test by using Wald's grouping method in the second step. Each firm's preranking beta for month t is estimated by using the market model for the previous five-year period. Based on each firm's preranking beta, stocks are formed into N (N = 2, 10, 20, 50, 100, and 200) beta-sorted portfolios in month t. Post-ranking beta of each beta-sorted portfolio in month t is the average of preranking betas of each portfolio. In the second step, the monthly cross-sectional regression coefficients for each portfolio, $\hat{\gamma}_{0t}$ and $\hat{\gamma}_{1t}$, are estimated through the ordinary least square. As shown in Panel A, parameters, $\bar{\hat{\gamma}}_0$ and $\bar{\hat{\gamma}}_1$, are the time-series averages of the month-by-month cross-sectional regression coefficient estimates by using average beta as post-ranking portfolio beta. We also provide another post-ranking beta which is estimated by market model using the previous five-year portfolio monthly returns and market returns. Panel B presents $\bar{\hat{\gamma}}_0$ and $\bar{\hat{\gamma}}_1$, which are the time-series averages of the month-by-month cross-sectional regression coefficient estimates by using portfolio beta from market model as the post-ranking beta. Coefficients are presented in percentages, and t-statistics are presented in the parentheses. The sample period is from January 1931 to December 2009. Results of subperiods are also provided

test. When we group cross-sectional observations into two groups in the second step, the corrected estimator of the price of the beta risk, $\hat{\gamma}_{1t}$, is 0.43% per month and has a t-statistics of 2.11, while the uncorrected estimator of the price of the beta risk has a smaller magnitude with t-statistics of 1.90. Therefore, Wald's two-group method can effectively reduce the errors-in-variables problem of estimated beta and give beta more explanatory power for excess return of the stock.

In addition to taking the average of the betas as post-ranking beta, we also provide another post-ranking beta used by Fama and French (1992). After grouping firms into N ($N = 2$, 10, 20, 50, 100, and 200) beta-sorted portfolios in month t, we run a time-series market model to regress the previous five-year portfolio monthly returns on market returns and obtain the post-ranking beta of the portfolio. Panel B of Table 8.2 presents the time-series averages of the month-by-month cross-sectional regression coefficient estimates by using the portfolio beta from the market model as the post-ranking beta. Similar to the results of Panel A of Table 8.2, two-group grouping method yields a higher price of the beta risk and supports the validation of the capital asset pricing model.

We further use the grouping method in both first step and second step to deal with measurement error of market return and estimated beta. Table 8.3 shows that only the two-group grouping method used in the second step can improve the price of the beta risk, which is market premium. It is consistent with the results from Tables 8.1 and 8.2.

8.2.3 Instrumental Variable Method for Testing Capital Asset Pricing Model

In this section, we will empirically use the instrumental method to correct measurement errors of market return and estimated beta in the capital asset pricing test. We also combine the grouping method and the instrumental variable method to test the capital asset pricing model.

Durbin's instrumental variable method can also deal with measurement errors of market return and estimated beta in the capital asset pricing test. To make the adjustment of the measurement errors problem of market return, we implement the instrumental variable method in the first step of capital asset pricing test. In the first step, the instrumental variables for 60 time-series observations of each stock are rank orders of their corresponding market returns. The estimated coefficients of the market model for each security can be estimated by the following:

$$\begin{bmatrix} \hat{\alpha}^{\mathrm{IV}} \\ \hat{\beta}^{\mathrm{IV}} \end{bmatrix} = (\mathbf{Z}'\mathbf{R_m})^{-1}\mathbf{Z}'\mathbf{R_i}, \tag{8.3}$$

where

$$\mathbf{Z}' = \begin{bmatrix} 1 & 1 & 1 & \cdots & 1 \\ 1 & 2 & 3 & \cdots & 60 \end{bmatrix},$$

$$\mathbf{R}'_{\mathbf{m}} = \begin{bmatrix} 1 & 1 & 1 & \cdots & 1 \\ R_{m,t1} & R_{m,t2} & R_{m,t3} & \cdots & R_{m,t60} \end{bmatrix}, \text{and}$$

$$\mathbf{R}'_{\mathbf{i}} = \begin{bmatrix} R_{i,t1} & R_{i,t2} & R_{i,t3} & \cdots & R_{i,t60} \end{bmatrix}.$$

The $\hat{\beta}^{\mathrm{IV}}$ is corrected by the instrumental variable method dealing with the measurement error of market return, and can be used in the second step of the capital asset pricing test.

Durbin (1954) provides t-statistics which can be used to test whether estimators obtained by OLS and Durbin's instrumental variable method are significantly different. The t-statistics is defined as

$$t = \frac{\hat{\beta}^{\mathrm{IV}} - \hat{\beta}^{\mathrm{OLS}}}{\text{s.e.}\left(\hat{\beta}^{\mathrm{OLS}}\right)\left(\frac{1}{\rho^2} - 1\right)^{1/2}},$$

where

$\hat{\beta}^{\mathrm{IV}}$	is the beta coefficient estimated by instrumental variable method,
$\hat{\beta}^{\mathrm{OLS}}$	is the beta coefficient estimated by ordinary least square,
$\text{s.e.}\left(\hat{\beta}^{\mathrm{OLS}}\right)$	is the standard error of beta coefficient estimated by the ordinary least square, and
ρ	is the correlation coefficient between the market return and the instrumental variable.

Table 8.3 Grouping method in both first step and second step for testing capital asset pricing model

Portfolio size (first step/second step)		$R_{it} = \gamma_{0t} + \gamma_{1t}\hat{\beta}_{it-1} + e_{it}$				
		$\bar{\hat{\gamma}}_0$		$\bar{\hat{\gamma}}_1$		$\overline{R^2}$
60	N	0.7715	(5.74)	0.3513	(1.90)	0.0329
2	2	0.7544	(6.28)	0.3661	(1.95)	–
2	10	0.8046	(6.41)	0.3169	(1.75)	0.5455
2	20	0.8096	(6.40)	0.3123	(1.73)	0.4373
2	50	0.8197	(6.48)	0.3028	(1.68)	0.2981
2	100	0.8177	(6.45)	0.3056	(1.70)	0.2076
2	200	0.8215	(6.47)	0.3054	(1.71)	0.1339
10	2	0.6881	(5.61)	0.4326	(2.15)	–
10	10	0.7505	(5.97)	0.3713	(1.90)	0.5664
10	20	0.7602	(6.00)	0.3618	(1.87)	0.4546
10	50	0.7675	(6.03)	0.3554	(1.85)	0.3164
10	100	0.7639	(6.00)	0.3585	(1.87)	0.2222
10	200	0.7663	(5.99)	0.3584	(1.86)	0.1453
15	2	0.7173	(5.78)	0.4046	(2.04)	–
15	10	0.7560	(5.91)	0.3668	(1.90)	0.5689
15	20	0.7648	(5.94)	0.3583	(1.87)	0.4573
15	50	0.7660	(5.93)	0.3570	(1.88)	0.3175
15	100	0.7711	(5.98)	0.3511	(1.85)	0.2228
15	200	0.7642	(5.89)	0.3567	(1.89)	0.1454
20	2	0.6857	(5.56)	0.4355	(2.16)	–
20	10	0.7538	(5.85)	0.3688	(1.92)	0.5666
20	20	0.7597	(5.85)	0.3627	(1.90)	0.4551
20	50	0.7630	(5.86)	0.3602	(1.90)	0.3139
20	100	0.7671	(5.89)	0.3542	(1.88)	0.2204
20	200	0.7656	(5.85)	0.3529	(1.88)	0.1438
30	2	0.6886	(5.49)	0.4320	(2.16)	–
30	10	0.7591	(5.83)	0.3631	(1.90)	0.5670
30	20	0.7625	(5.82)	0.3598	(1.90)	0.4544
30	50	0.7654	(5.84)	0.3575	(1.90)	0.3143
30	100	0.7682	(5.83)	0.3550	(1.89)	0.2208
30	200	0.7624	(5.80)	0.3615	(1.92)	0.1439

This table provides the corrected estimators of the capital asset pricing test by using Wald's grouping method in both first step and second step. In the first step, 60 time-series observations of each stock are divided into k groups based on their corresponding market returns (k = 2, 10, 15, 20, and 30). The market-adjusted preranking beta for each stock is estimated by the market model using k time-series observations. In the second step, based on each firm's preranking beta, portfolios are formed into N (N = 2, 10, 20, 50, 100, and 200) beta-sorted portfolios in month t. Post-ranking beta of each beta-sorted portfolio in month t is the average of preranking betas of each portfolio. The monthly cross-sectional regression coefficients for each portfolio, $\hat{\gamma}_{0t}$ and $\hat{\gamma}_{1t}$, are estimated through the ordinary least square. Parameters, $\bar{\hat{\gamma}}_0$ and $\bar{\hat{\gamma}}_1$, are the time-series averages of the month-by-month cross-sectional regression coefficient estimates by using average beta as post-ranking portfolio beta. Coefficients are presented in percentages, and t-statistics are presented in the parentheses. The sample period is from January 1931 to December 2009

Table 8.4 Instrumental variable method in the first step of capital asset pricing test

Time period	OBS	$\bar{\bar{\beta}}^{OLS}$	$\bar{\bar{\beta}}^{IV}$	95% significant (OBS)	95% significant (%)	99% significant (OBS)	99% significant (%)
1931–2009	1,760,755	1.0111	1.0157	139,854	7.94	36,836	2.09
1931–1950	138,676	1.0054	1.0065	18,255	13.16	6487	4.68
1951–1970	220,594	1.0299	1.0288	11,921	5.40	2254	1.02
1971–1990	596,591	1.0250	1.0283	42,693	7.16	10,302	1.73
1991–2009	804,894	0.9967	1.0043	66,985	8.32	17,793	2.21

This table provides corrected estimators of the market model by using Durbin's instrumental variable method. The uncorrected beta for each stock, $\hat{\beta}^{OLS}$, is estimated through the market model by least square method. The adjusted beta for each stock, $\hat{\beta}^{IV}$, is estimated through the Durbin's instrumental variable method. Parameters, $\bar{\bar{\beta}}^{OLS}$ and $\bar{\bar{\beta}}^{IV}$, are time-series and cross-sectional averages of uncorrected and corrected estimators. This table also presents the number of observations and the percentage of observations that are significant under the null hypothesis that the absolute value of $\hat{\beta}^{IV}$ is greater than the absolute value of $\hat{\beta}^{OLS}$ at 95 and 99% significance levels. The sample period runs from January 1931 to December 2009. Results of subperiods are also provided

Table 8.4 presents the time-series and cross-sectional averages of the estimated beta using ordinary least square and the instrumental variable method. During the period from 1931 to 2009, we find that only 7.94% (2.09%) of 1.76 million firm-years have corrected betas significantly different from uncorrected betas at the 95% (99%) significant level. Consistent with the result derived from the grouping method used in the first step of the capital asset pricing test, we find little evidence that the measurement error of the rate of return on the market index will lead to a significant bias on the beta estimation.

To make the adjustment of the measurement errors problem of estimated beta from the market model, we use the instrumental variable method in the second step of the capital asset pricing test. In the second step, the instrumental variables for cross-sectional observations in month t are rank orders of the estimated betas of the cross-sectional observations. Then the price of beta risk in month t can be estimated by the following:

$$\begin{bmatrix} \hat{\gamma}_{0,t} \\ \hat{\gamma}_{1,t} \end{bmatrix} = \left(\mathbf{Z}'\hat{\beta}\right)^{-1}\mathbf{Z}'\mathbf{R}_{\mathrm{i}}, \tag{8.4}$$

where

$$\mathbf{Z}' = \begin{bmatrix} 1 & 1 & 1 & \cdots & 1 \\ 1 & 2 & 3 & \cdots & N \end{bmatrix},$$

$$\boldsymbol{\beta}_{\mathrm{i}}' = \begin{bmatrix} 1 & 1 & 1 & \cdots & 1 \\ \beta_{i1,t} & \beta_{i2,t} & \beta_{i3,t} & \cdots & \beta_{iN,t} \end{bmatrix}, \text{and}$$

$$\mathbf{R}_{\mathrm{i}}' = [R_{iN,t} - R_{f,t} \quad R_{iN,t} - R_{f,t} \quad R_{iN,t} - R_{f,t} \quad \cdots \quad R_{iN,t} - R_{f,t}].$$

The t-statistics under the null hypothesis that the risk price estimated by the instrumental variable method is different from the risk price estimated by the ordinary least square is defined as

$$t = \frac{\hat{\gamma}_{1,t}^{IV} - \hat{\gamma}_{1,t}^{OLS}}{\text{s.e.}\left(\hat{\gamma}_{1,t}^{OLS}\right)\left(\frac{1}{\rho^2} - 1\right)^{1/2}}, \tag{8.5}$$

where

$\hat{\gamma}_{1,t}^{IV}$ is the beta coefficient estimated by instrumental variable method,

$\hat{\gamma}_{1,t}^{OLS}$ is the beta coefficient estimated by ordinary least square,

$\text{s.e.}\left(\hat{\gamma}_{1,t}^{OLS}\right)$ is the standard error of beta coefficient estimated by ordinary least square, and

ρ is the correlation coefficient between the beta risk and the instrumental variable.

Table 8.5 presents the results of the instrumental variable method for testing the capital asset pricing model. Uncorrected estimators, estimators corrected in the first step, estimators corrected in the second step, and estimators corrected in both first step and second step are presented in order. Results show that, after correcting by instrumental variables method in the second step, the corrected price of beta risk has been improved in both magnitude and significant levels. Moreover, comparing to the ordinary least square used in the second step, the instrumental variable method used in the second step yields a significant different risk premium in 15.83% (6.25%) to 32.92% (25.42%) of 948 months at the 95% (99%) significant level. It therefore indicates that the measurement error of the estimated beta might be a serious problem in the capital asset pricing model test and resulting different conclusions of the validation of the capital asset pricing model.

8.2.4 Applying Instrumental Variable Methods into Grouping Sample

We further use both the grouping method and the instrumental variable method to test the capital asset pricing model. We group time-series observations into k ($k = 2$, 10, 15, 20, and 30) groups in the first step to deal with the measurement error problem of market return. In the second step, we group cross-sectional observations into N ($N = 2$, 10, 20, 50, 100, and 200) groups and further use the instrumental variable method to correct the errors-in-variables problem of the estimated beta. Table 8.6 show results of applying instrumental variable methods into grouping data for testing the capital asset pricing model. Besides results using two-group grouping methods in the first step ($k = 2$), results combining grouping and instrumental variable methods all yield higher estimated market premium in both magnitude and significant levels than the uncorrected model.

8.2.5 Maximum Likelihood Method for Testing Capital Asset Pricing Model

As discussed in the previous section, Kim's (1995) maximum likelihood method can only deal with the errors-in-variables problem of the estimated beta from the market model, but the maximum likelihood method can test, besides the capital asset pricing model, the multifactor asset pricing model. In the empirical study, we use the maximum likelihood method to test both the capital asset pricing model and the multifactor asset pricing model which include the size factor. The corrected estimators of cross-sectional regression coefficients through the maximum likelihood method can be obtained by Eqs. (7.45), (7.46), and (7.47). Because the size factor is free from measurement error, the estimated coefficient of size factor in the multifactor asset pricing model will be overstated if a measurement error of the estimated beta exists. Therefore, we expect a relatively small and insignificant estimator of the size premium and a relatively large and significant estimator of the market premium.

Table 8.7 presents the time-series average of the cross-sectional regression estimators by the maximum likelihood method and by traditional ordinary least square method. After correcting the errors-in-variables problem by the maximum likelihood method, for the capital asset pricing model test, we can find that the estimated intercept decreases in both magnitude and significant level, and the estimated price of beta risk increases and turns to significant. For the multifactor asset pricing model, the estimated intercept and the estimated coefficient of the size factor decrease in both magnitude and significance level; and the estimated price of beta risk turns to significant. Our results therefore show that the evidence of failure in the capital asset pricing model or the finding of new risk factors might occur because of the errors-in-variables problem rather than model misspecification.

Table 8.5 Instrumental variable method for testing capital asset pricing model

Time period	Method (1st step/2nd step)		$R_{it} = \gamma_{0t} + \gamma_{1t}\hat{\beta}_{it-1} + e_{it}$												
			$\bar{\hat{\gamma}}_0$	*T*-stats.	Compare to OLS/OLS (95%/99%) (%)		Compare to IV/OLS (95%/99%) (%)		$\bar{\hat{\gamma}}_1$	*t*-stats.	Compare to OLS/OLS (95%/99%) (%)		Compare to IV/OLS (95%/99%) (%)		$\overline{R^2}$
1931–2009	OLS	OLS	0.7715	(5.74)					0.3513	(1.90)					0.0329
	IV	OLS	0.7884	(6.25)					0.3338	(1.75)					0.0331
	OLS	IV	0.7346	(5.77)	18.88	10.44			0.3877	(2.02)	21.84	14.35			–
	IV	IV	0.7491	(6.21)	26.37	18.04	18.57	11.08	0.3725	(1.90)	27.95	20.89	22.15	14.45	–
1931–1950	OLS	OLS	0.7655	(1.96)					1.0590	(1.82)					0.0587
	IV	OLS	0.6992	(2.02)					1.1224	(1.81)					0.0601
	OLS	IV	0.7022	(1.93)	17.08	10.83			1.1208	(1.86)	18.33	12.50			–
	IV	IV	0.6425	(1.96)	28.75	23.75	17.08	8.75	1.1774	(1.85)	29.58	25.42	18.33	10.83	–
1951–1970	OLS	OLS	0.8329	(5.09)					0.0466	(0.21)					0.0316
	IV	OLS	0.8709	(5.24)					0.0107	(0.05)					0.0310
	OLS	IV	0.8200	(5.16)	14.17	4.17			0.0605	(0.27)	15.83	6.25			–
	IV	IV	0.8505	(5.29)	19.58	9.58	13.75	8.33	0.0323	(0.15)	20.83	10.83	15.83	10.42	–
1971–1990	OLS	OLS	0.7575	(3.27)					−0.1441	(−0.52)					0.0192
	IV	OLS	0.7383	(3.24)					−0.1259	(−0.45)					0.0192
	OLS	IV	0.7292	(3.23)	19.17	12.92			−0.1177	(−0.41)	22.08	15.83			–
	IV	IV	0.7035	(3.16)	30.83	21.67	21.69	13.33	−0.0934	(−0.32)	32.92	23.75	23.33	15.42	–
1991–2009	OLS	OLS	0.7279	(1.07)					0.4486	(0.56)					0.0213
	IV	OLS	0.8482	(1.27)					0.3276	(0.45)					0.0217
	OLS	IV	0.6846	(1.08)	25.44	14.04			0.4925	(0.59)	31.58	23.25			–
	IV	IV	0.8026	(1.27)	26.32	17.11	21.93	14.04	0.3738	(0.49)	28.51	23.68	31.58	21.49	–

This table provides corrected estimators of the capital asset pricing test by using Durbin's instrumental variable approach in both first step and second step. In the first step, the instrumental variables for 60 time-series observations of each stock are ranking orders of their corresponding market returns. Then, the beta of each security can be estimated by the instrumental variable method. In the second step, the instrumental variables for cross-sectional observations in month t are the rank orders of the estimated betas of cross-sectional observations. The monthly cross-sectional regression coefficients in month t, $\hat{\gamma}_{0t}$ and $\hat{\gamma}_{1t}$, are estimated through the instrumental variable method. Parameters, $\bar{\hat{\gamma}}_0$ and $\bar{\hat{\gamma}}_1$, are the time-series averages of the month-by-month cross-sectional regression coefficient estimates. The t-statistics are provided under the null hypothesis that the estimator is different from zero. This table also presents the number of observations and the percentage of observations that are significant under the null hypothesis that the absolute value of corrected estimator is greater than the absolute value of the uncorrected estimator at 95 and 99% significance levels. Coefficients are presented in percentages, and t-statistics are presented in the parentheses. The sample period is from January 1931 to December 2009. Results of subperiods are also provided

Table 8.6 Applying instrumental variable methods into grouping sample for testing capital asset pricing model

Estimating method (first step/second step)		$R_{it} = \gamma_{0t} + \gamma_{1t}\hat{\beta}_{it-1} + e_{it}$							
		$\bar{\hat{\gamma}}_0$	*t*-stats.	Compare to OLS/OLS (95%/ 99%) (%)		$\bar{\hat{\gamma}}_1$	*t*-stats.	Compare to OLS/OLS (95%/ 99%) (%)	
60 (OLS)	*N* (OLS)	0.7715	(5.74)	–	–	0.3513	(1.90)	–	–
2 (OLS)	2 (IV)	0.7544	(6.28)	–	–	0.3661	(1.95)	–	–
2 (OLS)	10 (IV)	0.7773	(6.36)	8.44	0.32	0.3440	(1.88)	14.03	3.06
2 (OLS)	20 (IV)	0.7754	(6.34)	14.45	3.80	0.3459	(1.89)	18.46	7.17
2 (OLS)	50 (IV)	0.7777	(6.37)	17.19	7.07	0.3439	(1.88)	20.89	11.50
2 (OLS)	100 (IV)	0.7767	(6.35)	17.72	8.86	0.3456	(1.89)	21.31	12.66
2 (OLS)	200 (IV)	0.7795	(6.36)	18.04	9.07	0.3464	(1.91)	20.99	12.76
2 (OLS)	*N* (IV)	0.7777	(6.35)	18.35	10.13	0.3437	(1.88)	20.05	13.50
10 (OLS)	2 (IV)	0.6881	(5.61)	–	–	0.4326	(2.15)	–	–
10 (OLS)	10 (IV)	0.7253	(5.96)	12.66	0.32	0.3964	(2.02)	16.88	3.16
10 (OLS)	20 (IV)	0.7289	(6.00)	16.35	4.75	0.3928	(2.00)	20.36	8.65
10 (OLS)	50 (IV)	0.7316	(6.01)	17.83	8.02	0.3909	(2.00)	21.73	11.92
10 (OLS)	100 (IV)	0.7294	(5.99)	18.67	9.18	0.3927	(2.01)	21.31	12.13
10 (OLS)	200 (IV)	0.7304	(5.98)	18.57	9.18	0.3939	(2.00)	22.15	12.45
10 (OLS)	*N* (IV)	0.7315	(5.99)	18.57	11.39	0.3904	(2.00)	23.31	13.71
15 (OLS)	2 (IV)	0.7173	(5.78)	–	–	0.4046	(2.04)	–	–
15 (OLS)	10 (IV)	0.7304	(5.92)	10.34	0.21	0.3921	(2.02)	14.45	3.27
15 (OLS)	20 (IV)	0.7320	(5.93)	15.08	4.64	0.3908	(2.01)	19.20	8.44
15 (OLS)	50 (IV)	0.7316	(5.93)	16.77	7.91	0.3909	(2.01)	20.68	11.18
15 (OLS)	100 (IV)	0.7355	(5.96)	18.35	8.65	0.3862	(1.99)	20.89	11.92
15 (OLS)	200 (IV)	0.7334	(5.93)	18.46	8.97	0.3871	(2.00)	21.10	12.76
15 (OLS)	*N* (IV)	0.7323	(5.92)	18.78	10.55	0.3902	(2.01)	21.94	13.61
20 (OLS)	2 (IV)	0.6857	(5.56)	–	–	0.4355	(2.16)	–	–
20 (OLS)	10 (IV)	0.7283	(5.86)	11.18	0.63	0.3940	(2.03)	14.98	3.06
20 (OLS)	20 (IV)	0.7274	(5.85)	14.24	5.06	0.3948	(2.04)	17.19	8.12
20 (OLS)	50 (IV)	0.7279	(5.85)	16.88	8.02	0.3950	(2.04)	20.89	10.13
20 (OLS)	100 (IV)	0.7304	(5.87)	17.83	7.91	0.3904	(2.02)	21.73	10.86
20 (OLS)	200 (IV)	0.7281	(5.83)	18.57	8.86	0.3899	(2.02)	22.47	12.13
20 (OLS)	*N* (IV)	0.7274	(5.83)	18.78	10.13	0.3949	(2.04)	21.84	13.82
30 (OLS)	2 (IV)	0.6886	(5.49)	–	–	0.4320	(2.16)	–	–
30 (OLS)	10 (IV)	0.7275	(5.80)	11.81	0.21	0.3944	(2.04)	16.88	3.48
30 (OLS)	20 (IV)	0.7281	(5.79)	14.03	4.22	0.3939	(2.05)	17.30	8.54
30 (OLS)	50 (IV)	0.7303	(5.83)	16.88	7.59	0.3922	(2.04)	19.41	11.08
30 (OLS)	100 (IV)	0.7315	(5.82)	17.19	8.23	0.3913	(2.03)	19.94	11.71
30 (OLS)	200 (IV)	0.7259	(5.80)	17.30	8.97	0.3975	(2.05)	20.57	11.60
30 (OLS)	*N* (IV)	0.7304	(5.81)	18.46	10.34	0.3917	(2.04)	22.15	13.92
60 (OLS)	2 (IV)	0.6952	(5.55)	–	–	0.4253	(2.11)	–	–
60 (OLS)	10 (IV)	0.7332	(5.77)	10.02	0.21	0.3889	(2.02)	15.51	2.85
60 (OLS)	20 (IV)	0.7352	(5.80)	13.50	3.90	0.3869	(2.02)	17.41	7.91
60 (OLS)	50 (IV)	0.7335	(5.78)	16.56	6.96	0.3889	(2.03)	18.88	10.86

(continued)

Table 8.6 (continued)

Estimating method (first step/second step)		$R_{it} = \gamma_{0t} + \gamma_{1t}\hat{\beta}_{it-1} + e_{it}$							
		$\bar{\hat{\gamma}}_0$	*t*-stats.	Compare to OLS/OLS (95%/ 99%) (%)		$\bar{\hat{\gamma}}_1$	*t*-stats.	Compare to OLS/OLS (95%/ 99%) (%)	
60 (OLS)	100 (IV)	0.7348	(5.79)	17.19	8.33	0.3897	(2.03)	20.57	11.39
60 (OLS)	200 (IV)	0.7300	(5.75)	18.04	8.97	0.3948	(2.06)	21.31	12.24
60 (OLS)	*N* (IV)	0.7346	(5.77)	18.88	10.44	0.3877	(2.02)	21.84	14.35

This table provides the corrected estimators of the capital asset pricing test by combining grouping method and second step and instrumental variables method. In the first step, 60 time-series observations of each stock are divided into *k* groups based on their corresponding market returns (*k* = 2, 10, 15, 20, and 30). The market-adjusted preranking beta for each stock is estimated by the market model using *k* time-series observations. In the second step, based on each firm's preranking beta, portfolios are formed into *N* (N = 2, 10, 20, 50, 100, and 200) beta-sorted portfolios in month *t*. Post-ranking beta of each beta-sorted portfolio in month *t* is the average of preranking betas of each portfolio. The instrumental variables for cross-sectional beta-sorted portfolios in month t are the rank orders of estimated betas of beta-sorted portfolios. The monthly cross-sectional regression coefficients in month *t*, $\hat{\gamma}_{0t}$ and $\hat{\gamma}_{1t}$, are estimated through the instrumental variable method. $\bar{\hat{\gamma}}_0$ and $\bar{\hat{\gamma}}_1$, are the time-series averages of the month-by-month cross-sectional regression coefficient estimates. The *t*-statistics are provided under the null hypothesis that the estimator is different from zero. This table also presents the number of observations and the percentage of observations that are significant under the null hypothesis that the absolute value of corrected estimator is greater than the absolute value of uncorrected estimator at 95 and 99% significance levels. Coefficients are presented in percentages, and *t*-statistics are presented in the parentheses. The sample period runs from January 1931 to December 2009

8.2.6 Asset Pricing Model Tests with Individual Stocks

As discussed in Chap. 7, the grouping method in testing asset pricing model may have some limitations. For example, the grouping method may reduce statistical power in the first step of the asset pricing model test. That is, the grouping method is a trade-off between the bias and the variance of the first-pass estimator. In addition, the formation of portfolios for the second-step estimation may cause a loss of valuable information about cross-sectional behavior among individual securities. Ahn et al. (2009) demonstrate that different results may be obtained by using different portfolio grouping criteria. Shanken (1992) therefore uses individual stocks and applies maximum likelihood estimation to examine the asset pricing model. Kim (1995) also applies maximum likelihood method to mitigate errors-in-variables problem and therefore individual stock can be used in the examination of asset pricing model.

More recently, Cederburg and O'Doherty (2015) introduce a hierarchical Bayesian approach and uses US firm-level data to re-examine financial anomalies under capital asset pricing model. Their empirical results, supporting the capital asset pricing model, show that relations between firm characteristics and firm-level alphas are primarily confined to small stocks. Jegadeesh et al. (2018) propose an instrumental variable method allowing the use of individual stocks and small samples. They apply the instrumental variable method to test asset pricing models, including the standard CAPM, the Fama–French three- and five-factor models, the *q*-factor asset pricing model, and liquidity-adjusted CAPM. They find premiums on risk factors are all insignificant after controlling for asset characteristics.

8.3 Normality Test for Time-Series Estimators and Future Research

Since Fama and MacBeth (1973) introduced a rolling two-step procedure to test the capital asset pricing model, this procedure has become a standard in the empirical capital asset pricing test. Although the Fama–MacBeth procedure allows estimators to change over time, it still presents some problems in estimating and testing coefficients. Petersen (2009) points out the drawbacks of using OLS on the panel data analysis, stating that the residual may be

Table 8.7 Maximum likelihood method for testing capital asset pricing model

Time period	$R_{it} = \gamma_{0,t} + \gamma_{1,t}\hat{\beta}_{i,t-1} + e_{i,t}/R_{i,t} = \gamma_{0,t} + \gamma_{1,t}\hat{\beta}_{i,t-1} + \gamma_{2,t}\log(V_{i,t-1}) + e_{i,t}$											
	MLE corrected estimators						Uncorrected estimators					
	$\bar{\hat{\gamma}}_0$		$\bar{\hat{\gamma}}_1$		$\bar{\hat{\gamma}}_2$		$\bar{\hat{\gamma}}_0$		$\bar{\hat{\gamma}}_1$		$\bar{\hat{\gamma}}_2$	
1931–2009	0.1870	(1.16)	0.9534	(3.60)			0.7715	(5.74)	0.3513	(1.90)		
	0.8630	(2.52)	1.0706	(3.40)	−0.0696	(−2.96)	3.0689	(6.01)	0.0258	(0.34)	−0.1945	(−4.90)
1931–1950	−0.1172	(−0.26)	1.8187	(2.21)			0.7655	(2.25)	1.0590	(1.65)		
	0.6444	(0.64)	2.2536	(2.24)	−0.1116	(−1.70)	5.1474	(3.28)	0.1752	(0.76)	−0.3935	(−3.16)
1951–1970	0.6029	(3.95)	0.2896	(1.41)			0.8329	(3.91)	0.0466	(0.54)		
	1.2403	(3.29)	0.3000	(1.32)	−0.0536	(−1.7)	1.7041	(2.66)	0.0260	(0.23)	−0.0736	(−1.50)
1971–1990	0.5464	(2.46)	0.1204	(0.39)			0.7575	(2.57)	−0.1441	(−0.94)		
	0.8502	(1.29)	0.2058	(0.43)	−0.0350	(−0.81)	2.1360	(2.78)	−0.2638	(−2.21)	−0.1158	(−1.99)
1991–2009	−0.3088	(−0.28)	1.6183	(1.02)			0.7279	(1.10)	0.4486	(0.67)		
	0.7095	(0.49)	1.5465	(1.04)	−0.0787	(−0.68)	3.2995	(1.44)	0.1733	(0.55)	−0.1951	(−1.17)

This table provides the corrected estimators of the capital asset pricing test by using maximum likelihood method. The corrected estimators are the monthly cross-sectional regression coefficients through the maximum likelihood method, and the uncorrected estimators are those estimated through the ordinary least square. Parameters $\bar{\hat{\gamma}}_0$, $\bar{\hat{\gamma}}_1$, and $\bar{\hat{\gamma}}_2$ are the time-series averages of the month-by-month cross-sectional regression coefficients. Coefficients are presented in percentages, and t-statistics are presented in parentheses. The sample period runs from January 1931 to December 2009. Results of subperiods are also provided

correlated across firms or across time, resulting in biased standard errors of OLS estimators and wrong inference in testing. Although the Fama–MacBeth procedure adjusts the potential time-series correlation of the residual, Petersen (2009) shows that the Fama–MacBeth procedure cannot deal with the potential problem of the cross-sectional correlation of the residual. Moreover, to avoid the consideration of data mining, Graham et al. (2015) introduce a new multiple framework to test the various asset pricing models and suggest the hurdle of statistical significance cutoffs in asset pricing tests has better to be greater than 3.0. Harvey (2017) demonstrates that, besides reporting the usual *p*-value, reporting a Bayesianized *p*-value can reduce the impact of data mining and *p*-hacking.

If the time-series estimators in the second step are not distributed normally or symmetrically, inferences from *t*-test may be invalid. Tables 8.8, 8.9, 8.10, 8.11, 8.12 and 8.13 present the descriptive statistics and results of normality test of time-series estimators from various errors-in-variables estimation methods. We can find that most of estimators are not normally distributed. For example, in the Panel A of Table 8.9, the estimated slopes, $\hat{\gamma}_1$s, are positively skewed and with kurtosis greater than 3 for all different group sizes and time periods. The estimated intercepts, $\hat{\gamma}_0$s, are not skewed as much as the estimated slopes, but their kurtoses are greater than 3 for all different group sizes and time periods. We further use the Jarque–Bera (1987) test to check whether the time-series estimators departure from normality or not. For the various grouping methods and different time periods, the Jarque–Bera tests reject the null hypothesis that the time-series estimators are normally distributed at the 99% significance level. Therefore, using *t*-test may be not adequate because of the non-normal (or non-*t*) distribution of the time-series estimators in Fama–MacBeth procedure. The capital asset pricing test can be further improved by taking care of potential problems in Fama–MacBeth procedure. We will address this issue in future research.

8.4 The Investment Horizon of Beta Estimation

Issues of the data frequency and the investment horizon in estimating beta coefficient have been documented in the literature. In terms of the frequency of beta estimation, Lee (1976) derives a generalized functional form for CAPM with an assumption that the true investment horizon is finite and unobservable. Lee et al. (1990) explores the role of investment horizon for asset pricing. They find that some empirical CAPM tests ignore the discrepancy between the observed data periods and the true investment horizons and, therefore, reject CAPM. In addition, Gilbert et al. (2014) introduce uncertainty about the effect of systematic news on firm value into a frictionless model and show that additional factors in the asset pricing model are necessary at high frequencies, while the beta estimation is appropriate at low frequencies. Brennan and Zhang (2018) present an extended version of the capital asset pricing model which uses a probability distribution over different horizons to estimate parameters. They show that the probability distribution over horizon dates varies over time and the extended model outperforms standard CAPM and Fama-French 3-factor model.

Time aggregation on beta can also affect the specification of beta estimates. Cartwright and Lee (1987) show that the simple static market model is misspecified and it is found that

Table 8.8 Normality test for the time-series estimators of grouping method in the first step for testing capital asset pricing model

Portfolio size (second step)	$R_{it} = \gamma_{0t} + \gamma_{1t}\hat{\beta}_{it-1} + e_{it}$															
	$\hat{\gamma}_0$								$\hat{\gamma}_1$							
	Mean	Median	*t*-stats.	STD	SK	KU	J-B	*p*-value	Mean	Median	*t*-stats.	STD	SK	KU	J-B	*p*-value
1931–2009																
2	0.8174	1.0591	(6.42)	3.9186	−0.7898	5.2579	1191	$<10^{-4}$	0.3049	−0.1285	(1.70)	5.5236	3.4958	28.6309	34,310	$<10^{-4}$
10	0.7682	0.9579	(6.00)	3.9423	−0.6632	6.5984	1789	$<10^{-4}$	0.3542	−0.1874	(1.86)	5.8699	3.2088	25.0465	26,406	$<10^{-4}$
15	0.7667	0.9679	(5.90)	4.0023	−0.5028	7.8136	2452	$<10^{-4}$	0.3562	−0.1734	(1.89)	5.8176	3.2138	26.0927	28,525	$<10^{-4}$
20	0.7645	0.7925	(5.83)	4.0355	−0.4303	8.3373	2775	$<10^{-4}$	0.3582	−0.2242	(1.91)	5.7864	3.2009	26.2341	28,804	$<10^{-4}$
30	0.7693	0.9557	(5.82)	4.0730	−0.2575	9.1904	3347	$<10^{-4}$	0.3533	−0.1937	(1.90)	5.7396	3.1191	25.5840	27,391	$<10^{-4}$
60 (uncorrected)	0.7715	0.9751	(5.74)	4.1419	0.0323	11.6176	5331	$<10^{-4}$	0.3533	−0.1865	(1.90)	5.7060	3.0810	25.4766	27,138	$<10^{-4}$
1931–1950																
2	0.7457	0.7349	(2.22)	5.1959	−0.4418.	3.2779	115	$<10^{-4}$	1.0765	0.0177	(1.79)	9.3003	2.4845	11.1005	1547	$<10^{-4}$
10	0.6951	0.7457	(1.95)	5.5364	−0.2787	3.8937	155	$<10^{-4}$	1.1274	−0.2346	(1.84)	9.4777	2.5101	11.8340	1652	$<10^{-4}$
15	0.7188	0.7388	(1.96)	5.6900	−0.0924	4.9062	241	$<10^{-4}$	1.1044	−0.1455	(1.83)	9.3440	2.5612	12.7938	1899	$<10^{-4}$
20	0.7084	0.7311	(1.90)	5.7643	0.0046	5.3426	285	$<10^{-4}$	1.1151	−0.1745	(1.86)	9.2649	2.5677	13.0690	1972	$<10^{-4}$
30	0.7490	0.7376	(1.98)	5.8725	0.1936	5.9282	353	$<10^{-4}$	1.0758	−0.1978	(1.83)	9.1271	2.5260	13.0808	1966	$<10^{-4}$
60 (uncorrected)	0.7655	0.8183	(1.96)	6.0359	0.5325	7.8647	630	$<10^{-4}$	1.0590	−0.2646	(1.82)	9.0178	2.5328	13.4092	2055	$<10^{-4}$
1951–1970																
2	0.8757	1.0591	(5.03)	2.6956	−0.3974	1.3952	26	$<10^{-4}$	0.0025	0.0468	(0.01)	2.7639	0.1040	1.2244	15	0.0006
10	0.8618	0.9591	(5.22)	2.5575	−0.4000	1.3939	26	$<10^{-4}$	0.0195	−0.0368	(0.09)	3.3292	0.1896	1.5825	26	$<10^{-4}$
15	0.8529	0.9040	(5.21)	2.5382	−0.4034	1.3941	26	$<10^{-4}$	0.0287	−0.0434	(0.13)	3.3706	0.1971	1.4767	23	$<10^{-4}$
20	0.8464	0.9063	(5.18)	2.5308	−0.3959	1.3508	25	$<10^{-4}$	0.0343	−0.0356	(0.16)	3.3773	0.1971	1.4807	23	$<10^{-4}$
30	0.8413	0.9266	(5.16)	2.5255	−0.4073	1.3955	26	$<10^{-4}$	0.0387	−0.0493	(0.18)	3.3744	0.2153	1.5117	25	$<10^{-4}$
60 (uncorrected)	0.8329	0.8911	(5.09)	2.5372	−0.4133	1.4078	27	$<10^{-4}$	0.0466	−0.0719	(0.21)	3.3721	0.2065	1.4853	24	$<10^{-4}$
1971–1990																
2	0.7250	0.8681	(3.04)	3.6936	−1.1150	6.4619	467	$<10^{-4}$	−0.1122	−0.5854	(−0.46)	3.8064	2.0865	10.8070	1342	$<10^{-4}$
10	0.7365	0.9000	(3.20)	3.5663	−0.9801	7.1117	544	$<10^{-4}$	−0.1242	−0.7750	(−0.45)	4.3147	2.0744	10.5328	1282	$<10^{-4}$
15	0.7308	0.9072	(3.16)	3.5782	−0.8989	7.1008	537	$<10^{-4}$	−0.1174	−0.7247	(−0.42)	4.3084	2.0083	9.9631	1154	$<10^{-4}$
20	0.7477	0.9737	(3.22)	3.5935	−0.8991	6.9529	516	$<10^{-4}$	−0.1344	−0.6902	(−0.49)	4.2906	1.9778	9.5141	1062	$<10^{-4}$

(continued)

Table 8.8 (continued)

Portfolio size (second step)	$R_{it} = \gamma_{0t} + \gamma_{1t}\hat{\beta}_{it-1} + e_{it}$															
	$\hat{\gamma}_0$								$\hat{\gamma}_1$							
	Mean	Median	*t*-stats.	STD	SK	KU	J-B	*p*-value	Mean	Median	*t*-stats.	STD	SK	KU	J-B	*p*-value
30	0.7431	0.9524	(3.22)	3.5773	−0.8818	7.1104	537	$<10^{-4}$	−0.1300	−0.6477	(−0.47)	4.3067	2.0203	9.9863	1161	$<10^{-4}$
60 (uncorrected)	0.7575	0.9852	(3.27)	3.5862	−0.8615	7.0115	521	$<10^{-4}$	−0.1441	−0.7487	(−0.52)	4.2897	1.9925	9.7860	1116	$<10^{-4}$
1991–2009																
2	0.9285	1.2231	(3.80)	3.6866	−1.2514	4.4048	220	$<10^{-4}$	0.2502	0.1535	(1.09)	3.4648	1.4698	6.7713	518	$<10^{-4}$
10	0.7799	1.1856	(3.36)	3.5030	−1.4172	4.0301	231	$<10^{-4}$	0.3961	0.2641	(1.45)	4.1115	1.5569	5.9624	430	$<10^{-4}$
15	0.7643	1.2209	(3.27)	3.5305	−1.4179	3.9570	225	$<10^{-4}$	0.4119	0.2357	(1.52)	4.1009	1.5681	6.0652	443	$<10^{-4}$
20	0.7551	1.2179	(3.22)	3.5504	−1.4224	3.9107	222	$<10^{-4}$	0.4210	0.2705	(1.54)	4.1156	1.5950	6.1911	461	$<10^{-4}$
30	0.7425	1.2096	(3.15)	3.5640	−1.3708	3.7806	207	$<10^{-4}$	0.4328	0.2408	(1.57)	4.1591	1.6535	6.4821	503	$<10^{-4}$
60 (uncorrected)	0.7279	1.1679	(3.06)	3.5897	−1.3626	3.6530	197	$<10^{-4}$	0.4486	0.2423	(1.60)	4.2387	1.6377	6.3434	484	$<10^{-4}$

This table provides corrected estimators of the capital asset pricing test by using Wald's grouping method in the first step. 60 time-series observations of each stock are divided into *k* groups based on their corresponding market returns (*k* = 2, 10, 15, 20, and 30). The market-adjusted beta for each stock is estimated by the market model using *k* time-series observations. In the second step, the monthly cross-sectional regression coefficients for each stock, $\hat{\gamma}_{0t}$ and $\hat{\gamma}_{1t}$, are estimated through the ordinary least square. Parameters, $\bar{\hat{\gamma}}_0$ and $\bar{\hat{\gamma}}_1$, are time-series averages of the month-by-month cross-sectional regression coefficient estimates. The *t*-statistics, standard deviation, skewness, kurtosis, Jarque–Bera statistics, and the *p*-value of Jarque–Bera test are also presented. The sample period runs from January 1931 to December 2009. Results of subperiods are also provided

Table 8.9 Normality test for the time-series estimators of grouping method in the second step for testing capital asset pricing model

Panel A: post-ranking beta is average beta																
Portfolio size (second step)	$R_{it} = \gamma_{0t} + \gamma_{1t}\hat{\beta}_{it-1} + e_{it}$															
	$\hat{\gamma}_0$								$\hat{\gamma}_1$							
	Mean	Median	*t*-stats.	STD	SK	KU	J-B	*p*-value	Mean	Median	*t*-stats.	STD	SK	KU	J-B	*p*-value
1931–2009																
2	0.6952	0.9090	(5.55)	3.8578	−0.9606	11.4967	2997	$<10^{-4}$	0.4253	−0.1258	(2.11)	6.1978	3.1350	29.1202	28,502	$<10^{-4}$
10	0.7645	0.3580	(5.78)	4.0704	0.0191	14.4304	5161	$<10^{-4}$	0.3580	−0.1947	(1.89)	5.8395	2.9022	25.9639	22,161	$<10^{-4}$
20	0.7676	0.3547	(5.77)	4.0933	0.0652	14.6720	5382	$<10^{-4}$	0.3547	−0.2191	(1.89)	5.7815	2.9382	26.4223	23,034	$<10^{-4}$
50	0.7653	0.3575	(5.73)	4.1106	0.0467	14.7551	5459	$<10^{-4}$	0.3575	−0.1791	(1.91)	5.7608	2.9999	27.1943	24,544	$<10^{-4}$
100	0.7708	0.3541	(5.77)	4.1133	0.0411	14.8787	5574	$<10^{-4}$	0.3541	−0.1677	(1.90)	5.7435	3.0291	27.8303	25,803	$<10^{-4}$
200	0.7642	0.3611	(5.70)	4.1249	0.1094	14.5300	5253	$<10^{-4}$	0.3611	−0.1428	(1.95)	5.7165	2.9574	26.7231	23,612	$<10^{-4}$
N (uncorrected)	0.7715	0.3513	(5.74)	4.1419	0.0322	14.5490	5269	$<10^{-4}$	0.3513	−0.1865	(1.90)	5.7060	3.0763	28.3381	26,855	$<10^{-4}$
1931–1950																
2	0.5934	0.7642	(1.68)	5.4836	−0.7555	8.5972	336	$<10^{-4}$	1.2249	0.0903	(1.91)	9.9343	2.5329	15.8580	1910	$<10^{-4}$
10	0.7601	0.7692	(1.99)	5.9181	0.5108	10.6136	590	$<10^{-4}$	1.0643	0.2546	(1.79)	9.1990	2.3838	14.8789	1638	$<10^{-4}$
20	0.7802	0.7554	(2.03)	5.9561	0.5628	10.8135	623	$<10^{-4}$	1.0434	−0.3100	(1.77)	9.1123	2.4149	15.1373	1706	$<10^{-4}$
50	0.7662	0.7705	(1.98)	5.9935	0.5463	10.8185	623	$<10^{-4}$	1.0583	−0.2182	(1.80)	9.0933	2.4577	15.5042	1805	$<10^{-4}$
100	0.7698	0.7958	(1.99)	5.9844	0.5482	11.0118	654	$<10^{-4}$	1.0617	−0.2043	(1.81)	9.0718	2.4789	15.8541	1898	$<10^{-4}$
200	0.7526	0.7427	(1.94)	6.0122	0.6356	10.6486	601	$<10^{-4}$	1.0800	−0.2175	(1.86)	9.0097	2.4162	15.2744	1740	$<10^{-4}$
N (uncorrected)	0.7655	0.8183	(1.96)	6.0360	0.5291	10.6759	600	$<10^{-4}$	1.0590	−0.2646	(1.82)	9.0177	2.5172	16.1081	1972	$<10^{-4}$
1951–1970																
2	0.8012	0.8083	(5.01)	2.4798	−0.3792	4.1851	20	$<10^{-4}$	0.0795	−0.0810	(0.35)	3.5454	0.1720	4.6680	29	$<10^{-4}$
10	0.8204	0.9191	(5.03)	2.5281	−0.4072	4.3248	24	$<10^{-4}$	0.0593	0.0356	(0.27)	3.4248	0.1839	4.3016	18	0.0001
20	0.8178	0.8791	(5.01)	2.5276	−0.3951	4.3573	25	$<10^{-4}$	0.0616	−0.0361	(0.28)	3.3995	0.1816	4.3491	20	$<10^{-4}$
50	0.8317	0.9454	(5.11)	2.5235	−0.4032	4.3272	24	$<10^{-4}$	0.0481	−0.0710	(0.22)	3.3899	0.1981	4.3830	21	$<10^{-4}$
100	0.8323	0.8964	(5.10)	2.5273	−0.4106	4.3508	25	$<10^{-4}$	0.0483	−0.0655	(0.22)	3.3807	0.2043	4.4190	22	$<10^{-4}$
200	0.8327	0.8998	(5.05)	2.5262	−0.4180	4.3609	26	$<10^{-4}$	0.0565	−0.0617	(0.26)	3.3830	0.2058	4.4368	22	$<10^{-4}$
N (uncorrected)	0.8329	0.8911	(5.09)	2.5373	−0.4104	4.3529	25	$<10^{-4}$	0.0466	−0.0719	(0.21)	3.3719	0.2049	4.4306	22	$<10^{-4}$
1971–1990																
2	0.7235	0.8438	(3.19)	3.5105	−0.7323	8.8215	360	$<10^{-4}$	−0.1130	−0.6820	(−0.39)	4.5056	1.6187	10.0226	598	$<10^{-4}$
10	0.7524	0.9870	(3.30)	3.5328	−0.7493	9.8550	492	$<10^{-4}$	−0.1397	−0.7518	(−0.49)	4.3802	1.8710	11.8550	924	$<10^{-4}$

(continued)

Table 8.9 (continued)

Panel A: post-ranking beta is average beta																
Portfolio size (second step)	$R_{it} = \gamma_{0t} + \gamma_{1t}\hat{\beta}_{it-1} + e_{it}$															
	$\hat{\gamma}_0$								$\hat{\gamma}_1$							
	Mean	Median	*t*-stats.	STD	SK	KU	J-B	*p*-value	Mean	Median	*t*-stats.	STD	SK	KU	J-B	*p*-value
20	0.7549	0.9709	(3.29)	3.5563	−0.7869	9.7391	479	<10^{-4}	−0.1417	−0.7632	(−0.51)	4.3278	1.9073	12.1201	977	<10^{-4}
50	0.7841	0.9872	(3.24)	3.5727	−0.8379	9.7659	486	<10^{-4}	−0.1351	−0.6887	(−0.49)	4.3016	1.9340	12.2906	1013	<10^{-4}
100	0.7506	0.9913	(3.25)	3.5757	−0.8522	9.7847	489	<10^{-4}	−0.1373	−0.7164	(−0.50)	4.2874	1.9432	12.3446	1024	<10^{-4}
200	0.7578	0.9742	(3.28)	3.5773	−0.8629	9.7990	492	<10^{-4}	−0.1437	−0.7210	(−0.52)	4.2835	1.9463	12.3268	1021	<10^{-4}
N (uncorrected)	0.7575	0.9852	(3.27)	3.5863	−0.8562	9.8386	497	<10^{-4}	−0.1441	−0.7487	(−0.52)	4.2900	1.9805	12.5622	1071	<10^{-4}
1991–2009																
2	0.6612	1.1276	(1.06)	3.3097	−1.4391	6.9848	230	<10^{-4}	0.5142	0.2744	(0.60)	4.5408	1.4187	8.3334	347	<10^{-4}
10	0.7231	1.2016	(1.09)	3.5170	−1.4482	6.9201	226	<10^{-4}	0.4530	0.3431	(0.54)	4.4406	1.5292	8.7384	402	<10^{-4}
20	0.7150	1.1860	(1.07)	3.5355	−1.4136	6.7914	212	<10^{-4}	0.4611	0.3234	(0.56)	4.3899	1.5508	8.7507	406	<10^{-4}
50	0.7125	1.1896	(1.07)	3.5382	−1.4002	6.7428	208	<10^{-4}	0.4638	0.3716	(0.56)	4.3497	1.5914	8.9435	432	<10^{-4}
100	0.7285	1.2083	(1.08)	3.5616	−1.3822	6.6584	200	<10^{-4}	0.4485	0.3295	(0.55)	4.3230	1.6041	8.9745	437	<10^{-4}
200	0.7203	1.1373	(1.07)	3.5676	−1.3792	6.6347	198	<10^{-4}	0.4565	0.3529	(0.56)	4.3092	1.6052	8.9709	437	<10^{-4}
N (uncorrected)	0.7279	1.1679	(1.07)	3.5894	−1.3537	6.5472	189	<10^{-4}	0.4486	0.2423	(0.56)	4.2387	1.6270	9.1808	464	<10^{-4}

Panel B: Post-ranking beta is market model beta																
Portfolio size (second step)	$R_{it} = \gamma_{0t} + \gamma_{1t}\hat{\beta}_{it-1} + e_{it}$															
	$\hat{\gamma}_0$								$\hat{\gamma}_1$							
	Mean	Median	*t*-stats.	STD	SK	KU	J-B	*p*-value	Mean	Median	*t*-stats.	STD	SK	KU	J-B	*p*-value
1931–2009																
2	0.0656	0.3060	(0.50)	4.0470	−2.2081	22.0033	19,894	<10^{-4}	0.4237	−0.1164	(2.10)	6.2170	3.1620	26.6583	29,651	<10^{-4}
10	0.1360	0.3356	(1.11)	3.7740	−1.9746	17.6763	12,958	<10^{-4}	0.3548	−0.2103	(1.87)	5.8268	2.8990	23.0524	22,319	<10^{-4}
20	0.1382	0.3397	(1.14)	3.7324	−2.0444	18.4360	14,086	<10^{-4}	0.3523	−0.2136	(1.88)	5.7732	2.9596	23.8002	23,759	<10^{-4}
50	0.1320	0.3121	(1.10)	3.7028	−2.0830	18.9007	14,796	<10^{-4}	0.3591	−0.1718	(1.93)	5.7431	3.0144	24.3561	24,868	<10^{-4}
100	0.1444	0.3758	(1.21)	3.6790	−2.1465	20.3406	17,071	<10^{-4}	0.3473	−0.1553	(1.88)	5.7027	3.0672	25.7052	27,586	<10^{-4}
200	0.1401	0.3684	(1.17)	3.6889	−2.2452	21.6643	19,336	<10^{-4}	0.3539	−0.1307	(1.91)	5.7017	3.1309	26.8849	30,099	<10^{-4}
N (uncorrected)	0.7715	0.3513	(5.74)	4.1419	0.0322	14.5490	5269	<10^{-4}	0.3513	−0.1865	(1.90)	5.7060	3.0763	28.3381	26,855	<10^{-4}

(continued)

Table 8.9 (continued)

Panel B: Post-ranking beta is market model beta																
Portfolio size (second step)	$R_{it} = \gamma_{0t} + \gamma_{1t}\hat{\beta}_{it-1} + e_{it}$															
	$\hat{\gamma}_0$								$\hat{\gamma}_1$							
	Mean	Median	*t*-stats.	STD	SK	KU	J-B	*p*-value	Mean	Median	*t*-stats.	STD	SK	KU	J-B	*p*-value
1931–1950																
2	−0.2856	0.3514	(−0.74)	6.0171	−2.4040	15.8888	2756	$<10^{-4}$	1.2178	0.0708	(1.89)	9.9764	2.5634	13.3281	2039	$<10^{-4}$
10	−0.1168	0.4680	(−0.34)	5.3831	−2.3683	14.8406	2427	$<10^{-4}$	1.0533	−0.2662	(1.78)	9.1695	2.3985	12.1926	1717	$<10^{-4}$
20	−0.1032	0.4658	(−0.30)	5.3159	−2.4671	15.6826	2703	$<10^{-4}$	1.0383	−0.2936	(1.77)	9.0905	2.4534	12.6316	1836	$<10^{-4}$
50	−0.1315	0.5336	(−0.39)	5.2754	−2.5092	16.0728	2835	$<10^{-4}$	1.0688	−0.1624	(1.83)	9.0499	2.4903	12.8952	1911	$<10^{-4}$
100	−0.1019	0.5567	(−0.30)	5.2249	−2.6265	17.7353	3421	$<10^{-4}$	1.0400	−0.1306	(1.80)	8.9721	2.5533	13.8614	2182	$<10^{-4}$
200	−0.1117	0.6278	(−0.33)	5.2606	−2.7295	18.6922	3792	$<10^{-4}$	1.0595	−0.0969	(1.83)	8.9777	2.6090	14.5530	2390	$<10^{-4}$
N (uncorrected)	0.7655	0.8183	(1.96)	6.0360	0.5291	10.6759	600	$<10^{-4}$	1.0590	−0.2646	(1.82)	9.0177	2.5172	16.1081	1972	$<10^{-4}$
1951–1970																
2	0.0702	0.0923	(0.42)	2.5859	−0.0089	0.4367	1.9103	0.3848	0.0787	−0.0735	(0.34)	3.5431	0.1797	1.7637	32	$<10^{-4}$
10	0.0882	−0.0199	(0.55)	2.5007	−0.0337	0.1680	0.3276	0.8489	0.0598	−0.372	(0.27)	3.4224	0.1836	1.3558	20	$<10^{-4}$
20	0.0891	−0.0102	(0.56)	2.4818	−0.0410	0.1779	0.3837	0.8254	0.0587	−0.0360	(0.27)	3.4002	0.1799	1.3979	21	$<10^{-4}$
50	0.1020	0.0660	(0.64)	2.4698	−0.0538	0.1713	0.4093	0.8149	0.0460	−0.0783	(−0.21)	3.3883	0.1990	1.4176	22	$<10^{-4}$
100	0.1065	0.0275	(0.67)	2.4594	−0.0629	0.1725	0.4561	0.7961	0.0426	−0.1027	(0.20)	3.3800	0.2089	1.4521	23	$<10^{-4}$
200	0.1038	0.0104	(0.65)	2.4589	−0.0622	0.1822	0.4867	0.7840	0.0448	−0.0983	(0.21)	3.3766	0.2047	1.4670	23	$<10^{-4}$
N (uncorrected)	0.8329	0.8911	(5.09)	2.5373	−0.4104	4.3529	25	$<10^{-4}$	0.0466	−0.0719	(0.21)	3.3719	0.2049	4.4306	22	$<10^{-4}$
1971–1990																
2	0.3665	0.7067	(1.89)	3.0073	−0.8778	1.7958	63	$<10^{-4}$	−0.1137	−0.6736	(−0.39)	4.5098	1.6251	7.1699	620	$<10^{-4}$
10	0.3961	0.7056	(2.11)	2.9116	−1.0851	2.7544	123	$<10^{-4}$	−0.1409	−0.7504	(−0.50)	4.3789	1.8818	9.0698	964	$<10^{-4}$
20	0.4003	0.6343	(2.16)	2.8757	−1.1035	2.8742	131	$<10^{-4}$	−0.1447	−0.7621	(−0.52)	4.3269	1.9194	9.3421	1020	$<10^{-4}$
50	0.3931	0.6811	(2.14)	2.8488	−1.103	2.9099	133	$<10^{-4}$	−0.1373	−0.7162	(−0.50)	42.964	1.9472	9.5040	1055	$<10^{-4}$
100	0.3943	0.6609	(2.15)	2.8416	−1.1135	2.9649	138	$<10^{-4}$	−0.1384	−0.6958	(−0.50)	4.2845	1.9707	9.6411	1085	$<10^{-4}$
200	0.4000	0.6331	(2.18)	2.8376	−1.1163	2.9594	137	$<10^{-4}$	−0.1438	−0.6640	(−0.52)	4.2798	1.9792	9.6563	1089	$<10^{-4}$
N (uncorrected)	0.7575	0.9852	(3.27)	3.5863	−0.8562	9.8386	497	$<10^{-4}$	−0.1441	−0.7487	(−0.52)	4.2900	1.9805	12.5622	1071	$<10^{-4}$

(continued)

Table 8.9 (continued)

Panel B: Post-ranking beta is market model beta																
Portfolio size (second step)	$R_{it} = \gamma_{0t} + \gamma_{1t}\hat{\beta}_{it-1} + e_{it}$															
	$\hat{\gamma}_0$								$\hat{\gamma}_1$							
	Mean	Median	*t*-stats.	STD	SK	KU	J-B	*p*-value	Mean	Median	*t*-stats.	STD	SK	KU	J-B	*p*-value
1991–2009																
2	0.1137	0.1012	(0.47)	3.6630	−0.1619	2.6966	70	$<10^{-4}$	0.5164	0.2857	(1.71)	4.5518	1.4344	5.4971	365	$<10^{-4}$
10	0.1784	0.2765	(0.74)	3.6409	−0.2551	3.0995	94	$<10^{-4}$	0.4518	0.3431	(1.54)	4.4403	1.5403	5.9033	421	$<10^{-4}$
20	0.1680	0.2365	(0.70)	3.6104	−0.2793	3.0365	91	$<10^{-4}$	0.4623	0.3331	(1.59)	4.3928	1.5630	5.9030	424	$<10^{-4}$
50	0.1660	0.164	(0.70)	3.5755	−0.2744	2.9989	88	$<10^{-4}$	0.4624	0.3752	(1.61)	4.3519	1.6043	6.1051	452	$<10^{-4}$
100	0.1807	0.1736	(0.76)	3.5663	−0.2653	2.9428	85	$<10^{-4}$	0.4501	0.3221	(1.57)	4.3239	1.6163	6.1321	456	$<10^{-4}$
200	0.1699	0.1493	(0.72)	3.5565	−0.2636	2.9268	84	$<10^{-4}$	0.4602	0.3416	(1.61)	4.3096	1.6228	6.1507	459	$<10^{-4}$
N (uncorrected)	0.7279	1.1679	(1.07)	3.5894	−1.3537	6.5472	189	$<10^{-4}$	0.4486	0.2423	(0.56)	4.2387	1.6270	9.1808	464	$<10^{-4}$

This table provides corrected estimators of the capital asset pricing test by using Wald's grouping method in the second step. Each firm's preranking beta for month *t* is estimated by using the market model for the previous five-year period. Based on each firm's preranking beta, stocks are formed into *N* (*N* = 2, 10, 20, 50, 100, and 200) beta-sorted portfolios in month *t*. Post-ranking beta of each beta-sorted portfolio in month *t* is the average of preranking betas of each portfolio. In the second step, the monthly cross-sectional regression coefficients for each portfolio, $\hat{\gamma}_{0t}$ and $\hat{\gamma}_{1t}$, are estimated through the ordinary least square. As shown in Panel A, parameters, $\bar{\bar{\gamma}}_0$ and $\bar{\bar{\gamma}}_1$, are the time-series averages of the month-by-month cross-sectional regression coefficient estimates by using average beta as post-ranking portfolio beta. We also provide another post-ranking beta which is estimated by market model using the previous five-year portfolio monthly returns and market returns. Panel B presents $\bar{\bar{\gamma}}_0$ and $\bar{\bar{\gamma}}_1$, which are the time-series averages of the month-by-month cross-sectional regression coefficient estimates by using portfolio beta from market model as the post-ranking beta. The *t*-statistics, standard deviation, skewness, kurtosis, Jarque–Bera statistics, and the *p*-value of Jarque–Bera test are also presented. The sample period is from January 1931 to December 2009. Results of subperiods are also provided

Table 8.10 Normality test for the time-series estimators of grouping method in both first step and second step for testing capital asset pricing model

Portfolio size (first step/second step)		$R_{it} = \gamma_{0t} + \gamma_{1t}\hat{\beta}_{it-1} + e_{it}$															
		$\hat{\gamma}_0$								$\hat{\gamma}_1$							
		Mean	Median	*t*-stats.	STD	SK	KU	J-B	*p*-value	Mean	Median	*t*-stats.	STD	SK	KU	J-B	*p*-value
60	N	0.7715	0.3513	(5.74)	4.1419	0.0322	14.5490	5269	$<10^{-4}$	0.3513	−0.1865	(1.90)	5.7060	3.0763	28.3381	26,855	$<10^{-4}$
2	2	0.7544	0.9686	(6.28)	3.6980	−0.9485	4.8676	1078	$<10^{-4}$	0.3661	0.0027	(1.95)	5.7779	3.4428	27.9023	32,625	$<10^{-4}$
2	10	0.8046	0.9944	(6.41)	3.8640	−0.8097	5.3525	1235	$<10^{-4}$	0.3169	−0.0710	(1.75)	5.5918	3.3773	27.2252	31,080	$<10^{-4}$
2	20	0.8096	0.9971	(6.40)	3.8929	−0.8163	5.4583	1282	$<10^{-4}$	0.3123	−0.0964	(1.73)	5.5628	3.4707	28.5358	34,068	$<10^{-4}$
2	50	0.8197	1.0380	(6.48)	3.8948	−0.8048	5.2793	1203	$<10^{-4}$	0.3028	−0.0990	(1.68)	5.5532	3.4913	28.6289	34,300	$<10^{-4}$
2	100	0.8177	1.0297	(6.45)	3.9046	−0.7941	5.4654	1280	$<10^{-4}$	0.3056	−0.1106	(1.70)	5.5402	3.4907	28.7300	34,529	$<10^{-4}$
2	200	0.8215	1.0268	(6.47)	3.9102	−0.6595	5.5838	1300	$<10^{-4}$	0.3054	−0.1099	(1.71)	5.5050	3.3222	26.7102	29,925	$<10^{-4}$
10	2	0.6881	0.9066	(5.61)	3.7794	−0.9143	6.7673	1941	$<10^{-4}$	0.4326	−0.0589	(2.15)	6.1898	3.1446	24.9521	26,155	$<10^{-4}$
10	10	0.7505	0.9418	(5.97)	3.8743	−0.7599	6.2018	1610	$<10^{-4}$	0.3713	−0.1426	(1.90)	6.0032	3.1363	23.8710	24,062	$<10^{-4}$
10	20	0.7602	0.9736	(6.00)	3.9024	−0.7255	6.5393	1772	$<10^{-4}$	0.3618	−0.2057	(1.87)	5.9572	3.1674	24.5072	25,309	$<10^{-4}$
10	50	0.7675	0.9732	(6.03)	3.9170	−0.6606	6.5368	1757	$<10^{-4}$	0.3554	−0.1871	(1.85)	5.9050	3.1503	24.3149	24,921	$<10^{-4}$
10	100	0.7639	0.9536	(6.00)	3.9222	−0.6417	6.7538	1867	$<10^{-4}$	0.3585	−0.1751	(1.87)	5.9004	3.1616	24.3740	25,046	$<10^{-4}$
10	200	0.7663	0.9565	(5.99)	3.9373	−0.5917	7.1877	2096	$<10^{-4}$	0.3584	−0.1857	(1.86)	5.9233	3.2947	25.9964	28,410	$<10^{-4}$
15	2	0.7173	0.8816	(5.78)	3.8189	−0.7541	6.9316	1988	$<10^{-4}$	0.4046	−0.1763	(2.04)	6.1004	3.0150	23.8444	23,894	$<10^{-4}$
15	10	0.7560	0.9388	(5.91)	3.9398	−0.5202	7.5388	2288	$<10^{-4}$	0.3668	−0.1550	(1.90)	5.9316	3.0610	23.9971	24,227	$<10^{-4}$
15	20	0.7648	0.9412	(5.94)	3.9637	−0.5525	7.7186	2401	$<10^{-4}$	0.3583	−0.1634	(1.87)	5.9061	3.1529	25.1706	26,596	$<10^{-4}$
15	50	0.7660	0.9532	(5.93)	3.9766	−0.5018	7.8119	2450	$<10^{-4}$	0.3570	−0.1842	(1.88)	5.8578	3.1602	25.4049	27,072	$<10^{-4}$
15	100	0.7711	0.9330	(5.98)	39690	−0.4723	7.8185	2450	$<10^{-4}$	0.3511	−0.1600	(1.85)	5.8391	3.1562	25.3941	27,046	$<10^{-4}$
15	200	0.7642	0.9494	(5.89)	3.9943	−0.3196	8.3182	2749	$<10^{-4}$	0.3567	−0.1669	(1.89)	5.8256	3.1103	24.5127	25,263	$<10^{-4}$
20	2	0.6857	0.9354	(5.56)	3.790	−1.0306	8.0614	2735	$<10^{-4}$	0.4355	−0.1054	(2.16)	6.2091	3.2020	26.4980	29,355	$<10^{-4}$
20	10	0.7538	0.9615	(5.85)	3.9653	−0.4651	8.1102	2632	$<10^{-4}$	0.3688	−0.1877	(1.92)	5.9147	3.0675	24.2920	24,796	$<10^{-4}$
20	20	0.7597	0.9716	(5.85)	3.9989	−0.4466	8.4655	2862	$<10^{-4}$	0.3627	−0.2213	(1.90)	5.8727	3.1323	25.3288	26,891	$<10^{-4}$
20	50	0.7630	0.9618	(5.86)	4.0113	−0.4085	8.4061	2818	$<10^{-4}$	0.3602	−0.2035	(1.90)	5.8344	3.1444	25.4896	27,226	$<10^{-4}$

(continued)

Table 8.10 (continued)

Portfolio size (first step/second step)		$R_{it} = \gamma_{0t} + \gamma_{1t}\hat{\beta}_{it-1} + e_{it}$															
		$\hat{\gamma}_0$								$\hat{\gamma}_1$							
		Mean	Median	*t*-stats.	STD	SK	KU	J-B	*p*-value	Mean	Median	*t*-stats.	STD	SK	KU	J-B	*p*-value
20	100	0.7671	0.9712	(5.89)	4.0096	−0.4319	8.4215	2831	$<10^{-4}$	0.3542	−0.1922	(1.88)	5.8143	3.1522	25.6883	27,636	$<10^{-4}$
20	200	0.7656	0.9905	(5.85)	4.0284	−0.4057	8.8063	3089	$<10^{-4}$	0.3529	−0.1793	(1.88)	5.7850	3.1923	26.4796	29,306	$<10^{-4}$
30	2	0.6886	0.9325	(5.49)	3.8628	−0.8520	8.7303	3125	$<10^{-4}$	0.4320	−0.0655	(2.16)	6.1507	3.1231	26.1798	28,614	$<10^{-4}$
30	10	0.7591	0.9484	(5.83)	4.0101	−0.3540	8.7341	3033	$<10^{-4}$	0.3631	−0.1691	(1.90)	5.8792	2.9888	23.7827	23,753	$<10^{-4}$
30	20	0.7625	0.9323	(5.82)	4.0364	−0.2624	9.4485	3537	$<10^{-4}$	0.3598	−0.2175	(1.90)	5.8244	3.0491	24.6981	25,564	$<10^{-4}$
30	50	0.7654	0.9589	(5.84)	4.0375	−0.2771	9.0314	3234	$<10^{-4}$	0.3575	−0.1782	(1.90)	5.7963	3.0474	24.3976	24,979	$<10^{-4}$
30	100	0.7682	0.9519	(5.83)	4.0542	−0.2825	9.6367	3681	$<10^{-4}$	0.3550	−0.1728	(1.89)	5.7776	3.1396	26.0732	28,410	$<10^{-4}$
30	200	0.7624	0.9492	(5.80)	4.0451	−0.5007	8.4588	2866	$<10^{-4}$	0.3615	−0.1876	(1.92)	5.8070	3.2793	27.8442	32,323	$<10^{-4}$

This table provides the corrected estimators of the capital asset pricing test by using Wald's grouping method in both first step and second step. In the first step, 60 time-series observations of each stock are divided into *k* groups based on their corresponding market returns (k = 2, 10, 15, 20, and 30). The market-adjusted preranking beta for each stock is estimated by the market model using *k* time-series observations. In the second step, based on each firm's preranking beta, portfolios are formed into N (N = 2, 10, 20, 50, 100, and 200) beta-sorted portfolios in month *t*. Post-ranking beta of each beta-sorted portfolio in month *t* is the average of preranking betas of each portfolio. The monthly cross-sectional regression coefficients for each portfolio, $\hat{\gamma}_{0t}$ and $\hat{\gamma}_{1t}$, are estimated through the ordinary least square. Parameters, $\bar{\hat{\gamma}}_0$ and $\bar{\hat{\gamma}}_1$, are the time-series averages of the month-by-month cross-sectional regression coefficient estimates by using average beta as post-ranking portfolio beta. The *t*-statistics, standard deviation, skewness, kurtosis, Jarque–Bera statistics, and the *p*-value of Jarque–Bera test are also presented. The sample period is from January 1931 to December 2009

Table 8.11 Normality test for the time-series estimators of instrumental variable method for testing capital asset pricing model

Portfolio size (second step)		$R_{it} = \gamma_{0t} + \gamma_{1t}\hat{\beta}_{it-1} + e_{it}$															
		$\hat{\gamma}_0$								$\hat{\gamma}_1$							
		Mean	Median	*t*-stats.	STD	SK	KU	J-B	*p*-value	Mean	Median	*t*-stats.	STD	SK	KU	J-B	*p*-value
1931–2009																	
OLS	OLS	0.7715	0.9751	(5.74)	4.1419	0.0323	11.6176	5331	$<10^{-4}$	0.3513	−0.1865	(1.90)	5.7060	3.0810	25.4766	27,138	$<10^{-4}$
IV	OLS	0.7884	1.0153	(6.25)	3.8848	−0.7387	6.1838	1597	$<10^{-4}$	0.3338	−0.2106	(1.75)	5.8769	3.1975	25.0885	26,478	$<10^{-4}$
OLS	IV	0.7346	0.9651	(5.77)	3.9182	−0.2277	10.5700	4421	$<10^{-4}$	0.3877	−0.1400	(2.02)	5.9035	2.9429	23.7086	23,571	$<10^{-4}$
IV	IV	0.7491	0.9505	(6.21)	3.7142	−0.8820	6.1496	1617	$<10^{-4}$	0.3725	−0.0972	(1.90)	6.0219	3.1370	24.4284	25,126	$<10^{-4}$
1931–1950																	
OLS	OLS	0.7655	0.8183	(1.96)	6.0359	0.5325	7.8647	630	$<10^{-4}$	1.0590	−0.2646	(1.82)	9.0178	2.5328	13.4092	2055	$<10^{-4}$
IV	OLS	0.6992	0.8295	(2.02)	5.3686	−0.3974	3.7573	147	$<10^{-4}$	1.1224	−0.0719	(1.81)	9.6159	2.4145	11.1006	1465	$<10^{-4}$
OLS	IV	0.7022	0.7411	(1.93)	5.6387	0.2145	7.2885	533	$<10^{-4}$	1.1208	−0.1796	(1.86)	9.3516	2.4171	12.2636	1738	$<10^{-4}$
IV	IV	0.6425	0.7321	(1.96)	5.0911	−0.6460	3.8369	164	$<10^{-4}$	1.1774	0.0829	(1.85)	9.8341	2.3976	10.9053	1419	$<10^{-4}$
1951–1970																	
OLS	OLS	0.8329	0.8911	(5.09)	2.532	−0.4133	1.4078	27	$<10^{-4}$	0.0466	−0.0719	(0.21)	3.3721	0.2065	1.4853	24	$<10^{-4}$
IV	OLS	0.8709	0.9696	(5.24)	2.5767	−0.4014	1.4086	26	$<10^{-4}$	0.0107	−0.0147	(0.05)	3.2308	0.1453	1.3564	19	$<10^{-4}$
OLS	IV	0.8200	0.9098	(5.16)	2.4632	−0.4249	1.3590	26	$<10^{-4}$	0.0605	−0.0207	(0.27)	3.4456	0.1832	1.5923	27	$<10^{-4}$
IV	IV	0.8505	0.9557	(5.29)	2.4887	−0.4123	1.3537	25	$<10^{-4}$	0.0323	0.0290	(0.15)	3.3047	0.1321	1.4863	23	$<10^{-4}$
1971–1990																	
OLS	OLS	0.7575	0.9852	(3.27)	3.5862	−0.8615	7.0115	521	$<10^{-4}$	−0.1441	−0.7487	(−0.52)	4.2897	1.9925	9.7860	1116	$<10^{-4}$
IV	OLS	0.7383	0.9196	(3.24)	3.5257	−0.9337	7.0548	533	$<10^{-4}$	−0.1259	−0.7848	(−0.45)	4.3428	2.1676	11.7373	1566	$<10^{-4}$
OLS	IV	0.7292	0.9429	(3.23)	3.4965	−0.7100	636767	466	$<10^{-4}$	−0.1177	−0.6826	(−0.41)	4.4199	1.8064	8.6307	875	$<10^{-4}$
IV	IV	0.7035	0.8190	(3.16)	3.4440	−0.7816	6.7126	475	$<10^{-4}$	−0.0934	−0.8241	(−0.32)	4.4591	19758	10.3931	1236	$<10^{-4}$
1991–2009																	
OLS	OLS	0.7279	1.1679	(3.06)	3.5897	−1.3626	3.6530	197	$<10^{-4}$	0.4486	0.2423	(1.60)	4.2387	1.6377	6.3434	484	$<10^{-4}$
IV	OLS	0.8482	1.2893	(3.62)	3.5377	−1.3688	3.9409	219	$<10^{-4}$	0.3276	0.2481	(1.28)	3.8592	1.3603	5.3825	346	$<10^{-4}$
OLS	IV	0.6846	1.1694	(3.08)	3.3613	−1.4698	4.0906	241	$<10^{-4}$	0.4925	0.3489	(1.69)	4.3915	1.4366	5.3096	345	$<10^{-4}$
IV	IV	0.8026	1.2093	(3.62)	3.3504	−1.4084	40459	231	$<10^{-4}$	0.3738	0.3520	(1.41)	3.9990	1.2273	4.6723	265	$<10^{-4}$

This table provides corrected estimators of the capital asset pricing test by using Durbin's instrumental variable approach in both first step and second step. In the first step, the instrumental variables for 60 time-series observations of each stock are ranking orders of their corresponding market returns. Then, the beta of each security can be estimated by the instrumental variable method. In the second step, the instrumental variables for cross-sectional observations in month *t* are the rank orders of the estimated betas of cross-sectional observations. The monthly cross-sectional regression coefficients in month *t*, $\hat{\gamma}_{0t}$ and $\hat{\gamma}_{1t}$, are estimated through the instrumental variable method. Parameters, $\bar{\hat{\gamma}}_0$ and $\bar{\hat{\gamma}}_1$, are the time-series averages of the month-by-month cross-sectional regression coefficient estimates. The *t*-statistics are provided under the null hypothesis that the estimator is different from zero. This table also presents the number of observations and the percentage of observations that are significant under the null hypothesis that the absolute value of corrected estimator is greater than the absolute value of the uncorrected estimator at 95 and 99% significance levels. The *t*-statistics, standard deviation, skewness, kurtosis, Jarque–Bera statistics, and the p-value of Jarque–Bera test are also presented. The sample period is from January 1931 to December 2009. Results of subperiods are also provided

Table 8.12 Normality test for the time-series estimators of combining grouping and instrumental variable methods for testing capital asset pricing model

Portfolio size (second step)		$R_{it} = \gamma_{0t} + \gamma_{1t}\hat{\beta}_{it-1} + e_{it}$															
		$\hat{\gamma}_0$								$\hat{\gamma}_1$							
		Mean	Median	*t*-stats.	STD	SK	KU	J-B	*p*-value	Mean	Median	*t*-stats.	STD	SK	KU	J-B	*p*-value
60 (OLS)	*N* (OLS)	0.7715	0.9751	(5.74)	4.1419	0.0323	11.6176	5331	$<10^{-4}$	0.3513	−0.1865	(1.90)	5.7060	3.0810	25.4766	27,138	$<10^{-4}$
2 (OLS)	2 (IV)	0.7544	0.9686	(6.28)	3.6980	−0.9485	4.8676	1078	$<10^{-4}$	0.3661	0.0027	(1.95)	5.7779	3.4428	27.9023	32,625	$<10^{-4}$
2 (OLS)	10 (IV)	0.7773	0.9666	(6.36)	3.7643	−0.8480	5.3454	1242	$<10^{-4}$	0.3440	−0.0336	(1.88)	5.6262	3.4013	27.7582	32,263	$<10^{-4}$
2 (OLS)	20 (IV)	0.7754	0.9837	(6.34)	3.7675	−0.8670	5.4137	1276	$<10^{-4}$	0.3459	−0.0435	(1.89)	5.6271	3.4519	28.4483	33,850	$<10^{-4}$
2 (OLS)	50 (IV)	0.7777	0.9849	(6.37)	3.7590	−0.8689	5.3377	1245	$<10^{-4}$	0.3439	−0.0529	(1.88)	5.6318	3.4566	28.4505	33,860	$<10^{-4}$
2 (OLS)	100 (IV)	0.7767	0.9862	(6.35)	3.7664	−0.8530	5.5598	1336	$<10^{-4}$	0.3456	−0.0504	(1.89)	5.6253	3.4521	28.5723	34,130	$<10^{-4}$
2 (OLS)	200 (IV)	0.7795	0.9810	(6.36)	3.7746	−0.7611	5.5615	1313	$<10^{-4}$	0.3464	−0.0307	(1.91)	5.5889	3.2685	26.2061	28,815	$<10^{-4}$
2 (OLS)	*N* (IV)	0.7777	0.9859	(6.35)	3.7697	−0.8644	5.3926	1267	$<10^{-4}$	0.3437	−0.0490	(1.88)	5.6210	3.4411	28.2269	33,343	$<10^{-4}$
10 (OLS)	2 (IV)	0.6881	0.9066	(5.61)	3.7794	−0.9143	6.7673	1941	$<10^{-4}$	0.4326	−0.0589	(2.15)	6.1898	3.1446	24.9521	26,155	$<10^{-4}$
10 (OLS)	10 (IV)	0.7253	0.9150	(5.96)	3.7446	−0.8361	6.1869	1622	$<10^{-4}$	0.3964	−0.0833	(2.02)	6.0459	3.1136	23.8073	23,920	$<10^{-4}$
10 (OLS)	20 (IV)	0.7289	0.9168	(6.00)	3.7433	−0.8193	6.2651	1656	$<10^{-4}$	0.3928	−0.0859	(2.00)	6.0361	3.1030	23.7910	23,879	$<10^{-4}$
10 (OLS)	50 (IV)	0.7316	0.9248	(6.01)	3.7452	−0.8085	6.3314	1687	$<10^{-4}$	0.3909	−0.1079	(2.00)	6.0289	3.1033	23.8448	23,980	$<10^{-4}$
10 (OLS)	100 (IV)	0.7294	0.9226	(5.99)	3.7488	−0.7853	6.4241	1728	$<10^{-4}$	0.3927	−0.1061	(2.01)	6.0215	3.0814	23.5248	23,360	$<10^{-4}$
10 (OLS)	200 (IV)	0.7304	0.9250	(5.98)	3.7611	−0.7184	6.7624	1888	$<10^{-4}$	0.3939	−0.0916	(2.00)	6.0528	3.1788	24.6431	25,584	$<10^{-4}$
10 (OLS)	*N* (IV)	0.7315	0.9288	(5.99)	3.7585	−0.7959	6.4009	1718	$<10^{-4}$	0.3904	−0.0969	(2.00)	6.0167	3.0954	23.7758	23,843	$<10^{-4}$
15 (OLS)	2 (IV)	0.7173	0.8816	(5.78)	3.8189	−0.7541	6.9316	1988	$<10^{-4}$	0.4046	−0.1763	(2.04)	6.1004	3.0150	23.8444	23,894	$<10^{-4}$
15 (OLS)	10 (IV)	0.7304	0.9749	(5.92)	3.8019	−0.6702	7.2041	2121	$<10^{-4}$	0.3921	−0.1323	(2.02)	5.9829	3.0455	23.8723	23,976	$<10^{-4}$
15 (OLS)	20 (IV)	0.7320	0.9339	(5.93)	3.7977	−0.7049	7.3146	2192	$<10^{-4}$	0.3908	−0.1083	(2.01)	5.9878	3.0955	24.5805	25,380	$<10^{-4}$
15 (OLS)	50 (IV)	0.7316	0.9439	(5.93)	3.7996	−0.6796	7.3385	2200	$<10^{-4}$	0.3909	−0.1087	(2.01)	5.9771	3.0752	24.3635	24,941	$<10^{-4}$
15 (OLS)	100 (IV)	0.7355	0.9279	(5.96)	3.7966	−0.6133	7.4441	2248	$<10^{-4}$	0.3862	−0.1186	(1.99)	5.9648	3.0496	24.1721	24,549	$<10^{-4}$
15 (OLS)	200 (IV)	0.7334	0.9067	(5.93)	3.8071	−0.5055	7.5431	2288	$<10^{-4}$	0.3871	−0.0940	(2.00)	5.9574	3.0003	23.2734	22,818	$<10^{-4}$
15 (OLS)	*N* (IV)	0.7323	0.9105	(5.92)	3.8110	−0.6701	7.4034	2236	$<10^{-4}$	0.3902	−0.1120	(2.01)	5.9687	3.0795	24.4235	25,060	$<10^{-4}$
20 (OLS)	2 (IV)	0.6857	0.9354	(5.56)	3.7990	−1.0306	8.0614	2735	$<10^{-4}$	0.4355	−0.1054	(2.16)	6.2091	3.2020	26.4980	29,355	$<10^{-4}$

(continued)

Table 8.12 (continued)

Portfolio size (second step)		$R_{it} = \gamma_{0t} + \gamma_{1t}\hat{\beta}_{it-1} + e_{it}$															
		$\hat{\gamma}_0$								$\hat{\gamma}_1$							
		Mean	Median	*t*-stats.	STD	SK	KU	J-B	*p*-value	Mean	Median	*t*-stats.	STD	SK	KU	J-B	*p*-value
20 (OLS)	10 (IV)	0.7283	0.9603	(5.86)	3.8271	−0.6309	7.8264	2482	$<10^{-4}$	0.3940	−0.1240	(2.03)	5.9682	3.0481	24.2209	24,641	$<10^{-4}$
20 (OLS)	20 (IV)	0.7274	0.9563	(5.85)	3.8314	−0.6203	7.9357	2548	$<10^{-4}$	0.3948	−0.1323	(2.04)	5.9662	3.0693	24.5112	25,220	$<10^{-4}$
20 (OLS)	50 (IV)	0.7279	0.9394	(5.85)	3.8321	−0.5967	7.9406	2547	$<10^{-4}$	0.3950	−0.1286	(2.04)	5.9563	3.0570	24.4443	25,079	$<10^{-4}$
20 (OLS)	100 (IV)	0.7304	0.9575	(5.87)	3.8313	−0.6011	8.0347	2607	$<10^{-4}$	0.3904	−0.1370	(2.02)	5.9520	3.0533	24.4583	25,102	$<10^{-4}$
20 (OLS)	200 (IV)	0.7281	0.9732	(5.83)	3.8476	−0.5837	8.3284	2794	$<10^{-4}$	0.3899	−0.1284	(2.02)	5.9293	3.0624	24.8253	25,825	$<10^{-4}$
20 (OLS)	*N* (IV)	0.7274	0.9440	(5.83)	3.8425	−0.6027	7.9483	2553	$<10^{-4}$	0.3949	−0.1196	(2.04)	5.9513	3.0609	24.4430	25,080	$<10^{-4}$
30 (OLS)	2 (IV)	0.6886	0.9325	(5.49)	3.8628	−0.8520	8.7303	3125	$<10^{-4}$	0.4320	−0.0655	(2.16)	6.1507	3.1231	26.1798	28,614	$<10^{-4}$
30 (OLS)	10 (IV)	0.7275	0.9637	(5.80)	3.8622	−0.5175	8.5193	2909	$<10^{-4}$	0.3944	−0.1260	(2.04)	5.9451	2.9807	23.8061	23,790	$<10^{-4}$
30 (OLS)	20 (IV)	0.7281	0.9371	5.79	3.8701	−0.4531	8.9856	3222	$<10^{-4}$	0.3939	−0.1183	(2.05)	5.9317	2.9870	24.0275	24,214	$<10^{-4}$
30 (OLS)	50 (IV)	0.7303	0.9513	(5.83)	3.8549	−0.4743	8.6197	2970	$<10^{-4}$	0.3922	−0.1332	(2.04)	5.9314	2.9677	23.6712	23,524	$<10^{-4}$
30 (OLS)	100 (IV)	0.7315	0.9551	(5.81)	3.8732	−0.4563	9.3593	3493	$<10^{-4}$	0.3913	−0.1249	(2.03)	5.9250	3.0268	24.8690	25,877	$<10^{-4}$
30 (OLS)	200 (IV)	0.7259	0.9756	(5.80)	3.8562	−0.7139	8.4078	2873	$<10^{-4}$	0.3975	−0.1180	(2.05)	5.9757	3.1831	26.6797	29,717	$<10^{-4}$
30 (OLS)	*N* (IV)	0.7304	0.9487	(5.81)	3.8717	−0.4606	8.8953	3159	$<10^{-4}$	0.3917	−0.1351	(2.03)	5.9278	2.9896	24.0559	24,270	$<10^{-4}$
60 (OLS)	2 (IV)	0.6952	0.9090	(5.55)	3.8578	−0.9619	8.8469	3032	$<10^{-4}$	0.4253	−0.1258	(2.11)	6.1980	3.1402	26.2671	28,811	$<10^{-4}$
60 (OLS)	10 (IV)	0.7332	0.9698	(5.77)	3.9104	−0.2351	10.4373	4312	$<10^{-4}$	0.3889	−0.1376	(2.02)	5.9169	2.9103	23.2034	22,605	$<10^{-4}$
60 (OLS)	20 (IV)	0.7352	0.9699	(5.80)	3.9041	−0.2293	10.2787	4182	$<10^{-4}$	0.3869	−0.1471	(2.02)	5.9031	2.9058	23.1356	22,477	$<10^{-4}$
60 (OLS)	50 (IV)	0.7335	0.9602	(5.78)	3.9052	−0.2414	10.3191	4215	$<10^{-4}$	0.3889	−0.1413	(2.03)	5.9100	2.9225	23.3991	22,976	$<10^{-4}$
60 (OLS)	100 (IV)	0.7348	0.9681	(5.79)	3.9098	−0.2047	10.7416	4564	$<10^{-4}$	0.3897	−0.1334	(2.03)	5.9085	2.9191	23.6004	23,347	$<10^{-4}$
60 (OLS)	200 (IV)	0.7300	0.9595	(5.75)	3.9115	−0.1363	10.4038	4278	$<10^{-4}$	0.3948	−0.1158	(2.06)	5.8984	2.8589	22.5316	21,345	$<10^{-4}$
60 (OLS)	*N* (IV)	0.7346	0.9651	(5.77)	3.9182	−0.2277	10.5700	4421	$<10^{-4}$	0.3877	−0.1400	(2.02)	5.9035	2.9429	23.7086	23,571	$<10^{-4}$

This table provides the corrected estimators of the capital asset pricing test by combining grouping method and second step and instrumental variables method. In the first step, 60 time-series observations of each stock are divided into *k* groups based on their corresponding market returns (k = 2, 10, 15, 20, and 30). The market-adjusted preranking beta for each stock is estimated by the market model using *k* time-series observations. In the second step, based on each firm's preranking beta, portfolios are formed into *N* (N = 2, 10, 20, 50, 100, and 200) beta-sorted portfolios in month *t*. Post-ranking beta of each beta-sorted portfolio in month *t* is the average of preranking betas of each portfolio. The instrumental variables for cross-sectional beta-sorted portfolios in month t are the rank orders of estimated betas of beta-sorted portfolios. The monthly cross-sectional regression coefficients in month *t*, $\hat{\gamma}_{0t}$ and $\hat{\gamma}_{1t}$, are estimated through the instrumental variable method. $\bar{\hat{\gamma}}_0$ and $\bar{\hat{\gamma}}_1$, are the time-series averages of the month-by-month cross-sectional regression coefficient estimates. The *t*-statistics are provided under the null hypothesis that the estimator is different from zero. This table also presents the number of observations and the percentage of observations that are significant under the null hypothesis that the absolute value of corrected estimator is greater than the absolute value of uncorrected estimator at 95 and 99% significance levels. The *t*-statistics, standard deviation, skewness, kurtosis, Jarque–Bera statistics, and the *p*-value of Jarque–Bera test are also presented. The sample period runs from January 1931 to December 2009

Table 8.13 Normality test for the time-series estimators of maximum likelihood method for testing capital asset pricing model

	$R_{it} = \gamma_{0,t} + \gamma_{1,t}\hat{\beta}_{i,t-1} + e_{i,t}/R_{i,t} = \gamma_{0,t} + \gamma_{1,t}\hat{\beta}_{i,t-1} + \gamma_{2,t}\log(V_{i,t-1}) + e_{i,t}$									
	1931–2009		1931–1950		1951–1970		1971–1990		1991–2009	
$\bar{\bar{\gamma}}_0$										
Mean	0.1870	0.8630	−0.1172	0.6444	0.6029	1.2403	0.5464	0.8502	−0.3088	0.7095
Median	0.7518	1.1565	0.7851	1.6538	0.6846	0.8579	0.7082	1.3028	0.8975	0.6607
t-stats.	(1.16)	(2.52)	(0.26)	(0.64)	(3.95)	(3.29)	(2.46)	(1.29)	(−0.79)	(1.39)
STD	4.9779	10.5404	6.8577	15.6654	2.3623	5.8333	3.4444	10.2190	5.9016	7.6885
Skewness	−0.50425	−5.2711	−4.3889	−5.1935	−0.5363	0.3741	−0.2564	−3.9213	−4.9911	−0.8860
Kurtosis	49.1964	66.2496	33.4560	46.4568	1.3488	0.9071	6.8795	36.0691	34.6726	3.3848
J-B	99,619	177,756	11,964	22,661	30	14	476	13,625	12,367	139
p-value	$<10^{-4}$	$<10^{-4}$	$<10^{-4}$	$<10^{-4}$	$<10^{-4}$	0.0009	$<10^{-4}$	$<10^{-4}$	$<10^{-4}$	$<10^{-4}$
$\bar{\bar{\gamma}}_1$										
Mean	0.9534	1.0706	1.8187	2.2536	0.2896	0.3000	0.1204	0.2058	1.6183	1.5465
Median	0.0933	0.0939	0.3327	0.3614	0.1257	0.1338	−0.4360	−0.4525	0.3074	0.3114
t-stats.	(3.60)	(3.40)	(2.21)	(2.24)	(1.41)	(1.32)	(0.39)	(0.43)	(2.92)	(2.96)
STD	8.1650	9.6945	12.7519	15.6163	3.1825	3.5207	4.7505	7.3604	8.3701	7.8806
Skewness	6.8021	7.2029	5.3448	5.5658	0.3881	0.3755	3.1638	5.1154	5.1127	4.5775
Kurtosis	71.0500	82.7006	38.4210	42.3549	1.6528	2.5026	23.0004	46.2211	33.9865	27.0818
J-B	206710	278353	15904	19178	33	68	5691	22411	11967	7764
p-value	$<10^{-4}$	$<10^{-4}$	$<10^{-4}$	$<10^{-4}$	$<10^{-4}$	$<10^{-4}$	$<10^{-4}$	$<10^{-4}$	$<10^{-4}$	$<10^{-4}$
$\bar{\bar{\gamma}}_0$										
Mean		−0.0696		−0.1116		−0.0536		−0.0350		−0.0787
Median		−0.0660		−0.0835		−0.0497		−0.0864		−0.0465
t-stats.		(−2.96)		(−1.70)		(−1.70)		(−0.81)		(−1.95)
STD		0.7251		1.0198		0.4871		0.6722		0.6079
Skewness		0.5288		0.4414		−0.5248		1.8639		−0.4059
Kurtosis		24.9228		21.7571		1.3199		12.7973		3.6759
J-B		24579		4742		28		1777		135
p-value		$<10^{-4}$		$<10^{-4}$		$<10^{-4}$		$<10^{-4}$		$<10^{-4}$

This table provides the corrected estimators of the capital asset pricing test by using maximum likelihood method. The corrected estimators are the monthly cross-sectional regression coefficients through the maximum likelihood method, and the uncorrected estimators are those estimated through the ordinary least square. Parameters $\bar{\bar{\gamma}}_0$, $\bar{\bar{\gamma}}_1$, and c are the time-series averages of the month-by-month cross-sectional regression coefficients. Coefficients are presented in percentages, and *t*-statistics are presented in parentheses. The sample period runs from January 1931 to December 2009. Results of subperiods are also provided

temporal aggregation does affect the specification of a market model and beta estimates. Both dynamic and random coefficient market model can be used to determine the appropriate horizon for specifying the market model and estimating beta coefficients in capital asset pricing. Xiao et al. (2019) investigate the effects of temporal aggregation on beta estimates. They find that temporal aggregation affects both the specification of a market model and the stability of beta and R-square estimates. Campbell et al. (2018) use the first-order conditions of a long-term equity investor holding the aggregate equity market than overweighting value stocks and other equity portfolios. The low-frequency movements in equity volatility is associated with default risk and can explain the cross-section stock returns.

8.5 Conclusion

Numerous studies have shown that the market risk (β) has a very limited ability to explain cross-sectional average returns (e.g., Banz 1981; Carhart 1997; Chordia and Shivakumar 2006; Fama and French 1992; and Reinganum 1981). However, the failure of market risk in explaining cross-sectional stock returns might occur because of the errors-in-variables problem rather than model misspecification. In this chapter, we investigate the effect of the errors-in-variables problem on asset pricing test. When applying the two-step regression method to test the asset pricing model, the errors-in-variables problem underestimates the market beta, which suffers measurement error and overestimates other risk factors with no measurement error.

We apply three alternative correction methods for the errors-in-variables problem. The grouping method, the instrumental variable method, and the maximum likelihood method are discussed and used in the two-step capital asset pricing test. We use US firm-level data during the period 1931–2009 to test the capital asset pricing model. The grouping method, the instrumental method, and the maximum likelihood correction method are used in correcting the errors-in-variables problem in testing the capital asset pricing model. Empirical results support the role of the market beta in the capital asset pricing model after correcting the errors-in-variables problem.

Bibliography

Ahn, D. H., Conrad, J., & Dittmar, R. F. (2009). Basis assets. *Review of Financial Studies, 22*, 5133–5174.

Aït-Sahalia, Y, Parker, J. A., & Yogo, M. (2004). Luxury goods and the equity premium. *Journal of Finance, 59*, 2959–3004.

Acharya, V. V., & Pedersen, L. H. (2005). Asset pricing with liquidity risk. *Journal of Financial Economics, 77*, 375–410.

Bansal, R., Dittmar, R. F., & Lundblad, C. T. (2005). Consumption, dividends, and the cross section of equity returns. *Journal of Finance, 60*, 1639–1672.

Banz, R. W. (1981). The relationship between return and market value of common stocks. *Journal of Financial Economics, 9*, 3–18.

Black, F. (1972). Capital market equilibrium with restricted borrowing. *Journal of Business, 45*, 444–455.

Black, F., Jensen, M. C., & Scholes, M. (1972). The capital asset pricing model: Some empirical tests. In M. C. Jensen (Ed.). *Studies in the theory of capital markets*. Praeger.

Blume, M. E., & Friend, I. (1973). A new look at the capital asset pricing model. *Journal of Finance, 28*, 19–33.

Brennan M. J., & Zhang, Y. H. (2018). Capital asset pricing with a stochastic horizon. *Journal of Financial and Quantitative Analysis*, forthcoming.

Brennan, M. J. (1970). Taxes, market valuation and corporate financial policy. *National Tax Journal, 23*, 417–427.

Brennan, M. J. (1979). The pricing of contingent claims in discrete time models. *The Journal of Finance, 34*, 53–68.

Brennan, M. J., Wang, A. W., & Xia, Y. (2004). Estimation and test of a simple model of intertemporal capital asset pricing. *The Journal of Finance, 59*, 1743–1775.

Brown, S. J., & Warner, J. B. (1980). Measuring security price performance. *Journal of Financial Economics, 8*, 205–258.

Campbell, J. Y., & Vuolteenaho, T. (2004). Bad beta, good beta. *American Economic Review, 94*, 1249–1275.

Campbell, J. Y., Giglio, S., Polk, C., & Turley, R. (2018). An intertemporal CAPM with stochastic volatility, *Journal of Financial Economics*, *128*, 207–233.

Carhart, M. M. (1997). On persistence in mutual fund performance. *Journal of Finance, 52*, 57–82.

Cartwright, P. A., & Lee, C. F. (1987). Time aggregation and the estimation of the market model: empirical evidence, *Journal of Business and Economic Statistics, 5*, 131–143.

Cederburg, S., & O'Doherty, M. S. (2015). Asset-pricing anomalies at the firm level. *Journal of Econometrics, 186*, 113–128.

Chen, H. Y. (2011). Momentum strategies, dividend policy, and asset pricing test. Ph.D. dissertation, State University of New Jersey, Rutgers.

Chen, P. J., Chen, S. S., Lee, C. F., & Shih, Y. C. (2014). The evolution of capital asset pricing models. *Review of Quantitative Finance and Accounting, 42*(3), 415–448.

Cheng, P. L., & Grauer, R. R. (1980). An alternative test of the capital asset pricing model. *American Economic Review, 70*, 660–671.

Chordia, T., & Shivakumar, L. (2006). Earnings and price momentum. *Journal of Financial Economics, 80*, 627–656.

Cochrane, J. H. (1996). A cross-sectional test of an investment-based asset pricing model. *Journal of Political Economy, 104*, 572–621.

Durbin, J. (1954). Errors in variables. *Review of the International Statistical Institute, 22*, 23–32.

Dybvig, P. H., & Ross, S. A. (2003). Arbitrage, state prices and portfolio theory. In Constantinides, G., Stulz, R. M., & Harris, M. (Ed.). *Handbook of the economic of finance* (North Holland).

Fabozzi, F. J., & Francis, J. C. (1978). Beta as a random coefficient. *Journal of Financial and Quantitative Analysis, 13*, 101–116.

Fama, E. F. (1998). Market efficiency, long-term returns, and behavioral finance. *Journal of Financial Economics, 49*, 283–306.

Fama, E. F., & French, K. R. (1992). The cross-section of expected stock returns. *Journal of Finance, 47*, 427–465.

Fama, E. F., & MacBeth, J. D. (1973). Risk, return, and equilibrium: Empirical tests. *Journal of Political Economy, 81*, 607–636.

Gibbons, M. R. (1982). Multivariate tests of financial models: A new approach. *Journal of Financial Economics, 10*, 2–27.

Gilbert, T., Hrdlicka, C., Kalodimos, J., & Siegel, S. (2014). Daily data is bad for beta: Opacity and frequency-dependent betas. *The Review of Asset Pricing Studies, 4*, 78–117.

Graham, J. R., Harvey, C. R., & Puri, M. (2015). Capital allocation and delegation of decision-making authority within firms. *Journal of Financial Economics*, *115*(3), 449–470.

Gu, S., Kelly, B., & Xiu, D. (2018). *Empirical asset pricing via machine learning*. Working paper, University of Chicago.

Hansen, L. P., Heaton, J. C., & Li, N. (2008). Consumption strikes back? Measuring long-run risk. *Journal of Political Economy, 116*, 260–302.

Harvey, C. R., Liu, Y., & Zhu, H. (2016). … and the cross-section of expected returns. *Review of Financial Studies, 29*, 5–68.

Harvey, C. R. (2017). Presidential address: The scientific outlook in Financial Economics. *The Journal of Finance, 72*, 1399–1440.

Heaton, J., & Lucas, D. (2000). Portfolio choice and asset prices: The importance of entrepreneurial risk. *Journal of Finance, 55*, 1163–1198.

Jacobs, K., & Wang, K. Q. (2004). Idiosyncratic consumption risk and the cross section of asset returns. *The Journal of Finance, 59*, 2211–2252.

Jagannathan, R., & Wang, Z. (1993). *The CAPM is alive and well.* Staff report 165, Federal Reserve Bank of Minneapolis.

Jagannathan, R., & Wang, Z. (1996). The conditional CAPM and the cross-section of expected returns. *Journal of Finance, 51*, 3–53.

Jagannathan, R., & Wang, Z. (1998). An asymptotic theory for estimating beta-pricing models using cross-sectional regression. *Journal of Finance, 53*, 1285–1309.

Jagannathan, R., Skoulakis, G., & Wang, Z. (2009). The analysis of the cross section of security returns. In Aït-Sahalia, Y., & Hansen, L. (Ed.). *Handbook of financial econometrics* (North-Holland) Vol. 2, pp. 73–134.

Jarque, C. M., & Bera, A. K. (1987). A test for normality of observations and regression residuals. *International Statistical Review, 55*, 163–172.

Jegadeesh, N., Noh, J., Pukthuanghong, K., Roll, R., & Wang, J. L. (2018). Empirical tests of asset pricing models with individual assets: Resolving the errors-in-variables bias in risk premium estimation. *Journal of Financial Economics* (forthcoming).

Johnston, M. (1997). *Econometric method.* New York, NY: McGraw-Hill.

Kim, D. (1995). The errors in the variables problem in the cross-section of expected stock returns. *Journal of Finance, 50*, 1605–1634.

Kim, D. (1997). A reexamination of firm size, book-to-market, and earnings price in the cross-section of expected stock returns. *Journal of Financial and Quantitative Analysis, 32*, 463–489.

Kim, D. (2010). Issues related to the errors-in-variables problems in asset pricing tests. In C. F. Lee, A. C. Lee, & J. Lee (Ed.). *Handbook of quantitative finance and risk management*. Berlin: Springer.

Lee, C. F. (1973). *Errors-in-variables estimation procedures with applications to a capital asset pricing model*. State University of New York at Buffalo.

Lee, C. F., (1976). Investment horizon and the functional form of the capital asset pricing model, *The Review of Economics and Statistics, 58*, 356–363.

Lee, C. F. (1977). Performance measure, systematic risk, and errors-in-variables estimation method. *Journal of Economics and Business*, 122–127.

Lee, C. F. (1984). Random coefficient and errors-in-variables models for beta estimates: Methods and applications. *Journal of Business Research, 12*, 505–516.

Lee, C. F., & Chen, S. N. (1979). A random coefficient model for reexaming risk decomposition method and risk-return relationship test. *Financial Review, 14*, 65–65.

Lee, C. F., & Jen, F. C. (1978). Effects of measurement errors on systematic risk and performance measure of a portfolio. *Journal of Financial and Quantitative Analysis, 13*, 299–312.

Lee, C. F., Wu, C. & John Wei, K. C. (1990). The heterogeneous investment horizon and the capital asset pricing model: theory and implications. *Journal of Financial and Quantitative Analysis 25*, 361–376.

Lee, C. F., Wei, K. C., & Chen, H. Y. (2015). Multi-factor, multi-indicator approach to asset pricing: Methods and empirical evidence. In C. F. Lee & J. Lee (Ed.). *Handbook of financial econometrics and statistics*. Singapore: Springer.

Li, Q., Vassalou, M., & Xing, Y. (2006). Sector investment growth rates and the cross section of equity returns. *Journal of Business, 79*, 1637–1665.

Lintner, J. (1965). The valuation of risk assets and the selection of risky investments in stock portfolios and capital budgets. *Review of Economics and Statistics, 47*, 13.

Litzenberger, R. H., & Ramaswamy, K. (1979). The effects of personal taxes and dividends on capital asset prices: Theory and empirical evidence. *Journal of Financial Economics, 7*, 163–195.

MacKinlay, A. C., & Richardson, M. (1991). Using gerneralized methods of moments to test mean-variance efficiency. *Journal of Finance, 46*, 511–527.

Mossin, J. (1966). Equilibrium in a capital asset market. *Econometrica, 34*, 768–783.

Parker, J. A., & Julliard, C. (2005). Consumption risk and the cross section of expected returns. *Journal of Political Economy, 113*, 185–222.

Pástor, Ľ., & Stambaugh, R. F. (2003). Liquidity risk and expected stock returns. *Journal of Political Economy, 111*, 642–685.

Petersen, M. A. (2009). Estimating standard errors in finance panel data sets: Comparing approaches. *Review of Financial Studies, 22*, 435–480.

Petkova, R. (2006). *Do the Fama-French factors proxy for innovations in predictive variables? The Journal of Finance, 61*, 581–612.

Reinganum, M. R. (1981). Misspecification of capital asset pricing: Empirical anomalies based on earnings' yields and market values. *Journal of Financial Economics, 9*, 19–46.

Roll, R. (1969). Bias in fitting the Sharpe model to time series data. *Journal of Financial and Quantitative Analysis, 4*, 271–289.

Roll, R. (1977). A critique of the asset pricing theory's tests Part I: On past and potential testability of the theory. *Journal of Financial Economics, 4*, 129–176.

Shanken, J. (1992). On the estimation of beta-pricing models. *Review of Financial Studies, 5*, 1–33.

Sharpe, W. F. (1964). *Capital asset prices: A theory of market equilibrium under conditions of risk. Journal of Finance, 19*, 425–442.

Vassalou, M. (2003). News related to future GDP growth as a risk factor in equity returns. *Journal of Financial Economics, 68*, 47–73.

Wald, A. (1940). The fitting of straight lines if both variables are subject to error. *Annals Mathematical Statistics, 11*, 284, 300.

Xiao Y. Y., Tang, Y. S. & Lee, C. F. (2019). Impact of time aggregation on beta value and R square estimations under additive and multiplicative assumptions: Theoretical results and empirical evidence, Handbook of Financial Econometrics, Mathematics, Statistics, and Technology, World Scientific, Singapore, forthcoming.

Yogo M. (2006). A consumption-based explanation of expected stock returns. *Journal of Finance, 61*, 539–580.

Spurious Regression and Data Mining in Conditional Asset Pricing Models

9

Contents

C.-F. Lee et al., *Financial Econometrics, Mathematics and Statistics*,
https://doi.org/10.1007/978-1-4939-9429-8_9

Abstract
Based upon a pioneering paper entitled, "Spurious Regressions in Econometrics," by Granger and Newbold (J Econ 4: 111–120, 1974), this chapter investigates how the spurious regression phenomenon can affect asset pricing tests. The measure method used to examine this issue is by data mining methodology. Finally, potential solutions to the problems of spurious regression and data mining are discussed in some detail.

9.1 Introduction

The theoretical model of capital asset pricing model and methodologies to examine the capital asset pricing model have been discussed in the previous chapters. Based upon Ferson et al. (2003) and (2010), in this chapter, we discuss the issue of spurious regression proposed by Granger and Newbold (1974) and data mining in conditional asset pricing models. Section 9.2 defines the model to be used in this chapter. Section 9.3 discusses the issues of data mining and spurious regression in the simple predictive regression. Section 9.4 discusses the impact of spurious regression and data mining on conditional asset pricing. Section 9.5 describes the data. Section 9.6 presents the models used in the simulation experiments. Section 9.7 presents the simulation results for predictive regressions. Section 9.8 presents the simulation results for various forms of conditional asset pricing models. Section 9.9 discusses and evaluates solutions to the problems of spurious regression and data mining. Section 9.10 examines the robustness of results. Section 9.11 concludes the chapter.

9.2 Model Specification

In this section, we will briefly discuss the models to be used in this chapter. Predictive models for common stock returns have long been a staple of financial economics. Early studies, reviewed by Fama (1970), used such models to examine market efficiency. Stock returns are assumed to be predictable, based on lagged instrumental variables, in the current conditional asset pricing literature.

The simplest predictive model is a regression for the future stock return, r_{t+1}, on a lagged predictor variable:

$$r_{t+1} = \alpha + \beta Z_t + v_{t+1}. \tag{9.1}$$

Standard lagged variables include the levels of short-term interest rates, payout-to-price ratios for stock market indexes, and yield spreads between low-grade and high-grade bonds or between long- and short-term bonds. Table 9.1 surveys major studies that propose predictor variables. Many of these variables behave as persistent, or highly autocorrelated, time series. We study the finite-sample properties of stock return predictive regressions with persistent lagged regressors.

Regression models for stock or portfolio returns on contemporaneously measured market-wide factors have also long been a staple of financial economics. Such factor models are used in event studies (e.g., Fama et al. 1969), in tests of asset pricing theories such as the capital asset pricing model (CAPM, Sharpe 1964) and in other applications. For example, when the market return r_m is the factor, the regression model for the return r_{t+1} is:

Table 9.1 Common instrumental variables: sources, summary statistics, and OLS regression results

(1) Reference	(2) Predictor	(3) Period	(4) Obs	(5) ρ_Z	(6) σ_Z	(7) β	(8) t	(9) R^2	(10) HAC
Breen et al. (1989)	TB1y	5404–8612	393	0.97	0.0026	−2.49	−3.58	0.023	NW (5)
Campbell (1987)	Two-one	5906–7908	264	0.32	0.0006	11.87	2.38	0.025	NW (0)
	Six-one	5906–7908	264	0.15	0.0020	2.88	2.13	0.025	NW (0)
	Lag(two)-one	5906–7908	264	0.08	0.0010	9.88	2.67	0.063	NW (6)
Fama (1990)	ALLy–AAAy	5301–8712	420	0.97	0.0040	0.88	1.46	0.005	MA (0)
Fama and French (1988a)	Dividend yield	2701–8612	720	0.97	0.0013	0.40	1.36	0.007	MA (9)
Fama and French (1989)	AAAy–TB1y	2601–8612	732	0.92	0.0011	0.51	2.16	0.007	MA (9)
Keim and Stambaugh (1986)	UBAAy	2802–7812	611	0.95	0.0230	1.50	0.75	0.002	MA (9)
	UBAAy–TB1y	2802–7812	611	0.97	0.0320	1.57	1.48	0.007	MA (9)
Kothari and Shanken (1997)	DJBM	1927–1992	66	0.66	0.2270	0.28	2.63	0.078	MA (0)
Lettau and Ludvigson (2001)	"Cay"	52Q4–98Q4	184	0.79	0.0110	1.57	2.58	0.057	MA (7)
Pontiff and Schall (1998)	DJBM	2602–9409	824	0.97	0.2300	2.96	2.16	0.012	MA (9)
	SPBM	5104–9409	552	0.98	0.0230	9.32	1.03	0.001	MA (5)

$$r_{t+1} = \alpha + \beta r_{m,t+1} + u_{t+1}, \tag{9.2}$$

where $E(u_{t+1}) = E(u_{t+1} r_{m,t+1}) = 0$. The slope coefficients are the "betas," which measure the market-factor risk. When the returns are measured in excess of a reference asset like a risk-free Treasury-bill return, the intercepts are the "alphas," which measure the expected abnormal return. For example, when r_m is the market portfolio excess return, the CAPM implies that $\alpha = 0$, and the model is evaluated by testing that null hypothesis.

Recent work in conditional asset pricing allows for time-varying betas modeled as linear functions of lagged predictor variables, following Maddala (1977). Prominent examples include Shanken (1990), Cochrane (1996), Ferson and Schadt (1996), Jagannathan and Wang (1996), and Lettau and Ludvigson (2001). The time-varying beta coefficient is $\beta_t = b_0 + b_1 Z_t$ where Z_t is a lagged predictor variable. In some cases, the intercept or conditional alpha is also time-varying, as $\alpha_t = \alpha_0 + \alpha_1 Z_t$ (e.g., Christopherson et al. 1998). This results in the following regression model:

$$r_{t+1} = \alpha_0 + \alpha_1 Z_t + b_0 r_{m,t+1} + b_1 r_{m,t+1} Z_t + u_{t+1}, \tag{9.3}$$

where $E(u_{t+1}) = E(u_{t+1}[Z_t r_{m,t+1}]) = 0$. The conditional CAPM implies that $\alpha = 0$ and $\alpha_1 = 0$. This chapter also studies the finite-sample properties of asset pricing model regressions like (9.3) when there are persistent lagged regressors.

9.3 Spurious Regression and Data Mining in Predictive Regressions

In our analysis of regressions, like (9.1), that attempt to predict stock returns, we focus on two issues. The first is spurious regression, analogous to Yule (1926) and Granger and Newbold (1974). These studies warned that spurious relations may be found between the levels of trending time series that are actually independent. For example, given two independent random walks, it is likely that a regression of one on the other will produce a "significant" slope coefficient, evaluated by the usual *t*-statistics.

Stock returns are not highly autocorrelated, so you might think that spurious regression would not be an issue for stock returns. Thus, one may think that spurious regression problems are unlikely. However, the returns may be considered as the sum of an unobserved expected return, plus unpredictable noise. If the underlying expected returns are persistent time series, there is still a risk of spurious regression. Because the unpredictable noise represents a substantial portion of the variance of stock returns, the spurious regression will differ from the classical setting.

The second issue is "naïve data mining" as studied for stock returns by Lo and MacKinlay (1990), Foster et al. (1997), and others. If the standard instruments employed in the literature arise as the result of a collective search through the data, they may have no predictive power in the future. Stylized "facts" about the dynamic behavior of stock returns using these instruments (e.g., Cochrane 1999) could be artifacts of the sample. Such concerns are natural, given the widespread interest in predicting stock returns. Not all data mining is naïve. In fact, increasing computing power and data availability have allowed the development of some very sophisticated data mining (for statistical foundations, see Hastie et al. 2001).

We focus on spurious regression and the interaction between data mining and spurious regression bias. If the underlying expected return is not predictable over time, there is no spurious regression bias, even if the chosen regressor is highly autocorrelated. This is because, under the null hypothesis that there is no predictability, the autocorrelation of the regression errors the same as that of the left-hand-side asset returns. In this case, our analysis reduces to pure data mining as studied by Foster et al. (1997).

The spurious regression and data mining effects reinforce each other. If researchers have mined the data for regressors that produce high *t*-statistics in predictive regressions, then mining is more likely to uncover the spurious, persistent regressors. The standard regressors in the literature tend to be highly autocorrelated, as expected if the regressors result from this kind of a "spurious mining" process. For reasonable parameter values, all the regressions that we review from the literature are consistent with a spurious mining process, even when only a small number of instruments are considered in the mining.

While data mining amplifies the problem of spurious regressions, persistent lagged variables and spurious regression also magnify the impact of data mining. As a consequence, we show that standard corrections for data mining are inadequate in the presence of persistent lagged variables.

These results have profound potential implications for asset pricing regressions because the conditional asset pricing literature has, for the most part, used variables that were discovered based on predictive regressions like (9.1). It is important therefore to examine how data mining and spurious regression biases influence asset pricing regressions.

9.4 Spurious Regression, Data Mining, and Conditional Asset Pricing

The conditional asset pricing literature using regressions like (9.3) has evolved from the literature on pure predictive regressions. First, studies identified lagged variables that appear to predict stock returns. Later studies, beginning with Gibbons and Ferson (1985), used the same variables to study asset pricing models. Thus, it is reasonable to presume that data mining is directed at the simpler predictive regressions.

The question now is: How does this affect the validity of the subsequent asset pricing research that uses these variables in regressions like (9.3).

Table 9.2 summarizes representative studies that use the regression model (9.3). It lists the sample period, number of observations, and the lagged instruments employed. It also indicates whether the study uses the full model (9.3), with both time-varying betas and alphas, or restricted versions of the model in which either the time-varying betas or time-varying alphas are suppressed. Finally, the table summarizes the largest t-statistics for the coefficients α_1 and b_1 reported in each study. If we find that the largest t-statistics are insignificant in view of the joint effects of spurious regression and data mining, then none of the coefficients are significant. We return to this table later and revisit the evidence.

Table 9.2 Representative studies on conditional asset pricing models

(1) Reference	(2) Predictor	(3) Period	(4) Obs	(5) α_t	(6) β_t	(7) $\lvert\alpha_0/\alpha\rvert$	(8) α_1	(9) $t(\alpha_1)$	(10) b_1	(11) $t(b_1)$
Shanken (1990)	TB1y TB1vol	5301– 8212	360	Yes	Yes	NA	−0.48 −5.70	−1.17 −3.56	1.42 −8.40	5.92 −4.42
Cochrane (1996)	DY Term	47Q1– 93Q4	188	No	Yes	NA	None	None	−0.53 −0.31	−4.74 −1.76
Ferson and Schadt (1996)	TB1y DY Term Default	6801– 9012	276	No	Yes	0.72	None	None	NA	NA
Jagannathan and Wang (1996)	Default	6307– 9012	330	Yes	No	1.53	–65.7	–3.10	None	None
Christopherson et al. (1998)	TB1y DY Term	7901– 9012	144	Yes	Yes	0.77	–0.21 1.22 –0.21	–2.01 3.72 –1.85	NA	NA
Lettau and Ludvigson (2001)	"Cay"	63Q3– 98Q3	144	Yes	Yes	0.84	NA	NA	NA	NA
Petkova and Zhang (2005)	TB1y DY Term Default	2701– 0112	900	No	Yes	0.97	None	None	NA	NA

The first column indicates the published study. The second column specifies the lagged instruments used. The next two columns give the sample (*Period*) and the number of observations (*Obs*) on the stock returns. Columns five and six indicate whether the conditional model includes time-varying alpha (α_t) and the time-varying beta (β_t), respectively. The last five columns summarize the regression results. Column seven shows the ratio of an intercept (a pricing error) in the conditional model to that of the unconditional. Columns 8 and 9 report the point estimates of the time-varying alpha, α_1, and their corresponding t-statistics, $t(\alpha_1)$. Columns 10 and 11 report the point estimates of the time-varying beta coefficient, b_1, and their corresponding t-statistics, $t(b_1)$. For each predictor, the table reports regression estimates corresponding to the largest in absolute value t-statistics. The abbreviations in the table are as follows. *TB1y* and *TB1vol* are the yield and volatility on the one-month Treasury bill, respectively. *Three-one* is the difference between the lagged returns of a three-month and a one-month T-bill. The variable "*Cay*" is the linear function of consumption, asset wealth, and labor income. *DY* is the dividend yield of the CRSP index. *Term* is a spread between long-term and short-term bonds. *Default* is a spread between low-grade and high-grade corporate bonds. "None" stands parameter values not used in the study, "NA" stands for results not reported.

Using regression models like Eq. (9.3), the literature has produced a number of "stylized facts." First, studies typically find that the intercept is smaller in the "conditional" model (9.3) than in the "unconditional" model (9.2): $|\alpha| > |\alpha_0|$. The interpretation of these studies is that the conditional CAPM does a better job of "explaining" average returns than the unconditional CAPM. Examples with this finding include Cochrane (1996), Ferson and Schadt (1996), Ferson and Harvey (1997, 1999), Lettau and Ludvigson (2001), and Petkova and Zhang (2005). Second, studies typically find evidence of time-varying betas: The coefficient estimate for b_1 is statistically significant. Third, studies typically find that the conditional models fail to completely explain the dynamic properties of returns: The coefficient estimate for α_1 is significant, indicating a time-varying alpha. Our objective is to study the reliability of such inferences in the presence of persistent lagged instruments and data mining.

9.5 The Data

Table 9.1 surveys nine of the major studies that propose instruments for predicting stock returns. The table reports summary statistics for monthly data, covering various subperiods of 1926 through 1998. The sample size and period depends on the study and the variable, and the table provides the details. We attempt to replicate the data series that were used in these studies as closely as possible. The summary statistics are from our data. Note that the first-order autocorrelations of the predictor variables frequently suggest a high degree of persistence. For example, short-term Treasury-bill yields, monthly book-to-market ratios, the dividend yield of the S&P500, and some of the yield spreads have sample first-order autocorrelations of 0.97 or higher.

This table summarizes variables used in the literature to predict stock returns. The first column indicates the published study. The second column denotes the lagged instrument. The next two columns give the sample (Period) and the number of observations (Obs) on the stock returns. Columns five and six report the autocorrelation (ρ_Z) and the standard deviation of the instrument (σ_Z), respectively. The next three columns report regression results for Standard & Poors 500 excess return on a lagged instrument. The slope coefficient is β, the t-statistic is t, and the coefficient of determination is R^2. The last column (HAC) reports the method used in computing the standard errors of the slopes. The method of Newey–West (1987) is used with the number of lags given in parentheses. The abbreviations in the table are as follows. TB1y is the yield on the one-month Treasury bill. Two-one, Six-one, and Lag(two)-one are computed as the spreads on the returns of the two and one-month bills, six and one-month bills, and the lagged value of the two-month and current one-month bill. The yield on all corporate bonds is denoted as ALLy. The yield on AAA-rated corporate bonds is AAAy, and UBAAy is the yield on corporate bonds with a below BAA rating. The variable "Cay" is the linear function of consumption, asset wealth, and labor income. The book-to-market ratios for the Dow Jones Industrial Average and the S&P500 are, respectively, DJBM and SPBM.

Table 9.1 also summarizes regressions for the monthly return of the S&P500 stock index, measured in excess of the one-month Treasury-bill return from Ibbotson Associates, on the lagged instruments. These are OLS regressions using one instrument at a time. We report the slope coefficients, their t-ratios, and the adjusted R-squares. The R-squares range from less than one percent to more than seven percent, and eight of the 13 t-ratios are larger than 2.0. The t-ratios are based on the OLS slopes and Newey–West (1987) standard errors, where the number of lags is chosen based on the number of statistically significant residual autocorrelations.[1]

[1]Specifically, we compute 12 sample autocorrelations and compare their values with a cutoff at two approximate standard errors: $2\sqrt{T}$, where T is the sample size. The number of lags chosen is the minimum lag length at which no higher-order autocorrelation is larger than two standard errors. The number of lags chosen is indicated in the far right column.

The small R-squares in Table 9.1 suggest that predictability represents a tiny fraction of the variance in stock returns. However, even a small R-squared can signal economically significant predictability. For example, Kandel and Stambaugh (1990) and Fleming et al. (2001) find that optimal portfolios respond by a substantial amount to small R-squares in standard models. Studies combining several instruments in multiple regressions report higher R-squares. For example, Harvey (1989), using five instruments, reports adjusted R-squares as high as 17.9% for size portfolios. Ferson and Harvey (1991) report R-squares of 5.8–13.7% for monthly size and industry portfolio returns. These values suggest that the "true" R-squared, if we could regress the stock return on its time-varying conditional mean, might be substantially higher than we see in Table 9.1. To accommodate this possibility, we allow the true R-squared in our simulations to vary over the range from zero to 15%. For exposition, we focus on an intermediate value of 10%.

To incorporate data mining, we compile a randomly selected sample of 500 potential instruments, through which our simulated analyst sifts to mine the data for predictor variables. All the data come from the Web site Economagic.-com: Economic Time Series Page, maintained by Ted Bos. The sample consists of all monthly series listed on the main homepage of the site, except under the headings of LIBOR, Australia, Bank of Japan, and Central Bank of Europe. From the Census Bureau, we exclude building permits by region, state, and metro areas (more than 4000 series). From the Bureau of Labor Statistics, we exclude all noncivilian labor force data and state, city, and international employment (more than 51,000 series). We use the consumer price index (CPI) measures from the city average listings, but include no finer subcategories. The producer price index (PPI) measures include the aggregates and the two-digit subcategories. From the Department of Energy, we exclude data in Sect. 10, the International Energy series.

We first randomly select (using a uniform distribution) 600 out of the 10,866 series that were left after the above exclusions. From these 600, we eliminated series that mixed quarterly and monthly data and extremely sparse series, and took the first 500 from what remained.

Because many of the data are reported in levels, we tested for unit roots using an augmented Dickey–Fuller test (with a zero-order time polynomial). We could not reject the hypothesis of a unit root for 361 of the 500 series, and we replaced these series with their first differences. The 500 series are randomly ordered, and then permanently assigned numbers between one and 500. When a data miner in our simulations searches through, say 50 series, we use the sampling properties of the first 50 series to calibrate the parameters in the simulations.

We also use our sample of potential instruments to calibrate the parameters that govern the amount of persistence in the "true" expected returns in the model. On the one hand, if the instruments we see in the literature, summarized in Table 9.1, arise from a spurious mining process, they are likely to be more highly autocorrelated than the underlying "true" expected stock return. On the other hand, if the instruments in the literature are a realistic representation of expected stock returns, the autocorrelations in Table 9.1 may be a good proxy for the persistence of the true expected returns.[2] The mean autocorrelation of our 500 series is 15% and the median is two percent. Eleven of the 13 sample autocorrelations in Table 9.1 are higher than 15%, and the median value is 95%. We consider a range of values for the true autocorrelation based on these figures, as described below.

[2]There are good reasons to think that expected stock returns may be persistent. Asset pricing models like the consumption model of Lucas (1978) describe expected stock returns as functions of expected economic growth rates. Merton (1973) and Cox et al. (1985) propose real interest rates as candidate state variables, driving expected returns in intertemporal models. Such variables are likely to be highly persistent. Empirical models for stock return dynamics frequently involve persistent, autoregressive expected returns (e.g., Lo and MacKinlay 1988; Conrad and Kaul 1988; Fama and French 1988b; or Huberman and Kandel 1990).

9.6 The Models

9.6.1 Predictive Regressions

In the model for the predictive regressions, the data are generated by an unobserved latent variable, Z^*, as:

$$r_{t+1} = \mu + Z_t^* + u_{t+1}, \tag{9.4}$$

where u_{t+1} is white noise with variance, σ_u^2. We interpret the latent variable, Z_t^* as the deviations of the conditional mean return from the unconditional mean, μ, where the expectations are conditioned on an unobserved "market" information set at time t. The predictor variables follow an autoregressive process:

$$(Z_t^*, Z_t)' = \begin{Bmatrix} \rho^* & 0 \\ 0 & \rho \end{Bmatrix} (Z_{t-1}^*, Z_{t-1})' + (\varepsilon_t^*, \varepsilon_t)', \tag{9.5}$$

where Z_t is the measured predictor variable and ρ is the autocorrelation. The assumption that the true expected return is autoregressive (with parameter ρ^*) follows previous studies such as Lo and MacKinlay (1988), Conrad and Kaul (1988), Fama and French (1988b), and Huberman and Kandel (1990).

To generate the artificial data, the errors $(\varepsilon_t^*, \varepsilon_t)$ are drawn randomly as a normal vector with mean zero and covariance matrix, Σ. We build up the time series of the Z and Z^* through the vector autoregression Eq. (9.3), where the initial values are drawn from a normal with mean zero and variances, $\text{Var}(Z)$ and $\text{Var}(Z^*)$. The other parameters that calibrate the simulations are $\{\mu, \sigma_u^2, \rho, \rho^*, \text{and } \sum\}$.

We have a situation in which the "true" returns may be predictable, if Z_t^* could be observed. This is captured by the true R-squared, $\text{Var}(Z^*)/\left[\text{Var}(Z^*) + \sigma_u^2\right]$. We set $\text{Var}(Z^*)$ to equal the sample variance of the S&P500 return, in excess of a one-month Treasury-bill return, multiplied by 0.10. When the true R-squared of the simulation is 10%, the unconditional variance of the r_{t+1} that we generate is equal to the sample variance of the S&P500 return. When we choose other values for the true R-squared, these determine the values for the parameter σ_u^2. We set μ to equal the sample mean excess return of the S&P500 over the 1926 through 1998 period, or 0.71% per month.

The extent of the spurious regression bias depends on the parameters, ρ and ρ^*, which control the persistence of the measured and the true regressor. These values are determined by reference to Table 9.1 and from our sample of 500 potential instruments. The specifics differ across the special cases, as described below.

While the stock return could be predicted if Z_t^* could be observed, the analyst uses the measured instrument Z_t. If the covariance matrix $\sum$ is diagonal, Z_t and Z_t^* are independent, and the true value of δ in the regression (9.1) is zero.

To focus on spurious regression in isolation, we specialize Eq. (9.3) as follows. The covariance matrix $\sum$ is a 2×2 diagonal matrix with variances (σ^{*2}, σ^2). For a given value of ρ^* the value of σ^{*2} is determined as $\sigma^{*2} = (1 - \rho^{*2})\text{Var}(Z^*)$. The measured regressor has $\text{Var}(Z) = \text{Var}(Z^*)$. The autocorrelation parameters, $\rho^* = \rho$, are allowed to vary over a range of values. (We also allow ρ and ρ^* to differ from one another, as described below.)

Following Granger and Newbold (1974), we interpret a spurious regression as one in which the "t-ratios" in the regression (9.1) are likely to indicate a significant relation when the variables are really independent. The problem may come from the numerator or the denominator of the t-ratio: The coefficient or its standard error may be biased. As in Granger and Newbold, the problem lies with the standard errors.[3] The reason is simple to understand. When the null hypothesis that the regression slope $\delta = 0$ is true, the error term u_{t+1} of the regression Eq. (9.1) inherits

[3]While Granger and Newbold (1974) do not study the slopes and standard errors to identify the separate effects, our simulations designed to mimic their setting (not reported in the tables) confirm that their slopes are well behaved, while the standard errors are biased. Granger and Newbold use OLS standard errors, while we focus on the heteroscedasticity and autocorrelation-consistent standard errors that are more common in recent studies.

autocorrelation from the dependent variable. Assuming stationarity, the slope coefficient is consistent, but standard errors that do not account for the serial dependence correctly are biased.

Because the spurious regression problem is driven by biased estimates of the standard error, the choice of standard error estimator is crucial. In our simulation exercises, it is possible to find an efficient unbiased estimator, since we know the "true" model that describes the regression error. Of course, this will not be known in practice. To mimic the practical reality, the analyst in our simulations uses the popular autocorrelation-heteroscedasticity-consistent (HAC) standard errors from Newey and West (1987), with an automatic lag selection procedure. The number of lags is chosen by computing the autocorrelations of the estimated residuals, and truncating the lag length when the sample autocorrelations become "insignificant" at longer lags. (The exact procedure is described in Footnote 1, and modifications to this procedure are discussed below.)

This setting is related to Phillips (1986) and Stambaugh (1999). Phillips derives asymptotic distributions for the OLS estimators of the regression (9.1), in the case where $\rho = 1$, $u_{t+1} = 0$, and $\{\varepsilon_t^*, \varepsilon_t\}$ are general independent mean zero processes. We allow a nonzero variance of u_{t+1} to accommodate the large noise component of stock returns. We assume $\rho < 1$ to focus on stationary, but possibly highly autocorrelated, regressors.

Stambaugh (1999) studies a case where the errors $\{\varepsilon_t^*, \varepsilon_t\}$ are perfectly correlated, or equivalently, the analyst observes and uses the correct lagged stochastic regressor. A bias arises when the correlation between u_{t+1} and ε_{t+1}^* is not zero, related to the well-known small sample bias of the autocorrelation coefficient (e.g., Kendall 1954). In the pure spurious regression case studied here, the observed regressor Z_t is independent of the true regressor Z_t^*, and u_{t+1} is independent of ε_{t+1}^*. The Stambaugh bias is zero in this case. The point is that there remains a problem in predictive regressions, in the absence of the bias studied by Stambaugh, because of spurious regression.

9.6.2 Conditional Asset Pricing Models

The data in our simulations of conditional asset pricing models are generated according to:

$$\begin{aligned} r_{t+1} &= \beta_t r_m, u_{t+1}, \\ \beta_t &= 1 + Z_t^*, \\ r_{m,t+1} &= \mu + kZ_t^* + w_{t+1}. \end{aligned} \tag{9.6}$$

Our artificial analyst uses the simulated data to run the regression model (9.3), focusing on the t-statistics for the coefficients $\{\alpha_0, \alpha_1, b_0, b_1\}$. The variable Z_t^* in Eq. (9.6) is an unobserved latent variable that drives both expected market returns and time-varying betas. The term β_t in Eq. (9.6) is a time-varying beta coefficient. As Z_t^* has mean equal to zero, the expected value of beta is 1.0. When $k \neq 0$, there is an interaction between the time variation in beta and the expected market risk premium. A common persistent factor drives the movements in both expected returns and conditional betas. Common factors in time-varying betas and expected market premiums are important in asset pricing studies such as Chan and Chen (1988), Ferson and Korajczyk (1995), and Jagannathan and Wang (1996), and in conditional performance evaluation, as in Ferson and Schadt (1996). There is a zero intercept, or "alpha," in the data-generating process for r_{t+1}, consistent with asset pricing theory.

The market return data, $r_{m,t+1}$, are generated as follows. The parameter μ was described earlier. The variance of the error is $\sigma_w^2 = \sigma_{sp}^2 - k^2\text{Var}(Z^*)$, where $\sigma_{sp} = 0.057$ matches the S&P500 return and $\text{Var}(Z^*) = 0.055$ is the estimated average monthly variance of the market betas on 58 randomly selected stocks from CRSP over the period 1926–1997.[4] The predictor variables follow the autoregressive process (9.3).

9.7 Results for Predictive Regressions

9.7.1 Pure Spurious Regression

Table 9.3 summarizes the results for the case of pure spurious regression, with no data mining. We record the estimated slope coefficient in Eq. (9.1), its Newey–West t-ratio, and the coefficient of determination at each trial and summarize their empirical distributions. The experiments are run for two sample sizes, based on the extremes in Table 9.1. These are $T = 66$ and $T = 824$ in Panels A and B, respectively. In Panel C, we match the sample sizes to the studies in Table 9.1. In each case, 10,000 trials of the simulation are run; 50,000 trials on a subset of the cases produce similar results.

The table reports the 97.5% of the Monte Carlo distribution of 10,000 Newey–West t-statistics, the 95% for the estimated coefficients of determination, and the average estimated slopes from the regression

$$r_{t+1} = \alpha + \delta Z_t + v_{t+1},$$

where r_{t+1} is the excess return, Z_t is the predictor variable, and $t = 1,\ldots,T$. The parameter ρ^* is the autocorrelation coefficient of the predictors, Z_t^* and Z_t. The R^2 is the coefficient of determination from the regression of excess returns r_{t+1} on the unobserved, true instrument Z_t^*. Panel A depicts the results for $T = 66$ and Panel B for $T = 824$. Panel C gives the simulation results for the number of observations and the autocorrelations in Table 9.1. In Panel C, the true R^2 is set to 0.1. The theoretical critical R^2 is from the F-distribution.

The rows of Table 9.3 refer to different values for the true R-squares. The smallest value is 0.1%, where the stock return is essentially unpredictable, and the largest value is 15%. The columns of Table 9.3 correspond to different values of ρ^*, the autocorrelation of the true expected return, which runs from 0.00 to 0.99. In these experiments, we set $\rho = \rho^*$. The subpanels labeled critical t-statistic and critical estimated R^2 report empirical critical values from the 10,000 simulated trials, so that 2.5% of the t-statistics or five percent of the R-squares lie above these values.

The subpanels labeled mean δ report the average slope coefficients over the 10,000 trials. The mean estimated values are always small and very close to the true value of zero at the larger sample size. This confirms that the slope coefficient estimators are well behaved, so that bias due to spurious regression comes from the standard errors.

When $\rho^* = 0$, and there is no persistence in the true expected return, the table shows that spurious regression phenomenon is not a concern. This is true even when the measured regressor is highly persistent. (We confirm this with additional simulations, not reported in the tables, where we set $\rho^* = 0$ and vary ρ.) The logic is that when the slope in Eq. (9.1) is zero and $\rho^* = 0$, the regression error has no persistence, so the standard errors are well behaved. This implies that spurious regression is not a problem from the perspective of testing the null hypothesis that expected stock returns are unpredictable, even if a highly autocorrelated regressor is used.

[4]We calibrate the variance of the betas to actual monthly data by randomly selecting 58 stocks with complete CRSP data for January 1926 through December 1997. Following Fama and French (1997), we estimate simple regression betas for each stock's monthly excess return against the S&P500 excess return, using a series of rolling 5-year windows, rolling forward one month at a time. For each window, we also compute the standard error of the beta estimate. This produces a series of 805 beta estimates and standard error estimates for beta for each firm. We calibrate the variance of the true beta for each firm to equal the sample variance of the rolling beta estimates minus the average estimated variance of the estimator. Averaging the result across firms, the value of $\text{Var}(Z^*)$ is 0.0550. Repeating this exercise with firms that have data from January of 1926 through the end of 2004 increases the number of months used from 864 to 948 but decreases the number of firms from 58 to 46. The value of $\text{Var}(Z^*)$ in this case is 0.0549.

Table 9.3 Monte Carlo simulation results for regressions with a lagged predictor variable

Panel A: 66 observations						
R^2/ρ^*	0	0.5	0.9	0.95	0.98	0.99
Means: δ						
0.001	−0.0480	−0.0554	−0.0154	−0.0179	−0.0312	−0.0463
0.005	−0.0207	−0.0246	−0.0074	−0.0088	−0.0137	−0.0193
0.010	−0.0142	−0.0173	−0.0055	−0.0066	−0.0096	−0.0129
0.050	−0.0055	−0.0075	−0.0029	−0.0037	−0.0040	−0.0042
0.100	−0.0033	−0.0051	−0.0023	−0.0030	−0.0026	−0.0021
0.150	−0.0024	−0.0040	−0.0020	−0.0026	−0.0020	−0.0012
Critical t-statistics						
0.001	2.1951	2.3073	2.4502	2.4879	2.4746	2.4630
0.005	2.2033	2.3076	2.4532	2.5007	2.5302	2.5003
0.010	2.2121	2.3123	2.4828	2.5369	2.5460	2.5214
0.050	2.2609	2.3335	2.6403	2.7113	2.7116	2.6359
0.100	2.2847	2.3702	2.8408	2.9329	2.9043	2.7843
0.150	2.2750	2.3959	3.0046	3.1232	3.0930	2.9417
Critical estimated R^2						
0.001	0.0593	0.0575	0.0598	0.0599	0.0610	0.0600
0.005	0.0590	0.0578	0.0608	0.0607	0.0616	0.0604
0.010	0.0590	0.0579	0.0619	0.0623	0.0630	0.0612
0.050	0.0593	0.0593	0.0715	0.0737	0.0703	0.0673
0.100	0.0600	0.0622	0.0847	0.0882	0.0823	0.0766
0.150	0.0600	0.0649	0.0994	0.1032	0.0942	0.0850
Panel B: 824 observations						
R^2/ρ^*	0	0.5	0.9	0.95	0.98	0.99
Means: δ						
0.001	0.0150	0.0106	0.0141	0.0115	0.0053	−0.0007
0.005	0.0067	0.0049	0.0069	0.0055	0.0021	−0.0011
0.010	0.0048	0.0035	0.0052	0.0040	0.0014	−0.0012
0.050	0.0021	0.0017	0.0029	0.0021	0.0003	−0.0014
0.100	0.0015	0.0013	0.0023	0.0016	0.0001	−0.0014
0.150	0.0012	0.0011	0.0021	0.0014	−0.0000	−0.0014
Critical t-statistics						
0.001	1.9861	2.0263	2.0362	2.0454	2.0587	2.0585
0.005	1.9835	2.0297	2.0429	2.1123	2.1975	2.2558
0.010	1.9759	2.0279	2.0655	2.1479	2.3578	2.4957
0.050	1.9878	2.0088	2.2587	2.5685	3.1720	3.7095
0.100	1.9862	2.0320	2.3758	2.7342	3.6356	4.4528
0.150	2.0005	2.0246	2.4164	2.8555	3.8735	4.9151
Critical estimated R^2						

(continued)

Table 9.3 (continued)

Panel A: 66 observations

R^2/ρ^*	0	0.5	0.9	0.95	0.98	0.99
0.001	0.0046	0.0047	0.0047	0.0047	0.0049	0.0049
0.005	0.0046	0.0047	0.0048	0.0051	0.0056	0.0059
0.010	0.0046	0.0047	0.0050	0.0054	0.0065	0.0073
0.050	0.0046	0.0047	0.0066	0.0085	0.0132	0.0183
0.100	0.0047	0.0049	0.0084	0.0125	0.0220	0.0316
0.150	0.0046	0.0050	0.0104	0.0166	0.0308	0.0450

Panel C: Table 9.1 simulation

Obs	ρ^*	Critical theoretical R^2	Critical t-statistic	Critical estimated R^2
393	0.97	0.0098	3.2521	0.0311
264	0.32	0.0146	2.0645	0.0151
264	0.15	0.0146	2.0560	0.0151
264	0.08	0.0146	2.0318	0.0146
420	0.97	0.0092	3.2734	0.0304
720	0.97	0.0053	3.2005	0.0194
732	0.92	0.0053	2.3947	0.0103
611	0.95	0.0063	2.8843	0.0167
611	0.97	0.0063	3.2488	0.0219
66	0.66	0.0586	2.4221	0.0656
184	0.79	0.0209	2.2724	0.0270
824	0.97	0.0047	3.1612	0.0173
552	0.98	0.0070	3.6771	0.0293

Table 9.3 shows that spurious regression bias does not arise to any serious degree, provided ρ^* is 0.90 or less, and provided that the true R^2 is one percent or less. For these parameters, the empirical critical values for the t-ratios are 2.48 (T = 66, Panel A) and 2.07 (T = 824, Panel B). The empirical critical R-squares are close to their theoretical values. For example, for a five percent test with $T = (66, 824)$, the F-distribution implies critical R-squared values of (5.9%, 0.5%). The values in Table 9.3 when $\rho^* = 0.90$ and true R^2 = 1% are (6.2%, 0.5%); thus, the empirical distributions do not depart far from the standard rules of thumb.

Variables like short-term interest rates and dividend yields typically have first-order sample autocorrelations in excess of 0.95, as we saw in Table 9.1. We find substantial biases when the regressors are highly persistent. Consider the plausible scenario with a sample of T = 824 observations where $\rho = 0.98$ and true $R^2 = 10\%$. In view of the spurious regression phenomenon, an analyst who was not sure that the correct instrument is being used and who wanted to conduct a five percent, two-tailed t-test for the significance of the measured instrument, would have to use a t-ratio of 3.6. The coefficient of determination would have to exceed 2.2% to be significant at the five percent level. These cutoffs are substantially more stringent than the usual rules of thumb.

Panel C of Table 9.3 revisits the evidence from the literature in Table 9.1. The critical values for the t-ratios and R-squares are reported, along with the theoretical critical values for the R-squares, implied by the F-distribution. We set the true R-squared value equal to 10% and $\rho^* = \rho$ in each case. We find that seven of the 17 statistics in Table 9.1 that would be considered significant using the traditional standards are no longer significant in view of the spurious regression bias.

While Panels A and B of Table 9.3 show that spurious regression can be a problem in stock return regressions, Panel C finds that accounting for spurious regression changes the inferences about specific regressors that were found to be significant in previous studies. In particular, we question the significance of the term spread in Fama and French (1989), on the basis of either the t-ratio or the R-squared of the regression. Similarly, the book-to-market ratio of the Dow Jones index, studied by Pontiff and Schall (1998) fails to be significant with either statistic. Several other variables are marginal, failing on the basis of one but not both statistics. These include the short-term interest rate (Fama and Schwert 1977; using the more recent sample of Breen et al. 1989), the dividend yield (Fama and French 1988a), and the quality-related yield spread (Keim and Stambaugh 1986). All of these regressors would be considered significant using the standard cutoffs.

It is interesting to note that the biases documented in Table 9.2 do not always diminish with larger sample sizes; in fact, the critical t-ratios are larger in the lower right corner of the panels when $T = 824$ than when $T = 66$. The mean values of the slope coefficients are closer to zero at the larger sample size, so the larger critical values are driven by the standard errors. A sample as large as $T = 824$ is not by itself a cure for the spurious regression bias. This is typical of spurious regression with a unit root, as discussed by Phillips (1986) for infinite-sample sizes and nonstationary data.[5] It is interesting to observe similar patterns, even with stationary data and finite samples.

Phillips (1986) shows that the sample autocorrelation in the regression studied by Granger and Newbold (1974) converges in limit to 1.0. However, we find only mildly inflated residual autocorrelations (not reported in the tables) for stock return samples as large as $T = 2000$, even when we assume values of the true R^2 as large as 40%. Even in these extreme cases, none of the empirical critical values for the residual autocorrelations are larger than 0.5. Since $u_{t+1} = 0$ in the cases studied by Phillips, we expect to see explosive autocorrelations only when the true R^2 is very large. When R^2 is small the white noise component of the returns serves to dampen the residual autocorrelation. Thus, we are not likely to see large residual autocorrelations in stock return regressions, even when spurious regression is a problem. The residuals-based diagnostics for spurious regression, such as the Durbin–Watson tests suggested by Granger and Newbold, are not likely to be very powerful in stock return regressions. For the same reason, typical application of the Newey–West procedure, where the number of lags is selected by examining the residual autocorrelations, is not likely to resolve the spurious regression problem.

Newey and West (1987) show that their procedure is consistent for the standard errors when the number of lags used grows without bound as the sample size T increases, provided that the number of lags grows no faster than $T^{1/4}$. The lag selection procedure in Table 9.3 examines 12 lags. Even though no more than nine lags are selected for the actual data in Table 9.1, more lags would sometimes be selected in the simulations, and an inconsistency results from truncating the lag length.[6] However, in finite samples, an increase in the number of lags can

[5] Phillips derives asymptotic distributions for the OLS estimators of Eq. (9.1), in the case where $\rho = 1$, $u_{t+1} \equiv 0$. He shows that the t-ratio for diverges for large T, while $t(\delta)\sqrt{T}, \delta$, and the coefficient of determination converge to well-defined random variables. Marmol (1998) extends these results to multiple regressions with partially integrated processes and provides references to more recent theoretical literature. Phillips (1998) reviews analytical tools for asymptotic analysis when nonstationary series are involved.

[6] At very large sample sizes, a huge number of lags can control the bias. We verify this by examining samples as large as $T = 5000$, letting the number of lags grow to 240. With 240 lags, the critical t-ratio when the true

make things worse. When "too many" lags are used, the standard error estimates become excessively noisy, which thickens the tails of the sampling distribution of the t-ratios. This occurs for the experiments in Table 9.2. For example, letting the procedure examine 36 autocorrelations to determine the lag length (the largest number we find mentioned in published studies) the critical t-ratio in Panel A, for true $R^2 = 10\%$ and $\rho^* = 0.98$, increases from 2.9 to 4.8. Nine of the 17 statistics from Table 9.1 that are significant by the usual rules of thumb now become insignificant. The results calling these studies into question are therefore even stronger than before. Thus, simply increasing the number of lags in the Newey–West procedure is not likely to resolve the finite-sample, spurious regression bias.[7] We discuss this issue in more detail in Sect. 8.1.

We draw several conclusions about spurious regression in stock return predictive regressions. Given persistent expected returns, spurious regression can be a serious concern well outside the classic setting of Yule (1926) and Granger and Newbold (1974). Stock returns, as the dependent variable, are much less persistent than the levels of most economic time series. Yet, when the expected returns are persistent, there is a risk of spurious regression bias. The regression residuals may not be highly autocorrelated, even when spurious regression bias is severe. Given inconsistent standard errors, spurious regression bias is not avoided with large samples. Accounting for spurious regression bias, we find that seven of the 17 t-statistics and regression R-squares from previous studies of predictive regressions that would be significant by standard criteria are no longer significant.

$R^2 = 10\%$ and $\rho = 0.98$ falls from 3.6 in Panel B of Table 9.2 to a reasonably well-behaved value of 2.23.

[7]We conduct several experiments letting the number of lags examined be 24, 36, or 48, when $T = 66$ and $T = 824$. When $T = 66$, the critical t-ratios are always larger than the values in Table 9.2. When $T = 824$, the effects are small and of mixed sign. The most extreme reduction in a critical t-ratio, relative to Table 9.2, is with 48 lags, true $R^2 = 15\%$, and $\rho^* = 0.99$, where the critical value falls from 4.92 to 4.23.

9.7.2 Spurious Regression and Data Mining

We now consider the interaction between spurious regression and data mining in the predictive regressions, where the instruments to be mined are independent as in Foster et al. (1997). There are L-measured instruments over which the analyst searches for the "best" predictor, based on the R-squares of univariate regressions. In Eq. (9.5), Z_t becomes a vector of length L, where L is the number of instruments through which the analyst sifts. The error terms become an $L+1$ vector with a diagonal covariance matrix; thus, ε_t^* is independent of ε_t.

The persistence parameters in Eq. (9.5) become an $(L+1)$-square, diagonal matrix, with the autocorrelation of the true predictor equal to ρ^*. The value of ρ^* is either the average from our sample of 500 potential instruments, 15%, or the median value from Table 9.1, 95%. The remaining autocorrelations, denoted by the L-vector, ρ, are set equal to the autocorrelations of the first L instruments in our sample of 500 potential instruments.[8] When $\rho^* = 95\%$, we rescale the autocorrelations to center the distribution at 0.95 while preserving the range in the original data.[9] The simulations match the unconditional variances of the instruments, $\mathrm{Var}(Z)$, to the data. The first element of the covariance matrix Σ is equal to σ*2. For a typical i-th diagonal element of $\sum$, denoted by σ_i, the

[8]We calibrate the true autocorrelations in the simulations to the sample autocorrelations, adjusted for first-order finite-sample bias as: $\hat{\rho} + (1 + 3\hat{\rho})/T$, where $\hat{\rho}$ is the OLS estimate of the autocorrelation and T is the sample size.

[9]The transformation is as follows. In the 500 instruments, the minimum bias-adjusted autocorrelation is −0.571, the maximum is 0.999, and the median is 0.02. We center the transformed distribution about the median in Table 9.1, which is 0.95. If the original autocorrelation ρ is less than the median, we transform it to:

$$0.95 + (\rho - 0.02)\{(0.95 + 0.571)/(0.02 + 0.571)\}.$$

If the value is above the median, we transform it to:

$$0.95 + (\rho - 0.02)\{(0.999 + 0.95)/(0.999 - 0.02)\}.$$

elements of $\rho(Z_i)$ and $\text{Var}(Z_i)$ are matched to the data, and we set $\sigma_i^2 = \left[1 - \rho(Z_i)^2\right]\text{Var}(Z_i)$.

Table 9.4 summarizes the results. The columns correspond to different numbers of potential instruments, through which the analyst sifts to find the regression that delivers the highest sample R-squared. The rows refer to the different values of the true R-squared.

The table reports the 97.5% of the Monte Carlo distribution of 10,000 Newey–West t-statistics, the 95% for the estimated coefficients of determination, and the average estimated slopes from the regression

$$r_{t+1} = \alpha + \delta Z_t + v_{t+1},$$

where r_{t+1} is the excess return, Z_t is the predictor variable, and $t = 1, \ldots, T$. The R^2 is the coefficient of determination from the regression of excess returns r_{t+1} on the unobserved, true instrument Z_t^*, which has the autocorrelation ρ^*. The parameter L is the number of instruments mined, where the one with the highest estimated R^2 is chosen. Panels A and B depict the results for $T = 66$ and $T = 824$, respectively, when the autocorrelation of the true predictor, $\rho^* = 0.15$. Panels C and D depict the results for $T = 66$ and $T = 824$, respectively, when the autocorrelation of the true predictor, $\rho^* = 0.95$, the median autocorrelation in Table 9.1. In Panel E, the true R^2 is set to 0.1 and the original distribution of instruments is transformed so that their median autocorrelation is set at 0.95. The left-hand side of Panel E gives the critical L for the given number of observations and autocorrelation that is sufficient to generate critical t-statistics or R^2's in excess of the corresponding statistics in Table 9.1. The right-hand side of Panel E gives the critical L that is sufficient to generate critical t-statistics or R^2's in excess of the corresponding statistics in Table 9.1 when $\rho^* = 0.95$.

The rows with true $R^2 = 0$ refer to data mining only, similar to Foster et al. (1997). The columns where $L = 1$ correspond to pure spurious regression bias. We hold fixed the persistence parameter for the true expected return, ρ^*, while allowing ρ to vary depending on the measured instrument. When $L = 1$, we set $\rho = 15\%$. We consider two values for ρ^*, 15% or 95%.

Panels A and B of Table 9.4 show that when $L = 1$ (there is no data mining) and $\rho^* = 15\%$, there is no spurious regression problem, consistent with Table 9.2. The empirical critical values for the t-ratios and R-squared statistics are close to their theoretical values under normality. For larger values of L (there is data mining) and $\rho^* = 15\%$, the critical values are close to the values reported by Foster et al. (1997) for similar sample sizes.[10] There is little difference in the results for the various true R-squares. Thus, with little persistence in the true expected return there is no spurious regression problem, and no interaction with data mining.

Panels C and D of Table 9.4 tell a different story. When the underlying expected return is persistent ($\rho^* = 0.95$), there is a spurious regression bias. When $L = 1$, we have spurious regression only. The critical t-ratio in Panel C increases from 2.3 to 2.8 as the true R-squared goes from zero to 15%. The bias is less pronounced here than in Table 9.2, with $\rho = \rho^* = 0.95$, which illustrates that for a given value of ρ^*, spurious regression is worse for larger values of ρ.

Spurious regression bias interacts with data mining. Consider the extreme corners of Panel C. Whereas, with $L = 1$, the critical t-ratio increases from 2.3 to 2.8 as the true R-squared goes from zero to 15%, with $L = 250$, the critical t-ratio increases from 5.2 to 6.3 as the true R-squared is increased. Thus, data mining magnifies the effects of the spurious regression bias. When more instruments are examined, the more persistent ones are likely to be chosen, and the spurious regression problem is amplified. The slope coefficients are centered near zero, so the bias does not increase the average slopes of the selected regressors. Again, spurious regression works through the standard errors.

We can also say that spurious regression makes the data mining problem worse. For a given value

---

[10] Our sample sizes, T, are not the same as in Foster et al. (1997). When we run the experiments for their sample sizes, we closely approximate the critical values that they report.

Table 9.4 Monte Carlo simulation results of regressions with spurious regression and data mining, with independent regressors

Panel A: 66 observations; $\rho^* = 0.15$							
R^2/L	1	5	10	25	50	100	250
Means: δ							
0	−0.0004	0.0002	−0.0002	0.0004	−0.0001	0.0001	0.0005
0.001	−0.0114	0.0044	−0.0069	0.0208	−0.0078	0.0012	0.0162
0.005	−0.0050	0.0017	−0.0017	0.0113	−0.0014	−0.0031	0.0109
0.010	−0.0035	0.0008	−0.0014	0.0076	−0.0002	−0.0011	0.0098
0.050	−0.0014	0.0004	−0.0004	0.0018	−0.0023	−0.0013	0.0063
0.100	−0.0009	0.0006	−0.0004	0.0014	−0.0013	−0.0007	0.0044
0.150	−0.0007	0.0007	−0.0002	0.0009	−0.0010	−0.0010	0.0035
Critical t-statistics							
0	2.2971	3.2213	3.5704	4.1093	4.4377	4.8329	5.2846
0.001	2.2819	3.2105	3.5418	4.1116	4.4351	4.8238	5.2803
0.005	2.2996	3.2250	3.5466	4.1190	4.4604	4.7951	5.2894
0.010	2.2981	3.2109	3.5492	4.1198	4.4728	4.7899	5.2900
0.050	2.2950	3.2416	3.5096	4.0981	4.4036	4.8803	5.2527
0.100	2.3175	3.2105	3.5316	4.1076	4.4563	4.8772	5.2272
0.150	2.3040	3.2187	3.5496	4.0644	4.5090	4.8984	5.2948
Critical estimated R^2							
0	0.0594	0.0974	0.1153	0.1387	0.1548	0.1738	0.1944
0.001	0.0589	0.0969	0.1149	0.1386	0.1546	0.1739	0.1944
0.005	0.0591	0.0972	0.1151	0.1383	0.1545	0.1734	0.1948
0.010	0.0592	0.0967	0.1158	0.1386	0.1544	0.1733	0.1950
0.050	0.0596	0.0970	0.1163	0.1390	0.1557	0.1738	0.1955
0.100	0.0608	0.0969	0.1165	0.1392	0.1570	0.1738	0.1954
0.150	0.0612	0.0975	0.1165	0.1397	0.1577	0.1745	0.1967
Panel B: 824 observations; $\rho^* = 0.15$							
R^2/L	1	5	10	25	50	100	250
Means: δ							
0	0.0000	0.0000	0.0000	0.0000	−0.0001	−0.0002	0.0000
0.001	−0.0004	0.0032	−0.0017	0.0000	−0.0028	−0.0058	0.0015
0.005	−0.0002	0.0012	−0.0004	0.0000	−0.0020	−0.0031	0.0007
0.010	−0.0001	0.0009	−0.0004	−0.0003	−0.0015	−0.0020	0.0004
0.050	−0.0001	0.0005	0.0000	−0.0005	−0.0006	−0.0009	0.0004
0.100	0.0000	0.0005	−0.0001	−0.0003	−0.0001	−0.0002	0.0003
0.150	0.0000	0.0003	−0.0003	−0.0003	0.0001	−0.0002	0.0002
Critical t-statistics							
0	2.0283	2.5861	2.8525	3.1740	3.3503	3.5439	3.8045
0.001	2.0369	2.6000	2.8534	3.1785	3.3616	3.5443	3.7928

(continued)

Table 9.4 (continued)

Panel B: 824 observations; $\rho^* = 0.15$							
R^2/L	1	5	10	25	50	100	250
0.005	2.0334	2.6043	2.8565	3.1769	3.3625	3.5440	3.7906
0.010	2.0310	2.6152	2.8694	3.1782	3.3544	3.5477	3.7917
0.050	2.0272	2.6229	2.8627	3.1846	3.3450	3.5552	3.8039
0.100	2.0115	2.6304	2.8705	3.1807	3.3648	3.5673	3.8041
0.150	2.0044	2.6327	2.8618	3.1766	3.3691	3.5723	3.7965
Critical estimated R^2							
0	0.0047	0.0079	0.0096	0.0116	0.0130	0.0145	0.0166
0.001	0.0047	0.0079	0.0096	0.0116	0.0130	0.0145	0.0166
0.005	0.0047	0.0080	0.0096	0.0116	0.0129	0.0145	0.0166
0.010	0.0047	0.0080	0.0096	0.0115	0.0129	0.0145	0.0166
0.050	0.0047	0.0081	0.0096	0.0116	0.0130	0.0145	0.0167
0.100	0.0047	0.0081	0.0097	0.0117	0.0131	0.0146	0.0168
0.150	0.0047	0.0082	0.0096	0.0117	0.0130	0.0146	0.0168
Panel C: 66 observations; $\rho^* = 0.95$							
R^2/L	1	5	10	25	50	100	250
Means: δ							
0	−0.0005	0.0002	0.0006	−0.0001	−0.0006	−0.0003	0.0017
0.001	−0.0140	0.0069	0.0212	−0.0105	−0.0134	−0.0112	0.0557
0.005	−0.0060	0.0042	0.0082	−0.0068	−0.0024	−0.0033	0.0240
0.010	−0.0042	0.0031	0.0051	−0.0029	−0.0018	−0.0027	0.0145
0.050	−0.0016	0.0006	0.0035	−0.0023	−0.0016	−0.0019	0.0012
0.100	−0.0010	−0.0002	0.0021	−0.0013	−0.0017	−0.0005	0.0028
0.150	−0.0007	−0.0005	0.0015	−0.0008	−0.0011	−0.0001	0.0013
Critical t-statistics							
0	2.3446	3.3507	3.6827	4.1903	4.4660	4.9412	5.2493
0.001	2.3641	3.3547	3.6776	4.1756	4.5157	4.9201	5.2441
0.005	2.4030	3.3864	3.7013	4.1984	4.5625	4.9381	5.2760
0.010	2.3939	3.4197	3.7308	4.1952	4.6039	4.9718	5.3083
0.050	2.5486	3.5482	3.9676	4.4703	4.9512	5.2027	5.5539
0.100	2.6955	3.7336	4.1899	4.7485	5.2335	5.5027	5.9006
0.150	2.8484	3.9724	4.4329	4.9748	5.5547	5.8256	6.2563
Critical estimated R^2							
0	0.0579	0.0974	0.1140	0.1374	0.1515	0.1689	0.1885
0.001	0.0587	0.0981	0.1143	0.1376	0.1518	0.1692	0.1884
0.005	0.0596	0.0987	0.1153	0.1385	0.1530	0.1699	0.1895
0.010	0.0604	0.1002	0.1166	0.1402	0.1543	0.1711	0.1910
0.050	0.0691	0.1113	0.1307	0.1552	0.1711	0.1859	0.2057
0.100	0.0802	0.1265	0.1508	0.1774	0.1952	0.2099	0.2307
0.150	0.0911	0.1451	0.1728	0.2021	0.2209	0.2370	0.2587

Panel D: 824 observations; $\rho^* = 0.95$

R^2/L	1	5	10	25	50	100	250
Means: δ							
0	−0.0001	0.0000	0.0000	0.0000	0.0001	0.0002	0.0001
0.001	−0.0027	−0.0016	−0.0007	0.0005	0.0015	0.0072	0.0039
0.005	−0.0012	−0.0004	0.0003	0.0006	−0.0008	0.0029	0.0026
0.010	−0.0009	−0.0005	0.0000	0.0003	−0.0008	0.0013	0.0006
0.050	−0.0004	−0.0005	0.0001	−0.0002	0.0007	−0.0006	0.0001
0.100	−0.0003	−0.0002	−0.0001	−0.0003	0.0000	0.0001	−0.0004
0.150	−0.0003	0.0000	0.0000	−0.0002	0.0001	0.0002	−0.0002
Critical t-statistics							
0	1.9807	2.6807	2.8535	3.1579	3.3640	3.5673	3.8103
0.001	1.9989	2.6876	2.8758	3.1745	3.3702	3.5792	3.8252
0.005	2.0406	2.7588	2.9269	3.2218	3.4497	3.6493	3.9075
0.010	2.1108	2.8538	3.0150	3.3500	3.5548	3.7836	4.0351
0.050	2.4338	3.3118	3.6292	4.1202	4.3685	4.6795	4.9741
0.100	2.6274	3.6661	4.0003	4.5660	4.9129	5.2567	5.6937
0.150	2.7413	3.8720	4.2048	4.8481	5.2200	5.5846	6.0420
Critical estimated R^2							
0	0.0045	0.0080	0.0096	0.0113	0.0129	0.0145	0.0164
0.001	0.0046	0.0082	0.0097	0.0115	0.0130	0.0146	0.0167
0.005	0.0048	0.0086	0.0102	0.0121	0.0137	0.0153	0.0176
0.010	0.0050	0.0092	0.0108	0.0131	0.0146	0.0163	0.0187
0.050	0.0077	0.0145	0.0173	0.0216	0.0244	0.0273	0.0314
0.100	0.0113	0.0216	0.0264	0.0331	0.0374	0.0421	0.0482
0.150	0.0151	0.0293	0.0356	0.0446	0.0508	0.0568	0.0647

Panel E: Table 9.1 simulation

Obs	$\rho*$	Critical L (t-statistic)	Critical L (R^2)	$\rho*$	Critical L (t-statistic)	Critical L (R^2)
393	0.97	2	1	0.95	4	2
264	0.32	2	5	0.95	1	1
264	0.15	2	5	0.95	1	1
264	0.08	5	>500	0.95	1	10
420	0.97	1	1	0.95	1	1
720	0.97	1	1	0.95	1	1
732	0.92	1	1	0.95	1	1
611	0.95	1	1	0.95	1	1
611	0.97	1	1	0.95	1	1
66	0.66	2	2	0.95	1	2
184	0.79	2	7	0.95	1	3
824	0.97	1	1	0.95	1	2
552	0.98	1	1	0.95	1	1

of L, the critical t-ratios and R^2 values increase moving down the rows of Table 9.4. For example, with $L = 250$ and true $R^2 = 0$, we can account for pure data mining with a critical t-ratio of 5.2. However, when the true R-squared is 15%, the critical t-ratio rises to 6.3. The differences moving down the rows are even greater when $T = 824$, in Panel D. Thus, in the situations where the spurious regression bias is more severe, its impact on the data mining problem is also more severe.

Finally, Panel E of Table 9.4 revisits the studies from the literature in view of spurious regression and data mining. We report critical values for L, the number of instruments mined, sufficient to render the regression t-ratios and R-squares insignificant at the five percent level. We use two assumptions about persistence in the true expected returns: (i) ρ^* is set equal to the sample values from the studies, as in Table 9.1, or (ii) $\rho^* = 95\%$. With only one exception, the critical values of L are 10 or smaller. The exception is where the instrument is the lagged one-month excess return on a two-month Treasury bill, following Campbell (1987). This is an interesting example because the instrument is not very autocorrelated, at eight percent, and when we set $\rho^* = 0.08$ there is no spurious regression effect. The critical value of L exceeds 500. However, when we set $\rho^* = 95\%$ in this example, the critical value of L falls to 10, illustrating the strong interaction between the data mining and spurious regression effects.

9.8 Results for Conditional Asset Pricing Models

9.8.1 Cases with Small Amounts of Persistence

We first consider a special case of the model where we set $\rho^* = 0$ in the data-generating process for the market return and true beta, so that Z^* is white noise and $\sigma^2(\varepsilon^*) = \mathrm{Var}(Z^*)$. In this case, the predictable (but unobserved by the analyst) component of the stock market return and the betas follow white noise processes. We allow a range of values for the autocorrelation, ρ, of the measured instrument, Z, including values as large as 0.99. For a given value of ρ, we choose $\sigma^2(\varepsilon) = \mathrm{Var}(Z^*)(1 - \rho^2)$, so the measured instrument and the unobserved beta have the same variance. We find in this case that the critical values for all of the coefficients are well behaved. Thus, when the true expected returns and betas are not persistent, the use of even a highly persistent regressor does not create a spurious regression bias in the asset pricing regressions of Eq. (9.3).

It seems intuitive that there should be no spurious regression problem when there is no persistence in Z^*. Since the true coefficient on the measured instrument, Z, is zero, the error term in the regression is unaffected by the persistence in Z under the null hypothesis. When there is no spurious regression problem, there can be no interaction between spurious regression and data mining. Thus, standard corrections for data mining (e.g., White 2000) can be used without concern in these cases.

In our second experiment, the measured instrument and the true beta have the same degree of persistence, but their persistence is not extreme. We fix $\mathrm{Var}(Z) = \mathrm{Var}(Z^*)$ and choose, for a given value of $\rho^* = \rho$, $\sigma^2(\varepsilon) = \sigma^2(\varepsilon^*) = \mathrm{Var}(Z^*)(1 - \rho^2)$. For values of $\rho < 0.95$ and all values of the true predictive R-squared, R_p^2 the regressions seem generally well specified, even at sample sizes as small as $T = 66$. These findings are similar to the findings for the predictive regression (9.1). Thus, the asset pricing regressions (9.3) also appear to be well specified when the autocorrelation of the true predictor is below 0.95.

9.8.2 Cases with Persistence

Table 9.5 summarizes simulation results for a case that allows data mining and spurious regression. In this experiment, the true persistence parameter ρ^* is set equal to 0.95. The table summarizes the results for time-series samples of $T = 66$, $T = 350$, and $T = 960$. The number of variables over which the artificial agent searches in mining the data, ranges from one to 250. We focus on the two abnormal return coefficients, $\{\alpha_0, \alpha_1\}$ and on the time-varying beta coefficient, b_1.

Table 9.5 shows the results of 10,000 simulations of the estimates from the conditional asset pricing model, allowing for possible data mining of the lagged instruments. The regression model is:

$$r_{t+1} = \alpha_0 + \alpha_1 Z_t + b_0 r_{m,t+1} + b_1 r_{m,t+1} Z_t + u_{t+1}.$$

T is the sample size, L is the number of lagged instruments mined, R_p^2 is the true predictive R^2 in the artificial data-generating process.

Table 9.5 shows that the means of the coefficient α_0, the fixed part of the alpha, are close to zero, and they get closer to zero as the number of observations increases, as expected of a consistent estimator. The 5% critical t-ratios for α_0 are reasonably well specified at the larger sample sizes, although there is some bias at $T = 66$, where the critical values rise with the extent of data mining. Data mining has little effect on the intercepts at the larger sample sizes. Since the lagged instrument has a mean of zero, the intercept is the average conditional alpha. Thus, the issue of data mining for predictive variables appears to have no serious implications for measures of average abnormal performance in the conditional asset pricing regressions, provided $T > 66$. This justifies the use of such models for studying the cross section of average equity returns.

The coefficients α_1, which represent the time-varying part of the conditional alphas, present a different pattern. We would expect a data mining effect, given that the data are mined based on the coefficients on the lagged predictor in the simple predictive regression. The presence of the interaction term, however, would be expected to attenuate the bias in the standard errors, compared with the simple predictive regression. The table shows only a small effect of data mining on the α_1 coefficient, but a large effect on its t-ratio. The overall effect is the greatest at the smaller sample size ($T = 66$), where the critical t-ratios for the intermediate R_p^2 values (10% predictive R^2) vary from about 2.4 to 5.2 as the number of variables mined increases from one to 250. The bias diminishes with T, especially when the number of mined variables is small, and for $L = 1$ there is no substantial bias at $T = 360$ or $T = 960$ months.

The results on the α_1 coefficient are interesting in three respects. First, the critical t-ratios vary by only small amounts across the rows of the table. This indicates very little interaction between the spurious regression and data mining effects. Second, the table shows a smaller data mining effect than observed on the pure predictive regression. Thus, standard data mining corrections for predictive regressions will overcompensate in this setting. Third, the critical t-ratios for α_1 become smaller in Table 9.5 as the sample size is increased. This is just the opposite of what is found for the simple predictive regressions, where the inconsistency in the standard errors makes the critical t-ratios larger at larger sample sizes. Thus, the sampling distributions for time-varying alpha coefficients are not likely to be well approximated by simple corrections.[11]

Table 9.5 does not report the t-statistics for b_0, the constant part of the beta estimate. These are generally unbiased across all of the samples, except that the critical t-ratios are slightly inflated at the smaller sample size ($T = 66$) when data mining is not at issue ($L = 1$).

Finally, Table 9.5 shows results for the b_0 coefficients and their t-ratios, which capture the time-varying component of the conditional betas. Here, the average values and the critical t-ratios are barely affected by the number of variables mined. When $T = 66$ the critical t-ratios stay in a narrow range, from about 2.5 to 2.6, and they cluster closely around a value of 2.0 at the larger sample sizes. There are no discernible effects of data mining on the distribution of the

[11]We conducted some experiments in which we applied a simple local-to-unity correction to the t-ratios, dividing by the square root of the sample size. We found that this correction does not result in a t-ratio that is approximately invariant to the sample size.

Table 9.5 Simulating a conditional asset pricing model

	$T = 66$			$T = 350$			$T = 960$		
R_p^2	$L = 1$	$L = 25$	$L = 250$	$L = 1$	$L = 25$	$L = 250$	$L = 1$	$L = 25$	$L = 250$
Means: α_0									
0.001	0.000	0.001	0.000	0.000	0.000	0.000	0.000	0.000	0.000
0.005	0.000	0.000	0.000	0.000	0.000	0.000	0.000	0.000	0.000
0.01	−0.001	0.000	0.000	0.000	0.000	0.000	0.000	0.000	0.000
0.05	−0.001	−0.001	−0.001	0.000	−0.001	0.000	0.000	0.000	0.000
0.1	−0.002	−0.002	−0.002	0.000	−0.001	−0.001	0.000	0.000	0.000
0.15	−0.002	−0.002	−0.002	−0.001	−0.001	−0.001	0.000	0.000	0.000
Critical 5% t-statistics for α_0									
0.001	2.280	2.603	2.855	1.999	2.061	2.146	1.996	2.005	2.115
0.005	2.266	2.540	2.792	1.994	2.058	2.135	2.013	2.002	2.104
0.01	2.253	2.508	2.759	2.000	2.045	2.125	2.016	2.002	2.098
0.05	2.153	2.408	2.728	1.974	1.998	2.094	2.021	1.991	2.100
0.1	2.088	2.388	2.652	1.977	2.000	2.030	2.058	2.008	2.073
0.15	2.065	2.382	2.597	1.968	1.960	1.987	2.069	2.031	2.041
Means: α_1									
0.001	0.001	0.007	−0.003	−0.001	−0.002	0.003	−0.001	−0.002	−0.001
0.005	0.001	0.005	−0.001	−0.001	−0.001	0.002	−0.001	−0.002	−0.001
0.01	0.001	0.005	0.000	−0.001	−0.001	0.002	−0.001	−0.002	−0.001
0.05	0.001	0.005	−0.002	−0.001	−0.001	0.002	−0.001	−0.001	−0.002
0.1	0.001	0.005	0.000	−0.001	−0.001	0.002	−0.001	−0.001	0.000
0.15	0.001	0.004	0.000	−0.001	0.000	0.002	−0.001	−0.001	0.001
Critical 5% t-statistics for α_1									
0.001	2.392	3.992	5.305	2.023	3.240	3.891	1.910	3.097	3.748
0.005	2.390	3.961	5.252	2.024	3.206	3.874	1.905	3.092	3.719
0.01	2.387	3.912	5.198	2.025	3.198	3.872	1.902	3.091	3.712
0.05	2.412	3.924	5.237	2.039	3.172	3.855	1.902	3.062	3.706
0.1	2.426	3.912	5.163	2.036	3.159	3.837	1.912	3.040	3.690
0.15	2.423	3.913	5.086	2.024	3.155	3.785	1.909	3.024	3.629
Means: b_1									
0.001	−0.041	−0.017	0.026	−0.003	−0.002	0.011	0.010	0.008	−0.009
0.005	−0.038	−0.019	0.061	−0.003	0.001	0.002	0.009	0.007	−0.005
0.01	−0.038	−0.012	0.055	−0.003	0.000	0.004	0.008	0.004	−0.006
0.05	−0.039	−0.014	0.077	−0.003	0.006	−0.003	0.006	0.001	0.001
0.1	−0.040	−0.018	0.058	−0.003	0.010	0.003	0.005	0.000	0.004
0.15	−0.041	−0.015	0.062	−0.003	0.014	0.007	0.004	0.001	0.000
Critical 5% t-statistics for b_1									
0.001	2.576	2.534	2.634	2.122	2.098	2.175	1.996	2.013	2.075
0.005	2.574	2.579	2.611	2.116	2.138	2.210	2.022	2.036	2.075

(continued)

Table 9.5 (continued)

	T = 66			T = 350			T = 960		
R_p^2	L = 1	L = 25	L = 250	L = 1	L = 25	L = 250	L = 1	L = 25	L = 250
0.01	2.583	2.574	2.597	2.114	2.133	2.219	2.027	2.071	2.126
0.05	2.603	2.588	2.597	2.149	2.212	2.336	2.027	2.121	2.219
0.1	2.612	2.614	2.596	2.157	2.297	2.451	2.024	2.188	2.475
0.15	2.610	2.601	2.657	2.156	2.361	2.614	2.018	2.322	2.722

time-varying beta coefficients except when the R^2 values are very high. This is an important result in the context of the conditional asset pricing literature, which we characterize as having mined predictive variables based on the regression (9.1). Our results suggest that the empirical evidence in this literature for time-varying betas, based on the regression model (9.3), is relatively robust to the data mining.

9.8.3 Suppressing Time-Varying Alphas

Some studies in the conditional asset pricing literature use regression models with interaction terms, but without the time-varying alpha component (e.g., Cochrane 1996; Ferson and Schadt 1996; Ferson and Harvey 1999). Since the time-varying alpha component is the most troublesome term in the presence of spurious regression and data mining effects, it is interesting to ask if regressions that suppress this term may be better specified. Table 9.6 presents results for models in which the analyst runs regressions without the α_1 coefficient. The results suggest that the average alpha coefficient, α_0, and its t-statistic remain well specified regardless of data mining and potential spurious regression. Thus, once again we find little cause for concern about the inferences on average abnormal returns using the conditional asset pricing regressions, even though they use persistent, data mined lagged regressors.

The distribution of the average beta estimate, b_0, is not shown in Table 9.6. The results are similar to those obtained in a factor model regression where no lagged instrument is used. The coefficients and standard errors generally appear well specified. However, we find that the coefficient measuring the time-varying beta is somewhat more susceptible to bias than in the regression that includes α_1. The b_1 coefficient is biased, especially when T = 66, and its mean varies with the number of instruments mined. The critical t-ratios are inflated at the higher values of R_p^2 and when more instruments are mined.

Table 9.6 shows the results of 10,000 simulations of the estimates from the conditional asset pricing model with no time-varying alpha, allowing for the possibility of data mining for the lagged instruments. The regression model is:

$$r_{t+1} = \alpha_0 + b_0 r_{m,t+1} + b_1 r_{m,t+1} Z_t + u_{t+1}.$$

T is the sample size, L is the number of lagged instruments mined, R_p^2 is the true predictive R^2 in the artificial data-generating process.

These experiments suggest that including the time-varying alpha in the regression (9.3) helps "soak up" the bias so that it does not adversely affect the time-varying beta estimate. We conclude that if one is interested in obtaining good estimates of conditional betas, then in the presence of potential data mining and persistent lagged instruments, the time-varying alpha term should be included in the regression.

Table 9.6 Simulating a conditional asset pricing model with no time-varying alpha

	$T = 66$			$T = 350$			$T = 960$		
R_p^2	$L = 1$	$L = 25$	$L = 250$	$L = 1$	$L = 25$	$L = 250$	$L = 1$	$L = 25$	$L = 250$
Means: α_0									
0.001	0.000	0.000	−0.001	0.000	0.000	0.000	0.000	0.000	0.000
0.005	−0.001	−0.001	−0.001	0.000	0.000	0.000	0.000	0.000	0.000
0.01	−0.001	−0.001	−0.001	0.000	0.000	0.000	0.000	0.000	0.000
0.05	−0.002	−0.002	−0.002	0.000	0.000	−0.001	0.000	0.000	0.000
0.1	−0.002	−0.002	−0.003	0.000	−0.001	−0.001	0.000	0.000	0.000
0.15	−0.003	−0.003	−0.003	−0.001	−0.001	−0.001	0.000	0.000	−0.001
Critical 5% t-statistics for α_0									
0.001	2.165	2.146	2.123	1.951	1.988	1.947	1.990	1.965	1.934
0.005	2.138	2.114	2.089	1.949	1.981	1.929	1.983	1.961	1.922
0.01	2.132	2.102	2.054	1.944	1.966	1.918	1.992	1.956	1.931
0.05	2.042	2.020	1.986	1.907	1.921	1.917	1.983	1.970	1.921
0.1	1.984	1.956	1.929	1.896	1.910	1.885	1.988	1.962	1.904
0.15	1.949	1.922	1.893	1.900	1.884	1.829	1.993	1.965	1.869
Means: b_1									
0.001	0.007	−0.006	−0.043	−0.008	0.004	−0.001	−0.003	0.013	−0.012
0.005	0.008	−0.017	−0.060	−0.006	−0.003	0.000	−0.003	0.013	−0.008
0.01	0.008	−0.015	−0.064	−0.006	−0.003	−0.004	−0.003	0.010	−0.007
0.05	0.010	−0.031	−0.047	−0.003	−0.003	−0.003	−0.002	0.007	−0.001
0.1	0.010	−0.020	−0.035	−0.002	−0.005	−0.001	−0.002	0.000	0.001
0.15	0.010	−0.029	−0.042	0.000	−0.003	−0.003	−0.002	0.002	−0.002
Critical 5% t-statistics for b_1									
0.001	2.630	2.605	2.639	2.157	2.128	2.218	1.987	2.136	2.147
0.005	2.636	2.646	2.631	2.156	2.145	2.246	1.991	2.162	2.170
0.01	2.661	2.665	2.643	2.163	2.150	2.256	1.987	2.154	2.216
0.05	2.656	2.748	2.739	2.146	2.257	2.441	1.988	2.267	2.476
0.1	2.629	2.811	2.861	2.175	2.378	2.618	1.994	2.395	2.639
0.15	2.607	2.857	3.001	2.201	2.466	2.755	2.008	2.479	2.828

9.8.4 Suppressing Time-Varying Betas

There are examples in the literature where the regression is run with a linear term for a time-varying conditional alpha but no interaction term for a time-varying conditional beta (e.g., Jagannathan and Wang 1996). Table 9.7 considers this case.

The table shows the results of 10,000 simulations of the estimates from the conditional asset pricing model with no time-varying beta, allowing for the possibility of data mining for the lagged instruments. The regression model is:

Table 9.7 Simulating a conditional asset pricing model with no time-varying beta

	T = 66				T = 350			T = 960	
R_p^2	L = 1	L = 25	L = 250	L = 1	L = 25	L = 250	L = 1	L = 25	L = 250
Means: α_0									
0.001	0.000	0.000	0.000	0.000	0.000	0.000	0.000	0.000	0.000
0.005	0.000	0.000	0.000	0.000	0.000	0.000	0.000	0.000	0.000
0.01	−0.001	0.000	0.000	0.000	0.000	0.000	0.000	0.000	0.000
0.05	−0.001	−0.001	−0.001	0.000	0.000	0.000	0.000	0.000	0.000
0.1	−0.002	−0.001	−0.001	0.000	0.000	0.000	0.000	0.000	0.000
0.15	−0.002	−0.002	−0.002	−0.001	−0.001	0.000	0.000	0.000	0.000
Critical 5% t-statistics for α_0									
0.001	2.239	2.592	2.794	1.991	2.080	2.182	2.004	2.047	2.092
0.005	2.219	2.533	2.764	1.982	2.070	2.147	2.014	2.036	2.084
0.01	2.206	2.511	2.731	1.988	2.069	2.150	2.013	2.018	2.089
0.05	2.124	2.429	2.656	1.970	2.056	2.148	2.019	2.012	2.130
0.1	2.065	2.389	2.607	1.968	2.034	2.158	2.035	2.010	2.160
0.15	2.015	2.356	2.546	1.978	2.033	2.177	2.072	2.010	2.167
Means: α_1									
0.001	0.001	0.000	0.002	−0.001	0.001	0.000	0.000	−0.001	0.000
0.005	0.001	0.000	0.001	−0.001	0.001	0.001	0.000	−0.001	0.001
0.01	0.001	0.000	0.001	−0.001	0.001	0.001	0.000	−0.001	0.001
0.05	0.000	0.001	0.003	−0.001	0.001	0.001	0.000	−0.001	0.002
0.1	0.000	0.001	0.005	−0.001	0.000	0.000	0.000	0.000	0.001
0.15	0.000	0.000	0.003	−0.001	0.000	−0.001	0.000	−0.001	0.000
Critical 5% t-statistics for α_1									
0.001	2.307	4.015	5.298	2.068	3.172	3.912	2.020	3.073	3.794
0.005	2.311	4.005	5.169	2.066	3.149	3.890	2.017	3.061	3.761
0.01	2.312	3.998	5.142	2.061	3.148	3.891	2.017	3.058	3.754
0.05	2.322	3.968	5.040	2.054	3.156	3.885	2.003	3.029	3.739
0.1	2.323	3.908	5.041	2.056	3.138	3.849	2.012	3.001	3.713
0.15	2.323	3.901	5.075	2.056	3.124	3.783	2.005	3.003	3.659

$$r_{t+1} = \alpha_0 + \alpha_1 Z_t + b_0 r_{m,t+1} + u_{t+1}.$$

T is the sample size, L is the number of lagged instruments mined, R_p^2 is the true predictive R^2 in the artificial data-generating process.

First, the coefficient for the average beta in the regression with no b_1 term (not shown in the table) is reasonably well specified and largely unaffected by data mining on the lagged instrument. We find that the coefficients for alpha, α_0 and α_1, behave similarly to the corresponding coefficients in the full regression model (9.3). The estimates of the average alpha are reasonably well behaved and only mildly affected by the extent of data mining at smaller sample sizes. The bias in α_1 is severe. The bias leads the analyst to overstate the evidence for a time-varying alpha, and the bias is worse as the amount of data mining increases. Thus, the evidence in the literature for time-varying alphas, based on these asset pricing regressions, is likely to be overstated.

9.8.5 A Cross Section of Asset Returns

We extend the simulations to study a cross section of asset returns. We use five book-to-market (BM) quintile portfolios, equally weighted across the size dimension, as an illustration. The data are courtesy of Kenneth French. In these experiments, the cross section of assets features cross-sectional variation in the true conditional betas. Instead of $\ldots\beta_t = \beta_0 + \beta_1 Z_t^*$, the betas are $\beta_t = \beta_0 + \beta_1 Z_t^*$, where the coefficients β_0 and β_1 are the estimates obtained from regressions of each quintile portfolio's excess return on the market portfolio excess return and the product of the market portfolio with the lagged value of the dividend yield. The set of β_0's is $\{1.259, 1.180, 1.124, 1.118, 1.274\}$, the set of β_1's is $\{-1.715, 1.000, 3.766, 7.646, 8.970\}$.[12] The true predictive R-squared in the artificial data-generating process is set to 0.5%. This value matches the smallest R-squared from the regression of the market portfolio on the lagged dividend yield with a window of 60 months.

Table 9.8 shows simulation results for the conditional model with time-varying alphas and betas. The means of the b_0 and b_1 coefficients are shown in excess of their true values in the simulations. The critical t-statistics for both α_1 and b_1 are generally similar to the case where $R_p^2 =$ 0.5% in Table 9.5. As before, there is a large bias in the t-statistic for α_1 that increases with data mining but decreases somewhat with the sample size. The t-statistics for the time-varying betas are generally well specified.

[12]The β_1 coefficient for the BM2 portfolio is 1.0, replacing the estimated value of 0.047. When the β_1 coefficient is 0.047, the simulated return becomes nearly perfectly correlated with r_m and the simulation is uninformative. The dividend yield is demeaned and multiplied by 10. The dividend yield has the largest average sample correlation with the five BM portfolios among the standard instruments we examine.

The table shows the results of 10,000 simulations from a conditional asset pricing model, allowing for the possibility of data mining for the lagged instruments. The dependent variables are book-to-market quintile portfolios. The regression model is:

$$r_{t+1} = \alpha_0 + \alpha_1 Z_t + b_0 r_{m,t+1} + b_1 r_{m,t+1} Z_t + u_{t+1}.$$

T is the sample size, and L is the number of lagged instruments mined. The true predictive R^2 in the artificial data-generating process is 0.005.

We conduct additional experiments using the cross section of asset returns, where the conditional asset pricing regression suppresses either the time-varying alphas or the time-varying betas. The results are similar to those in Table 9.8. When the time-varying betas are suppressed, there is severe bias in α_1 that diminishes somewhat with the sample size. When time-varying alphas are suppressed there is a mild bias in b_1.

9.8.6 Revisiting Previous Evidence

In this section, we explore the impact of the joint effects of data mining and spurious regression bias on the asset pricing evidence based on regression (9.3). First, we revisit the studies listed in Table 9.2. Consider the models with both time-varying alphas and betas. If the data mining searches over 250 variables predicting the test asset return and $T = 350$, the 5% cutoff value to apply to the t-statistic for α_1 is larger than 3.8 in absolute value. For smaller sample sizes, the cutoff value is even higher. Note from Table 9.2 that the largest t-statistic for α_1 in Shanken (1990) with a sample size of 360 is −3.57 on the T-bill volatility, while the largest t-statistic for α_1 in Christopherson et al. (1998) with a sample size of 144 is 3.72 on the dividend yield. This means that the significance of the time-varying alphas in both of these studies is questionable. However, the largest t-statistic for b_1 in Shanken (1990) exceeds the empirical 5% cutoff, irrespective of spurious regression and data mining adjustments.

Table 9.8 Conditional asset pricing models with a cross section of returns

	$T = 66$			$T = 350$			$T = 960$		
BM quintile	$L = 1$	$L = 25$	$L = 250$	$L = 1$	$L = 25$	$L = 250$	$L = 1$	$L = 25$	$L = 250$
Means: α_0									
BM1 (low)	−0.002	−0.001	−0.001	−0.002	−0.002	−0.002	−0.002	−0.002	−0.002
BM2	0.000	0.000	0.000	0.000	0.000	0.000	0.000	0.000	0.000
BM3	0.000	0.002	0.001	0.002	0.002	0.002	0.002	0.002	0.003
BM4	0.001	0.003	−0.001	0.004	0.005	0.006	0.005	0.006	0.007
BM5 (high)	0.004	0.002	0.006	0.006	0.006	0.005	0.008	0.007	0.006
Critical 5% t-statistics for α_0									
BM1 (low)	2.157	2.593	2.705	1.691	1.847	1.914	1.526	1.618	1.651
BM2	2.297	2.523	2.742	1.916	2.067	2.156	1.960	2.034	2.056
BM3	2.296	2.681	2.916	2.093	2.218	2.380	2.143	2.206	2.301
BM4	2.343	2.686	3.060	2.059	2.253	2.399	2.132	2.233	2.368
BM5 (high)	2.369	2.659	3.099	2.135	2.233	2.326	2.269	2.284	2.359
Means: α_1									
BM1 (low)	0.002	0.000	0.001	−0.001	−0.004	0.002	0.000	0.000	0.000
BM2	0.000	0.000	0.002	0.001	0.000	0.001	0.000	0.000	0.001
BM3	0.009	0.006	−0.008	−0.001	−0.005	−0.008	0.001	0.007	0.001
BM4	0.019	−0.027	−0.004	0.007	−0.003	0.012	−0.003	−0.003	0.005
BM5 (high)	0.021	0.028	−0.068	0.005	0.028	0.027	0.002	−0.006	−0.009
Critical 5% t-statistics for α_1									
BM1 (low)	2.381	4.088	5.382	2.037	3.243	3.917	1.962	3.115	3.813
BM2	2.390	3.884	4.956	2.025	3.145	3.793	1.971	3.044	3.637
BM3	2.418	4.146	5.720	1.972	3.264	3.999	1.952	3.148	3.804
BM4	2.403	4.263	5.705	2.078	3.240	3.934	1.999	3.179	3.807
BM5 (high)	2.417	4.227	5.594	2.005	3.271	4.076	2.021	3.134	3.817
Means: b_1									
BM1 (low)	0.018	0.007	−0.054	−0.007	0.002	0.051	0.021	−0.001	0.004
BM2	−0.003	0.032	−0.015	0.000	−0.004	−0.009	−0.010	0.000	−0.002
BM3	0.108	0.050	0.128	0.066	−0.033	−0.042	−0.041	−0.004	0.034
BM4	−0.230	−0.050	−0.062	−0.016	−0.054	0.087	0.028	−0.068	0.032
BM5 (high)	−0.479	0.075	−0.389	0.041	0.113	−0.006	−0.136	0.032	0.058

(continued)

Table 9.8 (continued)

	$T = 66$			$T = 350$			$T = 960$		
BM quintile	$L = 1$	$L = 25$	$L = 250$	$L = 1$	$L = 25$	$L = 250$	$L = 1$	$L = 25$	$L = 250$
Critical 5% t-statistics for b_1									
BM1 (low)	2.612	2.522	2.548	2.065	2.168	2.181	2.045	2.115	2.061
BM2	2.521	2.591	2.559	2.146	2.126	2.149	2.056	2.056	2.067
BM3	2.573	2.501	2.585	2.083	2.159	2.089	2.103	2.023	2.061
BM4	2.624	2.556	2.552	2.048	2.114	2.091	2.130	2.004	2.035
BM5 (high)	2.628	2.640	2.536	2.177	2.149	2.068	2.084	2.050	1.982

This illustrates that the evidence for time-varying beta is robust to the joint effects of data mining and spurious regression bias, while the evidence for time-varying alphas is fragile.

Now consider the model with no time-varying alpha. If the data mining searches over 250 variables to predict the test asset return, the 5% cutoff value to apply to the t-statistic on b_1 is less than 3.5 in absolute value. Cochrane (1996) reports a t-statistic of −4.74 on the dividend yield in a time-varying beta, with a sample of $T = 186$. Thus, we find no evidence to doubt the inference that there is a time-varying beta. (However, the significance of the term premium in the time-varying beta, with a t-statistic of −1.76, is in doubt at the 10% level.)

Finally, consider the model with no time-varying beta. If the data mining searches over 25 variables to predict the test asset return, then the 5% cutoff value to apply to the t-statistic on α_1 is larger than 3.1 in absolute value. The largest t-statistic in Jagannathan and Wang (1996) with a sample size of 330 is 3.1. Therefore, their evidence for a time-varying alpha does not survive even with a modest amount of data mining.

We conclude that some aspects of the conditional asset pricing regression (9.3) are robust to data mining over persistent predictor variables, while others are not. The regression delivers reliable estimates of average abnormal returns and betas. However, the estimates of time-varying alphas may have vastly overstated statistical significance when standard tests are used.

9.9 Solutions to the Problems of Spurious Regression and Data Mining

9.9.1 Solutions in Predictive Regressions

The essential problem in dealing with the spurious regression bias is to get the right standard errors. We examine the Newey–West (1987) style standard errors that have been popular in recent studies. These involve a number of "lag" terms to capture persistence in the regression error. We use the automatic lag selection procedure described in footnote 1, and we compare it to a simple ordinary least squares (OLS) regression with no adjustment to the standard errors, and to a heteroscedasticity-only correction due to White (1980). Table 9.9 shows the critical t-ratios you would have to use in a 5%, two-tailed test, accounting for the possibility of spurious regression. Here we consider an extreme case with $\rho^* = 99\%$, because if we can find a solution that works in this case it should also work in most realistic cases. The critical t-ratios range from 2.24 to 6.12 in the first three columns. None of the approaches delivers the right critical value, which should be 1.96. The table shows that a larger sample size is no insurance against

Table 9.9 Possible solutions to the spurious regression problem: critical *t*-ratios

					Lagged return		Detrended (12)	
Observations	OLS	White	NW(auto)	NW(20)	OLS	NW(auto)	OLS	NW(auto)
60	2.24	2.36	2.71	3.81	2.19	2.67	2.06	2.46
350	4.04	4.10	3.87	3.77	3.74	3.73	2.28	2.21
2000	6.08	6.12	4.62	4.17	5.49	4.58	2.33	1.94

spurious regression. In fact, the problem is the worst at the largest sample size.

The Newey–West approach is consistent, which means that by letting the number of lags grow when you have longer samples, you should eventually get the right standard error and solve the spurious regression problem. So, the first potential solution we examine is simply to use more lags in the consistent standard errors. Unfortunately, it is hard to know how many lags to use. The reason is that in stock return regressions, the large unexpected part of stock returns is in the regression error, and this "noise" masks the persistence in the expected part of the return. If you use too few lags, the standard errors are biased and the spurious regression remains. The "White" example in column two is an illustration where the number of lags is zero. If you use too many lags, the standard errors will be inefficient and inaccurate, except in the largest sample sizes. We use simulations to evaluate the strategy of letting the number of lags grow large. We find that in realistic sample sizes, more lags do not help the spurious regression problem. The fourth column of Table 9.9 (denoted NW(20)) shows an example of this where 20 lags are used in monthly data. The critical *t*-ratios are still much larger than two. In the smaller sample size ($T = 60$), it is actually better to use the standard procedure, without any adjustments.

A second potential solution to the spurious regression problem is to include a lagged value of the dependent variable as an additional right-hand side variable in the regression. The logic of this approach is that the spurious regression problem is caused by autocorrelation in the regression residuals, which is inherited from the dependent variable. Therefore, logic suggests that putting a lagged dependent variable in the regression should "soak up" the autocorrelation, leaving a clean residual. The columns of Table 9.9 labeled "lagged return" evaluate this approach. It helps a little bit, compared with no adjustment, but the critical *t*-ratios are still much larger than two at the larger sample sizes. For a hypothetical monthly sample with 350 observations, a *t*-ratio of 3.7 is needed for significance. The reason that this approach does not work very well is the same reason that increasing the number of lags in the Newey–West method fails to work in finite samples. It is peculiar to stock return regressions, where the ex ante expected return may be persistent but the actual return includes a large amount of unpredictable noise. Spurious regression is driven in this case by persistence in the ex ante return, but the noise makes the lagged return a poor instrument for this persistence.[13]

Each cell contains the critical *t*-ratios at the 97.5% of 10,000 Monte Carlo simulations. OLS contains the critical *t*-ratios without any adjustment to the standard errors, in the White column the *t*-stats are formed using White's standard errors, the NW(auto) *t*-stats use Newey–West standard errors based on the automatic lag selection, the NW(20) *t*-stats use the Newey–West procedure with 20 lags. The regression model of stock returns in columns two-to-five has one independent variable, the lagged instrument; in columns six and seven—two independent variables, the lagged instrument and the lagged return; in the last two columns the only independent variable, the lagged instrument, is

[13]More formally, consider a case where the ex ante return is an AR(1) process, in Box–Jenkins notation. The realized return is distributed as an AR(1) plus noise, which is ARMA(1,1). Regressing the return on the lagged return, the residual may still be highly persistent.

stochastically detrended using a trailing 12-month moving average. The autocorrelation parameter of the ex ante expected return and the lagged predictor variable is set to 99%, and the ex ante return variance is 10% of the total return variance.

Of the various approaches we tried, the most practically useful insurance against spurious regression seems to be a form of "stochastic detrending" of the lagged variable, advocated by Campbell (1991). The approach is very simple. Just transform the lagged variable by subtracting off a trailing moving average of its own past values. Instead of regressing returns on Z_t, regress them on:

$$X_t = Z_t - \left(\frac{1}{\tau}\right) \sum_{j=1,\ldots,\tau} Z_{t-j}. \tag{9.7}$$

While different numbers of lags could be used in the detrending, Campbell uses 12 monthly lags, which seems natural for monthly data. We evaluate the usefulness of his suggestion in the last two columns of Table 9.9. With this approach, the critical t-ratios are less than 2.5 at all sample sizes, and much closer to 1.96 than any of the other examples. The simple detrending approach works pretty well. Detrending lowers the persistence of the transformed regressor, resulting in autocorrelations that are below the levels where spurious regression becomes a problem. Stochastic detrending can do this without destroying the information in the data about a persistent ex ante return, as would be likely to occur if the predictor variable is simply first differenced. Overall, we recommend stochastic detrending as a simple method for controlling the problem of spurious regression in stock returns.

9.9.2 Solutions in Conditional Asset Pricing Models

Since detrending works relatively well in simple predictive regressions, one would think of using it also in conditional asset pricing tests to correct the inflated t-statistics on the time-varying alpha coefficient. However, as we observed above, the bias in the t-statistic on a_1 is largely due to data mining rather than spurious regression. As a result, high t-statistics on a_1 for large number of data mining searches come not from the high autocorrelation of Z_t but rather from its high cross-correlation with the asset return. Therefore, simple detrending does not work in this case because chosen instruments may not necessarily have high persistency.

9.10 Robustness of the Asset Pricing Results

This section summarizes the results of a number of additional experiments. We extend the simulations of the asset pricing models to consider examples with more than a single lagged instrument. We consider asset pricing models with multiple factors, motivated by Merton's (1973) multiple-beta model. We also examine models where the data mining to select the lagged instruments focuses on predicting the market portfolio return instead of the test asset returns.

9.10.1 Multiple Instruments

The experiments summarized above focus on a single lagged instrument, while many studies in the literature use multiple instruments. We modify the simulations, assuming that the researcher mines two independent instruments with the largest absolute t-statistics and then uses both of them in the conditional asset pricing regression (9.3) with time-varying betas and alphas. (Thus, there are two a_1 coefficients and two b_1 coefficients.) These simulations reveal that the statistical behavior of both coefficients is similar to each other and similar to our results as reported in Table 9.5.

9.10.2 Multiple-Beta Models

We extend the simulations to study models with three state variables or factors. In building the three-factor model, we make the following assumptions. All three risk premiums are linear functions of one instrument, Z^*. The factors differ in their unconditional means and their disturbance terms, which are correlated with each other. The variance–covariance matrix of the disturbance terms matches that of the residuals from regressions of the three Fama and French (1993, 1996) factors on the lagged dividend yield. The true coefficients for the asset return on all three factors and their interaction terms with the correct lagged instrument, Z^*, are set to unity. Thus, the true conditional betas on each factor are equal to $1 + Z^*$. We find that the bias in the t-statistic for α_1 remains and is similar to the simulation in Table 9.5. There are no biases in the t-statistics associated with the b_1's for the larger sample sizes.

9.10.3 Predicting the Market Return

Much of the previous literature looked at more than one asset to select predictor variables. For the examples reported in the previous tables, the data mining is conducted by attempting to predict the excess returns of the tests assets. But a researcher might also choose instruments to predict the market portfolio return. We examine the sensitivity of the results to this change in the simulation design. The results for the conditional asset pricing model with both time-varying alphas and betas are re-examined. Recall that when the instrument is mined to predict the test asset return, there is an upward bias in the t-statistic for α_1. The bias increases with data mining and decreases somewhat with T. When the instruments are mined to predict the market, the bias in α_1 is small and is confined to the smaller sample size, $T = 66$. Mining to predict the market return has little impact on the sampling distribution of b_1.

9.10.4 Simulations Under the Alternative Hypothesis

Note that the return-generating process (9.6) does not include an intercept or alpha, consistent with asset pricing theory. Thus, the data are generated under the null hypothesis that an asset pricing model holds exactly. However, no asset pricing model is likely to hold exactly in reality. We therefore conduct experiments in which the data-generating process allows for a nonzero alpha. We modify Eq. (9.6) as follows:

$$\begin{aligned} r_{t+1} &= a_1 Z_t^* + \beta_t r_{m,t+1} + u_{t+1}, \\ \beta_t &= 1 + Z_t^* \\ r_{m,t+1} &= \mu + k Z_t^* + w_{t+1}. \end{aligned} \quad (9.8)$$

In the system (9.8), there is a time-varying alpha, proportional to Z_t^*. We set the coefficient $a_1 = k$ and estimate the model (9.3) again. With this modification, the bias in the time-varying alpha coefficient, α_1, is slightly worse at the larger R^2 values and larger values of L than it was before. The overall patterns, including the reduction in bias for larger values of T, are similar. We also run the model with no time-varying beta, and the results are similar to those reported above for that case.

9.11 Conclusion

Our results have distinct implications for tests of predictability and model selection. In tests of predictability, the researcher chooses a lagged variable and regresses future returns on the variable. The hypothesis is that the slope coefficient is zero. Spurious regression presents no problem from this perspective, because under the null hypothesis the expected return is not actually persistent. If this characterizes the academic studies of Table 9.1, the eight t-ratios larger than two suggest that ex ante stock returns are not constant over time.

The more practical problem is model selection. In model selection, the analyst chooses a lagged instrument to predict returns, for purposes such as implementing a tactical asset allocation strategy, active portfolio management, conditional performance evaluation or market timing. Here is where the spurious regression problem rears its ugly head. You are likely to find a variable that appears to work on the historical data, but will not work in the future. A simple form of stochastic detrending lowers the persistence of lagged predictor variables and can be used to reduce the risk of finding spurious predictive relations.

The pattern of evidence for the lagged variables in the academic literature is similar to what is expected under a spurious data mining process with an underlying persistent ex ante return. In this case, we would expect instruments to be discovered, then fail to work with fresh data. The dividend yield rose to prominence in the 1980s, but apparently fails to work for post-1990 data (Goyal and Welch 2003; Schwert 2003). The book-to-market ratio also seems to have weakened in recent data. When more data are available, new instruments appear to work (e.g., Lettau and Ludvigson 2001; Lee et al. 1999). Analysts should be wary that the new instruments, if they arise from the spurious mining process that we suggest, are likely to fail in future data, and thus fail to be practically useful.

We also study regression models for conditional asset pricing models in which lagged variables are used to model conditional betas and alphas. The conditional asset pricing literature has, for the most part used the same variables that were discovered based on simple predictive regressions, and our analysis characterizes the problem by assuming the data mining occurs in this way. Our results relate to several stylized facts that the literature on conditional asset pricing has produced.

Previous studies find evidence that the intercept, or average alpha, is smaller in a conditional model than in an unconditional model, suggesting, for example, that the conditional CAPM does a better job of explaining average abnormal returns. Our simulation evidence finds that the estimates of the average alphas in the conditional models are reasonably well specified in the presence of spurious regression and data mining, at least for samples larger than $T = 66$. Some caution should be applied in interpreting the common 60-month rolling regression estimator, but otherwise we take no issue with the stylized fact that conditional models deliver smaller average alphas.

Studies typically find evidence of time-varying betas based on significant interaction terms. Here again we find little cause for concern. The coefficient estimator for the interaction term is well specified in larger samples and largely unaffected by data mining in the presence of persistent lagged regressors. There is an exception when the model is estimated without a linear term in the lagged instrument. In this case, the coefficient measuring the time-varying beta is slightly biased. Thus, when the focus of the study is to estimate accurate conditional betas, we recommend that a linear term be included in the regression model.

Studies also find that even conditional models fail to explain completely the dynamic properties of stock returns. That is, the estimates indicate time-varying conditional alphas. We find that this result is the most problematic. The estimates of time variation in alpha inherit biases similar to, if somewhat smaller than, the biases in predictive regressions. We use our simulations to revisit the evidence of several prominent studies. Our analysis suggests that the evidence for time-varying alphas in the current literature should be viewed with some suspicion. Perhaps, the current generation of conditional asset pricing models do a better job of capturing the dynamic behavior of asset returns than existing studies suggest.

Finally, we think that our study, as summarized in this chapter, represents the beginning of what could be an important research direction at the nexus of econometrics and financial economics. The literature in this area has arrived at a good understanding of a number of econometric issues in asset pricing research; the two issues that we take on are only part of a much longer list that includes stochastic regressor bias, unit roots, cointegration, overlapping data, time aggregation, structural change, errors-in-variables, and many more. These issues have been discussed in Chaps. 7 and 8 of this book. But what is less understood is how these econometric issues interact with each other. We have seen that the interaction of data mining and spurious regression is likely to be a problem of practical importance. Many of other econometric issues also occur in combination in our empirical practice. We need to study these other interactions in future research.

Bibliography

Breen, W., Glosten, L. R., & Jagannathan, R. (1989). Economic significance of predictable variations in stock index returns. *Journal of Finance, 44,* 1177–1190.

Campbell, J. Y. (1987). Stock returns and the term structure. *Journal of Financial Economics, 18,* 373–400.

Campbell, J. Y. (1991). A variance decomposition for stock returns. *The Economic Journal, 101*(405), 157–179.

Chan, C. K., & Chen, N. F. (1988). An unconditional asset pricing test and the role of size as an instrumental variable for risk. *Journal of Finance, 43,* 309–325.

Christopherson, J. A., Ferson, W. E., & Glassman, D. (1998). Conditioning manager alpha on economic information: Another look at the persistence of performance. *Review of Financial Studies, 11,* 111–142.

Cochrane, J. H. (1996). A cross-sectional test of an investment-based asset pricing model. *Journal of Political Economy, 104,* 572–621.

Cochrane, J. H. (1999). *New facts in finance* (Working Paper). University of Chicago.

Conrad, J., & Kaul, G. (1988). Time-variation in expected returns. *Journal of Business, 61,* 409–425.

Cox, J. C., Ingersoll, J. E., Jr., & Ross, S. A. (1985). A theory of the term structure of interest rates. *Econometrica, 53,* 363–384.

Fama, E. F. (1970). Efficient capital markets: A review of theory and empirical work. *Journal of Finance, 25,* 383–417.

Fama, E. F. (1990). Stock returns, expected returns, and real activity. *The Journal of Finance, 45*(4), 1089–1108.

Fama, E. F., & French, K. R. (1988a). Dividend yields and expected stock returns. *Journal of Financial Economics, 22,* 3–25.

Fama, E. F., & French, K. R. (1988b). Permanent and temporary components of stock prices. *Journal of Political Economy, 96,* 246–273.

Fama, E. F., & French, K. R. (1989). Business conditions and expected returns on stocks and bonds. *Journal of Financial Economics, 25,* 23–49.

Fama, E. F., & French, K. R. (1993). Common risk factors in the returns on stocks and bonds. *Journal of Financial Economics, 33*(1), 3–56.

Fama, E. F., & French, K. R. (1996). Multifactor explanations of asset pricing anomalies. *The Journal of Finance, 51*(1), 55–84.

Fama, E. F., & French, K. R. (1997). Industry costs of equity. *Journal of Financial Economics, 43*(2), 153–193.

Fama, E. F., & Schwert, G. W. (1977). Asset returns and inflation. *Journal of Financial Economics, 5,* 115–146.

Fama, E. F., Fisher, L., Jensen, M. C., & Roll, R. (1969). The adjustment of stock prices to new information. *International Economic Review, 10,* 1–21.

Ferson, W. E., & Harvey, C. R. (1991). Sources of predictability in portfolio returns. *Financial Analysts Journal, 3,* 49–56.

Ferson, W. E., & Harvey, C. R. (1997). The fundamental determinants of international equity returns: A perspective on country risk and asset pricing. *Journal of Banking & Finance, 21,* 1625–1665.

Ferson, W. E., & Harvey, C. R. (1999). Conditioning variables and the cross-section of stock returns. *Journal of Finance, 54,* 1325–1360.

Ferson, W. E., & Korajczyk, R. (1995). Do arbitrage pricing models explain the predictability of stock returns? *Journal of Business, 68,* 309–349.

Ferson, W. E., & Schadt, R. (1996). Measuring fund strategy and performance in changing economic conditions. *Journal of Finance, 51,* 425–462.

Ferson, W., Sarkissian, S., & Simin, T. (2003). Spurious regression in financial economics? *Journal of Finance, 58,* 1393–1413.

Ferson, W., Sarkissian, S., & Simin, T. (2010). Spurious regression and data mining in conditional asset pricing models. In C. F. Lee, A. C. Lee, & J. Lee (Eds.), *Handbook of quantitative finance and risk management*. Singapore: Springer.

Fleming, J., Kirby, C., & Ostdiek, B. (2001). The economic value of volatility timing. *Journal of Finance, 61,* 329–352.

Foster, F. D., Smith, T., & Whaley, R. E. (1997). Assessing goodness-of-fit of asset pricing models: The distribution of the maximal R-squared. *Journal of Finance, 52,* 591–607.

Gibbons, M. R., & Ferson, W. E. (1985). Testing asset pricing models with changing expectations and an unobservable market portfolio. *Journal of Financial Economics, 14,* 216–236.

Goyal, A., & Welch, I. (2003). Predicting the equity premium with dividend ratios. *Management Science, 49,* 639–654.

Granger, C. W. J., & Newbold, P. (1974). Spurious regressions in econometrics. *Journal of Econometrics, 4,* 111–120.

Greene, W. H. (2017). *Econometric analysis* (8th ed.). New Jersey: Prentice Hall.

Harvey, C. R. (1989). Time-varying conditional covariances in tests of asset pricing models. *Journal of Financial Economics, 24,* 289–318.

Hastie, T., Tibshirani, R., & Friedman, J. (2001). *The elements of statistical learning*. Springer.

Huberman, G., & Kandel, S. (1990). Market efficiency and value line's record. *Journal of Business, 63,* 187–216.

Jagannathan, R., & Wang, Z. (1996). The conditional CAPM and the cross-section of expected returns. *Journal of Finance, 51,* 3–54.

Kandel, S., & Stambaugh, R. F. (1990). Expectations and volatility of consumption and asset returns. *Review of Financial Studies, 3,* 207–232.

Keim, D. B., & Stambaugh, R. F. (1986). Predicting returns in the bond and stock markets. *Journal of Financial Economics, 17,* 357–390.

Kendall, M. G. (1954). A note on the bias in the estimation of autocorrelation. *Biometrica, 41,* 403–404.

Kothari, S. P., & Shanken, J. (1997). Book-to-market time series analysis. *Journal of Financial Economics, 44,* 169–203.

Lee, C.-F., Lee, A. C., & Lee, J. (2010). *Handbook of quantitative finance and risk management*. New York: Springer.

Lee, C., Myers, J., & Swaminathan, B. (1999). What is the intrinsic value of the dow? *Journal of Finance, 54,* 1693–1742.

Lettau, M., & Ludvigson, S. (2001). Consumption, aggregate wealth and expected stock returns. *Journal of Finance, 56,* 815–849.

Lo, A. W., & MacKinlay, A. C. (1988). Stocks prices do not follow random walks. *Review of Financial Studies, 1,* 41–66.

Lo, A. W., & MacKinlay, A. C. (1990). Data snooping in tests of financial asset pricing models. *Review of Financial Studies, 3,* 431–467.

Lucas, R. E., Jr. (1978). Asset prices in an exchange economy. *Econometrica, 46,* 1429–1445.

Maddala, G. S. (1977). *Econometrics*. New York: McGraw-Hill.

Marmol, F. (1998). Spurious regression theory with non-stationary fractionally integrated processes. *Journal of Econometrics, 84,* 233–250.

Merton, R. C. (1973). An intertemporal capital asset pricing model. *Econometrica, 41,* 867–887.

Newey, W. K., & West, K. D. (1987). A simple, positive definite, heteroskedasticity and autocorrelation consistent covariance matrix. *Econometrica, 55,* 703–708.

Petkova, R., & Zhang, L. (2005). Is value riskier than growth? *Journal of Financial Economics, 78,* 187–202.

Phillips, P. C. B. (1986). Understanding spurious regressions in econometrics. *Journal of Econometrics, 33,* 311–340.

Phillips, P. C. B. (1998). New tools for understanding spurious regressions. *Econometrica, 66,* 1299–1326.

Pontiff, J., & Schall, L. (1998). Book-to-market as a predictor of market returns. *Journal of Financial Economics, 49,* 141–160.

Schwert, G. W. (2003). Anomalies and market efficiency. In G. M. Constantinides, M. Harris, & R. M. Stulz (Eds.), *Handbook of the economics of finance*. Amsterdam: North Holland.

Shanken, J. (1990). Intertemporal asset pricing: An empirical investigation. *Journal of Econometrics, 45,* 99–120.

Sharpe, W. F. (1964). Capital asset prices: A theory of market equilibrium under conditions of risk. *Journal of Finance, 19,* 425–442.

Stambaugh, R. S. (1999). Predictive regressions. *Journal of Financial Economics, 54,* 315–421.

White, H. (1980). A heteroskedasticity-consistent covariance matrix estimator and a direct test for heteroskedasticity. *Econometrica, 48*(4), 817–838.

White, H. (2000). A reality check for data snooping. *Econometrica, 68,* 1097–1126.

Yule, G. U. (1926). Why do we sometimes get nonsense correlations between time series? A study in sampling and the nature of time series. *Journal of the Royal Statistical Society, 89,* 1–64.

Part II
Time-Series Analysis and Its Applications

The purpose of Chap. 10 is to describe components of time-series analyses and to discuss alternative methods of economic and business forecasting in terms of time-series data. Specifically, we discuss a classical description of three time-series components, the moving average and seasonally adjusted time series, linear and log-linear time trend regressions, exponential smoothing and forecasting, autoregressive forecasting model, ARIMA model, and composite forecasting.

In Chap. 11, we attempt to achieve two goals. First, we present alternative theories for deriving optimal hedge ratios. We discuss various estimation methods and the relationship among lengths of hedging horizon, maturity of futures contract, data frequency, and hedging effectiveness. Moreover, we try to show how SAS program can be used to estimate hedge ratio in terms of ARCH method, GARCH method, EGARCH method, GJR-GARCH method, and TGARCH method.

10 Time Series: Analysis, Model, and Forecasting

Contents

This chapter is an update and expansion of Chap. 18 of the book entitled Statistics for Business and Financial Economics by Lee et al. (2013).

C.-F. Lee et al., *Financial Econometrics, Mathematics and Statistics*,
https://doi.org/10.1007/978-1-4939-9429-8_10

Abstract

In this chapter, we discuss traditional time-series analysis method, which includes moving-average method, autoregressive forecasting model, and ARIMA model. In addition, autoregressive conditional heteroscedasticity process is introduced. ARCH and GARCH models and models related to ARCH family are discussed in detail.

10.1 Introduction

Time-series analysis is one of most important tools in finance research. In this chapter, we will discuss traditional time-series analysis technique. In the next chapter, we will discuss the modern time-series analysis models which deal with heteroscedasticity and apply the estimation of the hedge ratio of futures contract.

We can incorporate both time-series and cross-sectional data and apply statistical analysis techniques in economic and business decision makings. Time-series data are any set of data from a quantifiable (or qualitative) event that are recorded *over time*. For example, we read newspapers everyday and can obtain the Dow Jones Industrial Average (DJIA) index over time. The series of DJIA index values, ordered through time, constitutes time-series data. Other types of time-series data are based on the rate of inflation, the consumer price index, the balance of trade, and the annual profit of a firm.

Cross-sectional data are observations made on individuals, groups of individuals, objects, or geographic areas *at a particular time*. For example, price per share for N firms in 1991 is a set of cross-sectional data. On the other hand, price per share for General Motors over time, P_t ($t = 1, 2, \ldots, T$), is a set of time-series data.

The purpose of this chapter is to describe components of time-series analyses and to discuss alternative methods of economic and business forecasting in terms of time-series data. Section 10.2 discusses a classical description of three time-series components. Section 10.3 discusses the moving average and seasonally adjusted time series. Section 10.4 will discuss linear and log linear time trend regressions. Section 10.5 will discuss exponential smoothing and forecasting. And autoregressive forecasting model will be discussed in Sect. 10.6. Section 10.7 will discuss ARIMA model. Section 10.8 will discuss ARCH and GARCH models. Section 10.9 will discuss composite forecasting. Finally, Sect. 10.10 concludes the chapter. Appendix 1 addresses the Holt–Winters forecasting model for seasonal series. Appendix 2 discusses composite forecasting.

10.2 The Classical Time-Series Component Model

Several factors result in the interdependence of time-series data over time; these factors are trend, seasonal, and business cycle factors. For example, the current earnings of a growing company tend to be greater than its earnings in the period just ended, and, of course, the expected earnings in the next period will be greater than the current earnings. Therefore, the correlation between any adjacent earnings is positive, and this is due to the trend factor. Seasonal factors also contribute to the interdependence of time-series data. Retail sales in the fourth quarter account for a major portion of total annual sales of department stores. This seasonal factor ensures that the sales volume

Table 10.1 Earnings per share of Johnson & Johnson

Year	EPS
2001	1.87
2002	2.2
2003	2.42
2004	2.87
2005	3.5
2006	3.76
2007	3.67
2008	4.62
2009	4.45
2010	4.85

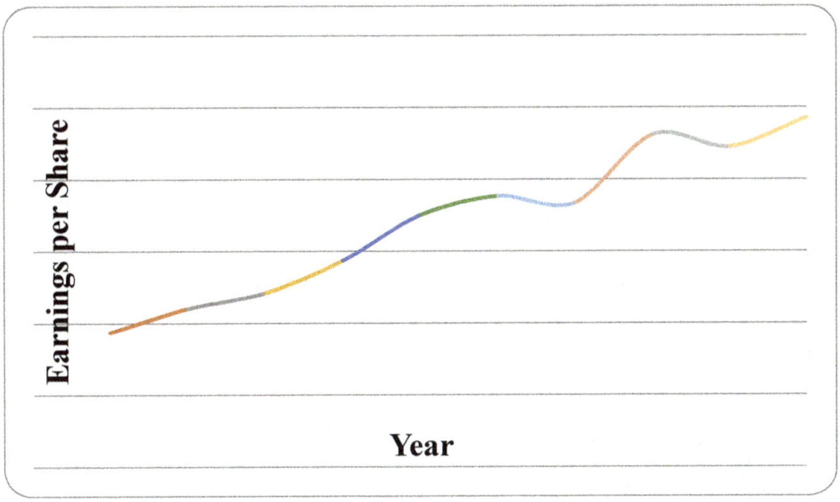

Fig. 10.1 Earnings per share of Johnson & Johnson

in the fourth quarter of each year is highly correlated with the fourth-quarter sales volume of any other year. The business cycle is another cause of interdependency in a time-series model. In short, it is traditionally assumed that the total variation in a time series is composed of four basic components: a **trend component**, a **seasonal component**, a **cyclical component**, and an **irregular component**. We will now discuss these four components in some detail.

The Trend Component

A trend is a pattern that exhibits a tendency either to grow or to decrease fairly steadily over time. For example, the earnings per share (EPS) of Johnson & Johnson exhibits two separate trends (or a quadratic trend) over time (see Table 10.1 and Fig. 10.1). One of the trends is from 2001 to 2007, the other from 2008 to 2010.

The Seasonal Component

The phenomenon of seasonality is common in the business world. Retailers can rely on greater sales volume in December than in any other month; stock returns are typically higher in January than in most other months—the "January effect."

Table 10.2 and Fig. 10.2 show earnings per share of IBM Corporation over a period of 44 quarters (first quarter 2000 to fourth quarter 2010). The table offers evidence of seasonal behavior for all quarters. The fourth-quarter figures tend to be relatively high, whereas those in the first quarter are relatively low. This seasonal behavior is quite clear in Fig. 10.2, where an obvious pattern almost repeats itself each year.

Table 10.2 Quarterly earnings per share of IBM Corporation

	Quarter			
Year	1	2	3	4
2000	0.85	1.95	3.06	4.58
2001	1.02	2.22	3.21	4.69
2002	0.75	1.01	2.01	3.13
2003	0.8	1.8	2.84	4.42
2004	0.81	1.84	2.77	4.48
2005	0.86	2.02	2.97	4.99
2006	1.09	2.4	3.87	6.15
2007	1.23	2.8	4.5	7.32
2008	1.67	3.67	5.79	9.07
2009	1.71	4.04	6.47	10.12
2010	2	4.64	7.49	11.69

The Cyclical Component and Business Cycles

Cyclical patterns are long-term oscillatory patterns that are unrelated to seasonal behavior. They are not necessarily regular but instead follow rather smooth patterns of upswings and

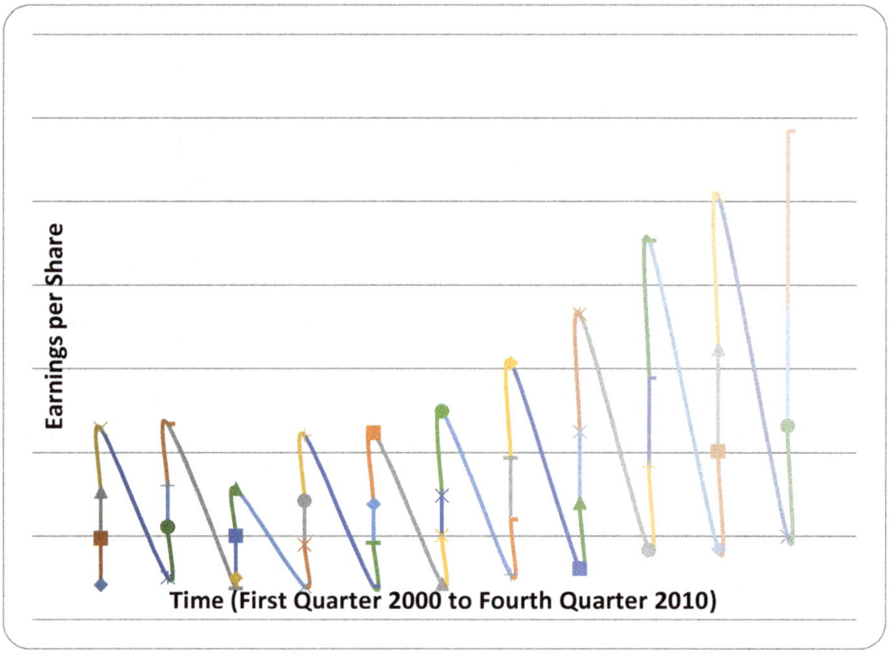

Fig. 10.2 Quarterly earnings per share of IBM

downswings, each swing lasting more than 2 or 3 years. Figure 10.3 demonstrates the cyclical pattern of the S&P 500 composite index during the period of 2000–2010, which will be discussed in detail in the next chapter. Figure 10.4 shows the cyclical patterns of monthly data of 3-month interest rates of return on Eurodollar deposits, US certificates of deposit (CDs), and Treasury bills during the period of 2000–2010.[1]

The National Bureau of Economic Research (NBER) and the US Department of Commerce have specified a number of time series as statistical business indicators of cyclical revivals and recessions. These time series have been classified into three groups.[2] The first group is the so-called **leading indicators**, such as the S&P index of the prices of 500 common stocks. These series have usually reached their cyclical turning points prior to the analogous turns in economic activity. The second group is the **coincident indicators**, such as unemployment rate, the index of industrial production, and GNP in current dollars. The third group is the **lagging indicators**, such as index of labor cost per unit of output in manufacturing, business expenditures, and new plant and equipment. A particular indicator series is considered a leading, a coinciding, or a lagging indicator of overall economic activity, depending on whether the cyclical component of the series exhibits a tendency to precede, match, or follow the cyclical behavior of the economy at large.

The Irregular Component

The last component of the variation in a time series is the irregular element introduced by the unexpected event. For example, the announcement of a takeover bid may cause the price of the target company's stock to jump up 20% or more in a single day. Fears of an outbreak of war in the Middle East and concerns about trade deficits and antitakeover legislation contributed to a spectacular decline in the stock market on

[1]A Eurodollar is any dollar on deposit outside the USA. In the bottom portion of Fig. 10.4, "spreads" are the differences between two different kinds of interest rates. For example, the Eurodollar rate is 0.40% higher than the US CD rate (0.27%) and T-Bill rate (0.14%) in November 2010.

[2]Index numbers are essential elements for these business indicators.

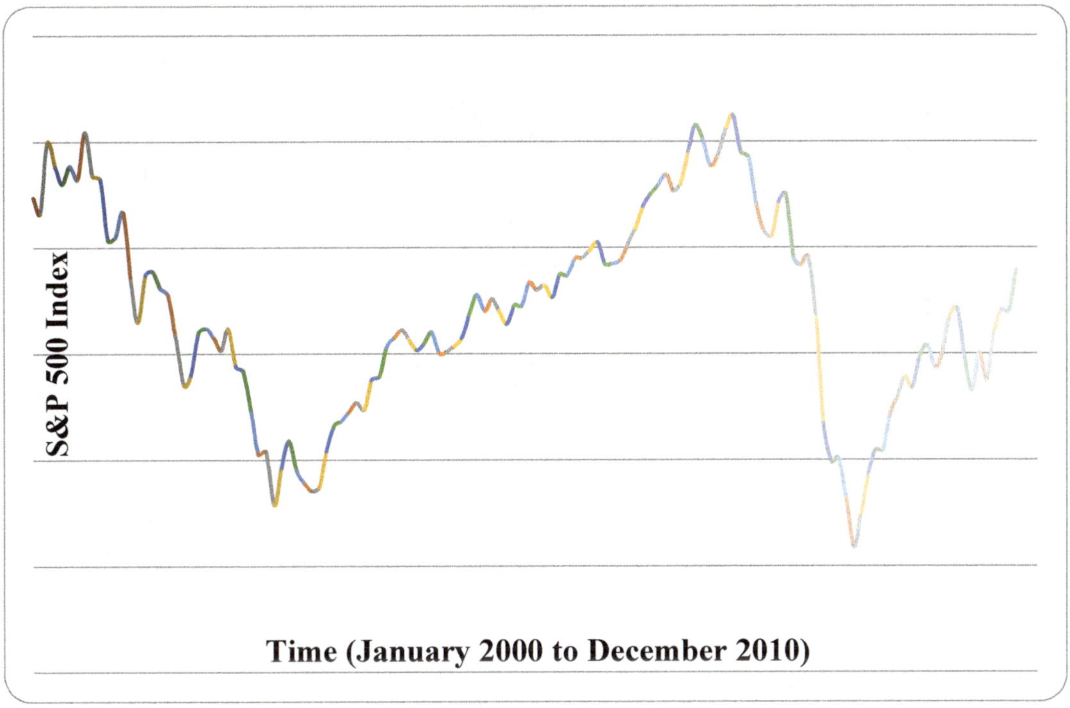

Fig. 10.3 S&P 500 composite index, January 2000 to December 2010

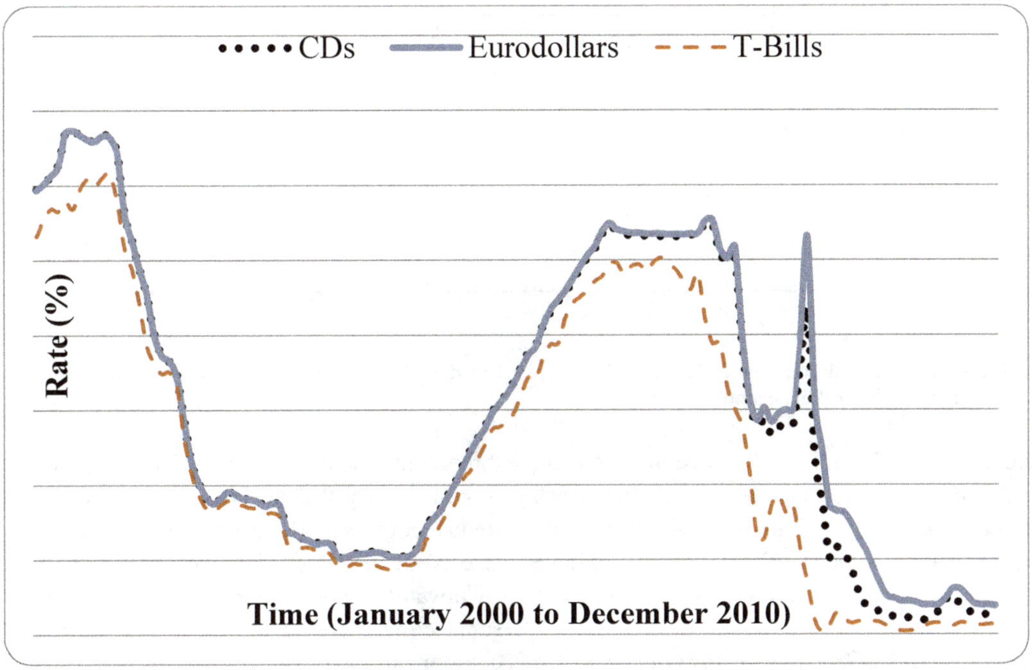

Fig. 10.4 Three-month rates on Eurodollar deposits, US CDs, and US T-Bills, 2000–2010 (monthly data)

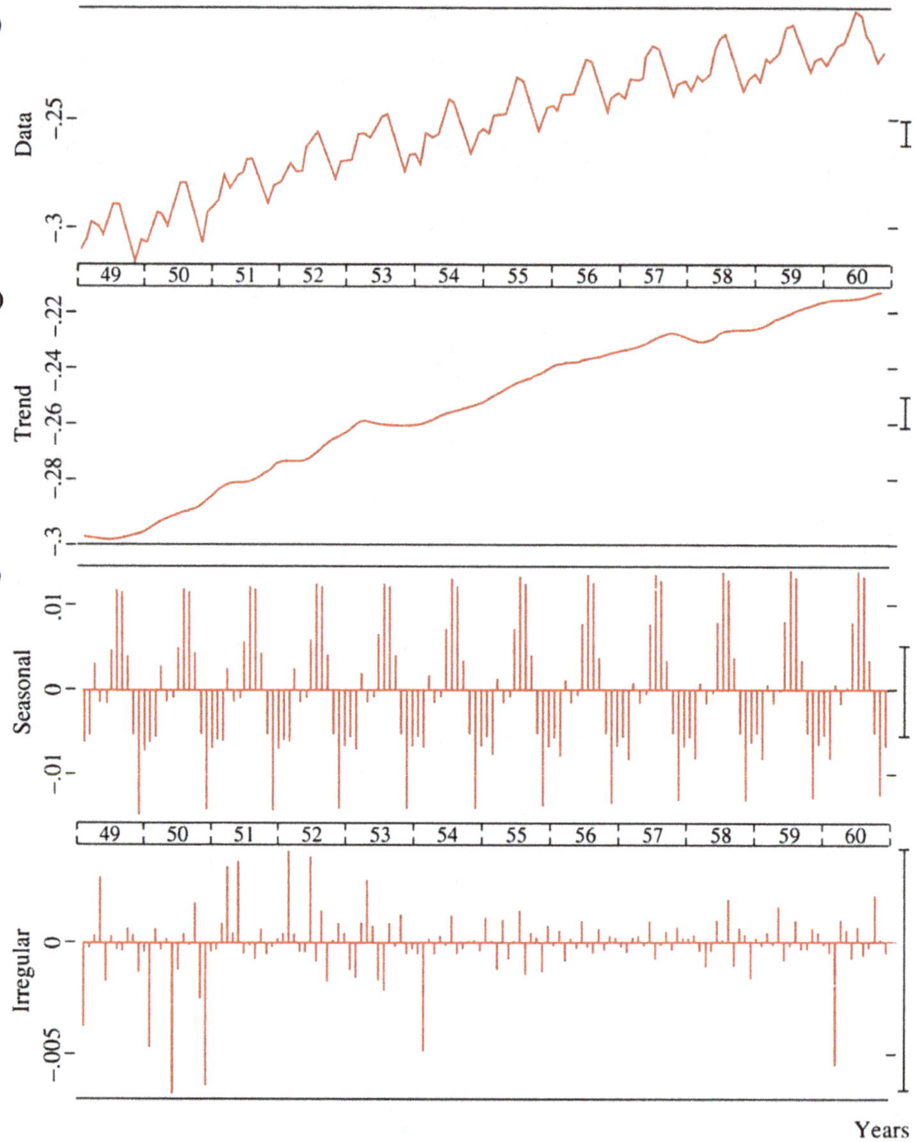

Fig. 10.5 Time-series decomposition. *Source* H. Levenback and J. P. Cleary (1984) *The Modem Forecaster* (New York: Lifetime Learning Publications), p. 50

October 19, 1987. And Iraq's invasion of Kuwait on August 3, 1990, caused worldwide stock markets to drop more than 10% within a week. These irregular elements arise suddenly and have a temporary impact on time-series behavior.

Example 10.1 (*Graphical Presentation of Time-Series Components*)
In Fig. 10.5, Levenbach and Cleary show how a set of time-series data can be broken down into three components. Figure 10.5a is a plot of the original series of data. Figure 10.5b presents the trend component (long-term trend plus cyclical effects) of the series. The data obviously exhibit an upward trend. Figure 10.5c presents the seasonal component of the data, and the irregular component appears in Fig. 10.5d.

Overall, a set of time-series data, x_t, can be described by using the additive model of Eq. (10.1) or the multiplicative model of Eq. (10.2).

$$x_t = T_t + C_t + S_t + I_t \quad (10.1)$$

$$x_t = T_t S_t C_t I_t \quad (10.2)$$

where

T_t = trend component
C_t = cyclical component
S_t = seasonal component
I_t = irregular component.

For long-term planning and decision making in terms of time-series components, business executives concentrate primarily on forecasting the trend movement. For intermediate-term planning—say, from about 2 to about 5 years—fluctuations in the business cycle are of critical importance too. For short-term planning, and for purposes of operational decisions and control, seasonal variations must also be taken into account. In the next four sections of this chapter, we will analyze these time-series factors (components) and see how they can be forecasted.

10.3 Moving Average and Seasonally Adjusted Time Series

In this section, we explain how the moving-average method is used to smooth time-series data. We also discuss how moving average and related techniques can be used to obtain seasonally adjusted time-series data.

Moving Averages

Moving averages are usually associated with data smoothing. Smoothing a time series reduces the effects of seasonality and irregularity. As a result, the smoothed data reveal more information about seasonal trends and business cycles. The most common moving-average method is the unweighted moving average, in which each value of the data carries the same weight in the smoothing process. For a time series $x_1, \ldots, x_n$ the formula for doing a three-term unweighted moving average is

$$z_t = \left(\frac{1}{3}\right)\sum_{i=0}^{2} x_{t-i} \quad (t = 3, \ldots, n) \quad (10.3)$$

Similarly, the k-term unweighted moving average is written

$$z_t = (1/k)\sum_{t=0}^{k-t} x_{t-i} \quad (t = k, \ldots, n) \quad (10.4)$$

Alternatively, the weighted moving average can be used to replace the unweighted moving average. A k-term weighted moving average can be defined as

$$z_t = \sum_{i=0}^{k-1} w_{t-i}x_{t-i} \quad (t = k, \ldots, n) \quad (10.5)$$

where $\sum_{i=0}^{k-1} w_{t-i} = 1$.

The w_{t-i}'s are known as weights and they sum to unity. If the w_{t-i}'s do not sum to unity, they can be normalized with a new set of weights (w^*_{t-i}) that sum to unity. The unweighted moving average is a special case of the weighted moving average with $w_i = 1/k$ for all i. An example of a weighted average calculation appears in Table 10.3. Here columns (1) and (2) represent observation value (x_{t-i}) and weight (w_i), respectively. Column (3) represents $x_{t-i}w_{t-i}$. From Table 10.3, we obtain

$$z_t = \sum_{i=0}^{3} x_{t-i}w_{t-i} = 0.0501$$

Table 10.3 Weighted average

(1) Observation value, x_{t-i}	(2) Weight, w_{t-i}	(3) $x_{t-i}w_{t-i}$
0.035	0.10	0.0035
0.002	0.30	0.0006
0.100	0.25	0.0250
0.060	0.35	0.0210
	1.00	0.0501 weighted average

One of the important applications of moving averages is to deseasonalize seasonal time-series data which will be discussed in the next section.

Seasonal Index and Seasonally Adjusted Time Series

In Sect. 10.2, we noted that many business and economic time series contain a strong seasonal component. This component generally needs to be removed for either monthly or quarterly data. This section demonstrates how the moving-average procedure is used to remove the seasonal component and to do related analysis.

Suppose we have a quarterly time series, x_t, with a seasonal component. Then we can apply Eq. (10.6), which is obtained by letting $k = 4$ in Eq. (10.4), to remove the seasonal component.

$$z_i = \left(\frac{1}{4}\right)\sum_{i=0}^{3} x_{t-i} \quad (t = 4, \ldots, n) \qquad (10.6)$$

Example 10.2 (Seasonally Adjusted Quarterly Earnings per Share of Johnson & Johnson)

For the data on quarterly earnings per share of J&J Corporation during the period of first quarter 2000 to fourth quarter 2010 given in Table 10.4, the first number in the series of the fourth-quarter moving average is

Table 10.4 Actual (x_t) and centered four-point moving-average (z_t^*) earnings per share of Johnson & Johnson from first quarter 2000 to fourth quarter 2010

(1)t	(2) Earnings per share, x_t	(3) Four-point moving average, z_t	(4) Centered four-point, moving average, z_t^*
1	0.86		
2	1.8	2.16	
3	2.68	2.195	2.1775
4	3.3	1.995	2.095
5	1.0	1.7025	1.84875
6	1.0	1.345	1.52375
7	1.51	1.245	1.295
8	1.87	1.2825	1.26375
9	0.6	1.3375	1.31
10	1.15	1.42	1.37875
11	1.73	1.445	1.4325
12	2.2	1.43	1.4375
13	0.7	1.4475	1.43875
14	1.09	1.5025	1.475
15	1.8	1.5375	1.52
16	2.42	1.6825	1.61
17	0.84	1.8475	1.765
18	1.67	1.96	1.90375
19	2.46	1.99	1.975
20	2.87	2.03	2.01
21	0.96	2.085	2.0575
22	1.83	2.2125	2.14875
23	2.68	2.25	2.23125
24	3.38	2.31	2.28
25	1.11	2.3925	2.35125
26	2.07	2.4875	2.44
27	3.01	2.4325	2.46
28	3.76	2.4025	2.4175
29	0.89	2.36	2.38125
30	1.95	2.3375	2.34875
31	2.84	2.4325	2.385
32	3.67	2.5575	2.495
33	1.27	2.7575	2.6575
34	2.45	2.995	2.87625
35	3.64	2.995	2.995
36	4.62	2.99	2.9925
37	1.27	2.99	2.99
38	2.43	2.9475	2.96875
39	3.64	3.04	2.99375
40	4.45	3.155	3.0975
41	1.64	3.28	3.2175
42	2.89	3.38	3.33
43	4.14		
44	4.85		

$$\frac{0.86 + 1.8 + 2.68 + 3.3}{4} = 2.16$$

and the second number is

$$\frac{1.8 + 2.68 + 3.3 + 1.0}{4} = 2.195$$

The Complete series appears in column (3) of Table 10.4.

This fourth-quarter moving-average time series is free from seasonally because it is always based on values such that each "season" is represented in each single observation of the new series (see Fig. 10.6). However, the location in time of the members of the series of moving averages does not correspond precisely with that of the members of the original series. Actually, the first fourth-quarter moving average would be centered midway between the second-quarter and third-quarter dates. Hence, the fourth-quarter moving-average series indicated in Eq. (10.6) should be rewritten either as

$$z_{t-0.5} = \left(\frac{1}{4}\right)\sum_{i=2}^{-t} x_{t-i} \quad (t = 3, 4, \ldots, n-2) \tag{10.7}$$

or

$$z_{t+0.5} = \left(\frac{1}{4}\right)\sum_{i=-1}^{2} x_{t+i} \quad (t = 2, 3, \ldots, n-2) \tag{10.7a}$$

Then the location-adjusted (centered) moving-average series can be written as

$$z_i^* = \frac{z_{t-0.5} + z_{t+0.5}}{2} \quad (t = 3, 4, \ldots, n-2) \tag{10.8}$$

when

$$t = 3, \; z_3^* = \frac{z_{2.5} + z_{3.5}}{2} = \frac{x_1 + 2x_2 + 2x_3 + 2x_4 + x_5}{8}$$

The location-adjusted moving averages, z_t^*, are given in column (4) of Table 10.4. Both x_t and z_t^* are presented in Fig. 10.6.

We can use the location-adjusted moving-average data obtained from Eq. (10.8) to calculate seasonally adjusted series if we assume that the seasonal pattern through time is very stable.

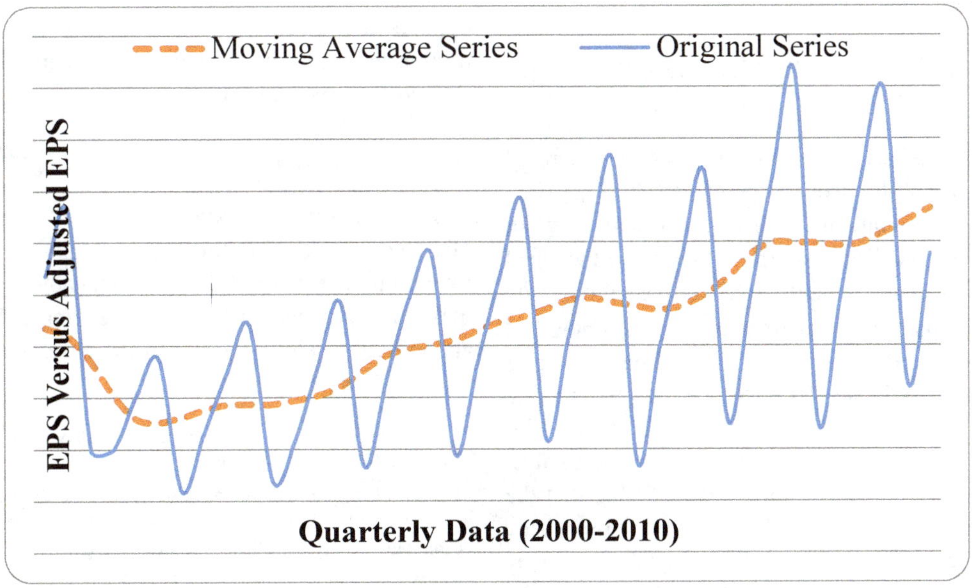

Fig. 10.6 Earnings per share versus moving-average EPS for Johnson & Johnson

To do this, we need first to divide original data (x_t) by the location-adjusted moving averages $\left(z_t^*\right)$ to obtain the **percentage of moving average**. That is,

$$\text{Percentage of moving average (PMA)} = 100\left(\frac{x_t}{z_t^*}\right) \quad (10.9)$$

The PMA of earnings per share for Johnson & Johnson is presented in column (4) of Table 10.5.

In our case, the first observation of PMA is

$$100\left(\frac{x_3}{z_3^*}\right) = 100\left(\frac{2.68}{2.1775}\right) = 123.0769$$

We assume that for any given quarter, in each year, the effect of seasonality is to raise or lower the observation by a constant proportionate amount (**seasonal index**) compared with what it would have been in the absence of seasonal influences. Then we use the so-called **seasonal index method** to remove the seasonal component.

Let us explore the logic of and procedure for calculating the seasonal index listed in column (5) of Table 10.5. By dividing z_t^* into x_t, we can explicitly write the percentage of moving average as

$$100\left(\frac{x_i}{z_t^*}\right) = \frac{100T_tC_tS_tI_t}{T_tC_t} = 100S_tI_t \quad (10.10)$$

The $100S_tI_t$ series for earnings per share of Johnson & Johnson is presented in Fig. 10.7. This series contains both seasonal and irregular components. The next step is to remove the effect of irregular movements from $100\left(x_t/z_t^*\right)$. We do this by taking the median of the percentage of moving-average figures for the same quarter as indicated in Table 10.6. The medians for the first through the fourth quarters are 47.400, 84.122, 121.152, and 149.278, respectively. The total of these medians is 401.953. It is desirable that the total of the four indexes be 400, in order that they average 100%, so we multiply each of them by an adjustment factor (400/401.953) to make the sum of the fourth-quarter seasonal indexes equal 400. The seasonal index is presented in column (5) of Table 10.5.[3] Dividing the seasonal index into the original quarterly data and multiplying the result by 100, we obtain the adjusted series presented in column (6) of Table 10.5 and in Fig. 10.8.

This seasonal index method of seasonal adjustment shows us one possible and simple way to solve the problem of eliminating the seasonal component. In practice, however, it generally can be solved by computer. Important government monthly and quarterly economic data such as consumer price indexes and employment and unemployment rates have strong seasonal components, and government agencies generally publish these data in both unadjusted and adjusted forms. The seasonal adjustment procedure used in official US government publications is the Census X-11 method which is based upon the moving-average method.[4] In the next section, we will look at time trend regression.

10.4 Linear and Log Linear Time Trend Regressions

If a time series is expected to change linearly overtime, the simple linear regression model defined in Eq. (10.11) can be used to relate the time series, x_t, to time t, and the least squares line is used to forecast future values of x_t.

$$x_t = \alpha + \beta t + \varepsilon_t \quad (10.11)$$

If the relationship between x_t and t is multiplicative instead of additive, then transforming x_t by taking the natural logarithm enables us to make the relationship linear. For example, let x_0 and x_t be the sales of a firm in the base year and

[3]The mean instead of the median can also be used to calculate the seasonal index.

[4]The X-11 model for decomposing time-series components can be found in Appendix 24.A of the book entitled "Financial Analysis, Planning and Forecasting: Theory and Application" by Lee, A. C, J. C. Lee, and C. F. Lee., 2nd ed. Singapore: World Scientific Publishing Company, 2009.

Table 10.5 Seasonal adjustment of earnings per share of Johnson & Johnson by the seasonal index method from first quarter 2000 to fourth quarter 2010

(1) Date	(2) EPS, x_t	(3) z_t^*	(4) 100 $\left(x_t/z_t^*\right)$	(5) Seasonal index	(6) Adjusted EPS [Col. (2) ÷ Col. (5)] × 100
2000					
1	0.86			47.1702	1.823185
2	1.8			83.71375	2.150184
3	2.68	2.1775	123.0769	120.5633	2.2229
4	3.3	2.095	157.5179	148.5528	2.221432
2001					
1	1	1.84875	54.0906	47.1702	2.119983
2	1	1.52375	65.62756	83.71375	1.194547
3	1.51	1.295	116.6023	120.5633	1.252455
4	1.87	1.26375	147.9723	148.5528	1.258812
2002					
1	0.6	1.31	45.80153	47.1702	1.27199
2	1.15	1.37875	83.40888	83.71375	1.373729
3	1.73	1.4325	120.7679	120.5633	1.434931
4	2.2	1.4375	153.0435	148.5528	1.480955
2003					
1	0.7	1.43875	48.65334	47.1702	1.483988
2	1.09	1.475	73.89831	83.71375	1.302056
3	1.8	1.52	118.4211	120.5633	1.492992
4	2.42	1.61	150.3106	148.5528	1.62905
2004					
1	0.84	1.765	47.59207	47.1702	1.780785
2	1.67	1.90375	87.7216	83.71375	1.994893
3	2.46	1.975	124.557	120.5633	2.040423
4	2.87	2.01	142.7861	148.5528	1.931973
2005					
1	0.96	2.0575	46.65857	47.1702	2.035183
2	1.83	2.14875	85.16579	83.71375	2.186021
3	2.68	2.23125	120.112	120.5633	2.2229
4	3.38	2.28	148.2456	148.5528	2.275285
2006					
1	1.11	2.35125	47.20893	47.1702	2.353181
2	2.07	2.44	84.83607	83.71375	2.472712
3	3.01	2.46	122.3577	120.5633	2.496615
4	3.76	2.4175	155.5326	148.5528	2.531087
2007					
1	0.89	2.38125	37.37533	47.1702	1.886785
2	1.95	2.34875	83.02288	83.71375	2.329366

(continued)

Table 10.5 (continued)

(1) Date	(2) EPS, x_t	(3) z_t^*	(4) $100\left(x_t/z_t^*\right)$	(5) Seasonal index	(6) Adjusted EPS [Col. (2) ÷ Col. (5)] × 100
3	2.84	2.385	119.0776	120.5633	2.35561
4	3.67	2.495	147.0942	148.5528	2.470502
2008					
1	1.27	2.6575	47.78928	47.1702	2.692378
2	2.45	2.87625	85.18036	83.71375	2.92664
3	3.64	2.995	121.5359	120.5633	3.019162
4	4.62	2.9925	154.386	148.5528	3.110005
2009					
1	1.27	2.99	42.47492	47.1702	2.692378
2	2.43	2.96875	81.85263	83.71375	2.902749
3	3.64	2.99375	121.5866	120.5633	3.019162
4	4.45	3.0975	143.6642	148.5528	2.995568
2010					
1	1.64	3.2175	50.97125	47.1702	3.476772
2	2.89	3.33	86.78679	83.71375	3.45224
3	4.14			120.5633	3.433882
4	4.85			148.5528	3.264833

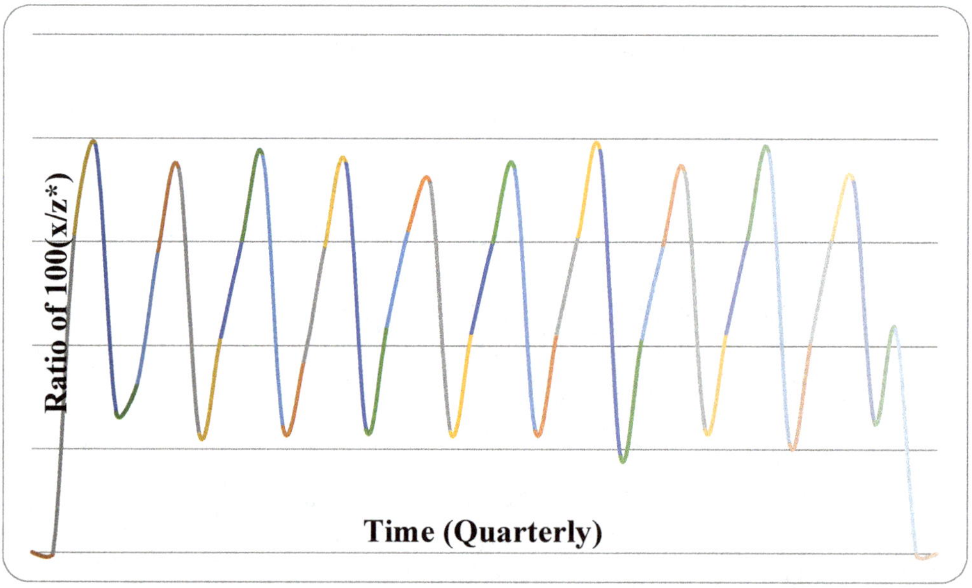

Fig. 10.7 Trend of $100\left(x_t/z_t^*\right)$ ratio for Johnson & Johnson

Table 10.6 Calculation of seasonal indexes of EPS for Johnson & Johnson Corporation

	Quarter				
Year	1	2	3	4	Sums
2000			123.077	157.518	
2001	54.091	65.628	116.602	147.972	
2002	45.802	83.409	120.768	153.043	
2003	48.653	73.898	118.421	150.311	
2004	47.592	87.722	124.557	142.786	
2005	46.659	85.166	120.112	148.246	
2006	47.209	84.836	122.358	155.533	
2007	37.375	83.023	119.078	147.094	
2008	47.789	85.180	121.536	154.386	
2009	42.475	81.853	121.587	143.664	
2010	50.971	86.787			
Median	47.400	84.122	121.152	149.278	401.953
Seasonal index	47.170	83.714	120.563	148.553	400.000

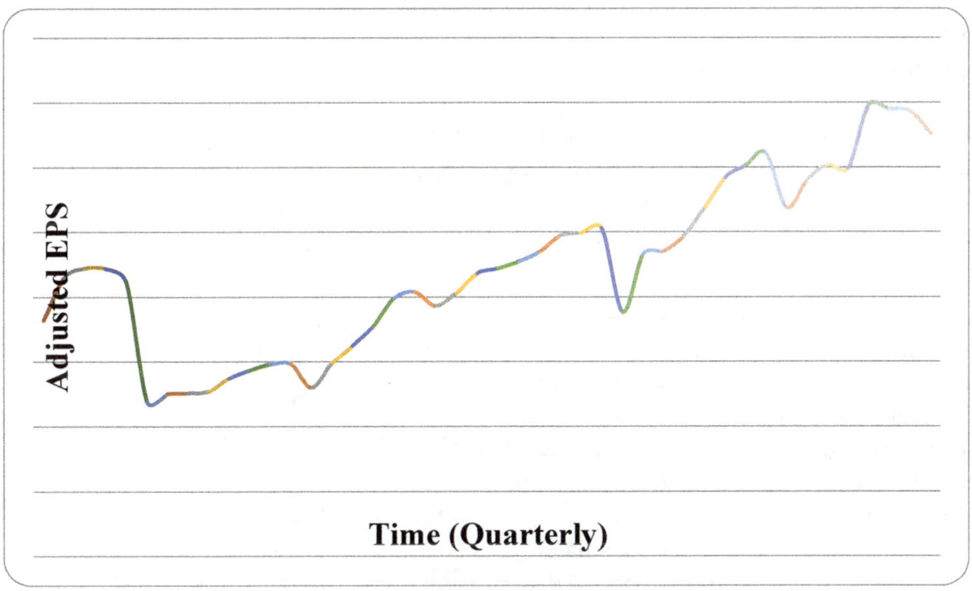

Fig. 10.8 Adjusted earnings per share (EPS) of Johnson & Johnson

in year t, respectively. Then the underlying relationship is

$$x_t = x_0 e^{gt}$$

where x_0 is the base-year sales figure, g is the growth rate, and t is the length of time in terms of number of periods. Then, via the natural logarithm transformation, we obtain

$$\begin{aligned}\log_e x_t &= \log_e(x_0 e^{gt}) \\ &= \log_e x_0 + gt\end{aligned} \tag{10.12}$$

where $\log_e$ is the natural logarithm operator. Equation (10.12) can be defined as a log linear regression model,[5]

$$\log_e x_t = \alpha' + \beta' t + \varepsilon'_t \tag{10.13}$$

where

$\alpha' = \log_e x_0$
$\beta' = g$ = growth rate of a firm's sales.

JNJ's annual sales data (1980–2010), presented in Table 10.7, are used to show how Eq. (10.12) can be employed to forecast JNJ's future sales, and Eq. (10.13) to estimate the growth rate of JNJ's historical sales.

Example 10.3 (*Forecasting Sales and Estimating Growth Rate*)

Suppose Johnson & Johnson Company is interested in forecasting its sales revenues for each of the next 6 years. The sales manager of the company would also like to estimate the historical growth rate of sales revenue.

To make forecasts and assess their reliability, we must construct a time-series model for the sales revenue data listed in Table 10.7. A plot of the data (Fig. 10.9) reveals a linearly increasing trend. Therefore, the linear time trend regression defined in Eq. (10.11) can be used to do forecasting. By the method of least squares, we obtain the least squares model in terms of sales (x_t) and time intervals (t) as

$$\hat{x}_t = \hat{\alpha} + \hat{\beta} t = -7965.026 + 2089.257t$$

With $R^2 = 0.903$.

This least squares line is shown in Fig. 10.9, and the result of straight-line model is given in Fig. 10.10. We can now forecast sales for years

Table 10.7 Johnson & Johnson's annual sales

Year	Sales, x_t (in millions)	t
1980	$4837.38	1
1981	$5399.00	2
1982	$5760.87	3
1983	$5972.87	4
1984	$6124.50	5
1985	$6421.30	6
1986	$7002.90	7
1987	$8012.00	8
1988	$9000.00	9
1989	$9757.00	10
1990	$11,232.00	11
1991	$12,447.00	12
1992	$13,753.00	13
1993	$14,138.00	14
1994	$15,734.00	15
1995	$18,842.00	16
1996	$21,620.00	17
1997	$22,629.00	18
1998	$23,657.00	19
1999	$27,471.00	20
2000	$29,139.00	21
2001	$33,004.00	22
2002	$36,298.00	23
2003	$41,862.00	24
2004	$47,348.00	25
2005	$50,434.00	26
2006	$53,194.00	27
2007	$61,035.00	28
2008	$63,747.00	29
2009	$61,897.00	30
2010	$61,587.00	31

[5] In this regression, we implicitly assume that x_t is lognormally distributed and that $\log_e x_t$ is normally distributed.

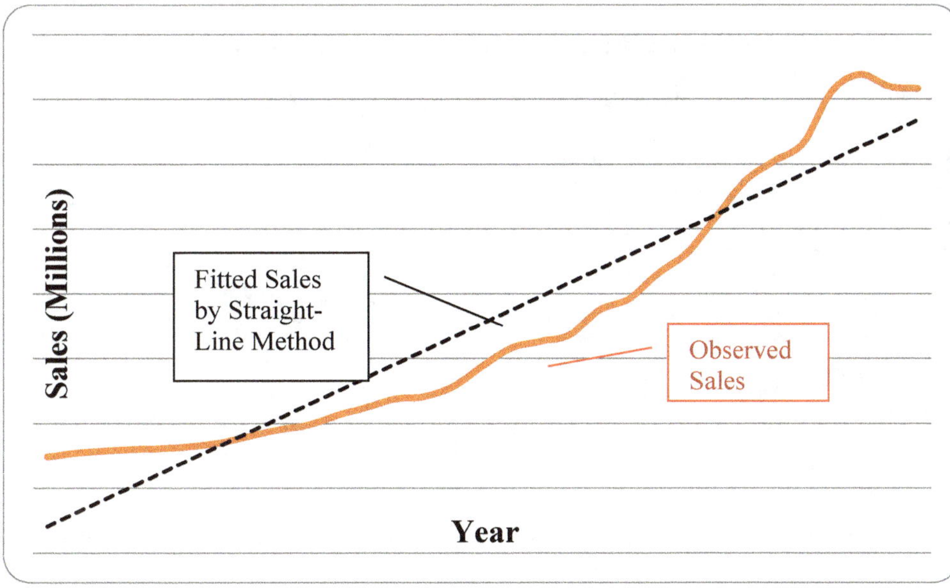

Fig. 10.9 JNJ's annual sales (1980–2010) and the linear trend regression

SUMMARY OUTPUT

Regression Statistics	
Multiple R	0.950
R Square	0.903
Adjusted R Square	0.899
Standard Error	6346.179
Observations	31.000

ANOVA

	df	*SS*	*MS*	*F*	*Significance F*
Regression	1.000	10825189941.84	10825189941.84	268.789	0.000
Residual	29.000	1167945584.055	40273985.65		
Total	30.000	11993135525.89			

	Coefficients	*Standard Error*	*t Stat*	*P-value*	*Lower 95%*	*Upper 95%*
Intercept	-7965.026	2335.910	-3.410	0.002	-12742.498	-3187.553
t	2089.257	127.434	16.395	0.000	1828.625	2349.890

Fig. 10.10 Least squares fit (straight-line method) to x_t = sales

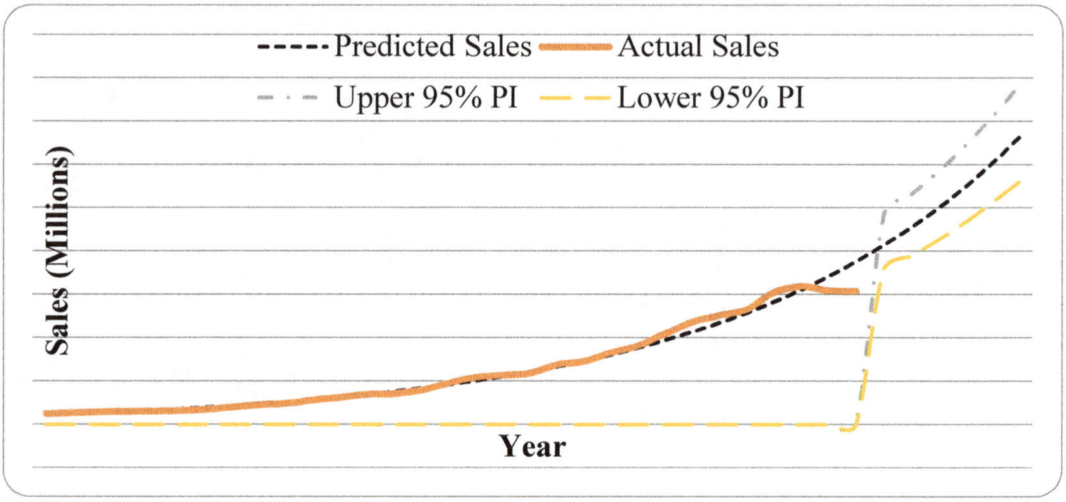

Fig. 10.11 Observed (years 1980–2010) and forecast (years 2011–2016) sales using log linear regression model

2011–2016 by log linear regression model defined in Eq. (10.13). The forecasts of sales by log linear regression model and the corresponding 95% prediction intervals are shown in Fig. 10.11. Although it is not easily perceptible in the figure, the prediction interval widens as we attempt to forecast further into the future. This agrees with the intuitive notion that short-term forecasts should be more reliable than long-term forecasts.

To estimate the growth rate for JNJ's sales during the period 1980–2010, we use data listed in Table 10.7 to fit the log linear regression of Eq. (10.13) and obtain

$$\log_e \hat{x}_t = \hat{\alpha}' + \hat{\beta}' t = 8.3005 + 0.0944t$$
$$(0.027)\,(0.001) \qquad R^2 = 0.993$$

Figures in parentheses are standard errors. This result implies that the estimated growth rate $g = \hat{\beta}' = 9.44\%$. In other words, the annual growth rate of JNJ's sales was 9.44% during the period 1980–2010.

10.5 Exponential Smoothing and Forecasting

Simple Exponential Smoothing and Forecasting

Smoothing techniques are often used to forecast future values of a time series. One problem that arises in using a moving average to forecast time series is that values at the ends of the series are lost, as shown in Sect. 10.3. Therefore, we must subjectively extend the graph of the moving average into the future. No exact calculation of a forecast is available, because generating the moving average at a future time period t requires that we know one or more future values of the series. A technique that leads to forecasts that can be explicitly calculated is called **exponential smoothing**. To use the exponential smoothing technique in forecasting, we need only past and current values of the time series.

To obtain an exponentially smoothed series, we first need to choose a weight α between

0 and 1, called the **exponential smoothing constant**. The exponentially smoothed series, denoted s_t, is then calculated as follows:

$$\begin{aligned} s_1 &= x_1 \\ s_2 &= \alpha x_2 + (1-\alpha)s_1 \\ s_3 &= \alpha x_3 + (1-\alpha)s_2 \\ &\vdots \\ s_i &= \alpha x_t + (1-\alpha)s_{t-1} \end{aligned} \tag{10.14}$$

We can see that the exponentially smoothed value at time t is simply a weighted average of the current time-series value x_t and the exponentially smoothed value at the previous time period, S_{t-1}. Then we can use s_t to do forecasting as follows:

$$\hat{x}_{t+1} = s_t = \alpha x_t + (1-\alpha)s_{t-1} \tag{10.15}$$

where $\hat{x}_{t+1}$ is the next period's forecast value. In other words, $\hat{x}_{t+1}$ is expressed in terms of the smoothing constant times x_t plus $(1-\alpha)$ times s_{t-1}.

If the manager of a company in 1990 ($t = 1$) knows only that current sales of his or her company equal $x_1 = 5000$ units and that current sales have been forecasted as $s_0 = 5100$ units, then he or she can use Eq. (10.15) to forecast 1991 sales. If we choose $\alpha = 0.30$ as a smoothing constant, then the sales for 1991 are forecasted in terms of Eq. (10.15) as

$$\begin{aligned} \hat{x}_2 = s_1 &= (0.30)(5000) + (1-0.30)(5100) \\ &= 5070 \text{ units} \end{aligned}$$

Rewriting Eq. (10.15) as

$$\hat{x}_{t+1} = s_t = s_{t-1} + \alpha(x_t - s_{t-1}) \tag{10.16}$$

implies that simple exponential smoothing is the weighted average of s_{t-1} and the forecast error $(x_t - s_{t-1})$ with weights of 1 and α, respectively. The term **exponential smoothing** refers to the fact that s_t can be expressed as a weighted average with exponentially decreasing weights, as we now illustrate.

We substitute the expressions for s_{t-1} and s_{t-2} into the expression for s_t as defined in Eq. (10.15) and obtain

$$\begin{aligned} s_{t-1} &= \alpha x_{t-1} + (1-\alpha)s_{t-2} \\ s_{t-2} &= \alpha x_{t-2} + (1-\alpha)s_{t-3} \end{aligned}$$

Repeatedly substituting s_{t-2} and s_{t-1} into Eq. (10.15) reveals that

$$\begin{aligned} s_t &= \alpha x_t + (1-\alpha)s_{t-1} \\ &= \alpha x_t + \alpha(1-\alpha)x_{t-1} + (1+\alpha)^2 s_{t-2} \\ &= \alpha x_t + \alpha(1-\alpha)x_{t-1} + \alpha(1-\alpha)^2 x_{t-2} + (1-\alpha)^3 s_{t-3} \end{aligned}$$

Continuous substitution for s_{t-k}, where $k = 2, 3, \ldots, t$, yields

$$s_t = \left[\alpha \sum_{k=0}^{t-1} (1-\alpha)^k x_{t-k}\right] + (1-\alpha)^t s_0 \quad (0<\alpha<1) \tag{10.17}$$

where s_0 is an initial estimate of the smoothed value.

The sum of weights approaches unity as t approaches infinity; hence, we use the term *average*.[6] The weights decrease geometrically with increasing k, so the most recent values of x_t are assigned the greatest weight. All the previous values of x_t are included in the expression for s_t. Because α is less than unity, the most remote values of x_t are associated with the smallest weights. The selection of α depends on the sensitivity of the response required by the model. For example, a small α is used to represent the small sensitivity of the response, and it implies that a single change will not affect the moving average much. The smaller the value of α, the slower the response. Note that the method discussed in this section is good only for short-term forecasting.

[6] Let $0 < \alpha \leq 1$, as $t \geq \infty$, $(1-\alpha)^t \geq 0$. Lety $= \alpha + \alpha(1-\alpha) + \alpha(1-\alpha)^2 + \cdots + \alpha(1-\alpha)^{t-1}$ (A)-
(B)Subtracting Eq. (B) from Eq. (A) yields $y = 1 - (1-\alpha)^t$. Because $\alpha < 1$, y approaches 1 if t approaches infinity. This implies that $\alpha + \alpha(1-\alpha) + \alpha(1-\alpha)^2 + \ldots = 1$.

In the next example, we draw on annual earnings per share (EPS) data for both Johnson & Johnson (J&J) and International Business Machines (IBM) to show how the simple exponential smoothing method defined in Eq. (10.15) can be used to do data analysis.

Example 10.4 (*Simple Exponential Smoothing of EPS for Both J&J and IBM*)

Consider the EPS for both J&J and IBM from 2000 to 2010 as shown in the second column of Table 10.8. Using $\alpha = 0.3$, we calculate the exponentially smoothed series presented in the third column of Table 10.8 as follows:

$$
\begin{aligned}
&\text{IBM}\\
s_{00} &= x_{00} = 4.58\\
s_{01} &= 0.3(4.45) + 0.7(4.58) = 4.541\\
&\vdots\\
s_{10} &= 0.3(11.69) + 0.7(7.734566)\\
&= 8.921196\\
&\text{JNJ}\\
s_{00} &= x_{00} = 3.45\\
s_{01} &= 0.3(1.87) + 0.7(3.45) = 2.976\\
&\vdots\\
s_{10} &= 0.3(4.85) + 0.7(3.939735)\\
&= 4.212814
\end{aligned}
$$

Table 10.8 Simple exponential smoothing ($\alpha = 0.3$) of EPS for J&J and IBM

t	x_t	s_t
IBM		
2000	4.58	4.58
2001	4.45	4.541
2002	2.1	3.8087
2003	4.4	3.98609
2004	5.03	4.299263
2005	4.96	4.497484
2006	6.2	5.008239
2007	7.32	5.701767
2008	9.07	6.712237
2009	10.12	7.734566
2010	11.69	8.921196
J&J		
2000	3.45	3.45
2001	1.87	2.976
2002	2.2	2.7432
2003	2.42	2.64624
2004	2.87	2.713368
2005	3.5	2.949358
2006	3.76	3.19255
2007	3.67	3.335785
2008	4.62	3.72105
2009	4.45	3.939735
2010	4.85	4.212814

We see from the table that the most recent estimates of smoothed EPS for J&J and IBM are

$$
\begin{aligned}
s_n &= s_{10} = 8.921196 \quad \text{(IBM)}\\
s_n &= s_{10} = 4.212814 \quad \text{(J\&J)}
\end{aligned}
$$

These values are then used as the forecast of EPS for both J&J and IBM for future years. The observed series and these forecasts for J&J and IBM are graphed in Figs. 10.12 and 10.13, respectively.

Finally, note that the choice of the smoothing constant (α) affects the precision of the forecast. In practice, we can try several different values to see which would have been most successful in predicting historical movement in the time series. For example, we might compute the smoothed series for values of α of 0.3, 0.4, 0.5, and 0.7 and calculate the forecast **mean squared error (MSE)** for these four different α-values.

$$\text{MSE} = \frac{\sum_{t=1}^{n} (x_t - \hat{x}_t)^2}{n} \tag{10.18}$$

where x_t and $\hat{x}_t$ are actual value and forecast value, respectively. The value of α for which this MSE is smallest is then used in the prediction of future values.

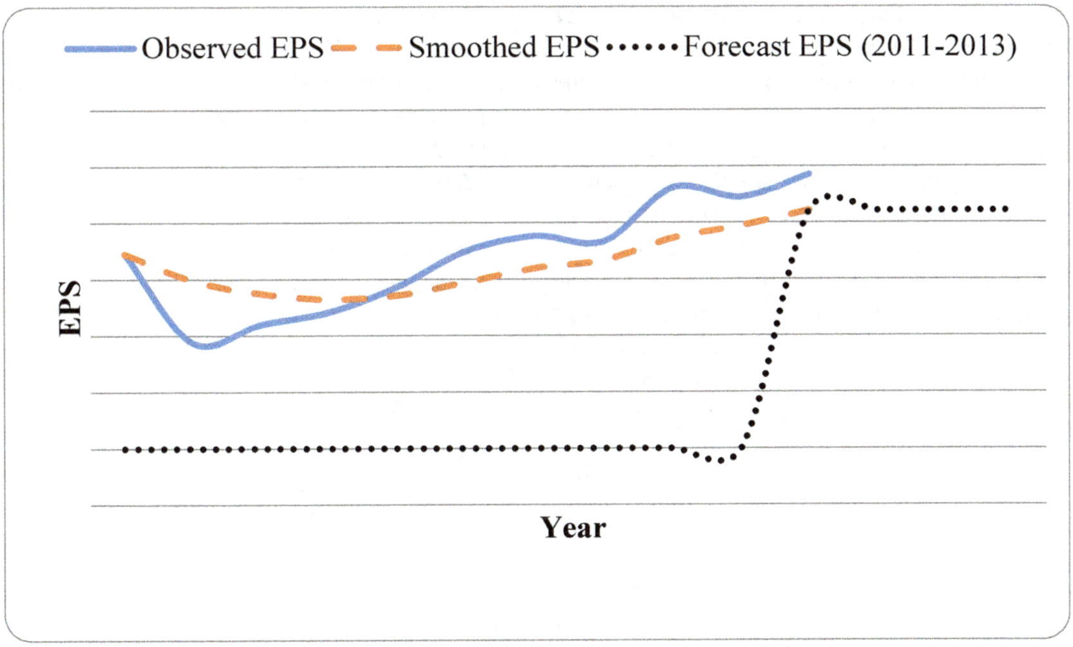

Fig. 10.12 Annual earnings per share of J&J (simple exponential smoothing)

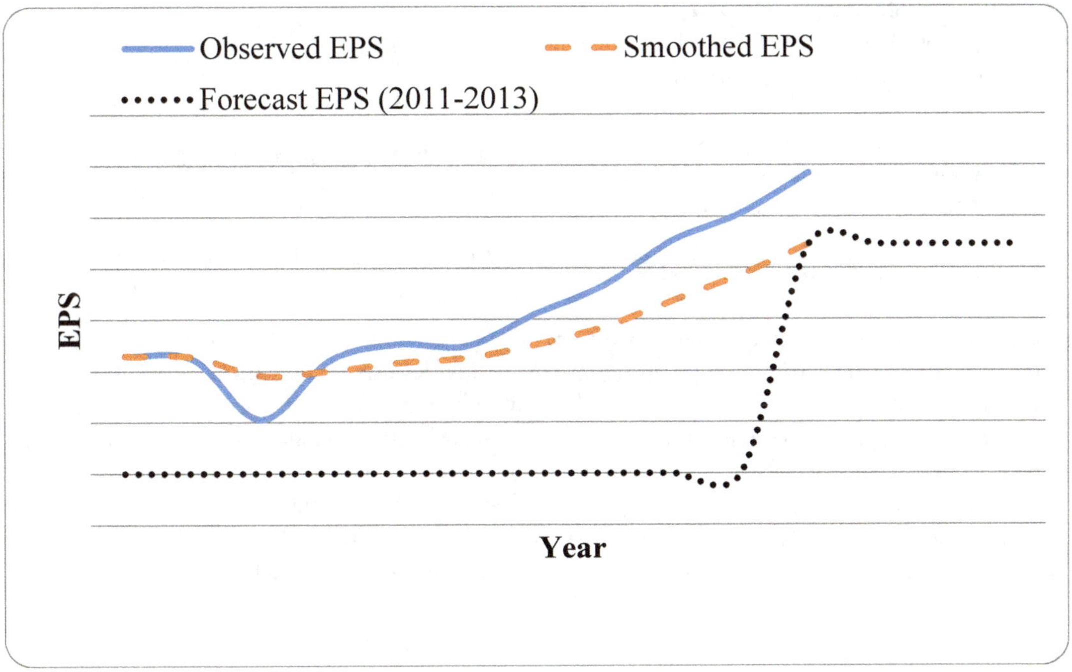

Fig. 10.13 Annual earnings per share of IBM (simple exponential smoothing)

The Holt–Winters Forecasting Model for Nonseasonal Series[7]

The simple exponential smoothing technique discussed in the previous section does not recognize the trend in the time series. In this section, we will generalize the simple exponential smoothing model defined in Eq. (10.15) by explicitly recognizing the trend in a time series. The **Holt–Winters forecasting model** consists of both an exponentially smoothed component (s_t) and a trend component (T_t). The trend component is used in calculating the exponentially smoothed value. Here s_t and T_t can be written as

$$s_t = \alpha x_t + (1 - \alpha)(s_{t-1} + T_{t-1}) \quad (10.19a)$$

$$T_t = \beta(s_t - s_{t-1}) + (1 - \beta)T_{t-1} \quad (10.19b)$$

where α and β are two smoothing constants, each of which is between 0 and 1. We estimate the trend component of the series by using a weighted average of the most recent change in the smoothed component [represented by $(s_t - s_{t-1})$] and the time trend estimate (represented by T_{t-1}) from the previous period. The procedure for calculating the Holt–Winters components is as follows:

1. Choose an exponential smoothing constant a between 0 and 1. Small values of α give less weight to the current values of the time series and more weight to the past. Large values of α give more weight to the current values of the series.
2. Choose a trend smoothing constant β between 0 and 1. Small values of β give less weight to the current changes in the level of the series and more weight to the past trend. Larger choices assign more weight to the most recent trend of the series.
3. Estimate the first observation of trend T_1 by one of the following two alternative methods.

Method I:
Let $T_1 = 0$. If there are a large number of observations in the time series, this method provides an adequate initial estimate for the trend.

Method 2:
Use the first five (or so) observations to estimate the initial trend by following the linear time trend regression line

$$x_t = a + bt + e_t$$

Then use the estimated slope $\hat{b}$ as the first trend observation; that is, $T_1 = \hat{b}$.

4. Calculate the components s_t and T_1 from the time series as follows:

$$\begin{aligned} s_1 &= x_1 \\ T_1 &= 0 \text{ or } \hat{b} \\ s_2 &= \alpha x_2 + (1 - \alpha)(s_1 + T_1) \\ T_2 &= \beta(s_2 - s_1) + (1 - \beta)T_t \\ &\vdots \\ s_t &= \alpha x_t + (1 - \alpha)(s_{t-1} + T_{t-1}) \\ T_t &= \beta(s_t - s_{t-1}) + (1 - \beta)T_{t-1} \end{aligned}$$

The data on earnings per share of J&J and IBM listed in Table 10.8 show how the forecasting model defined in Eqs. (10.19a) and (10.19b) can be used to do data analysis.

Example 10.5 (Using the Holt–Winters Model to Estimate the EPS of J&J and IBM)

Now let us use the Holt–Winters model to do the exponential smoothing for the EPS data for both J&J and IBM listed in Table 10.9. We begin by using the first five observations to estimate the first term of the trend component. The estimated slopes for the EPS of J&J and IBM are 0 and 1.275, respectively. Let $\alpha = 0.3$ and $\beta = 0.2$.

[7]The Holt–Winters forecasting model for seasonal series will be discussed in Appendix 2.

Table 10.9 EPS for IBM and J&J and their smoothed series in terms of the Holt–Winters forecasting model

t	x_t	s_t	T_t
IBM			
2000	4.58	4.58	0.085
2001	4.45	4.6005	0.0721
2002	2.1	3.90082	−0.08226
2003	4.4	3.992995	−0.04737
2004	5.03	4.270937	0.017693
2005	4.96	4.490041	0.057975
2006	6.2	5.043611	0.157094
2007	7.32	5.836494	0.284252
2008	9.07	7.005522	0.461207
2009	10.12	8.26271	0.620403
2010	11.69	9.725179	0.788816
J&J			
2000	3.45	3.45	0
2001	1.87	2.976	−0.0948
2002	2.2	2.67684	−0.13567
2003	2.42	2.504818	−0.14294
2004	2.87	2.514313	−0.11245
2005	3.5	2.731301	−0.04657
2006	3.76	3.007314	0.01795
2007	3.67	3.218685	0.056634
2008	4.62	3.678723	0.137315
2009	4.45	4.006227	0.175353
2010	4.85	4.382105	0.215458

Following the formula for the Holt–Winters components listed in Step 4, we calculate

J&J

$$s_1 = x_1 = 3.45$$
$$T_1 = 0$$
$$\begin{aligned} s_2 &= 0.3(1.87) + 0.7(3.45 + 0) \\ &= 2.976 \end{aligned}$$
$$\begin{aligned} T_2 &= 0.2(2.976 - 3.45) + 0.8(0) \\ &= -0.0948 \end{aligned}$$
$$\vdots$$

IBM

$$s_1 = x_1 = 4.58$$
$$T_1 = 0.085$$
$$\begin{aligned} s_2 &= 0.3(4.45) + 0.7(4.58 + 0.085) \\ &= 4.6005 \end{aligned}$$
$$\begin{aligned} T_2 &= 0.2(4.6005 - 4.58) + 0.8(0.085) \\ &= 0.0721 \end{aligned}$$
$$\vdots$$

The remaining calculations are carried out in precisely the same way. All s_t- and T_1-values for both J&J and IBM are given in Table 10.9.

How are these estimates of EPS level and trend used to forecast future observations? Given a series $x_1, x_2, \ldots, x_n$, the most recent EPS level and trend estimates are s_n and T_n, respectively. To do forecasting, we assume that the latest trend will continue from the most recent level. In general, standing at time n and looking m time periods into the future, we define the prediction for the m period ahead as

$$\hat{x}_{t+m} = s_t + mT_t \tag{10.20}$$

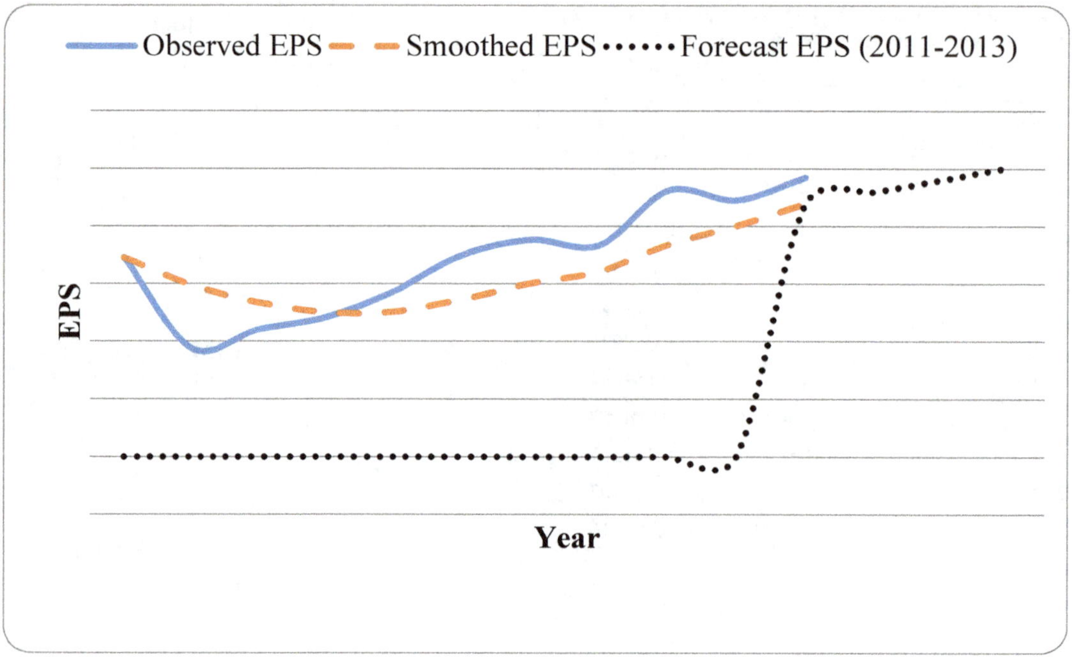

Fig. 10.14 Annual earnings per share of J&J with forecasts based on the Holt–Winters model

If $T_t = 0$, then this prediction reduces to the simple exponential smoothing prediction discussed in Example 10.3. On the basis of this formula and the information given in Table 10.9, we calculate the future predictions for both J&J and IBM as

J&J

$$s_{2011} = 4.382105 + 0.215458 = 4.597563$$

$$s_{2012} = 4.382105 + (2)(0.215458) = 4.813021$$

$$s_{2013} = 4.382105 + (3)(0.215458) = 5.028479$$

IBM

$$s_{2011} = 9.725179 + 0.788816 = 10.514$$

$$s_{2012} = 9.725179 + (2)(0.788816) = 11.30281$$

$$s_{2013} = 9.725179 + (3)(0.788816) = 12.09163$$

Figures 10.14 and 10.15 show the data series and three forecasts for J&J and IBM, respectively.

Finally, note that the choice of smoothing constants (α and β) affects the precision of a forecast. In practice, we can try several different values of α and β to see which would have been most successful in predicting historical movement in the time series. Again, the forecast mean squared error as defined in Eq. (10.18) can be used as a benchmark in deciding what values of α and β are appropriate for forecasting future observations.

10.6 Autoregressive Forecasting Model[8]

A time-series analysis always reveals some degree of correlation between elements. For example, a certain firm's current sales may be correlated with sales in the previous period and even with sales in several prior periods. Under

[8]It should be noted that the exponential smoothing model of Sect. 10.5 of the autoregressive models described herein are all special cases of **autoregressive integrated moving-average (ARIMA)** models developed by Box and Jenkins. The Box–Jenkins approach, however, is beyond the scope of this text.

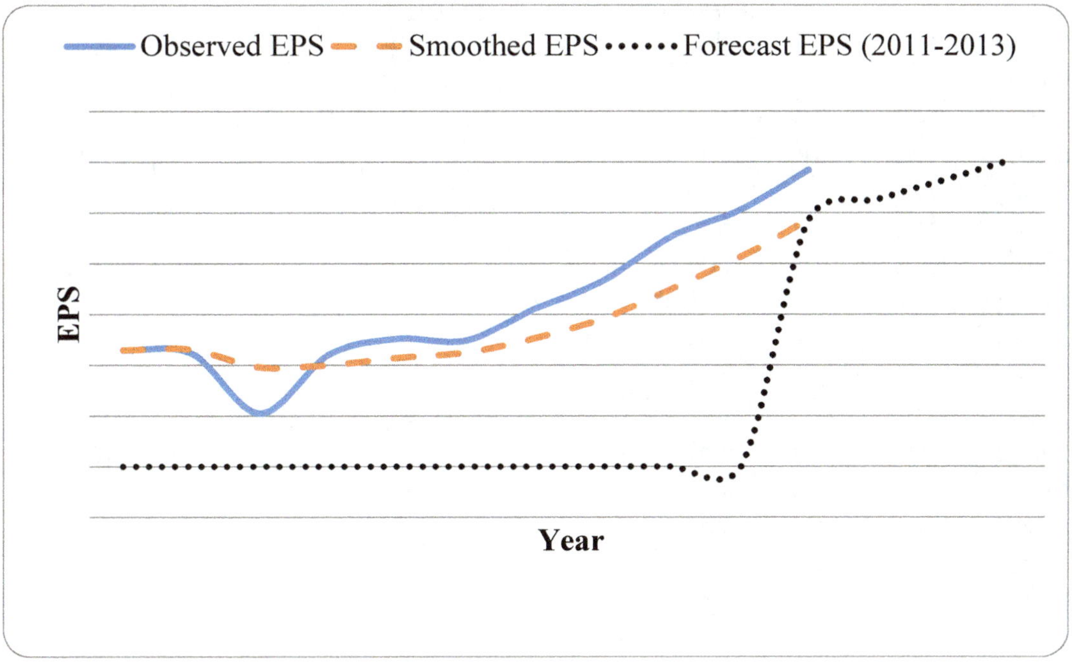

Fig. 10.15 Annual earnings per share of IBM with forecasts based on the Holt–Winters model

these circumstances, we can regress the time series x_t on some combination of its past values to derive a forecasting equation.

Suppose we attempt to predict the value of x_t by using previous observation. The prediction equation is

$$\hat{x}_t = a_0 + a_1 x_{t-1} \tag{10.21}$$

where α_0 and α_1 are the least squares regression estimates. This is called a first-order **autoregressive forecasting model**, AR(1). If the current value of a time series depends on the two most recent observations, we can use the model

$$\hat{x}_t = a_0 + a_1 x_{t-1} + a_2 x_{t-2} \tag{10.22}$$

where a_0, a_1, and a_2 are least squares regression estimates. This is called a second-order autoregressive model, AR(2). Generally, the autoregressive model of order p, AR(P), can be expressed as

$$\hat{x}_t = a_0 + a_1 x_{t-1} + a_2 x_{t-2} + \cdots + a_p x_{t-p} \tag{10.23}$$

where a_0, a_1, a_2, …, a_p are least squares regression estimates.

In the next example, quarterly data on Johnson & Johnson's sales are employed to show how the autoregressive model can be used in forecasting.

Example 10.6 (Sales Forecast for Johnson & Johnson)

Quarterly sales data for Johnson & Johnson from first quarter 2000 through fourth quarter 2010 are presented in Table 10.10 and Fig. 10.16.

Using the data in Table 10.10, we run the AR (1), AR(2), and AR(3) models.

$$\begin{aligned}\text{AR(1): Sales}_t &= 552.7913 + \underset{(0.026)}{0.9703}\ \text{sales}_{t-1}\\ R^2 &= 0.9719\end{aligned} \tag{10.24}$$

Table 10.10 Quarterly sales data for Johnson & Johnson (first quarter 2000 to fourth quarter 2010)

Quarter	S_t	S_{t-1}	S_{t-2}	S_{t-3}
2000Q1	7440			
2000Q2	7670	7440		
2000Q3	7438	7670	7440	
2000Q4	7298	7438	7670	7440
2001Q1	7855	7298	7438	7670
2001Q2	8179	7855	7298	7438
2001Q3	8058	8179	7855	7298
2001Q4	8225	8058	8179	7855
2002Q1	8743	8225	8058	8179
2002Q2	9073	8743	8225	8058
2002Q3	9079	9073	8743	8225
2002Q4	9403	9079	9073	8743
2003Q1	9821	9403	9079	9073
2003Q2	10,333	9821	9403	9079
2003Q3	10,454	10,333	9821	9403
2003Q4	11,254	10,454	10,333	9821
2004Q1	11,559	11,254	10,454	10,333
2004Q2	11,484	11,559	11,254	10,454
2004Q3	11,553	11,484	11,559	11,254
2004Q4	12,752	11,553	11,484	11,559
2005Q1	12,832	12,752	11,553	11,484
2005Q2	12,762	12,832	12,752	11,553
2005Q3	12,230	12,762	12,832	12,752
2005Q4	12,610	12,230	12,762	12,832
2006Q1	12,992	12,610	12,230	12,762
2006Q2	13,363	12,992	12,610	12,230
2006Q3	13,157	13,363	12,992	12,610
2006Q4	13,682	13,157	13,363	12,992
2007Q1	15,037	13,682	13,157	13,363
2007Q2	15,131	15,037	13,682	13,157
2007Q3	14,910	15,131	15,037	13,682
2007Q4	15,957	14,910	15,131	15,037
2008Q1	16,194	15,957	14,910	15,131
2008Q2	16,450	16,194	15,957	14,910
2008Q3	15,921	16,450	16,194	15,957
2008Q4	15,182	15,921	16,450	16,194
2009Q1	15,026	15,182	15,921	16,450
2009Q2	15,239	15,026	15,182	15,921
2009Q3	15,081	15,239	15,026	15,182
2009Q4	16,551	15,081	15,239	15,026
2010Q1	15,631	16,551	15,081	15,239
2010Q2	15,330	15,631	16,551	15,081
2010Q3	14,982	15,330	15,631	16,551
2010Q4	15,644	14,982	15,330	15,631

$$\begin{aligned} \text{AR(2): Sales}_t &= 586.6586 + \underset{(0.1623)}{0.9106}\ \text{sales}_{t-1} \\ &\quad + \underset{(0.1590)}{0.0580}\ \text{sales}_{t-2} \\ R^2 &= 0.9702 \end{aligned} \tag{10.25}$$

$$\begin{aligned} \text{AR(3): Sales}_t &= 737.5405 + \underset{(0.1616)}{0.8987}\ \text{sales}_{t-1} \\ &\quad - \underset{(0.2220)}{0.1082}\ \text{sales}_{t-2} + \underset{(0.1603)}{0.1697}\ \text{sales}_{t-3} \\ R^2 &= 0.9698 \end{aligned} \tag{10.26}$$

In Eqs. (10.24), (10.25), and (10.26), figures in parentheses under the coefficients are standard errors.

Table 10.10 makes it clear that the observations used to run AR(1), AR(2), and AR(3) are 43, 42, and 41, respectively. Therefore, by the central limit theorem, the parameter estimators divided by their standard errors approximate standard normal distributions.

From the standard error indicated in the parentheses and the parameter estimator, we can calculate the Z-statistic for each regression slope. Looking up these Z-statistics in Table 10.14 of Appendix 1 reveals that coefficients of sales_{t-1} in the AR(1), AR(2), and AR(3) model are significantly different from zero at the significance level of $\alpha = 0.05$. Hence, we conclude that the autoregressive processes can be used to forecast quarterly sales of Johnson & Johnson.

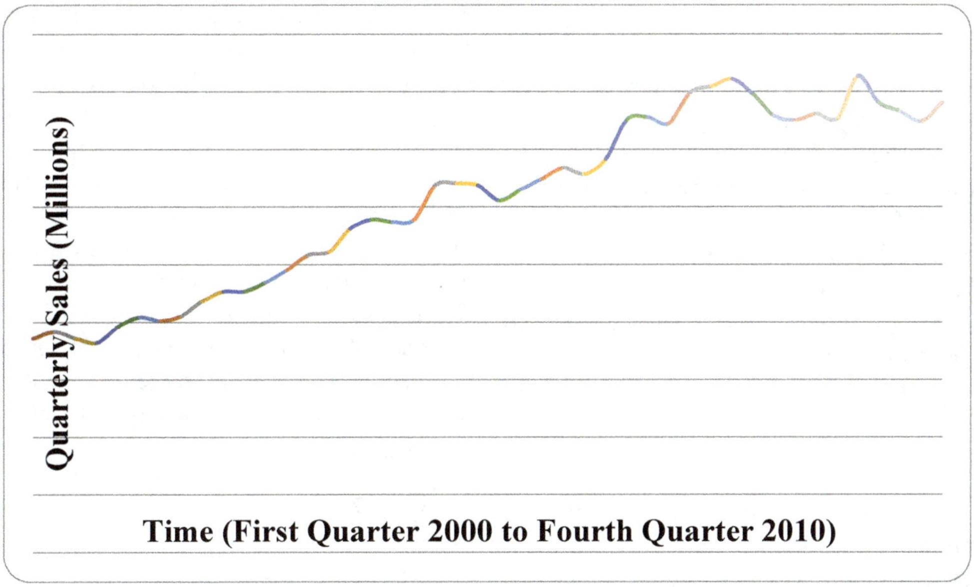

Fig. 10.16 Quarterly sales data for Johnson & Johnson

Substituting related quarterly sales data into the AR(1), AR(2), and AR(3) models, we obtain the following three alternative forecasted sales for the first quarter of 2011. Substituting $\text{sales}_{t-1} = 15{,}644$ into Eq. (10.24), we obtain the AR (1) forecast.

$$\begin{aligned}\text{Sales}_{2011Q1} &= 552.7913 + 0.9703(15,644)\\ &= 15,732\end{aligned}$$

Substituting $\text{sales}_{t-1} = 15{,}644$ and $\text{sales}_{t-2} = 14{,}982$ into Eq. (10.25), we obtain the AR (2) forecast.

$$\begin{aligned}\text{Sales}_{2011Q1} &= 586.6586 + 0.9106(15,644)\\ &\quad + 0.0580(14,982)\\ &= 15,700.72\end{aligned}$$

Substituting $\text{sales}_{t-1} = 15{,}644$, $\text{sales}_{t-2} = 14{,}982$, and $\text{sales}_{t-3} = 15{,}330$ into Eq. (10.26), we obtain the AR(3) forecast.

$$\begin{aligned}\text{Sales}_{2011Q1} &= 737.5405 + 0.8987(15,644)\\ &\quad - 0.1082(14,982) + 0.1697(15,330)\\ &= 15,777.44\end{aligned}$$

To determine which model we should choose, we can use the mean absolute relative prediction error (MARPE) to see which one gives us the smallest error.

$$\text{MARPE} = \frac{\left|\hat{S}_t - S_t\right|}{S_t} \tag{10.27}$$

where $\hat{S}_t$ represents the sales forecast for time period t and S_t represents actual reported sales for time period t.

10.7 ARIMA Models

In the last section, we have discussed autoregression model. This section will discuss the ARIMA model, which integrates autoregression model with moving-average model. The ARIMA model is one of a class of models used in forecasting time-series data. It is best to begin by examining the simplest of all possible time series, a purely random series. A series that is purely random is sometimes referred to as a white-noise or random-walk model. Mathematically, such a series can be described by the following equation:

$$y_t = a_t \quad (10.28)$$

in which the series a_t is assumed to have a mean of zero, to be unrelated to its past values, and to have a constant variance over time. Mathematically, these assumptions can be summarized as
(i) $E(a_t) = 0$,
(ii) $E(a_t, a_{t-i}) = 0$ for all t and $i \neq 0$
(iii) $\mathrm{Var}(a_t) = \sigma_a^2$ for all t.

Modifying Eq. (10.28) to allow the series to be concentrated around a nonzero mean δ, the series could now be described:

$$y_t = \delta + a_t. \quad (10.29)$$

Equation (10.29) is a model that can be used to represent many different time series in economics and finance. For example, in an efficient market a series of stock prices might be expected to randomly fluctuate around a constant mean. So the actual stock price observed in time period t would be equal to its average price plus some random shock in time period t.

The question now is how to model a purely random series. Fortunately, a theorem known as Wold's decomposition provides the answer. Wold's decomposition proves that any stationary time series (a series is stationary if it is centered around a constant mean) can be considered as a sum of self-deterministic components. This theorem states that a time series can be generated from a weighted average of past random shocks of infinite order. A model such as this is known as a moving average of infinite order and can be expressed by the following equation:

$$y_t = \delta + \Theta_1 a_{t-1} + \Theta_2 a_{t-2} + \ldots + \Theta_\infty a_{t-\infty} + a_t, \quad (10.30)$$

where δ = mean of the process; $a_{t-\infty}$ = random shock that occurred ∞ periods earlier; and Θ_∞ = parameter that relates the random shock ∞ periods earlier to the current value of y.

The moving-average model just discussed should not be confused with the moving-average concept previously discussed in this chapter. Previously, the term moving average was used to refer to an arithmetic average of stock prices over a specified number of days. The term moving average was used because the average was continually updated to include the most recent series of data. In time-series analysis, a moving-average process refers to a series generated by a weighted average of past random shocks.

Because it would be impossible to estimate a model of infinite order, in practice it is best to specify a model of finite order. A moving-average process of order q with zero mean would be expressed:

$$y_t = \Theta_1 a_{t-1} + \ldots + \Theta_q a_{t-q} + a_t. \quad (10.31)$$

A moving-average process is not the only way to model a stationary time series. Again, consider a moving-average process of infinite order:

$$y_t = \Theta_1 a_{t-1} + \Theta_2 a_{t-2} + \ldots + a_t. \quad (10.32)$$

Equation (10.32) can be rewritten in terms of the error term a_t:

$$a_t = y_t - \Theta_1 a_{t-1} - \Theta_2 a_{t-2} - \cdots. \quad (10.33)$$

Because Eq. (10.33) is recursive in nature, it is easy to generate an expression for a_{t-1}:

$$a_{t-1} = y_{t-1} - \Theta_1 a_{t-2} - \Theta_2 a_{t-3} - \cdots. \quad (10.34)$$

By substituting the expression for a_{t-1} into Eq. (10.33):

$$\begin{aligned} a_t &= y_t - \Theta_1 (y_{t-1} - \Theta_1 a_{t-2} - \Theta_2 a_{t-3} - \cdots) \\ &\quad - \Theta_2 a_{t-2} - \Theta_3 a_{t-3} - \cdots \\ &= y_t - \Theta_1 y_{t-1} + (\Theta_1^2 - \Theta_2) a_{t-2} \\ &\quad + (\Theta_1 \Theta_2 - \Theta_3) a_{t-3} + \cdots \end{aligned} \quad (10.35)$$

By generating expressions for $a_{t-2} a_{t-3}, \cdots$ and substituting them into Eq. (10.35) a_t can be expressed in terms of past values of y_t:

$$a_t = y_t + \phi_1 y_{t-1} + \phi_2 y_{t-2} + \ldots . \quad (10.36)$$

Rearranging Eq. (10.36) in terms of y_t:

$$y_t = -\phi_1 y_{t-1} - \phi_2 y_{t-2} - \ldots + a_t, \quad (10.37)$$

where $\phi_1 = \phi_1$ and $\phi_2 = \Theta_1^2 - \Theta_2$.

Equation (10.37) is known as an autoregressive process of infinite order. The term autoregressive refers to the fact that y_t is expressed in terms of its own past values $y_{t-1}, y_{t-2}, \ldots, y_{t-p}$.

Again, because it is impossible to estimate a model of infinite order, an approximate model of finite order is specified. An autoregressive process of order p can be expressed:

$$y_t = -\phi_1 y_{t-1} - \ldots - \phi_p y_{t-p} - \ldots + a_t. \quad (10.38)$$

Thus, a stationary time series can be expressed in two ways: (1) as a moving-average process in which the series can be represented as a weighted average of past random shocks, or (2) as an autoregressive process in which the series can be represented as a weighted average of its past values. A third possibility is that a series may involve some combination of the two processes. This process is referred to as a mixed autoregressive moving-average (ARMA) process.

An ARMA process of infinite order can be expressed:

$$y_t = \Theta_1 a_{t-1} + \Theta_2 a_{t-2} + \ldots - \phi_1 y_{t-1} - \phi_2 y_{t-2} - \ldots + a_t. \quad (10.39)$$

Again, an ARMA process of finite order must be specified in order to make estimation possible. An ARMA (p, q) process can be expressed:

$$y_t = \Theta_1 a_{t-1} + \ldots + \Theta_q a_{t-q} - \phi_1 y_{t-1} - \ldots - \phi_p y_{t-p} + a_t. \quad (10.40)$$

So far the discussion has focused on the estimation of stationary time series. Suppose the process of interest is not stationary. Fortunately, a nonstationary series can usually be made stationary by transforming the data in an appropriate manner. The most popular method of transforming a nonstationary series to a stationary one is by differencing the series. For example, suppose the series $y_1, y_2, \ldots, y_t$ is nonstationary. By differencing the series, a new series, $Z_1, Z_2, \ldots, Z_{t-1}$, is created. The new series can be defined as (i) $Z_1 = y_2 - y_1$ (ii) $Z_2 = y_3 - y_2 \ldots$ (iii) $Z_{t-1} = y_t - y_{t-1}$.

If the series Z_t is nonstationary, it may be necessary to differentiate the series Z_t.

The modeling of a series that has been differenced is referred to as an autoregressive integrated moving-average (ARIMA) process. A detailed discussion of ARIMA modeling is beyond the scope of this book; nevertheless, a brief outline of the ARIMA modeling procedure is in order (see Nelson 1973 or Nazem 1988, for details of the ARIMA procedure).

The ARIMA process uses the following three steps as (1) Identification, (2) Estimation, and (3) Forecasting.

The first step is to identify the appropriate model. Identification involves determining the degree of differencing necessary to make the series stationary and to determine the form (ARMA or ARIMA) and order of the process.

After a suitable model is identified, the parameters $\phi_1, \ldots, \phi_p, \Theta_1, \ldots, \Theta_q$ need to be estimated. The final step in the ARIMA process is to use the model for forecasting. Oftentimes, the adequacy of the model is checked by using the model to forecast within the sample. This allows a comparison of the forecasted values to the actual values. If the model is determined to be adequate, the model can be used to forecast future values of the series.

10.8 Autoregressive Conditional Heteroscedasticity

10.8.1 Autoregressive Conditional Heteroscedasticity (ARCH) Models

Some finance researchers are interested in forecasting financial time-series, such as stock returns and foreign exchange rates. In the time-series analysis, it is usually assumed that the error term follows a homoscedastic process. However, Engle (1982) found that, in inflation data, the variance of forecast errors in time-series models is not constant but varies from period to period. Specifically, he found autocorrelation in the variance of forecast errors and suggested an alternative model, autoregressive conditional heteroscedasticity (ARCH) model, to deal with this problem. In ARCH model, a regression model is expressed as:

$$y_t = a_0 + a_1 x_1 + \cdots + a_k x_k + \varepsilon_t \quad (10.41)$$

Assuming the disturbance term is conditionally heteroscedastic with respect to the information available at time $(t-1)$, the variance of disturbance term is

$$\mathrm{Var}[\varepsilon_t | \varepsilon_{t-1}] = \omega + \alpha_1 \varepsilon_{t-1}^2 \quad (10.42)$$

Equation (10.42) assumes that the variance of disturbance term is only serially correlated with the variance of disturbance term in the previous period. We therefore call such process as ARCH (1) model.

The autoregressive conditional heteroscedasticity model can be extend to a more general model with longer lags. An ARCH (q) model can be written as

$$\mathrm{Var}(\varepsilon_t) = \sigma_t^2 = \alpha_0 + \alpha_1 \varepsilon_{t-1}^2 + \cdots + \alpha_q \varepsilon_{t-q}^2 \quad (10.43)$$

10.8.2 Generalized Autoregressive Conditional Heteroscedasticity (GARCH) Model

The model of generalized autoregressive conditional heteroscedasticity (GARCH) is assumed that the conditional variance of disturbance term follows an autoregressive moving-average ARMA (p, q) process. A-GARCH (p, q) model can therefore be written as

$$y_t = a_0 + a_1 x_1 + \cdots + a_k x_k + \varepsilon_t \quad (10.44)$$

$$\begin{aligned}\mathrm{Var}(\varepsilon_t) = \sigma_t^2 &= \omega + \alpha_1 \varepsilon_{t-1}^2 + \cdots + \alpha_p \varepsilon_{t-p}^2 \\ &\quad + \beta_1 \sigma_{t-1}^2 + \cdots + \beta_p \sigma_{t-p}^2 \\ &= \omega + \sum_{i=1}^{q} \alpha_i \varepsilon_{t-i}^2 + \sum_{j=1}^{p} \beta_j \sigma_{t-j}^2\end{aligned} \quad (10.45)$$

10.8.3 The GARCH Universe

Hansen and Lunde (2005) present a family of GARCH-type models in terms of different density functions for the compounded daily return, r_t.[9] As Hansen and Lunde (2005) mention, the conditional variance (σ_t) is the main object of interest and the analysis includes a large number of parametric specifications for the conditional variance that are listed in Table 10.11. Hansen and Lunde (2005) use A-GARCH as Engle and Ng's (1993) model and H-GARCH as Hentshel's (1995) model. Duan (1997) and Loudon et al. (2000) show that some flexible specifications nest other specifications in GARCH-type models (e.g., H-GARCH and Aug-GARCH).

[9]Please see Hansen and Lunde (2005) for detailed discussion.

Table 10.11 Specifications for the conditional variance

Model
ARCH: $\sigma_t^2 = \omega + \sum_{i=1}^{q} \alpha_i \varepsilon_{t-i}^2$
GARCH: $\sigma_t^2 = \omega + \sum_{i=1}^{q} \alpha_i \varepsilon_{t-i}^2 + \sum_{j=1}^{p} \beta_j \sigma_{t-j}^2$
IGARCH: $\sigma_t^2 = \omega + \varepsilon_{t-1}^2 + \sum_{i=2}^{q} \alpha_i (\varepsilon_{t-i}^2 - \varepsilon_{t-1}^2) + \sum_{j=1}^{p} \beta_j (\sigma_{t-j}^2 - \varepsilon_{t-1}^2)$
Taylor/Schwert: $\sigma_t = \omega + \sum_{i=1}^{q} \alpha_i \lvert\varepsilon_{t-i}\rvert + \sum_{j=1}^{p} \beta_j \sigma_{t-j}$
A-GARCH: $\sigma_t^2 = \omega + \sum_{i=1}^{q} [\alpha_i \varepsilon_{t-i}^2 + r_i \varepsilon_{t-i}] + \sum_{j=1}^{p} \beta_j \sigma_{t-j}^2$
NA-GARCH: $\sigma_t^2 = \omega + \sum_{i=1}^{q} \alpha_i (\varepsilon_{t-i} + r_i \sigma_{t-i})^2 + \sum_{j=1}^{p} \beta_j \sigma_{t-j}^2$
V-GARCH: $\sigma_t^2 = \omega + \sum_{i=1}^{q} \alpha_i (e_{t-i} + r_i)^2 + \sum_{j=1}^{p} \beta_j \sigma_{t-j}^2$
Thr.-GARCH: $\sigma_t = \omega + \sum_{i=1}^{q} \alpha_i [(1 - r_i)\varepsilon_{t-i}^{+} - (1 + r_i)\varepsilon_{t-i}^{-}] + \sum_{j=1}^{p} \beta_j \sigma_{t-j}$
GJR-GARCH: $\sigma_t^2 = \omega + \sum_{i=1}^{q} [\alpha_i + r_i I_{(\varepsilon_{t-1} > 0)}] \varepsilon_{t-i}^2 + \sum_{j=1}^{p} \beta_j \sigma_{t-j}^2$
log-GARCH: $\log(\sigma_t) = \omega + \sum_{i=1}^{q} \alpha_i \lvert e_{t-i}\rvert + \sum_{j=1}^{p} \beta_j \log(\sigma_{t-j})$
EGARCH: $\log(\sigma_t^2) = \omega + \sum_{i=1}^{q} [\alpha_i e_{t-i} + r_i (\lvert e_{t-i}\rvert - E\lvert e_{t-i}\rvert)] + \sum_{j=1}^{p} \beta_j \log(\sigma_{t-j}^2)$
NGARCH[a]: $\sigma_t^\delta = \omega + \sum_{i=1}^{q} \alpha_i \lvert\varepsilon_{t-i}\rvert^\delta + \sum_{j=1}^{p} \beta_j \sigma_{t-j}^\delta$
A-PARCH: $\sigma^\delta = \omega + \sum_{i=1}^{q} \alpha_i [\lvert\varepsilon_{t-i}\rvert - r_i \varepsilon_{t-i}]^\delta + \sum_{j=1}^{p} \beta_j \sigma_{t-j}^\delta$
GQ-ARCH: $\sigma_t^2 = \omega + \sum_{i=1}^{q} \alpha_i \varepsilon_{t-i} + \sum_{i=1}^{p} \alpha_{ii} \varepsilon_{t-i}^2 + \sum_{i<j}^{p} \alpha_{ij} \varepsilon_{t-i} \varepsilon_{t-j} + \sum_{j=1}^{p} \beta_j \sigma_{t-j}^2$
H-GARCH: $\sigma_t^\delta = \omega + \sum_{i=1}^{q} \alpha_i \delta \sigma_{t-i}^\delta [\lvert e_t - \kappa\rvert - \tau(e_t - \kappa)]^v + \sum_{j=1}^{p} \beta_j \sigma_{t-j}^\delta$
Aug-GARCH[b]: $\sigma_t^2 = \begin{cases} \lvert\delta\phi_t - \delta + 1\rvert^{1/\delta} & \text{if } \delta \neq 0 \\ \exp(\phi_t - 1) & \text{if } \delta = 0 \end{cases}$ $\phi_t = \omega + \sum_{i=1}^{q} [\alpha_{1i} \lvert\varepsilon_{t-i} - \kappa\rvert^v + \alpha_{2i} \max(0, \kappa - \varepsilon_{t-i})^v] \phi_{t-j}$ $+ \sum_{i=1}^{q} [\alpha_{3i} f(\lvert\varepsilon_{t-i} - \kappa\rvert, v) + \alpha_{4i} f(\max(0, \kappa - \varepsilon_{t-i}), v)] \phi_{t-j}$ $+ \sum_{j=1}^{p} \beta_j \phi_{t-j}^2$

Source Hansen and Lunde (2005)

[a]This is the A-PARCH model without the leverage effect

[b]Here $f_{(x,v)} = (x^v - 1)/v$

10.9 Composite Forecasting

10.9.1 Composite Forecasting of Livestock Prices

Numerous approaches running from sophisticated multiple-equation regression techniques to rather naïve extrapolations or intuitive estimates are being utilized to produce forecasts. Bessler and Brandt (1979) examined three alternative procedures for forecasting time-dependent quarterly observations on hog, cattle, and broiler prices along with composite forecasts based on various linear combinations of these three procedures. The alternative methods for forecasting these prices are econometric models, time series (ARIMA), and expert opinion.

The results obtained by Bessler and Brandt for selected performance measures (mean squared error and turning points) applied to the forecasts of each method over the period 1976 quarter I through 1979 quarter II suggest that no method consistently outperformed or was outperformed by the other two methods. In terms of mean squared error, performance forecasts based on the ARIMA processes are lowest for hog and cattle prices, while the econometric model gives lowest mean squared error forecasts for broiler prices.

The mean forecast error is determined by taking the average of the difference between the summation of the overpredictions and the summation of the underpredictions. A negative sign would indicate that the average forecast series is above the mean of the actual series; a positive sign suggests an average forecast which is low. The mean absolute forecast error is simply the average of the absolute values of the forecast errors.

Composite forecasts based on the forecasts of the individual methods are formed using three procedures; minimum variance, adaptive weighting, and simple average composites. The empirical results from all three composite forecasting schemes generate performance levels that are at least as good as any of the individual forecasts and usually much better. In particular, the mean squared error of the best individual forecasting method is compared with that of the best composite for each of the three commodities. The composite forecast errors of the three commodities average 14% lower than the errors of the best individual forecasts.

The econometric model is essentially based on representations of the underlying economic behavioral system for a particular commodity. These representations attempt to identify and model the relevant supply-and-demand factors that together determine market price and quantity.

As an alternative to statistical models for forecasting, forecasts based upon expert opinions are available. These forecasts represent an accumulation of knowledge about the particular industry, commodity, or stock in question. In many respects, the forecasts of experts are like those of econometric- or ARIMA model forecasting in that they incorporate much of the same information from the same data sources. Expert opinions, however, are less restrictive or structures, in that the expert can change the weights assigned to different bits of information, or can select with relative ease the sources from which to draw the data. In addition, these expert forecasts are able to incorporate information that cannot, perhaps, be included in a more quantitative model in the form of data.

Recognizing that most forecasts contain some information that is not used in other forecasts, it seems possible that a combination of forecasts will quite often outperform any of the individual forecasts. Bessler and Brandt (1981) construct composite forecasts based upon composite weighting schemes. Bessler and Brandt use various tests or measures of performance to evaluate the price forecasts of econometric, ARIMA, expert opinion, and composite methods. Of the single-variable measures, they use the mean squared error, the mean forecast error, and the mean absolute forecast error. The mean squared error is a nonparametric statistic that provides a measure of the size of individual forecast errors from the actual values. Because the error is squared, large errors detract significantly from the performance of the method.

Performance indicators that track the movements of actual and forecast price series are called tracking measures. Examples of tracking measures are the number of turning points missed or falsely predicted compared with those correctly forecasted. Although these measures will not indicate which forecasting method most closely approximates the actual series, they are particularly useful when the forecaster is interested in knowing when a series is likely to turn upward or downward from its current pattern.

Bessler and Brandt's study does not find any specific forecasting method to be universally superior in terms of the performance measures. Although the ARIMA model performs best for two of three commodities, its performance is poorest for the third commodity in terms of the mean squared error criterion. The composite forecasting method's mean squared errors are lower than or nearly as low as the best of the individual methods. More important, in no case does a composite method of forecast generate errors that are as large as the worst of the individual methods.

The results of the performance evaluation suggest that forecasters should seriously consider using composite forecasting techniques. The idea that alternative forecasting methods use a variety of different information sources and means for assimilating the information and generating forecasts, a variety that can be captured by a composite forecast, is not only theoretically appealing but, in Bessler's and Brandt's study, somewhat empirically substantiated. Lee et al. (1986) use composite forecasting technique to forecast beta coefficient. It is found that composite forecasting performs better than both accounting beta forecasting and market beta forecasting in terms of mean squared error. Appendix 2 presents the composite forecasting method in detail.

10.9.2 Combined Forecasting of the Taiwan Weighted Stock Index

Ji et al. (2015, 2018) introduce various prediction models to forecast Taiwan Stock Exchange (TAIEX) index. Specifically, Ji et al. (2015, 2018) use intervention analysis integrated into the ARIMA-GARCH model, error correction model (ECM), intervention analysis integrated into the transfer function model, the simple average combination forecasting model, and the minimum error combination forecasting model to predict TAIEX index.

Empirical results show that, in terms of noncombination models, intervention analysis integrated into the transfer function model can obtain better predictive power than ECM and intervention analysis integrated into the ARIMA-GARCH model. Moreover, the minimum error combination forecasting model can improve prediction accuracy much better than noncombination models and also maintain robustness. Ji et al. (2015, 2018) find that the minimum error combination forecasting model has the best predictive power in terms of quantity index and the quality index.

10.10 Conclusion

In this chapter, we examined time-series component analysis and several methods of forecasting. The major components of a time series are the trend, cyclical, seasonal, and irregular components. To analyze these time-series components, we used the moving-average method to obtain seasonally adjusted time series. After investigating the analysis of time-series components, we discussed several forecasting models in detail. These forecasting models are linear time trend regression, simple exponential smoothing, the Holt–Winters forecasting model without seasonality, the Holt–Winters forecasting model with seasonality, and autoregressive forecasting.

Many factors determine the power of any forecasting model. They include the time horizon of the forecast, the stability of variance of data, and the presence of a trend, seasonal, or cyclical component. The model and method discussed in this chapter will be used in the next chapter to explore alternative time-series models.

Appendix 1: The Holt–Winters Forecasting Model for Seasonal Series

In this appendix, we will generalize the Holt–Winters forecasting model discussed in Sect. 10.5 to take into account the existence of seasonality. As in the nonseasonal case, we will use x_t, s_t, and T_t to denote, respectively, the observed value and the level and trend estimates at time t. F_t is used to denote the seasonal factor, so if the time series contains L periods per year, the seasonal factor for the corresponding period in the previous period will be F_{t-L}. The Holt–Winters method for seasonal series can be expressed by the following three equations:

$$s_t = \alpha\left(\frac{x_t}{F_{t-L}}\right) + (1-\alpha)(s_{t-1} + T_{t-1}) \quad (10.46)$$

$$T_t = \beta(s_t - s_{t-1}) + (1-\beta)T_{t-1} \quad (10.47)$$

$$F_t = \gamma\left(\frac{x_t}{s_t}\right) + (1-\gamma)F_{t-L}. \quad (10.48)$$

where α, β, and γ are smoothing constants whose values are set between 0 and 1.

In Eq. (10.46), the term $s_{t-1} + T_{t-1}$ represents an estimate of the level at time t, formed 1 time period earlier. This estimate is updated when the new observation x_t becomes available. However, here it is necessary to remove the influence of seasonality from that observation by deflating it by the latest available estimate, F_{t-L}, of the seasonal factor for that period. The updating equation for trend, Eq. (10.47), is identical to that used previously, Eq. (10.19b) in the text.

Finally, the seasonal factor is estimated by Eq. (10.48). The most recent estimate of the factor, available from the previous year, is F_{t-L}. However, dividing the new observation x_t by the level estimate s_t suggests a seasonal factor x_t/s_t. The new estimate of the seasonal factor is then a weighted average of these two quantities.

The procedure for forecasting via the Holt–Winters forecasting model for seasonal series is similar to that for nonseasonal series. Here the forecast for a particular month includes the effect of all three smoothing equations. The forecast for m periods ahead is

$$\hat{x}_{t+m} = (s_t + mT_t)(F_{t+m-L}) \quad (10.49)$$

If no seasonality exists—that is, if $F_{t+m-L} = 1$—then this equation reduces to Eq. (10.20) in the text.

We will use quarterly data listed in Table 10.4 in the text for Johnson & Johnson (J&J) during the period first quarter 2000 through fourth quarter 2010 to demonstrate how Eqs. (10.46), (10.47), (10.48), and (10.49) are used to do exponential smoothing and forecasting.

Example 10.7 (*The Holt–Winters Forecasting Model for J&J's Quarterly EPS*)

Table 10.4 and Fig. 10.6 in the text make it clear that Johnson & Johnson's quarterly EPS in the period 2000–2010 exhibited significant seasonality. The fourth-quarter EPS especially appeared to be considerably higher than those for the other three quarters.

The Holt–Winters forecasting model with seasonality is used to determine the smoothed value, s_t, and the predicted value, $\hat{x}$, for each time period. The smoothing constants are $\alpha = 0.2$, $\beta = 0.3$, and $\gamma = 0.3$.

First, we use the first three years of data to determine the seasonal indexes. Working with Eq. (10.9), we present the percentage of moving average (PMA) in terms of the first three years' data in column (4) of Table 10.12. Table 10.13 shows the procedure for calculating the seasonal index in terms of the first three years' data. These indexes are

Quarter 1 = 0.503 Quarter 2 = 0.751

Quarter 3 = 1.207 Quarter 4 = 1.539

and these are the four values of F_t in 1999.

The data from the first three years were seasonally adjusted to obtain d_t; see column (6) of Table 10.12. Drawing a least squares line through these 12 values by means of simple time trend linear regression produces

Table 10.12 Seasonal index and seasonally adjusted EPS for J&J in terms of the first 12 quarters' data

(1) Date	(2) x_t	(3) z_t^*	(4) x_t/z^*	(5) Seasonal index	(6) Seasonally adjusted EPS, d_t
2000					
1	0.86			0.503173	1.709154
2	1.8			0.750721	2.397696
3	2.68	2.1775	1.230769	1.207303	2.219824
4	3.3	2.095	1.575179	1.538804	2.144523
2001					
1	1	1.84875	0.540906	0.503173	1.987389
2	1	1.52375	0.656276	0.750721	1.332053
3	1.51	1.295	1.166023	1.207303	1.250722
4	1.87	1.26375	1.479723	1.538804	1.21523
2002					
1	0.6	1.31	0.458015	0.503173	1.192433
2	1.15	1.37875	0.834089	0.750721	1.531861
3	1.73			1.207303	1.432946
4	2.2			1.538804	1.429682

Table 10.13 Calculation of seasonal indexes of EPS for J&J

	Quarter				
Year	1	2	3	4	Sums
2000			1.231	1.575	
2001	0.541	0.656	1.166	1.480	
2002	0.458	0.834			
Median	0.499	0.745	1.198	1.527	3.970
Seasonal index	0.503	0.751	1.207	1.539	4.000

$$\hat{d}_t = 2.192966 - 0.08298t$$

The value $\hat{b} = 0.08298$ becomes the initial trend estimate of T_0. Finally, the initial smoothed value for fourth quarter 1999 is

$$\begin{aligned} s_0 &= [a + b(0)](\text{initial seasonal} \\ &\quad \text{index for fourth quarter}) \\ &= (2.192996)(1.539) = 3.374543 \end{aligned}$$

This estimate of s_0 becomes the forecast value for each of the quarters in 2000, as indicated in column (6) of Table 10.14.

The calculation of Table 10.14 in terms of $t = 10$ is shown as follows:

1. $x_{10} = 1.15$
2. Substituting related information into Eq. (10.46) yields

$$\begin{aligned} s_{10} &= 0.2\left(\frac{x_{10}}{F_{10-4}}\right) + 0.8(s_9 + T_9) \\ &= 0.2\left(\frac{x_{10}}{F_6}\right) + 0.8(s_9 + T_9) \\ &= 0.2\left(\frac{1.15}{0.678411}\right) + 0.8(1.042416 - 0.22779) \\ &= 0.99073 \end{aligned}$$

Table 10.14 Solution using Holt–Winters model with seasonality ($\alpha = 0.2$, $\beta = 0.3$, $\gamma = 0.3$)

t	x_t	T_t	F_t	s_t	$\hat{x}_t$	$x_t - \hat{x}_t$
			0.503173			
			0.750721			
			1.207303			
		−0.08298	1.538804	3.374543		
1	0.86	−0.17792	0.438941	2.975085	1.656227	−0.79623
2	1.8	−0.20189	0.724233	2.717271	2.09989	−0.29989
3	2.68	−0.21962	1.172438	2.456271	3.03683	−0.35683
4	3.3	−0.22515	1.523465	2.218224	3.441764	−0.14176
5	1	−0.20804	0.453593	2.050102	0.874843	0.125157
6	1	−0.23572	0.678411	1.749802	1.334081	−0.33408
7	1.51	−0.24929	1.129111	1.46885	1.775169	−0.26517
8	1.87	−0.24881	1.525832	1.221142	1.857959	0.012041
9	0.6	−0.22779	0.490191	1.042416	0.441041	0.158959
10	1.15	−0.17496	0.823116	0.99073	0.552653	0.597347
11	1.73	−0.13197	1.331536	0.959054	0.921098	0.808902
12	2.2	−0.09509	1.762796	0.950032	1.261986	0.938014
13	0.7	−0.0607	0.559727	0.969558	0.419086	0.280914
14	1.09	−0.03578	0.905841	0.991931	0.748093	0.341907
15	1.8	−0.01204	1.453671	1.035285	1.273149	0.526851
16	2.42	0.008934	1.898087	1.09316	1.803772	0.616228
17	0.84	0.032852	0.605039	1.181821	0.616872	0.223128
18	1.67	0.070587	1.007842	1.340457	1.100301	0.569699
19	2.46	0.087461	1.520538	1.467289	2.051194	0.408806
20	2.87	0.084899	1.885506	1.54621	2.951051	−0.08105
21	0.96	0.082233	0.601062	1.622222	0.986885	−0.02688
22	1.83	0.088911	1.023434	1.726716	1.717821	0.112179
23	2.68	0.085726	1.509804	1.805008	2.760729	−0.08073
24	3.38	0.079839	1.861778	1.871112	3.564991	−0.18499
25	1.11	0.073586	0.593272	1.930107	1.172642	−0.06264
26	2.07	0.07472	1.025748	2.007475	2.050647	0.019353
27	3.01	0.069407	1.49426	2.064483	3.143705	−0.13371
28	3.76	0.062548	1.837582	2.111027	3.97283	−0.21283
29	0.89	0.022143	0.546244	2.038891	1.289522	−0.39952
30	1.95	0.012544	1.006337	2.029037	2.1141	−0.1641
31	2.84	0.004085	1.46915	2.013386	3.050653	−0.21065
32	3.67	0.002868	1.833139	2.013415	3.707269	−0.03727
33	1.27	0.021389	0.565719	2.07802	1.101383	0.168617
34	2.45	0.041499	1.043702	2.166442	2.112714	0.337286
35	3.64	0.05768	1.51119	2.261878	3.243796	0.396204

(continued)

Table 10.14 (continued)

t	x_t	T_t	F_t	s_t	$\hat{x}_t$	$x_t - \hat{x}_t$
36	4.62	0.069723	1.870561	2.359699	4.252072	0.367928
37	1.27	0.058653	0.555249	2.392524	1.374369	−0.10437
38	2.43	0.051278	1.031013	2.426592	2.558299	−0.1283
39	3.64	0.047127	1.501008	2.464035	3.744531	−0.10453
40	4.45	0.039196	1.846676	2.484723	4.697282	−0.24728
41	1.64	0.064978	0.57719	2.609861	1.401404	0.238596
42	2.89	0.072672	1.042762	2.700485	2.757794	0.132206
43	4.14	0.071771	1.499056	2.770155	4.162532	−0.02253
44	4.85	0.058836	1.812537	2.798809	5.248116	−0.39812

3. Substituting related information into Eq. (10.47) yields

$$\begin{aligned} T_{10} &= 0.3(s_{10} - s_9) + 0.7T_9 \\ &= 0.3(0.99073 - 1.042416) + 0.7(-0.22779) \\ &= -0.17496 \end{aligned}$$

4. Substituting related information into Eq. (10.48) yields

$$\begin{aligned} F_{10} &= 0.3\left(\frac{x_{10}}{s_{10}}\right) + 0.7F_6 \\ &= 0.3\left(\frac{1.15}{0.99073}\right) + 0.7(0.678411) \\ &= 0.823116 \end{aligned}$$

Similarly, we can calculate all other values of s_t, T_t, and F_t, which are listed in columns (5), (3), and (4), respectively. Figure 10.17 presents actual data and smoothed data s_t.

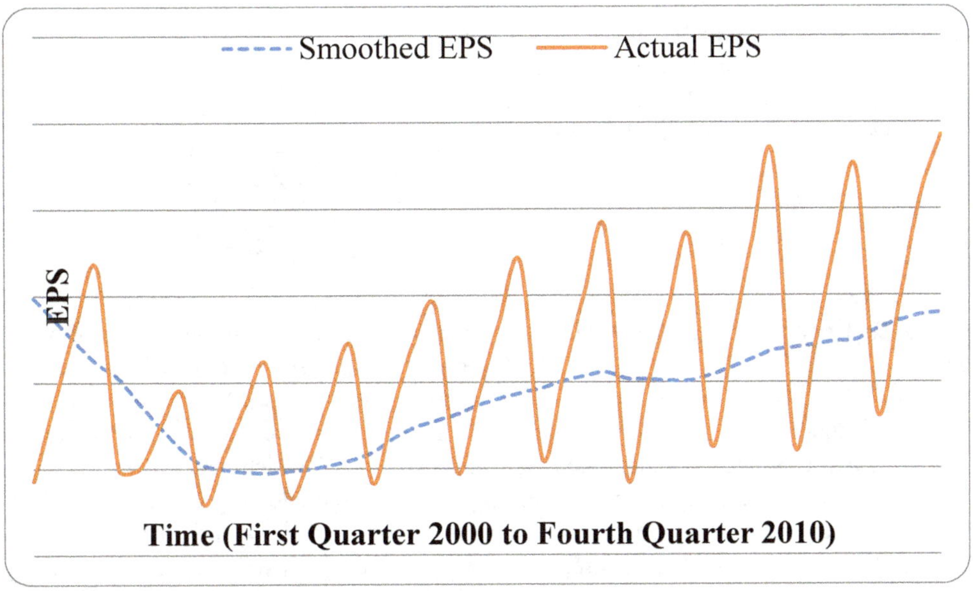

Fig. 10.17 Quarterly earnings per share of J&J (actual and smoothed EPS)

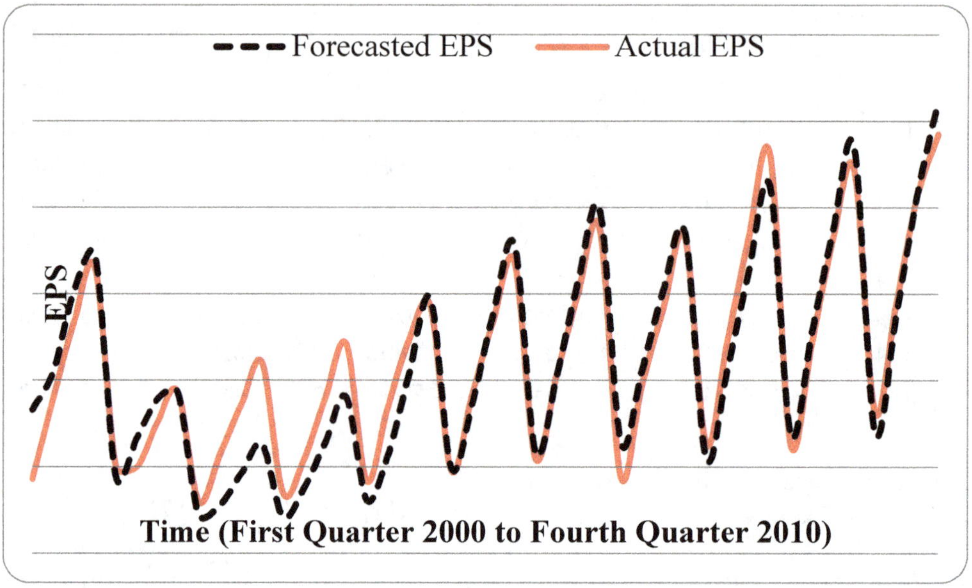

Fig. 10.18 Quarterly earnings per share of J&J (actual and forecasted EPS)

Using Eq. (10.49), we estimate $\hat{x}_{t+1}$ (t = 5, 6, …, 44); it is shown in column (6) of Table 10.14. For example,

$$\hat{x}_{11} = (s_{10} + T_{10})(F_7) = (0.99073 - 0.17496)(0.678411) = 0.921098$$

Figure 10.18 presents actual data and forecasted data ($\hat{x}_t$). If we let $m \geq 1$, then we can forecast future observations. For example, to forecast the EPS of J&J in the third quarter of 2002, we let m = 1. Finally, in the last column of Table 10.14, we present the residual in period t, $(x_t - \hat{x}_t)$.

Appendix 2: Composite Forecasting Method

Most forecasts contain some information that is independent of that contained in other forecasts; thus, a combination of the forecasts will, quite often, outperform any of the individual forecasts.

Nelson (1973) has shown that a composite forecast of unbiased forecasts is unbiased. For n individual unbiased forecasts $X_i (i = 1, 2, \ldots, n)$ with n weights a_i, each greater than or equal to zero and all weights summing to one, and a composite forecast X, the value of X is then given as

$$X = \sum_{i=1}^{n} a_i X_i \qquad \sum_{i=1}^{n} a_i = 1,\ a_i \geq 0.$$

The expected value of X is

$$E(X) = E\left(\sum_{i=1}^{n} a_i X_i\right) = \sum_{i=1}^{n} a_i E(X_i) = \sum_{i=1}^{n} a_i(\mu_x) = \mu_x$$

in which μ_x is the expected value of X_i. Therefore, the expected value of combination of n unbiased forecasts is itself unbiased.

If, however, a combination of n forecasts is formed, m of which are biased, the result is generally a biased composite forecast. By letting the expected value of the ith biased forecast be represented as $E(X_i) = \mu_x + \epsilon_i$, the composite bias can be represented as follows:

$$
\begin{aligned}
E(X) &= \sum_{i=1}^{n} a_i E(X_i) \\
&= \sum_{i=1}^{m} a_i(\mu + \in_i) + \sum_{i=m+1}^{n} a_i(\mu_x) \\
&= \mu_x + \sum_{i=1}^{m} a_i \in_i
\end{aligned}
$$

The composite of m-biased forecasts has a bias given by a combination of the individual forecast biases. This suggests that the composite of a biased forecast can be unbiased only if $\sum_{i=1}^{m} a_i \in_i = 0$. In particular, combining two forecasts, one with a positive bias and one with a negative bias, can, for proper choices of weights, result in an unbiased composite. However, for biased forecasts that do not balance each other, and assuming the assignment of zero weights to biased forecasts is not desired, numerous combinations of weights can be selected, each of which gives a composite that is unbiased.

The choice of weights can follow numerous approaches. These range from the somewhat naïve rule of thumb to more involved additive rules. One rule of thumb is that when several alternative forecasts are available but a history of performance on each is not, the user can combine all forecasts by finding their simple average.

Some additive rules may combine the econometric and ARIMA forecasts into a linear composite prediction of the form:

$$A_t = B_1(\text{Econometric})_t + B_2(\text{ARIMA})_t + \in_t, \tag{10.50}$$

where

A_t = actual value for period t;
B_1 and B_2 = fixed coefficients; and
$\in_t$ = composite prediction error.

Least squares fitting of (10.50) requires minimization of the sum of errors over values of B_1 and B_2 and, therefore, provides the minimum mean square-error linear composite prediction for the sample period. In the case that both the econometric model and ARIMA predictions are individually unbiased, then (10.50) can be rewritten as

$$A_t = B(\text{Econometric}) + (1 - B)(\text{ARIMA})_t + \in_t . \tag{10.51}$$

The least squares estimate of B in (10.51) is then given by

$$\hat{B} = \frac{\sum_{t=1}^{N} [(\text{ECM})_t - (\text{ARIMA})_t][A_t - (\text{ARIMA})_t]}{\sum_{t=1}^{N} [(\text{ECM})_t - (\text{ARIMA})_t]^2} \tag{10.52}$$

in which $(\text{ECM})_t$ and $(\text{ARIMA})_t$ represent forecasted values from econometric model and ARIMA model, respectively. Equation (10.52) is seen to be the coefficient of the regression of ARIMA prediction errors $[A_t - (\text{ARIMA})_t]$ on the difference between the two predictions. As would seem quite reasonable, the greater the ability of the difference between the two predictions to account for error committed by $(\text{ARIMA})_t$, the larger will be the weight given to $(\text{Econometric})_t$.

Composite predictions may be viewed as portfolios of predictions. If the econometric model's and ARIMA's errors are denoted by u_{1t}, and u_{2t}, respectively, then from (10.51) the composite prediction error is seen to be

$$\in_t = B(u_{1t}) + (1 - B)(u_{2t}). \tag{10.53}$$

The composite error is the weighted average of individual errors. The objective is to minimize the variance of the weighted average, given its expected value. In the case of prediction portfolios, the weighted average always has expectation zero if individual predictions are unbiased; or it may be given expectation zero by addition of an appropriate constant.

Minimizing composite error variance over a finite sample of observations leads to the estimate of B given by

$$\hat{B} = \frac{s_2^2 - s_{12}}{s_1^2 + s_2^2 - 2s_{12}}, \tag{10.54}$$

where s_1^2, s_2^2, and s_{12} are the sample variance of u_{1t}, the sample variance of u_{2t}, and the sample covariance of u_{1t} and u_{2t}, respectively. For large samples, or in the case that the variances $\mathrm{Var}(u_{1t})$ and $\mathrm{Var}(u_{2t})$ and the covariance $\mathrm{Cov}(u_{1t}u_{2t})$ are known, Eq. (10.54) becomes

$$B = \frac{\mathrm{Var}(u_{2t}) - \mathrm{Cov}(u_{1t}, u_{2t})}{\mathrm{Var}(u_{1t}) + \mathrm{Var}(u_{2t}) - 2\mathrm{Cov}(u_{1t}, u_{2t})}. \tag{10.55}$$

The minimum-variance weight is seen to depend on the covariance between individual errors as well as on their respective variance. Holding the covariance constant, the larger the variance of the ARIMA error relative to that of the econometric error, the larger the weight given to the econometric prediction.

Bibliography

Bessler, D. A., & Brandt, J. A. (1979). Composite forecasting of livestock prices: Analysis on combining alternative forecasting methods. *Purdue University Station Bulletin No. 265*.

Brandt, J., & Bessle, D. A. (1981). Composite forecasting: An application with U.S. hog prices. *American Journal of Agricultural Economics, 63*(1), 135–140.

Duan, J. (1997). Augmented GARCH (p, q) process and its diffusion limit. *Journal of Econometrics, 79,* 97–127.

Engle, R. F. (1982). Autoregressive conditional heteroscedasticity with estimates of the variance of United Kingdom inflations. *Econometrica, 50,* 987–1008.

Engle, R. F., & Ng, V. (1993). Measuring and testing the impact of news on volatility. *Journal of Finance, 48,* 1747–1778.

Greene, W. H. (2017). *Econometric analysis* (8th ed.). New Jersey: Prentice Hall.

Hansen, P. R., & Lunde, A. (2005). A forecast comparison of volatility models: Does anything beat a GARCH (1, 1)? *Journal of Applied Econometrics, 20,* 873–889.

Hentshel, L. (1995). All in the family: Nesting symmetric and asymmetric GARCH models. *Journal of Financial Economics, 39,* 71–104.

Ji, D. Y., Chen, H. Y., & Lee, C. F. (2015). Forecast performance of the Taiwan weighted stock index. *Review of Pacific Basin Financial Markets and Policies, 18,* 1550017-1–1550017-16.

Ji, D. Y., Chen, H. Y., & Lee, C. F. (2018). Forecast performance of the Taiwan weighted stock index: Update and expansion. In C.-F. Lee & J. Lee (Eds.), *Handbook of financial econometrics, mathematics, statistics, and technology*. World Scientific.

Lee, C.-F., Newbold, P., Finnerty, J. E., & Chu, C. C. (1986). On accounting-based, market-based and composite-based beta predictions: Methods and implications. *The Financial Review, 21*, 51–68.

Lee, C.-F., Lee, J. C., & Lee, A. C. (2013a). *Statistics for business and financial economics* (3rd ed.). New York: Springer.

Lee, C.-F., Finnerty, J., Lee, J., Lee, A. C., & Wort, D. (2013b). *Security analysis, portfolio management, and financial derivatives*. Singapore: World Scientific.

Lee, A. C., Lee, J. C., & Lee, C. F. (2017). *Financial analysis, planning and forecasting: Theory and application* (3rd ed.). Singapore: World Scientific.

Loudon, G. F., Watt, W. H., & Yadav, P. K. (2000). An empirical analysis of alternative parametric ARCH models. *Journal of Applied Econometrics, 15,* 117–136.

Nazem, S. M. (1988). Applied time series analysis for business and economic forecasting. Marcel Dekker, New York.

Nelson, C. R. (1973). Applied time series analysis for managerial forecasting. Holden-Day, New York.

Hedge Ratio and Time-Series Analysis

11

Contents

This chapter is an update and expansion of Chap. 32 of the book entitled Essentials of Excel, Excel VBA, SAS and Minitab for Statistical and Financial Analyses by Lee et al. (2016).

C.-F. Lee et al., *Financial Econometrics, Mathematics and Statistics*, https://doi.org/10.1007/978-1-4939-9429-8_11

Abstract

In this chapter, we theoretically develop alternative hedge ratio models. We then use one of the hedge ratio models and S&P index futures data to show how alternative time-series models can be used to estimate hedge ratio. Time-series models include OLS regression, ARCH model, GARCH model, etc.

11.1 Introduction

In this chapter, we first review the derivation and estimation of hedge ratio. We then apply various time-series models associated with conditional heteroscedasticity to estimate the hedge ratio. One of the best uses of derivative securities such as futures contracts is in hedging. Therefore, both academicians and practitioners have shown great interest in the issue of hedging with futures.

One of the main theoretical issues in hedging involves the determination of the optimal hedge ratio. However, the optimal hedge ratio depends on the particular objective function to be optimized. Many different objective functions are currently being used. For example, one of the most widely used hedging strategies is based on the minimization of the variance of the hedged portfolio (e.g., see Johnson 1960; Ederington 1979; Myers and Thompson 1989). This so-called minimum-variance (MV) hedge ratio is simple to understand and estimate. However, the MV hedge ratio completely ignores the expected return of the hedged portfolio. Therefore, this strategy is in general inconsistent with the mean-variance framework unless the individuals are infinitely risk-averse or the future price follows a pure martingale process (i.e., expected futures price change is zero).

Other strategies that incorporate both the expected return and risk (variance) of the hedged portfolio have been recently proposed (e.g., see Howard and D'Antonio 1984; Cecchetti et al. 1988; Hsin et al. 1994). These strategies are consistent with the mean-variance framework. However, it can be shown that if the futures price follows a pure martingale process, then the optimal mean-variance hedge ratio will be the same as the MV hedge ratio.

Another aspect of the mean-variance based strategies is that even though they are an improvement over the MV strategy, for them to be consistent with the expected utility maximization principle, either the utility function needs to be quadratic or the returns should be jointly normal. If neither of these assumptions is valid, then the hedge ratio may not be optimal with respect to the expected utility maximization principle. Some researchers have solved this problem by deriving the optimal hedge ratio based on the maximization of the expected utility (e.g., see Cecchetti et al. 1988; Lence 1995, 1996). However, this approach requires the use of specific utility function and specific return distribution.

Attempts have been made to eliminate these specific assumptions regarding the utility function and return distributions. Some of them involve the minimization of the mean extended-Gini (MEG) coefficient, which is consistent with the concept of stochastic dominance (e.g., see Cheung et al. 1990; Kolb and Okunev 1992, 1993; Lien and Luo 1993a; Shalit 1995; Lien and Shaffer 1999). Shalit (1995) shows that if the prices are normally distributed, then the

MEG-based hedge ratio will be the same as the MV hedge ratio.

Recently, hedge ratios based on the generalized semivariance (GSV) or lower partial moments have been proposed (e.g., see De Jong et al. 1997; Lien and Tse 1998, 2000; Chen et al. 2001). These hedge ratios are also consistent with the concept of stochastic dominance. Furthermore, these GSV-based hedge ratios have another attractive feature whereby they measure portfolio risk by the GSV, which is consistent with the risk perceived by managers, because of its emphasis on the returns below the target return (see Crum et al. 1981; Lien and Tse 2000). Lien and Tse (1998) show that if the futures and spot returns are jointly normally distributed and if the futures price follows a pure martingale process, then the minimum-GSV hedge ratio will be equal to the MV hedge ratio. Finally, Hung et al. (2006) has proposed a related hedge ratio that minimizes the value at risk associated with the hedged portfolio when choosing hedge ratio. This hedge ratio will also be equal to MV hedge ratio if the futures price follows a pure martingale process.

Most of the studies mentioned above [except Lence (1995, 1996)] ignore transaction costs as well as investments in other securities. Lence (1995, 1996) derives the optimal hedge ratio where transaction costs and investments in other securities are incorporated in the model. Using a CARA utility function, Lence finds that under certain circumstances the optimal hedge ratio is zero; i.e., the optimal hedging strategy is not to hedge at all.

In addition to the use of different objective functions in the derivation of the optimal hedge ratio, previous studies also differ in terms of the dynamic nature of the hedge ratio. For example, some studies assume that the hedge ratio is constant over time. Consequently, these static hedge ratios are estimated using unconditional probability distributions (e.g., see Ederington 1979; Howard and D'Antonio 1984; Benet 1992; Kolb and Okunev 1992, 1993; Ghosh 1993). On the other hand, several studies allow the hedge ratio to change over time. In some cases, these dynamic hedge ratios are estimated using conditional distributions associated with models such as ARCH (autoregressive conditional heteroscedasticity) and GARCH (generalized autoregressive conditional heteroscedasticity) (e.g., see Cecchetti et al. 1988; Baillie and Myers 1991; Kroner and Sultan 1993; Sephton 1993a). The GARCH based method has recently been extended by Lee and Yoder (2007) where regime-switching model is used. Alternatively, the hedge ratios can be made dynamic by considering a multiperiod model where the hedge ratios are allowed to vary for different periods. This is the method used by Lien and Luo (1993b).

When it comes to estimating the hedge ratios, many different techniques are currently being employed, ranging from simple to complex ones. For example, some of them use such a simple method as the ordinary least squares (OLS) technique (e.g., see Ederington 1979; Malliaris and Urrutia 1991; Benet 1992). However, others use more complex methods such as the conditional heteroscedastic (ARCH or GARCH) method (e.g., see Cecchetti et al. 1988; Baillie and Myers 1991; Sephton 1993a), the random coefficient method (e.g., see Grammatikos and Saunders 1983), the cointegration method (e.g., see Ghosh 1993; Lien and Luo 1993b; Chou, Fan, and Lee 1996), or the cointegration-heteroscedastic method (e.g., see Kroner and Sultan 1993). Recently, Lien and Shrestha (2007) has suggested the use of wavelet analysis to match the data frequency with the hedging horizon. Finally, Lien and Shresths (2010) also suggests the use of multivariate skew-normal distribution in estimating the minimum-variance hedge ratio.

It is quite clear that there are several different ways of deriving and estimating hedge ratios. In the chapter we review these different techniques and approaches and examine their relations.

The chapter is divided into six sections. In Sect. 11.2 alternative theories for deriving the optimal hedge ratios are reviewed. Various estimation methods are discussed in Sect. 11.3. Section 11.4 presents a discussion on the relationship among lengths of hedging horizon, maturity of futures contract, data frequency, and hedging effectiveness. In Sect. 11.5, we apply

SAS program to estimate hedge ratio by ARCH, GARCH, EGARCH, GJR-GARCH, and TGARCH models. Finally, in Sect. 11.6 we provide a summary and conclusion. In addition, Appendix 1 discusses theoretical models of the time-series analysis. Appendix 2 summarizes empirical models of the time-series analysis. Appendix 3 provides monthly data of S&P 500 index and its futures.

11.2 Alternative Theories for Deriving the Optimal Hedge Ratio

The basic concept of hedging is to combine investments in the spot market and futures market to form a portfolio that will eliminate (or reduce) fluctuations in its value. Specifically, consider a portfolio consisting of C_s units of a long spot position and C_f units of a short futures position.[1] Let S_t and F_t denote the spot and futures prices at time t, respectively. Since the futures contracts are used to reduce the fluctuations in spot positions, the resulting portfolio is known as the hedged portfolio. The return on the hedged portfolio, R_h, is given by:

$$R_h = \frac{C_s S_t R_s - C_f F_t R_f}{C_s S_t} = R_s - hR_f, \quad (11.1a)$$

where $h = \frac{C_f F_t}{C_s S_t}$ is the so-called hedge ratio, and $R_s = \frac{S_{t+1} - S_t}{S_t}$ and $R_f = \frac{F_{t+1} - F_t}{F_t}$ are so-called one-period returns on the spot and futures positions, respectively. Sometimes, the hedge ratio is discussed in terms of price changes (profits) instead of returns. In this case the profit on the hedged portfolio, ΔV_H, and the hedge ratio, H, are respectively given by:

$$\Delta V_H = C_s \Delta S_t - C_f \Delta F_t \quad \text{and} \quad H = \frac{C_f}{C_s}, \quad (11.1b)$$

where $\Delta S_t = S_{t+1} - S_t$ and $\Delta F_t = F_{t+1} - F_t$.

[1]Without loss of generality, we assume that the size of the futures contract is one.

The main objective of hedging is to choose the optimal hedge ratio (either h or H). As mentioned above, the optimal hedge ratio will depend on a particular objective function to be optimized. Furthermore, the hedge ratio can be static or dynamic. In Appendices 1 and 2, we will discuss the static hedge ratio and then the dynamic hedge ratio.

It is important to note that in the above setup, the cash position is assumed to be fixed and we only look for the optimum futures position. Most of the hedging literature assumes that the cash position is fixed, a setup that is suitable for financial futures. However, when we are dealing with commodity futures, the initial cash position becomes an important decision variable that is tied to the production decision. One such setup considered by Lence (1995, 1996) will be discussed in Appendix 3.

11.2.1 Static Case

We consider here that the hedge ratio is static if it remains the same over time. The static hedge ratios reviewed in this chapter can be divided into eight categories, as shown in Table 11.1. We will discuss each of them in the chapter.

11.2.1.1 Minimum-Variance Hedge Ratio

The most widely used static hedge ratio is the minimum-variance (MV) hedge ratio. Johnson (1960) derives this hedge ratio by minimizing the portfolio risk, where the risk is given by the variance of changes in the value of the hedged portfolio as follows:

$$\mathrm{Var}(\Delta V_H) = C_s^2 \mathrm{Var}(\Delta S) + C_f^2 \mathrm{Var}(\Delta F) - 2C_s C_f \mathrm{Cov}(\Delta S, \Delta F).$$

The MV hedge ratio, in this case, is given by:

$$H_J^* = \frac{C_f}{C_s} = \frac{\mathrm{Cov}(\Delta S, \Delta F)}{\mathrm{Var}(\Delta F)}. \quad (11.2a)$$

Alternatively, if we use definition (11.1a) and use $\mathrm{Var}(R_h)$ to represent the portfolio risk, then

Table 11.1 A list of different static hedge ratios

Hedge ratio	Objective function
• Minimum-variance (MV) hedge ratio	Minimize variance of R_h
• Optimum mean-variance hedge ratio	Maximize $E(R_h) - \frac{A}{2}\text{Var}(R_h)$
• Sharpe hedge ratio	Maximize $\frac{E(R_h)-R_F}{\sqrt{\text{Var}(R_h)}}$
• Maximum expected utility hedge ratio	Maximize $E[U(W_1)]$
• Minimum mean extended-Gini (MEG) coefficient hedge ratio	Minimize $\Gamma_v(R_h v)$
• Optimum mean-MEG hedge ratio	Maximize $E[R_h] - \Gamma_v(R_h v)$
• Minimum generalized semivariance (GSV) hedge ratio	Minimize $V_{\delta,\alpha}(R_h)$
• Maximum mean-GSV hedge ratio	Maximize $E[R_h] - V_{\delta,\alpha}(R_h)$
• Minimum VaR hedge ratio over a given time period τ	Minimize $Z_\alpha \sigma_h \sqrt{\tau} - E[R_h]\tau$

Notes

1. R_h = return on the hedged portfolio
   $E(R_h)$ = expected return on the hedged portfolio
   $\text{Var}(R_h)$ = variance of return on the hedged portfolio
   σ_h = standard deviation of return on the hedged portfolio
   Z_α = negative of left percentile at α for the standard normal distribution
   A = risk aversion parameter
   R_F = return on the risk-free security
   $E(U(W_1))$ = expected utility of end-of-period wealth
   $\Gamma_v(R_h v)$ = mean extended-Gini coefficient of R_h
   $V_{\delta,\alpha}(R_h)$ = generalized semivariance of R_h
2. With W_1 given by Eq. (11.17), the maximum expected utility hedge ratio includes the hedge ratio considered by Lence (1995, 1996)

the MV hedge ratio is obtained by minimizing $\text{Var}(R_h)$ which is given by:

$$\text{Var}(R_h) = \text{Var}(R_s) + h^2\text{Var}(R_f) - 2h\text{Cov}(R_s, R_f).$$

In this case, the MV hedge ratio is given by:

$$h_J^* = \frac{\text{Cov}(R_s, R_f)}{\text{Var}(R_f)} = \rho\frac{\sigma_s}{\sigma_f}, \qquad (11.2\text{b})$$

where ρ is the correlation coefficient between R_s and R_f, and σ_s and σ_f are standard deviations of R_s and R_f, respectively.

The attractive features of the MV hedge ratio are that it is easy to understand and simple to compute. However, in general the MV hedge ratio is not consistent with the mean-variance framework since it ignores the expected return on the hedged portfolio. For the MV hedge ratio to be consistent with the mean-variance framework, either the investors need to be infinitely risk-averse or the expected return on the futures contract needs to be zero.

11.2.1.2 Optimum Mean-Variance Hedge Ratio

Various studies have incorporated both risk and return in the derivation of the hedge ratio. For example, Hsin et al. (1994) derive the optimal hedge ratio that maximizes the following utility function:

$$\underset{C_f}{\text{Max}}\, V(E(R_h), \sigma; A) = E(R_h) - 0.5A\sigma_h^2, \qquad (11.3)$$

where A represents the risk aversion parameter. It is clear that this utility function incorporates both risk and return. Therefore, the hedge ratio based on this utility function would be consistent with the mean-variance framework. The optimal number of futures contract and the optimal hedge ratio are respectively given by:

$$h_2 = -\frac{C_f^* F}{C_s S} = -\left[\frac{E(R_f)}{A\sigma_f^2} - \rho\frac{\sigma_s}{\sigma_f}\right]. \qquad (11.4)$$

One problem associated with this type of hedge ratio is that in order to derive the optimum hedge ratio, we need to know the individual's risk aversion parameter. Furthermore, different individuals will choose different optimal hedge ratios, depending on the values of their risk aversion parameter.

Since the MV hedge ratio is easy to understand and simple to compute, it will be interesting and useful to know under what condition the above hedge ratio would be the same as the MV hedge ratio. It can be seen from Eqs. (11.2b) and (11.4) that if $A \to \infty$ or $E(R_f) = 0$, then h_2 would be equal to the MV hedge ratio h_J^*. The first condition is simply a restatement of the infinitely risk-averse individuals. However, the second condition does not impose any condition on the risk-averseness, and this is important. It implies that even if the individuals are not infinitely risk averse, then the MV hedge ratio would be the same as the optimal mean-variance hedge ratio if the expected return on the futures contract is zero (i.e., futures prices follow a simple martingale process). Therefore, if futures prices follow a simple martingale process, then we do not need to know the risk aversion parameter of the investor to find the optimal hedge ratio.

11.2.1.3 Sharpe Hedge Ratio

Another way of incorporating the portfolio return in the hedging strategy is to use the risk-return tradeoff (Sharpe measure) criteria. Howard and D'Antonio (1984) consider the optimal level of futures contracts by maximizing the ratio of the portfolio's excess return to its volatility:

$$\underset{C_f}{\text{Max}}\, \theta = \frac{E(R_h) - R_F}{\sigma_h}, \qquad (11.5)$$

where $\sigma_h^2 = \text{Var}(R_h)$ and R_F represents the risk-free interest rate. In this case the optimal number of futures positions, C_f^*, is given by:

$$C_f^* = -C_s \frac{\left(\frac{S}{F}\right)\left(\frac{\sigma_s}{\sigma_f}\right)\left[\frac{\sigma_s}{\sigma_f}\left(\frac{E(R_f)}{E(R_s)-R_F}\right) - \rho\right]}{\left[1 - \frac{\sigma_s}{\sigma_f}\left(\frac{E(R_f)\rho}{E(R_s)-R_F}\right)\right]}. \qquad (11.6)$$

From the optimal futures position, we can obtain the following optimal hedge ratio:

$$h_3 = -\frac{\left(\frac{\sigma_s}{\sigma_f}\right)\left[\frac{\sigma_s}{\sigma_f}\left(\frac{E(R_f)}{E(R_s)-R_F}\right) - \rho\right]}{\left[1 - \frac{\sigma_s}{\sigma_f}\left(\frac{E(R_f)\rho}{E(R_s)-R_F}\right)\right]}. \qquad (11.7)$$

Again, if $E(R_f) = 0$, then h_3 reduces to:

$$h_3 = \left(\frac{\sigma_s}{\sigma_f}\right)\rho, \qquad (11.8)$$

which is the same as the MV hedge ratio h_J^*.

As pointed out by Chen et al. (2001), the Sharpe ratio is a highly nonlinear function of the hedge ratio. Therefore, it is possible that Eq. (11.7), which is derived by equating the first derivative to zero, may lead to the hedge ratio that would minimize, instead of maximizing, the Sharpe ratio. This would be true if the second derivative of the Sharpe ratio with respect to the hedge ratio is positive instead of negative. Furthermore, it is possible that the optimal hedge ratio may be undefined as in the case encountered by Chen et al. (2001), where the Sharpe ratio monotonically increases with the hedge ratio.

11.2.1.4 Maximum Expected Utility Hedge Ratio

So far we have discussed the hedge ratios that incorporate only risk as well as the ones that incorporate both risk and return. The methods, which incorporate both the expected return and risk in the derivation of the optimal hedge ratio, are consistent with the mean-variance framework. However, these methods may not be consistent with the expected utility maximization principle unless either the utility function is quadratic or the returns are jointly normally distributed. Therefore, in order to make the hedge ratio consistent with the expected utility maximization principle, we need to derive the hedge ratio that maximizes the expected utility. However, in order to maximize the expected utility, we need to assume a specific utility function. For example, Cecchetti et al. (1988) derive the hedge ratio that maximizes the expected utility where the utility function is assumed to be the logarithm of terminal wealth. Specifically, they derive the optimal hedge ratio that maximizes the following expected utility function:

$$\int_{R_s}\int_{R_f} \log[1 + R_s - hR_f] f(R_s, R_f) dR_s dR_f,$$

where the density function $f(R_s, R_f)$ is assumed to be bivariate normal. A third-order linear bivariate ARCH model is used to get the conditional variance and covariance matrix, and a numerical procedure is used to maximize the objective function with respect to the hedge ratio.[2]

11.2.1.5 Minimum Mean Extended-Gini Coefficient Hedge Ratio

This approach of deriving the optimal hedge ratio is consistent with the concept of stochastic dominance and involves the use of the mean extended-Gini (MEG) coefficient. Cheung et al. (1990), Kolb and Okunev (1992), Lien and Luo (1993a), Shalit (1995), and Lien and Shaffer (1999) all consider this approach. It minimizes the MEG coefficient $\Gamma_v(R_h)$ defined as follows:

$$\Gamma_v(R_h) = -v\mathrm{Cov}\left(R_h, (1 - G(R_h))^{v-1}\right), \tag{11.9}$$

where G is the cumulative probability distribution and v is the risk aversion parameter. Note that $0 \le v < 1$ implies risk seekers, $v = 1$ implies risk-neutral investors, and $v > 1$ implies risk-averse investors. Shalit (1995) has shown that if the futures and spot returns are jointly normally distributed, then the minimum-MEG hedge ratio would be the same as the MV hedge ratio.

11.2.1.6 Optimum Mean-MEG Hedge Ratio

Instead of minimizing the MEG coefficient, Kolb and Okunev (1993) alternatively consider maximizing the utility function defined as follows:

$$U(R_h) = E(R_h) - \Gamma_v(R_h). \tag{11.10}$$

The hedge ratio based on the utility function defined by Eq. (11.10) is denoted as the M-MEG hedge ratio. The difference between the MEG and M-MEG hedge ratios is that the MEG hedge ratio ignores the expected return on the hedged portfolio. Again, if the futures price follows a martingale process (i.e., $E(R_f) = 0$), then the MEG hedge ratio would be the same as the M-MEG hedge ratio.

11.2.1.7 Minimum Generalized Semivariance Hedge Ratio

In recent years a new approach for determining the hedge ratio has been suggested (see De Jong et al. 1997; Lien and Tse 1998, 2000; Chen et al. 2001). This new approach is based on the relationship between the generalized semivariance (GSV) and expected utility as discussed by Fishburn (1977) and Bawa (1978). In this case the optimal hedge ratio is obtained by minimizing the GSV given below:

$$V_{\delta,\alpha}(R_h) = \int_{-\infty}^{\delta} (\delta - R_h)^{\alpha} dG(R_h), \quad \alpha > 0, \tag{11.11}$$

where $G(R_h)$ is the probability distribution function of the return on the hedged portfolio R_h. The parameters δ and α (which are both real numbers) represent the target return and risk aversion, respectively. The risk is defined in such a way that the investors consider only the returns below the target return (δ) to be risky. It can be shown (see Fishburn 1977) that $\alpha < 1$ represents a risk-seeking investor and $\alpha > 1$ represents a risk-averse investor.

The GSV, due to its emphasis on the returns below the target return, is consistent with the risk

[2] Lence (1995) also derives the hedge ratio based on the expected utility. We will discuss it later in Appendix 3.

perceived by managers (see Crum et al. 1981; Lien and Tse 2000). Furthermore, as shown by Fishburn (1977) and Bawa (1978), the GSV is consistent with the concept of stochastic dominance. Lien and Tse (1998) show that the GSV hedge ratio, which is obtained by minimizing the GSV, would be the same as the MV hedge ratio if the futures and spot returns are jointly normally distributed and if the futures price follows a pure martingale process.

11.2.1.8 Optimum Mean-Generalized Semivariance Hedge Ratio

Chen et al. (2001) extend the GSV hedge ratio to a Mean-GSV (M-GSV) hedge ratio by incorporating the mean return in the derivation of the optimal hedge ratio. The M-GSV hedge ratio is obtained by maximizing the following mean-risk utility function, which is similar to the conventional mean-variance based utility function (see Eq. 11.3):

$$U(R_{\mathrm{h}}) = E[R_{\mathrm{h}}] - V_{\delta,\alpha}(R_{\mathrm{h}}). \tag{11.12}$$

This approach to the hedge ratio does not use the risk aversion parameter to multiply the GSV as done in conventional mean-risk models (see Hsin et al. 1994, and Eq. 11.3). This is because the risk aversion parameter is already included in the definition of the GSV, $V_{\delta,\alpha}(R_{\mathrm{h}})$. As before, the M-GSV hedge ratio would be the same as the GSV hedge ratio if the futures price follows a pure martingale process.

11.2.1.9 Minimum Value-at-Risk Hedge Ratio

Hung et al. (2006) suggests a new hedge ratio that minimizes the value at risk of the hedged portfolio. Specifically, the hedge ratio h is derived by minimizing the following value at risk of the hedged portfolio over a given time period τ:

$$\mathrm{VaR}(R_{\mathrm{h}}) = Z_\alpha \sigma_{\mathrm{h}} \sqrt{\tau} - E[R_{\mathrm{h}}]\tau \tag{11.13}$$

The resulting optimal hedge ratio, which Hung et al. (2006) refer to as zero-VaR hedge ratio, is given by

$$h^{\mathrm{VaR}} = \rho \frac{\sigma_{\mathrm{s}}}{\sigma_{\mathrm{f}}} - E[R_{\mathrm{f}}] \frac{\sigma_{\mathrm{s}}}{\sigma_{\mathrm{f}}} \sqrt{\frac{1-\rho^2}{Z_\alpha^2 \sigma_{\mathrm{f}}^2 - E[R_{\mathrm{f}}]^2}} \tag{11.14}$$

It is clear that, if the futures price follows martingale process, the zero-VaR hedge ratio would be the same as the MV hedge ratio.

11.2.2 Dynamic Case

We have up to now examined the situations in which the hedge ratio is fixed at the optimum level and is not revised during the hedging period. However, it could be beneficial to change the hedge ratio over time. One way to allow the hedge ratio to change is by recalculating the hedge ratio based on the current (or conditional) information on the covariance (σ_{sf}) and variance (σ_{f}^2). This involves calculating the hedge ratio based on conditional information (i.e., $\sigma_{\mathrm{sf}}|\Omega_{t-1}$ and $\sigma_{\mathrm{f}}^2|\Omega_{t-1}$) instead of unconditional information. In this case, the MV hedge ratio is given by:

$$h_1|\Omega_{t-1} = -\frac{\sigma_{\mathrm{sf}}|\Omega_{t-1}}{\sigma_{\mathrm{f}}^2|\Omega_{t-1}}.$$

The adjustment to the hedge ratio based on new information can be implemented using such conditional models as ARCH and GARCH (to be discussed later) or using the moving window estimation method.

Another way of making the hedge ratio dynamic is by using the regime-switching GARCH model (to be discussed later) as suggested by Lee and Yoder (2007). This model assumes two different regimes where each regime is associated with different set of parameters and the probabilities of regime switching must also be estimated when implementing such method. Alternatively, we can allow the hedge ratio to change during the hedging period by considering multiperiod models, which is the approach used by Lien and Luo (1993b).

Lien and Luo (1993b) consider hedging with T periods' planning horizon and minimize the variance of the wealth at the end of the planning horizon, W_T. Consider the situation where $C_{s,t}$ is the spot position at the beginning of period t and the corresponding futures position is given by $C_{f,t} = -b_t C_{s,t}$. The wealth at the end of the planning horizon, W_T, is then given by:

$$\begin{aligned} W_T &= W_0 + \sum_{t=0}^{T-1} C_{s,t}[S_{t+1} - S_t - b_t(F_{t+1} - F_t)] \\ &= W_0 + \sum_{t=0}^{T-1} C_{s,t}[\Delta S_{t+1} - b_t \Delta F_{t+1}]. \end{aligned} \tag{11.15}$$

The optimal b_t's are given by the following recursive formula:

$$b_t = \frac{\mathrm{Cov}(\Delta S_{t+1}, \Delta F_{t+1})}{\mathrm{Var}(\Delta F_{t+1})} + \sum_{i=t+1}^{T-1} \left(\frac{C_{s,i}}{C_{s,t}}\right) \frac{\mathrm{Cov}(\Delta F_{t+1}, \Delta S_{i+1} + b_i \Delta F_{t+i})}{\mathrm{Var}(\Delta F_{t+1})}. \tag{11.16}$$

It is clear from Eq. (11.16) that the optimal hedge ratio b_t will change over time. The multiperiod hedge ratio will differ from the single-period hedge ratio due to the second term on the right-hand side of Eq. (11.16). However, it is interesting to note that the multiperiod hedge ratio would be different from the single-period one if the changes in current futures prices are correlated with the changes in future futures prices or with the changes in future spot prices.

11.2.3 Case with Production and Alternative Investment Opportunities

All the models considered in appendices 1 and 2 assume that the spot position is fixed or predetermined, and thus production is ignored. As mentioned earlier, such an assumption may be appropriate for financial futures. However, when we consider commodity futures, production should be considered in which case the spot position becomes one of the decision variables. In an important chapter, Lence (1995) extends the model with a fixed or predetermined spot position to a model where production is included. In his model Lence (1995) also incorporates the possibility of investing in a risk-free asset and other risky assets, borrowing, as well as transaction costs. We will briefly discuss the model considered by Lence (1995) below.

Lence (1995) considers a decision maker whose utility is a function of terminal wealth $U(W_1)$, such that $U' > 0$ and $U'' < 0$. At the decision date ($t = 0$), the decision maker will engage in the production of Q commodity units for sale at terminal date ($t = 1$) at the random cash price P_1. At the decision date, the decision maker can lend L dollars at the risk-free lending rate ($R_L - 1$) and borrow B dollars at the borrowing rate ($R_B - 1$), invest I dollars in a different activity that yields a random rate of return ($R_I - 1$) and sell X futures at futures price F_0. The transaction cost for the futures trade is f dollars per unit of the commodity traded to be paid at the terminal date. The terminal wealth (W_1) is therefore given by:

$$\begin{aligned} W_1 &= W_0 R \\ &= P_1 Q + (F_0 - F_1)X - f|X| \\ &\quad - R_B B + R_L L + R_I I, \end{aligned} \tag{11.17}$$

where R is the return on the diversified portfolio. The decision maker will maximize the expected utility subject to the following restrictions:

$$\begin{aligned} &W_0 + B \geq v(Q)Q + L + I, \\ &0 \leq B \leq k_B v(Q)Q, \quad k_B \geq 0, \\ &L \geq k_L F_0 |X|, \quad k_L \geq 0, I \geq 0 \end{aligned}$$

where $v(Q)$ is the average cost function, k_B is the maximum amount (expressed as a proportion of his initial wealth) that the agent can borrow, and k_L is the safety margin for the futures contract.

Using this framework, Lence (1995) introduces two opportunity costs: opportunity cost of alternative (suboptimal) investment (c_{alt}) and opportunity cost of estimation risk (e^{Bayes}).[3] Let R_{opt} be the return of the expected utility maximizing strategy and let R_{alt} be the return on a particular alternative (suboptimal) investment strategy. The opportunity cost of alternative investment strategy c_{alt} is then given by:

$$E\left[U(W_0 R_{\text{opt}})\right] = E[U(W_0 R_{\text{alt}} + c_{\text{alt}})]. \quad (11.18)$$

In other words, c_{alt} is the minimum certain net return required by the agent to invest in the alternative (suboptimal hedging) strategy rather than in the optimum strategy. Using the CARA utility function and some simulation results, Lence (1995) finds that the expected utility maximizing hedge ratios are substantially different from the minimum-variance hedge ratios. He also shows that under certain conditions, the optimal hedge ratio is zero; i.e., the optimal strategy is not to hedge at all.

Similarly, the opportunity cost of the estimation risk (e^{Bayes}) is defined as follows:

$$\begin{aligned} &E_\rho\left[E\left(U\left\{W_0\left[R_{\text{opt}}(\rho) - e_\rho^{\text{Bayes}}\right]\right\}\right)\right] \\ &\quad = E_\rho\left[E\left(U(W_0 R_{\text{opt}}^{\text{Bayes}})\right)\right], \end{aligned} \quad (11.19)$$

where $R_{\text{opt}}(\rho)$ is the expected utility maximizing return where the agent knows with certainty the value of the correlation between the futures and spot prices (ρ), $R_{\text{opt}}^{\text{Bayes}}$ is the expected utility maximizing return where the agent only knows the distribution of the correlation ρ, and $E_\rho[.]$ is the expectation with respect to ρ. Using simulation results, Lence (1995) finds that the opportunity cost of the estimation risk is negligible and thus the value of the use of sophisticated estimation methods is negligible.

[3]Our discussion of the opportunity costs is very brief. We would like to refer interested readers to Lence (1995) for a detailed discussion. We would also like to point to the fact that production can be allowed to be random as is done in Lence (1996).

11.3 Alternative Methods for Estimating the Optimal Hedge Ratio

In Sect. 11.2 we discussed different approaches to deriving the optimum hedge ratios. However, in order to apply these optimum hedge ratios in practice, we need to estimate these hedge ratios. There are various ways of estimating them. In this section we briefly discuss these estimation methods.

11.3.1 Estimation of the Minimum-Variance (MV) Hedge Ratio

11.3.1.1 OLS Method

The conventional approach to estimating the MV hedge ratio involves the regression of the changes in spot prices on the changes in futures price using the OLS technique (e.g., see Junkus and Lee 1985). Specifically, the regression equation can be written as:

$$\Delta S_t = a_0 + a_1 \Delta F_t + e_t, \quad (11.20)$$

where the estimate of the MV hedge ratio, H_j, is given by a_1. The OLS technique is quite robust and simple to use. However, for the OLS technique to be valid and efficient, assumptions associated with the OLS regression must be satisfied. One case where the assumptions are not completely satisfied is that the error term in the regression is heteroscedastic. This situation will be discussed later.

Another problem with the OLS method, as pointed out by Myers and Thompson (1989), is the fact that it uses unconditional sample moments instead of conditional sample moments, which use currently available information. They suggest the use of the conditional covariance and conditional variance in Eq. (11.21). In this case, the conditional version of the optimal hedge ratio (Eq. (11.21)) will take the following form:

$$H_J^* = \frac{C_f}{C_s} = \frac{\mathrm{Cov}(\Delta S, \Delta F)|\Omega_{t-1}}{\mathrm{Var}(\Delta F)|\Omega_{t-1}}. \qquad (11.21)$$

Suppose that the current information (Ω_{t-1}) includes a vector of variables (X_{t-1}) and the spot and futures price changes are generated by the following equilibrium model:

$$\Delta S_t = X_{t-1}\alpha + u_t,$$
$$\Delta F_t = X_{t-1}\beta + v_t.$$

In this case the maximum likelihood estimator of the MV hedge ratio is given by (see Myers and Thompson (1989)):

$$\hat{h}|X_{t-1} = \frac{\hat{\sigma}_{uv}}{\hat{\sigma}_v^2}, \qquad (11.22)$$

where $\hat{\sigma}_{uv}$ is the sample covariance between the residuals u_t and v_t, and $\hat{\sigma}_v^2$ is the sample variance of the residual v_t. In general, the OLS estimator obtained from Eq. (11.20) would be different from the one given by Eq. (11.22). For the two estimators to be the same, the spot and futures prices must be generated by the following model:

$$\Delta S_t = \alpha_0 + u_t, \quad \Delta F_t = \beta_0 + v_t.$$

In other words, if the spot and futures prices follow a random walk, then with or without drift, the two estimators will be the same. Otherwise, the hedge ratio estimated from the OLS regression (11.18) will not be optimal. We will show how SAS can be used to estimate the hedge ratio in terms of OLS method in Sect. 11.5.

An alternative way of estimating the MV hedge ratio involves the assumption that the spot price and futures price follows a multivariate skew-normal distribution as suggested by Lien and Shrestha (2010). The estimate of covariance matrix under skew-normal distribution can be different from the estimate of covariance matrix under the usual normal distribution resulting in different estimates of MV hedge ratio. Let Y be a k-dimensional random vector. Then Y is said to have skew-normal distribution if its probability density function is given as follows:

$$f_Y(y) = 2\phi_k(y, \Omega_Y)\Phi(\alpha^t y)$$

where α is a k-dimensional column vector, $\phi_k(y, \Omega_Y)$ is the probability density function of a k-dimensional standard normal random variable with zero mean and correlation matrix Ω_Y and $\Phi(\alpha^t y)$ is the probability distribution function of a one-dimensional standard normal random variable evaluated at $\alpha^t y$.

11.3.1.2 ARCH and GARCH Methods

Ever since the development of ARCH and GARCH models, the OLS method of estimating the hedge ratio has been generalized to take into account the heteroscedastic nature of the error term in Eq. (11.20). In this case, rather than using the unconditional sample variance and covariance, the conditional variance and covariance from the GARCH model are used in the estimation of the hedge ratio. As mentioned above, such a technique allows an update of the hedge ratio over the hedging period.

Consider the following bivariate GARCH model (see Cecchetti et al. 1988; Baillie and Myers 1991):

$$\begin{bmatrix} \Delta S_t \\ \Delta F_t \end{bmatrix} = \begin{bmatrix} \mu_1 \\ \mu_2 \end{bmatrix} + \begin{bmatrix} e_{1t} \\ e_{2t} \end{bmatrix} \Leftrightarrow \Delta Y_t = \mu + e_t,$$
$$e_t|\Omega_{t-1} \sim N(0, H_t), H_t = \begin{bmatrix} H_{11,t} & H_{12,t} \\ H_{12,t} & H_{22,t} \end{bmatrix},$$
$$\mathrm{vec}(H_t) = C + A\,\mathrm{vec}\left(e_{t-1}e_{t-1}'\right) + B\,\mathrm{vec}(H_{t-1}). \qquad (11.23)$$

The conditional MV hedge ratio at time t is given by $h_{t-1} = H_{12,t}/H_{22,t}$. This model allows the hedge ratio to change over time, resulting in a series of hedge ratios instead of a single hedge ratio for the entire hedging horizon. Equation (11.23) represents a GARCH model. This GARCH model will reduce to ARCH if B equal to zero.

The model can be extended to include more than one type of cash and futures contracts (see Sephton 1993a). For example, consider a portfolio that consists of spot wheat $(S_{1,t})$, spot canola (S_{2t}), wheat futures (F_{1t}) and canola futures (F_{2t}). We then have the following multivariate GARCH model:

$$\begin{bmatrix} \Delta S_{1t} \\ \Delta S_{2t} \\ \Delta F_{1t} \\ \Delta F_{2t} \end{bmatrix} = \begin{bmatrix} \mu_1 \\ \mu_2 \\ \mu_3 \\ \mu_4 \end{bmatrix} + \begin{bmatrix} e_{1t} \\ e_{2t} \\ e_{3t} \\ e_{4t} \end{bmatrix} \Leftrightarrow \Delta Y_t = \mu + e_t,$$
$$e_t|\Omega_{t-1} \sim N(0, H_t).$$

The MV hedge ratio can be estimated using a similar technique as described above. For example, the conditional MV hedge ratio is given by the conditional covariance between the spot and futures price changes divided by the conditional variance of the futures price change. We will show how SAS program can be used to estimate ratio in terms of ARCH and GARCH models in Sect. 11.5.

11.3.1.3 Regime-Switching GARCH Model

The GARCH model discussed above can be further extended by allowing regime switching as suggested by Lee and Yoder (2007). Under this model, the data generating process can be in one of two states or regime denoted by state variable $s_t = \{1, 2\}$, which is assumed to follow a first-order Markov process. The state transition probabilities are assumed to follow a logistic distribution where the transition probabilities are given by

$$\Pr(s_t = 1|s_{t-1} = 1) = \frac{e^{p_0}}{1 + e^{p_0}} \quad \text{and}$$
$$\Pr(s_t = 2|s_{t-1} = 2) = \frac{e^{q_0}}{1 + e^{q_0}}.$$

The conditional covariance matrix is given by

$$H_{t,s_t} = \begin{bmatrix} h_{1,t,s_t} & 0 \\ 0 & h_{2,t,s_t} \end{bmatrix} \begin{bmatrix} 1 & \rho_{t,s_t} \\ \rho_{t,s_t} & 1 \end{bmatrix} \begin{bmatrix} h_{1,t,s_t} & 0 \\ 0 & h_{2,t,s_t} \end{bmatrix}$$

where

$$h^2_{1,t,s_t} = \gamma_{1,s_t} + \alpha_{1,s_t} e^2_{1.t-1} + \beta_{1,s_t} h^2_{1,t-1}$$
$$h^2_{2,t,s_t} = \gamma_{2,s_t} + \alpha_{2,s_t} e^2_{2.t-1} + \beta_{2,s_t} h^2_{2,t-1}$$
$$\rho_{t,s_t} = (1 - \theta_{1,s_t} - \theta_{2,s_t})\rho + \theta_{1,s_t}\rho_{t-1} + \theta_{2,s_t}\phi_{t-1}$$
$$\phi_{t-1} = \frac{\sum_{j=1}^{2} \varepsilon_{1,t-j}\varepsilon_{2,t-j}}{\sqrt{\left(\sum_{j=1}^{2} \varepsilon^2_{1,t-j}\right)\left(\sum_{j=1}^{2} \varepsilon^2_{2,t-j}\right)}},$$
$$\varepsilon_{i,t} = \frac{e_{i,t}}{h_{it}}, \theta_1, \theta_2 \geq 0 \text{ and } \theta_1 + \theta_2 \leq 1$$

Once the conditional covariance matrix is estimated, the time-varying conditional MV hedge ratio is given by the ratio of the covariance between the spot and futures returns to the variance of the futures return.

11.3.1.4 Random Coefficient Method

There is another way to deal with heteroscedasticity. This involves use of the random coefficient model as suggested by Grammatikos and Saunders (1983). This model employs the following variation of Eq. (11.20):

$$\Delta S_t = \beta_0 + \beta_t \Delta F_t + e_t, \quad (11.24)$$

where the hedge ratio $\beta_t = \bar{\beta} + v_t$ is assumed to be random. This random coefficient model can, in some cases, improve the effectiveness of hedging strategy. However, this technique does not allow for the update of the hedge ratio over time even though the correction for the randomness can be made in the estimation of the hedge ratio.

11.3.1.5 Cointegration and Error Correction Method

The techniques described so far do not take into consideration the possibility that spot price and futures price series could be nonstationary. If these series have unit roots, then this will raise a different issue. If the two series are cointegrated as defined by Engle and Granger (1987), then the regression Eq. (11.20) will be misspecified and an error correction term must be included in the equation. Since the arbitrage condition ties the spot and futures prices, they cannot drift far apart in the long run. Therefore, if both series follow a random walk, then we expect the two series to be cointegrated in which case we need to estimate the error correction model. This calls for the use of the cointegration analysis.

The cointegration analysis involves two steps. First, each series must be tested for a unit root (e.g., see Dickey and Fuller 1981; Phillips and Perron 1988). Second, if both series are found to have a single unit root, then the cointegration test must be performed (e.g., see Engle and Granger 1987; Johansen and Juselius 1990; Osterwald-Lenum 1992).

If the spot price and futures price series are found to be cointegrated, then the hedge ratio can be estimated in two steps (see Ghosh 1993; Chou et al. 1996). The first step involves the estimation of the following cointegrating regression:

$$S_t = a + bF_t + u_t. \tag{11.25}$$

The second step involves the estimation of the following error correction model:

$$\Delta S_t = \rho u_{t-1} + \beta \Delta F_t + \sum_{i=1}^{m} \delta_i \Delta F_{t-i} + \sum_{j=1}^{n} \theta_i \Delta S_{t-j} + e_j, \tag{11.26}$$

where u_t is the residual series from the cointegrating regression. The estimate of the hedge ratio is given by the estimate of β. Some researchers (e.g., see Lien and Luo 1993b) assume that the long-run cointegrating relationship is $(S_t - F_t)$, and estimate the following error correction model:

$$\Delta S_t = \rho(S_{t-1} - F_{t-1}) + \beta \Delta F_t + \sum_{i=1}^{m} \delta_i \Delta F_{t-i} + \sum_{j=1}^{n} \theta_i \Delta S_{t-j} + e_j. \tag{11.27}$$

Alternatively, Chou et al. (1996) suggest the estimation of the error correction model as follows:

$$\Delta S_t = \alpha \hat{u}_{t-1} + \beta \Delta F_t + \sum_{i=1}^{m} \delta_i \Delta F_{t-i} + \sum_{j=1}^{n} \theta_i \Delta S_{t-j} + e_j, \tag{11.28}$$

where $\hat{u}_{t-1} = S_{t-1} - (a + bF_{t-1})$; i.e., the series $\hat{u}_t$ is the estimated residual series from Eq. (11.25). The hedge ratio is given by β in Eq. (11.27).

Kroner and Sultan (1993) combine the error correction model with the GARCH model considered by Cecchetti et al. (1988) and Baillie and Myers (1991) in order to estimate the optimum hedge ratio. Specifically, they use the following model:

$$\begin{bmatrix} \Delta \log_e(S_t) \\ \Delta \log_e(F_t) \end{bmatrix} = \begin{bmatrix} \mu_1 \\ \mu_2 \end{bmatrix} + \begin{bmatrix} \alpha_s(\log_e(S_{t-1}) - \log_e(F_{t-1})) \\ \alpha_f(\log_e(S_{t-1}) - \log_e(F_{t-1})) \end{bmatrix} + \begin{bmatrix} e_{1t} \\ e_{2t} \end{bmatrix}, \tag{11.29}$$

where the error processes follow a GARCH process. As before, the hedge ratio at time $(t-1)$ is given by $h_{t-1} = H_{12,t}/H_{22,t}$.

11.3.2 Estimation of the Optimum Mean-Variance and Sharpe Hedge Ratios

The optimum mean-variance and Sharpe hedge ratios are given by Eqs. (11.4) and (11.7) respectively. These hedge ratios can be estimated simply by replacing the theoretical moments by their sample moments. For example, the expected returns can be replaced by sample average returns, the standard deviations can be replaced by the sample standard deviations, and the correlation can be replaced by sample correlation.

11.3.3 Estimation of the Maximum Expected Utility Hedge Ratio

The maximum expected utility hedge ratio involves the maximization of the expected utility. This requires the estimation of distributions of the changes in spot and futures prices. Once the distributions are estimated, one needs to use a numerical technique to get the optimum hedge ratio. One such method is described in Cecchetti et al. (1988) where an ARCH model is used to estimate the required distributions.

11.3.4 Estimation of Mean Extended-Gini (MEG) Coefficient-Based Hedge Ratios

The MEG hedge ratio involves the minimization of the following MEG coefficient:

$$\Gamma_v(R_h) = -vCov\left(R_h, (1 - G(R_h))^{v-1}\right).$$

In order to estimate the MEG coefficient, we need to estimate the cumulative probability density function $G(R_h)$. The cumulative probability density function is usually estimated by ranking the observed return on the hedged portfolio. A detailed description of the process can be found in Kolb and Okunev (1992), and we briefly describe the process here.

The cumulative probability distribution is estimated by using the rank as follows:

$$G(R_{h,i}) = \frac{Rank(R_{h,i})}{N},$$

where N is the sample size. Once we have the series for the probability distribution function, the MEG is estimated by replacing the theoretical covariance by the sample covariance as follows:

$$\Gamma_v^{\text{sample}}(R_h) = -\frac{v}{N}\sum_{i=1}^{N}(R_{h,i} - \bar{R}_h)\left((1 - G(R_{h,i}))^{v-1} - \Theta\right), \tag{11.30}$$

where

$$\bar{R}_h = \frac{1}{N}\sum_{i=1}^{N} R_{h,i} \quad \text{and}$$

$$\Theta = \frac{1}{N}\sum_{i=1}^{N}(1 - G(R_{h,i}))^{v-1}.$$

The optimal hedge ratio is now given by the hedge ratio that minimizes the estimated MEG. Since there is no analytical solution, the numerical method needs to be applied in order to get the optimal hedge ratio. This method is sometimes referred to as the empirical distribution method.

Alternatively, the instrumental variable (IV) method suggested by Shalit (1995) can be used to find the MEG hedge ratio. Shalit's method provides the following analytical solution for the MEG hedge ratio:

$$h^{IV} = \frac{\text{Cov}\left(S_{t+1}, [1 - G(F_{t+1})]^{v-1}\right)}{\text{Cov}\left(F_{t+1}, [1 - G(F_{t+1})]^{v-1}\right)}.$$

It is important to note that for the IV method to be valid, the cumulative distribution function of the terminal wealth (W_{t+1}) should be similar to the cumulative distribution of the futures price (F_{t+1}); i.e., $G(W_{t+1}) = G(F_{t+1})$. Lien and Shaffer (1999) find that the IV-based hedge ratio (h^{IV}) is significantly different from the minimum-MEG hedge ratio.

Lien and Luo (1993a) suggest an alternative method of estimating the MEG hedge ratio. This method involves the estimation of the cumulative distribution function using a nonparametric kernel function instead of using a rank function as suggested above.

Regarding the estimation of the M-MEG hedge ratio, one can follow either the empirical distribution method or the nonparametric kernel method to estimate the MEG coefficient. A numerical method can then be used to estimate the hedge ratio that maximizes the objective function given by Eq. (11.10).

11.3.5 Estimation of Generalized Semivariance (GSV) Based Hedge Ratios

The GSV can be estimated from the sample by using the following sample counterpart:

$$V_{\delta,\alpha}^{\text{sample}}(R_{\text{h}}) = \frac{1}{N}\sum_{i=1}^{N}(\delta - R_{h,i})^{\alpha}U(\delta - R_{\text{h},i}), \tag{11.31}$$

where

$$U(\delta - R_{\text{h},i}) = \begin{cases} 1 & \text{for } \delta \geq R_{\text{h},i} \\ 0 & \text{for } \delta < R_{\text{h},i} \end{cases}.$$

Similar to the MEG technique, the optimal GSV hedge ratio can be estimated by choosing the hedge ratio that minimizes the sample GSV, $V_{\delta,\alpha}^{\text{sample}}(R_{\text{h}})$. Numerical methods can be used to search for the optimum hedge ratio. Similarly, the M-GSV hedge ratio can be obtained by minimizing the mean-risk function given by Eq. (11.12), where the expected return on the hedged portfolio is replaced by the sample average return and the GSV is replaced by the sample GSV.

One can instead use the kernel density estimation method suggested by Lien and Tse (2000) to estimate the GSV, and numerical techniques can be used to find the optimum GSV hedge ratio. Instead of using the kernel method, one can also employ the conditional heteroscedastic model to estimate the density function. This is the method used by Lien and Tse (1998).

11.4 Hedging Horizon, Maturity of Futures Contract, Data Frequency, and Hedging Effectiveness

In this section we discuss the relationship among the length of hedging horizon (hedging period), maturity of futures contracts, data frequency (e.g., daily, weekly, monthly, or quarterly), and hedging effectiveness.

Since there are many futures contracts (with different maturities) that can be used in hedging, the question is whether the minimum-variance (MV) hedge ratio depends on the time to maturity of the futures contract being used for hedging. Lee et al. (1987) find that the MV hedge ratio increases as the maturity is approached. This means that if we use the nearest to maturity futures contracts to hedge, then the MV hedge ratio will be larger compared to the one obtained using futures contracts with a longer maturity.

Aside from using futures contracts with different maturities, we can estimate the MV hedge ratio using data with different frequencies. For example, the data used in the estimation of the optimum hedge ratio can be daily, weekly, monthly, or quarterly. At the same time, the hedging horizon could be from a few hours to more than a month. The question is whether a relationship exists between the data frequency used and the length of the hedging horizon.

Malliaris and Urrutia (1991) and Benet (1992) utilize Eq. (11.20) and weekly data to estimate the optimal hedge ratio. According to Malliaris and Urrutia (1991), the ex ante hedging is more effective when the hedging horizon is one week compared to a hedging horizon of four weeks. Benet (1992) finds that a shorter hedging horizon (four weeks) is more effective (in ex ante test) compared to a longer hedging horizon (eight-weeks and twelve-weeks). These empirical results seem to be consistent with the argument that when estimating the MV hedge ratio, the hedging horizon's length must match the data frequency being used.

There is a potential problem associated with matching the length of the hedging horizon and the data frequency. For example, consider the case where the hedging horizon is three months (one quarter). In this case we need to use quarterly data to match the length of the hedging horizon. In other words, when estimating Eq. (11.20) we must employ quarterly changes in spot and futures prices. Therefore, if we have five years' worth of data, then we will have 19 nonoverlapping price changes, resulting in a sample size of 19. However, if the hedging horizon is one week, instead of three months, then we will end up with approximately 260 nonoverlapping price changes (sample size of 260) for the same five years' worth of data. Therefore, the matching method is associated with a reduction in sample size for a longer hedging horizon.

One way to get around this problem is to use overlapping price changes. For example, Geppert (1995) utilizes k-period differencing for a k-period hedging horizon in estimating the regression-based MV hedge ratio. Since Geppert (1995) uses approximately 13 months of data for estimating the hedge ratio, he employs overlapping differencing in order to eliminate the reduction in sample size caused by differencing. However, this will lead to correlated observations instead of independent observations and will require the use of a regression with autocorrelated errors in the estimation of the hedge ratio.

In order to eliminate the autocorrelated errors problem, Geppert (1995) suggests a method based on cointegration and unit-root processes. We will briefly describe his method. Suppose that the spot and futures prices, which are both unit-root processes, are cointegrated. In this case the futures and spot prices can be described by the following processes (see Stock and Watson 1988; Hylleberg and Mizon 1989):

$$S_t = A_1 P_t + A_2 \tau_t, \tag{11.32a}$$

$$F_t = B_1 P_t + B_2 \tau_t, \tag{11.32b}$$

$$P_t = P_{t-1} + w_t, \tag{11.32c}$$

$$\tau_t = \alpha_1 \tau_{t-1} + v_t, \quad 0 \le |\alpha_1| < 1, \tag{11.32d}$$

where P_t and τ_t are permanent and transitory factors that drive the spot and futures prices and w_t and v_t are white noise processes. Note that P_t follows a pure random walk process and τ_t follows a stationary process. The MV hedge ratio for a k-period hedging horizon is then given by (see Geppert 1995):

$$H_J^* = \frac{A_1 B_1 k \sigma_w^2 + 2A_2 B_2 \left(\frac{(1-\alpha^k)}{1-\alpha^2}\right) \sigma_v^2}{B_1^2 k \sigma_w^2 + 2B_2^2 \left(\frac{(1-\alpha^k)}{1-\alpha^2}\right) \sigma_v^2}. \tag{11.33}$$

One advantage of using Eq. (11.33) instead of a regression with nonoverlapping price changes is that it avoids the problem of a reduction in sample size associated with nonoverlapping differencing.

An alternative way of matching the data frequency with the hedging horizon is by using the wavelet to decompose the time series into different frequencies as suggested by Lien and Shrestha (2007). The decomposition can be done without the loss of sample size (see Lien and Shrestha (2007) for detail). For example, the daily spot and future returns series can be decomposed using the maximal overlap discrete wavelet transform (MODWT) as follows:

$$R_{s,t} = B_{J,t}^{s} + D_{J,t}^{s} + D_{J-1,t}^{s} + \cdots + D_{1,t}^{s}$$
$$R_{f,t} = B_{J,t}^{f} + D_{J,t}^{f} + D_{J-1,t}^{f} + \cdots + D_{1,t}^{f}$$

where $D_{j,t}^{s}$ and $D_{j,t}^{f}$ are the spot and futures returns series with changes on the time scale of length 2^{j-1} days respectively.[4] Similarly, $B_{J,t}^{s}$ and $B_{J,t}^{2}$ represents spot and futures returns series corresponding to time scale of 2^J days and longer. Now, we can run the following regression to find the hedge ratio corresponding to hedging horizon equal to 2^{j-1} days:

$$D_{j,t}^{s} = \theta_{j,0} + \theta_{j,1} D_{j,t}^{f} + e_j \tag{11.34}$$

where the estimate of the hedge ratio is given by the estimate of $\theta_{j,1}$.

11.5 Empirical Results of Hedge Ratio Estimation

In this section, we empirically show how SAS program can be used to estimate the hedge ratio in terms of OLS method, ARCH model, GARCH model, E-GARCH model, GJR-GARCH model, and GARCH model. Prices and monthly changes of prices for S&P 500 index and its futures during January 2000 to July 2015 are used in the following empirical studies. Sample data are presented in Appendix 3.

[4]For example, represents daily time scale and represents 8-day time scale.

11.5.1 OLS Method

We use Eq. (11.20) to conduct OLS method and estimate the hedge ratio. By using SAS regression procedure, we obtain the following program code and empirical results.

```
proc reg data = sp500monthly;
model C_spot = C_future;
ods output ParameterEstimates = estimates_ols;
run;
```

Variable	DF	Estimate	Std. Err	*t*-value	Probt
Intercept	1	0.10961	0.28169	0.39	0.6976
C_future	1	0.98795	0.00535	184.51	<0.0001

C_spot—the change of the monthly spot prices from Jan. 1st, 2000 to July 31st, 2015.
C_future—the change of the monthly futures prices from Jan. 1st, 2000 to July 31st, 2015.

11.5.2 ARCH GARCH

Based upon prices and monthly price changes of S&P 500 index and its futures in Appendix 3, we use ARCH and GARCH method in term of Eq. (11.23) to estimate hedge ratio in this section.

```
proc autoreg data = sp500monthly;
model C_spot = C_future/nlag=2 garch=(q=1,p=1) maxit=50;
ods output ParameterEstimates=estimate_autoreg;
run;
```

Variable	DF	Estimate	Std. Err	*t*-value	Probt
Intercept	1	0.0563	0.009424	5.97	<0.0001
C_future	1	0.9997	0.000461	2167.28	<0.0001
AR1	1	0.9944	0.007702	129.12	<0.0001
AR2	1	1.0042	0.015	66.86	<0.0001
ARCH0	1	0.005013	0.005881	0.85	0.394
ARCH1	1	0.8427	0.1154	7.3	<0.0001
GARCH1	1	0.589	0.0371	15.88	<0.0001

C_spot—the change of the monthly spot prices from Jan. 1st, 2000 to July 31st, 2015.
C_future—the change of the monthly futures prices from Jan. 1st, 2000 to July 31st, 2015.

11.5.3 EGARCH

We use EGARCH method discussed in the previous chapter to estimate hedge ratio. Based upon prices and monthly price changes of S&P 500 index and its futures in Appendix 3, we obtain the following program code and empirical results.

```
proc autoreg data= sp500monthly ;
  model C_spot = C_future / nlag=2 garch=( q=1, p=1 , type = exp) ;
run;
```

Variable	DF	Estimate	Std. Err	*t*-value	Probt
Intercept	1	0.0517	0.0149	3.46	0.0005
C_future	1	1.0002	0.000474	2111.78	<.0001
AR1	1	0.9964	0.0102	97.31	<.0001
AR2	1	1.0096	0.0168	60.02	<.0001
EARCH0	1	0.0985	0.0240	4.11	<.0001
EARCH1	1	0.7371	0.1016	7.25	<.0001
EGARCH1	1	0.9516	0.0173	55.09	<.0001
THETA	1	0.2286	0.0860	2.66	0.0079

11.5.4 GJR-GARCH

We here apply GJR-GARCH method to estimate hedge ratio. The specification of the conditional variance for GJR-GARCH is presented in Table 10.11 of previous chapter. We use prices and monthly price changes of S&P 500 index and its futures in Appendix 3 to run a GJR-GARCH model as the following program code obtain the empirical results.

Parameter	Estimate	Approx. Std Err	*t*-value	Approx. Pr > |*t*|
arch0	0.077656	0.0534	1.45	0.1476
arch1	0.726043	0.1619	4.48	<.0001
garch1	0.662838	0.0418	15.85	<.0001
phi	−0.68261	0.2411	−2.83	0.0051

```
*no intercept
/* mean model */
 C_spot = intercept ;
```

Parameter	Estimate	Approx. Std. Err	*t*-value	Approx. Pr > |*t*|
intercept	6.826198	3.1803	2.15	0.0332
arch0	205.353	243.5	0.84	0.4002
arch1	0.279427	0.1445	1.93	0.0546
garch1	0.74021	0.1328	5.57	<0.0001
phi	−0.21296	0.1964	−1.08	0.2798

```
proc model data = sp500monthly ;
  parms arch0 .1 arch1 .2 garch1 .75 phi .1;
  /* mean model */
  C_spot = C_future ;
  /* variance model */
  if zlag(resid.C_spot) > 0 then
    h.C_spot = arch0 + arch1*xlag(resid.C_spot**2,mse.C_spot) + garch1*xlag(h.C_spot,mse.C_sp
  else
    h.C_spot = arch0 + arch1*xlag(resid.C_spot**2,mse.C_spot) + garch1*xlag(h.C_spot,mse.C_sp
        phi*xlag(resid.C_spot**2,mse.C_spot) ;
  /* fit the model */
  fit C_spot / method = marquardt fiml ;
 run ;
```

11.5.5 TGARCH

We use TGARCH method to estimate hedge ratio. The specification of the conditional variance for TGARCH is presented in Table 10.11 of previous chapter. We use prices and monthly price changes of S&P 500 index and its futures in Appendix 3 to run a TGARCH model as the following program code obtain the empirical results.

Parameter	Estimate	Approx. Std. Err	*t*-value	Approx. Pr > |t|
arch0	5.258712	0.2910	18.07	<0.0001
arch1_plus	5.849029	0.3921	14.92	<0.0001
arch1_minus	4.788963	0.2495	19.20	<0.0001
garch1	0.658774	0.0155	42.51	<0.0001

```
*no intercept
/* mean model */
 C_spot = intercept;
```

Parameter	Estimate	Approx. Std. Err	*t*-value	Approx. Pr > |t|
intercept	4.608663	0.6367	7.24	<0.0001
arch0	5.298545	0.5150	10.29	<0.0001
arch1_plus	4.995944	0.5565	8.98	<0.0001
arch1_minus	5.021968	0.5207	9.64	<0.0001
garch1	0.761879	0.0110	69.33	<0.0001

11.6 Conclusion

In this chapter we have reviewed various approaches to deriving the optimal hedge ratio, as summarized in Appendix 1. These approaches can be divided into the mean-variance based approach, the expected utility maximizing approach, the mean extended-Gini coefficient-based approach, and the generalized semivariance-based approach. All these approaches will lead to the same hedge ratio as the conventional minimum-variance (MV) hedge ratio if the futures price follows a pure martingale process and if the futures and spot prices are jointly normal. However, if these conditions do not hold, then the hedge ratios based on the various approaches will be different.

The MV hedge ratio is the most understood and most widely used hedge ratio. Since the statistical properties of the MV hedge ratio are well known, statistical hypothesis testing can be performed with the MV hedge ratio. For example, we can test whether the optimal MV hedge ratio is the same as the naïve hedge ratio. Since the MV hedge ratio ignores the expected return, it will not be consistent with the mean-variance analysis unless the futures price follows a pure martingale process. Furthermore, if the martingale and normality condition do not hold, then the MV hedge ratio will not be consistent with the expected utility maximization principle. Following the MV hedge ratio is the mean-variance hedge ratio. Even if this hedge ratio incorporates the expected return in the derivation of the optimal hedge ratio, it will not be consistent with the expected maximization principle unless either the normality condition holds or the utility function is quadratic.

In order to make the hedge ratio consistent with the expected utility maximization principle, we can derive the optimal hedge ratio by maximizing the expected utility. However, to implement such approach, we need to assume a specific utility function and we need to make an assumption regarding the return distribution. Therefore, different utility functions will lead to different optimal hedge ratios. Furthermore, analytic solutions for such hedge ratios are not known and numerical methods need to be applied.

New approaches have recently been suggested in deriving optimal hedge ratios. These include the mean-Gini coefficient-based hedge ratio, semivariance-based hedge ratios and value-at-risk-based hedge ratios. These hedge ratios are consistent with the second-order stochastic dominance principle. Therefore, such hedge ratios are very general in the sense that they are consistent with the expected utility maximization principle and make very few assumptions on the utility function. The only requirement is that the marginal utility be positive and the second derivative of the utility function be negative. However, both of these hedge ratios do not lead to a unique hedge ratio. For example, the mean-Gini coefficient-based hedge ratio depends on the risk aversion parameter (ν) and the semivariance-based hedge ratio depends on the risk aversion parameter (α) and target return (δ). It is important to note, however, that the semivariance-based hedge ratio has some appeal in the sense that the semivariance as a measure of risk is consistent with the risk perceived by individuals. The same argument can be applied to value-at-risk-based hedge ratio.

So far as the derivation of the optimal hedge ratio is concerned, almost all of the derivations do not incorporate transaction costs. Furthermore, these derivations do not allow investments in securities other than the spot and corresponding futures contracts. As shown by Lence (1995), once we relax these conventional assumptions, the resulting optimal hedge ratio can be quite different from the ones obtained under the conventional assumptions. Lence's (1995) results are based on a specific utility function and some other assumption regarding the return distributions. It remains to be seen if such results hold for the mean extended-Gini coefficient-based as well as semivariance-based hedge ratios.

In this chapter we have also reviewed various ways of estimating the optimum hedge ratio, as summarized in Appendix 2. As far as the estimation of the conventional MV hedge ratio is concerned, there are a large number of methods that have been proposed in the literature. These methods range from a simple regression method to complex cointegrated heteroscedastic methods with regime-switching, and some of the estimation methods include a kernel density function method as well as an empirical distribution method. Except for many of mean-variance based hedge ratios, the estimation involves the use of a numerical technique. This has to do with the fact that most of the optimal hedge ratio formulae do not have a closed-form analytic expression. Again, it is important to mention that based on his specific model, Lence (1995) finds that the value of complicated and sophisticated estimation methods is negligible. It remains to be seen if such a result holds for the mean extended-Gini coefficient-based as well as semivariance-based hedge ratios.

In this chapter, we have also discussed about the relationship between the optimal MV hedge ratio and the hedging horizon. We feel that this relationship has not been fully explored and can be further developed in the future. For example, we would like to know if the optimal hedge ratio approaches the naïve hedge ratio when the hedging horizon becomes longer.

The main thing we learn from this review is that if the futures price follows a pure martingale process and if the returns are jointly normally distributed, then all different hedge ratios are the same as the conventional MV hedge ratio, which is simple to compute and easy to understand. However, if these two conditions do not hold, then there are many optimal hedge ratios (depending on which objective function one is trying to optimize) and there is no single optimal hedge ratio that is distinctly superior to the remaining ones. Therefore, further research needs to be done to unify these different approaches to the hedge ratio. In addition, by using different time-series techniques such as ARCH, GARCH, and other techniques, we perform empirical results for hedge ratio.

For those who are interested in research in this area, we would like to finally point out that one requires a good understanding of financial economic theories and econometric methodologies. In addition, a good background in data analysis and computer programming would also be helpful. Overall, this chapter has carefully discussed how alternative hedge ratios can be derived theoretically. In the final section of this chapter, we also applied five time-series methods to estimate hedge ratio by using the monthly data of the S&P 500 index and its futures.

Appendix 1: Theoretical Models

References	Return definition and objective function	Summary
Johnson (1960)	Ret_1 O_1	The chapter derives the minimum-variance hedge ratio. The hedging effectiveness is defined as E_1, but no empirical analysis is done
Hsin et al. (1994)	Ret_2 O_2	The chapter derives the utility function-based hedge ratio. A new measure of hedging effectiveness E_2 based on a certainty equivalent is proposed. The new measure of hedging effectiveness is used to compare the effectiveness of futures and options as hedging instruments
Howard and D'Antonio (1984)	Ret_2 O_3	The chapter derives the optimal hedge ratio based on maximizing the Sharpe ratio. The proposed hedging effectiveness E_3 is based on the Sharpe ratio
Cecchetti et al. (1988)	Ret_2 O_4	The chapter derives the optimal hedge ratio that maximizes the expected utility function: $\int_{R_s}\int_{R_f}\log[1+R_s(t)-h(t)R_f(t)]f_t(R_s,R_f)dR_s dR_f$, where the density function is assumed to be bivariate normal. A third-order linear bivariate ARCH model is used to get the conditional variance and covariance matrix. A numerical procedure is used to maximize the objective function with respect to hedge ratio. Due to ARCH, the hedge ratio changes over time. The chapter uses certainty equivalent (E_2) to measure the hedging effectiveness
Cheung et al. (1990)	Ret_2 O_5	The chapter uses mean-Gini (v = 2, not mean extended-Gini coefficient) and mean-variance approaches to analyze the effectiveness of options and futures as hedging instruments
Kolb and Okunev (1992)	Ret_2 O_5	The chapter uses mean extended-Gini coefficient in the derivation of the optimal hedge ratio. Therefore, it can be considered as a generalization of the mean-Gini coefficient method used by Cheung et al. (1990)
Kolb and Okunev (1993)	Ret_2 O_6	The chapter defines the objective function as O_6, but in terms of wealth (W) $U(W)=E[W]-\Gamma_v(W)$ and compares with the quadratic utility function $U(W)=E[W]-m\sigma^2$. The chapter plots the EMG efficient frontier in W and $\Gamma_v(W)$ space for various values of risk aversion parameters (v)
Lien and Luo (1993b)	Ret_1 O_9	The chapter derives the multiperiod hedge ratios where the hedge ratios are allowed to change over the hedging period. The method suggested in the chapter still falls under the minimum-variance hedge ratio
Lence (1995)	O_4	This chapter derives the expected utility maximizing hedge ratio where the terminal wealth depends on the return on a diversified portfolio that consists of the production of a spot commodity, investment in a risk-free asset, investment in a risky asset, as well as borrowing. It also incorporates the transaction costs
De Jong et al. (1997)	Ret_2 O_7 (also uses O_1 and O_3)	The chapter derives the optimal hedge ratio that minimizes the generalized semivariance (GSV). The chapter compares the GSV hedge ratio with the minimum-variance (MV) hedge ratio as well as the Sharpe hedge ratio. The chapter uses E_1 (for the MV hedge ratio), E_3 (for the Sharpe hedge ratio) and E_4 (for the GSV hedge ratio) as the measures of hedging effectiveness

(continued)

References	Return definition and objective function	Summary
Chen et al. (2001)	Ret_1 O_8	The chapter derives the optimal hedge ratio that maximizes the risk-return function given by $U(R_h) = E[R_h] - V_{\delta,\alpha}(R_h)$. The method can be considered as an extension of the GSV method used by De Jong et al. (1997)
Hung et al. (2006)	Ret_2 O_{10}	The chapter derives the optimal hedge ratio that minimizes the value at risk for a hedging horizon of length τ given by $Z_\alpha \sigma_h \sqrt{\tau} - E[R_h]\tau$

Notes

A. Return Model:

(Ret_1)

$\Delta V_H = C_s \Delta P_s + C_f \Delta P_f \Rightarrow \text{hedge ratio} = H = \frac{C_f}{C_s}$,

C_s = units of spot commodity and C_f = units of futures contract

(Ret_2)

$$R_h = R_s + hR_f, \quad R_s = \frac{S_t - S_{t-1}}{S_{t-1}}, \qquad \text{(a)}$$

$$R_f = \frac{F_t - F_{t-1}}{F_{t-1}} \Rightarrow \text{hedge ratio} : h = \frac{C_f F_{t-1}}{C_s S_{t-1}}$$

$$R_f = \frac{F_t - F_{t-1}}{S_{t-1}} \Rightarrow \text{hedge ratio} : h = \frac{C_f}{C_s}. \qquad \text{(b)}$$

B. Objective Function:

(O_1) Minimize
$\mathrm{Var}(R_h) = C_s^2\sigma_s^2 + C_f^2\sigma_f^2 + 2C_sC_f\sigma_{sf}$
or
$\mathrm{Var}(R_h) = \sigma_s^2 + h^2\sigma_f^2 + 2h\sigma_{sf}$

(O_2) Maximize
$E(R_h) - \frac{A}{2}\mathrm{Var}(R_h)$

(O_3) Maximize
$\frac{E(R_h) - R_F}{\mathrm{Var}(R_h)}$ (Sharpe ratio),
R_F = risk-free interest rate

(O_4) Maximize
$E[U(W)], \quad U(.) = \text{utility function}$,
W = terminal wealth

(O_5) Minimize
$\Gamma_v(R_h), \quad \Gamma_v(R_h) = -v\mathrm{Cov}\left(R_h, (1 - F(R_h))^{v-1}\right)$

(O_6) Maximize
$E[R_h] - \Gamma_v(R_h v)$

(O_7) Minimize
$V_{\delta,\alpha}(R_h) = \int_{-\infty}^{\delta} (\delta - R_h)^\alpha dG(R_h), \alpha > 0$

(O_8) Maximize
$U(R_h) = E[R_h] - V_{\delta,\alpha}(R_h)$

(O_9) Minimize
$\mathrm{Var}(W_t) = \mathrm{Var}\left(\sum_{t=1}^{T} C_{st}\Delta S_t + C_{ft}\Delta F_t\right).$

(O_{10}) Minimize
$Z_\alpha \sigma_h \sqrt{\tau} - E[R_h]\tau$

C. Hedging Effectiveness:

(E_1) $e = \left(1 - \frac{\mathrm{Var}(R_h)}{\mathrm{Var}(R_s)}\right)$

(E_2) $e = R_h^{ce} - R_{ss}^{ce}, R_h^{ce}(R_s^{ce})$ = certainty equivalent return of hedged(unhedged)portfolio

(E_3) $e = \frac{\frac{(E[R_h] - R_F)}{\mathrm{Var}(R_h)}}{\frac{(E[R_s] - R_F)}{\mathrm{Var}(R_s)}}$ or

$e = \frac{(E[R_h] - R_F)}{\mathrm{Var}(R_h)} - \frac{(E[R_s] - R_F)}{\mathrm{Var}(R_s)}$

(E_4) $e = 1 - \frac{V_{\delta,\alpha}(R_h)}{V_{\delta,\alpha}(R_s)}.$

Appendix 2: Empirical Models

References	Commodity	Summary
Ederington (1979)	GNMA futures (1/1976-12/1977), Wheat (1/1976-12/1977), Corn (1/1976-12/1977), T-bill futures (3/1976-12/1977) [weekly data]	The chapter uses the Ret_1 definition of return and estimates the minimum-variance hedge ratio (O_1). E_1 is used as a hedging effectiveness measure. The chapter uses nearby contracts (3–6 months, 6–9 months and 9–12 months) and a hedging period of 2 and 4 weeks. OLS (M_1) is used to estimate the parameters. Some of the hedge ratios are found not to be different from zero and the hedging effectiveness increases with the length of hedging period. The hedge ratio also increases (closer to unity) with the length of hedging period
Grammatikos and Saunders (1983)	Swiss franc, Canadian dollar, British pound, DM, Yen (1/1974-6/1980) [weekly data]	The chapter estimates the hedge ratio for the whole period and moving window (2-year data). It is found that the hedge ratio changes over time. Dummy variables for various subperiods are used, and shifts are found. The chapter uses a random coefficient (M_3) model to estimate the hedge ratio. The hedge ratio for Swiss franc is found to follow a random coefficient model. However, there is no improvement in effectiveness when the hedge ratio is calculated by correcting for the randomness
Junkus and Lee (1985)	Three stock index futures for Kansas City Board of Trade, New York Futures Exchange, and Chicago Mercantile Exchange (5/82-3/83) [daily data]	The chapter tests the applicability of four futures hedging models: a variance-minimizing model introduced by Johnson (1960), the traditional one to one hedge, a utility maximization model developed by Rutledge (1972), and a basis arbitrage model suggested by Working (1953). An optimal ratio or decision rule is estimated for each model, and measures for the effectiveness of each hedge are devised. Each hedge strategy performed best according to its own criterion. The Working decision rule appeared to be easy to use and satisfactory in most cases. Although the maturity of the futures contract used affected the size of the optimal hedge ratio,, there was no consistent maturity effect on performance. Use of a particular ratio depends on how closely the assumptions underlying the model approach a hedger's real situation
Lee et al. (1987)	S&P 500, NYSE, Value Line (1983) [daily data]	The chapter tests for the temporal stability of the minimum-variance hedge ratio. It is found that the hedge ratio increases as maturity of the futures contract nears. The chapter also performs a functional form test and finds support for the regression of rate of change for discrete as well as continuous rates of change in prices

(continued)

References	Commodity	Summary
Cecchetti et al. (1988)	Treasury bond, Treasury bond futures (1/1978-5/1986) [monthly data]	The chapter derives the hedge ratio by maximizing the expected utility. A third-order linear bivariate ARCH model is used to get the conditional variance and covariance matrix. A numerical procedure is used to maximize the objective function with respect to the hedge ratio. Due to ARCH, the hedge ratio changes over time. It is found that the hedge ratio changes over time and is significantly less (in absolute value) than the minimum-variance (MV) hedge ratio (which also changes over time). E_2 (certainty equivalent) is used to measure the performance effectiveness. The proposed utility maximizing hedge ratio performs better than the MV hedge ratio
Cheung et al. (1990)	Swiss franc, Canadian dollar, British pound, German mark, Japanese yen (9/1983-12/1984) [daily data]	The chapter uses mean-Gini coefficient ($v = 2$) and mean-variance approaches to analyze the effectiveness of options and futures as hedging instruments. It considers both mean-variance and expected return mean-Gini coefficient frontiers. It also considers the minimum-variance (MV) and minimum mean-Gini coefficient hedge ratios. The MV and minimum mean-Gini approaches indicate that futures is a better hedging instrument. However, the mean-variance frontier indicates futures to be a better hedging instrument whereas the mean-Gini frontier indicates options to be a better hedging instrument
Baillie and Myers (1991)	Beef, Coffee, Corn, Cotton, Gold, Soybean (contracts maturing in 1982 and 1986) [daily data]	The chapter uses a bivariate GARCH model (M_2) in estimating the minimum-variance (MV) hedge ratios. Since the models used are conditional models, the time series of hedge ratios are estimated. The MV hedge ratios are found to follow a unit-root process. The hedge ratio for beef is found to be centered around zero. E_1 is used as a hedging effectiveness measure. Both in-sample and out-of-sample effectiveness of the GARCH-based hedge ratios is compared with a constant hedge ratio. The GARCH-based hedge ratios are found to be significantly better compared to the constant hedge ratio
Malliaris and Urrutia (1991)	British pound, German mark, Japanese yen, Swill franc, Canadian dollar (3/1980–12/1988) [weekly data]	The chapter uses regression autocorrelated errors model to estimate the minimum-variance (MV) hedge ratio for the five currencies. Using overlapping moving windows, the time series of the MV hedge ratio and hedging effectiveness are estimated for both ex post (in-sample) and ex ante (out-of-sample) cases. E_1 is used to measure the hedging effectiveness for the ex post case whereas average return is used to measure the hedging effectiveness. Specifically, the average return close to zero is used to indicate

(continued)

References	Commodity	Summary
		a better performing hedging strategy. In the ex post case, the four-week hedging horizon is more effective compared to the one-week hedging horizon. However, for the ex ante case the opposite is found to be true
Benet (1992)	Australian dollar, Brazilian cruzeiro, Mexican peso, South African rand, Chinese yuan, Finish markka, Irish pound, Japanese yen (8/1973–12/1985) [weekly data]	This chapter considers direct and cross-hedging, using multiple futures contracts. For minor currencies, the cross-hedging exhibits a significant decrease in performance from ex post to ex ante. The minimum-variance hedge ratios are found to change from one period to the other except for the direct hedging of Japanese yen. On the ex ante case, the hedging effectiveness does not appear to be related to the estimation period length. However, the effectiveness decreases as the hedging period length increases
Kolb and Okunev (1992)	Corn, Copper, Gold, German mark, S&P 500 (1989) [daily data]	The chapter estimates the mean extended-Gini (MEG) hedge ratio (M_9) with v ranging from 2 to 200. The MEG hedge ratios are found to be close to the minimum-variance hedge ratios for a lower level of risk parameter v (for v from 2 to 5). For higher values of v, the two hedge ratios are found to be quite different. The hedge ratios are found to increase with the risk aversion parameter for S&P 500, Corn, and Gold. However, for Copper and German mark, the hedge ratios are found to decrease with the risk aversion parameter. The hedge ratio tends to be more stable for higher levels of risk
Kolb and Okunev (1993)	Cocoa (3/1952 to 1976) for four cocoa-producing countries (Ghana, Nigeria, Ivory Coast, and Brazil) [March and September data]	The chapter estimates the Mean-MEG (M-MEG) hedge ratio (M_{12}). The chapter compares the M-MEG hedge ratio, minimum-variance hedge ratio, and optimum mean-variance hedge ratio for various values of risk aversion parameters. The chapter finds that the M-MEG hedge ratio leads to reverse hedging (buy futures instead of selling) for v less than 1.24 (Ghana case). For high-risk aversion parameter values (high v) all hedge ratios are found to converge to the same value
Lien and Luo (1993a)	S&P 500 (1/1984–12/1988) [weekly data]	The chapter points out that the mean extended-Gini (MEG) hedge ratio can be calculated either by numerically optimizing the MEG coefficient or by numerically solving the first-order condition. For $v = 9$ the hedge ratio of -0.8182 is close to the minimum-variance (MV) hedge ratio of -0.8171. Using the first-order condition, the chapter shows that for a large v the MEG hedge ratio converges to a constant. The empirical result shows that the hedge ratio decreases with the risk aversion parameter v. The chapter finds that the MV and MEG hedge ratio (for low v) series (obtained

(continued)

References	Commodity	Summary
		by using a moving window) are more stable compared to the MEG hedge ratio for a large v. The chapter also uses a nonparametric Kernel estimator to estimate the cumulative density function. However, the kernel estimator does not change the result significantly
Lien and Luo (1993b)	British pound, Canadian dollar, German mark, Japanese yen, Swiss franc (3/1980–12/1988), MMI, NYSE, S&P (1/1984–12/1988) [weekly data]	This chapter proposes a multiperiod model to estimate the optimal hedge ratio. The hedge ratios are estimated using an error correction model. The spot and futures prices are found to be cointegrated. The optimal multiperiod hedge ratios are found to exhibit a cyclical pattern with a tendency for the amplitude of the cycles to decrease. Finally, the possibility of spreading among different market contracts is analyzed. It is shown that hedging in a single market may be much less effective than the optimal spreading strategy
Ghosh (1993)	S&P futures, S&P index, Dow Jones Industrial Average, NYSE composite index (1/1990–12/1991) [daily data]	All the variables are found to have a unit root. For all three indices the same S&P 500 futures contracts are used (cross-hedging) . Using the Engle–Granger two-step test, the S&P 500 futures price is found to be cointegrated with each of the three spot prices: S&P 500, DJIA, and NYSE. The hedge ratio is estimated using the error correction model (ECM) (M_4). Out-of-sample performance is better for the hedge ratio from the ECM compared to the Ederington model
Sephton (1993a)	Feed wheat, Canola futures (1981–82 crop year) [daily data]	The chapter finds unit roots on each of the cash and futures (log) prices, but no cointegration between futures and spot (log) prices. The hedge ratios are computed using a four-variable GARCH(1,1) model. The time series of hedge ratios are found to be stationary. Reduction in portfolio variance is used as a measure of hedging effectiveness. It is found that the GARCH-based hedge ratio performs better compared to the conventional minimum-variance hedge ratio
Sephton (1993b)	Feed wheat, Feed barley, Canola futures (1988/89) [daily data]	The chapter finds unit roots on each of the cash and futures (log) prices, but no cointegration between futures and spot (log) prices. A univariate GARCH model shows that the mean returns on the futures are not significantly different from zero. However, from the bivariate GARCH canola is found to have a significant mean return. For canola the mean-variance utility function is used to find the optimal hedge ratio for various values of the risk aversion parameter. The time series of the hedge ratio (based on bivariate GARCH model) is found to be stationary. The benefit

(continued)

References	Commodity	Summary
		in terms of utility gained from using a multivariate GARCH decreases as the degree of risk aversion increases
Kroner and Sultan (1993)	British pound, Canadian dollar, German mark, Japanese yen, Swiss franc (2/1985–2/1990) [weekly data]	The chapter uses the error correction model with a GARCH error (M_5) to estimate the minimum-variance (MV) hedge ratio for the five currencies. Due to the use of conditional models, the time series of the MV hedge ratios are estimated. Both within-sample and out-of-sample evidence shows that the hedging strategy proposed in the chapter is potentially superior to the conventional strategies
Hsin et al. (1994)	British pound, German mark, Yen, Swiss franc (1/1986–12/1989) [daily data]	The chapter derives the optimum mean-variance hedge ratio by maximizing the objective function O_2. The hedging horizons of 14, 30, 60, 90, and 120 calendar days are considered to compare the hedging effectiveness of options and futures contracts. It is found that the futures contracts perform better than the options contracts
Shalit (1995)	Gold, Silver, Copper, Aluminum (1/1977–12/1990) [daily data]	The chapter shows that if the prices are jointly normally distributed, the mean extended-Gini (MEG) hedge ratio will be same as the minimum-variance (MV) hedge ratio. The MEG hedge ratio is estimated using the instrumental variable method.. The chapter performs normality tests as well as the tests to see if the MEG hedge ratios are different from the MV hedge ratios. The chapter finds that for a significant number of futures contracts the normality does not hold and the MEG hedge ratios are different from the MV hedge ratios
Geppert (1995)	German mark, Swiss franc, Japanese yen, S&P 500, Municipal Bond Index (1/1990–1/1993) [weekly data]	The chapter estimates the minimum-variance hedge ratio using the OLS as well as the cointegration methods for various lengths of hedging horizon. The in-sample results indicate that for both methods the hedging effectiveness increases with the length of the hedging horizon. The out-of-sample results indicate that in general the effectiveness (based on the method suggested by Malliaris and Urrutia (1991)) decreases as the length of the hedging horizon decreases. This is true for both the regression method and the decomposition method proposed in the chapter. However, the decomposition method seems to perform better than the regression method in terms of both mean and variance
De Jong et al. (1997)	British pound (12/1976–10/1993), German mark (12/1976–10/1993), Japanese yen (4/1977–10/1993) [daily data]	The chapter compares the minimum-variance, generalized semivariance and Sharpe hedge ratios for the three currencies. The chapter computes the out-of-sample hedging effectiveness using nonoverlapping 90-day periods where the first 60 days are used to

(continued)

References	Commodity	Summary
		estimate the hedge ratio and the remaining 30 days are used to compute the out-of-sample hedging effectiveness. The chapter finds that the naïve hedge ratio performs better than the model based hedge ratios
Lien and Tse (1998)	Nikkei Stock Average (1/1989–8/1996) [daily data]	The chapter shows that if the rates of change in spot and futures prices are bivariate normal and if the futures price follows a martingale process, then the generalized semivariance (GSV) (referred to as lower partial moment) hedge ratio will be same as the minimum-variance (MV) hedge ratio. A version of the bivariate asymmetric power ARCH model is used to estimate the conditional joint distribution, which is then used to estimate the time-varying GSV hedge ratios. The chapter finds that the GSV hedge ratio significantly varies over time and is different from the MV hedge ratio
Lien and Shaffer (1999)	Nikkei (9/86–9/89), S&P (4/82–4/85), TOPIX (4/90–12/93), KOSPI (5/96–12/96), Hang Seng (1/87–12189), IBEX (4/93–3/95) [daily data]	This chapter empirically tests the ranking assumption used by Shalit (1995). The ranking assumption assumes that the ranking of futures prices is the same as the ranking of the wealth. The chapter estimates the mean extended-Gini (MEG) hedge ratio based on the instrumental variable (IV) method used by Shalit (1995) , and the true MEG hedge ratio. The true MEG hedge ratio is computed using the cumulative probability distribution estimated employing the kernel method instead of the rank method. The chapter finds that the MEG hedge ratio obtained from the IV method to be different from the true MEG hedge ratio. Furthermore, the true MEG hedge ratio leads to a significantly smaller MEG coefficient compared to the IV-based MEG hedge ratio
Lien and Tse (2000)	Nikkei Stock Average (1/1988–8/996) [daily data]	The chapter estimates the generalized semivariance (GSV) hedge ratios for different values of parameters using a nonparametric kernel estimation method. The kernel method is compared with the empirical distribution method. It is found that the hedge ratio from one method is not different from the hedge ratio from another. The Jarque–Bera (1987) test indicates that the changes in spot and futures prices do not follow normal distribution
Chen et al. (2001)	S&P 500 (4/1982–12/1991) [weekly data]	The chapter proposes the use of the M-GSV hedge ratio. The chapter estimates the minimum-variance (MV), optimum mean-variance, Sharpe, mean extended-Gini (MEG), generalized semivariance (GSV), mean-MEG (M-MEG), and mean-GSV (M-GSV) hedge ratios. The Jarque–Bera

(continued)

References	Commodity	Summary
		(1987) Test and D'Agostino (1971) D Statistic indicate that the price changes are not normally distributed. Furthermore, the expected value of the futures price change is found to be significantly different from zero. It is also found that for a high level of risk aversion, the M-MEG hedge ratio converges to the MV hedge ratio whereas the M-GSV hedge ratio converges to a lower value
Hung et al. (2006)	S&P 500 (01/1997–12/1999) [daily data]	The chapter proposes minimization of value at risk in deriving the optimum hedge ratio. The chapter finds cointegrating relationship between the spot and futures returns and uses bivariate constant correlation GARCH(1,1) model with error correction term. The chapter compares the proposed hedge ratio with MV hedge ratio and hedge ratio (HKL hedge ratio) proposed by Hsin, Kuo and Lee (1994). The chapter finds the performance of the proposed hedge ratio to be similar to the HKL hedge ratio. Finally, the proposed hedge ratio converges to the MV hedge ratio for high risk-averse levels
Lee, and Yoder (2007)	Nikkei 225 and Hang Send index futures (01/1989–12/2003) [weekly data]	The chapter proposes regime-switching time-varying correlation GARCH model and compares the resulting hedge ratio with constant correlation GARCH and time-varying correlation GARCH. The proposed model is found to outperforms the other two hedge ratio in both in-sample and out-of-sample for both contracts.
Lien and Shrestha (2007)	23 different futures contracts (sample period depends on contracts) [daily data]	This chapter proposes wavelet base hedge ratio to compute the hedge ratios for different hedging horizons (1-day, 2-day, 4-day, 8-day, 16 day, 32-day, 64-day, 128-day; and 256-day and longer). It is found that the wavelet based hedge ratio and the error correction based hedge ratio are larger than MV hedge ratio. The performance of wavelet based hedge ratio improves with the length of hedging horizon
Lien and Shrestha (2010)	22 different futures contracts (sample period depends on contracts) [daily data]	The chapter proposes the hedge ratio based on skew-normal distribution (SKN hedge ratio). The chapter also estimates the semivariance (lower partial moment (LPM)) hedge ratio and MV hedge ratio among other hedge ratios. SKN hedge ratios are found to be different from the MV hedge ratio based on normal distribution. SKN hedge ratio performs better than LPM hedge ratio for long hedger especially for the out-of-sample cases

Notes

A. Minimum-Variance Hedge Ratio:
A.1. OLS:

(M$_1$): $\Delta S_t = a_0 + a_1 \Delta F_t + e_t$: Hedge ratio $= a_1$
$R_s = a_0 + a_1 R_f + e_t$: Hedge ratio $= a_1$

A.2. Multivariate Skew-Normal:

(M$_2$): The return vector $Y = \begin{bmatrix} R_s \\ R_f \end{bmatrix}$ is assumed to have skew-normal distribution with covariance matrix V:

$$\text{Hedge ration} = H_{skn} = \frac{V(1,2)}{V(2,2)}$$

A.3. ARCH/GARCH:

(M$_3$): $\begin{bmatrix} \Delta S_t \\ \Delta F_t \end{bmatrix} = \begin{bmatrix} \mu_1 \\ \mu_2 \end{bmatrix} + \begin{bmatrix} e_{1t} \\ e_{2t} \end{bmatrix}$,
$e_t|\Omega_{t-1} \sim N(0, H_t), \quad H_t = \begin{bmatrix} H_{11,t} & H_{12,t} \\ H_{12,t} & H_{22,t} \end{bmatrix}$,

$$\text{Hedge ratio} = H_{12,t}/H_{22,t}$$

A.4. Regime-Switching GARCH:

(M$_4$):The transition probabilities are given by:

$$\Pr(s_t = 1|s_{t-1} = 1) = \frac{e^{p_0}}{1+e^{p_0}} \text{ and}$$
$$\Pr(s_t = 2|s_{t-1} = 2) = \frac{e^{q_0}}{1+e^{q_0}}$$

The GARCH model: Two-series GARCH model with first series as return on futures.

$$H_{t,s_t} = \begin{bmatrix} h_{1,t,s_t} & 0 \\ 0 & h_{2,t,s_t} \end{bmatrix} \begin{bmatrix} 1 & \rho_{t,s_t} \\ \rho_{t,s_t} & 1 \end{bmatrix} \begin{bmatrix} h_{1,t,s_t} & 0 \\ 0 & h_{2,t,s_t} \end{bmatrix}$$
$$h^2_{1,t,s_t} = \gamma_{1,s_t} + \alpha_{1,s_t} e^2_{1.t-1} + \beta_{1,s_t} h^2_{1,t-1},$$
$$h^2_{2,t,s_t} = \gamma_{2,s_t} + \alpha_{2,s_t} e^2_{2.t-1} + \beta_{2,s_t} h^2_{2,t-1}$$
$$\rho_{t,s_t} = (1 - \theta_{1,s_t} - \theta_{2,s_t})\rho + \theta_{1,s_t}\rho_{t-1} + \theta_{2,s_t}\phi_{t-1}$$
$$\phi_{t-1} = \frac{\sum_{j=1}^{2} \varepsilon_{1,t-j}\varepsilon_{2,t-j}}{\sqrt{\left(\sum_{j=1}^{2} \varepsilon^2_{1,t-j}\right)\left(\sum_{j=1}^{2} \varepsilon^2_{2,t-j}\right)}},$$
$$\varepsilon_{i,t} = \frac{e_{i,t}}{h_{it}}, \quad \theta_1, \theta_2 \geq 0 \text{ and } \theta_1 + \theta_2 \leq 1,$$

$$\text{Hedge ration} = \frac{H_{t,s_t}(1,2)}{H_{t,s_t}(2,2)}$$

A.5. Random Coefficient:

(M$_5$): $\Delta S_t = \beta_0 + \beta_t \Delta F_t + e_t$
$\beta_t = \bar{\beta} + v_t, \quad \text{Hedge ration} = \bar{\beta}$

A.6. Cointegration and Error Correction:

(M$_6$):
$S_t = a + bF_t + u_t$

$$\Delta S_t = \rho u_{t-1} + \beta \Delta F_t + \sum_{i=1}^{m} \delta_i \Delta F_{t-i}$$
$$+ \sum_{j=1}^{n} \theta_i \Delta S_{t-j} + e_j, \quad EC \text{ Hedge ration} = \beta$$

A.7. Error Correction with GARCH:

(M7): $\begin{bmatrix} \Delta \log_e(S_t) \\ \Delta \log_e(F_t) \end{bmatrix} = \begin{bmatrix} \mu_1 \\ \mu_2 \end{bmatrix}$
$+ \begin{bmatrix} \alpha_s(\log_e(S_{t-1}) - \log_e(F_{t-1})) \\ \alpha_f(\log_e(S_{t-1}) - \log_e(F_{t-1})) \end{bmatrix} + \begin{bmatrix} e_{1t} \\ e_{2t} \end{bmatrix}$,
$e_t|\Omega_{t-1} \sim N(0, H_t), H_t = \begin{bmatrix} H_{11,t} & H_{12,t} \\ H_{12,t} & H_{22,t} \end{bmatrix}$

$$\text{Hedge ration} = h_{t-1} = H_{12,t}/H_{22,t}$$

A.8. Common Stochastic Trend:

$$(\mathrm{M}_8):\quad S_t = A_1P_t + A_2\tau_t,\ F_t = B_1P_t + B_2\tau_t,$$
$$P_t = P_{t-1} + w_t,$$
$$\tau_t = \alpha_1\tau_{t-1} + v_t,\quad 0 \le |\alpha_1| < 1,$$

Hedge ration for k-period investment horizon

$$= H_J^* = \frac{A_1B_1k\sigma_w^2 + 2A_2B_2\left(\frac{(1-\alpha^k)}{1-\alpha^2}\right)\sigma_v^2}{B_1^2k\sigma_w^2 + 2B_2^2\left(\frac{(1-\alpha^k)}{1-\alpha^2}\right)\sigma_v^2}.$$

B. Optimum Mean-Variance Hedge Ratio:

(M_9): Hedge ratio $= h_2 = -\frac{C_f^*F}{C_sS} = -\left[\frac{E(R_f)}{A\sigma_f^2} - \rho\frac{\sigma_s}{\sigma_f}\right]$, where the moments $E[R_f], \sigma_s$ and σ_f are estimated by sample moments.

C. Sharpe Hedge Ratio:

(M_{10}): *Hedge ratio* =

$$h_3 = -\frac{\left(\frac{\sigma_s}{\sigma_f}\right)\left[\frac{\sigma_s}{\sigma_f}\left(\frac{E(R_f)}{E(R_s)-i}\right)-\rho\right]}{\left[1-\frac{\sigma_s}{\sigma_f}\left(\frac{E(R_f)\rho}{E(R_s)-i}\right)\right]},$$ where the moments and correlation are estimated by their sample counterparts.

D. Mean-Gini Coefficient-Based Hedge Ratios:

(M_{11}): The hedge ratio is estimated by numerically minimizing the following mean extended-Gini coefficient, where the cumulative probability distribution function is estimated using the rank function:

$$\widehat{\Gamma}_v(R_h) = -\frac{v}{N}\sum_{i=1}^{N}\left(R_{h,i} - \overline{R}_h\right)\left(\left(1 - G(R_{h,i})\right)^{v-1} - \Theta\right).$$

(M_{12}): The hedge ratio is estimated by numerically solving the first-order condition, where the cumulative probability distribution function is estimated using the rank function.

(M_{13}): The hedge ratio is estimated by numerically solving the first-order condition, where the cumulative probability distribution function is estimated using the kernel-based estimates.

(M_{14}): The hedge ratio is estimated by numerically maximizing the following function:

$$U(R_h) = E(R_h) - \Gamma_v(R_h)$$

where the expected values and the mean extended-Gini coefficient are replaced by their sample counterparts and the cumulative probability distribution function is estimated using the rank function.

E. Generalized Semivariance-Based Hedge Ratios:

(M_{15}): The hedge ratio is estimated by numerically minimizing the following sample generalized hedge ratio:

$$V_{\delta,\alpha}^{\text{sample}}(R_h) = \frac{1}{N}\sum_{i=1}^{N}(\delta - R_{h,i})^\alpha U(\delta - R_{h,i}),$$

$$\text{where } U(\delta - R_{h,i}) = \begin{cases} 1 & \text{for } \delta \ge R_{h,i} \\ 0 & \text{for } \delta < R_{h,i} \end{cases}.$$

(M_{16}): The hedge ratio is estimated by numerically maximizing the following function:

$$U(R_h) = R_h - V_{\delta,\alpha}^{\text{sample}}(R_h).$$

F. Minimum Value-at-Risk Hedge Ratio:

(M_{17}): The hedge ratio is estimated by minimizing the following value at risk:

$$\mathrm{VaR}(R_h) = Z_\alpha\sigma_h\sqrt{\tau} - E[R_h]\tau$$

The resulting hedge ratio is given by

$$h^{\mathrm{VaR}} = \rho\frac{\sigma_s}{\sigma_f} - E[R_f]\frac{\sigma_s}{\sigma_f}\sqrt{\frac{1-\rho^2}{Z_\alpha^2\sigma_f^2 - E[R_f]^2}}$$

Appendix 3: Monthly Data of S&P 500 Index and Its Futures

(January 2000–June 2015)

Date	SPOT	FUTURES	C_spot	C_future
2000M01	1394.46	1401	−28.04	−29
2000M02	1366.42	1372	132.16	143.3
2000M03	1498.58	1515.3	−46.15	−55.3
2000M04	1452.43	1460	−31.83	−37.8
2000M05	1420.6	1422.2	34	45.9
2000M06	1454.6	1468.1	−23.77	−29.2
2000M07	1430.83	1438.9	86.85	82.3
2000M08	1517.68	1521.2	−81.17	−67.5
2000M09	1436.51	1453.7	−7.11	−13.5
2000M10	1429.4	1440.2	−114.45	−118.7
2000M11	1314.95	1321.5	5.33	13.5
2000M12	1320.28	1335	45.73	37.9
2001M01	1366.01	1372.9	−126.07	−130.9
2001M02	1239.94	1242	−79.61	−72.8
2001M03	1160.33	1169.2	89.13	85.1
2001M04	1249.46	1254.3	6.36	3.1
2001M05	1255.82	1257.4	−31.4	−25.7
2001M06	1224.42	1231.7	−13.19	−16.4
2001M07	1211.23	1215.3	−77.65	−80.2
2001M08	1133.58	1135.1	−92.64	−91.4
2001M09	1040.94	1043.7	18.84	17
2001M10	1059.78	1060.7	79.67	79.3
2001M11	1139.45	1140	8.63	9.2
2001M12	1148.08	1149.2	−17.88	−18.8
2002M01	1130.2	1130.4	−23.47	−23.5
2002M02	1106.73	1106.9	40.66	42.3
2002M03	1147.39	1149.2	−70.47	−72
2002M04	1076.92	1077.2	−9.78	−9.7
2002M05	1067.14	1067.5	−77.33	−77.4
2002M06	989.81	990.1	−78.19	−78.6
2002M07	911.62	911.5	4.45	4.6
2002M08	916.07	916.1	−100.79	−101.1
2002M09	815.28	815	70.48	70.4
2002M10	885.76	885.4	50.55	50.6
2002M11	936.31	936	−56.49	−57.1
2002M12	879.82	878.9	−24.12	−24.2
2003M01	855.7	854.7	−14.55	−13.8

(continued)

Date	SPOT	FUTURES	C_spot	C_future
2003M02	841.15	840.9	7.03	6.1
2003M03	848.18	847	68.74	69.1
2003M04	916.92	916.1	46.67	47.2
2003M05	963.59	963.3	10.91	10
2003M06	974.5	973.3	15.81	16
2003M07	990.31	989.3	17.7	18.4
2003M08	1008.01	1007.7	−12.04	−13.6
2003M09	995.97	994.1	54.74	55.4
2003M10	1050.71	1049.5	7.49	8.3
2003M11	1058.2	1057.8	53.72	52.8
2003M12	1111.92	1110.6	19.21	19.3
2004M01	1131.13	1129.9	13.81	14.7
2004M02	1144.94	1144.6	−18.73	−19.7
2004M03	1126.21	1124.9	−18.91	−18.8
2004M04	1107.3	1106.1	13.38	14.2
2004M05	1120.68	1120.3	20.16	20.1
2004M06	1140.84	1140.4	−39.12	−39.3
2004M07	1101.72	1101.1	2.52	3
2004M08	1104.24	1104.1	10.34	10.8
2004M09	1114.58	1114.9	15.62	15.4
2004M10	1130.2	1130.3	43.62	43.8
2004M11	1173.82	1174.1	38.1	39.6
2004M12	1211.92	1213.7	−30.65	−32
2005M01	1181.27	1181.7	22.33	22.4
2005M02	1203.6	1204.1	−23.01	−20.2
2005M03	1180.59	1183.9	−23.74	−25.4
2005M04	1156.85	1158.5	34.65	33.8
2005M05	1191.5	1192.3	−0.17	3.2
2005M06	1191.33	1195.5	42.85	41.3
2005M07	1234.18	1236.8	−13.85	−15.4
2005M08	1220.33	1221.4	8.48	12.9
2005M09	1228.81	1234.3	−21.8	−24.5
2005M10	1207.01	1209.8	42.47	41.3
2005M11	1249.48	1251.1	−1.19	3.7
2005M12	1248.29	1254.8	31.79	28.8
2006M01	1280.08	1283.6	0.58	−1.2
2006M02	1280.66	1282.4	14.17	20.9
2006M03	1294.83	1303.3	15.78	12.6
2006M04	1310.61	1315.9	−40.52	−43.8
2006M05	1270.09	1272.1	0.11	7.3

(continued)

Date	SPOT	FUTURES	C_spot	C_future
2006M06	1270.2	1279.4	6.46	2.4
2006M07	1276.66	1281.8	27.16	23.8
2006M08	1303.82	1305.6	32.03	39.8
2006M09	1335.85	1345.4	42.09	37.8
2006M10	1377.94	1383.2	22.69	19.7
2006M11	1400.63	1402.9	17.67	25.5
2006M12	1418.3	1428.4	19.94	14.6
2007M01	1438.24	1443	−31.42	−34.1
2007M02	1406.82	1408.9	14.04	22.3
2007M03	1420.86	1431.2	61.51	57.2
2007M04	1482.37	1488.4	48.25	44.5
2007M05	1530.62	1532.9	−27.27	−17.5
2007M06	1503.35	1515.4	−48.08	−53.5
2007M07	1455.27	1461.9	18.72	14.8
2007M08	1473.99	1476.7	52.76	61.4
2007M09	1526.75	1538.1	22.63	16.8
2007M10	1549.38	1554.9	−68.24	−71.2
2007M11	1481.14	1483.7	−12.79	−6.5
2007M12	1468.35	1477.2	−89.8	−97.6
2008M01	1378.55	1379.6	−47.92	−48.3
2008M02	1330.63	1331.3	−7.93	−7.3
2008M03	1322.7	1324	62.89	62
2008M04	1385.59	1386	14.79	14.6
2008M05	1400.38	1400.6	−120.38	−119.5
2008M06	1280	1281.1	−12.62	−14
2008M07	1267.38	1267.1	15.45	15.5
2008M08	1282.83	1282.6	−116.47	−113.6
2008M09	1166.36	1169	−197.61	−201.7
2008M10	968.75	967.3	−72.51	−72
2008M11	896.24	895.3	7.01	4.8
2008M12	903.25	900.1	−77.37	−77.6
2009M01	825.88	822.5	−90.79	−88.3
2009M02	735.09	734.2	62.78	60.6
2009M03	797.87	794.8	74.94	75.2
2009M04	872.81	870	46.33	48.1
2009M05	919.14	918.1	0.18	−2.6
2009M06	919.32	915.5	68.16	68.9
2009M07	987.48	984.4	33.14	35.3
2009M08	1020.62	1019.7	36.46	33.2
2009M09	1057.08	1052.9	−20.89	−19.9

(continued)

Date	SPOT	FUTURES	C_spot	C_future
2009M10	1036.19	1033	59.44	61.8
2009M11	1095.63	1094.8	19.47	15.9
2009M12	1115.1	1110.7	−41.23	−40.3
2010M01	1073.87	1070.4	30.62	33
2010M02	1104.49	1103.4	64.94	61.8
2010M03	1169.43	1165.2	17.26	18.2
2010M04	1186.69	1183.4	−97.28	−94.9
2010M05	1089.41	1088.5	−58.7	−61.9
2010M06	1030.71	1026.6	70.89	71.7
2010M07	1101.6	1098.3	−52.27	−50
2010M08	1049.33	1048.3	91.87	88.4
2010M09	1141.2	1136.7	42.06	43
2010M10	1183.26	1179.7	−2.71	−0.1
2010M11	1180.55	1179.6	77.09	73.4
2010M12	1257.64	1253	28.48	29.4
2011M01	1286.12	1282.4	41.1	43.7
2011M02	1327.22	1326.1	−1.39	−5.1
2011M03	1325.83	1321	37.78	38.7
2011M04	1363.61	1359.7	−18.41	−15.8
2011M05	1345.2	1343.9	−24.56	−28.4
2011M06	1320.64	1315.5	−28.36	−27.1
2011M07	1292.28	1288.4	−73.39	−70.7
2011M08	1218.89	1217.7	−87.47	−91.7
2011M09	1131.42	1126	121.88	123.3
2011M10	1253.3	1249.3	−6.34	−3.3
2011M11	1246.96	1246	10.64	6.6
2011M12	1257.6	1252.6	54.81	55.6
2012M01	1312.41	1308.2	53.27	56.2
2012M02	1365.68	1364.4	42.79	38.8
2012M03	1408.47	1403.2	−10.56	−9.6
2012M04	1397.91	1393.6	−87.58	−84.4
2012M05	1310.33	1309.2	51.83	47.2
2012M06	1362.16	1356.4	17.16	18.2
2012M07	1379.32	1374.6	27.26	30.5
2012M08	1406.58	1405.1	34.09	29.1
2012M09	1440.67	1434.2	−28.51	−27.4
2012M10	1412.16	1406.8	4.02	7.6
2012M11	1416.18	1414.4	10.01	5.7
2012M12	1426.19	1420.1	71.92	73.2
2013M01	1498.11	1493.3	16.57	20

(continued)

Date	SPOT	FUTURES	C_spot	C_future
2013M02	1514.68	1513.3	54.51	49.4
2013M03	1569.19	1562.7	28.38	29.5
2013M04	1597.57	1592.2	33.17	36.8
2013M05	1630.74	1629	−24.46	−29.7
2013M06	1606.28	1599.3	79.44	81.2
2013M07	1685.72	1680.5	−52.75	−49.2
2013M08	1632.97	1631.3	48.58	43
2013M09	1681.55	1674.3	74.99	76.7
2013M10	1756.54	1751	49.27	53.1
2013M11	1805.81	1804.1	42.55	37
2013M12	1848.36	1841.1	−65.77	−64.5
2014M01	1782.59	1776.6	76.86	81
2014M02	1859.45	1857.6	12.88	7
2014M03	1872.33	1864.6	11.62	13.3
2014M04	1883.95	1877.9	39.62	43.6
2014M05	1923.57	1921.5	36.66	30.9
2014M06	1960.23	1952.4	−29.56	−27.6
2014M07	1930.67	1924.8	72.7	76.6
2014M08	2003.37	2001.4	−31.09	−35.9
2014M09	1972.28	1965.5	45.77	45.9
2014M10	2018.05	2011.4	49.51	54.9
2014M11	2067.56	2066.3	−8.66	−13.9
2014M12	2058.9	2052.4	−63.91	−64
2015M01	1994.99	1988.4	109.51	114.4
2015M02	2104.5	2102.8	−36.61	−42
2015M03	2067.89	2060.8	17.62	18.1
2015M04	2085.51	2078.9	21.88	27.1
2015M05	2107.39	2106	−44.28	−51.6
2015M06	2063.11	2054.4	40.73	44

Bibliography

Anderson T. W. (1971). *The statistical analysis of time series analysis*. New York: Wiley.

Baillie, R. T., & Myers, R. J. (1991). Bivariate Garch estimation of the optimal commodity futures hedge. *Journal of Applied Econometrics, 6,* 109–124.

Bawa, V. S. (1978). Safety-first, stochastic dominance, and optimal portfolio choice. *Journal of Financial and Quantitative Analysis, 13,* 255–271.

Benet, B. A. (1992). Hedge period length and ex-ante futures hedging effectiveness: The case of foreign-exchange risk cross hedges. *Journal of Futures Markets, 12,* 163–175.

Cecchetti, S. G., Cumby, R. E., & Figlewski, S. (1988). Estimation of the optimal futures hedge. *Review of Economics and Statistics, 70,* 623–630.

Chen, S. S., Lee, C. F., & Shrestha, K. (2001). On a mean-generalized semivariance approach to determining the hedge ratio. *Journal of Futures Markets, 21,* 581–598.

Cheung, C. S., Kwan, C. C. Y., & Yip, P. C. Y. (1990). The hedging effectiveness of options and futures: a mean-Gini approach. *Journal of Futures Markets, 10,* 61–74.

Chou, W. L., Fan, K. K., & Lee, C. F. (1996). Hedging with the Nikkei index futures: The conventional model versus the error correction model. *Quarterly Review of Economics and Finance, 36,* 495–505.

Crum, R. L., Laughhunn, D. L., & Payne, J. W. (1981). Risk-seeking behavior and its implications for financial models. *Financial Management, 10,* 20–27.

D'Agostino, R. B. (1971). An omnibus test of normality for moderate and large size samples. *Biometrika, 58,* 341–348.

De Jong, A., De Roon, F., & Veld, C. (1997). Out-of-sample hedging effectiveness of currency futures for alternative models and hedging strategies. *Journal of Futures Markets, 17,* 817–837.

Dickey, D. A., & Fuller, W. A. (1981). Likelihood ratio statistics for autoregressive time series with a unit root. *Econometrica, 49,* 1057–1072.

Ederington, L. H. (1979). The hedging performance of the new futures markets. *Journal of Finance, 34,* 157–170.

Engle, R. F., & Granger, C. W. (1987). Co-integration and error correction: Representation, estimation and testing. *Econometrica, 55,* 251–276.

Fishburn, P. C. (1977). Mean-risk analysis with risk associated with below-target returns. *American Economic Review, 67,* 116–126.

Geppert, J. M. (1995). A statistical model for the relationship between futures contract hedging effectiveness and investment horizon length. *Journal of Futures Markets, 15,* 507–536.

Ghosh, A. (1993). Hedging with stock index futures: Estimation and forecasting with error correction model. *Journal of Futures Markets, 13,* 743–752.

Grammatikos, T., & Saunders, A. (1983). Stability and the hedging performance of foreign currency futures. *Journal of Futures Markets, 3,* 295–305.

Greene, W. H. (2017). *Econometric analysis* (8th ed.). New Jersey: Prentice Hall.

Hamilton, James D., Time Series Analysis. Princeton University Press, 1994.

Hansen, P. R., & Lunde, A. (2005). A forecast comparison of volatility models: does anything beat a garch (1,1)? *Journal of Applied Econometrics, 20,* 873–889.

Howard, C. T., & D'Antonio, L. J. (1984). A risk-return measure of hedging effectiveness. *Journal of Financial and Quantitative Analysis, 19,* 101–112.

Hsin, C. W., Kuo, J., & Lee, C. F. (1994). A new measure to compare the hedging effectiveness of foreign currency futures versus options. *Journal of Futures Markets, 14,* 685–707.

Hung, J. C., Chiu, C. L., & Lee, M. C. (2006). Hedging with zero-value at risk hedge ratio. *Applied Financial Economics, 16,* 259–269.

Hylleberg, S., & Mizon, G. E. (1989). Cointegration and error correction mechanisms. *Economic Journal, 99,* 113–125.

Jarque, C. M., & Bera, A. K. (1987). A test for normality of observations and regression residuals. *International Statistical Review, 55,* 163–172.

Johansen, S., & Juselius, K. (1990). Maximum likelihood estimation and inference on cointegration—With applications to the demand for money. *Oxford Bulletin of Economics and Statistics, 52,* 169–210.

Johnson, L. L. (1960). The theory of hedging and speculation in commodity futures. *Review of Economic Studies, 27,* 139–151.

Junkus, J. C., & Lee, C. F. (1985). Use of three index futures in hedging decisions. *Journal of Futures Markets, 5,* 201–222.

Kolb, R. W., & Okunev, J. (1992). An empirical evaluation of the extended mean-Gini coefficient for futures hedging. *Journal of Futures Markets, 12,* 177–186.

Kolb, R. W., & Okunev, J. (1993). Utility maximizing hedge ratios in the extended mean Gini framework. *Journal of Futures Markets, 13,* 597–609.

Kroner, K. F., & Sultan, J. (1993). Time-varying distributions and dynamic hedging with foreign currency futures. *Journal of Financial and Quantitative Analysis, 28,* 535–551.

Lee, C. F., Bubnys, E. L., & Lin, Y. (1987). Stock index futures hedge ratios: Test on horizon effects and functional form. *Advances in Futures and Options Research, 2,* 291–311.

Lee, H. T., & Yoder, J. (2007). Optimal hedging with a regime-switching time-varying correlation GARCH model. *Journal of Futures Markets, 27,* 495–516.

Lee, Cheng Few, Lee, John, Chang, Jow-Ran, & Tai, Tzu. (2016). *Essentials of Excel, Excel VBA, SAS and Minitab for statistical and financial analyses*. New York: Springer.

Lence, S. H. (1995). The economic value of minimum-variance hedges. *American Journal of Agricultural Economics, 77,* 353–364.

Lence, S. H. (1996). Relaxing the assumptions of minimum variance hedging. *Journal of Agricultural and Resource Economics, 21,* 39–55.

Lien, D., & Luo, X. (1993a). Estimating the extended mean-Gini coefficient for futures hedging. *Journal of Futures Markets, 13,* 665–676.

Lien, D., & Luo, X. (1993b). Estimating multiperiod hedge ratios in cointegrated markets. *Journal of Futures Markets, 13,* 909–920.

Lien, D., & Shaffer, D. R. (1999). Note on estimating the minimum extended Gini hedge ratio. *Journal of Futures Markets, 19,* 101–113.

Lien, D., & Shrestha, K. (2007). An empirical analysis of the relationship between hedge ratio and hedging horizon using wavelet analysis. *Journal of Futures Markets, 27,* 127–150.

Lien, D., & Shrestha, K. (2010). Estimating optimal hedge ratio: A multivariate skew-normal distribution. *Applied Financial Economics, 20,* 627–636.

Lien, D., & Tse, Y. K. (1998). Hedging time-varying downside risk. *Journal of Futures Markets, 18,* 705–722.

Lien, D., & Tse, Y. K. (2000). Hedging downside risk with futures contracts. *Applied Financial Economics, 10,* 163–170.

Malliaris, A. G., & Urrutia, J. L. (1991). The impact of the lengths of estimation periods and hedging horizons on the effectiveness of a hedge: Evidence from foreign currency futures. *Journal of Futures Markets, 3,* 271–289.

Myers, R. J., & Thompson, S. R. (1989). Generalized optimal hedge ratio estimation. *American Journal of Agricultural Economics, 71,* 858–868.

Osterwald-Lenum, M. (1992). A note with quantiles of the asymptotic distribution of the maximum likelihood cointegration rank test statistics. *Oxford Bulletin of Economics and Statistics, 54,* 461–471.

Phillips, P. C. B., & Perron, P. (1988). Testing unit roots in time series regression. *Biometrika, 75,* 335–346.

Rutledge, D. J. S. (1972). Hedgers' demand for futures contracts: A theoretical framework with applications to the United States soybean complex. *Food Research Institute Studies, 11,* 237–256.

Sephton, P. S. (1993a). Hedging wheat and canola at the Winnipeg commodity exchange. *Applied Financial Economics, 3,* 67–72.

Sephton, P. S. (1993b). Optimal hedge ratios at the Winnipeg commodity exchange. *Canadian Journal of Economics, 26,* 175–193.

Shalit, H. (1995). Mean-Gini hedging in futures markets. *Journal of Futures Markets, 15,* 617–635.

Stock, J. H., & Watson, M. W. (1988). Testing for common trends. *Journal of the American Statistical Association, 83,* 1097–1107.

Tsay, R. S. (2010). *Analysis of financial time series* (3rd edn.).

Working, H. (1953). Hedging reconsidered. *Journal of Farm Economics, 35,* 544–556.

Part III

Statistical Distributions, Option Pricing Model and Risk Management

Statistical distributions such as binomial distribution, multinomial distribution, normal distribution, log-normal distribution, Poisson distribution, central chi-square distribution, non-central chi-square distribution, copula distribution, nonparametric distribution, and other distributions are important in finance research. In Chap. 12, we will discuss how binomial and multinomial distribution can be used to derive the option pricing model. In Chap. 13, we show how to use two alternative binomial option pricing model approaches to derive Black–Scholes option pricing model.

In Chap. 14, we will discuss how normal and log-normal distribution can be used to derive the option pricing model. In Chap. 15, we will show how copula distribution can be used to do credit risk analysis. In Chap. 16, we will show how multivariate analysis such as factor analysis and discriminant analysis can be used to do financial rating analysis. In Part IV, we will continue to discuss how statistics distribution can be used to derive the option pricing model. In addition, we will also show how Itô's calculus can be used to derive the option pricing model.

The Binomial, Multinomial Distributions, and Option Pricing Model

12

Contents

Abstract

In this chapter, we show how to use binomial and mutinomial distributions to derive option pricing models. In addition, we show how the Black and Scholes option pricing model is a limited case of binomial and multinomial option pricing model. Finally, a lattice framework of option pricing model is discussed in some detail.

12.1 Introduction

Based upon Lee et al. (2013) and Lee and Lee (2010a, b), we show how the binomial and multinomial distribution can be used to derive the call option pricing model. We define the basic definition of option. In addition, we define and examine the simple binomial option pricing model. Specifically, we derive the generalized n-period binominal option pricing model.

C.-F. Lee et al., *Financial Econometrics, Mathematics and Statistics*,
https://doi.org/10.1007/978-1-4939-9429-8_12

Instead of two possible movements for the stock price, as considered by Cox et al. (1979) and Rendleman and Barter (1979), it is natural to extend it to the situation in which there are $k + 1$ possible price movements. We therefore present the extension proposed by Madan et al. (1989). More details are provided and should be helpful to the readers. We also derive the multinomial option pricing model and the Black and Scholes model as a limiting case.

We will discuss binomial distribution in Sect. 12.2. The simple binomial option pricing model will be discussed in Sect. 12.3. The generalized binomial option pricing model is demonstrated in Sect. 12.4. Section 12.5 will show multiple option pricing models. A lattice framework for option pricing will be shown in Sect. 12.6. Finally, Sect. 12.7 concludes the chapter.

12.2 Binomial Distribution

Binomial Distribution

If n trials of a Bernoulli process are observed, then the total number of successes in the n trials is a random variable, and the associated probability distribution is known as a **binomial distribution**. The number of successes, the number of trials, and the probability of success on a trial are the three pieces of information we need to generate a binomial distribution.

To develop the binomial distribution, assume that each of the n trials of an experiment will generate one of two outcomes, a success, S, or a failure, F. Suppose the trials generate x successes and $(n - x)$ failures. The probability of success on a particular trial is p, and the probability of failure is $(1 - p)$. Thus, the probability of obtaining a specific sequence of outcomes is

$$p^x(1-p)^{n-x} \tag{12.1}$$

Equation (12.1) presents the joint probability of x successes and $(n - x)$ failures occurring simultaneously. Because the n trials are independent of each other, the probability of any particular sequence of outcomes is, by the multiplication rule of probabilities, equal to the product of the probabilities for the individual outcomes.

Probability Function

There are several ways in which x successes can be arranged among $(n - x)$ failures. Therefore, the probability of x successes in n trials for a binomial random variable X is

$$\begin{aligned} P(X=x) &= \binom{n}{x} p^x(1-p)^{n-x} \\ &= \frac{n!}{x!(n-x)!} p^x(1-p)^{n-x}, \\ x = 0, 1, \ldots, n, \end{aligned} \tag{12.2}$$

where

$$\binom{n}{x} = n \text{ combinations taken } x \text{ at a time}$$

$$n! = n(n-1)(n-2)(n-3) \tag{12.1}$$

The symbol $n!$ is read "n factorial." When $n = 0$, then $n! = 0! = 1$. Equation (12.2) is the **binomial probability function**, which gives the probability of x successes in n trials: Using this formula, we can evaluate a binomial probability.

Example 12.1 (Probability Distribution for JNJ Stock)

Suppose that the price of a share of stock in Johnson & Johnson company in the future will either go up (U) or come down (D) in 1 d with the probabilities 0.40 and 0.60, respectively. Calculate the probability of each possible outcome of the stock price 4 days later.[1]

Using the outcome tree approach, we find the possible outcomes e and probabilities $p(e)$ indicated in Table 12.1.

[1] Assume that the price movement of JNJ stock today is completely independent of its movement in the past.

Table 12.1 Probability distribution of JNJ stock 4 days later

Outcome, *e*	Probability, *p*(*e*)	
e_1	(*UUUU*)	(0.4)(0.4)(0.4)(0.4) = 0.0256
e_2	(*UUUD*)	(0.4)(0.4)(0.4)(0.6) = 0.0384
e_3	(*UUDU*)	(0.4)(0.4)(0.6)(0.4) = 0.0384
e_4	(*UUDD*)	(0.4)(0.4)(0.6)(0.6) = 0.0576
e_5	(*UDUU*)	(0.4)(0.6)(0.4)(0.4) = 0.0384
e_6	(*UDUD*)	(0.4)(0.6)(0.4)(0.6) = 0.0576
e_7	(*UDDU*)	(0.4)(0.6)(0.6)(0.4) = 0.0576
e_8	(*UDDD*)	(0.4)(0.6)(0.6)(0.6) = 0.0864
e_9	(*DUUU*)	(0.6)(0.4)(0.4)(0.4) = 0.0384
e_{10}	(*DUUD*)	(0.6)(0.4)(0.4)(0.6) = 0.0576
e_{11}	(*DUDU*)	(0.6)(0.4)(0.6)(0.4) = 0.0576
e_{12}	(*DUDD*)	(0.6)(0.4)(0.6)(0.6) = 0.0864
e_{13}	(*DDUU*)	(0.6)(0.6)(0.4)(0.4) = 0.0576
e_{14}	(*DDUD*)	(0.6)(0.6)(0.4)(0.6) = 0.0864
e_{15}	(*DDDU*)	(0.6)(0.6)(0.6)(0.4) = 0.0864
e_{16}	(*DDDD*)	(0.6)(0.6)(0.6)(0.6) = 0.1296

The probability of JNJ stock going up three times and coming down once is the sum of the probabilities associated with e_2, e_3, e_5, and e_9: 0.0384 + 0.0384 + 0.0384 + 0.0384 = 0.1536.

Alternatively, this probability can be calculated in terms of the binomial combination formula (Eq. 12.2).

$$\binom{4}{3}(0.4)^3(0.6) = \frac{4!}{(4-1)!1!}(0.0384)$$
$$= 0.1536$$

Hence, the binomial combination formula can be used to replace the diagram for calculating such a probability.

Example 12.2 (*Probability Function of Insurance Sales*)

Assume that an insurance sales agent believes that the probability of her making a sale is 0.20. She makes five contacts and, eager to leave nothing to chance, calculates a binomial distribution.

$$P(0\text{ success}) = \frac{5!}{0!5!}0.2^0 0.8^5 = 0.3277$$

$$P(1\text{ success}) = \frac{5!}{1!4!}0.2^1 0.8^4 = 0.4096$$

$$P(2\text{ successes}) = \frac{5!}{2!3!}0.2^2 0.8^3 = 0.2048$$

$$P(3\text{ successes}) = \frac{5!}{3!2!}0.2^3 0.8^2 = 0.0512$$

$$P(4\text{ successes}) = \frac{5!}{4!1!}0.2^4 0.8^1 = 0.0064$$

$$P(5\text{ successes}) = \frac{5!}{5!0!}0.2^5 0.8^0 = 0.0003$$

Alternatively, these numbers can be calculated by the MINITAB program as shown here

```
MTB > SET INTO Cl
DATA > 0 1 2 3 4 5
DATA > END
MTB > PDF C1;
SUBC > BINOMIAL 5 0.2.
```

Probability Density Function
Binomial with $n = 5$ and $p = 0.200000$

x	$P(X = x)$
0.00	0.3277
1.00	0.4096
2.00	0.2048
3.00	0.0512
4.00	0.0064
5.00	0.0003

Figure 12.1 gives the probability distribution for this sales agent's successes. Because the events of the sales agent's number of successes are mutually exclusive, the probability that she has 3 or more successes is equal to P(3 successes) + P(4 successes) + P(5 successes) = 0.0512 + 0.0064 + 0.0003 = 0.0579.

Example 12.3 (Cumulative Probability Distribution for Insurance Sales)
Suppose the sales agent we met in Example 12.2 wants to determine the probability of making between 1 and 4 sales.

$$P(1\,\text{success}) + P(2\,\text{successes}) + P(3\,\text{successes}) + P(4\,\text{successes}) = 0.672$$

Unless the number of trials n is very small, it is easier to determine binomial probabilities by using Table 12.2. All three variables listed in

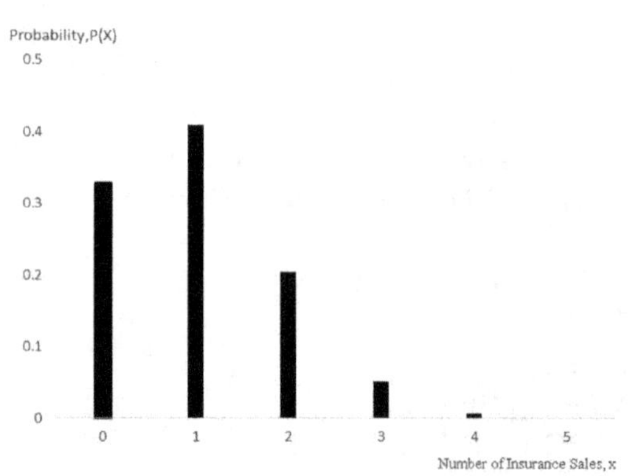

Fig. 12.1 Binomial probability distribution for Example 8.2 ($n = 5$, $p = 0.2$)

Eq. (12.2) (n, p, and x) appear in the binomial distribution table extracted from the National Bureau of Standard Tables. Using probabilities from this table, we can calculate both individual probabilities and cumulative probabilities.

The individual probabilities drawn for Example 12.3 from the binomial table are listed in Table 12.2. These probabilities are identical to those we found with Eq. (12.2).

The cumulative binomial function can be defined as

$$B(n,p) = \sum_{x=0}^{n} \binom{n}{x} p^x (1-p)^{n-x} \tag{12.3}$$

Using Table 12.2, we can calculate the cumulative probabilities for the sales agent having two or more successes.

$$\begin{aligned} P(X \geq 2 | n = 5, p = 0.2) &= P(X = 2) + P(X = 3) \\ &\quad + P(X = 4) + P(X = 5) \\ &= \sum_{x=2}^{5} \binom{n}{x} 0.2^x 0.8^{5-x} \\ &= 0.2048 + 0.0512 \\ &\quad + 0.0064 + 0.0003 = 0.2627 \end{aligned}$$

In a nationwide poll of 2052 adults by the American Association of Retired Persons (*USA Today,* August 8, 1985), approximately 40% of those surveyed described the current version of the federal income tax system as fair. Suppose we randomly sample 20 of the 2052 adults surveyed and record x as the number who think the federal income tax system is fair. To a reasonable degree of approximation, x is a binomial random variable. The probability that x is less than or equal to 10 can be defined as

Table 12.2 Part of binomial table ($n = 5$, $p = 0.2$)

x	$P(x)$
5	0.0003
4	0.0064
3	0.0512
2	0.2048
1	0.4096
0	0.3277

$$
\begin{aligned}
&P(X \le 10|n = 20, p = 0.4) \\
&= \sum_{x=1}^{10} \binom{n}{x} (0.4)^x (0.6)^{20-x} \\
&= 0 + 0.005 + 0.0031 + 0.0123 \\
&\quad + 0.0350 + 0.0746 + 0.1659 + 0.1797 \\
&\quad + 0.1597 + 0.1171 = 0.8725
\end{aligned}
$$

Another situation that requires the use of a binomial random variable is **lot acceptance sampling**, where we must decide, on the basis of *sample* information about the quality of the lot, whether to accept a lot (batch) of goods delivered from a manufacturer. It is possible to calculate the probability of accepting a shipment with any given proportion of defectives in accordance with Eq. (12.2).

Example 12.4 (Cumulative Probability Distribution: A Shipment of Calculator Chips)

A shipment of 800 calculator chips arrives at Century Electronics. The contract specifies that Century will accept this lot if a sample of size 20 drawn from the shipment has no more than one defective chip. What is the probability of accepting the lot by applying this criterion if, in fact, 5% of the whole lot (40 chips) turns out to be defective? What if 10% of the lot is defective?

This is a binomial situation where there are n = 20 trials, and p = the probability of success (chip is defective) = 0.05. The shipment is accepted if the number of defectives is either 0 or 1, so the probability of the shipment being accepted is

$$
\begin{aligned}
P(\text{shipment accepted}) &= P(X \le 1) \\
&= P(0) + P(1)
\end{aligned}
$$

Using Table 12.2 (n = 20, p = 0.05), we obtain $P(0)$ = 0.3585 and $P(1)$ = 0.3774. Hence, the probability that Century Electronics accepts delivery is

$$
\begin{aligned}
P(\text{shipment accepted}) &= 0.3585 + 0.3774 \\
&= 0.7359
\end{aligned}
$$

Similarly, if 10% of the items in the shipment are defective (that is, if p = 0.10), then

$$
\begin{aligned}
P(\text{shipment accepted}) &= 0.1216 + 0.2702 \\
&= 0.3918
\end{aligned}
$$

This implies that the higher the proportion of defectives in the shipment, the less likely is acceptance of the delivery. And that's as it should be.

Mean and Variance

The *expected value* (mean) of the binomial distribution is simply the number of trials times the probability of a success.

$$\mu = np \tag{12.4}$$

The variance of the binomial distribution is equal to

$$\sigma^2 = np\,(1 - p) \tag{12.5}$$

Thus, the standard deviation of the binomial distribution is $\sqrt{np\,(1-p)}$ The derivation of Eqs. (12.4) and (12.5) can be found in Appendix 6A of Lee et al. (2013).

Example 12.5 (Probability Distribution of Insurance Sales)

In the insurance sales case we discussed in Examples 12.3 and 12.4, the expected number of sales can be calculated in terms of Eq. (12.4) as np = 5(0.20) = 1. The variance of the distribution can be calculated in terms of Eq. (12.5) as $np(1 - p)$ = 5(0.2)(0.8) = 0.8. Thus, the expected number of sales by the sales agent is equal to 1, and the standard deviation is $\sqrt{0.8} = 0.894$.

12.3 The Simple Binomial Option Pricing Model

Before discussing the binomial option model, we must recognize its two major underlying assumptions. First, the binomial approach assumes that trading takes place in discrete time—that is, on a period-by-period basis. Second, it is assumed that the stock price (the price of the underlying asset) can take on only two possible values each period; it can go up or go down.

Say we have a stock whose current price per share S can advance or decline during the next period by a factor of either u (up) or d (down). This price either will increase by the proportion $u-1 \geq 0$ or will decrease by the proportion $1-d, \quad 0<d<1$. Therefore, the value S in the next period will be either uS or dS. Next, suppose that a call option exists on this stock with a current price per share of C and an exercise price per share of X and that the option has one period left to maturity. This option's value at expiration is determined by the price of its underlying stock and the exercise price X. The value is either

$$C_u = \text{Max}(0, uS - X) \tag{12.6}$$

Or

$$C_d = \text{Max}(0, dS - X) \tag{12.7}$$

Why is the call worth Max (0, $uS - X$) if the stock price us uS? The option holder is not obliged to purchase the stock at the exercise price of X, so she or he will exercise the option only when it is beneficial to do so. This means the option can never have a negative value. When is it beneficial for the option holder to exercise the option? When the price per share of the stock is greater than the price per share at which he or she can purchase the stock by using the option, which is the exercise price, X. Thus, if the stock price uS exceeds the exercise price X, the investor can exercise the option and buy the stock. Then, he or she can immediately sell it for uS, making a profit of $uS - X$ (ignoring commission). Likewise, if the stock price declines to dS, the call is worth $\text{Max}(0, dS - X)$.

Also for the moment, we will assume that the risk-free interest rate for both borrowing and lending is equal to r percent over the one time period and that the exercise price of the option is equal to X.

To intuitively grasp the underlying concept of option pricing, we must set up a *risk-free portfolio*—a combination of assets that produces the same return in every state of the world over our chosen investment horizon. The investment horizon is assumed to be one period (the duration of this period can be any length of time, such as an hour, a day, a week, etc.). To do this, we buy h share of the stock and sell the call option at its current price of C. Moreover, we choose the value of h such that our portfolio will yield the same payoff whether the stock goes up or down.

$$h(uS) - C_u = h(dS) - C_d \tag{12.8}$$

By solving for h, we can obtain the number of shares of stock we should buy for each call option we sell.

$$h = \frac{C_u - C_d}{(u-d)S} \tag{12.9}$$

Here, h is called the hedge ratio. Because our portfolio yields the same return under either of the two possible states for the stock, it is without risk and therefore should yield the risk-free rate of return, r percent, which is equal to the risk-free borrowing and lending rate, the condition must be true; otherwise, it would be possible to earn a risk-free profit without using any money. Therefore, the ending portfolio value must be equal to $(1 + r)$ times the beginning portfolio value, $hS - C$.

$$\begin{aligned}(1+r)(hS - C) &= h(uS) - C_u \\ &= h(dS) - C_d\end{aligned} \tag{12.10}$$

Note that S and C represent the beginning values of the stock price and the option price, respectively.

Setting $R = 1 + r$, rearranging to solve for C, and using the value of h from Eq. (12.9), we get

$$C = \left[\left(\frac{R-d}{u-d}\right)C_u + \left(\frac{u-R}{u-d}\right)C_d\right]/R \tag{12.11}$$

where $d < r < u$. To simplify this equation, we set

$$p = \frac{R-d}{u-d} \quad \text{so} \quad 1-p = \left\{\frac{u-R}{u-d}\right\} \qquad (12.12)$$

Thus, we get the option's value with one period to expiration:

$$C = [pC_u + (1-p)C_d]/R \qquad (12.13)$$

This is the binomial call option valuation formula in its most basic form. In other words, this is the binomial valuation formula with one period to expiration of the option.

To illustrate the model's qualities, let us plug in the following values, while assuming the option has one period to expiration. Let

$$X = \$100$$
$$S = \$100$$
$$U = (1.10). \quad \text{So} \quad uS = \$110$$
$$D = (0.90). \quad \text{So} \quad dS = \$90$$
$$R = 1 + r = 1 + 0.07 = 1.07$$

First, we need to determine the two possible option values at maturity, as indicated in Table 12.3.

Next we calculate the value of p as indicated in Eq. (12.12).

Table 12.3 Possible option value at maturity

Today				
Stock (S)	Option (C)	Next Period (Maturity)		
		uS = $110	C_u =	Max (0,uS – X)
			=	Max (0,110 – 100)
$100	C		=	Max (0,10)
			=	$ 10
		dS = $ 90	C_d =	Max (0,dS – X)
			=	Max (0,90 – 100)
			=	Max (0, –10)
			=	$0

$$p = \frac{1.07 - 0.90}{1.10 - 0.90} = 0.85$$

$$\text{so} \quad 1 - p = \frac{1.10 - 1.07}{1.10 - 0.90} = 0.15$$

Solving the binomial valuation equation as indicated in Eq. (12.13), we get

$$C = [0.85(10) + 0.15(0)]/1.07 = \$7.94$$

The correct value for this particular call option today, under the specified conditions, is \$7.94. If the call option does not sell for \$7.94, it will be possible to earn arbitrage profits. That is, it will be possible for the investor to earn a risk-free profit while using none of his or her own money. Clearly, this type of opportunity cannot continue to exist indefinitely.

12.4 The Generalized Binomial Option Pricing Model

Suppose we are interested in the case where there is more than one period until the option expires. We can extend the one-period binomial model to consideration of two or more periods. Because we are assuming that the stock follows a binomial process, from one period to the next it can only go up by a factor of u or go down by a factor of d. After one period the stock's price is either uS or dS. Between the first and second periods, the stock's price can once again go up by u or down by d, so the possible prices for the stock two periods from now are uuS, udS, and ddS. This process is demonstrated in tree diagram from (Fig. 12.2) in Example 12.6 later.

Note that the option's price at expiration, two periods from now, is a function of the same relationship that determined its expiration price in the one-period model, more specifically, the call option's maturity value is always

$$C_T = \text{Max}\,[0, S_T - X] \tag{12.14}$$

where T designated the maturity date of the option.

To derive the option's price with two periods to go (T = 2), it is helpful as an intermediate step to derive the value of C_u and C_d with one period to expiration when the stock price is either uS or dS, respectively.

$$C_u = [pC_{uu} + (1-p)C_{ud}]/R \tag{12.15}$$

$$C_d = [pC_{du} + (1-p)C_{dd}]/R \tag{12.16}$$

Equation (12.15) tells us that if the value of the option after one period is C_u, the option will be worth either C_{uu} (if the stock price goes up) or C_{ud} (if stock price goes down) after one more period (at its expiration date). Similarly, Eq. (12.16) shows that the value of the option is C_d after one period, and the option will be worth either C_{du} or C_{dd} at the end of the second period. Replacing C_u and C_d in Eq. (12.13) with their expressions in Eqs. (12.15) and (12.16), respectively, we can simplify the resulting equation to yield the two-period equivalent of the one-period binomial pricing formula, which is

$$C = \left[p^2 C_{uu} + 2p(1-p)C_{ud} + (1-p)^2 C_{dd}\right]/R^2 \tag{12.17}$$

In Eq. (12.17), we used the fact that $C_{ud} = C_{du}$ because the price will be the same in either case.

Following Eqs. (12.15) and (12.16), we can obtain

$$C_{uu} = [pC_{uuu} + (1-p)C_{uud}]/R \tag{12.18}$$

$$C_{dd} = [pC_{ddu} + (1-p)C_{ddd}]/R \tag{12.19}$$

$$C_{ud} = [pC_{udu} + (1-p)C_{udd}]/R \tag{12.20}$$

Substituting C_{uu}, C_{ud} and C_{dd} in Eq. (12.17), we can obtain the **three-period option pricing model**

$$C = [p^3 C_{uuu} + 3p^2(1-p)C_{uud} + 3p(1-p)^2 C_{udd} + (1-p)^3 C_{ddd}]/R^3 \tag{12.21}$$

Equation (12.21) can also be written in term of a generalized form as follows:

$$C = \frac{1}{R^3}\sum_{k=0}^{3}\frac{3!}{k!(3-k)!}p^k(1-p)^{3-k}\,\text{Max}\left[0, u^k d^{3-k} S - K\right] \tag{12.22}$$

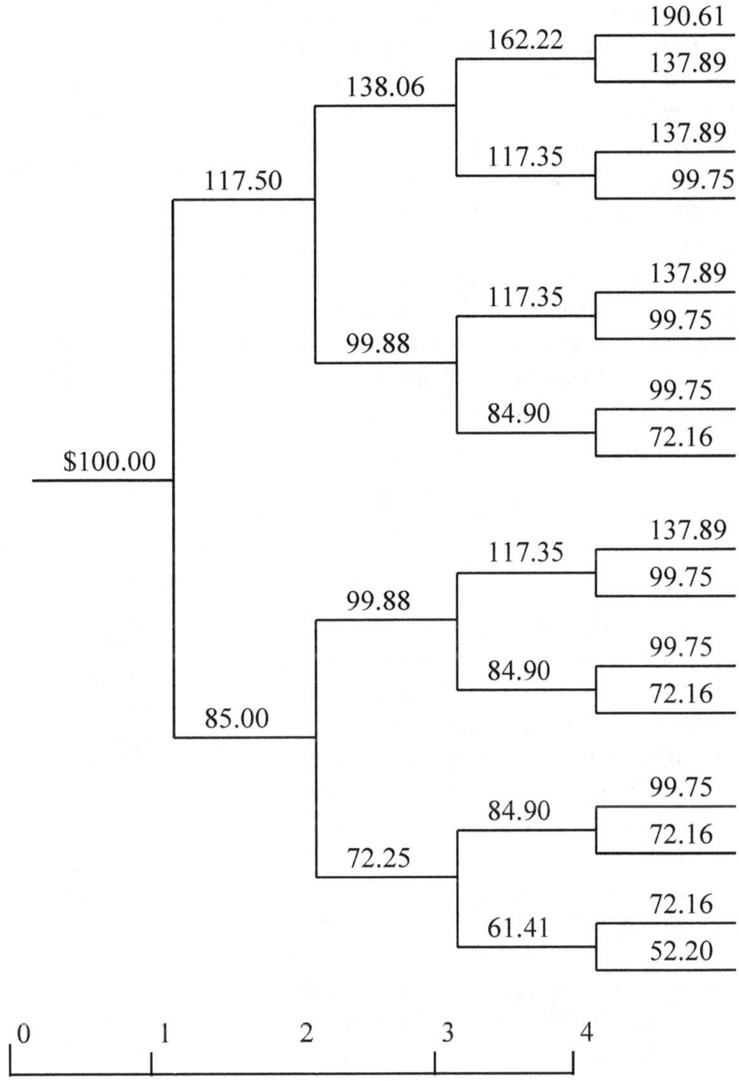

Fig. 12.2 Price path of underlying stock. *Source* Rendelman and Bartter (1979)

Following Eqs. (12.18), (12.19), and (12.20), we can obtain

$$C_{uuu} = [pC_{uuuu} + (1-p)C_{uuud}]/R \quad (12.23)$$

$$C_{ddd} = [pC_{dddu} + (1-p)C_{dddd}]/R \quad (12.24)$$

$$C_{uud} = [pC_{uudu} + (1-p)C_{uudd}]/R \quad (12.25)$$

$$C_{udd} = [pC_{uddu} + (1-p)C_{uddd}]/R \quad (12.26)$$

Substituting *Cuuu*, *Cuud*, *Cudd*, and *Cddd* in Eq. (12.21), we can obtain the **four-period option pricing model**

$$\begin{aligned} C = {} & [p^4 C_{uuuu} + 4p^3(1-p)C_{uuud} \\ & + 6p^2(1-p)^2 C_{uudd} + 4p(1-p)^3 C_{uddd} \\ & + (1-p)^4 C_{dddd}]/R^4 \end{aligned} \quad (12.27)$$

Equation (12.27) can also be written in term of a generalized form as follows:

$$C = \frac{1}{R^4}\sum_{k=0}^{4} \frac{4!}{k!(4-k)!} p^k (1-p)^{4-k} \operatorname{Max}\left[0, u^k d^{4-k} S - K\right] \quad (12.28)$$

We know the values of the parameters S and X. If we assume that R, u, and d will remain constant over time, the possible maturity values for the option can be determined exactly. Thus, deriving the option's fair value with two periods to maturity is a relatively simple process of working backward from the possible maturity values.

Finally, using this same procedure of going from a one-period model, two-period model, three-period model and to a four-period model, we can extend the binomial approach to its more generalized form, with n-period maturity:

$$C = \frac{1}{R^n}\sum_{k=0}^{n}\frac{n!}{k!(n-k)!}p^k(1-p)^{n-k} \operatorname{Max}\left[0, u^k d^{n-k}S - X\right] \tag{12.29}$$

To actually get this form of the binomial model, we could extend the two-period model to three periods, then from three periods to four periods, and so on. Equation (12.29) would be the result of these efforts. To show how Eq. (12.29) can be used to assess a call option's value, we modify the example as follows: $S = \$100$, $X = \$100$, $R = 1.07$, $n = 3$, $u = 1.1$, and $d = 0.90$.

First, we calculate the value of p from Eq. (12.12) as 0.85, so $1 - p$ is 0.15. Next we calculate the four possible ending values for the call option after three periods in terms of Max $[0, u^k d^{n-k}S - X]$

$$C_1 = \operatorname{Max}\left[0, (1.1)^3(0.90)^0(100) - 100\right] = 33.10$$

$$C_2 = \operatorname{Max}\left[0, (1.1)^2(0.90)(100) - 100\right] = 8.90$$

$$C_3 = \operatorname{Max}\left[0, (1.1)(0.90)^2(100) - 100\right] = 0$$

$$C_4 = \operatorname{Max}\left[0, (1.1)^0(0.90)^3(100) - 100\right] = 0$$

Now we insert these numbers (C_1, C_2, C_3, and C_4) into the model and sum the terms.

$$\begin{aligned} C &= \frac{1}{(1.07)^3}\left[\frac{3!}{0!3!}(0.85)^0(0.15)^3 \times 0 \right. \\ &\quad + \frac{3!}{1!2!}(0.85)^1(0.15)^2 \times 0 \\ &\quad + \frac{3!}{2!1!}(0.85)^2(0.15)^1 \times 8.90 \\ &\quad \left. + \frac{3!}{3!0!}(0.85)^3(0.15)^0 \times 33.10\right] \\ &= \frac{1}{1.225}\left[0 + 0 + \frac{3 \times 2 \times 1}{2 \times 1 \times 1}(0.7225)(0.15)(8.90) \right. \\ &\quad \left. + \frac{3 \times 2 \times 1}{3 \times 2 \times 1 \times 1} \times (0.61413)(1)(33.10)\right] \\ &= \frac{1}{1.225}[(0.32513 \times 8.90) + (0.61413 \times 33.10)] \\ &= \$18.96 \end{aligned}$$

As this example suggests, working out a multiple-period problem by hand with this formula can become laborious as the number of periods increases. Fortunately, programming this model into a computer is not too difficult.

Now let us derive a binomial option pricing model in terms of the cumulative binomial density function. As a first step, we can rewrite Eq. (12.29) as

$$\begin{aligned} C &= S\left[\sum_{k=m}^{n}\frac{n!}{k!(n-K)!}p^K(1-p)^{n-k}\frac{u^k d^{n-k}}{R^n}\right] \\ &\quad - \frac{X}{R^n}\left[\sum_{k=m}^{n}\frac{n!}{k!(n-k)!}p^k(1-p)^{n-k}\right] \end{aligned} \tag{12.30}$$

This formula is identical to Eq. (12.29) except that we have removed the Max operator. In order to remove the Max operator, we need to make $u^k d^{n-k}S - X$ positive, which we can do by changing the counter in the summation from $k = 0$ to $k = m$. What is m? It is the minimum number of upward stock movements necessary for the option to terminate "in the money" (that is, $u^k d^{n-k}S - X > 0$). How can we interpret Eq. (12.30)? Consider the second term in brackets; it is just a cumulative binomial distribution with parameters of n and p. Likewise, via a small algebraic manipulation we can show that the first term in the brackets is also a cumulative binomial distribution. This can be done by defining $P' \equiv (u/R)p$ and $1 - P' \equiv (d/R)(1 - p)$. Thus

$$P^k(1-p)^{n-k}\frac{u^k d^{n-k}}{R^n} = p'^k(1-p')^{n-k}$$

Therefore, the first term in brackets of (12.30) is also a cumulative binomial distribution with parameters of n and p'. Using the definition of cumulative binomial function, we can write the binomial call option model as

$$C = SB_1(n, p', m) - \frac{X}{R^n}B_2(n, p, m) \quad (12.31)$$

where

$$B_1(n, p', m) = \sum_{k=m}^{n} C_k^n p'^k(1-p')^{n-k}$$

$$B_2(n, p, m) = \sum_{k=m}^{n} C_k^n p^k(1-p)^{n-k}$$

and m is the minimum amount of time the stock has to go up for the investor to finish *in the money* (that is, for the stock price to become larger than the exercise price).

In this chapter, we showed that by employing the definition of a call option and by making some simplifying assumptions, we could use the binomial distribution to find the value of a call option. In the following sections, we will show how the binomial distribution is related to the normal distribution and how this relationship can be used to derive one of the most famous valuation equations in finance, the Black–Scholes option pricing model.

Example 12.6 (A Decision Tree Approach to Analyzing Future Stock Price)

By making some simplifying assumptions about how a stock's price can change from one period to the next, it is possible to forecast the future price of the stock by means of a decision tree. To illustrate this point, let us consider the following example.

Suppose the price of Company A's stock is currently \$100. Now let us assume that from one period to the next, the stock can go up by 17.5% or go down by 15%. In addition, let us assume that there is a 50% chance that the stock will go up and a 50% chance that the stock will go down. It is also assumed that the price movement of a stock (or of the stock market) today is completely independent of its movement in the past; in other words, the price will rise or fall today by a random amount. A sequence of these random increases and decreases is known as a **random walk**.

Given this information, we can lay out the paths that the stock's price may take. Figure 12.2 shows the possible stock prices for company A for four periods.

Note that in period 1, there are two possible outcomes: The stock can go up in value by 17.5% to \$117.50 or down by 15% to \$85.00. In period 2, there are four possible outcomes. If the stock went up in the first period, it can go up again to \$138.06 or down in the second period to \$99.88. Likewise, if the stock went down in the first period, it can go down again to \$72.25 or up in the second period to \$99.88. Using the same argument, we can trace the path of the stock's price for all four periods.

If we are interested in forecasting the stock's price at the end of period 4, we can find the average price of the stock for the 16 possible outcomes that can occur in period 4.

$$\bar{P} = \frac{\sum_{i=1}^{16} P_i}{16} = \frac{190.61 + 137.89 + \cdots + 52.20}{16}$$
$$= \$105.09$$

We can also find the standard deviation for the stock's return.

$$\sigma_P = \left[\frac{(190.61 - 105.09)^2 + \cdots + (52.20 - 105.09)^2}{16}\right]^{1/2}$$
$$= \$34.39$$

$\bar{P}$ and σ_P can be used to predict the future price of stock A.

12.5 Multinomial Option Pricing Model

12.5.1 Derivation of the Option Pricing Model

Suppose that the stock price follows a multiplicative multinomial process over discrete periods. The rate of return on the stock over each period can have $(k + 1)$ possible values: $f_i - 1$ with probability q_i, $i = 1, \ldots k+1$. Thus, if the current stock price is S, the stock price at the end of the period will be one of $f_i S'$ s. We can represent this movement with the following diagram:

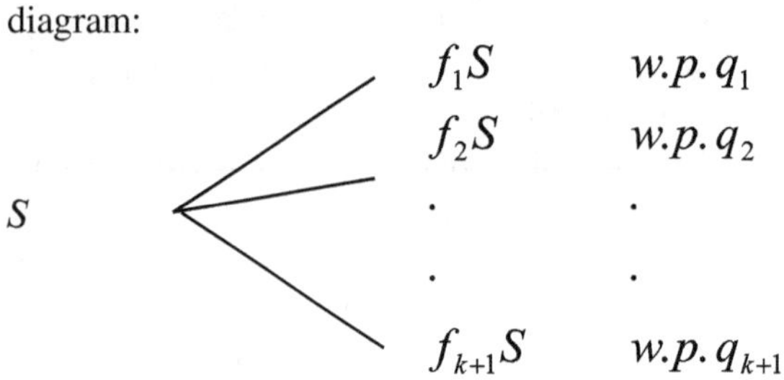

where $w.p.$ denotes with probability.

We need the following definitions and notations in our presentation. Let $\underset{\sim}{X} = (x_1, x_2, \ldots x_k, x_{k+1})^T$, and $\underset{\sim}{q} = (q_1, q_2, \ldots q_k, q_{k+1})^T$, $0 < q_j < 1, j = 1, 2, \ldots k+1$. $\underset{\sim}{X}$ is said to have the multinomial distribution, or $\underset{\sim}{X} \sim \text{Mult}\,(n, \underset{\sim}{q})$, if and only if the joint probability density is

$$
\begin{aligned}
f(\underset{\sim}{X}) &= f(x_1, x_2, \ldots, x_{k+1}) \\
&= \frac{n!}{x_1! x_2! \ldots x_{k+1}!} q_1^{x_1} q_{2x_2} \ldots q_{k+1}^{x_{k+1}} \qquad (12.32) \\
&= \begin{bmatrix} n \\ x_1, x_2, \ldots x_{k+1} \end{bmatrix} \prod_{j=1}^{k+1} q_j^{x_j}
\end{aligned}
$$

for all $0 \le x_j \le n$, where $\sum_{j=1}^{k+1} x_j = \underset{\sim}{1}^T \underset{\sim}{X} = n$ and $\sum_{j=1}^{k+1} q_j = \underset{\sim}{1}^T \underset{\sim}{q} = 1$, where

$\underset{\sim (k+1)}{1}$: $(k+1)$-dimensional vector of unit entries. Let

C = the current value of the n-period option price;
K = the option exercise price;
S^* = the stock price at the end of the n-period;
n = the number of periods to maturity; and
$\hat{r}$ = one plus the riskless rate per period.

The following theorem gives the option price for the multinomial case. This theorem extends the option price formula from the binomial case to the multinomial case.

Theorem 12.1 *The current value of the n-period multinomial option price C is given by*

$$
C = \sum_{\underset{\sim}{X} \in A} \begin{bmatrix} n \\ x_1, x_2, \ldots, x_{k+1} \end{bmatrix} \left\{ S \prod_{j=1}^{k+1} \left(\frac{f_j q_j}{\hat{r}} \right)^{x_j} - K \hat{r}^{-n} \prod_{j=1}^{k+1} q_j \right\} \qquad (12.33)
$$

where $A = \left\{ \underset{\sim}{X} \mid \underset{\sim}{X} = (x_1, x_2, \ldots, x_{k+1})^T, A = x_j \in N \cup \{0\}, A = \underset{\sim}{1}^T \underset{\sim}{X} = n, S* > K \right\}$

and S and S are the current stock price and the stock price at the end of the n-period, respectively.*

Proof Since

$$
\begin{aligned}
S* &= f_1^{x_1} \cdot f_2^{x_{21}} \cdot f_{k+1}^{x_{k+1}} \cdot S = \left(\prod_{j=1}^{k+1} f_j^{x_j} \right) \cdot S,\ C \\
&= \hat{r}^{-n} E[\max\,(S* - K, 0)],
\end{aligned}
$$

and

$$A = \left\{ \underset{\sim}{X} | \underset{\sim}{1}^T \underset{\sim}{X} = n, S* > K \right\},$$

we have

$$\begin{aligned} C &= \sum_{\underset{\sim}{X} \in A} \left\{ \begin{bmatrix} n \\ x_1, x_2, \ldots, x_{k+1} \end{bmatrix} (S * - K) \cdot \prod_{j=1}^{k+1} q_j^{x_j} \right\} / \hat{r}^n \\ &= \sum_{\underset{\sim}{X} \in A} \left\{ \begin{bmatrix} n \\ x_1, x_2, \ldots, x_{k+1} \end{bmatrix} \left(S \prod_{j=1}^{k+1} f_i^{x_j} - K \right) \cdot \prod_{j=1}^{k+1} q_j^{x_j} \right\} / \hat{r}^n \\ &= \sum_{\underset{\sim}{X} \in A} \left(\begin{bmatrix} n \\ x_1, x_2, \ldots, x_{k+1} \end{bmatrix} \left\{ S \prod_{j=1}^{k+1} (f_i q_j)^{x_j} - K \prod_{j=1}^{k+1} q_j^{x_j} \right\} \right) / \hat{r}^n \\ &= \sum_{\underset{\sim}{X} \in A} \left(\begin{bmatrix} n \\ x_1, x_2, \ldots, x_{k+1} \end{bmatrix} \left\{ S \prod_{j=1}^{k+1} \left(\frac{f_i q_j}{\hat{r}} \right)^{x_j} - K \hat{r}^{-n} \prod_{j=1}^{k+1} q_j^{x_j} \right\} \right) \end{aligned}$$

Now, let $\underset{\sim}{\upsilon} = (\upsilon_1, \upsilon_2, \ldots \upsilon_k, \upsilon_{k+1})^T$, $\upsilon_j = \frac{f_i q_j}{\hat{r}}$.

If investors were risk-neutral, that is,

$$\sum_{j=1}^{k+1} (f_j S) \cdot q_j = \hat{r} S,$$

or

$$\sum_{j=1}^{k+1} f_j q_j = \hat{r}, \text{ implying } \sum_{j=1}^{k+1} \frac{f_i q_j}{\hat{r}} = \sum_{j=1}^{k+1} \upsilon_j = \underset{\sim}{1}^T \underset{\sim}{\upsilon} = 1,$$

then $\underset{\sim}{\upsilon}$ and $\underset{\sim}{q}$ are both the probability vectors in multinomial distribution.

So we can rewrite Eq. (12.33) as

$$C = S P_{\underset{\sim}{\upsilon}}(A_\upsilon) - K \hat{r}^{-n} P_{\underset{\sim}{q}}(A_q), \tag{12.34}$$

where

$$P_{\underset{\sim}{\upsilon}}(A_\upsilon) = \sum_{\underset{\sim}{X} \in A} \left\{ \begin{bmatrix} n \\ x_1, x_2, \ldots, x_{k+1} \end{bmatrix} \prod_{j=1}^{k+1} \upsilon_j^{x_j} \right\} \tag{12.35}$$

$$P_{\underset{\sim}{q}}(A_q) = \sum_{\underset{\sim}{X} \in A} \left\{ \begin{bmatrix} n \\ x_1, x_2, \ldots, x_{k+1} \end{bmatrix} \prod_{j=1}^{k+1} q_j^{x_j} \right\}. \tag{12.36}$$

12.5.2 The Black and Scholes Model as a Limiting Case

We next consider the limiting result. If t is the fixed length of time to expiration, and n is the number of periods of length l prior to expiration, then $l \equiv \frac{t}{n}$.

As trading takes more and more frequently, l gets closer and closer to zero. We must then adjust the interval-dependent variable $\hat{r}$ and f_j's in such a way that we obtain empirically realistic results as l becomes smaller when $n \to \infty$.

Let r denote one plus the interest rate over a fixed length of time while $\hat{r}$ means one plus the interest rate over a period of length l.

Suppose that $\hat{r}^n = r^t$ for all n. Then, $\hat{r} = r^{\frac{t}{n}}$, which shows how $\hat{r}$ must depend on n for the total return over time t to be independent of n.

Denote $\underset{\sim}{\eta}^T = (\eta_1, \eta_2, \ldots \eta_{k+1}) = (\ln f_1, \ln f_2, \ldots, \ln f_{k+1})$.

If $\eta_j = h \cdot \xi_j$, where $h = \sqrt{\frac{t}{n}}\sigma + o(\frac{1}{\sqrt{n}})$, $\sum_{j=1}^{k+1} q_j \xi_j^2 = 1$, then $\ln(\frac{S*}{S}) = \sum_{j=1}^{k+1} x_j \log f_j = \underset{\sim}{\eta}^T \underset{\sim}{x}$.

Let $\underset{\sim}{\eta}^T(n \underset{\sim}{q}) = n \underset{\sim}{\eta}^T \underset{\sim}{q} \equiv \mu t$, and $\left[\sum_{j=1}^{k+1} q_j \eta_j^2\right] \cdot n = nh^2 \sum_{j=1}^{k+1} q_j \xi_j^2 \equiv \sigma^2 t + \frac{\mu^2 t^2}{n}$, then $\underset{\sim}{\eta}^T \underset{\sim}{q} = \mu \cdot \frac{t}{n}$.

Since $\underset{\sim}{\eta}^T \underset{\sim}{q} = h \underset{\sim}{\xi}^T \underset{\sim}{q} = \mu \frac{t}{n}$, $h = O\left(\frac{1}{\sqrt{n}}\right)$, we have $\underset{\sim}{\xi}^T \underset{\sim}{q} = O\left(\frac{1}{\sqrt{n}}\right)$.

For any vector $\underset{\sim}{y}$, let $\underset{\sim}{\hat{y}}$ denote the truncated vector obtained from $\underset{\sim}{y}$ by deleting its last entry. That is, if $\underset{\sim}{y}^T = (y_1, y_2, \ldots y_{p-1}, y_p)$, then $\underset{\sim}{\hat{y}}^T = (y_1, y_2, \ldots y_{p-1}, y_p)$.

Bhattacharya and Rao (1976) showed that if $\underset{\sim}{X} \sim \text{Mult}(n, \underset{\sim}{q})$, then for large n, $\underset{\sim}{\hat{X}} \overset{D}{\sim} N_k\left(n\,\underset{\sim}{\hat{q}}, n\underset{\sim}{\sum}{}_q^*\right)$, where $\underset{\sim}{\sum}{}_q^* = \Delta(\underset{\sim}{\hat{q}}) - \underset{\sim}{\hat{q}}\,\underset{\sim}{\hat{q}}^T$, and $\Delta(\underset{\sim}{\hat{q}})$ is a diagonal matrix, that is,

$$\Delta(\underset{\sim}{\hat{q}}) = (b_{ij}), b_{ij} = \begin{cases} q_i, & i = j \\ 0, & i \neq j \end{cases}$$

We next observe that

$$\underset{\sim}{\hat{X}} \in A_q \quad \text{iff} \quad X* = f_1^{x_1} f_2^{x_2} \ldots f_{k+1}^{x_{k+1}} \cdot S > K,$$

or

$$\underset{\sim}{\hat{X}} \in A_q \quad \text{iff} \quad \ln\left(f_1^{x_1} f_2^{x_2} \ldots f_{k+1}^{x_{k+1}}\right) = \sum_{j=1}^{k+1} \ln(f_j) x_j = \underset{\sim}{\hat{\eta}}^T \underset{\sim}{\hat{X}} + \eta_{k+1} x_{k+1} > \ln\frac{K}{S},$$

or

$$\underset{\sim}{\hat{X}} \in A_q \quad \text{iff} \quad Z^* = \frac{\underset{\sim}{\hat{\eta}}^T \underset{\sim}{\hat{X}} - n\underset{\sim}{\hat{\eta}}^T \underset{\sim}{\hat{q}}}{\left[n\,\underset{\sim}{\hat{\eta}}^T \underset{\sim}{\sum}{}_q^* \underset{\sim}{\hat{\eta}}\right]^{\frac{1}{2}}} > -\frac{\ln\frac{S}{K} + \eta_{k+1} x_{k+1} + n\underset{\sim}{\hat{\eta}}^T \underset{\sim}{\hat{q}}}{\left[n\,\underset{\sim}{\hat{\eta}}^T \underset{\sim}{\sum}{}_q^* \underset{\sim}{\hat{\eta}}\right]^{\frac{1}{2}}},$$

where $Z^* \sim N(0, 1)$ is a random variable and the right-hand side is also a random variable because it is a function of x_{k+1}.

Madan et al. (1989) used the random variable $(\underset{\sim}{\hat{\eta}} - \eta_{k+1}\underset{\sim}{1})^T(\underset{\sim}{\hat{X}} - n\,\underset{\sim}{\hat{q}})$ to solve the problem:

Since,

$$\begin{aligned}
\left(\underset{\sim}{\hat{\eta}} - \eta_{k+1}\underset{\sim}{1}\right)^T \left(\underset{\sim}{\hat{X}} - n\underset{\sim}{\hat{q}}\right)
&= \underset{\sim}{\hat{\eta}}^T \underset{\sim}{\hat{X}} - n\underset{\sim}{\hat{\eta}}^T \underset{\sim}{\hat{q}} - \eta_{k+1} \underset{\sim}{1}^T \underset{\sim}{\hat{X}} + n\eta_{k+1} \underset{\sim}{1}^T \underset{\sim}{\hat{q}} \\
&= \underset{\sim}{\hat{\eta}}^T \underset{\sim}{\hat{X}} - n\underset{\sim}{\hat{\eta}}^T \underset{\sim}{\hat{q}} - \eta_{k+1}(n - x_{k+1}) + n\eta_{k+1}(1 - q_{k+1}) \\
&= \underset{\sim}{\hat{\eta}}^T \underset{\sim}{\hat{X}} - n\underset{\sim}{\hat{\eta}}^T \underset{\sim}{\hat{q}} + \eta_{k+1} x_{k+1} - n\eta_{k+1} q_{k+1}
\end{aligned}$$

we have

$$\underset{\sim}{\hat{\eta}}^T \underset{\sim}{\hat{X}} - n\,\underset{\sim}{\hat{\eta}}^T \underset{\sim}{\hat{q}} = (\underset{\sim}{\hat{\eta}} - \eta_{k+1}\,\underset{\sim}{1})^T (\underset{\sim}{\hat{X}} - n\,\underset{\sim}{\hat{q}}) + n\eta_{k+1} q_{k+1} - \eta_{k+1} x_{k+1}.$$

Hence,

$$\begin{aligned}
\underset{\sim}{\hat{X}} \in A_q \quad \text{iff} \quad & \underset{\sim}{\hat{\eta}}^T \underset{\sim}{\hat{X}} - n\,\underset{\sim}{\hat{\eta}}^T \underset{\sim}{\hat{q}} > -\ln\frac{S}{K} - \eta_{k+1} x_{k+1} - n\,\underset{\sim}{\hat{\eta}}^T \underset{\sim}{\hat{q}} \\
\text{iff} \quad & \left(\underset{\sim}{\hat{\eta}} - \eta_{k+1}\,\underset{\sim}{1}\right)^T \left(\underset{\sim}{\hat{X}} - n\,\underset{\sim}{\hat{q}}\right) > -\ln\frac{S}{K} - \eta_{k+1} x_{k+1} - n\,\underset{\sim}{\hat{\eta}}^T \underset{\sim}{\hat{q}} - n\eta_{k+1} q_{k+1} + \eta_{k+1} x_{k+1} \\
& = -\ln\frac{S}{K} - n\,\underset{\sim}{\hat{\eta}}^T \underset{\sim}{\hat{q}}.
\end{aligned}$$

Thus, $P_{\underset{\sim}{q}}(A_q) = P_{\underset{\sim}{q}}\left(\left(\underset{\sim}{\hat{\eta}} - \eta_{k+1}\,\underset{\sim}{1}\right)^T \left(\underset{\sim}{\hat{X}} - n\,\underset{\sim}{\hat{q}}\right). > -\left(\ln\frac{S}{K} + n\,\underset{\sim}{\hat{\eta}}^T \underset{\sim}{\hat{q}}\right)\right)$.

Consequently, $\left(\underset{\sim}{\hat{\eta}} - \eta_{k+1}\,\underset{\sim}{1}\right)^T \left(\underset{\sim}{\hat{X}} - n\,\underset{\sim}{\hat{q}}\right) \overset{D}{\rightarrow} N\left(0, n\left(\underset{\sim}{\hat{\eta}} - \eta_{k+1}\,\underset{\sim}{1}\right)^T \underset{\sim}{\sum}{}_q^* \left(\underset{\sim}{\hat{\eta}} - \eta_{k+1}\,\underset{\sim}{1}\right)\right)$.

We need the following lemma.

Lemma 12.1

(a) *If* $\sum_{\underset{\sim}{q}} = \Delta(\underset{\sim}{q}) - \underset{\sim}{q}\,\underset{\sim}{q}^T$,

then $\left(\underset{\sim}{\hat{\eta}} - \eta_{k+1}\underset{\sim}{1}\right)^T \sum^*_{\underset{\sim}{q}}\left(\underset{\sim}{\hat{\eta}} - \eta_{k+1}\underset{\sim}{1}\right)$

$= \left(\underset{\sim}{\eta} - \eta_{k+1}\underset{\sim}{1}\right)^T \sum_{\underset{\sim}{q}}\left(\underset{\sim}{\eta} - \eta_{k+1}\underset{\sim}{1}\right)$

(b) *if* $\sum_{\underset{\sim}{\upsilon}} = \Delta(\underset{\sim}{\upsilon}) - \underset{\sim}{\upsilon}\,\underset{\sim}{\upsilon}^T$, $\sum^*_{\underset{\sim}{\hat{\upsilon}}} = \Delta(\underset{\sim}{\hat{\upsilon}}) - \underset{\sim}{\hat{\upsilon}}\,\underset{\sim}{\hat{\upsilon}}^T$,

then

$(\underset{\sim}{\hat{\eta}} - \eta_{k+1}\underset{\sim}{1})^T \sum^*_{\underset{\sim}{\upsilon}}(\underset{\sim}{\hat{\eta}} - \eta_{k+1}\underset{\sim}{1})$

$= (\underset{\sim}{\eta} - \eta_{k+1}\underset{\sim}{1})^T \sum_{\underset{\sim}{\upsilon}}(\underset{\sim}{\hat{\eta}} - \eta_{k+1}\underset{\sim}{1})$.

If

$$Z_q = \frac{\ln\frac{S}{K} + n\underset{\sim}{\eta}^T\underset{\sim}{q}}{\left[n\left(\underset{\sim}{\eta} - \eta_{k+1}\underset{\sim}{1}\right)^T \sum_{\underset{\sim}{q}}\left(\underset{\sim}{\eta} - \eta_{k+1}\underset{\sim}{1}\right)\right]^{\frac{1}{2}}},$$

then

$$P_{\underset{\sim}{q}}(A_q) = P_q\left(\frac{\left(\underset{\sim}{\hat{\eta}} - \eta_{k+1}\underset{\sim}{1}\right)^T\left(\underset{\sim}{\hat{X}} - n\,\underset{\sim}{\hat{q}}\right)}{\left[n\left(\underset{\sim}{\hat{\eta}} - \eta_{k+1}\underset{\sim}{1}\right)^T \sum_{\underset{\sim}{q}}\left(\underset{\sim}{\hat{\eta}} - \eta_{k+1}\underset{\sim}{1}\right)\right]^{\frac{1}{2}}} > -Z_q\right)$$
$$\approx 1 - N(-Z_q) = N(Z_q).$$

Similarly, if

$$Z_\upsilon = \frac{\ln\frac{S}{K} + n\,\underset{\sim}{\eta}^T\underset{\sim}{\upsilon}}{\left[n(\underset{\sim}{\eta} - \eta_{k+1}\underset{\sim}{1})^T \sum_{\underset{\sim}{\upsilon}}(\underset{\sim}{\eta} - \eta_{k+1}\underset{\sim}{1})\right]^{\frac{1}{2}}},$$

then $P_{\underset{\sim}{\upsilon}}(A_\upsilon) \approx N(Z_\upsilon)$

Theorem 12.2 *If* $\hat{r}^{-1}\underset{\sim}{f}^T\underset{\sim}{q} = r^{-\frac{t}{n}}\underset{\sim}{f}^T\underset{\sim}{q} = 1$ *and* $\eta_j = \ln f_j = h\cdot\xi_j$, $h = \sqrt{\frac{t}{n}}\sigma + o\left(\frac{1}{\sqrt{n}}\right)$ *for all* j, $\sum_{j=1}^{k+1} q_j\xi_j^2 = 1$.

Then,

(1) $\underset{\sim}{\eta}^T\underset{\sim}{q} = \frac{\left(\ln r - \frac{\sigma^2}{2}\right)t}{n} + o\left(\frac{1}{n}\right))$

(2) $\underset{\sim}{\eta}^T\underset{\sim}{\upsilon} = \frac{\left(\ln r + \frac{\sigma^2}{2}\right)t}{n} + o\left(\frac{1}{n}\right)$

(3)

$\lim_{n\to\infty} n\left(\underset{\sim}{\hat{\eta}} - \eta_{k+1}\underset{\sim}{1}\right)^T \sum^*_{\underset{\sim}{q}}\left(\underset{\sim}{\hat{\eta}} - \eta_{k+1}\underset{\sim}{1}\right) =$

$\lim_{n\to\infty} n(\underset{\sim}{\eta} - \eta_{k+1}\underset{\sim}{1})^T \sum_{\underset{\sim}{q}}(\underset{\sim}{\eta} - \eta_{k+1}\underset{\sim}{1}) = \sigma^2 t$

(4) $\lim_{n\to\infty} n\left(\underset{\sim}{\eta} - \eta_{k+1}\underset{\sim}{1}\right)^T \sum_{\underset{\sim}{\upsilon}}\left(\underset{\sim}{\eta} - \eta_{k+1}\underset{\sim}{1}\right) =$

$\sigma^2 t$

With the assumptions of Theorem 12.2, *let*

$$d_1 = \lim_{n\to\infty} Z_\upsilon = \frac{\ln\frac{S}{K}}{(\sigma^2 t)^{\frac{1}{2}}} + \frac{\left(\ln r + \frac{\sigma^2}{2}\right)t}{(\sigma^2 t)^{\frac{1}{2}}}$$
$$= \frac{\ln\frac{S}{K}}{\sigma\sqrt{t}} + \left(\frac{\ln r}{\sigma} + \frac{\sigma}{2}\right)\sqrt{t}$$

$$d_2 = \lim_{n\to\infty} Z_q = \frac{\ln\frac{S}{K}}{(\sigma^2 t)^{\frac{1}{2}}} + \frac{\left(\ln r - \frac{\sigma^2}{2}\right)t}{(\sigma^2 t)^{\frac{1}{2}}}$$
$$= d_1 - \sigma\sqrt{t}$$

Then, $C \to SN(d_1) - Kr^{-t}N(d_2)$, *which is the Black–Scholes formula.*

Here, we use the fact that $P_{\underset{\sim}{\upsilon}}(A_\upsilon) \to N(d_1)$, *and* $P_{\underset{\sim}{q}}(A_q) \to N(d_2)$.

12.6 A Lattice Framework for Option Pricing

12.6.1 Modification of the Two-State Approach for a Single-State Variable

In this section, we will present a lattice framework for option pricing proposed by Boyle (1989). We will first discuss a modification of the CRR lattice binomial approach for option pricing in which there is only a simple state variable. We will discuss the case with two-state variables later.

Instead of the two-jump process used in CRR, we will use a three-jump process. Let us consider an asset (S) with a lognormal distribution of returns. Over a small time interval h, this distribution is approximated by a three-point jump process such that the expected return on the asset is the risk-free rate, and the variance of the approximating distribution is equal to the variance of the corresponding lognormal distribution.

Let

T = time to option maturity (in years);
X = exercise price of option;
R = the risk-free annual interest rate;
σ^2 = the variance of the annual rate of return on the underlying asset (yearly);
$h = \frac{T}{n}$: length of one-time step;
Su = value of asset after an up jump;
S = value of asset after a horizontal jump;
Sd = value of asset after a down jump.

Let S_n be a random variable whose distribution is as follows:

Nature of jump	Probability asset	Price (value of S_n)
Up	p_1	Su
Horizontal	p_2	S
Down	p_3	Sd

We further assume that $ud = 1$.

It is trivial to see that

$$E(S_n) = \left[p_1 u + p_2 + p_3 \frac{1}{u}\right] S = MS,$$

$$\begin{aligned}\mathrm{Var}(S_n) &= p_1\left(S^2u^2 - S^2M^2\right) + p_2\left(S^2 - S^2M^2\right) \\ &\quad + p_3\left(S^2\frac{1}{u^2} - S^2M^2\right) = S^2 V,\end{aligned}$$

$$M = p_1 u + p_2 + \frac{p_3}{u},$$

$$V = p_1\left(u^2 - M^2\right) + p_2\left(1 - M^2\right) + p_3\left(\frac{1}{u^2} - M^2\right).$$

For the approximation to the lognormal distribution, we impose three conditions:

(i) The probabilities are positive and sum to one.

(ii) The mean of the discrete distribution, MS, is equal to the lognormal distribution, in a risk-neutral world, i.e,

$$SM = S\exp(rh).$$

(iii) The variance of the discrete distribution, S^2V, is equal to the variance of the lognormal distribution $S^2V = S^2M^2[\exp(\sigma^2 h) - 1]$.

Before proceeding further, we note that if $\log\left(\frac{S_n}{S}\right)$ is normally distributed with mean $h\mu$ and variance $h\sigma^2$, then

$$E\left(\frac{S_n}{S}\right) = e^{h\mu + h\sigma^2/2},$$

which is equal to e^{rh} in a risk-neutral world. Hence, $h(\mu + \sigma^2/2) = hr$. Furthermore,

$$\begin{aligned}\mathrm{Var}\left(\frac{S_n}{S}\right) &= e^{2h\mu}e^{h\sigma^2}\left(e^{h\sigma^2} - 1\right) \\ &= e^{2h\left(\mu + \frac{\sigma^2}{2}\right)}\left(e^{h\sigma^2} - 1\right) \\ &= e^{2hr}\left(e^{h\sigma^2} - 1\right).\end{aligned}$$

By setting $E\left(\frac{S_n}{S}\right) = M$ and $\mathrm{Var}\left(\frac{S_n}{S}\right) = V$, we obtain the conditions (ii) and (iii).

It is noted that although M and V have been specified, they will be forced to take on values $\exp(rh)$ and $M^2[\exp(\sigma^2 h) - 1]$, respectively, in the approximation.

The above three conditions are:

$$p_1 + p_2 + p_3 = 1, \tag{12.37}$$

$$p_1 Su + p_2 S + p_3 \frac{S}{u} = SM, \quad (12.38)$$

$$p_1 (S^2u^2 - S^2M^2) + p_2 (S^2 - S^2M^2) + p_3 \left(\frac{S^2}{u^2} - S^2M^2\right) = S^2V. \quad (12.39)$$

Even though there are three equations in three unknowns (assuming the stretch parameter u is specified), we need only two equations as p_1, p_2 and p_3 are linearly dependent. From (12.37), we have $p_2 = 1 - p_1 - p_2$. Substituting this into (12.38) and (12.39) we have:

$$p_1(u - 1) + p_3\left(\frac{1}{u} - 1\right) = M - 1, \quad (12.40)$$

$$p_1(u^2 - 1) + p_3\left(\frac{1}{u^2} - 1\right) = V + M^2 - 1. \quad (12.41)$$

From (12.40), (12.41), we have:

$$p_1(u, M, V) = \frac{(V + M^2 - M)u - (M - 1)}{(u - 1)(u^2 - 1)}, \quad (12.42)$$

$$p_3(u, M, V) = \frac{u^2(V + M^2 - M) - u^3(M - 1)}{(u - 1)(u^2 - 1)}, \quad (12.43)$$

$$p_2 = 1 - p_1 - p_2. \quad (12.44)$$

From (12.42), (12.43), we see that $u \neq 1$, in fact, $u > 1$. Also $p_i(u, M, V)$ indicates that the probability p_i is a function of u, M, V.

Recall that in CRR, u is set to be

$$u = \exp(\sigma\sqrt{h}). \quad (12.45)$$

However, it is noted by Boyle that p_2 will result in negative values for many realistic parameter values. Thus, it is proposed to consider

$$u = \exp(\lambda\sigma\sqrt{h}). \quad (12.46)$$

where $\lambda \geq 1$ and is called the jump amplitude. Table 12.4 shows a list of jump amplitudes and jump probabilities.

Before moving on to the two assets extension, it is noted that for a range of parameter values, the accuracy of the three-jump method with 5 time intervals was comparable to that of the CRR method with 20 time intervals, as claimed by Boyle.

12.6.2 A Lattice Model for Valuation of Options on Two Underlying Assets

We next consider the extension to the two assets case. Assume that the two assets are bivariate

Table 12.4 Jump amplitudes and jump probabilities $\sigma = 0.2, r = 0.1, T = 1.0, n = 20$

λ	u	p_1	p_3	p_3
1.00	1.0457	0.5539	0.4646	−0.0184
1.10	1.0504	0.4610	0.3798	0.1592
1.20	1.0511	0.3900	0.3156	0.2943
1.30	1.0599	0.3346	0.2659	0.3995
1.40	1.0646	0.2904	0.2266	0.4829
1.50	1.0694	0.2547	0.1951	0.5502
1.60	1.0742	0.2253	0.1694	0.6053
1.70	1.0790	0.2008	0.1482	0.6510
1.80	1.0838	0.1802	0.1305	0.6892
1.90	1.0887	0.1627	0.1156	0.7216
2.00	1.0936	0.1477	0.1030	0.7493

lognormally distributed. Let $S_{1,n}$ and $S_{2,n}$ be the two discrete random variables at the end of the interval corresponding to the two assets with current values S_1 and S_2, respectively. As before, we assume that the option matures after time T with exercise price X. Also $T = nh$. The means, variances, and covariance of the two assets at the end of this interval are obtained from the log-normal distribution as follows:

Let $Y_i = \log \frac{S_{i,n}}{S_i}$,

$$\begin{pmatrix} Y_1 \\ Y_2 \end{pmatrix} \sim N\left(\begin{pmatrix} \mu_{y_1} \\ \mu_{y_2} \end{pmatrix}, \begin{pmatrix} \sigma_{y_{11}} & \sigma_{y_{12}} \\ \sigma_{y_{21}} & \sigma_{y_{22}} \end{pmatrix}\right)$$

where

$$\mu_{y_i} = \mu_i h,\ \sigma_{y_{ij}} = \sigma_{ij} h.$$

Since the marginal distribution of Y_i is normal, we have as in the one asset case

$$E(S_{i,n}) = S_i M_i,$$

$$M_i = e^{rh},$$

$$\begin{aligned} \mathrm{Var}(S_{i,n}) &= S_i^2 M_i^2 \exp\left[\left(\sigma_i^2 h - 1\right)\right] \\ &= S_i^2 V_i. \end{aligned}$$

where

$$V_i = M_i^2 \exp\left[\left(\sigma_i^2 h - 1\right)\right]$$

By the property of the moment-generating function for the bivariate normal distribution, we have

$$\begin{aligned} E(S_{1,n} S_{2,n}) &= S_1 S_2 E(e^{y_1 + y_2}) \\ &= S_1 S_2 E\left(e^{(1,1)\begin{pmatrix} y_1 \\ y_2 \end{pmatrix}} \right) \\ &= S_1 S_2 e^{\mu_1 h + \mu_2 h + \frac{h}{2}\left(\sigma_1^2 + \sigma_2^2 + 2\sigma_{12}\right)} \\ &= S_1 S_2 M_1 M_2 \exp[\rho \sigma_1 \sigma_2 h], \end{aligned}$$

$$\begin{aligned} \mathrm{Cov}(S_{1,n}, S_{2,n}) &= E(S_{1,n} S_{2,n}) \\ &\quad - [E(S_{1,n})][E(S_{2,n})] \\ &= S_1 S_2 M_1 M_2 [\exp(\rho \sigma_1 \sigma_2 h) - 1] \\ &= S_1 S_2 R, \end{aligned}$$

$$\begin{aligned} \mathrm{Var}(S_{i,n}) &= E\left(S_{i,n}^2\right) - [E(S_{i,n})]^2 \\ &= S_i^2 M_i^2 \left[\exp\left(\sigma_i^2 h - 1\right)\right] \\ &= S_i^2 V_i \end{aligned}$$

where

$V_i = M_i^2 \left[\exp\left(\sigma_i^2 h - 1\right)\right].$

Thus, we have

	Asset 1	Asset 2
Mean	$S_1 M_1$	$S_2 M_2$
Variance	$S_1^2 V_1$	$S_1^2 V_2$

where

$$\begin{aligned} M_i &= \exp(rh) \\ V_i &= M_i^2 \left[\exp\left(\sigma_i^2 h - 1\right)\right], \\ i &= 1, 2. \end{aligned}$$

Furthermore,

$$E(S_{1,n} S_{2,n}) = S_1 S_2 M_1 M_2 [\exp(\rho \sigma_1 \sigma_2 h)] \tag{12.47}$$

where ρ is the correlation coefficient between the two assets.

Thus, the covariance between the two assets is

$$\begin{aligned} \mathrm{Cov}(S_{1,n}, S_{2,n}) &= S_1 S_2 M_1 M_2 [\exp(\rho \sigma_1 \sigma_2 h) - 1] \\ &= S_1 S_2 R \end{aligned} \tag{12.48}$$

In order to obtain an efficient algorithm, a five-point jump process is natural and is given below.

Event	Probability	Value of assets given that event occurred	
		Asset 1	Asset 2
E_1	p_1	S_1u_1	S_2u_2
E_2	p_2	S_1u_1	S_2d_2
E_3	p_3	S_1d_1	S_2d_2
E_4	p_4	S_1d_1	S_2u_2
E_5	p_5	S_1	S_2

where $u_1d_1 = u_2d_2 = 1$ and $\sum_{i=1}^{5} p_i = 1$.

In order to obtain the probabilities and the stretch parameters, we will first find the probabilities in terms of u_1, u_2 and other variables. By equating the means and variances of the discrete and lognormal distributions, we have

$$(p_1+p_2)S_1u_1+p_5S_1+(p_3+p_4)S_1d_1 = S_1M_1, \quad (12.49)$$

$$(p_1+p_4)S_2u_2+p_5S_2+(p_2+p_3)S_2d_2 = S_2M_2, \quad (12.50)$$

$$(p_1+p_2)\left(S_1^2u_1^2 - S_1^2M_1^2\right)+p_5\left(S_1^2 - S_1^2M_1^2\right) + (p_3+p_4)\left(S_1^2d_1^2 - S_1^2M_1^2\right) = S_1^2V_1. \quad (12.51)$$

$$(p_1+p_4)\left(S_2^2u_2^2 - S_2^2M_2^2\right)+p_5\left(S_2^2 - S_2^2M_2^2\right) + (p_2+p_3)\left(S_2^2d_2^2 - S_2^2M_2^2\right) = S_2^2V_2. \quad (12.52)$$

Furthermore,

$$\sum_{i=1}^{5} p_i = 1. \quad (12.53)$$

Note that there is a strong correspondence between the three Eqs. (12.37), (12.38), and (12.39) and the triplet (12.53), (12.50), and (12.52). If we regard (p_1+p_2), (p_3+p_4), and p_5 as the new probabilities, we can solve these probabilities as before. They are given in Eqs. (12.42), (12.43), and (12.44) with appropriate modifications. Likewise, we can solve for (p_1+p_4), (p_2+p_3), and p_5. Thus, we can write:

$$p_1+p_2 = f(u_1,M_1,V_1) = f_1, \quad (12.54)$$

$$p_3+p_4 = g(u_1,M_1,V_1) = g_1, \quad (12.55)$$

$$p_1+p_4 = f(u_2,M_2,V_2) = f_2, \quad (12.56)$$

$$p_2+p_3 = g(u_2,M_2,V_2) = g_2. \quad (12.57)$$

Obviously, $f_1+g_1 = f_2+g_2$, which gives the relationship to be satisfied by u_1 and u_2.

From (12.47), we obtain the following equation by equating the expected value of the product of the assets under approximating discrete and lognormal distributions,

$$(p_1u_1u_2+p_2u_2d_2+p_3d_1d_2+p_4d_1u_2+p_5)S_1S_2 = RS_1S_2. \quad (12.58)$$

From Eqs. (12.54), (12.55), (12.56), (12.57), and (12.58), we obtain the following:

As in the one asset case, set

$$u_i = \exp\left(\lambda_i\sigma_i\sqrt{h}\right).$$

Table 12.5 shows a list of jump amplitudes and jump probabilities.

$$\begin{aligned}
p_1 &= \frac{u_1u_2(R-1) - f_1\left(u_1^2-1\right) - f_2\left(u_2^2-1\right) + (f_2+g_2)(u_1u_2-1)}{\left(u_1^2-1\right)\left(u_2^2-1\right)},\\
p_2 &= \frac{f_1\left(u_1^2-1\right)u_2^2 + f_2\left(u_2^2-1\right) - (f_2+g_2)(u_1u_2-1) - u_1u_2(R-1)}{\left(u_1^2-1\right)\left(u_2^2-1\right)},\\
p_3 &= \frac{u_1u_2(R-1) - f_1\left(u_1^2-1\right)u_2^2 + g_2\left(u_2^2-1\right)u_1^2 + (f_2+g_2)\left(u_1u_2-u_2^2\right)}{\left(u_1^2-1\right)\left(u_2^2-1\right)},\\
p_4 &= \frac{f_1\left(u_1^2-1\right) + f_2\left(u_2^2-1\right)u_1^2 - (f_2+g_2)(u_1u_2-1) - u_1u_2(R-1)}{\left(u_1^2-1\right)\left(u_2^2-1\right)}.
\end{aligned} \tag{12.59}$$

We next present some numerical examples as given by Boyle (1988) which will illustrate the operation of the methods presented in the section. Both Johnson (1981) and Stulz (1982) have derived a number of exact results for European options when there are two underlying assets. These results can be used to check the accuracy of the approach. In particular, expressions exist for a European call option on the maximum of two assets and for a European put option on the minimum of two assets. Stulz and Johnson (1985) has obtained a generalization of these results for the case involving a European call option on the maximum or minimum of several assets. His results could be used as benchmarks to check extensions of our method to situations involving three or more assets.

The advantage of the present approach is that it permits early exercise and thus can be used to value American options and, in particular, American put options. The approach presented in Sect. 12.3 can also be modified to handle dividends and other payouts.

For our numerical solutions, we assume $S_1 = 40$, $S_2 = 40$, $\sigma_1 = 0.20$, $\sigma_2 = 0.30$, $\rho = 0.5$, $r = 5$ percent per annum effective = 0.048790 continuously, $T = 7$ months = 0.5833333 years, and exercise prices = 35, 40, 45. We display in Table 12.6 the results for three types of options: European call options on the maximum of the two assets; European put

Table 12.5 Jump amplitudes and jump probabilities for 5-jump process $\sigma_1 = 0.2, \sigma_2 = 0.25, r = 0.1, n = 20, T = 1, \rho = 0.5$

λ	u_1	u_2	p_1	p_2	p_3	p_4	p_5
1.00	1.0457	1.0574	0.4201	0.1337	0.3448	0.1198	−0.0184
1.10	1.0504	1.0633	0.3499	0.1111	0.2814	0.0984	0.1592
1.20	1.0551	1.0692	0.2962	0.0938	0.2334	0.0822	0.2943
1.30	1.0599	1.0752	0.2543	0.0803	0.1963	0.0696	0.3995
1.40	1.0646	1.0812	0.2208	0.0696	0.1670	0.0597	0.4829
1.50	1.0694	1.0873	0.1937	0.0609	0.1434	0.0517	0.5502
1.60	1.0742	1.0934	0.1715	0.0538	0.1243	0.0451	0.6053
1.70	1.0790	1.0995	0.1529	0.0479	0.1085	0.0397	0.6510
1.80	1.0838	1.1056	0.1373	0.0429	0.0953	0.0352	0.6892
1.90	1.0887	1.1118	0.1240	0.0387	0.0842	0.0314	0.7216
2.00	1.0936	1.1180	0.1126	0.0351	0.0748	0.0282	0.7493

Table 12.6 Different types of options involving two assets: comparison of lattice approach of this section with accurate results

	Number of steps			
Exercise price	10	20	50	Accurate value
European call on the maximum of the two assets				
35	9.404	9.414	9.419	9.420
40	5.466	5.477	5.483	5.488
45	2.817	2.790	2.792	2.795
European put on the minimum of the two assets				
35	1.425	1.394	1.392	1.387
40	3.778	3.790	3.795	3.798
45	7.475	7.493	7.499	7.500
American put on the minimum of the two assets				
35	1.450	1.423	1.423	
40	3.870	3.885	3.892	
45	7.645	7.674	7.689	

options on the minimum of the two assets; and American put options on the minimum of two assets. Since we assume no dividend payments, the value of the European call on the maximum of the two assets will be identical to the value of an American call option with the same specifications. The agreement between the numbers obtained using the lattice approximation and the accurate values is quite reasonable, especially when 50 time steps are used. In this case, the maximum difference is 0.005, and the accuracy would be adequate for most applications. For this set of parameter values, the American put is not much more valuable than its European counterpart.

12.7 Conclusion

In this chapter, we demonstrate how an option pricing model can be derived in a less mathematically fashion using binominal option pricing model. We have first discussed the basic concepts of call options, and we show how the decision trees can be used to derive binominal call option pricing model. The binominal call option pricing model can be used to derive Black–Scholes option pricing model (1973) as shown by Rendleman and Barter (RB 1979) and Cox et al. (CRR 1979). In addition, we try to extend binomial option pricing model to multinomial option pricing model. We derive the multinomial option pricing model and apply it to the limiting case of Black and Scholes model. We also introduce a lattice framework for option pricing model. To apply the lattice model, we present the modified models of the two-state approach on single-state variable and two-state variables. The relationship between binomial option pricing model and Black and Scholes (1973) option pricing model will be discussed in Chaps. 13 and 21 in detail.

Bibliography

Bhattacharya, R. N., & Rao, R. R. (1976). *Normal approximation and asymptotic expansions*. New York: Wiley.

Black, F., & Scholes, M. (1973). The pricing of options and corporate liabilities. *Journal of Political Economy, 3,* 637–659.

Boyle, P. (1988). A lattice framework for option pricing with two state variables. *Journal of Financial and Quantitative Analysis, 23,* 1–12.

Boyle, P. P. (1989). The quality option and the timing option in futures contracts. *Journal of Finance, 44,* 101–103.

Cox, J. C., & Rubinstein, M. (1985). *Option markets*. Englewood Cliffs, NJ: Prentice-Hall.

Cox, J. C., Ross, S. A., & Rubinstein, M. (1979). Option pricing: A simple approach. *Journal of Financial Economics, 7,* 229–263.

Hull, J. (2014). *Options, futures, and other derivatives* (9th ed.). Prentice Hall.

Jarrow, R., & Turnbull, S. (1999). *Derivatives securities* (2nd ed.). South-Western College Publishing.

Johnson, H. (1981). The *pricing of complex options*. Unpublished manuscript.

Johnson, H. (1987). Options on the maximum or the minimum of several assets. *Journal of Financial and Quantitative Analysis, 22,* 277–283.

Lee, J. C. (2001). Using microsoft excel and decision trees to demonstrate the binomial option pricing model. *Advances in Investment Analysis and Portfolio Management, 8,* 303–329.

Lee, C. F., & Chen, Y. (2016). Alternative methods to derive option pricing models: Review and comparison. *Review of Quantitative Finance and Accounting, 47*(2), 417–451.

Lee, C. F., & Lee, J. C. (2010a). Multinomial option pricing model. *Handbook of quantitative finance and risk management* (pp. 399–406). New York: Springer.

Lee, C. F., & Lee, J. C. (2010b). Applications of the binomial distribution to evaluate call options. *Handbook of quantitative finance and risk management* (pp. 393–397). New York: Springer.

Lee, C. F., & Lee, A. C. (2013). *Encyclopedia of finance* (2nd ed.). New York, NY: Springer.

Lee, J. C., Lee, C. F., Wang, R. S., & Lin, T. I. (2004). On the limit properties of binomial and multinomial option pricing models: Review and integration. In *Advances in quantitative analysis of finance and accounting new series volume 1* (Vol. 1). Singapore: World Scientific.

Lee, C. F., Lee, A. C., & Lee, J. (2010). *Handbook of quantitative finance and risk management*. New York: Springer.

Lee, C. F., Lee, J. C., & Lee, A. C. (2013). *Statistics for business and financial economics*. Singapore: World Scientific Publishing Company.

MacBeth, J., & Merville, L. (1979). An empirical examination of the Black–Scholes call option pricing model. *The Journal of Finance, 34,* 1173–1186.

Madan, D. B., Milne, F., & Shefrin, H. (1989). The multinomial option pricing model and its Brownian and Poisson limits. *The Review of Financial Studies, 2* (2), 251–265.

Rendelman, R. J., Jr., & Bartter, B. J. (1979). Two-state option pricing. *Journal of Finance, 34*(5), 1093–1110.

Stulz, R. (1982). Options on the minimum or the maximum of two risky assets: analysis and applications. *Journal of Financial Economics, 10*(2), 161–185.

Stulz, R., & Johnson, H. (1985). An analysis of secured debt. *Journal of Financial Economics, 14*(4), 501–521.

13 Two Alternative Binomial Option Pricing Model Approaches to Derive Black–Scholes Option Pricing Model

Contents

Abstract

Based upon the models derived in the previous chapters, based on the binomial model, we present two alternative approaches to derive the Black and Scholes model. These two approaches are developed by Rendleman and Barter (Journal of Finance 24:1093–1110, 1979) and Cox et al. (Journal of Financial Economics 7:229–263, 1979). The relative advantages between these two methods are discussed in some detail.

13.1 Introduction

Based upon the binomial option pricing model derived in the last chapter, the main purpose of this chapter is to review two famous binomial option pricing models: Rendleman and Barter (RB 1979) and Cox, Ross, and Rubinstein (CRR 1979). First, we will give an alternative detailed derivation of the two models and show that the limiting results of the two models both lead to the celebrated Black–Scholes formula. Then we will

This chapter draws upon the paper by Lee and Lin (2010) which was published in the Handbook entitled, Quantitative Finance and Risk Management (2010) Chap. 26, edited by Lee et al. (2010).

C.-F. Lee et al., *Financial Econometrics, Mathematics and Statistics*,
https://doi.org/10.1007/978-1-4939-9429-8_13

make comparisons of the two different approaches and analyze the advantages of each approach.

In the following of this chapter, we will discuss the two-state option pricing model of Rendleman and Barter in Sect. 13.2. And the binomial option pricing model of CRR will be demonstrated in Sect. 13.3. We will show comparison of the two approaches in Sect. 13.4. Finally, Sect. 13.5 concludes Appendix presents the proof of binomial theorem.

13.2 The Two-State Option Pricing Model of Rendleman and Bartter

In Rendleman and Barter (1979), a stock price can either advance or decline during the next period. Let H_T^+ and H_T^- represent the returns per dollar invested in the stock if the price rises (the + state) or falls (the − state), respectively, from time $T-1$ to time T. And V_T^+ and V_T^- the corresponding end-of-period values of the option.

Let R be the riskless interest rate, Rendleman and Barter (1979) show that the price of the option can be represented as a recursive form

$$P_{T-1} = \frac{V_-^+(1+R-H_T^-)+V_T^-(H_T^+-1-R)}{(H_T^+-H_T^-)(1+R)}$$

that can be applied at any time $T-1$ to determine the price of the option as a function of its value at time T.

13.2.1 The Discrete-Time Model

From the above equation, the value of a call option at maturing date $T-1$ is given by

$$W_{T-1} = \frac{W_T^+(1+R-H^-)+W_T^-(H^+-(1+R))}{(H^+-H^-)(1+R)} \tag{13.1}$$

Similarly,

$$W_{T-2} = \frac{W_{T-1}^+(1+R-H^-)+W_{T-1}^-(H^+-(1+R))}{(H^+-H^-)(1+R)} \tag{13.2}$$

Substituting (13.1) into (13.2) can get,

$$\begin{aligned} W_{T-2} = {} & \frac{(W_T^{++}(1+R-H^-)+W_T^{+-}(H^+-(1+R)))(1+R-H^-)}{(H^+-H^-)^2(1+R)^2} \\ & + \frac{(W_T^{-+}(1+R-H^-)+W_T^{--}(H^+-(1+R)))(H^++(1+R))}{(H^+-H^-)^2(1+R)^2} \end{aligned} \tag{13.3}$$

Noting that $W_T^{+-}=W_T^{-+}$, so (13.3) can be simplified as:

$$W_{T-2} = \frac{(W_T^{++}(1+R-H^-)^2+2W_T^{+-}(H^+-(1+R)))(1+R-H^-)+W_T^{--}(H^+-(1+R))^2}{(H^+-H^-)^2(1+R)^2} \tag{13.4}$$

We can use this recursive form to get W_0:

Since after T periods, there are $\binom{T}{0}$ ways that a sequence of (T) pluses can occur, $\binom{T}{1}$ ways that ($T-1$) pluses can occur, $\binom{T}{2}$ ways that ($T-2$) pluses can occur, and so on…

Hence, by binomial theorem, W_0 can be represented as:

$$W_0 = \frac{1}{[(H^+ - H^-)(1+R)]^T} \times \begin{bmatrix} \binom{T}{0} W_T^{+\cdots+}(1+R-H^-)^T \\ \times(H^+ - (1+R))^0 \\ +\binom{T}{1} W_T^{+\cdots-}(1+R-H^-)^{T-1} \\ \times(H^+ - (1+R))^1 \\ +\binom{T}{2} W_T^{+\cdots--}(1+R-H^-)^{T-2} \\ \times(H^+ - (1+R))^2 \\ \vdots \\ +\binom{T}{T-1} W_T^{+--\cdots-}(1+R-H^-)^1 \\ \times(H^+ - (1+R))^{T-1} \\ +\binom{T}{T} W_T^{-\cdots-}(1+R-H^-)^0 \\ \times(H^+ - (1+R))^T \end{bmatrix} \tag{13.5}$$

Next to determine the value of the option at maturity. Suppose that stock increases i times and declines $(T-i)$ times, then the price of the stock will be $S_0 H^{+^i} H^{-^{T-i}}$ on the expiration date. So the option will be exercised if

$$S_0 H^{+^i} H^{-^{T-i}} > X$$

The maturity value of the option will be

$$W_T = S_0 H^{+^i} H^{-^{T-i}} - X \tag{13.6}$$

Let a denote the minimum integer value of i in (13.6) for which the inequality is satisfied.

$$a = 1 + \mathrm{INT}\left[\frac{\ln(X/S_0) - T\ln(H^-)}{\ln H^+ - \ln H^-}\right] \tag{13.7}$$

where INT[·] is the integer operator.

i.e., taking natural logarithm of RHS of (13.6),

$$\ln S_0 + i\ln H^+ + (T-i)\ln H^- = \ln X$$

$$\Rightarrow i\ln H^+ - i\ln H^- = \ln\left(\frac{X}{S_0}\right) - T\ln H^-$$

$$\Rightarrow i = \frac{\ln\left(\frac{X}{S_0}\right) - T\ln H^-}{\ln H^+ - \ln H^-}$$

Hence, the maturing value of the option is given by

$$\begin{aligned} W_T &= S_0 H^{+^i} H^{-^{T-i}} - X \ldots \text{if} \ldots i \geq a \\ W_T &= 0 \ldots \text{if} \ldots i < a \end{aligned} \tag{13.8}$$

Substituting (13.8) into (13.5), then the generalized option pricing equation for the discrete time is

$$W_0 = \frac{\sum_{i=a}^{T}\binom{T}{i}(S_0 H^{+^i} H^{-^{T-i}} - X)(1+R-H^-)^i (H^+ - (1+R))^{T-i}}{(H^+ - H^-)^T (1+R)^T} \tag{13.9}$$

13.2.2 The Continuous Time Model

For (13.9), we can write is as:

$$
\begin{aligned}
W_0 &= \frac{\sum_{i=a}^{T}\binom{T}{i}(S_0 H^{+^i} H^{-^{T-i}})(1+R-H^-)^i(H^+-(1+R))^{T-i} - \sum_{i=a}^{T}\binom{T}{i}X(1+R-H^-)^i(H^+-(1+R))^{T-i}}{(H^+-H^-)^T(1+R)^T} \\
&= \frac{S_0\sum_{i=a}^{T}\binom{T}{i}[H^+(1+R-H^-)]^i[H^-(H^+-1-R)]^{T-i}}{(H^+-H^-)^T(1+R)^T} \\
&\quad - \frac{X\sum_{i=a}^{T}\binom{T}{i}(1+R-H^-)^i(H^+-(1+R))^{T-i}}{(H^+-H^-)^T(1+R)^T} \\
&= S_0\sum_{i=a}^{T}\binom{T}{i}\left[\frac{(1+R-H^-)H^+}{(1+R)(H^+-H^-)}\right]^i\left[\frac{(H^+-1-R)H^-}{(H^+-H^-)(1+R)}\right]^{T-i} \\
&\quad - \frac{X}{(1+R)^T}\sum_{i=a}^{T}\binom{T}{i}\left[\frac{(1+R-H^-)}{(H^+-H^-)}\right]^i\left[\frac{(H^+-1-R)}{(H^+-H^-)}\right]^{T-i}
\end{aligned}
\tag{13.10}
$$

Since $\frac{(1+R-H^-)H^+}{(1+R)(H^+-H^-)} + \frac{(H^+-1-R)H^-}{(H^+-H^-)(1+R)} = 1$, $\frac{(1+R-H^-)}{(H^+-H^-)} + \frac{(H^+-1-R)}{(H^+-H^-)} = 1$, therefore, can interpret it as "pseudo probability."

Let $\varphi = \frac{(1+R-H^-)H^+}{(1+R)(H^+-H^-)}$ and $\phi = \frac{(1+R-H^-)}{(H^+-H^-)}$, we can restate (13.10) as:

$$W_0 = S_0 B(a,T,\varphi) - \frac{X}{(1+R)^T}B(a,T,\phi) \tag{13.11}$$

where $B(a,T,(\cdot))$ is the cumulative binomial probability function, the number of successes will fall between a and T after T trials.

As $T \to \infty$,

$$W_0 \sim S_0 N(Z_1, Z_1') - \frac{X}{(1+R)^T}N(Z_2, Z_2') \tag{13.12}$$

where $N(Z, Z')$is the probability of a normally distributed random variable with zero mean and variance 1 taking values between a lower limit Z and a upper limit Z'. And by the property of binomial pdf,

$$
\begin{aligned}
Z_1 &= \frac{a-T\varphi}{\sqrt{T\varphi(1-\varphi)}}, \quad Z_1' = \frac{T-T\varphi}{\sqrt{T\varphi(1-\varphi)}} \\
Z_2 &= \frac{a-T\phi}{\sqrt{T\phi(1-\phi)}}, \quad Z_2' = \frac{T-T\phi}{\sqrt{T\phi(1-\phi)}}
\end{aligned}
$$

Thus,

$$
\begin{aligned}
W_0 = S_0&\left(\lim_{T\to\infty} Z_1, \lim_{T\to\infty} Z_1'\right) \\
&- \frac{X}{\lim_{T\to\infty}(1+R)^T}N\left(\lim_{T\to\infty} Z_2, \lim_{T\to\infty} Z_2'\right)
\end{aligned}
\tag{13.13}
$$

Let $1+R = e^{r/T}$, then $\lim_{T\to\infty}(1+R)^T = e^r$. And since $\lim_{T\to\infty} Z_1' = \lim_{T\to\infty} Z_2' = \infty$, the remaining things to be determined are $\lim_{T\to\infty} Z_1$ and $\lim_{T\to\infty} Z_2$.

From Eqs. (13.10) and (13.11) of the text in the Rendleman and Barter (1979),

$$\begin{aligned} H^+ &= e^{\mu/T + (\sigma/\sqrt{T})\sqrt{\frac{(1-\theta)}{\theta}}} \\ H^- &= e^{\mu/T - (\sigma/\sqrt{T})\sqrt{\frac{\theta}{(1-\theta)}}} \end{aligned} \tag{13.10}$$

substituting H^+ and H^- into (13.7), so

$$Z_1 = \frac{a - T\varphi}{\sqrt{T\varphi(1-\varphi)}} = \frac{1 + \mathrm{INT}\left[\frac{\ln\left(\frac{X}{S_0}\right) - \mu + \sigma\sqrt{T}\sqrt{\frac{\theta}{1-\theta}}}{\frac{\sigma}{\sqrt{T\theta(1-\theta)}}}\right] - T\varphi}{\sqrt{T\varphi(1-\varphi)}}$$

In the limit, the term $1 + \mathrm{INT}[\cdot]$ will be simplified to $[\cdot]$. So,

$$\begin{aligned} Z_1 &\sim \frac{\ln\left(\frac{X}{S_0}\right) - \mu}{\frac{\sigma}{\sqrt{T\theta(1-\theta)}}}\frac{1}{\sqrt{T\varphi(1-\varphi)}} + \frac{\sigma\sqrt{T}\sqrt{\frac{\theta}{1-\theta}}}{\frac{\sigma}{\sqrt{T\theta(1-\theta)}}}\frac{1}{\sqrt{T\varphi(1-\varphi)}} \\ &\sim \frac{\ln\left(\frac{X}{S_0}\right) - \mu}{\sigma\sqrt{\frac{\varphi(1-\varphi)}{\theta(1-\theta)}}} + \frac{\sigma\sqrt{T}\sqrt{\frac{\theta}{1-\theta}}}{\frac{\sigma}{\sqrt{T\theta(1-\theta)}}}\frac{1}{\sqrt{T\varphi(1-\varphi)}} - \frac{T\varphi}{\sqrt{T\varphi(1-\varphi)}} \\ &\sim \frac{\ln\left(\frac{X}{S_0}\right) - \mu}{\sigma\sqrt{\frac{\varphi(1-\varphi)}{\theta(1-\theta)}}} + \frac{\sigma\sqrt{T}\sqrt{\frac{\theta}{1-\theta}}}{\frac{\sigma\sqrt{T\varphi(1-\varphi)}}{\sqrt{T\theta(1-\theta)}}} - \frac{T\varphi}{\sqrt{T\varphi(1-\varphi)}} \\ &\sim \frac{\ln\left(\frac{X}{S_0}\right) - \mu}{\sigma\sqrt{\frac{\varphi(1-\varphi)}{\theta(1-\theta)}}} + \frac{\sqrt{T}\sqrt{\frac{\theta}{1-\theta}}\sqrt{T\theta(1-\theta)}}{\sqrt{T\varphi(1-\varphi)}} - \frac{T\varphi}{\sqrt{T\varphi(1-\varphi)}} \\ &\sim \frac{\ln\left(\frac{X}{S_0}\right) - \mu}{\sigma\sqrt{\frac{\varphi(1-\varphi)}{\theta(1-\theta)}}} + \frac{\sqrt{T}(\theta-\varphi)}{\sqrt{\varphi(1-\varphi)}} \end{aligned} \tag{13.14}$$

Substituting H^+, H^-, and $1 + R = e^{r/T}$ in φ,

$$\begin{aligned} \varphi &= \frac{(1+R-H^-)H^+}{(1+R)(H^+ - H^-)} \\ &= \frac{\left[e^{\frac{r}{T}} - e^{\frac{\mu}{T} - (\sigma/\sqrt{T})\sqrt{\frac{\theta}{1-\theta}}}\right] e^{\frac{\mu}{T} + (\sigma/\sqrt{T})\sqrt{\frac{1-\theta}{\theta}}}}{e^{\frac{r}{T}}\left[e^{\frac{\mu}{T} + (\sigma/\sqrt{T})\sqrt{\frac{1-\theta}{\theta}}} - e^{\frac{\mu}{T} - (\sigma/\sqrt{T})\sqrt{\frac{\theta}{1-\theta}}}\right]} \\ &= \frac{\left[e^{\frac{r}{T}} - e^{\frac{\mu}{T} - (\sigma/\sqrt{T})\sqrt{\frac{\theta}{1-\theta}}}\right] e^{(\sigma/\sqrt{T})\sqrt{\frac{1-\theta}{\theta}}}}{e^{\frac{r}{T}}\left[e^{(\sigma/\sqrt{T})\sqrt{\frac{1-\theta}{\theta}}} - e^{(\sigma/\sqrt{T})\sqrt{\frac{\theta}{1-\theta}}}\right]} \\ &= \frac{\left[e^{(\sigma/\sqrt{T})\sqrt{\frac{1-\theta}{\theta}}} - e^{\frac{\mu}{T} - \frac{r}{T} - (\sigma/\sqrt{T})\sqrt{\frac{\theta}{1-\theta}} + (\sigma/\sqrt{T})\sqrt{\frac{1-\theta}{\theta}}}\right]}{\left[e^{(\sigma/\sqrt{T})\sqrt{\frac{1-\theta}{\theta}}} - e^{(\sigma/\sqrt{T})\sqrt{\frac{\theta}{1-\theta}}}\right]} \\ &= \frac{\left[e^{(\sigma/\sqrt{T})\sqrt{\frac{1-\theta}{\theta}}}\right]}{\left[e^{(\sigma/\sqrt{T})\sqrt{\frac{1-\theta}{\theta}}} - e^{(\sigma/\sqrt{T})\sqrt{\frac{\theta}{1-\theta}}}\right]} - \frac{\left[e^{\frac{\mu}{T} - \frac{r}{T} + (\sigma/\sqrt{T})\left(\sqrt{\frac{1-\theta}{\theta}} - \sqrt{\frac{\theta}{1-\theta}}\right)}\right]}{\left[e^{(\sigma/\sqrt{T})\sqrt{\frac{1-\theta}{\theta}}} - e^{(\sigma/\sqrt{T})\sqrt{\frac{\theta}{1-\theta}}}\right]} \end{aligned}$$

Now expanding in Taylor's series in T,

$$\begin{aligned} &= \frac{\sigma\sqrt{T}\sqrt{\frac{\theta}{1-\theta}} + o\left(\frac{1}{T}\right)}{\sigma\sqrt{T}\left[\left(\sqrt{\frac{1-\theta}{\theta}} + \sqrt{\frac{\theta}{1-\theta}}\right) + \frac{1}{2}\frac{\sigma}{T}\left(\frac{1-\theta}{\theta} - \frac{\theta}{1-\theta}\right)\right] + o\left(\frac{1}{T}\right)} \\ &\quad + \frac{r - \mu - \frac{1}{2}\sigma^2\left(\frac{\theta}{1-\theta}\right) + \sigma^2 + O\left(\frac{1}{T}\right)}{\sigma\sqrt{T}\left[\left(\sqrt{\frac{1-\theta}{\theta}} + \sqrt{\frac{\theta}{1-\theta}}\right) + \frac{1}{2}\frac{\sigma}{T}\left(\frac{1-\theta}{\theta} - \frac{\theta}{1-\theta}\right)\right] + o\left(\frac{1}{T}\right)} \\ &= \frac{1}{\sqrt{T}}\frac{r - \mu - \frac{1}{2}\sigma^2\left(\frac{\theta}{1-\theta}\right) + \sigma^2}{\sigma\left(\sqrt{\frac{1-\theta}{\theta}} + \sqrt{\frac{\theta}{1-\theta}}\right) + \frac{1}{2}\left(\frac{\sigma^2}{\sqrt{T}}\right)\left(\frac{1-\theta}{\theta} - \frac{\theta}{1-\theta}\right) + o\left(\frac{1}{T}\right)} \\ &\quad + \frac{\sqrt{\frac{\theta}{1-\theta}}}{\left[\left(\sqrt{\frac{1-\theta}{\theta}} + \sqrt{\frac{\theta}{1-\theta}}\right) + \frac{1}{2}\left(\frac{\sigma}{\sqrt{T}}\right)\left(\frac{1-\theta}{\theta} - \frac{\theta}{1-\theta}\right)\right] + o\left(\frac{1}{T}\right)} \\ &\quad + o\left(\frac{1}{T}\right) \end{aligned}$$

where

$$\begin{aligned}H^{+} &= 1+\left(\frac{\mu}{T}+\frac{\sigma}{\sqrt{T}}\sqrt{\frac{1-\theta}{\theta}}\right)\\&\quad+\frac{1}{2!}\left(\frac{\mu}{T}+\frac{\sigma}{\sqrt{T}}\sqrt{\frac{1-\theta}{\theta}}\right)^{2}\\&\quad+\frac{1}{3!}\left(\frac{\mu}{T}+\frac{\sigma}{\sqrt{T}}\sqrt{\frac{1-\theta}{\theta}}\right)^{3}+\cdots\\&= 1+\frac{\mu}{T}+\frac{\sigma}{\sqrt{T}}\sqrt{\frac{1-\theta}{\theta}}\\&\quad+\frac{\sigma^2}{2T}\left(\frac{1-\theta}{\theta}\right)+o\left(\frac{1}{T}\right)\end{aligned}$$

and

$$\begin{aligned}H^{-} &= 1+\left(\frac{\mu}{T}-\frac{\sigma}{\sqrt{T}}\sqrt{\frac{\theta}{1-\theta}}\right)\\&\quad+\frac{1}{2!}\left(\frac{\mu}{T}-\frac{\sigma}{\sqrt{T}}\sqrt{\frac{\theta}{1-\theta}}\right)^{2}\\&\quad+\frac{1}{3!}\left(\frac{\mu}{T}-\frac{\sigma}{\sqrt{T}}\sqrt{\frac{\theta}{1-\theta}}\right)^{3}+\cdots\\&= 1+\frac{\mu}{T}+\frac{\sigma}{\sqrt{T}}\sqrt{\frac{\theta}{1-\theta}}+\frac{\sigma^2}{2T}\left(\frac{\theta}{1-\theta}\right)\\&\quad+o\left(\frac{1}{T}\right)\end{aligned}$$

where $o(\frac{1}{T})$ denotes a function tending to zero more rapidly than $\frac{1}{T}$. (When we expanding in Taylor's series in T, the rest of the terms tending to zero more rapidly than $\frac{1}{T}$ so regard them as a function $o(\frac{1}{T})$.) Hence,

$$\begin{aligned}\lim_{T\to\infty}\varphi &= \frac{\sqrt{\frac{\theta}{1-\theta}}}{\sqrt{\frac{1-\theta}{\theta}}+\sqrt{\frac{\theta}{1-\theta}}}=\frac{\sqrt{\frac{\theta}{1-\theta}}}{\frac{1-\theta+\theta}{\sqrt{\theta(1-\theta)}}}\\&= \sqrt{\frac{\theta}{1-\theta}}\sqrt{\theta(1-\theta)}\\&= \sqrt{\theta^2}\\&= \theta\end{aligned}$$

and,

$$\begin{aligned}&\lim_{T\to\infty}\sqrt{T}(\theta-\varphi)\\&= \lim_{T\to\infty}\sqrt{T}\left[\theta-\frac{1}{\sqrt{T}}\frac{r-\mu-\frac{1}{2}\sigma^2\left(\frac{\theta}{1-\theta}\right)+\sigma^2}{\sigma\left(\sqrt{\frac{1-\theta}{\theta}}+\sqrt{\frac{\theta}{1-\theta}}\right)+\frac{1}{2}\left(\frac{\sigma^2}{\sqrt{T}}\right)\left(\frac{1-\theta}{\theta}-\frac{\theta}{1-\theta}\right)+o\left(\frac{1}{T}\right)}\right.\\&\qquad\left.-\frac{\sqrt{\frac{\theta}{1-\theta}}}{\left[\left(\sqrt{\frac{1-\theta}{\theta}}+\sqrt{\frac{\theta}{1-\theta}}\right)+\frac{1}{2}\left(\frac{\sigma}{\sqrt{T}}\right)\left(\frac{1-\theta}{\theta}-\frac{\theta}{1-\theta}\right)\right]+o\left(\frac{1}{T}\right)}-o\left(\frac{1}{T}\right)\right]\\&= \lim_{T\to\infty}\left[\sqrt{T}\theta-\frac{r-\mu-\frac{1}{2}\sigma^2\left(\frac{\theta}{1-\theta}\right)+\sigma^2}{\sigma\left(\sqrt{\frac{1-\theta}{\theta}}+\sqrt{\frac{\theta}{1-\theta}}\right)+\frac{1}{2}\left(\frac{\sigma^2}{\sqrt{T}}\right)\left(\frac{1-\theta}{\theta}-\frac{\theta}{1-\theta}\right)+o\left(\frac{1}{T}\right)}\right.\\&\qquad\left.-\frac{\sqrt{T}\sqrt{\frac{\theta}{1-\theta}}}{\left[\left(\sqrt{\frac{1-\theta}{\theta}}+\sqrt{\frac{\theta}{1-\theta}}\right)+\frac{1}{2}\left(\frac{\sigma}{\sqrt{T}}\right)\left(\frac{1-\theta}{\theta}-\frac{\theta}{1-\theta}\right)\right]+o\left(\frac{1}{T}\right)}-\sqrt{T}o\left(\frac{1}{T}\right)\right]\\&= \lim_{T\to\infty}\frac{\sigma\sqrt{T}\theta\left[\left(\sqrt{\frac{1-\theta}{\theta}}+\sqrt{\frac{\theta}{1-\theta}}\right)+\frac{1}{2}\left(\frac{\sigma^2}{\sqrt{T}}\right)\left(\frac{1-\theta}{\theta}-\frac{\theta}{1-\theta}\right)+o\left(\frac{1}{T}\right)\right]-\sqrt{T}\sqrt{\frac{\theta}{1-\theta}}}{\sigma\left(\sqrt{\frac{1-\theta}{\theta}}+\sqrt{\frac{\theta}{1-\theta}}\right)+\frac{1}{2}\left(\frac{\sigma^2}{\sqrt{T}}\right)\left(\frac{1-\theta}{\theta}-\frac{\theta}{1-\theta}\right)+o\left(\frac{1}{T}\right)\sigma}\\&\qquad-\frac{r-\mu-\frac{1}{2}\sigma^2\left(\frac{\theta}{1-\theta}\right)+\sigma^2}{\sigma\left(\sqrt{\frac{1-\theta}{\theta}}+\sqrt{\frac{\theta}{1-\theta}}\right)+\frac{1}{2}\left(\frac{\sigma^2}{\sqrt{T}}\right)\left(\frac{1-\theta}{\theta}-\frac{\theta}{1-\theta}\right)+o\left(\frac{1}{T}\right)}-\sqrt{T}o\left(\frac{1}{T}\right)\\&= \frac{\frac{1}{2}\theta\sigma^2\left(\frac{1-\theta}{\theta}-\frac{\theta}{1-\theta}\right)-\left(r-\mu-\frac{1}{2}\sigma^2\left(\frac{\theta}{1-\theta}\right)+\sigma^2\right)}{\sigma\left(\sqrt{\frac{1-\theta}{\theta}}+\sqrt{\frac{\theta}{1-\theta}}\right)}\\&= \frac{\frac{1}{2}\theta\sigma^2\left(\frac{(1-\theta)^2-\theta^2}{\theta(1-\theta)}\right)-r+\mu+\frac{1}{2}\sigma^2\left(\frac{\theta}{1-\theta}\right)-\sigma^2}{\sigma\frac{1}{\sqrt{\theta(1-\theta)}}}\\&= \frac{\frac{1}{2}\sigma^2\left(\frac{1-2\theta}{1-\theta}\right)-r+\mu+\frac{1}{2}\sigma^2\left(\frac{\theta}{1-\theta}\right)-\sigma^2}{\sigma\frac{1}{\sqrt{\theta(1-\theta)}}}\\&= \frac{\frac{1}{2}\sigma^2\left(\frac{1-\theta}{1-\theta}\right)-\frac{1}{2}\sigma^2\left(\frac{\theta}{1-\theta}\right)-r+\mu+\frac{1}{2}\sigma^2\left(\frac{\theta}{1-\theta}\right)-\sigma^2}{\sigma\frac{1}{\sqrt{\theta(1-\theta)}}}\end{aligned}$$

After canceling terms,

$$\lim_{T\to\infty}\sqrt{T}(\theta-\varphi)=\frac{-\sqrt{\theta(1-\theta)}\left(r-\mu+\frac{1}{2}\sigma^2\right)}{\sigma}$$

Now substituting $\lim\limits_{T\to\infty}\varphi$ for φ and $\lim\limits_{T\to\infty}\sqrt{T}(\theta-\varphi)$ for $\sqrt{T}(\theta-\varphi)$ into (2.14),

$$\begin{aligned}\lim_{T\to\infty}Z_1 &= \frac{\ln\left(\frac{X}{S_0}\right)-\mu}{\sigma\sqrt{\frac{\theta(1-\theta)}{\theta(1-\theta)}}}\\&\quad-\frac{\sqrt{\theta(1-\theta)}\left(r-\mu+\frac{1}{2}\sigma^2\right)}{\sigma\sqrt{\theta(1-\theta)}}\\&= \frac{\ln\left(\frac{X}{S_0}\right)-r-\frac{1}{2}\sigma^2}{\sigma}\end{aligned}$$

Similarly,

$$\lim_{T\to\infty}Z_2=\frac{\ln\left(\frac{X}{S_0}\right)-r+\frac{1}{2}\sigma^2}{\sigma}$$

Since $N(Z,\infty) = N(-\infty, -Z)$, let $D_1 = -\lim_{T\to\infty} Z_1$, $D_2 = -\lim_{T\to\infty} Z_2$, the continuous time version of the two-state model is obtained:

$$w_0 = S_0 N(-\infty, D_1) - Xe^{-r} N(-\infty, D_2)$$

$$D_1 = \frac{\ln\left(\frac{X}{S_0}\right) + r + \frac{1}{2}\sigma^2}{\sigma}$$

$$D_2 = D_1 - \sigma$$

The above equation is identical to the Black–Scholes model.

13.3 The Binomial Option Pricing Model of CRR

In this section, we will concentrate on the limiting behavior of the binomial option pricing model proposed by Cox et al. (1979).

13.3.1 The Binomial Option Pricing Formula of CRR

Let S be the current stock price, K the option exercise price, and $R - 1$ the riskless rate. It is assumed that the stock follows a binomial process, from one period to the next it can only go up by a factor of U with probability P or go down by a factor of d with probability $(1 - p)$. After n periods to maturity, CRR showed that the option price C is:

$$C = \frac{1}{R^n} \sum_{k=0}^{n} \frac{n!}{k!(n-k)!} p^k (1-p)^{n-k} \quad \text{Max}\left[0, u^k d^{n-k} S - K\right]. \tag{13.15}$$

An alternative expression for C, which is easier to evaluate, is

$$\begin{aligned} C &= S\left[\sum_{k=m}^{n} \frac{n!}{k!(n-k)!} p^k (1-p)^{n-k} \frac{u^k d^{n-k}}{R}\right] \\ &\quad - \frac{K}{R^n}\left[\sum_{k=m}^{n} \frac{n!}{k!(n-k)!} p^k (1-p)^{n-k}\right] \\ &= SB(m; n, p') - \frac{K}{R^n} B(m; n, p). \end{aligned} \tag{13.16}$$

where $B(m; n, p) = \sum_{k=m}^{n} {}_nC_k p^k (1-p)^{n-k}$ and m is the minimum number of upward stock movements necessary for the option to terminate in the money, i.e., m is the minimum value of k in (13.15) such that $u^m d^{n-m} S - X > 0$.

13.3.2 Limiting Case

We now show that the binomial option pricing formula as given in Eq. (13.16) will converge to the celebrated Black–Scholes option pricing model. The Black–Scholes formula is

$$C = SN(d_1) - e^{-rt} XN(d_1 - \sigma\sqrt{t}) \tag{13.17}$$

where

$$d_1 = \frac{\ln\left(\frac{S}{XR^{-t}}\right)}{\sigma\sqrt{t}} + \frac{\sigma\sqrt{t}}{2} \tag{13.18}$$

σ^2 = the variance of stock rate of return
t = the fixed length of calendar time to expiration date, such that $h = \frac{t}{n}$.

We wish to show that Eq. (13.16) will coincide with Eq. (13.17) when $n \to \infty$.

In order to show the limiting result that the binomial option pricing formula converges to the continuous version of Black–Scholes option pricing formula, we suppose that h represents the lapsed time between successive stock price changes. Thus, if t is the fixed length of calendar time to expiration, and n is the total number of periods each with length h, then $h = \frac{t}{n}$. As the trading frequency increases, h will get closer to zero. When $h \to 0$, this is equivalent to $n \to \infty$.

Let $\hat{R}$ be one plus the interest rate over a trading period of length h. Then, we will have

$$\hat{R}^T = R^t \quad (13.19)$$

for any choice of n. Thus, $\hat{R} = R^{\frac{t}{n}}$, which shows that $\hat{R}$ must depend on n for the total return over elapsed time t to be independent of n. Also, in the limit, $(1+r)^{-n}$ tends to e^{-rt} as $n \to \infty$.

Let S^* be the stock price at the end of the nth period with the initial price S. If there are j up periods, then

$$\ln\frac{S^*}{S} = j\ \ln\ u + (n-j)\ln\ d = j\ \ln\left(\frac{u}{d}\right) + n\ \ln\ d \quad (13.20)$$

where j is the number of upward moves during the n periods.

Since j is the realization of a binomial random variable with probability of a success being q, we have expectation of log (S^*/S)

$$E\left(\ln\frac{S^*}{S}\right) = \left[q\ \ln\left(\frac{u}{d}\right) + \ln\ d\right]n = \hat{\mu}n, \quad (13.21)$$

and its variance

$$\mathrm{Var}\left(\mathrm{lv}\frac{S^*}{S}\right) = \left[\ln\left(\frac{u}{d}\right)\right]^2 q(1-q)n = \hat{\sigma}^2 n. \quad (13.22)$$

Since we divide up our original longer time period t into many shorter subperiods of length h so that $t = hn$, our procedure calls for making n longer, while keeping the length t fixed. In the limiting process, we would want the mean and the variance of the continuously compounded log rate of return of the assumed stock price movement to coincide with that of actual stock price as $n \to \infty$. Let the actual values denoted as $\hat{\mu}\ n$ and $\hat{\sigma}^2 n$ respectively. Then we want to choose u, d, and q in such a manner that $\hat{\mu}\ n \to \mu\ t$ and $\hat{\sigma}^2 n \to \sigma^2 t$ as $n \to \infty$. It can be shown that if we set

$$u = e^{\sigma\sqrt{\frac{t}{n}}},$$
$$d = e^{-\sigma\sqrt{\frac{t}{n}}}, \quad (13.23)$$
$$q = \frac{1}{2} + \frac{1}{2}\left(\frac{\mu}{\sigma}\right)\sqrt{\frac{t}{n}},$$

then $\hat{\mu}\ n \to \mu\ t$ and $\hat{\sigma}^2 n \to \sigma^2 t$ as $n \to \infty$. In order to proceed further, we need the following version of the central limit theorem.

Lyapounov's Condition. *Suppose* $X_1, X_2, \cdots$ *are independent and uniformly bounded with* $E(X_i) = 0$, $Y_n = X_1 + \cdots + X_n$, *and* $s^2 = E(Y_n^2) = \mathrm{Var}(Y_n)$.

If $\lim_{n\to\infty} \sum_{k=1}^{n} \frac{1}{s_n^{2+\delta}} E|X_k|^{2+\delta} = 0$ *for some* $\delta > 0$, *then the distribution of* $\frac{Y_n}{s_n}$ *converges to the standard normal as* $n \to \infty$.

Theorem 1 *If*

$$\frac{p|\ln u - \hat{\mu}|^3 + (1-p)|\ln d - \hat{\mu}|^3}{\hat{\sigma}^3\sqrt{n}} \to 0 \text{ as } n \to \infty \quad (13.24)$$

then

$$Pr\left[\frac{\ln\left(\frac{S^*}{S}\right) - \hat{\mu}n}{\hat{\sigma}\sqrt{n}} \le z\right] \to N(z) \quad (13.25)$$

where N(z) is the cumulative standard normal distribution function.

Proof See Appendix.

It is noted that the condition (13.24) is a special case of the Lyapounov's condition which is stated as follows. When $\delta = 1$, we have the condition (13.24).

This theorem says that when the fixed length t is divided into many subperiods, the log rate of return will approach to the normal distribution when the number of subperiods approached infinity. For this theorem to hold, the condition stated in Eq. (13.24) has to be satisfied. We next show that this condition is indeed satisfied.

We will next show that the binomial option pricing model as given in Eq. (13.16) will indeed coincide with the Black–Scholes option pricing formula as given in Eq. (13.17). Observe that $\hat{R}^{-n}$ is always equal to R^{-t}, as evidenced from Eq. (13.19). Thus, comparing the two option pricing formulae given in Eqs. (13.16) and (13.17), we see that there are apparent similarities. In order to show the limiting result, we need to show that as $n \to \infty$,

$$B(m; n, p') \to N(x) \text{ and } B(m; n, p) \to N(x - \sigma\sqrt{t}).$$

In this section, we will only show the second convergence result, as the same argument will hold true for the first convergence. From the definition of $B(n, pm)$, it is clear that

$$\begin{aligned} 1 - B(m; n, p) &= Pr(j \le m - 1) \\ &= Pr\left(\frac{j - np}{\sqrt{np(1-p)}} \le \frac{m - 1 - np}{\sqrt{np(1-p)}}\right). \end{aligned} \tag{13.26}$$

Recall that we consider a stock to move from S to uS with probability p and dS with probability $(1 - p)$. During the fixed calendar period of $t = nh$ with n subperiods of length h, if there are j up moves, then

$$\ln\frac{S^*}{S} = j \ln\left(\frac{u}{d}\right) + n \ln d \tag{13.27}$$

The mean and variance of the continuously compounded rate of return for this stock are $\hat{\mu}_P$ and $\hat{\sigma}_P^2$ where

$$\hat{\mu}_P = p \ln\left(\frac{u}{d}\right) + \ln d \text{ and } \hat{\sigma}_P^2 = \left[\ln\left(\frac{u}{d}\right)\right]^2 p(1-p).$$

From Eq. (13.27) and the definitions for $\hat{\mu}_P$ and $\hat{\sigma}_P^2$, we have

$$\frac{j - np}{\sqrt{np(1-p)}} = \frac{\ln\left(\frac{S^*}{S}\right) - \hat{\mu}_P n}{\hat{\sigma}_P\sqrt{n}}. \tag{13.28}$$

Also, from the binomial option pricing formula we have

$$\begin{aligned} m - 1 &= \frac{\ln\left(\frac{X}{Sd^n}\right)}{\ln\left(\frac{u}{d}\right)} - \varepsilon \\ &= \left[\ln\frac{X}{S} - n \ln d\right] / \ln\frac{u}{d} - \varepsilon, \end{aligned}$$

where is a real number between 0 and 1.

From the definitions of $\hat{\mu}_P$ and $\hat{\sigma}_P^2$, it is easy to show that

$$\frac{m - 1 - np}{\sqrt{np(1-p)}} = \frac{\ln\left(\frac{X}{S}\right) - \hat{\mu}_P n - \varepsilon \ln\left(\frac{u}{d}\right)}{\hat{\sigma}_P\sqrt{n}}.$$

Thus, from Eq. (13.26), we have

$$\begin{aligned} &1 - B(m; n, p) \\ &= Pr\left(\frac{\ln\frac{S^*}{S} - \hat{\mu}_P n}{\hat{\sigma}_P\sqrt{n}} \le \frac{\ln\frac{X}{S} - \hat{\mu}_P n - \varepsilon\ln\left(\frac{u}{d}\right)}{\hat{\sigma}_P\sqrt{n}}\right). \end{aligned} \tag{13.29}$$

We will now check the condition given by Eq. (13.24) in order to apply the central limit theorem. Now recall that

$$p = \frac{\hat{r} - d}{u - d},$$

with $\hat{r} = r^{\frac{t}{n}}$, and d and u are given in Eq. (13.23).

We have

$$\begin{aligned} p &= \frac{e^{\frac{t}{n}\log r} - e^{-\sigma\sqrt{\frac{t}{n}}}}{e^{\sigma\sqrt{\frac{t}{n}}} - e^{-\sigma\sqrt{\frac{t}{n}}}} \\ &= \frac{1 + \frac{t}{n}\ln r - \left[1 - \sigma\sqrt{\frac{t}{n}} + \frac{1}{2}\sigma^2\frac{t}{n}\right] + O\left(n^{-\frac{3}{2}}\right)}{1 + \sigma\sqrt{\frac{t}{n}} - \left[1 - \sigma\sqrt{\frac{t}{n}}\right] + O\left(n^{-\frac{3}{2}}\right)} \\ &= \frac{1}{2} + \frac{1}{2}\left[\frac{\ln r - \frac{1}{2}\sigma^2}{\sigma}\right]\sqrt{\frac{t}{n}} + O\left(n^{-1}\right). \end{aligned} \tag{13.30}$$

Hence, the condition given by Eq. (13.10) is satisfied because

$$\frac{p|\ln u - \hat{\mu}_p|^3 + (1-p)|\ln d - \hat{\mu}_p|^3}{\hat{\sigma}_p^3\sqrt{n}} = \frac{(1-p)^2+p^2}{\sqrt{np(1-p)}} \to 0, \quad \text{as}\, n \to \infty.$$

Finally, in order to apply the central limit theorem, we have to evaluate $\hat{\mu}_p n$, $\hat{\sigma}_p^2 n$, and $\log\left(\frac{u}{d}\right)$ as $n \to \infty$. It is clear that

$$\hat{\mu}_p n \to \left(\ln r - \frac{1}{2}\sigma^2\right)t,\ \hat{\sigma}_p^2\sqrt{n} \to \sigma^2 t \,\text{and}\, \ln\left(\frac{u}{d}\right) \to 0.$$

Hence, in order to evaluate the asymptotic probability in Eq. (13.26), we have

$$\frac{\ln\left(\frac{X}{S}\right) - \hat{\mu}_p n - \varepsilon \ln\left(\frac{u}{d}\right)}{\hat{\sigma}_p\sqrt{n}} \to z = \frac{\ln\left(\frac{X}{S}\right) - \left(\ln r - \frac{1}{2}\sigma^2\right)t}{\sigma\sqrt{t}}.$$

Using the fact that $1 - N(z) = N(-z)$, we have, as $n \to \infty$

$$B(m; n, p) \to N(-z) = N(x - \sigma\sqrt{t}).$$

Similar argument holds for $B(m; n, p')$, and hence we completed the proof that the binomial option pricing formula as given in Eq. (13.16) includes the Block–Scholes option pricing formula as a limiting case.

13.4 Comparison of the Two Approaches

From the results of last two sections, we show that both RB and CRR models lead to the celebrated Black–Scholes formula. The following table shows the comparisons of the necessary mathematical and statistical knowledge and assumptions for the two models.

Model	Rendleman and Barter (1979)	Cox et al. (1979)
Mathematical and probability theory knowledge	Basic algebra	Basic algebra
	Taylor expansion	Taylor expansion
	Binomial theorem	Binomial theorem
	Central limit theorem	Central limit theorem
	Properties of binomial distribution	Properties of binomial distribution Lyapounov's condition
Assumption	1. The distribution of returns of the stock is stationary over time and the stock pays no dividends. (Discrete-time model)	The stock follows a binomial process from one period to the next it can only go up by a factor of "u" with probability "p" or go down by a factor of "d" with probability "$1 - p$"
	2. The mean and variance of logarithmic returns of the stock are held constant over the life of the option. (Continuous time model)	In order to apply the central limit theorem, "u", "d", and "p" are needed to be chosen

(continued)

Model	Rendleman and Barter (1979)	Cox et al. (1979)
Advantage and disadvantage	1. Readers who have undergraduate level training in mathematics and probability theory can follow this approach	1. Readers who have advanced level knowledge in probability theory can follow this approach; but for those who do not, CRR approach may be difficult to follow
	2. The approach of RB is intuitive. But the derivation is more complicated and tedious than the approach of CRR	2. The assumption on the parameters "*u*", "*d*", "*p*" makes CRR approach more restricted than RB approach

Hence, like we indicate in the table, CRR is easy to follow if one has the advanced level knowledge in probability theory but the assumptions on the model parameters make its applications limited. On the other hand, RB model is intuitive and does not require higher-level knowledge in probability theory. However, the derivation is more complicated and tedious.

13.5 Conclusion

This chapter can help to understand the statistical aspects of option pricing models for economics and finance professions. Also, it gives important financial and economic intuitions for readers in statistics professions. Therefore, by showing two alternative binomial option pricing models approaches to derive the Black–Scholes model, this chapter is useful for understanding the relationship between the two important optional pricing models and the Black–Scholes formula.

For readers who are interested in the binomial option pricing model, they can compare the two different approaches and find the best one which fits their interests and is easier to follow.

Appendix: The Binomial Theorem

The Binomial Theorem

$$(x+y)^n = \sum_{k=0}^{n} \binom{n}{k} x^k y^{n-k}$$

Lindberg–Levy Central Limit Theorem

If $x_1, \ldots, x_n$ are a random sample from a probability distribution with finite mean μ and finite variance σ^2 and $\bar{x} = \left(\frac{1}{n}\right) \sum_{i=1}^{n} x_i$, then

$$\sqrt{n}(\bar{x}_n - \mu) \longrightarrow dN\left[0, \sigma^2\right]$$

Proof of Theorem 1

Since

$$\begin{aligned} & p|\ln u - \hat{\mu}|^3 \\ & = p\left|\ln u - p \ln\frac{u}{d} - \ln d\right|^3 \\ & = p(1-p)^3\left|\ln\frac{u}{d}\right|^3 \end{aligned}$$

And

$$\begin{aligned} & (1-p)|\ln d - \hat{\mu}|^3 \\ & = (1-p)\left|\ln d - p \ln\frac{u}{d} - \ln d\right|^3 \\ & = p^3(1-p)\left|\ln \frac{u}{d}\right|^3, \end{aligned}$$

We have

$$\begin{aligned} & p|\ln u - \hat{\mu}|^3 + (1-p)|\ln d - \hat{\mu}|^3 \\ & = p(1-p)\left[(1-p)^2 - p^2\right]\left|\ln\frac{u}{d}\right|^3 \end{aligned}$$

.

Thus

$$\frac{p|\ln u-\hat{\mu}|^3+(1-p)|\ln d-\hat{\mu}|^3}{\hat{\sigma}^3\sqrt{n}} = \frac{p(1-p)[(1-p)^2-p^2]\left|\ln\frac{u}{d}\right|^3}{\left(\sqrt{p(1-p)}\ln\left(\frac{u}{d}\right)\right)^3\sqrt{n}} = \frac{(1-p)^2+p^2}{\sqrt{np(1-p)}}\rightarrow 0 \quad \text{as}\, n\rightarrow\infty.$$

Hence, the condition for the theorem to hold as stated in Eq. (13.24) is satisfied.

Bibliography

Amram, M., & Kulatilaka, N. (2001). *Real options*. USA: Oxford University Press.

Banz, R., & Miller, M. (1978). Prices for state contingent claims: Some estimates and applications. *Journal of Business, 51,* 653–672.

Bartter, B. J., & Rendleman, R. J., Jr. (1979). Fee-based pricing of fixed rate bank loan commitments. *Financial Management, 8.*

Bhattachayra, M. (1980). Empirical properties of the Black-Scholes formula under ideal conditions. *Journal of Financial and Quantitative Analysis, 15,* 1081–1105.

Bhattacharya, R. N., & Rao, R. R. (1976). *Normal approximations and asymptotic expansions*. New York: Wiley.

Black, F. (1972). Capital market equilibrium with restricted borrowing. *Journal of Business, 45,* 444–455.

Black, F., & Scholes, M. (1973). The pricing of options and corporate liabilities. *Journal of Political Economy, 31,* 637–654.

Bookstaber, R. M. (1981). *Option pricing and strategies in investing, reading*. MA: Addison-Wesley.

Brennan, M. J., & Schwartz, E. S. (1977). The valuation of American put options. *Journal of Financial Economics, 32,* 449–462.

Cox, J. c., & Rubinstein, M. (1985). *Option markets*. Englewood Cliffs, NJ: Prentice-Hall.

Cox, J., Ross, S. A., & Rubinstein, M. (1979). Option pricing: A simplified approach. *Journal of Financial Economics, 7,* 229–263.

Cox, J., & Ross, S. A. (1976). The valuation of options for alternative stochastic processes. *Journal of Financial Economics, 3,* 145–166.

Duffie, D., & Huang, C. (1985). Implementing arrow debreu equilibria by continuous trading of few long lived securities. *Econometrica, 6,* 1337–1356.

Finnerty, J. (1978). The chicago board options exchange and market efficiency. *Journal of Financial and Quantitative Analysis, 13,* 29–38.

Galai, D., & Masulis, R. W. (1976). The option pricing model and the risk factor of stock. *Journal of Financial Economics, 3,* 53–81.

Geske, R. (1979). The valuation of compound options. *Journal of Financial Economics, 7,* 63–81.

Harrison, J. M., & Kreps, D. M. (1979). Martingales and arbitrage in multiperiod securities markets. *Journal of Economic Theory, 20,* 381–408.

Harrison, J. M., & Pliska, S. R. (1981). Martingales and stochastic integrals in the theory of continuous trading. *Stochastic Process and their Applications, 11,* 215–260.

Hull, J. (2014). *Options, futures, and other derivatives* (9th ed.). Englewood Cliffs, NJ: Prentice-Hall.

Jarrow, R., & Turnbull, S. (1999). *Derivatives securities* (2nd ed.). South-Western College Pub.

Jones, E. P. (1984). Option arbitrage and strategy with large price changes. *Journal of Financial Economics, 13,* 91–113.

Liaw, K. T., & Moy, R. L. (2000). *The Irwin guide to stocks, bonds, futures, and options*. New York: McGraw-Hill Companies.

Lee, C. F., & Lee, A. C. (2013). *Encyclopedia of finance* (2nd ed.). New York: Springer.

Lee, C. F., Lee, A. C. & Lee, J. (2010). *Handbook of quantitative finance and risk management*. New York: Springer.

Lee, C. F., & Lin, C. S. M. (2010). Two alternative binomial option pricing model approaches to derive Black-Scholes option pricing model. In C. F. Lee, A. C. Lee, & J. Lee (Eds.), *Handbook of quantitative finance and risk management* (pp. 409–419). Singapore: Springer.

Lee, J. C., Lee, C. F., Wang, R. S. & Lin, T. I. (2004). On the limit properties of binomial and multinomial option pricing models: Review and integration. In *Advances in quantitative analysis of finance and accounting New Series Volume 1*. Singapore: World Scientific.

MacBeth, J., & Merville, L. (1979). An empirical examination of the Black-Scholes call option pricing model. *The Journal of Finance, 34,* 1173–1186.

Madan, D. B., & Milne, F. (1987). *The multinomial option pricing model and its limits*. Department of Econometrics working papers, University of Sydney, Sydney, NSW 2066.

Madan, D. B., & Milne, F. (1988a). *Arrow debreu option pricing for long-tailed return distributions*. Department of Econometrics working papers, University of Sydney, Sydney, NSW 2066.

Madan, D. B., & Milne, F. (1988b). *Option pricing when share prices are subject to predictable jumps*. Department of Econometrics working papers, University of Sydney, Sydney, NSW 2066.

McDonald, R. L. (2005). *Derivatives markets* (2nd ed.). Boston, Massachusetts: Addison Wesley.

Merton, R. C. (1973). The theory of rational option pricing. *Bell Journal of Economics and Management Science, 4,* 141–183.

Merton, R. C. (1976). Option pricing when underlying stock returns are discontinuous. *Journal of Financial Economics, 3,* 125–144.

Merton, R. C. (1977). On the pricing of contingent claims and the Modigliani-Miller theorem. *Journal of Financial Economics, 5,* 241–250.

Milne, F., & Shefrin, H. (1987). A critical examination of limiting forms of the binomial option pricing theory. Department of Economics, Australian National University, Canberra, ACT 2601.

Øksendal, B. (2007). *Stochastic differential equations: An introduction with applications* (6th ed.). Berlin, Heidelberg: Springer.

Parkinson, M. (1977). Option pricing: The American put. *Journal of Business, 50,* 21–36.

Rendleman, R. J., Jr., & Barter, B. J. (1979). Two-state option pricing. *Journal of Finance, 24,* 1093–1110.

Rendleman, R. J., Jr., & Barter, B. J. (1980). The pricing of options on debt securities. *Journal of Financial and Quantitative Analysis, 15,* 11–24.

Rubinstein, M. (1976). The valuation of uncertain income streams and the pricing of options. *Bell Journal of Economics and Management Science, 7,* 407–425.

Ritchken, P. (1987). *Option: Theory, strategy, and applications*. Glenview, IL: Scott, Foresman.

Shreve, S. E. (2004). *Stochastic calculus for finance I: The binomial asset pricing model*. New York: Springer.

Shreve, S. E. (2010). *Stochastic calculus for finance II: Continuous-time model*. New York: Springer.

Summa, J. F., & Lubow, J. W. (2001). *Options on futures*. New York: Wiley.

Trennepohl, G. (1981). A comparison of listed option premia and Black-Scholes model prices: 1973–1979. *Journal of Financial Research,* 11–20.

Zhang, P. G. (1998). *Exotic options: A guide to second generation options* (2nd ed.). World Scientific Pub Co Inc.

Normal, Lognormal Distribution, and Option Pricing Model

14

Contents

This chapter draws upon (1) Chapter 27 of Handbook of Quantitative Finance and Risk Management (2010) by Lee, Lee and Lee; and (2) Chapter 19 of the book entitled Security Analysis, Portfolio Management, and Financial Derivatives (2013) by Lee, Finnerty, Lee, Lee, and Wort.

C.-F. Lee et al., *Financial Econometrics, Mathematics and Statistics*,
https://doi.org/10.1007/978-1-4939-9429-8_14

Abstract

In this chapter, we discuss univariate and multivariate normal and lognormal distribution. The relationship between these distributions is discussed. Then we use these two statistical distributions to derive European options and American options with and without dividend payment. Excel programs used to evaluate these two kinds of option pricing models are also discussed in detail.

14.1 Introduction

The normal (or Gaussian) distribution is the most important distribution in probability and statistics. One of the justifications for using the normal distribution is the central limit theorem. Also, most of the statistical theory is based on the normality assumption. However, in finance research, the lognormal distribution is playing a more important role. Part of the reason is the fact that in finance, we are dealing with random quantities which are positive in nature. Hence, taking the natural logarithm is quite reasonable. Also, empirical data quite often support the assumption of the lognormality for random quantity such as the stock price movements.

We will discuss the normal distribution in Sect. 14.2 and the lognormal distribution in Sect. 14.3. Section 14.4 will discuss the relationship between normal distribution and lognormal distribution. Multivariate normal and lognormal distributions will be shown in Sect. 14.5. Section 14.6 will demonstrate the normal distribution as an application to the binomial and Poisson distributions. Section 14.7 will present applications of the lognormal distribution in option pricing. The bivariate normal density function will be presented in Sect. 14.8. American call options will be discussed in Sect. 14.9. Section 14.10 will discuss price bounds for options. Finally, Sect. 14.11 concludes the chapter. In addition, Appendix 1 will present Microsoft Excel program for calculating cumulative bivariate normal density function. Appendix 2 will discuss Microsoft Excel program for calculating the American call options.

14.2 The Normal Distribution

A random variable X is said to be normally distributed with mean μ and variance σ^2 if it has the pdf

$$f(x) = \frac{1}{\sqrt{2\pi}\sigma} e^{-\frac{1}{2}\left(\frac{x-\mu}{\sigma}\right)^2}, \quad \sigma > 0. \tag{14.1}$$

The normal distribution is symmetric around μ, which is the mean and the mode of the distribution. It is easy to see that the pdf of $Z = \frac{X-\mu}{\sigma}$ is

$$g(z) = \frac{1}{\sqrt{2\pi}} e^{-\frac{z^2}{2}} \tag{14.2}$$

which is the pdf of the standard normal and is independent of the parameters μ and σ^2.

The CDF of Z is

$$P(Z \le z) = N(z) \tag{14.3}$$

which has been well tabulated. Also, software package or system such S-plus will provide the value $N(z)$ instantly. For a discussion of some approximations for $N(z)$, the reader is referred to

Johnson and Kotz (1970). For the CDF of X we have

$$P(X \leq x) = P\left(\frac{X-\mu}{\sigma} \leq \frac{x-\mu}{\sigma}\right) = N\left(\frac{x-\mu}{\sigma}\right). \tag{14.4}$$

The normal distribution as given in Eq. (14.1) is very important in practice. It is very useful in describing phenomena such as the scores of tests and heights of students in a certain school. It is useful serving as an approximation for distribution such as binomial and Poisson. It is also quite useful in studying the option pricing.

We next discuss some properties of the normal distribution. If X is normally distributed with mean μ and variance σ^2, then the mgf of X is

$$M_x(t) = e^{\mu t + t^2\sigma^2/2}, \tag{14.5}$$

which is useful in deriving the moment of X. Equation (14.5) is also useful in deriving the moments of the lognormal distribution. From Eq. (14.5), it is easy to verify that

$E(X) = \mu$ and $\text{Var}(X) = \sigma^2$.

If $X_1, \cdots, X_n$ are independent, normally distributed random variable, then any linear function of these variables is also normally distributed. In fact, if X_i is normally distributed with mean μ_i and variance σ_i^2, then $\sum_{i=1}^n a_i X_i$ is normally distributed with mean $\sum_{i=1}^n a_i \mu_i$ and variance $\sum_{i=1}^n a_i^2 \sigma_i^2$, when $a_i's$ are constants. If X_1 and X_2 are independent, and each is normally distributed with mean 0 and variance σ^2, then $(X_1^2 - X_2^2)/(X_1^2 + X_2^2)$ is also normally distributed.

14.3 The Lognormal Distribution

A random variable X is said to be lognormality distributed with parameters μ and σ^2 if

$$Y = \log X \tag{14.6}$$

is normally distributed with mean μ and variance σ^2. It is clear that X has to be a positive random variable. This distribution is quite useful in studying the behavior of stock prices.

For the lognormal distribution as described above, the pdf is

$$g(x) = \frac{1}{\sqrt{2\pi}\sigma x} e^{-\frac{1}{2\sigma^2}(\log x - \mu)^2}, \quad x > 0. \tag{14.7}$$

The lognormal distribution is sometimes called the antilognormal distribution, because it is the distribution of the random variable X which is the antilog of the normal random variable Y. However, "lognormal" is most commonly used in the literature. When applied to economic data, especially production function, it is often called the Cobb–Douglas distribution.

We next discuss some properties of the lognormal distribution, as defined in Eq. (14.6). The rth moment of X is

$$\mu_r' = E(X^r) = E(e^{rY}) = e^{\mu r + \frac{r^2\sigma^2}{2}}. \tag{14.8}$$

It is noted that we have utilized the fact the mgf of the normal random variable Y with mean μ and variance σ^2 is $M_Y(t) = e^{\mu t + \frac{t^2\sigma^2}{2}}$. Thus, $E(e^{rY})$ is simply $M_Y(r)$, which is the right-hand size of Eq. (14.8). From Eq. (14.8) we have

$$E(X) = e^{\mu + \frac{\sigma^2}{2}}, \tag{14.9}$$

$$\text{Var}(X) = e^{2\mu} e^{\sigma^2}[e^{\sigma^2} - 1]. \tag{14.10}$$

It is noted that the moment sequence $\{\mu_r'\}$ does not belong only to lognormal distribution. Thus, the lognormal distribution cannot be defined by its moments.

The CDF of X is

$$P(X \leq x) = P(\log X \leq \log x) = N\left(\frac{\log x - \mu}{\sigma}\right), \tag{14.11}$$

because log$X$ is normally distributed with mean μ and variance σ^2.

The distribution of X is unimodal with the mode at

$$\mathrm{mode}(X) = e^{(\mu-\sigma^2)}. \tag{14.12}$$

Let x_α be the $(100)\alpha$ percentile for the lognormal distribution and z_α be the corresponding percentile for the standard normal, then

$$\begin{aligned} P(X \le x_\alpha) &= P\left(\frac{\log X - \mu}{\sigma} \le \frac{\log x_\alpha - \mu}{\sigma}\right) \\ &= N\left(\frac{\log x_\alpha - \mu}{\sigma}\right). \end{aligned} \tag{14.13}$$

Thus $z_\alpha = \frac{\log x_\alpha - \mu}{\sigma}$, implying that

$$x_\alpha = e^{\mu + \sigma z_\alpha}. \tag{14.14}$$

Thus, the percentile of the lognormal distribution can be obtained from the percentile of the standard normal.

From Eq. (14.13), we also see that

$$\mathrm{median}(X) = e^{\mu}, \tag{14.15}$$

as $z_{0.5} = 0$. Thus, $\mathrm{median}(X) > \mathrm{mod}\, e(X)$. Hence, the lognormal distribution is not symmetric.

14.4 The Lognormal Distribution and Its Relationship to the Normal Distribution

By comparing the pdf of normal distribution given in Eq. (14.1) and the pdf of lognormal distribution given in Eq. (14.7), we know that

$$f(x) = \frac{f(y)}{x}. \tag{14.16}$$

In addition, from Eq. (14.6), it is easy to see that

$$dx = x dy. \tag{14.17}$$

The CDF for the lognormal distribution can be expressed as

$$\begin{aligned} F(a) &= \Pr(X \le a) = \Pr(\log X \le \log a) \\ &= \Pr\left(\frac{\log X - \mu}{\sigma} \le \frac{\log a - \mu}{\sigma}\right) \\ &= N(d) \end{aligned} \tag{14.18}$$

where

$$d = \frac{\log a - \mu}{\sigma} \tag{14.19}$$

and $N(d)$ is the CDF of the standard normal distribution which can be obtained from normal table. $N(d)$ can also be obtained from S-plus or other software package. Alternatively, the value of $N(d)$ can be approximated by the following formula:

$$N(d) = a_0 e^{-\frac{d^2}{2}}(a_1 t + a_2 t^2 + a_3 t^3) \tag{14.20}$$

where

$$\begin{aligned} t &= \frac{1}{1 + 0.33267d}, \\ a_0 &= 0.3989423, \\ a_1 &= 0.4361936, \\ a_2 &= -0.1201676, \\ a_3 &= 0.9372980. \end{aligned}$$

In case we need $\Pr(X > a)$, then we have

$$\begin{aligned} \Pr(X > a) &= 1 - \Pr(X \le a) = 1 - N(d) \\ &= N(-d). \end{aligned} \tag{14.21}$$

Since for any h, $E(X^h) = E(e^{hY})$, the hth moment of X, the following moment-generating function of Y, which is normally distributed with mean μ and variance σ^2,

$$M_Y(t) = e^{\mu t + \frac{1}{2}t^2\sigma^2}. \quad (14.22)$$

For example,

$$\mu_X = E(X) = E(e^Y) = M_Y(1) = e^{\mu + \frac{1}{2}t^2\sigma^2}.$$

$$E(X^h) = E(e^{hY}) = M_Y(h) = e^{\mu h + \frac{1}{2}t^2\sigma^2}. \quad (14.23)$$

Hence,

$$\begin{aligned}\sigma_X^2 &= E(X^2) - (EX)^2 = e^{2\mu + 2\sigma^2} - e^{2\mu + \sigma^2} \\ &= e^{2\mu + \sigma^2}(e^{\sigma^2} - 1)\end{aligned} \quad (14.24)$$

Thus, fractional and negative moments of a lognormal distribution can be obtained from Eq. (14.23).The mean of a lognormal random variable can be defined as

$$\int_0^\infty xf(x)\mathrm{d}x = e^{\mu + \frac{\sigma^2}{2}}. \quad (14.25)$$

If the lower bound a is larger than 0; then the partial mean of x can be shown as

$$\int_0^\infty xf(x)\mathrm{d}x = \int_{\log(a)}^\infty f(y)e^y\mathrm{d}y = e^{\mu + \frac{\sigma^2}{2}}N(d) \quad (14.26)$$

where

$$d = \frac{\mu - \log(a)}{\sigma} + \sigma.$$

This implies that partial mean of a lognormal variable is the mean of x times an adjustment term, $N(d)$.

14.5 Multivariate Normal and Lognormal Distributions

The normal distribution with the pdf given in Eq. (14.1) can be extended to the p-dimensional case. Let $\boldsymbol{X} = (X_1, \ldots, X_p)'$ be a $p \times 1$ random vector. Then, we say that $\boldsymbol{X} \sim N_p(\boldsymbol{\mu}, \Sigma)$ if it has the pdf

$$f(\boldsymbol{x}) = (2\pi)^{-p/2}|\Sigma|^{-1/2}\exp\left[-\frac{1}{2}(\boldsymbol{x} - \boldsymbol{\mu})'\Sigma^{-1}(\boldsymbol{x} - \boldsymbol{\mu})\right]. \quad (14.27)$$

In Eq. (14.27), $\boldsymbol{\mu}$ is the mean vector and Σ is the covariance matrix which is symmetric and positive definite. The moment-generating function of $\boldsymbol{X}$ is

$$M_x(t) = E\left(e^{t'x}\right) = e^{t'\mu + \frac{1}{2}t'\Sigma t} \quad (14.28)$$

where $t = (t_1, \ldots, t_p)'$ is a $p \times 1$ vector of real values.

From Eq. (14.28), it can be shown that $E(\boldsymbol{X}) = \boldsymbol{\mu}$ and $\mathrm{Cov}(\boldsymbol{X}) = \Sigma$.

If $\mathbf{C}$ is a $q \times p$ matrix of rank $q \le p$. Then, $\mathbf{CX} \sim N_q(\mathbf{C}\mu, \mathbf{C}\Sigma\mathbf{C}')$. Thus, linear transformation of a normal random vector is also a multivariate normal random vector.

Let $\mathbf{X} = \begin{pmatrix} \mathbf{X}^{(1)} \\ \mathbf{X}^{(2)} \end{pmatrix}$, $\mu = \begin{pmatrix} \mu^{(1)} \\ \mu^{(2)} \end{pmatrix}$, and $\Sigma = \begin{pmatrix} \Sigma_{11} & \Sigma_{12} \\ \Sigma_{21} & \Sigma_{22} \end{pmatrix}$, where $\mathbf{X}^{(i)}$ and $\mu^{(i)}$ are $p_i \times 1$, $p_1 + p_2 = p$, and Σ_{ij} is $p_i \times p_j$.

Then, the marginal distribution of $X^{(i)}$ is also a multivariate normal with mean vector $\mu^{(i)}$ and covariance matrix Σ_{ii}, i.e., $\mathbf{X}^{(i)} \sim N_{p_i}(\mu^{(i)}, \Sigma_{ii})$. Furthermore, the conditional distribution of $\mathbf{X}^{(1)}$ given $\mathbf{X}^{(2)} = x^{(2)}$, where $x^{(2)}$ is a known vector, is normal with mean vector $\mu_{1\cdot2}$ and covariance matrix $\Sigma_{11\cdot2}$

where

$$\mu_{1\cdot2} = \mu^{(1)} + \Sigma_{12}\Sigma_{22}^{-1}\left(x^{(2)} - \mu^{(2)}\right) \quad (14.29)$$

and

$$\Sigma_{11\cdot2} = \Sigma_{11} - \Sigma_{12}\Sigma_{22}^{-1}\Sigma_{21}, \quad (14.30)$$

i.e., $\mathbf{X}^{(1)}|\mathbf{X}^{(2)} = x^{(2)} \sim N_{p_1}(\mu_{1\cdot2}, \Sigma_{11\cdot2})$.

We next consider a bivariate version of correlated lognormal distribution.

Let $\begin{pmatrix} Y_1 \\ Y_2 \end{pmatrix} = \begin{pmatrix} \log(X_1) \\ \log(X_2) \end{pmatrix} \sim N\left(\begin{pmatrix} \mu_1 \\ \mu_2 \end{pmatrix}, \begin{pmatrix} \sigma_{11} & \sigma_{12} \\ \sigma_{21} & \sigma_{22} \end{pmatrix}\right)$.

The joint pdf of X_1 and X_2 can be obtained from the joint pdf of Y_1 and Y_2 by observing that

$$dx_1 dx_2 = x_1 x_2 dy_1 dy_2 \tag{14.31}$$

which is an extension of Eq. (14.17) to the bivariate case.

Hence, the joint pdf of X_1 and X_2 is

$$g(x_1, x_2) = \frac{1}{2\pi|\Sigma|x_1 x_2} \times \exp\left\{-\frac{1}{2}[(\log x_1, \log x_2) - \mu]' \times \Sigma^{-1}[(\log x_1, \log x_2) - \mu]\right\}. \tag{14.32}$$

From the property of the multivariate normal distribution, we have $Y_i \sim N(\mu_i, \sigma_{ii})$. Hence, X_i is lognormal with

$$E(X_i) = e^{\mu_i + \frac{\sigma_{ii}}{2}}, \tag{14.33}$$

$$\mathrm{Var}(X_i) = e^{2\mu_i} e^{\sigma_{ii}}[e^{\sigma_{ii}} - 1]. \tag{14.34}$$

Furthermore, by the property of the moment generating for the bivariate normal distribution, we have

$$\begin{aligned} E(X_1 X_2) &= E\left(e^{Y_1 + Y_2}\right) \\ &= e^{\mu_1 + \mu_2 + \frac{1}{2}(\sigma_{11} + \sigma_{22} + 2\sigma_{12})} \\ &= E(X_1)E(X_2) \cdot \exp(\rho\sqrt{\sigma_{11}\sigma_{22}}). \end{aligned} \tag{14.35}$$

Thus, the covariance between X_1 and X_2 is

$$\begin{aligned} \mathrm{Cov}(X_1, X_2) &= E(X_1 X_2) - E(X_1)E(X_2) \\ &= E(X_1)E(X_2) \cdot (\exp(\rho\sqrt{\sigma_{11}\sigma_{22}}) - 1) \\ &= \exp\left(\mu_1 + \mu_2 + \frac{1}{2}(\sigma_{11} + \sigma_{22})\right) \cdot (\exp(\rho\sqrt{\sigma_{11}\sigma_{22}}) - 1) \end{aligned} \tag{14.36}$$

From the property of conditional normality of Y_1 given $Y_2 = y_2$, we also see that the conditional distribution of Y_1 given $Y_2 = y_2$ is lognormal.

The extension to the p-variate lognormal distribution is trivial. Let $\boldsymbol{Y} = (Y_1, \ldots, Y_p)'$ where $Y_i = \log X_i$. If $\boldsymbol{Y} \sim N_p(\boldsymbol{\mu}, \Sigma)$, where $\boldsymbol{\mu} = (\mu_1 \ldots \mu_p)'$ and $\Sigma = (\sigma_{ij})$.

The joint pdf of $X_1, \ldots, X_p$, can be obtained from the following theorem.

Theorem 14.1 (Anderson 2003)

Let the pdf of $Y_1, \ldots, Y_p$ *be* $f(y_1, \ldots, y_p)$, *consider the p-valued functions*

$$x_i = x_i(y_1, \ldots, y_p), \quad i = 1, \ldots, p. \tag{14.37}$$

We assume that the transformation from the y-space to the x-space is one-to-one with the inverse transformation

$$y_i = y_i(x_1, \ldots, x_p), \quad i = 1, \ldots, p. \tag{14.38}$$

Let the random variables $X_1, \ldots, X_p$ *be defined by*

$$X_i = x_i(Y_1, \ldots, Y_p), \quad i = 1, \ldots, p. \tag{14.39}$$

Then the pdf of $X_1, \ldots, X_p$ *is*

$$g(x_1, \ldots, x_p) = f(y_1(x_1, \ldots, x_p), \ldots, y_p(x_1, \ldots, x_p))\, J(x_1, \ldots, x_p) \tag{14.40}$$

where $J(x_1, \ldots, x_p)$ *is the Jacobian of transformations*

$$J(x_1, \ldots, x_p) = \mathrm{mod}\begin{vmatrix} \frac{\partial y_1}{\partial x_1} & \cdots & \frac{\partial y_1}{\partial x_p} \\ \vdots & \ddots & \vdots \\ \frac{\partial y_p}{\partial x_1} & \cdots & \frac{\partial y_p}{\partial x_p} \end{vmatrix}, \tag{14.41}$$

where "mod" means a modulus or absolute value.

Applying the above theorem with $f(y_1, \ldots, y_p)$ *being a p-variate normal, and*

$$J(x_1,\ldots,x_p)=\mathrm{mod}\begin{vmatrix}\frac{1}{x_1} & 0 & \cdots & 0\\ 0 & \frac{1}{x_2} & \cdots & 0\\ \vdots & & \ddots & \vdots\\ 0 & \cdots & & \frac{1}{x_p}\end{vmatrix}=\prod_{i=1}^{p}\frac{1}{x_i}, \quad (14.42)$$

we have the following joint pdf of $X_1,\ldots,X_p$

$$g(x_1,\ldots,x_p)=(2\pi)^{-p/2}|\Sigma|^{-\frac{p}{2}}\left(\prod_{i=1}^{p}\frac{1}{x_i}\right)$$
$$\times\exp\left\{-\frac{1}{2}\left[(\log x_1,\ldots,\log x_p)-\mu\right]'\Sigma^{-1}\left[(\log x_1,\ldots,\log x_p)-\mu\right]\right\}. \quad (14.43)$$

It is noted that when $p=2$, *Eq. (14.43) reduces to the bivariate case given in Eq. (14.32).*

Then the first two moments are

$$E(X_i)=e^{\mu_i+\frac{\sigma_{ii}}{2}}, \quad (14.44)$$

$$\mathrm{Var}(X_i)=e^{2\mu_i}e^{\sigma_{ii}}\left[e^{\sigma_{ii}}-1\right]. \quad (14.45)$$

$$\mathrm{Cor}(X_i,X_j)=\exp\left(\mu_i+\mu_j+\frac{1}{2}(\sigma_{ii}+\sigma_{jj})\right)\cdot\left(\exp\left(\rho_{ij}\sqrt{\sigma_{ii}\sigma_{jj}}\right)-1\right) \quad (14.46)$$

where ρ_{ij} *is the correlation between* Y_i *and* Y_j.

For more details concerning properties of the multivariate lognormal distribution, the reader is referred to Johnson and Kotz (1972).

14.6 The Normal Distribution as an Application to the Binomial and Poisson Distributions

The cumulative normal density function tells us the probability that a random variable Z will be less than some value x. Note in Fig. 14.1 that $P(Z > x)$ is simply the area under the normal curve from $-\infty$ up to point x.

One of the many applications of the cumulative normal distribution function is in valuing stock options. A call option gives the option holder the right to purchase, at a specified price known as the exercise price, a specified number of shares of stock during a given time period.

A call option is a function of the following five variables:

1. Current price of the firm's common stock (S),
2. Exercise price (or strike price) of the option (X),
3. Term to maturity in years (T),
4. Variance of the stock's price (σ^2),
5. Risk-free rate of interest (r).

The binomial option pricing model defined in Eq. (14.22) can be written as

$$C=S\left[\sum_{k=m}^{T}\frac{T!}{k!(T-k)!}p'^k(1-p')^{T-k}\right]-\frac{X}{(1+r)^T}\left[\sum_{k=m}^{T}\frac{T!}{k!(T-k)!}p^k(1-p)^{T-k}\right]$$
$$=SB(T,p',m)-\frac{X}{(1+r)^T}B(T,p,m), \quad (14.47)$$

and $C=0$ if $m<T$,

where

S = Current price of the stock,
T = term to maturity in years,

Fig. 14.1 The cumulative normal density function

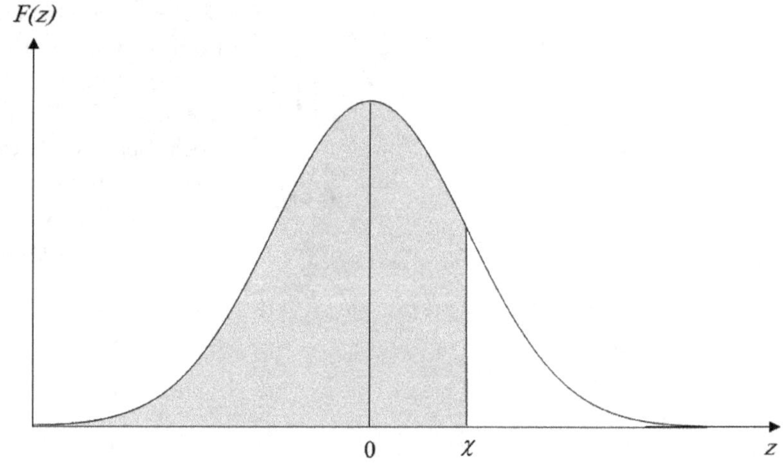

m = minimum number of upward movements in stock price that is necessary for the option to terminate "in-the-money",
$p = \frac{R-d}{u-d}$ and $1 - p = \frac{u-R}{u-d}$,
X = option exercise price (or strike price),
$R = 1 + r = 1$ + risk-free rate of return,
$u = 1$ + percentage of price increase,
$d = 1$ + percentage of price decrease,
$p' = \left(\frac{u}{R}\right)p$,
$B(n, p, m) = \sum_{k=m}^{n} {}_nC_k p^k (1-p)^{n-k}$.

By a form of the central limit theorem, when $T \to \infty$, the option price C converges to C below[1]

$$C = SN(d_1) - XR^{-T}N(d_2) \qquad (14.48)$$

C = Price of the call option,
$d_1 = \frac{\log\left(\frac{S}{Xr^{-t}}\right)}{\sigma\sqrt{t}} + \frac{1}{2}\sqrt{t}$,
$d_2 = d_1 - \sigma\sqrt{t}$,
$N(d)$ is the value of the cumulative standard normal distribution,
t is the fixed length of calendar time to expiration, and

h is the elapsed time between successive stock price changes and $T = ht$.

If future stock price is constant over time, then $\sigma^2 = 0$. It can be shown that both $N(d_1)$ and $N(d_2)$ are equal to 1 and that Eq. (14.48) becomes

$$C = S - Xe^{-rT}. \qquad (14.49)$$

Alternatively, Eqs. (14.48) and (14.49) can be understood in terms of the following steps:

Step 1: The future price of the stock is constant over time.

Because a call option gives the option holder the right to purchase the stock at the exercise price X, the value of the option, C, is just the current price of the stock less the present value of the stock's purchase price. Mathematically, the value of the call option is

$$C = S - \frac{X}{(1+r)^T}. \qquad (14.50)$$

Note that Eq. (14.50) assumes discrete compounding of interest, whereas Eq. (14.49) assumes continuous compounding of interest. To adjust Eq. (14.50) for continuous compounding, we substitute e^{-rT} for $\frac{1}{(1+r)^T}$ to get

[1]See Rubinstein (1976), "Option Pricing: A Simplified Approach" for details.

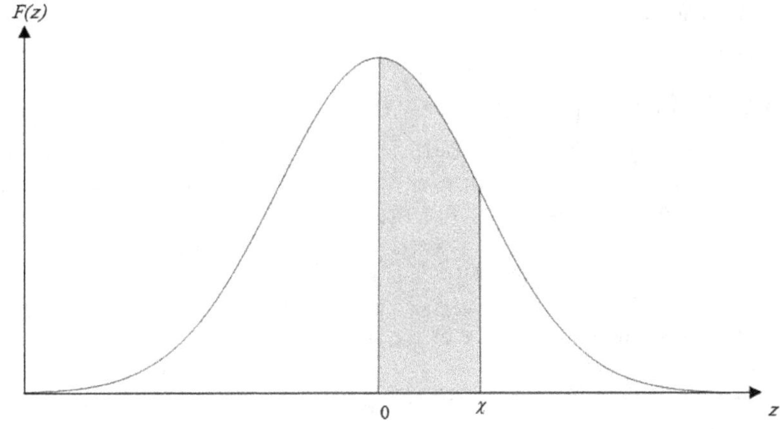

Fig. 14.2 The probability of the normal density function

$$C = S - Xe^{-rT}. \tag{14.51}$$

Step 2: Assume the price of the stock fluctuates over time (S_t).

In this case, we need to adjust Eq. (14.49) for the uncertainty associated with that fluctuation. We do this by using the cumulative normal distribution function. In deriving Eq. (14.48), we assume that S_t follows a lognormal distribution, as discussed in Sect. 14.3.

The adjustment factors $N(d_1)$ and $N(d_2)$ in the Black–Scholes option valuation model are simply adjustments made to Eq. (14.49) to account for the uncertainty associated with the fluctuation of the price of the stock.

Equation (14.48) is a continuous option pricing model. Compare this with the binomial option price model given in Eq. (14.47), which is a discrete option pricing model. The adjustment factors $N(d_1)$ and $N(d_2)$ are cumulative normal density functions. The adjustment factors $B(T,p',m)$ and $B(T,p,m)$ are complementary binomial distribution functions.

We can use Eq. (14.48) to determine the theoretical value, as of November 29, 1991, of one of IBM's options with maturity on April 1992. In this case, we have $X = \$90$, $S = \$92.5$, $\sigma = 0.2194$, $r = 0.0435$, and $T = \frac{5}{12} = 0.42$ (in years).[2] Armed with this information, we can calculate the estimated d_1 and d_2.

$$\begin{aligned} x &= \frac{\{\ln(\frac{92.5}{90}) + [(0.0435) + \frac{1}{2}(0.2194)^2](0.42)\}}{(0.2194)(0.42)^{\frac{1}{2}}} \\ &= 0.392, \end{aligned}$$

$$x - \sigma\sqrt{t} = x - (0.2194)(0.42)^{\frac{1}{2}} = 0.25.$$

In Eq. (14.45), $N(d_1)$ and $N(d_2)$ are the probabilities that a random variable with a standard normal distribution takes on a value less than d_1 and a value less than d_2, respectively. The values for $N(d_1)$ and $N(d_2)$ can be found by using the tables for the standard normal distribution, which provide the probability that a variable Z is between 0 and x (see Fig. 14.2).

To find the cumulative normal density function, we need to add the probability that Z is less than zero to the value given in the standard normal distribution table. Because the standard normal distribution is symmetric around zero, we know that the probability that Z is less than zero is 0.5, so $P(Z<x) = P(Z<0) + P(0<Z<x) = 0.5 +$ value from table.

[2]Values of $X = \$90$, $S = \$92.5$, $r = 0.0435$ were obtained from Section C of the Wall Street Journal on December 2, 1991. And $\sigma = 0.2194$ is estimated in terms of monthly rate of return during the period January 1989 to November 1991.

We can now compute the values of $N(d_1)$ and $N(d_2)$.

$$N(d_1) = P(Z<d_1) = P(Z<0) + P(0<Z<d_1) \\ = P(Z<0.392) = 0.5 + 0.1517 = 0.6517$$

$$N(d_2) = P(Z<d_2) = P(Z<0) + P(0<Z<d_2) \\ = P(Z<0.25) = 0.5 + 0.0987 = 0.5987$$

Then, the theoretical value of the option is

$$C = (92.5)(0.6517) - [(90)(0.5987)]/e^{(0.0435)(0.42)} \\ = 60.282 - 53.883/1.0184 = \$7.373$$

and the actual price of the option on November 29, 1991 was $7.75.

14.7 Applications of the Lognormal Distribution in Option Pricing

To derive the Black–Scholes formula it is assumed that there are no transaction costs, no margin requirements, and no taxes; that all shares are infinitely divisible; and that continuous trading can be accomplished. It is also assumed that the economy is risk neutral and the stock price follows a lognormal distribution. Denote the current stock price by S and the stock price at the end of jth period by S_j.

Then, $\frac{S_j}{S_{j-1}} = \exp[K_j]$ is a random variable with a lognormal distribution where K_j is the rate of return in jth period and is a random variable with normal distribution. Let K_t have the expected value μ_k and variance σ_k^2 for each j. Then, $K_1 + K_2 + \cdots + K_t$ is a normal random variable with expected value $t\mu_k$ and variance $t\sigma_k^2$. Thus, we can define the expected value (mean) of $\frac{S_t}{S} = \exp[K_1 + K_2 + \cdots + K_t]$ as

$$E\left(\frac{S_t}{S}\right) = \exp\left[t\mu_k + \frac{t\sigma_k^2}{2}\right]. \tag{14.52}$$

Under the assumption of a risk-neutral investor, the expected return $E(\frac{S_t}{S})$ is assumed to be $\exp(rt)$ (where r is the riskless rate of interest). In other words,

$$\mu_k = r - \frac{\sigma_k^2}{2}. \tag{14.53}$$

In the risk-neutral assumptions of Cox and Ross (1976) and Rubinstein (1976), the call option price C can be determined by discounting the expected value of the terminal option price by the riskless rate of interest:

$$C = \exp[-rt]E[\mathrm{Max}(S_T - X, 0)] \tag{14.54}$$

where T is the time of expiration and X is the striking price.

Note that

$$\begin{aligned} \mathrm{Max}(S_T - X, 0) &= \left(S\left(\frac{S_T}{S} - \frac{X}{S}\right)\right), \quad &\text{for } \frac{S_T}{S} > \frac{X}{S} \\ &= 0, \quad &\text{for } \frac{S_T}{S} < \frac{X}{S} \end{aligned} \tag{14.55}$$

Equations (14.54) and (14.55) say that the value of the call option today will be either $S_t - X$ or 0, whichever is greater. If the price of stock at time t is greater than the exercise price, the call option will expire in-the-money. This simply means that an investor who owns the call option will exercise it. The option will be exercised regardless of whether the option holder would like to take physical possession of the stock. If the investor would like to own the stock, the cheapest way to obtain the stock is by exercising the option. If the investor would not like to own the stock, he or she will still exercise the option and immediately sell the stock in the market. Since the price the investor paid (X) is lower that the price he or she can sell the stock for (S_t), the investor realizes an immediate profit of $S_t - X$. If the price of the stock (S_t) is less than the exercise price (X), the option expires out-of-the-money. This occurs because in purchasing shares of the stock, the investor will find it cheaper to purchase the stock in the market than to exercise the option.

Let $X = \frac{S_T}{S}$ be lognormally distributed with parameters $\mu = tr - \frac{t\sigma_k^2}{2}$ and $\sigma^2 = t\sigma_k^2$. Then,

$$C = \exp[-rt]E[\mathrm{Max}(S_t - X)]$$
$$= \exp[-rt]\int_{\frac{X}{S}}^{\infty} S\left[x - \frac{X}{S}\right]g(x)\mathrm{d}x$$
$$= \exp[-rt]S\int_{\frac{X}{S}}^{\infty} xg(x)\mathrm{d}x - \exp[-rt]S\frac{X}{S}\int_{\frac{X}{S}}^{\infty} g(x)\mathrm{d}x \tag{14.56}$$

where $g(x)$ is the probability density function of $X_t = \frac{S_t}{S}$.

Substituting $\mu = tr - t\sigma_k^2/2, \sigma^2 = t\sigma_k^2$ and $a = \frac{X}{S}$ into Eqs. (14.18) and (14.26), we obtain

$$\int_{\frac{X}{S}}^{\infty} xg(x)\mathrm{d}x = e^{rt}N(d_1) \tag{14.57}$$

$$\int_{\frac{X}{S}}^{\infty} g(x)\mathrm{d}x = N(d_2) \tag{14.58}$$

where

$$d_1 = \frac{tr - \frac{t}{2}\sigma_k^2 - \log\left(\frac{X}{S}\right)}{\sqrt{t}\sigma_k} + \sqrt{t}\sigma_k = \frac{\log\left(\frac{S}{X}\right) + \left(r + \frac{1}{2}\sigma_k^2\right)t}{\sqrt{t}\sigma_k} \tag{14.59}$$

and

$$d_2 = \frac{\log\left(\frac{S}{X}\right) + \left(r - \frac{1}{2}\sigma_k^2\right)t}{\sqrt{t}\sigma_k} = d_1 - \sqrt{t}\sigma_k \tag{14.60}$$

Substituting Eq. (14.58) into Eq. (14.56), we obtain

$$C = SN(d_1) - X\exp[-rt]N(d_2), \tag{14.61}$$

This is Eq. (14.48) defined in Sect. 14.6.

We have defined a put option as a contract conveying the right to sell a designated security at a stipulated price. It can be shown that the relationship between a call option (C) and a put option (P) can be defined as

$$C + Xe^{-rt} = P + S \tag{14.62}$$

Substituting Eq. (14.33) into (14.34), we obtain the put option formula as

$$P = Xe^{-rt}N(-d_2) - SN(-d_1) \tag{14.63}$$

where S, C, r, t, d_1 and d_2 are identical to those defined in the call option model.

14.8 The Bivariate Normal Density Function

In correlation analysis, we assume a population where both X and Y vary jointly. It is called a joint distribution of two variables. If both X and Y are normally distributed, then we call this known distribution a bivariate normal distribution.

We can define the PDF of the normally distributed random variables X and Y as

$$f(X) = \frac{1}{\sigma_X\sqrt{2\pi}}\exp\left[\frac{-(X - \mu_X)}{2\sigma_X^2}\right], \quad -\infty < X < \infty, \tag{14.64}$$

$$f(Y) = \frac{1}{\sigma_Y\sqrt{2\pi}}\exp\left[\frac{-(Y - \mu_Y)}{2\sigma_Y^2}\right], \quad -\infty < Y < \infty, \tag{14.65}$$

where μ_X and μ_Y are population means for X and Y, respectively; σ_X and σ_Y are population standard deviations of X and Y, respectively; $\pi = 3.1416$; and exp represents the exponential function.

If ρ represents the population correlation between X and Y, then the PDF of the bivariate normal distribution can be defined as

$$f(X, Y) = \frac{1}{2\pi\sigma_X\sigma_Y\sqrt{1 - \rho^2}}\exp(-q/2), \quad -\infty < X < \infty, -\infty < Y < \infty, \tag{14.66}$$

where $\sigma_X > 0, \sigma_Y > 0$, and $-1 < \rho < 1$,

$$q = \frac{1}{1-\rho^2}\left[\left(\left(\frac{X-\mu_X}{\sigma_X}\right)^2 - 2\rho\left(\frac{X-\mu_X}{\sigma_X}\right)\left(\frac{Y-\mu_Y}{\sigma_Y}\right) + \left(\frac{X-\mu_Y}{\sigma_Y}\right)^2\right)\right].$$

It can be shown that the conditional mean of Y, given X, is linear in X and given by

$$E(Y|X) = \mu_Y + \rho\left\{\frac{\sigma_Y}{\sigma_X}\right\}(X - \mu_X). \quad (14.67)$$

It is also clear that given X, we can define the conditional variance of Y as

$$\sigma(Y|X) = \sigma_Y^2(1-\rho^2). \quad (14.68)$$

Equation (14.67) can be regarded as describing the population linear regression line. For example, if we have a bivariate normal distribution of heights of brothers and sisters, we can see that they vary together and there is no cause-and-effect relationship. Accordingly, a linear regression in terms of the bivariate normal distribution variable is treated as though there were a two-way relationship instead of an existing causal relationship. It should be noted that regression implies a causal relationship only under a "prediction" case.

Equation (14.66) represents a joint PDF for X and Y. If $\rho = 0$, then Eq. (14.66) becomes

$$f(X, Y) = f(X)f(Y). \quad (14.69)$$

This implies that the joint PDF of X and Y is equal to the PDF of X times the PDf of Y. We also know that both X and Y are normally distributed. Therefore, X is independent of Y.

Example 14.1 Using a mathematics aptitude test to predict grade in statistics

Let X and Y represent scores in a mathematics aptitude test and numerical grade in elementary statistics, respectively. In addition, we assume that the parameters in Eq. (14.66) are

$$\mu_X = 550 \quad \sigma_X = 40 \quad \mu_Y = 80 \quad \sigma_Y = 4$$
$$\rho = 0.7.$$

Substituting this information into Eqs. (14.67) and (14.68), respectively, we obtain

$$\begin{aligned} E(Y|X) &= 80 + 0.7(4/40)(X - 550) \\ &= 41.5 + 0.07X, \end{aligned} \quad (14.70)$$

$$\sigma^2(Y|X) = (16)(1 - 0.49) = 8.16. \quad (14.71)$$

If we know nothing about the aptitude test score of a particular student (say, john), we have to use the distribution of Y to predict his elementary statistics grade.

$$95\% \text{ interval} = 80 \pm (1.96)(4) = 80 \pm 7.84.$$

That is, we predict with 95% probability that John's grade will fall between 87.84 and 72.16.

Alternatively, suppose we know that John's mathematics aptitude score is 650. In this case, we can use Eqs. (14.70) and (14.71) to predict John's grade in elementary statistics.

$$E(Y|X = 650) = 41.5 + (0.07)(650) = 87,$$

and

$$\sigma^2(Y|X) = (16)(1 - 0.49) = 8.16.$$

We can now base our interval on a normal probability distribution with a mean of 87 and a standard deviation of 2.86.

$$95\% \text{ interval} = 87 \pm (1.96)(2.86) = 87 \pm 5.61.$$

That is, we predict with 95% probability that John's grade will fall between 92.61 and 81.39.

Two things have happened to this interval. First, the center has shifted upward to take into account the fact that John's mathematics aptitude score is above average. Second, the width of the interval has been narrowed from $87.84 - 72.16 = 15.68$ grade points to

92.61 − 81.39 = 11.22 grade points. In this sense, the information about John's mathematics aptitude score has made us less uncertain about his grade in statistics.

14.9 American Call Options

14.9.1 Price American Call Options by the Bivariate Normal Distribution

An option contract which can be exercised only on the expiration date is called European call. If the contract of a call option can be exercised at any time of the option's contract period, then this kind of call option is called American call.

When a stock pays a dividend, the American call is more complex. Following Whaley (1981), the valuation equation for American call option with one known dividend payment can be defined as

$$C(S,T,X) = S^x\left[N1(b1) + N2(a1,-b1;\sqrt{t/T})\right] -Xe^{-rt}\left[N1(b2)e^{r(T-t)} + N2(a2,-b2;\sqrt{t/T})\right] + De^{-rt}N1(b2), \tag{14.72a}$$

where

$$a1 = \frac{\ln\left(\frac{S^x}{X}\right) + \left(r + \frac{1}{2}\sigma^2\right)T}{\sigma\sqrt{T}}, \quad a2 = a1 - \sigma\sqrt{T}, \tag{14.72b}$$

$$b1 = \frac{\ln\left(\frac{S^x}{S_t^*}\right) + \left(r + \frac{1}{2}\sigma^2\right)t}{\sigma\sqrt{t}}, \quad b2 = b1 - \sigma\sqrt{t}, \tag{14.72c}$$

$$S^x = S - De^{-rt}, \tag{14.73}$$

S^x represents the correct stock net price of the present value of the promised dividend per share (D);

t represents the time dividend to be paid.

S_t^* is the exdividend stock price for which

$$C(S_t^*, T-t) = S_t^* + D - X \tag{14.74}$$

S, X, r, σ^2, T have been defined previously in this chapter.

Both $N_1(b_1)$ and $N_2(b_2)$ are cumulative univariate normal density function; $N_2(a,b;\rho)$ is the cumulative bivariate normal density function with upper integral limits, a and b, and correlation coefficient, $\rho = -\sqrt{t/T}$.

American call option on a nondividend-paying stock will never optimally be exercised prior to expiration. Therefore, if no dividend payments exist, Eqs. (14.72a–14.72c) will reduce to European Option pricing model with no dividend payment.

We have shown how the cumulative univariate normal density function can be used to evaluate the European call option in previous sections of this chapter. If a common stock pays a discrete dividend during the option's life, the American call option valuation equation requires the evaluation of a cumulative bivariate normal density function. While there are many available approximations for the cumulative bivariate normal distribution, the approximation provided here relies on Gaussian quadratures. The approach is straightforward and efficient, and its maximum absolute error is 0.00000055.

Following Eq. (14.66), the probability that x' is less than a and that y' is less than b for the standardized cumulative bivariate normal distribution

$$P(X' < a, Y' < b) = \frac{1}{2\pi\sqrt{1-\rho^2}} \int_{-\infty}^{a}\int_{-\infty}^{b} \exp\left[\frac{2x'^2 - 2\rho x'y' + y'^2}{2(1-\rho^2)}\right] dx'dy',$$

where $x' = \frac{x-\mu_x}{\sigma_x}$, $y' = \frac{y-\mu_y}{\sigma_y}$ and p is the correlation between the random variables x' and y'.

The first step in the approximation of the bivariate normal probability $N_2(a,b;\rho)$ is as follows:

$$\varphi(a, b; \rho) \approx 0.31830989\sqrt{1-\rho^2}\sum_{i=1}^{5}\sum_{j=1}^{5} w_i w_j f(x'_i, x'_j), \quad (14.75)$$

where

$$f(x'_i, x'_j) = \exp[a_1(2x'_i - a_1) + b_1(2x'_j - b_1) + 2\rho(x'_i - a_1)(x'_j - b_1)].$$

The pairs of weights, (w) and corresponding abscissa values (x') are

I, j	W	x′
1	0.24840615	0.10024215
2	0.39233107	0.48281397
3	0.21141819	1.0609498
4	0.033246660	1.7797294
5	0.00082485334	2.6697604

This portion is based upon Appendix 14.1 of Stoll H. R. and R. E Whaley. *Futures and Options*. Cincinnati, OH: South Western Publishing, 1993

and the coefficients a_1 and b_1 are computed using

$$a_1 = \frac{a}{\sqrt{2(1-\rho^2)}} \text{ and } b_1 = \frac{b}{\sqrt{2(1-\rho^2)}}.$$

The second step in the approximation involves computing the product $ab\rho$; if $ab\rho \le 0$, compute the bivariate normal probability, $N_2(a, b; \rho)$, using the following rules:

$$\begin{aligned}
&(1)\ \text{If } a \le 0, b \le 0, \text{ and } \rho \le 0,\\
&\quad \text{then } N_2(a,b;\rho) = \varphi(a,b;\rho);\\
&(2)\ \text{If } a \le 0, b \ge 0, \text{ and } \rho > 0,\\
&\quad \text{then } N_2(a,b;\rho) = N_1(a) - \varphi(a,-b;-\rho);\\
&(3)\ \text{If } a \ge 0, b \le 0, \text{ and } \rho > 0,\\
&\quad \text{then } N_2(a,b;\rho) = N_1(b) - \varphi(-a,b;-\rho);\\
&(4)\ \text{If } a \ge 0, b \ge 0, \text{ and } \rho \le 0,\\
&\quad \text{then } N_2(a,b;\rho) = N_1(a) + N_1(b) - 1 + \varphi(-a,-b;\rho).
\end{aligned} \quad (14.76)$$

If $ab\rho > 0$, compute the bivariate normal probability, $N_2(a, b; \rho)$, as

$$N_2(a, b; \rho) = N_2(a, 0; \rho_{ab}) + N_2(b, 0; \rho_{ab}) - \delta, \quad (14.77)$$

where the values of $N_2(\bullet)$ on the right-hand side are computed from the rules, for $ab\rho \le 0$

$$\rho_{ab} = \frac{(\rho a - b)\text{Sgn}(a)}{\sqrt{a^2 - 2\rho ab + b^2}}, \quad \rho_{ba} = \frac{(\rho b - a)\text{Sgn}(b)}{\sqrt{a^2 - 2\rho ab + b^2}}, \quad \delta = \frac{1 - \text{Sgn}(a) \times \text{Sgn}(b)}{4}$$

and

$$\text{Sgn}(x) = \begin{cases} 1 & x \ge 0 \\ -1 & x < 0 \end{cases},$$

$N_1(d)$ is the cumulative univariate normal probability.

14.9.2 Pricing an American Call Option: An Example

An American call option whose exercise price is \$48 has an expiration time of 90 days. Assume the risk-free rate of interest is 8% annually, the underlying price is \$50, the standard deviation of the rate of return of the stock is 20%, and the stock pays a dividend of \$2 exactly for 50 days. (a) What is the European call value? (b) Can the early exercise price predicted? (c) What is the value of the American call?

(a) The current stock net price of the present value of the promised dividend is

$$S^x = 50 - 2e^{-0.08(50/365)} = 48.0218.$$

The European call value can be calculated as

$$C = (48.0218)N(d_1) - 48e^{-0.08(90/365)}N(d_2),$$

where

$$d_1 = \frac{[\ln(48.208/48) + (0.08 + 0.5(0.20)^2)(90/365)]}{0.20\sqrt{90/365}}$$
$$= 0.25285$$
$$d_2 = 0.292 - 0.0993 = 0.15354.$$

From standard normal table, we obtain

$$N(0.25285) = 0.5 + 0.3438 = 0.599809$$
$$N(0.15354) = 0.5 + 0.3186 = 0.561014.$$

So the European call value is

$$\begin{aligned} C &= (48.516)(0.599809) \\ &\quad - 48(0.980)(0.561014) \\ &= 2.40123. \end{aligned}$$

(b) The present value of the interest income that would be earned by deferring exercise until expiration is

$$\begin{aligned} X(1 - e^{-r(T-t)}) &= 48(1 - e^{-0.08(90-50)/365}) \\ &= 48(1 - 0.991) = 0.432. \end{aligned}$$

Since $d = 2 < 0.432$, therefore, the early exercise is not precluded.

(c) The value of the American call is now calculated as

$$\begin{aligned} C &= 48.208\left[N_1(b_1) + N_2\left(a_1, -b_1; \sqrt{50/90}\right)\right] \\ &\quad - 48e^{-0.08(90/365)}\left[N_1(b_2)e^{-0.08(40/365)} + N_2\left(a_2, -b_2; -\sqrt{50/90}\right)\right] \\ &\quad + 2e^{-0.08(50/365)}N_1(b_2) \end{aligned} \tag{14.78}$$

since both b_1 and b_2 depend on the critical exdividend stock price S_t^*, which can be determined by

$$C(S_t^*, 40/365; 48) = S_t^* + 2 - 48.$$

By using trial and error, we find that $S_t^* = 46.9641$. An Excel program used to calculate this value is presented in Table 14.1.

Substituting $S^x = 48.208$, $X = \$48$ and S_t^* into Eqs. (14.72b) and (14.72c), we can calculate a_1, a_2, b_1, and b_2

$$a_1 = d_1 = 0.25285.$$
$$a_1 = d_1 = 0.25285.$$
$$b_1 = \frac{\ln\left(\frac{48.208}{46.9641}\right) + \left(0.08 + \frac{0.2^2}{2}\right)\left(\frac{50}{365}\right)}{(0.20)\sqrt{50/365}}$$
$$= 0.4859$$
$$b_2 = 0.485931 - 0.074023 = 0.4119.$$

In addition, we also know $\rho = -\sqrt{50/90} = -0.7454$.

From the above information, we now calculate related normal probability as follows:

$$N_1(b_1) = N_1(0.4859) = 0.6865$$
$$N_1(b_2) = N_1(0.7454) = 0.6598$$

Table 14.1 Calculation of S_t^* (critical exdividend stock price)

S^* (critical exdividend stock price)	46	46.962	46.963	46.9641	46.9	47
X (exercise price of option)	48	48	48	48	48	48
r (risk-free interest rate)	0.08	0.08	0.08	0.08	0.08	0.08
Volalitity of stock	0.2	0.2	0.2	0.2	0.2	0.2
$T-t$ (expiration date − exercise date)	0.10959	0.10959	0.10959	0.10959	0.10959	0.10959
d_1	−0.4773	−0.1647	−0.1644	−0.164	−0.1846	−0.1525
d_2	−0.5435	−0.2309	−0.2306	−0.2302	−0.2508	−0.2187
D (divent)	2	2	2	2	2	2
c (value of European call option to buy one share)	0.60263	0.96319	0.96362	0.9641	0.93649	0.9798
p (value of European put option to sell one share)	2.18365	1.58221	1.58164	1.58102	1.61751	1.56081
$C(S_t^*, T-t; X) - St^* - D + X$	0.60263	0.00119	0.00062	2.3E −06	0.03649	−0.0202

I[a]		Column C*
2		
3	S^* (critical exdividend stock price)	46
4	X (exercise price of option)	48
5	r (risk-free interest rate)	0.08
6	volatility of stock	0.2
7	$T-t$ (expiration date-exercise date)	=(90-50)/365
8	d_1	=(LN(C3/C4) + (C5 + C6^2/2)*(C7))/(C6*SQRT (C7))
9	d_2	=(LN(C3/C4) + (C5-C6^2/2)*(C7))/(C6*SQRT(C7))
10	D (divent)	2
11		
12	c (value of European call option to buy one share)	=C3*NORMSDIST(C8)-C4*EXP(-C5*C7)* NORMSDIST(C9)
13	p (value of European put option to sell one share)	=C4*EXP(-C5*C7)*NORMSDIST(-C9)- C3*NORMSDIST(-C8)
14		
15	$C(S_t^*, T-t; X)-S_t^*-D+X$	=C12-C3-C10 + C4

[a]The table shows the number of the row and the column in the excel sheet

Following Eq. (14.77), we now calculate the value of $N_2(0.25285, -0.4859; -0.7454)$ and $N_2(0.15354, -0.4119; -0.7454)$ as follows:

Since $ab\rho < 0$ for both cumulative bivariate normal density function, therefore, we can use Equation
$N_2(a, b; \rho) = N_2(a, 0; \rho_{ab}) + N_2(b, 0; \rho_{ba}) - \delta$
to calculate the value of both $N_2(a,\ b;\ \rho)$ as follows:

$$\rho_{ab} = \frac{[(-0.7454)(0.25285) + 0.4859](1)}{\sqrt{(0.25285)^2 - 2(0.7454)(0.25285)(-0.4859) + (0.4859)^2}} = 0.87002$$

$$\rho_{ba} = \frac{[(-0.7454)(-0.4859) - 0.25285](1)}{\sqrt{(0.25285)^2 - 2(0.7454)(0.25285)(-0.4859) + (0.4859)^2}} = -0.31979$$

$$\delta = (1 - (1)(-1))/4 = 1/2$$

$$\begin{aligned} N_2(0.292, -0.4859; -0.7454) &= N_2(0.292, 0.0844) + N_2(-0.5377, 0.0656) - 0.5 \\ &= N_1(0) + N_1(-0.5377) - \Phi(-0.292, 0; -0.0844) - \Phi(-0.5377, 0; -0.0656) \\ &\quad - 0.5 = 0.07525 \end{aligned}$$

Using a Microsoft Excel programs presented in Appendix 1, we obtain

$$N_2(0.1927, -0.4119; -0.7454) = 0.06862.$$

Then, substituting the related information into Eq. (14.78), we obtain $C = \$3.08238$ and all related results are presented in Appendix 2.

14.10 Price Bounds for Options

14.10.1 Options Written on Nondividend-Paying Stocks

To derive the lower price bounds and the put–call parity relations for options on nondividend-paying stocks, simply set the cost-of-carry rate, b, equal to the riskless rate of interest, r. Note that, the only cost of carrying the stock is interest.

The lower price bounds for the European call and put options are

$$c(S, T; X) \geq \max[0, S - Xe^{-rT}], \qquad (14.79a)$$

and

$$p(S, T; X) \geq \max[0, Xe^{-rT} - S], \qquad (14.79b)$$

respectively, and the lower price bounds for the American call and put options are

$$C(S, T; X) \geq \max[0, S - Xe^{-rT}], \qquad (14.80a)$$

and

$$P(S, T; X) \geq \max[0, Xe^{-rT} - S], \qquad (14.80b)$$

respectively. The put–call parity relation for nondividend-paying European stock options is

$$c(S, T; X) - p(S, T; X) = S - Xe^{-rT}, \qquad (14.81a)$$

and the put–call parity relation for American options on nondividend-paying stocks is

$$S - X \leq C(S, T; X) - P(S, T; X) \leq S - Xe^{-rT}. \tag{14.81b}$$

For nondividend-paying stock options, the American call option will not rationally be exercised early, while the American put option may be done so.

14.10.2 Options Written on Dividend-Paying Stocks

If dividends are paid during the option's life, the above relations must reflect the stock's drop in value when the dividends are paid. To manage this modification, we assume that the underlying stock pays a single dividend during the option's life at a time that is known with certainty. The dividend amount is D and the time to exdividend is t.

If the amount and the timing of the dividend payment are known, the lower price bound for the European call option on a stock is

$$c(S, T; X) \geq \max[0, S - De^{-rt} - Xe^{-rT}]. \tag{14.82a}$$

In this relation, the current stock price is reduced by the present value of the promised dividend. Because a European-style option cannot be exercised before maturity, the call option holder has no opportunity to exercise the option while the stock is selling cum dividend. In other words, to the call option holder, the current value of the underlying stock is its observed market price less the amount that the promised dividend contributes to the current stock value, that is, $S - De^{-rt}$. To prove this pricing relation, we use the same arbitrage transactions, except we use the reduced stock price $S - De^{-rt}$ in place of S. The lower price bound for the European put option on a stock is

$$p(S, T; X) \geq \max[0, Xe^{-rT} - S - De^{-rt}]. \tag{14.82b}$$

Again, the stock price is reduced by the present value of the promised dividend. Unlike the call option case, however, this serves to increase the lower price bound of the European put option. Because the put option is the right to sell the underlying stock at a fixed price, a discrete drop in the stock price, such as that induced by the payment of a dividend, serves to increase the value of the option. An arbitrage proof of this relation is straightforward when the stock price, net of the present value of the dividend is used in place of the commodity price.

The lower price bounds for American stock options are slightly more complex. In the case of the American call option, for example, it may be optimal to exercise just prior to the dividend payment because the stock price falls by an amount D when the dividend is paid. The lower price bound of an American call option expiring at the exdividend instant would be 0 or $S - Xe^{-rt}$, whichever is greater. On the other hand, it may be optimal to wait until the call option's expiration to exercise. The lower price bound for a call option expiring normally is (14.82a). Combining the two results, we get

$$C(S, T; X) \geq \max[0, S - Xe^{-rt}, S - De^{-rt} - Xe^{-rT}] \tag{14.83a}$$

The last two terms on the right-hand side of (14.83a) provide important guidance in deciding whether to exercise the American call option early, just prior to the exdividend instant. The second term in the squared brackets is the present value of the early exercise proceeds of the call. If the amount is less than the lower price bound of the call that expires normally, that is, if

$$S - Xe^{-rt} \leq S - De^{-rT} - Xe^{-rt}, \tag{14.84}$$

the American call option will not be exercised just prior to the exdividend instant.

To see why, simply rewrite (14.84) so it reads

$$D < X[1 - e^{-r(T-t)}]. \tag{14.85}$$

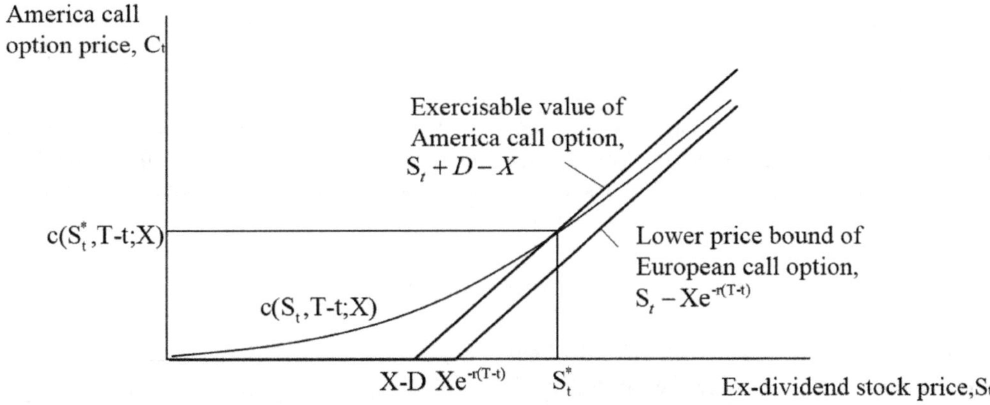

Fig. 14.3 American call option price as a function of the exdividend stock price immediately prior to the exdividend instant. Early exercise may be optimal

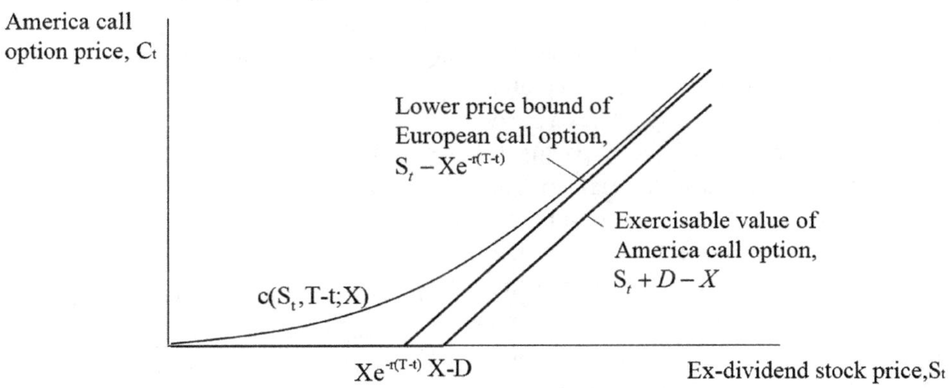

Fig. 14.4 American call option price as a function of the exdividend stock price immediately prior to the exdividend instant. Early exercise will not be optimal

In other words, the American call will not be exercised early if the dividend captured by exercising prior to the exdividend date is less than the interest implicitly earned by deferring exercise until expiration.

Figure 14.3 depicts a case in which early exercise could occur at the exdividend instant, t. Just prior to exdividend, the call option may be exercised yielding proceeds $S_t + D - X$, where S_t, is the exdividend stock price. An instant later, the option is left unexercised with value c(S_t,T − t; X), where c(.) is the European call option formula. Thus, if the exdividend stock price, S_t is above the critical exdividend stock price where the two functions intersect, S_t^*, the option holder will choose to exercise his or her option early just prior to the exdividend instant. On the other hand, if $S_t \leq S_t^*$, the option holder will choose to leave her position open until the option's expiration.

Figure 14.4 depicts a case in which early exercise will not occur at the exdividend instant, t. Early exercise will not occur if the functions, $S_t + D - X$ and $c(S_t, T - t; X)$ do not intersect, as is depicted in Fig. 14.4. In this case, the lower boundary condition of the European call, $S_t - Xe^{-r(T-t)}$, lies above the early exercise proceeds, $S_t + D - X$, and hence the call option will not be exercised early. Stated explicitly, early exercise is not rational if

$$S_t + D - X < S_t - Xe^{-r(T-t)}.$$

This condition for no early exercise is the same as (14.84), where S_t is the exdividend stock price and where the investor is standing at the exdividend instant, t. The condition can also be written as

$$D < X[1 - e^{-r(T-t)}]. \quad (14.85)$$

In words, if the exdividend stock price decline—the dividend—is less than the present value of the interest income that would be earned by deferring exercise until expiration, early exercise will not occur. When condition (14.85) is met, the value of the American call is simply the value of the corresponding European call.

The lower price bound of an American put option is somewhat different. In the absence of a dividend, an American put may be exercised early. In the presence of a dividend payment, however, there is a period just prior to the exdividend date when early exercise is suboptimal. In that period, the interest earned on the exercise proceeds of the option is less than the drop in the stock price from the payment of the dividend. If t_n represents a time prior to the dividend payment at time t, early exercise is suboptimal, where $(X - S)e^{-r(t-t_n)}$ is less than $(X - S + D)$. Rearranging, early exercise will not occur between t_n and t if[3]

$$t_n > t - \frac{\ln\left(1 + \frac{D}{X-S}\right)}{r}. \quad (14.86)$$

Early exercise will become a possibility again immediately after the dividend is paid. Overall, the lower price bound of the American put option is

$$P(S, T; X) \geq \max[o, X - (S - De^{-rt})]. \quad (14.83b)$$

Put–call parity for European options on dividend-paying stocks also reflects the fact that the current stock price is deflated by the present value of the promised dividend, that is

$$c(S, T; X) - p(S, T; X) = S - De^{-rt} - Xe^{-rT}. \quad (14.87)$$

That the presence of the dividend reduces the value of the call and increases the value of the put is again reflected here by the fact that the term on the right-hand side of (14.87) is smaller than it would be if the stock paid no dividend.

Put–call parity for American options on dividend-paying stocks is represented by a pair of inequalities, that is

$$S - De^{-rt} - X \leq C(S, T; X) - P(S, T; X) \leq S - De^{-rt} - Xe^{-rT}. \quad (14.88)$$

To prove the put–call parity relation (14.88), we consider each inequality in turn. The left-hand side condition of (14.88) can be derived by considering the values of a portfolio that consists of buying a call, selling a put, selling the stock, and lending $X + De^{-rt}$ with no risk. Table 14.2 contains these portfolio values.

In Table 14.2, it can be seen that, if all of the security positions stay open until expiration, the terminal value of the portfolio will be positive, independent of whether the terminal stock price is above or below the exercise price of the options. If the terminal stock price is above the exercise price, the call option is exercised, and the stock acquired at exercise price X is used to deliver, in part, against the short stock position. If the terminal stock price is below the exercise price, the put is exercised. The stock received in the exercise of the put is used to cover the short stock position established at the outset. In the event, the put is exercised early at time T, the investment in the riskless bonds is more than sufficient to cover the payment of the exercise

[3] It is possible that the dividend payment is so large that early exercise prior to the dividend payment is completely precluded. For example, consider the case where $X = 50$, $S = 40$, $D = 1$, $t = 0.25$, and $r = 0.10$. Early exercise is precluded if $r = 0.25 - \ln[1 - 1/(50 - 40)]/0.10 = -0.7031$. Because the value is negative, the implication is that there is no time during the current dividend period (i.e., from 0 to t) where it will not pay the American put option holder to wait until the dividend is paid to exercise his option.

Table 14.2 Arbitrage transactions for establishing put–call parity for American stock options $S - De^{-rt} - X \leq C(S, T; X) - P(S, T; X)$

		ExDividend day(t)	Put exercised early (γ)	Put exercised normally(T)	
Position	Initial Value		Intermediate value	Terminal value $S_T \leq X$	$S_T > X$
Buy American call	$-C$		C_γ	0	$S_T - X$
Sell American put	P		$-(X - S_\gamma)$	$-(\tilde{X} - S_T)$	0
Sell stock	S	$-D$	$-S_\gamma$	$-S_T$	$-S_T$
Lend $D\ e^{-rt}$	$-De^{-rt}$	D		Xe^{rT}	Xe^{rT}
Lend X	$-X$		$Xe^{r\gamma}$		
Net portfolio value	$-C + P + S - De^{-rt} - X$	0	$C_\gamma + X(e^{r\gamma} - 1)$	$X(e^{rT} - 1)$	$X(e^{rT} - 1)$

price to the put option holder, and the stock received from the exercise of the put is used to cover the stock sold when the portfolio was formed. In addition, an open call option position that may still have value remains.

In other words, by forming the portfolio of securities in the proportions noted above, we have formed a portfolio that will never have a negative future value. If the future value is certain to be non-negative, the initial value must be nonpositive, or the left-hand inequality of (14.88) holds.

The right-hand side of (14.88) may be derived by considering the portfolio used to prove European put–call parity. Table 14.2 contains the arbitrage portfolio transactions. In this case, the terminal value of the portfolio is certain to equal zero, should the option positions stay open until that time. In the event, the American call option holder decides to exercise the call option early, the portfolio holder uses his long stock position to cover his stock obligation on the exercised call and uses the exercise proceeds to retire his outstanding debt. After these actions are taken, the portfolio holder still has an open long put position and cash in the amount of $X\left[1 - e^{-r(T-t)}\right]$. Since the portfolio is certain to have non-negative outcomes, the initial value must be nonpositive or the right-hand inequality of (14.88) must hold.

14.11 Conclusion

In this chapter, we first introduced univariate and multivariate normal distribution and lognormal distribution. Then, we showed how normal distribution can be used to approximate binomial distribution. We used the concepts normal and lognormal distributions to derive Black–Scholes formula under the assumption that investors are risk neutral. The theory of American call stock option pricing model for one dividend payment is also presented. Moreover, the evaluations of stock option models without dividend payment and with dividend payment are discussed, respectively. Finally, we provided an excel program for evaluating American option pricing model with one dividend payment.

Appendix 1: Microsoft Excel Program for Calculating Cumulative Bivariate Normal Density Function

```
Option Explicit

Public Function Bivarncdf(a As Double, b As Double, rho As Double) As Double


    Dim rho_ab As Double, rho_ba As Double
    Dim delta As Double

    If (a <= 0 And b >= 0 And rho > 0) Then
        Bivarncdf = Application.WorksheetFunction.NormSDist(a) - Phi(a, -b, -rho)
    End If

    If (a >= 0 And b <= 0 And rho > 0) Then
        Bivarncdf = Application.WorksheetFunction.NormSDist(b) - Phi(-a, b, -rho)
    End If

    If (a >= 0 And b >= 0 And rho <= 0) Then
        Bivarncdf          =          Application.WorksheetFunction.NormSDist(a)          +
Application.WorksheetFunction.NormSDist(b) - 1 + Phi(-a, -b, rho)
    End If
  Else

    rho_ab = ((rho * a - b) * IIf(a >= 0, 1, -1)) / Sqr(a ^ 2 - 2 * rho * a * b + b ^ 2)
    rho_ba = ((rho * b - a) * IIf(b >= 0, 1, -1)) / Sqr(a ^ 2 - 2 * rho * a * b + b ^ 2)
    delta = (1 - IIf(a >= 0, 1, -1) * IIf(b >= 0, 1, -1)) / 4
    Bivarncdf = Bivarncdf(a, 0, rho_ab) + Bivarncdf(b, 0, rho_ba) - delta

  End If
End Function

Public Function Phi(a As Double, b As Double, rho As Double) As Double

  Dim a1 As Double, b1 As Double
  Dim w(5) As Double, x(5) As Double
  Dim i As Integer, j As Integer
  Dim doublesum As Double

  a1 = a / Sqr(2 * (1 - rho ^ 2))
  b1 = b / Sqr(2 * (1 - rho ^ 2))

  w(1) = 0.24840615
  w(2) = 0.39233107
  w(3) = 0.21141819
  w(4) = 0.03324666
  w(5) = 0.00082485334
```

```
    x(1) = 0.10024215
    x(2) = 0.48281397
    x(3) = 1.0609498
    x(4) = 1.7797294
    x(5) = 2.6697604

    doublesum = 0

    For i = 1 To 5
        For j = 1 To 5
        doublesum = doublesum + w(i) * w(j) * Exp(a1 * (2 * x(i) - a1) + b1 * (2 * x(j) - b1) + 2 * rho * (x(i) - a1)
* (x(j) - b1))
        Next j
    Next i

    Phi = 0.31830989 * Sqr(1 - rho ^ 2) * doublesum

End Function
```

Appendix 2: Microsoft Excel Program for Calculating the American Call Options

Number[a]	A	B	C*
1	Option Pricing Calculation		
2			
3	S (current stock price)=	50	
4	S_t* (critical exdividend stock price)=	46.9641	
5	S(current stock price NPV of promised dividend)=	48.0218	=B3-B11*EXP(-B7*B10)
6	X (exercise price of option)=	48	
7	r(risk-free interest rate)=	0.08	
8	σ(volatility of stock)=	0.2	
9	T(expiration date)=	0.24658	
10	t(exercise date)=	0.13699	
11	D(Dividend)=	2	
12	d_1(nondividend-paying)=	0.65933	=(LN(B3/B6) + (B7 + 0.5*B8^2)*B9)/(B8*SQRT(B9))
13	d_2(nondividend-paying)=	0.56001	=B12-B8*SQRT(B9)
14	d_1*(critical exdividend stock price)=	-0.16401	=(LN(B4/B6) + (B7 + 0.5*B8^2)*(B9-B10))/(B8*SQRT (B9-B10))
15	d_2*(critical exdividend stock price)=	-0.23022	=B14-B8*SQRT(B9-B10)
16	d_1(dividend-paying)=	0.25285	=(LN(B5/B6) + (B7 + 0.5*B8^2)*(B9))/(B8*SQRT(B9))

(continued)

Number[a]	A	B	C*
17	d_2(dividend-paying)=	0.15354	=B16-B8*SQRT(B9)
18	a_1=	0.25285	=(LN((B3-B11*EXP(-B7*B10))/ B6) + (B7 + 0.5*B8 ^2)*(B9))/(B8*SQRT(B9))
19	a_2=	0.15354	=B18-B8*SQRT(B9)
20	b_1=	0.48593	=(LN((B3-B11*EXP(-B7*B10))/ B4) + (B7 + 0.5*B8 ^2)*(B10))/(B8*SQRT(B10))
21	b_2=	0.41191	=B20-B8*SQRT(B10)
22			
23	$C(S_t^*,T\text{-}t;X)$=	0.9641	=B4*NORMSDIST(B14)-B6*EXP(-B7*(B9-B10)) *NORMSDIST(B15)
24	$C(S_t^*,T\text{-}t;X)\text{-}S_t^*\text{-}D + X$=	2.3E-06	=B23-B4-B11 + B6
25			
26	$N_1(a_1)$=	0.59981	=NORMSDIST(B18)
27	$N_1(a_2)$=	0.56101	=NORMSDIST(B19)
28	$N_1(b_1)$=	0.68649	=NORMSDIST(B20)
29	$N_1(b_2)$=	0.6598	=NORMSDIST(B21)
30	$N_1(\text{-}b_1)$=	0.31351	=NORMSDIST(-B20)
31	$N_1(\text{-}b_2)$=	0.3402	=NORMSDIST(-B21)
32	ρ=	-0.74536	=-SQRT(B10/B9)
33	$a = a_1;b = \text{-}b_1$		
34	$\Phi(a,\text{-}b;\text{-}\rho)$=	0.20259	=phi(-B20,0,-B37)
35	$\Phi(\text{-}a,b;\text{-}\rho)$=	0.04084	=phi(-B18,0,-B36)
36	ρab=	0.87002	=((B32*B18-(-B20))*IF(B18 <=0,1,-1))/ SQRT(B 18^2-2*B32*B18*-B20 + (-B20)^2)
37	ρba=	-0.31979	=((B32*-B20-(B18))*IF(-B20 <=0,1,-1))/ SQRT (B18^2-2*B32*B18*-B20 + (-B20)^2)
38	$N_2(a,0;\rho ab)$=	0.45916	=bivarncdf(B18,0,B36)
39	$N_2(b,0;\rho ba)$=	0.11092	=bivarncdf(-B20,0,B37)
40	δ=	0.5	=(1-IF(B18 <=0,1,-1)*IF(-B20 <=0,1,-1))/4
41	$a = a_2;b = b_2$		
42	$\Phi(a,\text{-}b;\text{-}\rho)$=	0.24401	=phi(-B21,0,-B45)
43	$\Phi(\text{-}a,b;\text{-}\rho)$=	0.02757	=phi(-B19,0,-B44)
44	ρab=	0.94558	=((B32*B19-(-B21))*IF(B19 <=0,1,-1))/ SQRT (B19^2-2*B32*B19*-B21 + (-B21)^2)
45	ρba=	-0.48787	=((B32*-B21-(B19))*IF(-B21 <=0,1,-1))/ SQRT (B19^2-2*B32*B19*-B21 + (-B21)^2)
46	$N_2(a,0;\rho ab)$=	0.47243	=bivarncdf(B19,0,B44)

(continued)

Number[a]	A	B	C*
47	$N_2(b,0;\rho ba)$=	0.09619	=bivarncdf(-B21,0,B45)
48	δ=	0.5	=(1-IF(B19 <=0,1,-1)*IF(-B21 <=0,1,-1))/4
49			
50	$N_2(a_1,-b_1;\rho)$=	0.07007	=bivarncdf(B18,-B20,B32)
51	$N_2(a_2,-b_2;\rho)$=	0.06862	=bivarncdf(B19,-B21,B32)
52			
53	***c*(value of European call option to buy one share)**	2.40123	=B5*NORMSDIST(B16)-B6*EXP(-B7*B9)*NO RMSDIST(B17)
54	***p*(value of European put option to sell one share)**	1.44186	=-B5*NORMSDIST(-B16) + B6*EXP(-B7*B9)*NO RMSDIST(-B17)
55	***c*(value of American call option to buy one share)**	3.08238	=(B3-B11*EXP(-B7*B10))*(NORMSDIST (B20) +bivarncdf(B18,-B20,-SQRT(B10/B9)))-B6*EXP (-B7*B9)*(NORMSDIST(B21)*EXP(B7*(B9-B10)) + bivarncdf(B19,-B21,-SQRT (B10/B9))) + B11*EX P(-B7*B10)*NORMSDIST(B21)

[a]The table above shows the number of the row and the column in the excel sheet

Bibliography

Anderson, T. W. (2003). *An introduction to multivariate statistical analysis* (3rd ed.). New York: Wiley.

Cox, J. C., & Ross, S. A. (1976). A survey of some new results in financial option pricing theory. *Journal of Finance*, *31*(2), 383–402.

Johnson, N. L., & Kotz, S. (1970). *Continuous univariate Distributions-I Distributions in statistics*. New York: Wiley.

Johnson, N. L., & Kotz, S. (1972). *Distributions in statistics: Continuous multivariate distribution*. New York: Wiley.

Lee, C.-F., Lee, J. C., & Lee, A. C. (2010a). Normal, lognormal distribution and option pricing model. In C. F. Lee, A. C. Lee, & J. Lee (Eds.), *Handbook of quantitative finance and risk management*. Singapore: Springer.

Lee, C. F., Finnerty, J., Lee, J., Lee, A., & Wort, D. (2013). *Security analysis and portfolio management, and financial derivatives*. Singapore: World Scientific.

Lee, C.-F., Lee, A. C., & Lee, J. (2010). *Handbook of quantitative finance and risk management*. New York: Springer.

Rubinstein, M. (1976). The valuation of uncertain income streams and the pricing of options. *The Bell Journal of Economics*, 407–425.

Whaley, R. E. (1981). On the valuation of American call options on stocks with known dividends. *Journal of Financial Economics*, *9*(2), 207–211.

Copula, Correlated Defaults, and Credit VaR

15

Contents

Abstract

Almost every financial institution devotes a lot of attention and energy to credit risk. The default correlations of credit assets have a fatal influence on credit risk. How to model default correlation correctly has become a prerequisite for the effective management of credit risk. In this chapter, we provide a new approach to estimating future credit risk on target portfolio based on the framework of CreditMetrics™ by J. P. Morgan. However, we adopt the perspective of factor copula and then bring the principal component analysis concept into factor structure to construct a more appropriate dependence structure among credits. In order to examine the proposed method, we use real market data instead of virtual ones. We also develop a tool for risk analysis that is convenient to use, especially for banking loan businesses. The results indicate that people assume dependence structures are normally distributed, which could lead to underestimated risks. On the other hand, our proposed method captures better features of risks, including conspicuous fat-tail effects, even though the factors appear normally distributed.

This chapter draws upon the paper by Chang and Chen (2010) which was published as Chap. 46 of Handbook of Quantitative Finance and Risk Management (2010) edited by Lee et al.

C.-F. Lee et al., *Financial Econometrics, Mathematics and Statistics*,
https://doi.org/10.1007/978-1-4939-9429-8_15

15.1 Introduction

Credit risk is a risk that generally refers to counterparty fails to fulfill its contractual obligations. The history of financial institutions has shown that many banking association failures were due to credit risk. For the integrity and regularity, financial institutions attempt to quantify credit risk as well as market risk. Credit risk has great influence on all financial institutions as long as they have contractual agreements. The evolution of measuring credit risk has been progressed for a long time. Many credit risk measure models were published, such as CreditMetrics by J. P. Morgan, CreditRisk+ by Credit Suisse. On the other side, New Basel Accords (Basel II Accords) which are the recommendation on banking laws and regulations construct a standard to promote greater stability in financial system. Basel II Accords allowed banks to estimate credit risk by using either a standardized model or an internal model approach, based on their own risk management system. The former approach is based on external credit ratings provided by external credit assessment institutions. It describes the weights, which fall into five categories for banks and sovereigns, and four categories for corporations. The latter approach allows banks to use their internal estimation of creditworthiness, subject to regulatory. How to build a credit risk measurement model after banking has constructed internal customer credit rating? How to estimate their default probability and default correlations? This thesis attempts to implement a credit risk model tool which links to internal banking database and gives the relevant reports automatically. The developed model facilitates banks to boost their risk management capability.

The dispersion of the credit losses, however, critically depends on the correlations between default events. Several factors such as industry sectors and corporation sizes will affect correlations between every two default events. The CreditMetrics^{TM} model (1997) issued from J. P. Morgan proposed a binomial normal distribution to describe the correlations (dependence structures). In order to describe the dependence structure between two default events in detail, we adopt copula function instead of binomial normal distribution to express the dependence structure.

When estimating credit portfolio losses, both the individual default rates of each firm and joint default probabilities across all firms need to be considered. These features are similar to the valuation process of collateralized debt obligation (CDO). A CDO is a way of creating securities with widely different risk characteristics from a portfolio of debt instrument. The estimating process is almost the same between our goal and CDO pricing. We focus on how to estimate risks. Most CDO pricing literature adopted copula functions to capture the default correlations. Li (2000) extended Sklar's issue (1959) that a copula function can be applied to solve financial problems of default correlation. Li (2000) pointed out that if the dependence structure were assumed to be normally distributed through binomial normal probability density function, the joint transformation probability would be consistent with the result from using a normal copula function. But this assumption is too strong. It has been discovered that most financial data have skew or fat-tail phenomenon. Bouye et al. (2000) and Embrechts et al. (1999) pointed out that the estimating VaR would be underestimated if the dependence structure was described by normal copula comparing to actual data. Hull and White (2004) combined factor analysis and copula functions as a factor copula concept to investigate reasonable spread of CDO. How to find a suitable correlation to describe the dependence structure between every two default events and to speed up the computational complexity is our main object.

This chapter aims to (1) construct an efficient model to describe the dependence structure; (2) use this constructed model to analyze overall credit, marginal, and industrial risks; and (3) build up an automatic tool for banking system to analyze its internal credit risks. Section 15.2 presents the methodology of the copula function to analyze credit risks. Section 15.3 discusses experimental results. Finally, Sect. 15.4 concludes this chapter.

15.2 Methodology

15.2.1 CreditMetrics

This chapter adopts the main framework of CreditMetrics and calculates credit risks by using real commercial bank loans. The calculating data set for this chapter is derived from a certain commercial bank in Taiwan. Although there may be some conditions which are different from the situations proposed by CreditMetrics, the calculating process by CreditMetrics can still be appropriately applied to this chapter. For instance, number of rating degrees in CreditMetrics adopting S&P's rating category are 7, i.e., AAA to C, but in this loans data set, there are 9 degrees instead. The following is the introduction to CreditMetrics model framework.

This model can be roughly divided into three components, i.e., value at risk due to credit, exposures, and correlations, respectively, as shown in Fig. 15.1. In this section, these three components and how does this model work out on credit risk valuation will be briefly introduced. For more further detail, refer to CreditMetrics technique document.

15.2.1.1 Value at Risk Due to Credit

The process of valuing value at risk due to credit can be decomposed into three steps. For simplicity, we assumed there is only one stand-alone instrument which is a corporation bond. (The bond property is similar to loan they both receive certain amount of cash flow every period and receive principal at the maturity.) This bond has five-year maturity and pays an annual coupon at the rate of 5% to express the calculation process if necessary. Some modifications to fit real situations will be considered later. In Step 1, CreditMetrics assumed that all risks of one portfolio due to credit rating changes, no matter defaulting or rating migrating. It is significant to estimate not only the likelihood of default but also the chance of migration to move toward any possible credit quality state at the risk horizon. Therefore, a standard system that evaluated "rating changing" under a certain horizon of time is necessary. This information is represented more concisely in transition matrix. Transition matrix can be calculated by observing the historical pattern of rating change and default. They have been published by S&P and Moody's rating agencies, or calculated by private banking internal rating systems. Besides, the transition matrix should be estimated for the same time interval (risk horizon) which can be defined by user demand, usually one-year period. Table 15.1 is an example to represent one-year transition matrix.

In the transition matrix table, AAA level is the highest credit rating and D is the lowest and also represents default occurs. According to the above transition matrix table, a company which stays in AA level at the beginning of the year has the

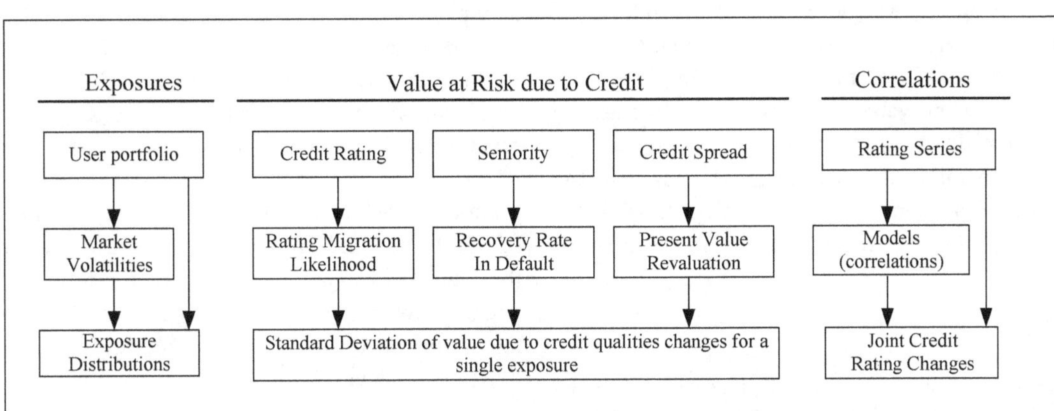

Fig. 15.1 Structure of CreditMetrics model

Table 15.1 One-year transition matrix

Initial rating	Rating at year-end (%)							
	AAA	AA	A	BBB	BB	B	CCC	D
AAA	90.81	8.33	0.68	0.06	0.12	0	0	0
AA	0.70	90.65	7.79	0.64	0.06	0.14	0.02	0
A	0.09	2.27	91.05	5.52	0.74	0.26	0.01	0.06
BBB	0.02	0.33	5.95	86.93	5.30	1.17	0.12	0.18
BB	0.03	0.14	0.67	7.73	80.53	8.84	1.00	1.06
B	0	0.11	0.24	0.43	6.48	83.46	4.07	5.20
CCC	0.22	0	0.22	1.30	2.38	11.24	64.86	19.79

Source Morgan*'s* (1997)

Table 15.2 Recovery rates by seniority class

Class		Recovery rate of Taiwan debt business research using TEJ data	
		Mean (%)	Standard deviation (%)
Loan	Secured	55.38	35.26
	Unsecured	33.27	30.29
Corporation bond	Secured	67.99	26.13
	Unsecured	36.15	37.17

Source Da-Bai Shen, Yong-Kang Jing, Jia-Cian Tsia (2003), Research of Taiwan recovery rate with TEJ Data Bank

probability of 0.64% to go down to BBB level at the end of the year. By the same way, a company which stays in CCC level at the beginning of the year has the probability of 2.38% to go up to BB level at the end of the year. In this chapter, the transition matrix is to be seen as an external data.[1]

In Step 1, we describe the likelihood of migration move to any possible quality states (AAA to CCC) at the risk horizon. Step 2 is valuation. The value at the risk horizon must be determined. According to different states, the valuation falls into two categories. First, in the event of a default, recovery rate of different seniority class is needed. Second, in the event of up (down) grades, the change in credit spread that results from the rating migration must be estimated too.

In default category, Table 15.2 gives recovery rates by seniority class which this chapter adopts to revaluate instruments. For instance, if the holding bond (five-year maturity and pays an annual coupon at the rate of 5%) is unsecured and the default occurs, the recovery value will be estimated using its mean value which is 36.15%.

In rating migration category, the action of revaluation is equal to determine the cash flows which result from holding the instrument (corporation bond position). Assuming a face value of \$100, the bond pays \$5 (an annual coupon at the rate of 5%) each at the end of the next four years. Now, the calculating process to describe the value V of the bond assuming the bond upgrades to level A by the formula below:

[1]We do not focus on how to model probability of default (PD) but focus on how to establish the dependence structure. The one-year transition matrix is a necessary input to our model.

Table 15.3 One-year forward zero curves by credit rating category

Category	Year 1	Year 2	Year 3	Year 4
AAA	3.60	4.17	4.73	5.12
AA	3.65	4.22	4.78	5.17
A	3.72	4.32	4.93	5.32
BBB	4.10	4.67	5.25	5.63
BB	5.55	6.02	6.78	7.27
B	6.05	7.02	8.03	8.52
CCC	15.05	15.02	14.03	13.52

Source Morgan's (1997)

$$V = 5 + \frac{5}{(1+3.72\%)} + \frac{5}{(1+4.32\%)^2} + \frac{5}{(1+4.93\%)^3} + \frac{105}{(1+5.32)^4} = 108.66$$

The discount rate in above formula comes from the forward zero curves shown in Table 15.3, which is derived from CreditMetrics technical document. This chapter does not focus on how to calculate forward zero curves. It is also seen as an external input data.

Step 3 is to estimate the volatility of value due to credit quality changes for this stand-alone exposure (level A, corporation bond). From Step 1 and Step 2, the likelihood of all possible outcomes and distribution of values within each outcome are known. CreditMetrics used two measures to calculate the risk estimate. One is standard deviation, and the other is percentile level. Besides these two measures, this chapter also embraces marginal VaR which denotes the increment VaR due to adding one new instrument in the portfolio.

15.2.1.2 Exposures

As discussed above, the instrument is limited to corporation bonds. CreditMetrics is allowed following generic exposure types.

1. Noninterest bearing receivables,
2. Bonds and loans,
3. Commitments to lend,
4. Financial letters of credit,
5. Market-driven instruments (swap, forwards, etc.).

The exposure type this chapter aims at is loans. The credit risk calculation process of loans is similar to bonds as previous example. The only difference is that loans do not pay coupons. Instead, loans receive interests. But the CreditMetrics model can definitely fit the goal of this chapter to estimate credit risks on banking loans business.

15.2.1.3 Correlations

In most circumstances, there is usually more than one instrument in a target portfolio. Now, multiple exposures are taken into consideration. In order to extend the methodology to a portfolio of multiple exposures, estimating the contribution to risk brought by the effect of nonzero credit quality correlations is necessary. Thus, the estimation of joint likelihood in the credit quality comovement is the next problem to be resolved. There are many academic papers which address the problems of estimating correlations within a credit portfolio. For example, Gollinger and Morgan (1993) used time series of default likelihood to correlate default likelihood, and Steveson and Fadil (1995) correlated the default experience across 33 industry groups. On the other hand, CreditMetrics proposed a method to estimate default correlation. They have several assumptions.

A. A firm's asset value is the process which drives its credit rating changes and default.

B. The asset returns are normally distributed.
C. Two asset returns are correlated and bivariate normally distributed, and multiple asset returns are correlated and multivariate normally distributed.

According to assumption A, individual threshold of one firm can be calculated. For a two exposure portfolio, which credit ratings are level B and level AA, and standard deviations of returns are σ and σ', respectively; it only remains to specify is the correlation ρ between two asset returns. The covariance matrix for the bivariate normal distribution:

$$\Sigma = \begin{pmatrix} \sigma^2 & \rho\sigma\sigma' \\ \rho\sigma\sigma' & \sigma'^2 \end{pmatrix}$$

Then, the joint probability of comovement that both two firms stay in the same credit rating can be described by the following formula:

$$Pr\{Z_{\mathrm{BB}} < R_1 < Z_{\mathrm{B}}, Z'_{\mathrm{AAA}} < R_2 < Z'_{\mathrm{AA}}\} = \int_{Z_{\mathrm{BB}}}^{Z_{\mathrm{B}}} \int_{Z'_{\mathrm{AAA}}}^{Z'_{\mathrm{AA}}} f(r, r'; \Sigma)(\mathrm{d}r')\mathrm{d}r$$

where Z_{BB}, Z_{B}, Z'_{AAA}, Z'_{AA} are the thresholds. Figure 15.2 gives a concept of the probability calculation. These three assumptions regarding estimating the default correlation are too strong, especially assuming the multiple asset returns are multinormally distributed. In the next session, a better way of using copula to examine the default correlation is proposed.

15.2.2 Copula Function

Consider a portfolio consists of m credits. Marginal distribution of each individual credit risks (defaults occur) can be constructed by using either the historical approach or the market implicit approach (derived credit curve from market information). But the question is—how to describe the joint distribution or comovement between these risks (default correlation)? In a sense, every joint distribution function for a vector of risk factors implicitly contains both a description of the marginal behavior of individual risk factors and a description of their dependence structure. The simplest way is assuming the dependence structure to be mutual independence among the credit risks. However, the independent assumption of the credit risks is obviously not realistic.

Undoubtedly, the default rate for a group of credits tends to be higher when the economy is in a recession and lower in a booming. This implies that each credit is subject to the same factors from macroeconomic environment and that there exists some form of dependence among these credits. The copula approach provides a way of isolating the description of the dependence structure. That is, the copula provides a solution to specify a joint distribution of risks, with given marginal distributions. Of course, this problem is no unique solution. There are many different techniques in statistics which can specify a joint distribution with given marginal distributions and a correlation structure. In the following section, the copula function is briefly introduced.

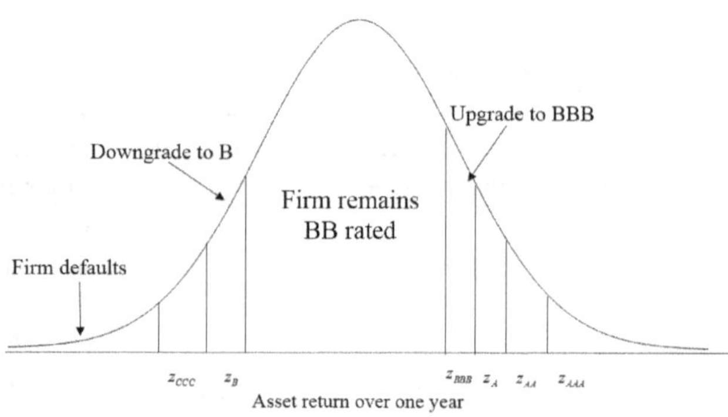

Fig. 15.2 Distribution of asset returns with rating change thresholds

15.2.2.1 Copula Function

A m-dimensional copula is a distribution function on $[0, 1]^m$ with standard uniform marginal distributions.

$$C(u) = C(u_1, u_2, \ldots, u_m) \tag{15.1}$$

C is called a copula function.

The copula function C is a mapping of the form $C : [0, 1]^m \rightarrow [0, 1]$, i.e., a mapping of the m-dimensional unit cube $[0, 1]^m$ such that every marginal distribution is uniform on the interval $[0, 1]$. The following two properties must hold

1. $C(u_1, u_2, \ldots, u_m, \Sigma)$ is increasing in each component u_i.
2. $C(1, \ldots, 1, u_i, 1, \ldots, 1, \Sigma) = u_i$ for all $i \in \{1, \ldots, m\}$, $u_i \in [0, 1]$.

15.2.2.2 Sklar's Theorem

Sklar (1959) underlined applications of the copula. Let $F(\cdot)$ be a m-dimensional joint distribution function with marginal distribution F_1, $F_2, \ldots, F_m$. There exits a copula C : $[0, 1]^m \rightarrow [0, 1]$ such that,

$$F(x_1, x_2, \ldots, x_m) = C(F_1(x_1), F_2(x_2), \ldots F_m(x_m)) \tag{15.2}$$

If the margins are continuous, then C is unique.

For any $x_1, \ldots, x_m$ in $\Re = [-\infty, \infty]$ and X has joint distribution function F, then

$$\begin{aligned} &F(x_1, x_2, \ldots, x_m) = \\ &\quad F_2(X_2) \leq F_2(x_2), \ldots, F_m(X_m) < F_m(x_m)] \end{aligned} \tag{15.3}$$

According to (15.2), the distribution function of $(F_1(X_1), F_2(X_2), \ldots, F_m(X_m))$ is a copula. Let $x_i = F_i^{-1}(u_i)$, then

$$\begin{aligned} &C(u_1, u_2, \ldots, u_m) \\ &\quad = F(F_1^{-1}(u_1), F_2^{-1}(u_2), \ldots, F_m^{-1}(u_m)) \end{aligned} \tag{15.4}$$

This gives an explicit representation of C in terms of F and its margins.

15.2.2.3 Copula of *F*

Li (2000) used the copula function conversely. The copula function links univariate marginals to their full multivariate distribution. For m uniform random variables, $U_1, U_2, \ldots, U_m$, the joint distribution function C is defined as

$$\begin{aligned} &C(u_1, u_2, \ldots, u_m, \Sigma) \\ &\quad = Pr[U_1 \leq u_1, U_2 \leq u_2, \ldots, U_m \leq u_m] \end{aligned} \tag{15.5}$$

where Σ is correlation matrix of $U_1, U_2, \ldots, U_m$.

As mentioned above, we can obtain univariate marginal distribution functions $F_1(x_1), F_2(x_2), \ldots, F_m(x_m)$. The same as above, let $x_i = F_i^{-1}(u_i)$, the joint distribution function F can be described as follows:

$$\begin{aligned} &F(x_1, x_2, \ldots, x_m) \\ &\quad = C(F_1(x_1), F_2(x_2), \ldots, F_m(x_m), \Sigma) \end{aligned} \tag{15.6}$$

The joint distribution function F is defined by using a copula.

The property can be easily shown as follows:

$$\begin{aligned} &C(F_1(x_1), F_2(x_2), \ldots F_m(x_m), \Sigma) \\ &\quad = Pr[U_1 \leq F_1(x_1), U_2 \leq F_2(x_2), \ldots, U_m \leq F_m(x_m)] \\ &\quad = Pr[F_1^{-1}(U_1) \leq x_1, F_2^{-1}(U_2) \leq x_2, \ldots, F_m^{-1}(U_m) \leq x_m] \\ &\quad = Pr[X_1 \leq x_1, X_2 \leq x_2, \ldots, X_m \leq x_m] \\ &\quad = F(x_1, x_2, \ldots, x_m) \end{aligned}$$

The marginal distribution of X_i is

$$\begin{aligned} &C(F_1(+\infty), F_2(+\infty), \ldots, F_i(x_i), \ldots, F_m(+\infty), \Sigma) \\ &\quad = Pr[X_1 \leq +\infty, X_2 \leq +\infty, \ldots, X_i \leq x_i, \ldots, X_m \leq +\infty] \\ &\quad = Pr[X_i \leq x_i] \\ &\quad = F_i(x_i) \end{aligned} \tag{15.7}$$

Li showed that with given marginal functions, we can construct the joint distribution through some copulas accordingly. But what kind of copula should be chosen corresponding to realistic joint distribution of a portfolio? For example, the CreditMetrics chose Gaussian copula to construct multivariate distribution.

By (15.6), this Gaussian copula is given by

$$\begin{aligned}C^{\mathrm{Ga}}(u,\Sigma) &= Pr(\Phi(X_1)\le u_1,\Phi(X_2)\le u_2,\ldots,\Phi(X_m)\le u_m,\Sigma)\\ &= \Phi_\Sigma(\Phi^{-1}(u_1),\Phi^{-1}(u_2),\ldots,\Phi^{-1}(u_m))\end{aligned} \tag{15.8}$$

where Φ denotes the standard univariate normal distribution, Φ^{-1} denotes the inverse of a univariate normal distribution, and Φ_Σ denotes multivariate normal distribution. In order to easily describe the construction process, we only discuss two random variables u_1 and u_2 to demonstrate the Gaussian copula.

$$C^{\mathrm{Ga}}(u_1,u_2,\rho) = \int_{-\infty}^{\Phi^{-1}(u_1)}\int_{-\infty}^{\Phi^{-1}(u_2)} \frac{1}{2\pi\sqrt{(1-\rho^2)}}\exp\left\{-\frac{v_1^2-2\rho v_1 v_2+v_2^2}{2(1-\rho^2)}\right\}dv_2dv_1 \tag{15.9}$$

where ρ denotes the correlation of u_1 and u_2.

Equation (15.9) is also equivalent to the bivariate normal copula which can be written as follows:

$$C(u_1,u_2,\rho) = \Phi_2(\Phi^{-1}(u_1),\Phi^{-1}(u_2)) \tag{15.10}$$

Thus, with given individual distribution (e.g., credit migration over one year's horizon) of each credit asset within a portfolio, we can obtain the joint distribution and default correlation of this portfolio through copula function. In our methodology, we do not use copula function directly. In the next session, we bring in the concept of factor copula for further improvement to form the default correlation. Using factor copula has two advantages. One is to avoid constructing a high-dimensional correlation matrix. If there are more and more instruments ($N > 1000$) in our portfolio, we need to store N-by-N correlation matrix, and scalability is one problem. The other advantage is to speed up the computation time because of the lower dimension.

15.2.3 Factor Copula Model

In this section, copula models having a factor structure will be introduced. It is called factor copula because this model describes dependence structure between random variables not from the perspective of a certain copula form, such as Gaussian copula, but from the factors model. Factor copula models have been broadly used to assess price of collateralized debt obligation (CDO) and credit default swap (CDS). The main concept of factor copula model is: Under a certain macro environment, credit default events are independent of each other. And the main causes affect default events come from potential market economic conditions. This model provides another way to avoid dealing with multivariate normal distribution (high-dimensional) simulation problem.

Continuing the above example—a portfolio is consisted of m credits. First, we consider the simplest example which contains only one factor. Define V_i is the asset value of ith credit under single factor copula model. Then, this ith credit asset value can be expressed by one factor M (mutual factor) chosen from macroeconomic factors and one error term ε_i.

$$V_i = r_i M + \sqrt{1-r_i^2}\,\varepsilon_i \tag{15.11}$$

where r_i is weight of M, and the mutual factor M is independent of ε_i.

Let the marginal distribution of $V_1, V_2,\ldots, V_m$ is F_i, $i = 1, 2,\ldots, m$. Then, the m-dimensional copula function can be written as

$$\begin{aligned}C(u_1,u_2,\ldots,u_m) &= F\left(F_1^{-1}(u_1),F_2^{-1}(u_2),\ldots,F_m^{-1}(u_m)\right)\\ &= Pr\left(V_1\le F_1^{-1}(u_1),V_2\le F_2^{-1}(u_2),\ldots,V_m\le F_m^{-1}(u_m)\right)\end{aligned} \tag{15.12}$$

F is the joint cumulative distribution function of $V_1, V_2,\ldots, V_m$.

It has been known that M and ε_i are independent of each other; according to iterated expectation theorem, (15.12) can be written as

$$
\begin{aligned}
&C(u_1, u_2, \ldots, u_m) \\
&= E\{Pr(V_1 \le F_1^{-1}(u_1), V_2 \le F_2^{-1}(u_2), \ldots, V_m \le F_m^{-1}(u_n))|M\} \\
&= E\left\{\prod_{i=1}^{m} Pr(r_iM + \sqrt{1-r_i^2}\varepsilon_i \le F_i^{-1}(u_i))|M\right\} \\
&= E\left\{\prod_{i=1}^{m} F_{\varepsilon,i}(\frac{F_i^{-1}(u_i) - r_iM}{\sqrt{1-r_i^2}})|M\right\} \\
&= \int\left(\prod_{i=1}^{m} F_{\varepsilon,i}\left(\frac{F_i^{-1}(u_i) - r_iM}{\sqrt{1-r_i^2}}\right)\right)g(M)\mathrm{d}M
\end{aligned}
\tag{15.13}
$$

Using above formula, m-dimensional copula function can be derived. Moreover, according to (15.13), the joint cumulative distribution F can also be derived

$$
\begin{aligned}
&F(t_1, t_2, \ldots, t_m) \\
&= \int\left(\prod_{i=1}^{m} F_{\varepsilon,i}\left(\frac{(F_i^{-1}(F_{T,i}(t_i)) - r_iM)}{\sqrt{1-r_i^2}}\right)\right)g(M)\mathrm{d}M
\end{aligned}
\tag{15.14}
$$

Let $F_i(t_i) = Pr(T_i \le t_i)$ represent i-credit default probability (default occurs before time t_i), where F_i is marginal cumulative distribution. We note here that CDX pricing cares about when the default time T_i occurs. Under the same environment (systematic factor M) Andersen and Sidenius (2004), the default probability $Pr(T_i \le t_i)$ will equal to $Pr(V_i \le c_i)$, which represents that the probability asset value V_i is below its threshold c_i. Then, joint default probability of these m credits can be described as follows:

$$
\begin{aligned}
&F(c_1, c_2, \ldots, c_m) \\
&\quad = Pr(V_1 \le c_1, V_2 \le c_2, \ldots, V_m \le c_m)
\end{aligned}
$$

Now, we bring the concept of principal component analysis (PCA). People use PCA to reduce the high dimensions or multivariable problems. If someone would like to explain one thing (or some movement of random variables), he has to gather interpreting variables related to those variables movements or their correlation. Once the kinds of interpreting variables are too huge or complicated, it becomes harder to explain those random variables and will produce complex problems. Principal component analysis provides a way to extract approximate interpreting variables to cover maximum variance of variables. Those representative variables may not be "real" variables. Virtual variables are allowed and depend on the explaining meaning. We do not talk about PCA calculation processes, and the further detail could refer to Jorion (2000). Based on factor model, the asset value of m credits with covariance matrix Σ can be described as follows:

$$
V_i = r_{i1}y_1 + r_{i2}y_2 + \ldots + r_{im}y_m + \varepsilon_i \tag{15.15}
$$

where y_i are common factors between these m credits, and r_{ij} is the weight (factor loading) of each factor. The factors are independent of each other. The question is: How to determinate those y_i factors and their loading? We use PCA to derive the factor loading. Factor loadings are based on listed price of those companies in the portfolio to calculate their dependence structure. The experimental result will be shown in the next section. We note here that the dependence structure among assets has been absorbed into factor loadings (Fig. 15.3).

15.3 Experimental Results

The purpose of this chapter is to estimate credit risk by using principal component analysis to construct dependence structure without giving any assumptions to specify formulas of copula. In other words, the data were based on itself to describe the dependence structure.

15.3.1 Data

In order to analyze credit VaR empirically through the proposed method, this investigation adopts the internal loan account data, loan application data, and customer information data from a commercial bank on current market in Taiwan. For reliability of data authenticities, instead of virtual ones, we apply the data in Taiwan market to the model. This also means now the portfolio pool contains only the loans of listed companies

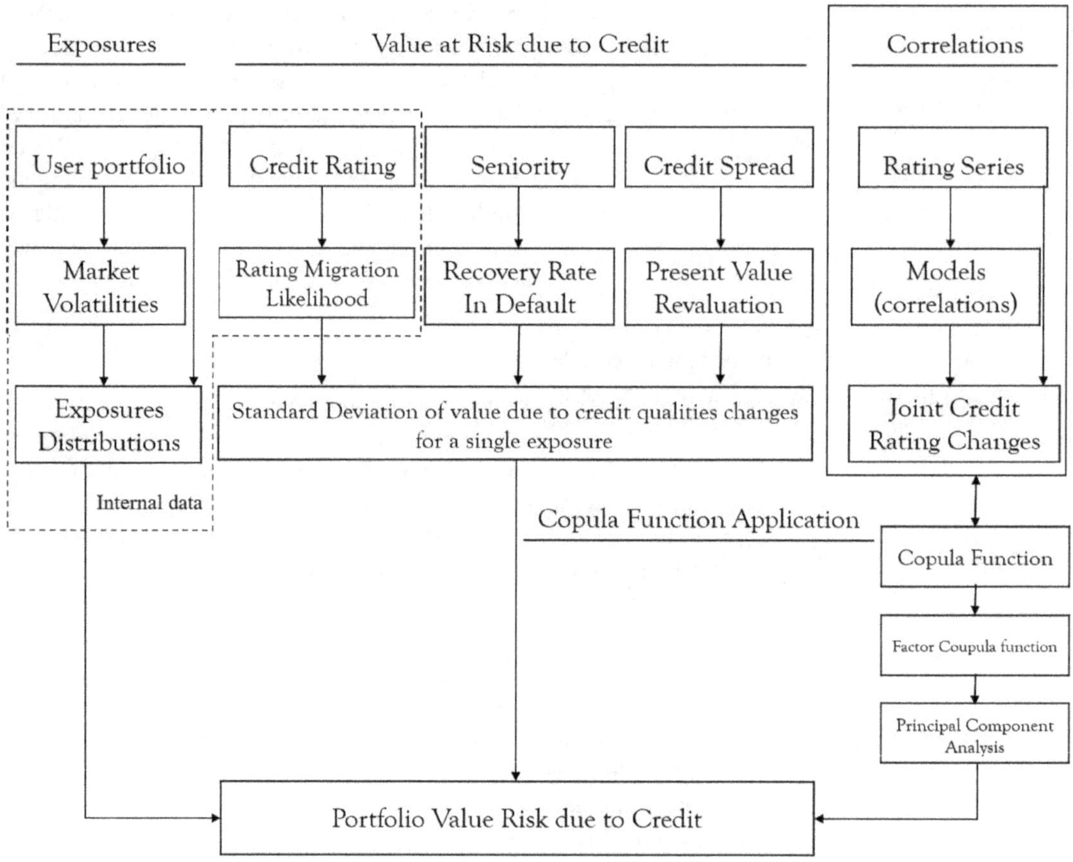

Fig. 15.3 Architecture of the proposed model

and does not contain the loan of unlisted companies. According to the period of these data, we can estimate two future portfolio values. They are values on 2003 and 2004, respectively.

All requirement data are downloaded automatically from database system to workspace for computations. Before going to the detail of the experiments, the relevant data and experimental environment are introduced as follows.

15.3.1.1 Requirements of Data Input

1. Commercial bank internal data: These internal data contain nearly 40,000 entries' customer's data, 50,000 entries' loans data and 3000 entries' application data. These data contain maturity dates, outstanding amount, credit ratings, interest rate for lending, market type, etc., up to December 31, 2004.
2. One-year period transition matrix: The data were extracted from Yang (2005), who used the same commercial bank history data to estimate a transition matrix which obeyed Markov Chain (Table 15.4).
3. Zero forward rate: Refer to Yang (2005), based on computed transition matrix to estimate the term structure of credit spreads. Furthermore, they added corresponding risk-free interest rate to calculate zero forward rates from discounting zero spot rates (Table 15.5).
4. Listed share prices at exchange market and overcounter market: We collected weekly listed share prices of all companies at exchange and overcounter market in Taiwan from January 1, 2000 to December 31, 2003, in total three years data, through Taiwan Economic Journal Data Bank (TEJ).

Table 15.4 One-year transition matrix (commercial data)

Initial rating	Rating at year-end (%)									
	1	2	3	4	5	6	7	8	9	D
1	100	0	0	0	0	0	0	0	0	0
2	3.53	70.81	8.90	5.75	6.29	1.39	0.19	2.74	0.06	0.34
3	10.76	0.03	72.24	0.24	10.39	5.78	0.31	0.09	0.06	0.10
4	1.80	1.36	5.85	57.13	18.75	11.31	2.45	0.32	0.70	0.33
5	0.14	0.44	1.58	2.39	75.47	16.97	1.49	0.61	0.49	0.42
6	0.09	0.06	0.94	2.44	13.66	70.58	6.95	1.68	0.76	2.81
7	0.05	0.05	0.27	3.72	3.75	14.49	66.39	8.05	0.12	3.11
8	0.01	0	0.03	0.45	0.21	1.34	2.00	77.10	0.44	18.42
9	0	0	0.02	0.09	1.46	1.80	1.36	3.08	70.06	22.13
D	0	0	0	0	0	0	0	0	0	100

Source Yang (2005)

Table 15.5 One-year forward zero curves by credit rating category (commercial data)

Yield (%)		1 year	2 year	3 year	4 year	5 year	6 year	7 year	8 year	9 year
Credit rating	1	1.69	2.08	2.15	2.25	2.41	2.53	2.58	2.62	2.7
	2	2.57	2.88	3.19	3.44	3.72	3.94	4.07	4.18	4.33
	3	2.02	2.41	2.63	2.85	3.11	3.32	3.45	3.56	3.71
	4	2.6	2.93	3.28	3.59	3.91	4.17	4.34	4.48	4.65
	5	2.79	3.1	3.48	3.81	4.15	4.42	4.6	4.75	4.93
	6	4.61	5.02	5.16	5.31	5.51	5.67	5.76	5.83	5.93
	7	6.03	6.16	6.56	6.83	7.07	7.23	7.28	7.31	7.36
	8	22.92	23.27	22.54	21.91	21.36	20.78	20.15	19.52	18.94
	9	27.51	27.82	26.4	25.17	24.09	23.03	21.97	20.97	20.08

Source Yang (2005)

15.3.2 Simulation

In this section, the simulation procedure of analyzing banking VaR is briefly introduced. There are two main methods of experiments: A and B. A is the method that this chapter proposed which uses factor analysis to explain the dependence structure and to simulate the distribution of future values. B is contrast set which is traditionally and popularly used in most applications such as CreditMetrics. We call it the multinormal (normal/Gaussian copula) simulation method.

Both of these two methods need three input data tables—credit transition matrix, forward zero curves, and share prices of each corporation in portfolio pool. The detail of normal copula method procedure does not mention here, and readers can refer to technical documentation of CreditMetrics. Now, the process of factor analysis method is shown as follows:

1. Extracting the data entries that do not mature under given date, from database system including credit ratings, outstanding amounts, interest rates, etc.
2. According to the input transition matrix, we can calculate standardized thresholds for each credit rating.

3. Using the share prices of those corporations in the portfolio pool to calculate equities correlations.
4. Using principal component analysis to obtain each factor loading for all factors under the assumption that these factors obey some distributions (e.g., standard normal distribution) to simulate their future asset value and future possible credit ratings.
5. According to possible credit ratings, to discount the outstanding amounts by their own forward zero curves to evaluate future value distributions.
6. Displaying the analysis results.

15.3.3 Discussion

15.3.3.1 Tool and Interface Preview

For facility and convenience, this chapter uses MATLAB and MySQL to construct an application tool to help tool users analyze future portfolio value more efficiently. Following is this tool's interactive interfaces.

Basic Information of Experimental Data: (Pie Chart)

The noncomputing data are extracted, and pie charts give user the basic view of loan information. These charts present the proportion of each composition of three key elements—loan amount of companies, industry, and credit rating. These charts can also assist users construct an overview of concerned portfolio.

Pie chart of loan amount weight in terms of enterprise, industry, and credit rating is shown in Fig. 15.4.

Information According to Experimental Data: (Statistic Numbers)

Besides graphic charts, the second part demonstrates a numerical analysis. First part is the extraction of the company data which has maturity more than the given months, and second part is to extract the essential data of top weighting companies. Part I and II extract data without any computation; the only thing that has been done is to sort or remove some useless data (Figs. 15.5 and 15.6).

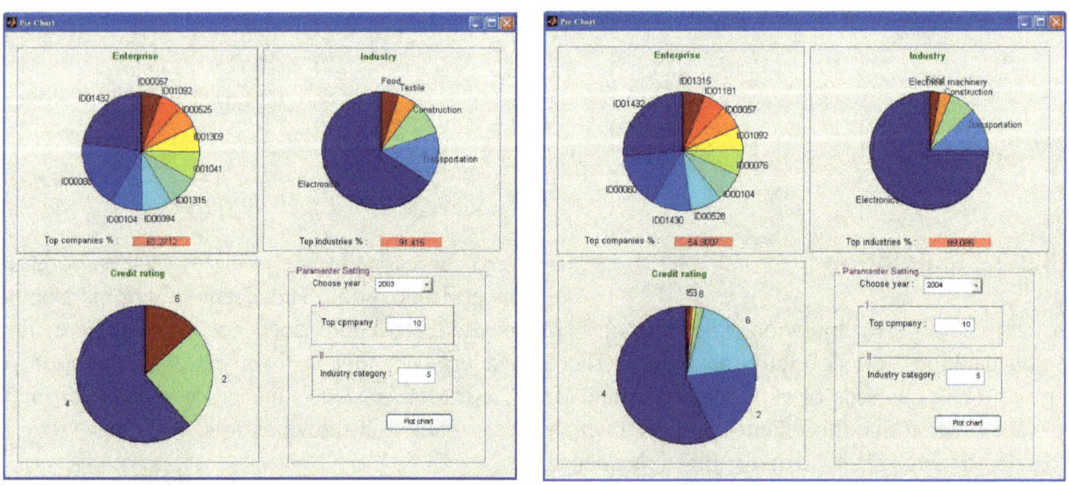

At the end of 2003 At the end of 2004

Fig. 15.4 Interface of part I

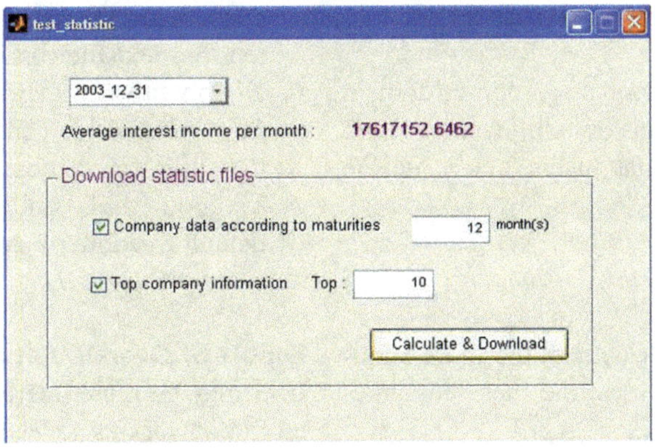

Fig. 15.5 Interface of part II

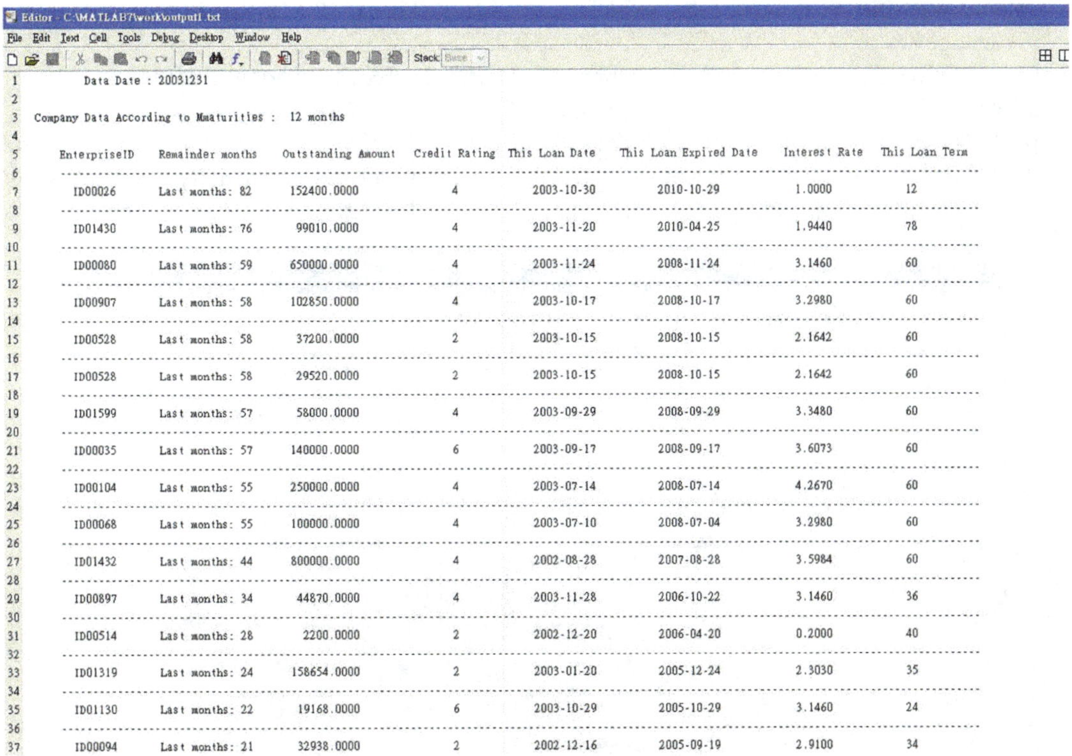

Editor - C:\MATLAB7\work\output1.txt

Data Date : 20031231

Company Data According to Mmaturities : 12 months

EnterpriseID	Remainder months	Outstanding Amount	Credit Rating	This Loan Date	This Loan Expired Date	Interest Rate	This Loan Term
ID00026	Last months: 82	152400.0000	4	2003-10-30	2010-10-29	1.0000	12
ID01430	Last months: 76	99010.0000	4	2003-11-20	2010-04-25	1.9440	78
ID00080	Last months: 59	650000.0000	4	2003-11-24	2008-11-24	3.1460	60
ID00907	Last months: 58	102850.0000	4	2003-10-17	2008-10-17	3.2980	60
ID00528	Last months: 58	37200.0000	2	2003-10-15	2008-10-15	2.1642	60
ID00528	Last months: 58	29520.0000	2	2003-10-15	2008-10-15	2.1642	60
ID01599	Last months: 57	58000.0000	4	2003-09-29	2008-09-29	3.3480	60
ID00035	Last months: 57	140000.0000	6	2003-09-17	2008-09-17	3.6073	60
ID00104	Last months: 55	250000.0000	4	2003-07-14	2008-07-14	4.2670	60
ID00068	Last months: 55	100000.0000	4	2003-07-10	2008-07-04	3.2980	60
ID01432	Last months: 44	800000.0000	4	2002-08-28	2007-08-28	3.5984	60
ID00897	Last months: 34	44870.0000	4	2003-11-28	2006-10-22	3.1460	36
ID00514	Last months: 28	2200.0000	2	2002-12-20	2006-04-20	0.2000	40
ID01319	Last months: 24	158654.0000	2	2003-01-20	2005-12-24	2.3030	35
ID01130	Last months: 22	19168.0000	6	2003-10-29	2005-10-29	3.1460	24
ID00094	Last months: 21	32938.0000	2	2002-12-16	2005-09-19	2.9100	34

Fig. 15.6 Companies' data downloads from part II interface

Set Criteria and Derive Fundamental Experimental Result

This portion is the core of the proposed tool; it provides several functions of computations. Here are the parameters that users must decide themselves:

1. Estimated year.
2. Confidence level.
3. Simulation times. Of course, the more simulation time user chooses, the more computational time will need.
4. Percentage of explained factors which is defined for PCA method. Using the eigenvalues of given normalized assets (equities) values, we can determinate the explained percentage.
5. This function gives user the option to estimate all or portion of the companies of portfolio pool. The portion part is sorted according to the loan amount of each enterprise. User can choose multiple companies they are most concerned. The computational result is written to a text file for further analysis.
6. Distribution of factors. This is defined for PCA method too. There are two distributions that user can choose—standard normal distribution and student t-distribution. The default freedom of student t-distribution is set as one (Fig. 15.7).

Report of Overall VaR Contributor

User may be more interested in the detail of risk profile at various levels. In this part, industries are discriminated from nineteen sections and credits are discriminated from nine level. This allows user to see where the risk is concentrated visually (Fig. 15.8).

15.3.3.2 Experimental Result and Discussion

Table 15.6 represents the experimental results. For objectivity, all simulations times are set to 100,000 times which is large enough to obtain

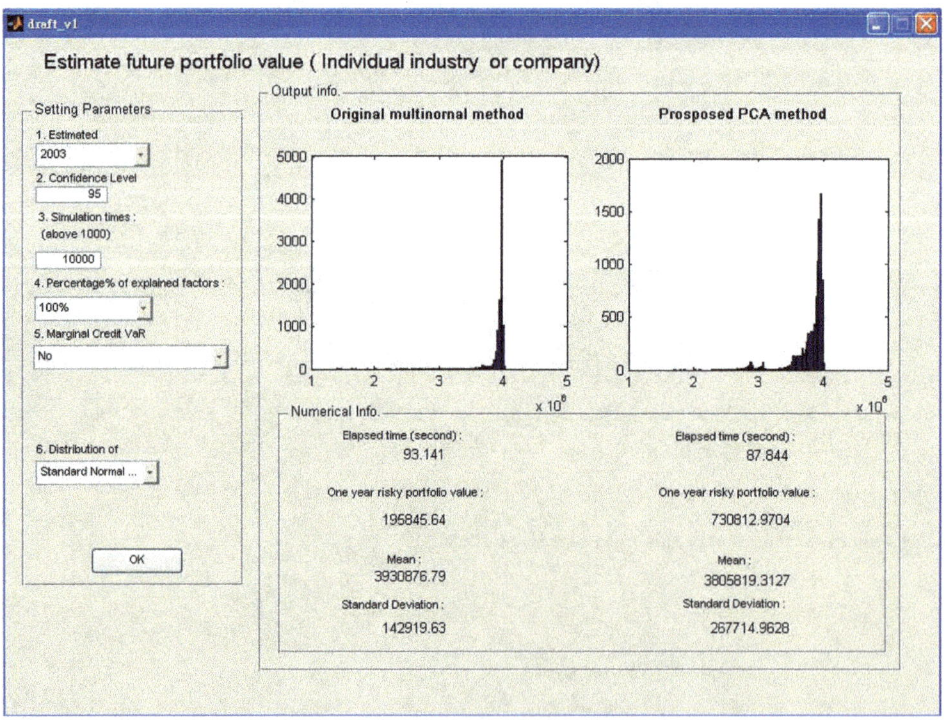

Fig. 15.7 Interface of part III

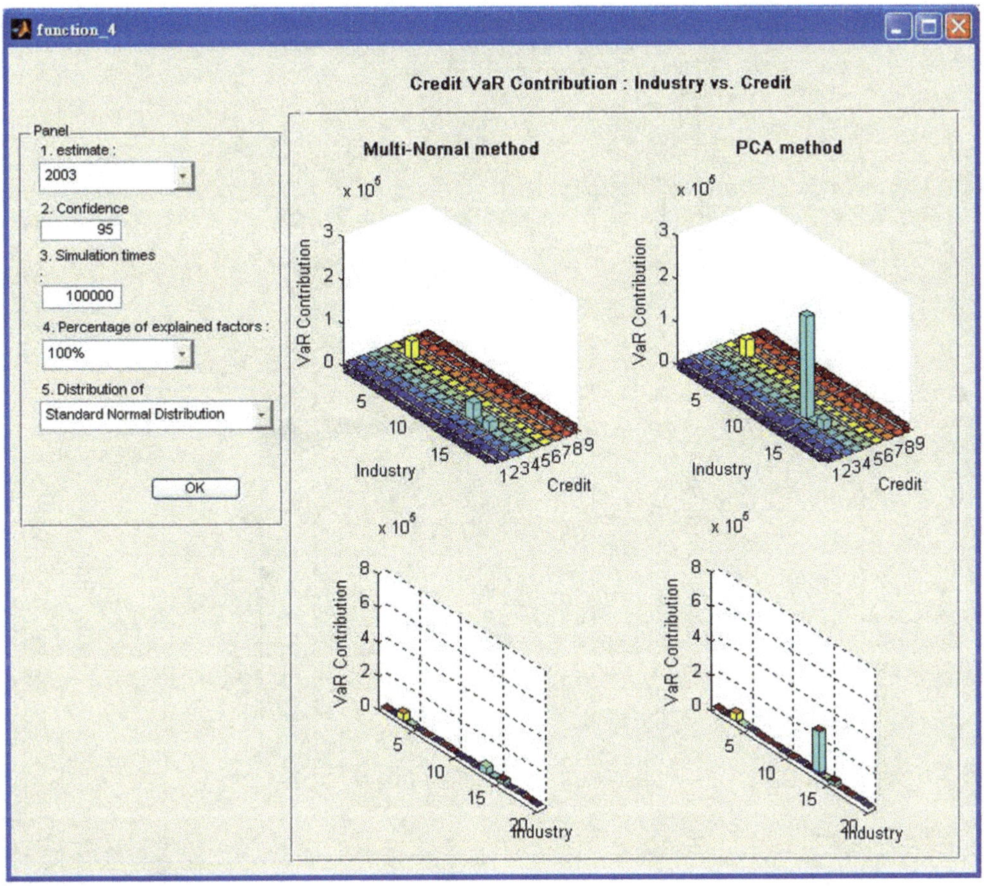

Fig. 15.8 VaR contribution of individual credits and industries

stable numerical results.[2] Based on the provided data, the one-year future portfolio value of listed corporations on 2003 and 2004 can be estimated. In other words, for instance, we stand on t_0 = January 1, 2002, to estimate the portfolio value at the time t_T = December 31, 2003. Similarly, we stand on t_0 = January 1, 2003, to estimate the portfolio value at the time t_T = December 31, 2004. The following tables list the experimental results of factor copula methods of different factor distributions and comparison with multinormal method by CreditMetrics. The head of the tables is parameter setting, and the remained fields are experimental results. We note here to the Eq. (15.15)

$$V_i = r_{i1}y_1 + r_{i2}y_2 + \ldots + r_{im}y_m + \varepsilon_i$$

the distribution of factors y_1, $y_2 \ldots y_m$ listed in following table are standard normally distributed and student t-distributed (assumes freedoms are 2, 5 and 10).

There are some messages that can be derived from above table. First, obviously, risk of future portfolio value by multinormal method is less than that of the proposed method. Risk amount of the proposed method is 3–5 times over multinormal method. This result corresponds to most research that copula function can capture the fat-tail phenomenon which prevails over practical market more adequately. Second, the distribution of future portfolio value by the proposed method is more diversified than multinormal method which concentrated on nearly 400,000

[2]We have examined the simulation times, 100,000 times is enough to have a stable computational result.

Table 15.6 Experimental result of estimated portfolio value at the end of 2003

Estimate year: 2003				
Parameter setting:				
Simulation time: *100,000*		Percentage of explained factors: *100.00%*		
Involved listed enterprise number: *40*		Loan account entries: *119*		
Result:				
Factor distribution assumption: Normal distribution $F \sim N(0, 1)$				
	Credit VaR 95%	Credit VaR 99%	Portfolio mean	Portfolio s.d.
Multinormal	192,113.4943	641,022.0124	3,931,003.1086	136,821.3770
PCA	726,778.6308	1,029,766.9285	3,812,565.6170	258,628.5713

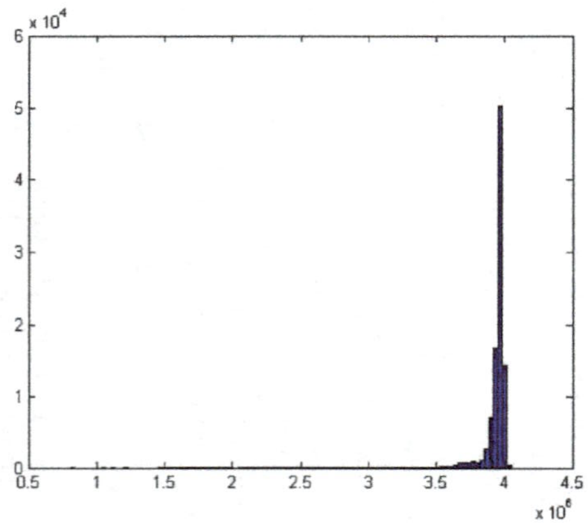

Multinormal method

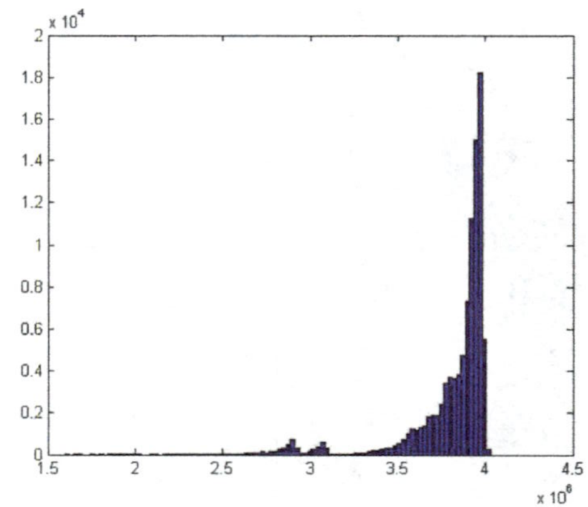

PCA method

Factor distribution assumption: student t-distribution, freedom = (2)				
	Credit VaR 95%	Credit VaR 99%	Portfolio mean	Portfolio s.d.
Multinormal	191,838.2019	620,603.6273	3,930,460.5935	136,405.9177
PCA	1,134,175.1655	1,825,884.8901	3,398,906.5097	579,328.2159

(continued)

Table 15.6 (continued)

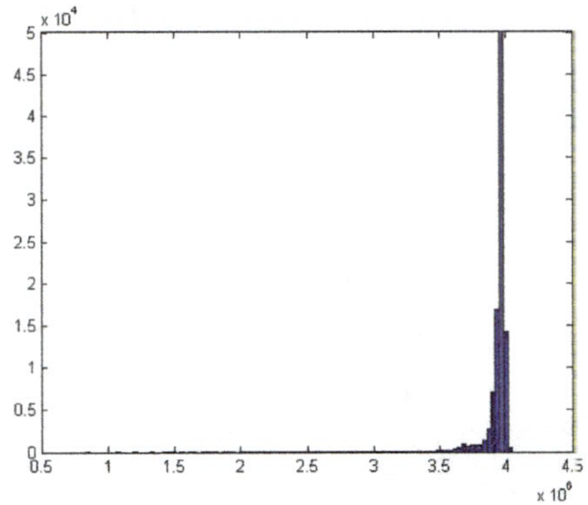

Multinormal method

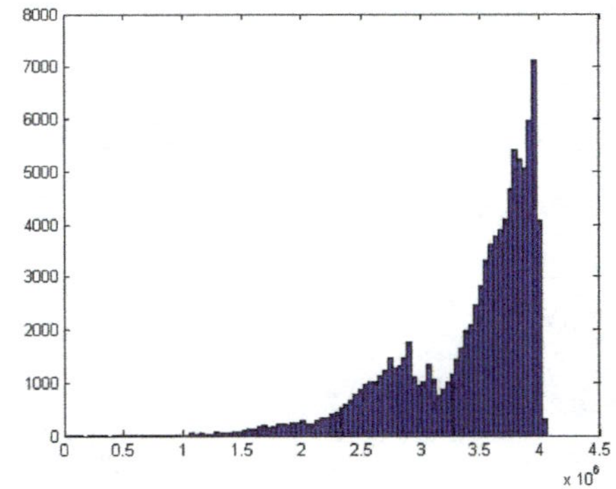

PCA method

Factor distribution assumption: student t-distribution, freedom = (5)

	Credit VaR 95%	Credit VaR 99%	Portfolio mean	Portfolio s.d.
Multinormal	192,758.7482	610,618.5048	3,930,923.6708	135,089.0618
PCA	839,129.6162	1,171,057.2562	3,728,010.5847	337,913.7886

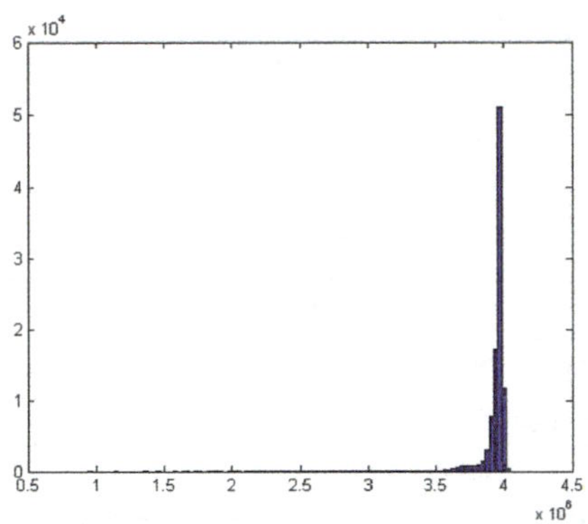

Multinormal method

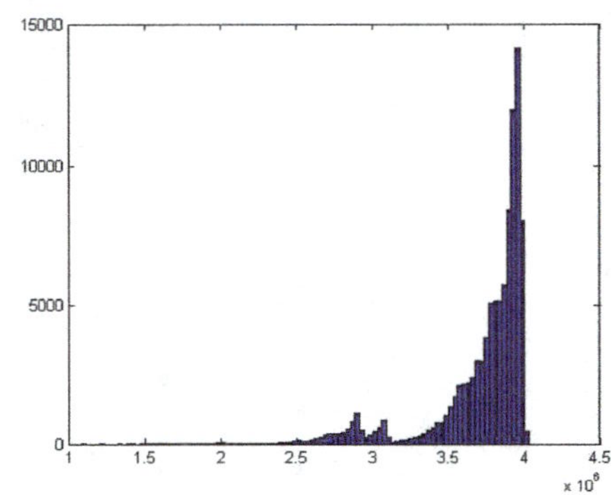

PCA method

(continued)

Table 15.6 (continued)

Factor distribution assumption: student t-distribution, freedom = (10)

	Credit VaR 95%	Credit VaR 99%	Portfolio mean	Portfolio s.d.
Multinormal	192,899.0228	600,121.1074	3,930,525.7612	137,470.3856
PCA	773,811.8411	1,080,769.3036	3,779,346.2750	291,769.4291

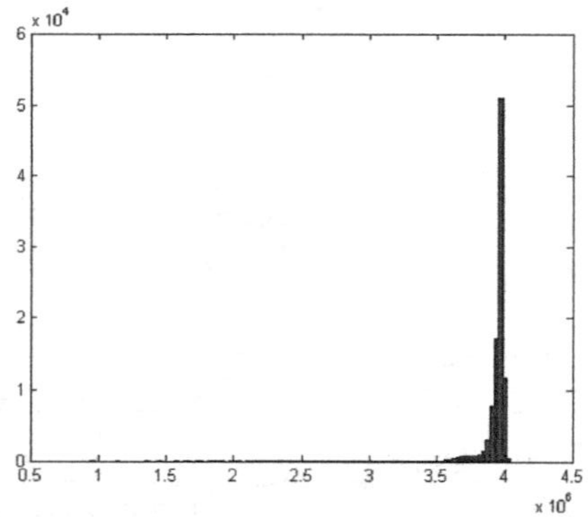

Multinormal method

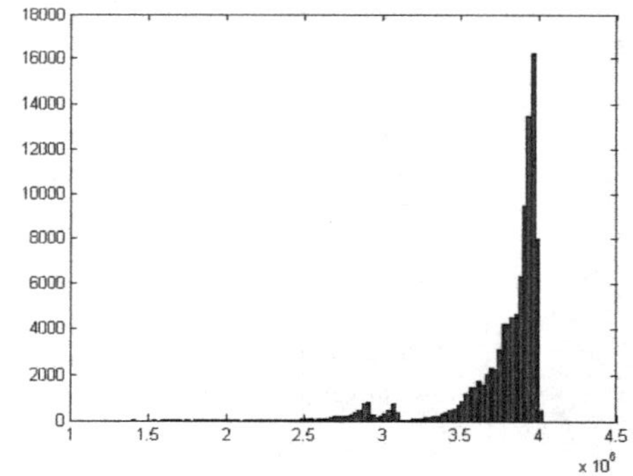

PCA method

Table 15.7 CPU time for factor computation (simulation time: 100,000 year: 2003)

Method	Explained ratio (s)				
	100%	90–95%	80–85%	70–80%	Below 60%
Multinormal	2.5470	2.6090	2.2350	2.2500	2.3444
PCA	1.2030	0.7810	0.7030	0.6720	0.6090

Table 15.8 Individual credit VaR of top 5 industries

Industry	Credit VaR	
	Multinormal method	PCA method
(No. 1) Electronics	40,341	252,980
(No. 2) Plastic	42,259	42,049
(No. 3) Transportation	22,752	22,391
(No. 4) Construction	7011	7007
(No. 5) Textile	2884	2765

Table 15.9 Estimate portfolio value at the end of 2004 with different explained levels

95% Confidence level, $F \sim (0, 1)$					
	100%	90–95%	80–85%	70–80%	60–70%
Multinormal	208,329.40	208,684.38	209,079.72	208,686.22	207,710.63
PCA	699,892.33	237,612.60	200,057.74	187,717.73	183,894.43

with 50,000 times while the proposed method with 17,000 times. Third, it is very clear to see that risks with factors using student t-distribution to simulate is more than with normal distribution, and the risk amount tends toward the same, while the degree of freedom becomes larger. Fourth, the mean of portfolio of the proposed method is smaller than of multinormal method, but the standard deviation of the proposed method is much more than multinormal method. It shows the overall possible portfolio values by the proposed method have the trend to become less worth and also fluctuate more rapidly.

The above discrepancies between two methods give us some inferences. First, the proposed method provides another way to estimate more actual credit risks of portfolio containing risky credits through market data, and this method captures fat-tail event more notably. Second, the computation time of the proposed method is shorter than multinormal method. Following Table 15.7, when using fully explained factors, computation by the proposed method is still faster than that of the multinormal method. The computation time decreases as the required explained ratio sets lower; this means less number of factors are used for the expected explained level. Third, Table 15.8 which retrieves individual Credit VaR contribution to whole portfolio from 19 industries shows the main risk comes from electronics industry. Based on the commercial data, we can find out that among its loan account entries, the electronics industry customers have the proportion of exceeding half of loan entries (63/119). The Credit VaR of electronics industry computed by the proposed method is six times more than by multinormal method. This effect reveals that the multinormal method lacks the ability to catch concentrative risks. On the contrary, based on the factor structure, the mutual factor loadings extracted by the correlation among companies express more actual risks. Forth, for finite degree of freedom, the t-distribution has fatter tails than Gaussian distribution and is known to generate tail dependence in the joint distribution.

Table 15.9 shows the impact on risks amount by using different factor numbers. According to

Table 15.9, the risks decrease as the explained level decreases; this is a trade-off between time consuming and afforded risk amount. Most research and reports say 80% explained level is large enough to be accepted.

15.4 Conclusion

Credit risks and default correlation issues have been probed in recent research, and many solutions have been proposed. We take another view to examine credit risks and derivative tasks. In our perspective, the loan credits in target portfolio like the widely different risk characteristics from a portfolio of debt instruments, their properties, and behavior are the same in the main.

In this chapter, we propose a new approach which connects the principal component analysis and copula functions to estimate credit risks of bank loan businesses. The advantage of this approach is that we do not need to specify particular copula functions to describe dependence structure among credits. On the contrary, we use a factor structure which covers market factor and idiosyncratic factor and the computed risks have heavy tail phenomenon. Another benefit is to reduce the difficulties to estimate parameters which copula functions will use. This approach provides another way and has better performance than the conventional method such as assuming the dependence structures are normally distributed.

In order to describe the risk features and other messages that bank policymakers may like to know, we wrote a tool for risk estimation and result display. It contains basic data information preview which just downloads data from database and does some statistical analyses. It also provides different parameter setting, uses Monte Carlo simulation to calculate credit VaR, and finally gives an overview of individual credit VaR contributions. The experimental results are consistent with previous studies that the risk will be underestimated compared with real risks if people assume dependence structure is normally distributed. In addition, the aforementioned approach and tool still have some rooms to be improved such as recovery rate estimations, how to chosen distributions of factors, and more friendly user interface.

Bibliography

Andersen, L., Sidenius, J. (2004). Extensions to the gaussian copula: Random recovery and random factor loadings. *Journal of Credit Risk, 1*(1), 29–70.

Bouye, E., Durrleman, V., Nikeghbali, A., Riboulet, G., & Roncalli, T. (2000). *Copulas for finance—A reading guide and some application*. Groupe de Recherche.

Chang, J. R., & Chen, A. C. (2010). Copula, correlated defaults, and credit VaR. In *Handbook of quantitative finance and risk management* (pp. 697–711). New York: Springer.

Embrechts, P., McNeil, A., & Straumann, D. (1999). *Correlation and dependence in risk management: Properties and pitfalls*. ETHZ Zentrum, Zurich: Mimeo.

Gollinger, T. L., & Morga, J. B. (1993). Calculation of an efficient frontier for a commercial loan portfolio. *Journal of Portfolio Management*, 39–46.

Gupton, G. M., Finger, C. C., & Bhatia, M. (1997). *CreditMetrics–Technical document*. New York: Morgan Guaranty Trust Company.

Hull, J., & White, A. (2004). Valuing of a CDO and an n-th to default CDS without Monte Carlo simulation. *Journal of Derivatives, 12*(2), 8–48.

Jorion, P. (2000). *Value at risk*. New York: McGraw Hill.

Lee, C.-F., Lee, A. C., & Lee, J. (2010). *Handbook of quantitative finance and risk management*. New York: Springer.

Li, D. X. (2000). On default correlation: A copula function approach. *Journal of Fixed Income, 9*, 43–54.

Shen, D. B., Jing, Y. K., & Tsia, J. C. (2003). Research of Taiwan recovery rate with TEJ Data Bank. Working Paper.

Sklar, A. (1959). Functions de repartition an n dimension et leurs marges. *Publications de l'Institut Statistique de l'Université de Paris, 8*, 229–231.

Stevenson, B. G., & Fadil, M. W. (1995). Modern portfolio theory: Can it work for commercial loans? *Commercial Lending Review, 10*(2), 4–12.

Yang, T.-C. (2005). *The pricing of credit risk of portfolio base on listed corporation in Taiwan market* (Master thesis). National Tsing Hua University, Taiwan.

Multivariate Analysis: Discriminant Analysis and Factor Analysis

16

Contents

Abstract

In this chapter, we discuss two multivariate analysis models, which include discriminant analysis and factor analysis. In addition, we discuss principal component analysis. The derivations of both discriminant analysis and principal component analysis are presented in Appendices 1 and 2. It is also shown that two groups of discriminant analysis can be analyzed in terms of dummy regression analysis.

16.1 Introduction

Financial ratios are widely used in all financial analysis and planning. Banks use a firm's current and quick ratio to determine acceptability, for commercial loans; the leverage ratio is used as a proxy for a firm's capital measure in predicting bankruptcy and to analyze the impact of leverage

This chapter draws upon Chap. 3 of the Book Entitled, Financial Analysis, Planning and Forecasting by Lee et al. (2017).

C.-F. Lee et al., *Financial Econometrics, Mathematics and Statistics*,
https://doi.org/10.1007/978-1-4939-9429-8_16

on the market value of a firm. Furthermore, for financial planning and forecasting, firm managers use activity ratios, that is, the asset turnover ratio and the inventory turnover ratio, to determine the total amount of assets required to sustain a level of activity. In financial analysis and planning determination, lenders or managers need to measure a customer's (either an individual's or a firm's) short-term or long-term financial position. The well-known statistical techniques of factor analysis and discriminant analysis can be used in such instances to identify important financial ratios and to construct an overall financial indicator, that is, a "financial *z*-score."

Although the measurement of financial *z-scores* is a compromise between theory and practice, *z*-scores have been used extensively by practitioners and academicians in credit analysis, financial distress determination, and bankruptcy prediction. Factor analysis has been used to determine important financial ratios and in testing other finance-related issues. Two-group discriminant analysis and *k*-group discriminant analysis have been applied to bond-rating analysis as well. Other multivariate analysis techniques gaining wide acceptance in both investment analysis and financial management are principal components and cluster analysis.[1]

The theory and methodology of factor analysis and discriminant analysis are explored in this chapter. In Sect. 16.2, the linear algebra needed for factor analysis, discriminant analysis, and portfolio analysis is reviewed in accordance with the basic concepts of algebra. In Sect. 16.3, the theory and methodology of two-group discriminant analysis will be explored in accordance with both the dummy regression method and the analysis-of-variance (eigenvalue) method. Section 16.4 will discuss the theory and methodology of *k-group* discriminant analysis. In Sect. 16.5, the theory and the methodology of principal component and factor analysis are investigated. Finally, in Sect. 16.6, the results of this chapter are summarized. In addition, Appendix 1 discusses the relationship between discriminant analysis and dummy regression analysis. Appendix 2 discusses principle-component analysis.

16.2 Important Concepts of Linear Algebra

In performing financial analysis and planning, the most important concepts of linear algebra that are needed are: (i) linear combination and its distribution, (ii) operation of vectors and matrices, and (iii) the linear equation system and its solution.

Linear Combination and its Distribution

If $x_1, x_2, \ldots x_n$ are one set of variables, then a linear combination of these variables is:

$$Y = a_1x_1 + a_2x_2 + \cdots + a_nx_n. \tag{16.1}$$

In financial analysis, $x_1, x_2, \ldots x_n$ can be used to represent amounts of *i* products ($i = 1, 2, \ldots, n$) to be purchased. The a_i coefficients ($i = 1, 2, \ldots, n$) can be used to represent the net profit of producing one unit of product *i*; *Y* can be used to represent the total profit of a firm. The variables of linear combination will be used as an objective function (the function to, be minimized or maximized) for (i) portfolio analysis, (ii) linear programming in performing capital rationing, and (iii) financial analysis, planning, and forecasting.[2] In both factor analysis and discriminant analysis, the variables used to obtain a linear combination are generally random instead of deterministic. Hence, the distribution of a linear combination is required for performing empirical analysis.

[1]Principal component analysis is one of the major factor-analytic techniques for summarizing multivariate data; cluster analysis is an appropriate set of techniques for summarizing multivariate data; cluster analysis is an appropriate set of techniques to partition the data set into homogeneous subsets of objects. See Green and Tull (1978) for detail.

[2]For the detailed discussion of linear programming in performing capital rationing and financial analysis, planning, and forecasting, please check Chaps. 12, 21, 22, and 23 of Lee et al. (2017).

Linear discriminant analysis, which will be discussed in Sects. 16.3 and 16.4, is a linear combination of a set of random variables.

In calculating the financial z-score, Eq. (16.1) can be rewritten as:

$$\tilde{z} = a_1\tilde{x}_1 + a_2\tilde{x}_2 + \cdots + a_m\tilde{x}_m, \qquad (16.1')$$

where the $\tilde{x}_i$'s $(i = 1, 2, \ldots, m)$ represent the related financial ratios; $\tilde{z}$ is the financial z-score. The financial ratio discussed in Chap. 2, which is used to compute the financial z-score, can be either normally or log-normally distributed. If the $\tilde{x}_i$'s are normally distributed, then Anderson (2003) and others show that i is normally distributed. If $\bar{x}_i$, σ_i^2 are the mean and the variance *for* $\tilde{x}_i$, respectively, and ρ_{ij} is the correlation coefficient between $\tilde{x}_i$ and $\tilde{x}_j$, then the mean and variance of $\tilde{z}$ can be defined as,

$$\bar{z} = \sum_{i=1}^{m} a_i\bar{x}_i, \qquad (16.2a)$$

$$\sigma_{\tilde{z}}^2 = \sum_{i=1}^{m} a_i^2\sigma_i^2 + 2\sum\sum_{i>j} a_i a_j \rho_{ij}\sigma_i\sigma_j \qquad (16.2b)$$

where the symbol $\sum\sum_{i>j}$ denotes summation over all possible pairs of i and j values in the range *from* 1 through m, with the restriction that i is at least one greater than j. If $i = 2$,

$$\sigma_z^2 = a_1^2\sigma_1^2 + a_2^2\sigma_2^2 + 2a_1a_2\rho_{12}\sigma_1\sigma_2 \quad (\rho_{12}\sigma_1\sigma_2 = \sigma_{12}) \qquad (16.2b')$$

If $i = 3$, then:

$$\sigma_z^2 = a_1^2\sigma_1^2 + a_2^2\sigma_2^2 + a_3^2\sigma_3^2 + 2a_1a_2\rho_{12}\sigma_1\sigma_2 + 2a_1a_3\rho_{32}\sigma_1\sigma_{32} + 2a_3a_2\rho_{23}\sigma_3\sigma_2 \qquad (16.2b'')$$

Vectors, Matrices, and Their Operations

In estimating financial z-scores, we need time-series ratio data in order to estimate the coefficients a_1, a_2, …, a_m. Under this circumstance, the X_i's (i = 1, 2, …, m) are vectors. If the current ratio (X_1) and the leverage ratio (X_2) are the only two ratios to be used in estimating z, then time-series financial ratio data can be written in terms of vectors as

$$X_1 = \begin{bmatrix} X_{11} \\ X_{12} \\ \vdots \\ X_{1n} \end{bmatrix} \text{ and } X_2 = \begin{bmatrix} X_{21} \\ X_{22} \\ \vdots \\ X_{2n} \end{bmatrix},$$

where X_{ij} represents the ratio i in time period j. Vector X_1 and vector X_2 can be used to formulate a matrix of the ratios used in computing the financial z-score:

$$X = \begin{bmatrix} X_{11} & X_{12} \\ X_{12} & X_{22} \\ \vdots & \vdots \\ X_{1n} & X_{n2} \end{bmatrix}$$

A matrix is a rectangular array of numbers. Matrix X is a $n \times 2$ matrix because it has n rows and two columns.

To represent all observations of financial ratios for either factor or discriminant analysis, Matrix $\boldsymbol{X}$ can be generalized as

$$X = \begin{bmatrix} X_{11} & X_{21} & \cdots & X_{11} \\ X_{21} & X_{22} & \cdots & X_{1m} \\ \vdots & \vdots & \cdots & X_{2m} \\ X_{n1} & X_{12} & \cdots & X_{nm} \end{bmatrix} \qquad (16.3)$$

Now X is a $n \times m$ matrix. A computer generally uses this type of matrix to store ratio information for performing related analyses.

In portfolio analysis, the variance of a portfolio can be written as Eq. (16.2b″). In vector and matrix notation, Eq. (16.2b″) can be written as

$$\sigma^2 = \underset{1\times 3}{\overset{A'}{[a_1a_2a_3]}} \ \underset{3\times 3}{\overset{B}{\begin{bmatrix} \sigma_1^2 & \sigma_{12} & \sigma_{13} \\ \sigma_{21} & \sigma_2^2 & \sigma_{23} \\ \sigma_{31} & \sigma_{32} & \sigma_3^2 \end{bmatrix}}} \ \underset{3\times 1}{\overset{A}{\begin{bmatrix} a_1 \\ a_2 \\ a_3 \end{bmatrix}}}, \qquad (16.2b''')$$

where B is a 3×3 covariance matrix, and σ_{ij} represents the covariance between x_i and x_j.

In Eq. (16.2b‴), A is a 3 × 1 coefficient vector and A' is the transposition of A. A *transpose* of a matrix A is defined to be a matrix obtained by interchanging the corresponding rows and columns *of* A, that is, first with first, second with second, and so on. Here, we have only one row and three columns, so taking the transpose of A is a fairly simple operation. For example, if

$$B = \begin{bmatrix} 3 \\ 5 \\ 6 \end{bmatrix},$$

then $B' = (3, 5, 6)$. Similarly, if

$$A = \begin{bmatrix} a_1 \\ a_2 \\ a_3 \end{bmatrix},$$

then $A' = (a_1, a_2, a_3)$.

The multiplicative rule of matrices and vectors is used to show that Eq. (16.2b‴) is indeed equal to Eq. (16.2b″). The rule for matrix multiplication requires "row–column multiplication."

In order for any two matrices to be multiplied together, the number of columns in one must be equal to the number of rows in another. For example:

$$\underset{(1\times 3)}{A = [2 \quad 1 \quad 0]}, \quad B = \underset{(3\times 2)}{\begin{bmatrix} 2 & 0 \\ 0 & 3 \\ -1 & 0 \end{bmatrix}}.$$

These two matrices can be multiplied together. An element in the ith row and jth column of the product AB is obtained by multiplying the ith row of A by the jth column of B. Therefore:

$$AB = [2 \quad 1 \quad 0] \begin{bmatrix} 2 & 0 \\ 0 & 3 \\ -1 & 0 \end{bmatrix} = [4 \quad 3].$$

We obtained the first element of AB from (2 × 2) + (1 × 0) + (−1 × 0) = 4. To see how this is done for Eq. (16.2b‴),

Step 1: Multiply A' by B; we then have

$$C = [(a_1\sigma_1^2 + a_2\sigma_{21} + a_3\sigma_{31}), (a_1\sigma_{12} + a_2\sigma_2^2 + a_3\sigma_{32}), (a_1\sigma_{31} + a_2\sigma_{23} + a_3\sigma_3^2)].$$

Step 2: Multiply C by A, and we get:

$$\begin{aligned}\sigma^2 &= a_1(a_1\sigma_1^2 + a_2\sigma_{21} + a_3\sigma_{31}) + a_2(a_1\sigma_{12} \\ &\quad + a_2\sigma_2^2 + a_3\sigma_{32}) + a_3(a_1\sigma_{31} + a_2\sigma_{23} + a_3\sigma_3^2) \\ &= a_1^2\sigma_1^2 + a_2^2\sigma_2^2 + a_3^2\sigma_3^2 + 2(a_1a_2\sigma_{12} \\ &\quad + a_1a_3\sigma_{13} + a_2a_3\sigma_{23}).\end{aligned}$$

Linear Equation System and its Solution

A general linear equation system can be defined as

$$\begin{aligned} a_{11}X_1 + \cdots + a_{1m}X_m &= b_1 \\ &\vdots \\ a_{n1}X_1 + \cdots + a_{nm}X_m &= b_n. \end{aligned} \tag{16.4}$$

In matrix formulation, Eq. (16.4) can be written as $AX = b$, or

$$\begin{bmatrix} a_{11} \cdots & a_{1m} \\ \vdots & \vdots \\ a_{n1} \cdots & a_{nm} \end{bmatrix} \begin{bmatrix} X_1 \\ \vdots \\ X_m \end{bmatrix} = \begin{bmatrix} b_1 \\ \vdots \\ b_n \end{bmatrix}.$$

The solution of Eq. (16.4) can be obtained either by the derivation of the inversion of A or by using Cramer's rule.[3]

In order to take the inverse of a matrix, the matrix must be a nonsingular square matrix. That is, the number of rows must equal the number of columns. Then, the following condition must be satisfied:

$$A^{-1}A = I,$$

where I is the identity matrix.

[3]The procedure of inversion will be discussed in the example in this section.

An identity matrix is a matrix such that:

$$I = \begin{bmatrix} 1 & 0 & \cdots & 0 \\ 0 & 1 & & 0 \\ \vdots & & & \vdots \\ 0 & & \cdots & 1 \end{bmatrix}.$$

If the inversion of coefficient matrix A is used, then, the solution of X can be defined as

$$X = A^{-1}b, \qquad (16.5)$$

A^{-1} can generally be obtained by using a computer program. Alternatively, Cramer's rule can be used to obtain the solution as defined in Eq. (16.6):

$$X_i = \frac{|\hat{A}_i|}{|A|} \qquad (16.6)$$

where $\hat{A}_i$ is the matrix obtained from A by replacing the ith column with the constant vector. Both $|\hat{A}_i|$ and $|A|$ represent the determinants. A determinant of a matrix is the value of a matrix. An example is:

$$A = \begin{bmatrix} a_{11} & a_{12} \\ a_{21} & a_{22} \end{bmatrix}, \quad |A| = a_{11}a_{22} - a_{12}a_{21}.$$

If both $\hat{A}_i$ and A are square matrices, then the determinants of $\hat{A}_i$ and A are unique numbers associated with these matrices. Numerically, if

$$A = \begin{bmatrix} 2 & 8 \\ 7 & 6 \end{bmatrix},$$

then $|A| = (2)(6) - (7)(8) = -44$.

Equation (16.4) is a general equation system. If all elements of $(b_1 \ldots b_n)$ are zero, then it is a homogeneous equation system. A special case of homogeneous equation systems can be defined as

$$(A - \lambda I)X = 0, \qquad (16.7)$$

where A and X are identical to those defined in Eq. (16.4), A is an unknown scalar quantity, and I is an $n \times m$ matrix with all unity elements in the diagonal elements. There is a trivial solution $(X = 0)$ and a nontrivial solution $\left((A - \lambda I)X = \hat{0}\right)$ for this set of homogeneous equations.

Tatsuoka (1988), Moore (1968), Anton (2004), and others show that the condition of Eq. (16.7) poses a nontrivial solution that is

$$|A - \lambda I| = 0, \qquad (16.8)$$

which is called the *characteristic* equation of matrix A.

Conceptually, the existence of this characteristic equation can be justified as follows:

If $|A - \lambda I| \neq 0$, then $A - \lambda I$ is not a singular matrix, and hence, it possesses an inverse; then premultiplying both sides of Eq. (16.7) by $(A - \lambda I)^{-1}$ will yield $X = (A - \lambda I)^{-1}0 = 0$; that is, the trivial solution is the only solution of the equation. We therefore conclude that in order for a set of homogeneous equations to possess a nontrivial solution, there must exist a characteristic solution as defined in Eq. (16.8).

To obtain the nontrivial solution, we should first find the unknown scalar quantity A. The scalar A is called an eigenvalue of A, and X is said to be an eigenvector corresponding to A. One of the meanings of the word "eigen" in German is "proper"; eigenvalues are also called proper values, characteristic values; or latent roots, by some writers.

A numerical example is now used to show how eigenvalue can be calculated. The eigenvalue is the characteristic root associated with $|A - \lambda I| = 0$. To find the eigenvalue of the matrix

$$A = \begin{bmatrix} 3 & 2 & 4 \\ 2 & 0 & 2 \\ 4 & 2 & 3 \end{bmatrix},$$

we note that the characteristic equation of A is

$$\begin{bmatrix} 3-\lambda & 2 & 4 \\ 2 & -\lambda & 2 \\ 4 & 2 & 3-\lambda \end{bmatrix} = 0.$$

This determinantal equation can be easily reduced to $(1 + \lambda)^2 = 0$ or $(\lambda - 8) = 0$. This implies that there are three roots for A: $\lambda_1 = -1$, $\lambda_2 = -1$, and $\lambda_3 = 8$. This concept and this method of eigenvalues and the eigenvector are useful in understanding discriminant analysis and principal component analysis.

As an example, we will show how the eigenvector associated with $\lambda_3 = 8$ can be calculated:

(a) To find $A - 8I$:

$$A - 8I = \begin{bmatrix} 3-\lambda & 2 & 4 \\ 2 & -\lambda & 2 \\ 4 & 2 & 3-\lambda \end{bmatrix}.$$

(b) To find the adjoint of the above matrix:

(i) Calculate the cofactor of $A - 8I$:

Cofactor of $A - 8I$

$$= \begin{bmatrix} (-1)^{1+1}(36) & (-1)^{1+2}(-18) & (-1)^{1+3}(36) \\ (-1)^{2+1}(-18) & (-1)^{2+2}(9) & (-1)^{2+3}(-18) \\ (-1)^{3+1}(36) & (-1)^{3+2}(-18) & (-1)^{3+3}(36) \end{bmatrix}$$

$$= \begin{bmatrix} 36 & 18 & 36 \\ 18 & 9 & 18 \\ 36 & 18 & 36 \end{bmatrix}.$$

(ii) Adjoint of $A - 8I = \begin{bmatrix} 36 & 18 & 36 \\ 18 & 9 & 18 \\ 36 & 18 & 36 \end{bmatrix}$.

Note that all these columns are proportional; this fact offers a partial check on the calculation. (In this case, the first and third columns are identical, but this will not be true in general.) Before carrying out the last, step, the reader should verify that:

$$(A - 8I)\begin{bmatrix} 36 \\ 18 \\ 36 \end{bmatrix} = 0 \text{ or that } A\begin{bmatrix} 36 \\ 18 \\ 36 \end{bmatrix} = 8I\begin{bmatrix} 36 \\ 18 \\ 36 \end{bmatrix}.$$

The eigenvector is one vector solution of x corresponding to the eigenvalue 8.

(c) In order to have x satisfying the limit norm condition, $x'x = 1$, we divided each element of

$$\begin{bmatrix} 36 \\ 18 \\ 36 \end{bmatrix}$$

by $\sqrt{36^2 + 18^2 + 36^2} = 54$, and obtain

$$x' = \left(\tfrac{36}{54} \quad \tfrac{18}{54} \quad \tfrac{36}{54}\right) = [0.6667 \quad 0.3333 \quad 0.6667]$$

This example has shown how the solution of a special kind of homogeneous equation system [as indicated in Eq. (16.7)] can be solved. This kind of equation system differs in two respects from the regular equation system of Eq. (16.4). First, the vector of constraint on the right-hand side is a null vector. Second, the matrix of coefficients on the left-hand side, $A - \lambda I$, itself involves an unknown scalar quantity A. Therefore, in order to solve x, we should first solve λ by using the characteristic equation of matrix A' as indicated in Eq. (16.8).

Alternatively, x can be solved by the equation system as:

$$\begin{aligned} -5x_1 + 2x_2 + 4x_3 &= 0, \\ 2x_1 - 8x_2 + 2x_3 &= 0, \\ 4x_1 + 2x_2 - 5x_3 &= 0. \end{aligned}$$

The solution of this equation system is: $2x_1 = x_2 = 2x_3$. Hence, ($2d$, d, $2d$) is the eigenvector associated with the eigenvalue $\lambda = 8$ (where d is a constant). This implies that one element in the eigenvector is arbitrary. We may normalize the vector by setting its length at unity, $x_1^2 + x_2^2 + x_3^2 = 1$, and obtain the normalized solution as indicated above.

The concepts and methods of solving for eigenvalue and eigenvectors discussed in this section are important to understanding both discriminant analysis and factor analysis, to be discussed in later sections in this chapter.

Now the inversion of matrix A is discussed, by the definition

$$A^{-1} = \frac{1}{|A|}\,(\text{Adjoint A}); \tag{16.9}$$

by using the previous example, we have

$$|A| = \begin{vmatrix} 3 & 2 & 4 \\ 2 & 0 & 2 \\ 4 & 2 & 3 \end{vmatrix} = (0+16+16)-(0+12+12) = 8.$$

Cofactor of A

$$= \begin{bmatrix} \begin{matrix}1+1\\(-1)(-4)\end{matrix} & \begin{matrix}1+2\\(-1)(-2)\end{matrix} & \begin{matrix}1+3\\(-1)(4)\end{matrix} \\ \begin{matrix}2+1\\(-1)(-2)\end{matrix} & \begin{matrix}2+2\\(-1)(-7)\end{matrix} & \begin{matrix}2+3\\(-1)(-2)\end{matrix} \\ \begin{matrix}3+1\\(-1)(4)\end{matrix} & \begin{matrix}3+2\\(-1)(-4)\end{matrix} & \begin{matrix}3+3\\(-1)(-4)\end{matrix} \end{bmatrix} = \begin{bmatrix} -4 & 2 & 4 \\ 2 & -7 & 2 \\ 4 & 4 & -4 \end{bmatrix}.$$

$$\text{Adjoint of } A = \begin{bmatrix} -4 & 2 & 4 \\ 2 & -7 & 4 \\ 4 & 2 & -4 \end{bmatrix},$$

And therefore

$$A^{-1} = \begin{bmatrix} -\frac{1}{2} & \frac{1}{4} & \frac{1}{2} \\ \frac{1}{4} & -\frac{7}{8} & \frac{1}{2} \\ \frac{1}{2} & \frac{1}{4} & -\frac{1}{2} \end{bmatrix}.$$

16.3 Two-Group Discriminant Analysis

Following Eq. (16.1), a linear two-group discriminant function can be defined as

$$Y_i = a_1 x_{1i} + a_2 x_{2i} + \cdots + a_m x_{mi}, \tag{16.10}$$

where Y_i is used to indicate two alternative options, and $x_{1i}, x_{2i}, \ldots, x_{mi}$ are explanatory variables; Y_i is a binary variable. In credit analysis, Y_i can be used to represent good and bad accounts; in corporate bankruptcy analysis, to represent bankrupt and nonbankrupt firms; and in banking analysis, to present the problem and nonproblem banks. Two different methods can be used to estimate the coefficients of Eq. (16.10). These two methods are the dummy regression method or the eigenvalue method.

It is important for readers to understand the relationship between the logic of two-group discriminant analysis and the multiple-regression technique to estimate related discriminant function parameters.

The purposes of discriminant analysis are (1) to test for mean group differences and to describe the overlaps among the groups, and (2) to construct a classification scheme based upon a set of m variables in order to assign previously unclassified observations to appropriate groups. For example, in a study of corporate bankruptcy, Altman (1968) used data from a sample of failed firms and a sample of existing firms to determine whether, on average, bankrupt firms had significantly different financial ratios prior to failure than did solvent firms. When his statistical tests indicated significant differences between the two groups, Altman then developed a classification rule that used financial ratios to predict potential corporate failures.

Following Tatsuoka (1988), Johnston and Dinardo (1996), and Eisenbeis and Avery (1972), the basic equation of discriminant analysis as derived in Appendix 1 can be defined as

$$\begin{aligned} &(B - EC)A = 0, \\ &D' = [\overline{X}_{1,1} - \overline{X}_{1,2} \ldots, \overline{X}_{m,1} - \overline{X}_{m,2}], \end{aligned} \tag{16.11}$$

where

B = DD', between-group variance;
C = Within-group variance;
A = Coefficient vector representing the coefficients of Eq. (16.8);
E = Ratio of the weighted between-group variance to the pooled within variance.

Table 16.1 Roster of liquidity and leverage ratios for two groups with two predictors and a "dummy" criterion variable Y

Group 1			Group 2		
$[N_1 = 6]$			$[N_2 = 8]$		
x_{1i}	x_{2i}	Y_i	x_{1i}	x_{2i}	Y_i
2.0	0.50	1	1.8	0.35	0
1.8	0.48	1	1.9	0.34	0
2.3	0.49	1	1.7	0.42	0
3.1	0.41	1	1.5	0.49	0
1.9	0.43	1	2.2	0.36	0
2.5	0.44	1	2.8	0.38	0
			1.6	0.55	0
			1.4	0.56	0
$\sum x_{1i} = 13.6$ $\sum x_{1i}^2 = 32$ $\sum x_{1i}x_{2i} = 6.179$			$\sum x_{2i} = 2.75$ $\sum x_{2i}^2 = 1.2671$		
$\sum x_{1i} = 14.9$ $\sum x_{1i}^2 = 29.19$ $\sum x_{1i}x_{2i} = 6.245$			$\sum x_{2i} = 3.45$ $\sum x_{2i}^2 = 1.5447$		

Since Eq. (16.11) is similar to Eq. (16.7), the characteristic equation associated with Eq. (16.10) is

$$(C^{-1}B - EI)A = 0 \tag{16.12}$$

In order to use the linear discriminant function for empirical analysis, one must estimate the coefficients of Eq. (16.11). To illustrate the computation of two-group discriminant functions as a multiple regression equation and the eigenvalue method of discriminant analysis, we shall use a numerical example as indicated in Table 16.1. Table 16.1 shows the number scores of two groups on two predictor variables, X_1 [liquidity ratio], and X_2 [leverage ratio], and on a dummy variable Y. All members of group 1 are assigned $Y = 1$ and all members of group 2 are given $Y = 0$.

There are two alternative methods, the dummy regression and the eigenvalue method, to estimate the discriminant function.

(i) Dummy Regression Method:

If $m = 2$, then Eq. (16.10) can be written as

$$Y_i = a_1X_{1i} + a_2X_{2i}. \tag{16.13}$$

This equation can be rewritten as

$$y_i = a_1x_{1i} + a_2x_{2i} \tag{16.14}$$

where $y_i = Y_i - \overline{Y}$, $x_{1i} = a_1X_{1i} + a_2\overline{X}_1$, and $x_{2i} = X_{2i} - \overline{X}_2$.

The equation system used to solve a_1 and a_2 can be defined as

$$\mathrm{Var}(x_{1i})a_1 + \mathrm{Cov}(x_{1i}, x_{2i})a_2 = \mathrm{Cov}(x_{1i}, y_i) \tag{16.15a}$$

$$\mathrm{Cov}(x_{1i}, x_{2i})a_1 + \mathrm{Var}(x_{2i})a_2 = \mathrm{Cov}(x_{2i}, y_i) \tag{16.15b}$$

Following the data listed in Table 16.1, $\mathrm{Var}(x_{1i})$, $\mathrm{Var}(x_{2i})$, $\mathrm{Cov}(x_{1i}, x_{2i})$, $\mathrm{Cov}(x_{1i}, y_i)$, and $\mathrm{Cov}(x_{2i}, y_i)$ are calculated as follows:

$$\mathrm{Var}(x_{1i}) = \frac{\sum X_{1i}^2}{n} - \left(\frac{\sum X_{1i}}{n}\right)^2$$
$$= \frac{32 + 29.19}{n} - \left(\frac{13.6 + 14.9}{n}\right)^2$$
$$= \frac{61.19}{14} - \left(\frac{28.5}{14}\right)^2$$
$$= 4.3707 - 4.144 = 0.2267;$$

$$\mathrm{Var}(x_{2i}) = \frac{\sum X_{2i}^2}{n} - \left(\frac{\sum X_{2i}}{n}\right)^2$$
$$= \frac{1.2671 + 1.5447}{14} - \left(\frac{2.75 + 3.45}{14}\right)^2$$
$$= \frac{2.8122}{14} - \left(\frac{62}{14}\right)^2$$
$$= 02008 - 0196 = 0.0048;$$

$$\mathrm{Cov}(x_{1i}, x_{2i}) = \frac{12.424}{14} - (2.0357)(0.4428)$$
$$= 0.8874 - 0.9014$$
$$= -0.014;$$

$$\mathrm{Cov}(x_{1i}, y_i) = \frac{\sum X_{1i} Y_i}{n} - (\overline{X}_1)(\overline{Y})$$
$$\frac{13.6}{14} - (2.0357)(0.4285)$$
$$= 0.9714 - 0.8722$$
$$= 0.0992;$$

$$\mathrm{Cov}(x_{2i}, y_i) = \frac{\sum X_{2i} Y_i}{n} - (\overline{X}_2)(\overline{Y})$$
$$\frac{2.75}{14} - (0.4428)(0.4288)$$
$$= 0.0.1964 - 0.1897$$
$$= 0.00668.$$

Following Cramer's rule, a_1 and a_2 can be estimated as:

$$a_1 = \frac{\begin{vmatrix} 0.0992 & -0.0140 \\ 0.0066 & 0.0048 \end{vmatrix}}{\begin{vmatrix} 0.2267 & -0.0140 \\ -0.0140 & 0.0048 \end{vmatrix}}$$
$$= \frac{0.00047616 + 0.0000924}{0.0010886 - 0.000196}$$
$$= \frac{0.00056886}{0.0008926} = 0.63697;$$

$$a_2 = \frac{\begin{vmatrix} 0.2267 & 0.00992 \\ -0.0140 & 0.0066 \end{vmatrix}}{\begin{vmatrix} 0.2267 & -0.0140 \\ -0.0140 & 0.0048 \end{vmatrix}}$$
$$= \frac{0.00149 + 0.00138}{0.00108 - 0.00019}$$
$$= \frac{0.00288}{0.00089} = 3.2359.$$

Normalizing the regression coefficient by dividing a_2 into a_1, we obtain

$$\begin{bmatrix} \frac{a_1}{a_2} \\ 1 \end{bmatrix} = \begin{bmatrix} \frac{0.63697}{3.2359} \\ 1 \end{bmatrix} = \begin{bmatrix} 0.1968 \\ 1 \end{bmatrix}$$

(ii) Eigenvalue Method:

For the eigenvalue method, the elements of C and B can be calculated as:

$$C_{11} = 32 - \frac{(13.6)^2}{6} + 29.19 - \frac{(14.9)^2}{8} = 2.612;$$

$$C_{22} = 1.2671 - \frac{(2.75)^2}{6} + 1.5447 - \frac{(3.45)^2}{8} = 0.0636;$$

$$C_{12} = 6.179 - \frac{(13.6)(2.75)}{6} + 6.245 - \frac{(14.9)(3.45)}{8} = 0.2350;$$

$$B_{11} = 6\left[\frac{13.6}{6} - 2.0357\right]^2 + 8\left[\frac{14.9}{8} - 2.0357\right]^2 = 0.5601;$$

$$B_{22} = 6\left[\frac{2.75}{6} - 0.4428\right]^2 + 8\left[\frac{3.45}{8} - 0.4428\right]^2 = 0.0025;$$

$$B_{12} = 6\left(\frac{13.6}{6} - 2.0357\right)\left(\frac{2.75}{6} - 0.4428\right) + 8\left(\frac{14.9}{8} - 2.0357\right)\left(\frac{3.45}{8} - 0.4428\right) = 0.03753.$$

Following the above-mentioned information, the matrices C and B can be written as:

$$C = \begin{bmatrix} 2.612 & -0.2350 \\ -0.2350 & 0.0636 \end{bmatrix} \text{ and } B = \begin{bmatrix} 0.5601 & 0.03753 \\ 0.03753 & 0.0025 \end{bmatrix}.$$

Based upon the matrix inversion and multiplication rules, we can obtain:

$$C^{-1}B = \begin{bmatrix} 0.4007 & 0.0268 \\ 0.0268 & 0.1384 \end{bmatrix},$$

and the matrix to be substituted in Eq. (16.12). The characteristic equation, $\left|C^{-1}B - EI\right| = 0$, is found to be $E^2 - 0.5391E = 0$, whose single nonzero root is $E_1 = 0.5391$. The adjoint of $C^{-1}B - E_1I$ is:

$$\text{Adjoint of } C^{-1}B - E_1I = \begin{bmatrix} -0.4007 & -0.0268 \\ -0.2071 & -0.1384 \end{bmatrix}.$$

To find the adjoint of a matrix, take the transpose of the cofactor. The cofactor of $C^{-1}B - E_1I$ is:

$$C^{-1}B - E_1I = \begin{bmatrix} \begin{matrix}1+1\\(-1)(-0.4007)\end{matrix} & \begin{matrix}1+2\\(-1)(2.071)\end{matrix} \\ \begin{matrix}2+1\\(-1)(0.0268)\end{matrix} & \begin{matrix}2+2\\(-1)(-0.1384)\end{matrix} \end{bmatrix}$$

$$\text{Cofactor of} = \begin{bmatrix} -0.4007 & -2.071 \\ -0.0268 & -0.1384 \end{bmatrix}$$

and the transpose is:

$$\begin{bmatrix} -0.4007 & -0.0268 \\ -2.071 & -0.1384 \end{bmatrix};$$

hence the eigenvector of $C^{-1}B$, with larger elements set equal to unity, is

$$A_1 = \begin{bmatrix} -0.4007/-2.071 \\ 1 \end{bmatrix} = \begin{bmatrix} 0.1935 \\ 1 \end{bmatrix},$$

when the vector of regression weights obtained earlier is similarly rescaled, we find

$$a = \begin{bmatrix} 0.63697065/3.2359 \\ 1 \end{bmatrix} = \begin{bmatrix} 0.1968 \\ 1 \end{bmatrix}.$$

This agrees, within rounding errors, with the A_1 just obtained by the general method of discriminant analysis mentioned above.

Alternatively, the nonzero root, $E_1 = 0.5391$, can be substituted into Eq. (16.11), yielding:

$$\begin{bmatrix} -0.4007 - 0.5391 & -0.0268 \\ -2.071 & -0.1384 - 0.5391 \end{bmatrix}\begin{bmatrix} a_1 \\ a_2 \end{bmatrix} = 0 \quad (16.16)$$

Equation (16.12) implies that:

$$-0.1384a_1 + 0.0268a_2 = 0;$$
$$2.071a_1 - 0.4007a_2 = 0;$$

that is:

$$2.071a_1 - 0.4007a_2 = 0$$

In this case, one element in the characteristic (eigen) vector is arbitrary. We may normalize the vector by setting its length at unity, that is, by making $a_1^2 + a_2^2 = 1$. When this is combined with the fact that $a_1 = 0.1936a_2$, it gives these results: $a_1 = 0.1901$ and $a_2 = 0.9819$. From these figures, we have

$$a = \begin{bmatrix} 0.1907/0.9818 \\ 1 \end{bmatrix} = \begin{bmatrix} 0.1935 \\ 1 \end{bmatrix}.$$

This result is identical to the result obtained by solving the eigenvector.

16.4 k-Group Discriminant Analysis

The two-group discriminant analysis theory we have discussed can be readily generalized to the k-group case. Assume there are k samples ($G = 1, 2, \ldots, k$) of size N. The linear discrimination function is similar to that defined in Eq. (16.10), but we now have k groups instead of two. The k-group discriminant function can be defined as

$$Y_j = a_{j1}X_1 + a_{j2}X_2 + \cdots + a_{jm}X_m \quad (j = 1, 2, \ldots k) \tag{16.17}$$

The k-group analogy to the two-group case of maximizing the ratio E is to find the set of $(m \times 1)$ vectors $A_1, A_2, \ldots A_m$, that maximizes the ratio

$$E_j = A'BA/A'CA \tag{16.18}$$

where

$$A' = [A_1, A_2, \ldots A_m],$$

B is the matrix of the weighted among-group deviation sums of squares of X, and
C is the matrix of pooled within-group deviation sums of squares of X.

Following Tatsuoka (1988), and Appendix 1, the optimum discriminant function can be defined as Eq. (16.19):

$$(C^{-1}B - EI)A = 0 \tag{16.19}$$

In sum, the theory and methodology of two-group discriminant function analysis can be easily extended to k-group analysis. Both eigenvalues and eigenvectors can be solved in accordance with the procedure discussed in Sect. 16.3.

Following Tatsuoka (1988), the discriminant functions associated with Eq. (16.13) can be explicitly defined as

$$Y_1 = a_{11}X_1 + a_{12}X_2 + \cdots + a_{1m}X_m \tag{16.20a}$$
$$Y_2 = a_{21}X_1 + a_{22}X_2 + \cdots + a_{2m}X_m \tag{16.20b}$$
$$Y_3 = a_{31}X_1 + a_{32}X_2 + \cdots + a_{3m}X_m \tag{16.20c}$$
$$\vdots \quad \vdots \quad \vdots \quad \vdots$$
$$Y_r = a_{r1}X_1 + a_{r2}X_2 + \cdots + a_{rm}X_m, \tag{16.20r}$$

where $r < k$. These equations can be used to perform r-group discriminant analysis. The implications of these equations need further explanation.

To obtain the equations indicated in Eqs. (16.20a)–(16.20r), we need r discriminant-criterion values, denoted as $E_1, E_2, \ldots E_r$ in descending I order of magnitude, and r associated eigenvectors $A_1, A_2, \ldots A_r$. The eigenvectors are determined only up to an arbitrary multiplier, because if A satisfies Eq. (16.19) for some E, it is clear that dA also satisfies the equation for some E (where d is an arbitrary constant). It is customary to choose the multiplier for each eigenvector in one of two ways: (i) so that its norm will be unity (that is, $A_p' A_p = 1$, for each p), or (ii) so that its largest element will be unity.

From the fact that eigenvalues E_r are, by definition, the values assumed by the discriminant criterion for linear combinations using the elements of corresponding eigenvectors A_m as combining weights, it is clear that the eigenvector provides a set of weights such that the transformed variable [as indicated in Eq. (16.20 a)] has the largest discriminant criterion, E_1, achievable by any linear combination of the m predictor variables.

It is clear that the weights of the linear combination as indicated in Eq. (16.20b) are the elements of A_2. It can be shown that Y_2 has the discriminant criterion value E_2, which is the largest achievable by any linear combination of the X's that is uncorrelated with Y_1. Similarly, Y_3, as indicated in Eq. (16.20c), has the largest discriminant criterion value (E_3) among all linear combinations of the X's that are uncorrelated with Y_1 and Y_2; and so on, until Y_r, using the elements of A_r as weights, has the largest possible discriminant criterion value among linear combinations that are uncorrelated with all the preceding linear combinations $Y_1, Y_2, \ldots, Y_{r-1}$. The linear combinations $Y_1, Y_2, \ldots, Y_{r-1}$ are called the first, second, …, rth (linear) discriminant functions for optimally differentiating among the k given groups. Pinches and Mingo (1973) used four discriminant functions in analyzing industrial bond ratings.

We have discussed the theory and methodology for fitting a linear discriminant function (LDF) over the analysis sample. The estimated LDF can be used to investigate the group difference. After the LDF is estimated, the estimated LDF can also be used to do either descriptive or predictive analyses. As argued by Joy and Tollefson (1975), predictive analysis (ex post discrimination) refers to cross validation: classifying members of a time-coincident holdout or validation sample. Joy and Tollefson (1975) argued that prediction requires intertemporal validation (testing predictive results over time), whereas explanation requires only cross validation. Classification procedures are needed to do either explanation or prediction.

Following Eisenbeis and Avery (1972) and Altman and Eisenbeis (1968), the discriminant function coefficients are derived either (a) to minimize the expected overall error rate R or (b) to minimize the overall costs of misclassification. Using Altman's (1968) bankruptcy analysis for an example, Rand C can be defined as

$$R = q_1 p(1|2) + q_2 p(2|1) \tag{16.21a}$$

$$R = q_1 p(1|2) C_{12} + q_2 p(2|1) C_{21}, \tag{16.21b}$$

where

q_1 = Prior probability of being classified as bankrupt,

q_2 = Prior probability of being classified as nonbankrupt,

$p(2|1)$ = Conditional probability of being classified as nonbankrupt when, in fact, the firm is bankrupt,

$p(1|2)$ = Conditional probability of being classified as bankrupt when, in fact, the firm is nonbankrupt,

C_{12} = Cost of classifying a bankrupt firm as nonbankrupt,

C_{21} = Cost of classifying a nonbankrupt firm as bankrupt.

If $p(1|2) = 0.20$, $p(2|1) = 0.40$, $q_1 = 0.01$, and $q_2 = 0.99$, then

$$R = (0.01)(0.20) + (0.99)(0.40) = 0.3962,$$
$$C = 0.0002\, C_{12} + 0.3960\, C_{21}.$$

Application of the above-mentioned classification procedure will be discussed further in the next chapter.

16.5 Factor Analysis and Principal Component Analysis

Factor analysis was developed by psychologists and has only recently been j applied in marketing, finance, and accounting. Anderson (2003), Tatsuoka (1988), Green and Tull (1978), and Churchill and Iacobucci (2004) have argued that factor, analysis is one of the more popular "analyses of interdependence" techniques. In studies of interdependence, all variables have equal footing, and the analysis is concerned with the whole set of relationships among variables that characterize the objects. In factor analysis, two key concepts—factor scores and factor loadings—should first be explained.

Factor Score

A factor score is simply a linear combination (or linear composite) of the original variables; it can be defined as

$$f_i = b_1Y_1 + b_2Y_2 + \cdots + b_jY_j + b_pY_p, \quad (16.22)$$

where f_i is the ith factor score, $Y_j (j = 1, 2, \ldots, p)$ are original variables. For example, Johnson and Dinardo (1996) used factor analysis to classify 61 financial ratios into eight groups (factors). In Eq. (16.22) the coefficients (weights) $b_1, b_2, \ldots, b_p$, are parameters to be estimated.

Factor Loadings

A factor loading is defined simply as the correlation (across objects) of a set of factors with the original variables. In Johnson's case (1979), he used 306 primary manufacturing firms and 159 retailing firms to calculate eight factors. For manufacturing firms, he used 306 original variables for each factor; for retailing firms, he used 159 original variables for each factor. By using the simple correlation coefficient formula, he calculated 61 factor loadings for each industry.

Following Anderson (2003), the basic model of factor analysis can be defined as

$$Y = \mu + \beta f + U, \quad (16.23)$$

where f is an m-component vector of (nonobservable) factor scores, μ is a fixed vector of means, and U is a vector of (nonobservable) errors (or errors plus specific factors).

The $p \times m$ matrix β consists of factor loadings ($m < p$). When f is random, we assume $E(f) = 0$, $E(U) = 0$, $E(f'f) = M$, $E(UU') = \sum$, a diagonal matrix, and $fU' = 0$. Then $E(Y) = \mu$, and the covariance matrix of the observable Y is

$$E(Y - \mu)(Y - \mu)' = \beta M \beta' + \Sigma. \quad (16.24)$$

There are two alternative models in estimating the factor score: the principal component method and the maximum likelihood method. The principal component method for extracting factors or calculating the coefficient matrix to meet the foregoing statistical assumption can be found in both Johnson and Dinardo (1996) and Tatsuoka (1988). The maximum likelihood method can be found in Lawley (1940), Lawley and Maxwell (1963), and Joreskog (1967). The principal component method is discussed in Appendix 2.

16.6 Conclusion

In this chapter, method and theory of both discriminant analysis and factor analysis needed for determining useful financial ratios, predicting corporate bankruptcy, determining bond rating, and analyzing the relationship between bankruptcy avoidance and merger are discussed in detail. Important concepts of linear algebra-linear combination and matrix operations—required to understand both discriminant and factor analysis are discussed.

Appendix 1: Relationship Between Discriminant Analysis and Dummy Regression Analysis

Data composed of two samples of size N_1 and N_2 for two-group discriminant analysis must meet the following assumptions: (1) that the groups being investigated are discrete and identifiable; (2) that each observation in each group can be described by a set of measurements on m characteristics or variables; and (3) that these m variables have a multivariate normal distribution in each population. In vector notation the nth observation can be represented as an $m \times 1$ column vector of the form

$$X'_n = (X_{1n}, X_{2n}, \cdots, X_{mn}),$$

where $n = 1, \ldots, N_1$, or $n = 1, \ldots, N_2$.

Under these assumptions, the linear discriminant function can be defined as Eq. (16.10) of the chapter text (repeated here for convenience)

$$Y_i = a_1 X_{1i} + aX_{2i} + \cdots + a_m X_{mi} \qquad (16.25)$$

The a_i's were then chosen to maximize the ratio of the weighted between-group variance to the pooled within-group variance. Ladd (1966) has proposed a discriminant criterion E, as defined as Eq. (16.26) or Eq. (16.18) of the text, to determine the coefficients $a_1, a_2, \ldots, a_m$:

$$E \frac{A'DD'A}{A'CA} = \frac{A'BA}{A'CA} \qquad (16.26)$$

where

$A' = [a_1, a_2, \cdots, a_m]$;
$D' = [\overline{X}_{1,1} - \overline{X}_{1,2}, \overline{X}_{2,1} - \overline{X}_{2,2}, \overline{X}_{m,1} - \overline{X}_{m,2}]$;
C = Within-group variable matrix;
DD' = Between-group variable matrix.

There are two alternative methods that can be used to derive the basic equation of discriminant analysis as defined in Eq. (16.27) later in this Appendix.

(a) Subsequent Vector Derivation Method

Symbolically, we may find the derivative of E with the respect to the column vector of A and equate the result to the $(m \times 1)$ vector [see Tatsuoka (1971, pp. 160, 161)]. The vector equation thus obtained is

$$\frac{\partial E}{\partial A} = \frac{2[(BA)(A'CA) - (A'BA)(CA)]}{(A'CA)^2} = 0$$

Dividing both numerator and denominator of the middle member by $A'CA$ and using the definition of E in Eq. (16.26), this equation reduces to

$$\frac{2[BA - ECA]}{A'CA} = 0,$$

which is equivalent to

$$(B - EC)A = 0. \qquad (16.27)$$

(b) Long-Hand Method

If $m = 2$, then Eq. (16.26) can be rewritten as

$$E = \frac{b_{11}a_1^2 + b_{22}a_2^2 + 2b_{12}a_1a_2}{c_{11}a_1^2 + c_{22}a_2^2 + 2c_{12}a_1a_2}, \qquad (16.26a)$$

where b_{11}, b_{22}, and b_{12} are elements of B, and c_{11}, c_{21}, and c_{12} are elements of C.

Taking the partial derivative of E with respect to a_1, we obtain:

$$\begin{aligned}\frac{\partial E}{\partial a_1} &= [(2b_{11}a_1 + 2b_{12}a_2)(C_{11}a_1^2 + C_{22}a_2^2 \\ &\quad + 2C_{12}a_1a_2) - (b_{11}a_1^2 + b_{22}a_2^2 + 2b_{12}a_1a_2) \\ &\quad (2C_{11}a_1 + 2C_{12}a_2)] \times (C_{11}a_1^2 + C_{22}a_2^2 \\ &\quad + 2C_{12}a_1a_2)^{-2} = 2[(b_{11}a_1 + b_{12}a_2) \\ &\quad - E(C_{11}a_1 + C_{12}a_2)] \\ &\quad \times (C_{11}a_1^2 + C_{22}a_2^2 + 2C_{12}a_1a_2)^{-1}.\end{aligned}$$

Setting this equation equal to zero and simplifying, we get:

$$b_{11}a_1 + b_{12}a_2 = E(C_{11}a_1 + C_{12}a_2).$$

Using vector notation, we have:

$$[b_{11}, b_{12}]A = E[C_{11}, C_{12}]A. \quad (16.28)$$

Similarly, it can be shown that, upon equating $\partial E/\partial a_2$ to zero and simplifying, we get:

$$[b_{21}, b_{22}]A = E[C_{21}, C_{22}]A. \quad (16.29)$$

It is evident that Eqs. (16.28) and (16.29) can be written as a single matrix equation:

$$\begin{bmatrix} b_{11} & b_{12} \\ b_{21} & b_{22} \end{bmatrix} A = E \begin{bmatrix} C_{11} & C_{12} \\ C_{21} & C_{22} \end{bmatrix} A$$

or

$$(B - EC)A = 0. \quad (16.27)$$

Equation (16.27) can be used to formulate the characteristic equation for solving eigenvector A as indicated in the text.

If $A*$ maximizes the function shown in Eq. (16.25), then so does any $A** = KA*$, where K is a scalar. Substituting $A**$ for $A*$ simply multiplies both the numerator and denominator by K^2. Because the coefficients themselves are not unique, there are several methods of calculating the discriminant function. Johnston and Dinardo shows (1996), for example, that the vector $A*$ that maximizes the ratio E is proportional to the vector A (that is, $A = KA*$), which maximizes $G = A'DD'A$ subject to the constraint that $L = A'CA$, where L is an arbitrary constant.

Let λ be a Lagrange multiplier and define

$$F = A'DD'A - \lambda[A'CA - L]. \quad (16.30)$$

Setting the derivatives of F with respect to A equal to zero yields

$$\frac{\partial F}{\partial A} = 0 = 2DD'a - 2\lambda CA, \quad (16.31)$$

where $D'A$ is a scalar, say H. Hence, Eq. (16.31) can be rewritten as $(\frac{\lambda}{H})CA = D$, and thus a solution is

$$A\left(\frac{\lambda}{H}\right) = C^{-1}D = A_1, \text{ say}, \quad (16.32)$$

which is proportional to A. It can be seen that Al is a solution to Eq. (16.27):

$$(DD' - \lambda C)A_1 = (DD' - \lambda C)A\left(\frac{\lambda}{H}\right) = 0.$$

Alternatively, a new objective function can be developed as follows:

From Eq. (16.27) we have

$$BA = ECA. \quad (16.33)$$

Add EBA to both sides of Eq. (16.33), obtaining

$$(1 + E)BA = E(B + C)A$$

or

$$(B - E'S)A = 0, \quad (16.34)$$

where $E' = E/(1 + E)$. This implies that Eq. (16.34) can be used as an alternative discriminant function of Eq. (16.27). Under this circumstance, an alternative objective function can be defined as

$$E' = \frac{A'DD'A}{A'SA},$$

where $S = [S_{ij}] = \textit{the } m \times m$ matrix of S_{ij}, and S_{ij} is the sum of cross products of the deviations of X_i, and X_j about the overall means.

Following the same procedure mentioned above, we can obtain

$$A\left(\frac{\lambda}{H}\right) = S^{-1}D = A_2. \quad (16.35)$$

Ladd (1966) has shown that A_2 is proportional to A_1.

These results imply that the parameters of a two-group discriminant function can be estimated by using the related data of S and D.

If we let $Y = 1$ for observations in group 1 and $Y = 0$ for those in group 2, then, following the multiple regression technique, Ladd (1966) showed that the regression coefficient vector is

$$A = S^{-1}MD, \tag{16.36}$$

where $M = [N_1N_2]/[N_1 + N_2]$, and N_1 and N_2 are total observations in group 1 and group 2, respectively.

A comparison of Eq. (16.35) with Eq. (16.36) shows that the parameters obtained from the eigenvalue method differ from those of the dummy regression method only by a constant M.

The implications of Eqs. (16.35) and (16.36) can be discussed as follows: If $m = 2$, then Eq. (16.25) can be defined as

$$Y_i = a_1X_{1i} + a_2X_{2i}. \tag{16.25$'$}$$

Equation (16.25′) can be written in derived form as

$$y_i = a_1x_{1i} + a_2x_{2i}. \tag{16.37}$$

where $y_i = Y_i - \overline{Y}$, $x_{1i} = X_{1i} - \overline{X}_1$, $X_{2i} = X_{2i} - \overline{X}_2$. The equation system used to solve a_1 and a_2 can therefore be defined as

$$\left(\sum x_{1i}^2\right)a_1 + \left(\sum x_{1i}x_{2i}\right)a_2 = \sum x_{1i}y_i, \tag{16.38a}$$

$$\left(\sum x_{1i}x_{2i}\right)a_1 + \left(\sum x_{2i}^2\right)a_2 = \sum x_{2i}y_i. \tag{16.38b}$$

Since Y_i is a dichotomous variable, we write

$$\begin{aligned}\sum X_1V_1y = \sum X_1Y - N\overline{X}, \overline{Y} = N_1\overline{X}_{11} \\ - \left(\frac{N_1\overline{X}_{11} + N_2\overline{X}_{12}}{N_1 + N_2}\right)\left(\frac{NN_1}{N_1 + N_2}\right) \\ = \left(\frac{N_1N_2}{N}\right)(\overline{X}_{11} - \overline{X}_{12}) = \left(\frac{N_1N_2}{N}\right)(D_1)\end{aligned} \tag{16.39}$$

Similarly, we can show that

$$\sum x_2y = \left(\frac{N_1N_2}{N}\right)(\dot{X}_{21} - \dot{X}_{22}) = \left(\frac{N_1N_2}{N}\right)(D_2) \tag{16.40}$$

From Eqs. (16.38a), (16.38b), (16.39), and (16.40), we obtain

$$A = S^{-1}MD,$$

where
$A' = (a_1, a_2), D' = (D_1, D_2), M = N_1N_2/N$, and

$$S = \begin{bmatrix} \sum x_{1i}^2 & \sum x_{1i}x_{2i} \\ \sum x_{2i}x_{1i} & \sum x_{2i}^2 \end{bmatrix}.$$

Appendix 2: Principal Component Analysis

Following Anderson (2003), principal components are linear combinations of random or statistical variables that have special properties in terms of variances. For example, the first principal component is the normalized linear combination (that is, the sum of squares of the coefficients being one) with maximum variance. In effect, transforming the original vector variable to the vector of principal components amounts to a rotation of coordinate axes to a new coordinate system that has inherent statistical properties.

The principal components turn out to be the eigenvectors of the covariance matrix, as discussed in the chapter text. Thus the study of principal components can be considered as putting into statistical terms the usual developments of eigenroots and eigenvectors (for positive and semidefinite matrices).

From the point of view of statistical theory, the set of principal components yields a convenient set of coordinates, and the accompanying variances of the components characterize their statistical properties. In statistical practice, the method of principal components is used to find the linear combinations with large variance. In many empirical studies, the number of variables under consideration is too large to handle. A way of reducing the number of variables to be treated is to discard the linear combinations that have small variances and to study only those with large variances.

For example, a physical anthropologist may make dozens of measurements of each of a number of individuals, measurements such as ear length, ear breadth, facial length, facial breadth, and so forth. He may be interested in describing and analyzing how individuals differ in these kinds of physiological characteristics. Eventually he will want to "explain" these differences, but first he wants to know what measurements or combinations of measurements show considerable variation; that is, which should need further study. The principal component approach provides a set of linearly combined measurements. It may be that most of the variation from individual to individual resides in two linear combinations; if so, the anthropologist can direct his study to these two quantities. Other linear combinations may vary so little from one person to the next that study of them will tell the researcher little about individual variation.

Using Johnson's (1979) financial ratio study as an example, we have data matrix Y of 306 observations on 61 variables.

$$Y = \begin{bmatrix} Y_{11} & \cdots & Y_{1n} \\ Y_{21} & & Y_{2n} \\ \vdots & & \vdots \\ Y_{p1} & \cdots & Y_{pn} \end{bmatrix} \quad \begin{pmatrix} n = 1, 2, \ldots, 306 \\ p = 1, 2, \ldots, 61 \end{pmatrix}$$

Following Johnston and Dinardo (1996), the procedure of estimating principal components can be described as follows: we express the observations of Y's as deviations from the sample means, for we are concerned with studying the variation in the data.

The method of principal components may be approached in a number of ways. One is to ask how many dimensions or how much independence there really is in the set of p variables. More explicitly, we consider the transformation of the Y's to a net set of variables that will be pairwise uncorrelated and of which the first will have the maximum possible variance, the second the maximum possible variance among those uncorrelated with the first, and so forth. Let

$$f_{1t} = b_{11}y_{1t} + b_{21}y_{2t} + \cdots + b_{p1}y_{pt} \quad (t = 1, \ldots, n)$$

denote the first new variable. In matrix form

$$f_1 = yb_1, \tag{16.41}$$

where f_1, is an n-element vector and b_1 a p-element vector. The sum of squares of f_1, is

$$f_1'f_1 = f_1'Y'Yf_1. \tag{16.42}$$

We wish to choose b_1 to maximize $f_1'f_1$. Clearly, some constraint must be imposed on b_1, or $f_1'f_1$, could be made infinitely large, so let us normalize by setting

$$b_1'b_1 = 1. \tag{16.43}$$

The problem now is to maximize Eq. (16.42) subject to Eq. (16.43). Define ϕ as

$$\phi = b_1'Y'Yb_1 - \lambda_1(b_1'b_1 - 1),$$

where λ_1, is a Lagrange multiplier. Thus

$$\frac{\partial\phi}{\partial b_1} = 2Y'Yb_1 - 2\lambda_1 b_1.$$

Setting $\partial\phi/\partial b_1 = 0$ gives

$$(Y'Y)b_1 = \lambda_1 b_1. \tag{16.44}$$

Thus b_1 is a latent (eigen) vector of $X'X$ corresponding to the root of λ_1. From Eqs. (16.42) and (16.44) we see that

$$f_1'f_1 = \lambda_1 b_1'b_1 = \lambda_1,$$

and so we must choose λ_1, as the largest latent root of $Y'Y$. The $Y'Y$ matrix, in the absence of perfect collinearity, will be positive definite $(Y'Y > 0)$ and thus have positive latent roots. The first principal component of Y is then f_1.

Now define $f_2 = Yb_2$. We wish to choose b_2 to maximize $b_2'Y'Yb_2$ subject to $b_2'b_2 = 1$ and $b_1'b_2 = 0$. The reason for the second condition is that f_2 is to be uncorrelated with f_1. The covariation between them is given by

$$\begin{aligned} b_2'Y'Yb_2 &= \lambda_1 b_1'b_2 \\ &= 0 \quad \text{if and only if } b_1'b_2 = 0 \end{aligned}$$

Define

$$\phi = b_2'Y'Yb_2 - \lambda_2\left(b_2'b_2 - 1\right) - \mu\left(b_1'b_2\right),$$

where λ_2 and μ are Lagrange multipliers:

$$\frac{\partial \phi}{\partial b_2} = 2Y'Yb_2 - 2\lambda_2 b_2 - \mu b_1 = 0.$$

Premultiply by b_1',

$$2b_1'Y'Yb_2 - \mu = 0.$$

But from $(Y'Y)b_1 = \lambda_1 b_1$,

$$b_2'(Y'Y)b_1 = \lambda_1 Y_1'Y_1 = 0.$$

Thus $\mu = 0$, and we have

$$(Y'Y)b_2 = \lambda_2 b_2, \tag{16.45}$$

and λ_2 should obviously be chosen as the second largest latent root of $Y'Y$. We can proceed in this way for each of the p roots of $Y'Y$, and assemble the resultant vectors in the orthogonal matrix:

$$B = \begin{bmatrix} b_1 & b_2 & \dots & b_p \end{bmatrix}. \tag{16.46}$$

The p principal components of Y are then given by the $n \times p$ matrix F,

$$F = YB. \tag{16.47}$$

Moreover,

$$F'F = B'Y'YB = \Lambda = \begin{bmatrix} \lambda_1 & 0 & \cdots & 0 \\ 0 & \lambda_2 & \cdots & 0 \\ \vdots & \vdots & & \vdots \\ 0 & 0 & \cdots & \lambda_p \end{bmatrix}, \tag{16.48}$$

showing that the principal components are indeed pairwise uncorrelated and that their variances are given by

$$f_i'f_k = \lambda_i \quad (i = 1, \ldots, p). \tag{16.49}$$

This appendix has already shown the major concepts and procedures of principal component analysis. For further detail, we suggest that readers consult Anderson (2003), Johnston and Dinardo (1996), and Stuart (1982).

Bibliography

Altman, E. I. (1968). Financial ratios, discriminant analysis, and the prediction of corporate bankruptcy. *Journal of Finance, 23,* 589–609.

Altman, E. I., & Eisenbeis, R. A. (1968). Financial applications of discriminant analysis: A clarification. *Journal of Financial and Quantitative Analysis, 13,* 185–195.

Altman, E. I., Haldeman, R., & Narayanan, P. (1977). ZETA analysis, a new model for bankruptcy classification. *Journal of Banking and Finance,* 29–54.

Anderson, T. W. (2003). *An introduction to multivariate statistical analysis* (3rd ed.). New York: Wiley.

Anton, H. (2004). *Elementary linear algebra* (9th ed.). New York: Wiley.

Churchill, G. A., Jr., & Iacobucci, D. (2004). *Marketing research: Methodological foundations* (9th ed.). Mason, OH: South-Western College Publishing.

Eisenbeis, R. A., & Avery, R. B. (1972). *Discriminant analysis and classification procedure.* Lexington, MA: Lexington Books.

Green, P. E. (1988). *Research for marketing decisions* (5th ed.). Englewood Cliffs: Prentice-Hall.

Green, P. E., & Tull, D. S. (1978). *Research for marketing decisions.* New Jersey: Prentice-Hall.

Johnson, W. B. (1979). The cross-sectional stability of financial patterns. *Journal of Financial and Quantitative Analysis, 14,* 1035–1048.

Johnston, J., & Dinardo, J. (1996). *Econometrics methods* (4th ed.). New York: McGraw-Hill.

Joreskog, K. G. (1967). Some contributions to maximum-likelihood factor analysis. *Psychometrika, 32,* 443–482.

Joy, O. M. (1978). Some clarifying comments on discriminant analysis. *Journal of Financial and Quantitative Analysis, 13,* 197–200.

Joy, O. M., & Tollefson, J. O. (1975). On the financial applications of discriminant analysis. *Journal of Financial and Quantitative Analysis, 10,* 723–739.

Ladd, G. W. (1966). Linear probability functions and discriminant functions. *Econometrica, 34,* 873–885.

Lawley, D. N. (1940). The estimation of factor loadings by the method of maximum likelihood. *Proceedings of the Royal Society of Edinburg, Series A, 62,* 64–82.

Lawley, D. N., & Maxwell, A. E. (1963). *Factor analysis as a statistical method.* London: Butterworth.

Lee, A. C., Lee, J. C., & Lee, C. F. (2017). *Financial analysis, planning and forecasting: Theory and application* (3rd ed.). Singapore: World Scientific.

Moore, J. T. (1968). *Elements of linear algebra and matrix theory.* New York: McGraw-Hill Book Company.

Penman, S. H. (2006). *Financial statement analysis and security valuation* (3rd ed.). New York: McGraw-Hill/Irwin.

Pinches, G. E., & Mingo, K. A. (1973). A multivariate analysis of industrial bond ratings. *The Journal of Finance, 28*(1), 1–18.

Stuart, M. (1982). A geometric approach to principal-components analysis. *The American Statistician, 36,* 365–367.

Tatsuoka, M. M. (1971). *Multivariate analysis.* New York: Wiley.

Tatsuoka, M. M. (1988). *Multivariate analysis: Techniques for educational and psychological research* (2nd ed.). New York: Wiley.

Part IV

Statistics, Itô's Calculus and Option Pricing Model

In Chap. 17, we will show how characteristic function and noncentral chi-square can be used to analyze stochastic volatility–option pricing model. In Chap. 18, we will discuss alternative methods to estimate implied variance. In Chap. 19, we will show the numerical valuation of Asian options with higher moments in the underlying distribution. Both European and American options will be discussed in this chapter. In Chap. 20, we will first review Itô Lemma and stochastic differential equation, and then we will show how this mathematical technique can be used to derive the option pricing model. In Chap. 21, we will discuss the relationship between binomial option pricing model and Black–Scholes option pricing model. In addition, we also show how to use stochastic calculus to derive Black–Scholes model in detail. In Chap. 22, we will show how to use noncentral chi-square distribution to derive constant elasticity of variance option pricing model. In Chap. 23, we will discuss option pricing and hedging performance under stochastic volatility and stochastic interest rates. Finally, in Chap. 24, we will show how nonparametric distribution can be used to derive option bounds. Some empirical studies or option bounds are also provided.

17 Stochastic Volatility Option Pricing Models

Contents

Abstract

In this chapter, we assume that the volatility of option pricing model is stochastic instead of deterministic. We apply such assumption to the nonclosed-form solution developed by Scott (J Finance Quant Anal 22:419–438, 1987) and the closed-form solution of Heston (Rev Financ Stud 6:327–343, 1993). In both cases, we consider a model in which the variance of stock-price returns varies according to an independent diffusion process. For the closed-form option pricing model, the results are expressed in terms of the characteristic function.

This chapter draws upon the paper by Lee and Lee (2010) which was published as Chap. 32 of Handbook of Quantitative Finance and Risk Management (2010) edited by Lee et al.

17.1 Introduction

The variance of stock returns plays a vital role in option pricing, as evidenced in the Black–Scholes pricing formula. Also, it has long been observed that stock price changes over time. Thus, the assumption of a constant variance of stock-price returns does not seem to be reasonable in option price.

In this chapter, we will consider the situation in which the variance of stock-price returns is not a constant. Instead, we will assume the variance of stock-price returns is random and follows some distributions. Emphasis will be placed on the nonclosed-form solution developed by Scott

C.-F. Lee et al., *Financial Econometrics, Mathematics and Statistics*,
https://doi.org/10.1007/978-1-4939-9429-8_17

(1987) and the closed-form solution of Heston (1993). In both cases, we consider a model in which the variance of stock-price returns varies according to an independent diffusion process.

In the following of this chapter, we will discuss nonclosed-form type of option pricing model in Sect. 17.2. We will review the characteristic function in Sect. 17.3. Closed-form type of option pricing model will be presented in Sect. 17.4. Finally, Sect. 17.5 concludes. In addition, Appendix will demonstrate the market price of the risk.

17.2 Nonclosed-Form Type of Option Pricing Model

In the Black–Scholes model, the variance of stock returns was assumed to be constant. However, empirical studies show that volatility seems to change day by day. Its behavior looks like performing a random walk. Thus, it is intuitional to assume the following stochastic processes for stock prices with random variance:

$$\begin{aligned} dS &= \alpha S dt + \sigma S dz_1 \\ d\sigma &= \beta(\bar{\sigma} - \sigma)dt + \gamma dz_2 \end{aligned}, \tag{17.1}$$

where dz_1 and dz_2 are Wiener's processes. Since both stock price and volatility are assumed random variables, we need the following generalization of *Itô's* lemma.

Lemma 17.1 Generalization of *Itô's* Lemma
Let f be a function of variables x_1, x_2, ..., x_n and time t. Suppose that x_i's follow Itô's process:

$$dx_i = a_i dt + b_i dz_i, \quad i = 1, 2, \ldots, n,$$

where dz_i's are Wiener's processes, instantaneous correlation coefficients between dz_i and dz_j is r_{ij}, and a_i's and b_i's are functions of x_1, x_2, ..., x_n and t. Then

$$df = \left(\sum_i \frac{\partial f}{\partial x_i} a_i + \frac{\partial f}{\partial t} + \frac{1}{2}\sum_{i,j} \frac{\partial^2 f}{\partial x_i \partial x_j} b_i b_j \rho_{ij}\right) dt + \sum_i \frac{\partial f}{\partial x_i} b_i dz_i$$

Proof A Taylor series expansion of f gives

$$\Delta f = \sum_i \frac{\partial f}{\partial x_i}\Delta x_i + \frac{\partial f}{\partial t}\Delta t + \frac{1}{2}\sum_{i,j}\frac{\partial^2 f}{\partial x_i \partial x_j}\Delta x_i \Delta x_j + \frac{1}{2}\sum_i \frac{\partial^2 f}{\partial x_i \partial t}\Delta x_i \Delta t + \frac{1}{2}\frac{\partial^2 f}{\partial t^2}(\Delta t)^2 + R_2$$

A discrete form for dx can be written as

$$\Delta x_i = a_i \Delta t + b_i \sqrt{\Delta t}\varepsilon_i,$$

where e_i's are standard normal variables. Then

$$\Delta x_i \Delta x_j = a_i a_j (\Delta t)^2 + a_i b_j (\Delta t)^{\frac{3}{2}}\varepsilon_j + a_j b_i (\Delta t)^{\frac{3}{2}}\varepsilon_i + b_i b_j \Delta t \varepsilon_i \varepsilon_j.$$

The first term on the right is of order $(dt)^2$ and the second and third terms are distributed with mean 0 and a standard deviation of order $(dt)^{3/2}$. As Dt tends to 0,

$$\lim_{\Delta t \to 0} \Delta x_i \Delta x_j = \rho_{ij} b_i b_j,$$

where r_{ij} is the correlation coefficient of e_i and e_j.

Similarly, the terms, $\Delta x_i \Delta t$, are all distributed with mean 0 and a standard deviation of order $(dt)^{3/2}$, so they are comparably small with Dt as Dt→0.

It can be seen that all the terms in R_2 are at most as large as $(Dt)^{3/2}$, thus we can collect all terms of order Dt and write down the differential df as:

$$\begin{aligned} df &= \sum_i \frac{\partial f}{\partial x_i}(a_i dt + b_i dz_i) + \frac{\partial f}{\partial t}dt + \frac{1}{2}\sum_{i,j}\frac{\partial^2 f}{\partial x_i \partial x_j} b_i b_j \rho_{ij} dt \\ &= \left[\sum_i \frac{\partial f}{\partial x_i} a_i + \frac{\partial f}{\partial t} + \frac{1}{2}\sum_{i,j}\frac{\partial^2 f}{\partial x_i \partial x_j} b_i b_j \rho_{ij}\right] dt + \sum_i \frac{\partial f}{\partial x_i} b_i dz_i \end{aligned}$$

This completes the proof.

A call option price on the stock shall have the form $C(S, s, t)$, a function of the stock price, volatility, and time to expiration. With the introduction of a random variance and thus two sources of uncertainty, a portfolio with only one option and one stock is not sufficient for creating a riskless investment strategy. A portfolio consisting of one stock and two options having different expiration dates is required, or it should be

$$C(\cdot,\cdot,\tau_1) + w_2 C(\cdot,\cdot,\tau_2) + w_3 S.$$

Now we may use *Itô's* lemma above to derive the stochastic differential:

$$\begin{aligned} \mathrm{d}C &= C_1 \mathrm{d}S + C_2 \mathrm{d}\sigma + C_3 \mathrm{d}\tau \\ &\quad + \frac{1}{2} C_{11} (\mathrm{d}S)^2 + \frac{1}{2} H_{22} (\mathrm{d}\sigma)^2 + \frac{1}{2} C_{33} (\mathrm{d}\tau)^2 \\ &\quad + C_{12} \mathrm{d}S \mathrm{d}\sigma + C_{13} \mathrm{d}S \mathrm{d}\tau + C_{23} \mathrm{d}\sigma \mathrm{d}\tau \\ &= \left[C_1 \alpha S + C_2 \beta(\bar{\sigma} - \sigma) - C_3 + \frac{1}{2} C_{11} \sigma^2 S^2 \right. \\ &\quad \left. + \frac{1}{2} C_{22} \gamma^2 + \frac{1}{2} C_{12} \sigma S \gamma \rho \right] \mathrm{d}t + C_1 \sigma S \mathrm{d}z_1 + C_2 \gamma \mathrm{d}z_2 \end{aligned} \tag{17.2}$$

where r is the instantaneous correlation between $\mathrm{d}z_1$ and $\mathrm{d}z_2$, and the subscripts on C indicate partial derivatives and $\tau = T - t$ so that $\mathrm{d}\tau = -\mathrm{d}t$.

To eliminate the risk from the portfolio, we observe that the coefficients for dz_1 and dz_2 in the return of the portfolio, $\mathrm{d}C(\cdot,\cdot,\tau_1) + w_2 \mathrm{d}C(\cdot,\cdot,\tau_2) + w_3 \mathrm{d}S$, are respectively

$$C_1(\cdot,\cdot,\tau_1)\sigma S + w_2 C_1(\cdot,\cdot,\tau_2)\sigma S + w_3 \sigma S$$

and

$$C_2(\cdot,\cdot,\tau_1)\gamma + w_2 C_2(\cdot,\cdot,\tau_2)\gamma.$$

Set both the two quantities to be 0; we get

$$\hat{w}_2 = -C_2(\cdot,\cdot,\tau_1)/C_2(\cdot,\cdot,\tau_2).$$

and

$$\begin{aligned} \hat{w}_3 = &-C_1(\cdot,\cdot,\tau_1) + C_2(\cdot,\cdot,\tau_1) \\ &C_1(\cdot,\cdot,\tau_2)/C_2(\cdot,\cdot,\tau_2). \end{aligned}$$

Then the return of the portfolio becomes:

$$\begin{aligned} &\mathrm{d}C(\cdot,\cdot,\tau_1) + \hat{w}_2 \mathrm{d}C(\cdot,\cdot,\tau_2) + \hat{w}_3 \mathrm{d}S \\ &\quad = \left\{ -C_3(\tau_1) + \frac{1}{2} C_{11}(\tau_1)\sigma^2 S^2 \right. \\ &\quad + \frac{1}{2} C_{12}(\tau_1)\sigma S \gamma \delta + \frac{1}{2} C_{22}(\tau_1)\gamma^2 \\ &\quad - \frac{C_2(\tau_1)}{C_2(\tau_2)} \left[-C_3(\tau_2) + \frac{1}{2} C_{11}(\tau_2)\sigma^2 S^2 \right. \\ &\quad \left. \left. + \frac{1}{2} C_{12}(\tau_2)\sigma S \gamma \rho + \frac{1}{2} C_{22}(\tau_2)\gamma^2 \right] \right\} \mathrm{d}t \end{aligned}$$

Since it is a riskless return, it should be equal to the risk-free rate, that is,

$$\begin{aligned} &\mathrm{d}C(\cdot,\cdot,\tau_1) + \hat{w}_2 \mathrm{d}C(\cdot,\cdot,\tau_2) + \hat{w}_3 \mathrm{d}S \\ &\quad = r[C(\cdot,\cdot,\tau_1) + \hat{w}_2 C(\cdot,\cdot,\tau_2) + \hat{w}_3 S]\mathrm{d}t. \end{aligned}$$

Substituting $\hat{w}_2$ and $\hat{w}_3$ into the above two differential equations, we get

$$\begin{aligned} &-C_3(\tau_1) + \tfrac{1}{2} C_{11}(\tau_1)\sigma^2 S^2 + \tfrac{1}{2} C_{12}(\tau_1)\sigma S \gamma \delta \\ &\quad + \tfrac{1}{2} C_{22}(\tau_1)\gamma^2 - \tfrac{C_2(\tau_1)}{C_2(\tau_2)} \left[-C_3(\tau_2) + \tfrac{1}{2} C_{11}(\tau_2)\sigma^2 S^2 \right. \\ &\quad \left. + \tfrac{1}{2} C_{12}(\tau_2)\sigma S \gamma \rho + \tfrac{1}{2} C_{22}(\tau_2)\gamma^2 \right] \\ &= r[C(\cdot,\cdot,\tau_1) - C_2(\cdot,\cdot,\tau_1)C(\cdot,\cdot,\tau_2)/C_2(\cdot,\cdot,\tau_2) \\ &\quad + (-C_1(\cdot,\cdot,\tau_1) + C_2(\cdot,\cdot,\tau_1) \\ &\quad C_1(\cdot,\cdot,\tau_2)/C_2(\cdot,\cdot,\tau_2))S] \end{aligned},$$

implying

$$\begin{aligned} &C_3(\tau_1) - \frac{1}{2} C_{11}(\tau_1)\sigma^2 S^2 - \frac{1}{2} C_{12}(\tau_1)\sigma S \gamma \delta \\ &\quad - \frac{1}{2} C_{22}(\tau_1)\gamma^2 + C(\tau_1)r - C_1(\tau_1)Sr \\ &\quad - \frac{C_2(\tau_1)}{C_2(\tau_2)} [C_3(\tau_2) - \frac{1}{2} C_{11}(\tau_2)\sigma^2 S^2 \\ &\quad - \frac{1}{2} C_{12}(\tau_2)\sigma S \gamma \rho - \frac{1}{2} C_{22}(\tau_2)\gamma^2 + C(\tau_2)r \\ &\quad - C_1(\tau_2)Sr] = 0 \end{aligned} \tag{17.3}$$

It is easily observed that the solutions of the following equation are also solutions of (17.3):

$$\begin{aligned} &C_3 - \frac{1}{2} C_{11}\sigma^2 S^2 - \frac{1}{2} C_{12}\sigma S \gamma \rho - \frac{1}{2} C_{22}\gamma^2 \\ &\quad + Cr - C_1 Sr = 0. \end{aligned}$$

But further investigations show that a more general solution can be obtained from

$$C_3 - \frac{1}{2}C_{11}\sigma^2 S^2 - \frac{1}{2}C_{12}\sigma S\gamma\rho - \frac{1}{2}C_{22}\gamma^2 + Cr - C_1 Sr - C_2 b^* = 0$$

where b^* is an arbitrary function or constant. This means that a unique solution for the option price function in the random variance model cannot be obtained only through arbitrage. Alternative view on this problem is that we may form a duplicating portfolio for an option in this model which contains the stock, the riskless bond, and another option. We cannot determine the price of a call option without knowing the price of any other call on the same stock. However, this is just what we are trying to do with the equation. In other words, only with a predetermined market price of the risk we can obtain a unique solution for the option price function.

In Appendix, some aspects to the market price of the risk are shown.

To derive a unique option pricing function, the following equation based on the arbitrage pricing theory by Ross (1976) is introduced,

$$E\left(\frac{dC}{C}\right) = \left[r + \lambda_1 \frac{C_1}{C}\sigma S + \lambda_2 \frac{C_2}{C}\gamma\right] dt \equiv \left[r + \frac{C_1 S}{C}(\alpha - r) + \frac{C_2}{C}\lambda_2^*\right] dt$$

where $(a–r)$ is the risk premium on the stock and l_2^* is the risk premium associated with ds. Equating these expressions, we may get:

$$\begin{aligned}&\left(C_1\alpha S + C_2\beta(\bar{\sigma} - \sigma) - C_3 + \tfrac{1}{2}C_{11}\sigma^2 S^2 + \tfrac{1}{2}C_{22}\gamma^2 + \tfrac{1}{2}C_{12}\sigma S\gamma\delta\right)/C \\ &= \left[r + \tfrac{C_1 S}{C}(\alpha - r) + \tfrac{C_2}{C}\lambda_2^*\right]\end{aligned},$$

or

$$C_3 - \frac{1}{2}C_{11}\sigma^2 S^2 - \frac{1}{2}C_{12}\sigma S\gamma\delta - \frac{1}{2}C_{22}\gamma^2 + Cr - C_1 Sr - C_2[\beta(\bar{\sigma} - \sigma) - \lambda^*] = 0 \tag{17.4}$$

This equation with the boundary conditions has a unique solution, and it is easy to show that this solution also satisfies Eq. (17.3). Now it is seen that the expected return on the stock does not influence the option price but the expected change and the risk premium associated with the volatility do.

The solution for the option price function follows from Lemma 4 of Cox et al. (1985):

$$C(S, \sigma, t; r, K) = \widehat{E}(e^{-rt}\max\{0, S_t - K\}|S_0, \sigma_0), \tag{17.5}$$

where $\widehat{E}$ represents a risk-adjusted expectation. For the risk adjustment, we reduce the mean parameter of dP and dS by the corresponding risk premiums. For the stock return, a is replaced by the risk-free rate, r, and for the standard deviation, $[\beta(\bar{\sigma} - \sigma) - \lambda^*]$ is used in place of $\beta(\bar{\sigma} - \sigma)$. Following Karlin and Taylor (1981), the backward equation for the function can be derived and it can be shown that it solves the equation in (17.4) with the adjustments on the dS and ds processes.

The option pricing function in (17.5) is a general solution to this random variance model. For operational issues, parameters of the s process, the risk premium l^*, and the instantaneous correlation coefficients between the stock return and ds are all required. With these parameters and the current value of s given, we do simulations to compute the option prices.

The model can be simplified with l^* and d set as 0. The risk premium could be zero if, for example, the volatility risk of the stock can be diversifiable or if ds is uncorrelated with the marginal utility of wealth. Now we develop the distribution of the stock-price function at expiration with $l^* = 0$ and $d = 0$. Consider the following solution conditional on the process $\{s_s$: $0 < s < t\}$:

$$S_t = S_0 \exp\left\{\int_0^t \left(r - \frac{1}{2}\sigma_{(s)}^2\right) ds + \int_0^t \sigma_{(s)} dz_1(s)\right\}.$$

We have

$$d\exp\left(\int_0^t\left(r-\frac{\sigma^2}{2}\right)\mathrm{d}s\right)=\exp\left(\int_0^t\left(r-\frac{\sigma^2}{2}\right)\mathrm{d}s\right)\cdot d\left(\int_0^t\left(r-\frac{\sigma^2}{2}\right)\mathrm{d}s\right)=\exp\left(\int_0^t\left(r-\frac{\sigma^2}{2}\right)\mathrm{d}s\right)\left(r-\frac{\sigma^2}{2}\right)\mathrm{d}t$$

and

$$d\exp\left(\int_0^t\sigma\mathrm{d}z_1\right)=\frac{1}{2}\exp\left(\int_0^t\sigma\mathrm{d}z_1\right)\cdot d^2\left(\int_0^t\sigma\mathrm{d}z_1\right)+\exp\left(\int_0^t\sigma\mathrm{d}z_1\right)d\left(\int_0^t\sigma\mathrm{d}z_1\right)=\frac{1}{2}\sigma^2\exp\left(\int_0^t\sigma\mathrm{d}z_1\right)\mathrm{d}t+\sigma\exp\left(\int_0^t\sigma\mathrm{d}z_1\right)\mathrm{d}z_1$$

So we see that

$$\begin{aligned}\mathrm{d}S_t&=S_0\exp\left\{\int_0^t\left(r-\frac{1}{2}\sigma_{(s)}^2\right)\mathrm{d}s\right\}\cdot d\exp\left\{\int_0^t\sigma_{(s)}\mathrm{d}z_1(s)\right\}+S_0\exp\left\{\int_0^t\sigma_{(s)}\mathrm{d}z_1(s)\right\}\cdot d\exp\left\{\int_0^t\left(r-\frac{1}{2}\sigma_{(s)}^2\right)\mathrm{d}s\right\}\\&=S_0\exp\left\{\int_0^t\left(r-\frac{1}{2}\sigma_{(s)}^2\right)\mathrm{d}s\right\}\left[\frac{1}{2}\sigma^2\exp\left(\int_0^t\sigma\mathrm{d}z_1\right)\mathrm{d}t+\sigma\exp\left(\int_0^t\sigma\mathrm{d}z_1\right)\mathrm{d}z_1\right]\\&\quad+S_0\exp\left\{\int_0^t\sigma_{(s)}\mathrm{d}z_1(s)\right\}\exp\left(\int_0^t\left(r-\frac{\sigma^2}{2}\right)\mathrm{d}s\right)\left(r-\frac{\sigma^2}{2}\right)\mathrm{d}t=rS_t\mathrm{d}t+\sigma S_t\mathrm{d}\sigma\end{aligned}$$

That is, S_t is a solution of the stochastic equation for dS in (17.1).

Furthermore, we note that $\int_0^t\sigma_{(s)}\mathrm{d}z_1(s)$ can be characterized as the Riemann sum:

$$\sum_{i=0}^{n}\sigma_{(t_i)}(z_{t_{i+1}}-z_{t_i})$$

with a proper partition $\{t_0=0,\ \ldots,\ t_n=t\}$ for which $\max(t_{i-1}-t_i)\to 0$ as $n\to\infty$. Since Z_1 is a Brownian motion, each piece of the path is independently normally distributed. So we can see that the Riemann sum is distributed as a normal distribution with mean 0 and variance $\sum_{i=0}^{n}\sigma_{(t_i)}^2(z_{t_{i+1}}-z_{t_i})$. As n is large enough, the sum approaches the integral and thus

$$\int_0^t\sigma_{(s)}\mathrm{d}z_1(s)\sim N(0,\mathrm{V}).$$

with $V=\int_0^t\sigma_{(s)}^2\mathrm{d}s$. Then the distribution of stock price conditional on S_0 and $\{s_s\}$ is lognormal,

$$\ln(S_t/S_0)|S_0,\{\sigma_s\}\sim N\left(rt-\frac{V}{2},V\right).$$

The formula for option price becomes

$$C_t|S_0,\{\sigma_s\} = \mathrm{e}^{-rt}E(\max\{0,S_t-K\}|S_0,\{\sigma_s\}) \\ = S_0N(d_1) - K\mathrm{e}^{-rt}N(d_2),$$

where

$$d_1 = \frac{\ln(S_0/K)+rt+V/2}{\sqrt{V}} \text{ and } d_2 = d_1 - \sqrt{V}.$$

This result is essentially the Black–Scholes formula with s^2t replaced by V. To find the unconditional expectation of the option price, we should integrate this formula over the distribution of V,

$$C(S_0,\sigma_0,t;r,K,\beta,\bar{\sigma},\gamma) = \int_0^\infty [S_0N(d_1) - K\mathrm{e}^{-rt}N(d_2)]\mathrm{d}F(V;t,\sigma_0,\beta,\bar{\sigma},\gamma) \quad (17.6)$$

This integral converges because F is a distribution function and the function inside the bracket is bounded given the values of S_0, K, r, and t. This finishes the problem.

What are left is the estimation of the parameters in the ds processes and the current values of s. A common approach in the empirical literature is to use actual option prices to infer the values of s_0. Statistical techniques about point estimation can be applied here. One method is to determine the unconditional distribution of stock returns as a function of a, b, $\bar{\sigma}$, and g. Then apply the maximum likelihood method to get the estimates. Another approach is to use the method of moments, and this is illustrated in Scott (1987).

17.3 Review of Characteristic Function

One of the most important problems in statistics is the determination of cumulative distribution functions (CDF) of measurable functions of random variables. The characteristic function is one possible way in deriving the CDF of interest. It is also useful in generating moments and cumulants of the distribution.

The characteristic function $\varphi(t)$ of a random variable X having CDF $F(x)$ is defined as

$$\phi_x(t) = E(\mathrm{e}^{itx}) = \int_{-\infty}^{\infty} \mathrm{e}^{itx}\mathrm{d}F(x), \quad (17.7)$$

where $i = \sqrt{-1}$ and t is a real number. The function $\varphi_x(t)$ is sometimes referred as the characteristic function corresponding to $F(x)$ or more briefly, the characteristic function of X. Since $\mathrm{e}^{itx} = \cos tx + i\sin tx$, $\cos tx$ and $\sin tx$ are both integrable over $(-\infty,\infty)$ for any real t, then $\varphi_x(t)$ is a complex number whose real and imaginary parts are finite for every value of t.

It is noted that the moment-generating function $M_x(t)$ and $\varphi_x(t)$ are related by the following,

$$M_x(t) = E(\mathrm{e}^{tx}) = \phi\left(\frac{t}{i}\right). \quad (17.8)$$

If the rth moment μ_r' exists, we can differentiate Eq. (17.7) h times, $0<h\le r$, with respect to t and obtain

$$\mu_h' = \frac{\phi^{(h)}(0)}{i^h} \quad 0<h\le r. \quad (17.9)$$

Once the characteristic function of a measurable function of random variables is obtained, it is usually of interest to obtain the CDF and the pdf. The answer is provided in the following theorem of Levy (1925).

Theorem 1 (Levy) *Let X be a random variable having characteristic function $\varphi_x(t)$ and CDF $F(x)$. Then if $F(x)$ is continuous at $X = x \pm \delta$, $\delta > 0$, we have*

$$F(x+\delta) - F(x-\delta) = \lim_{A\to\infty}\frac{1}{\pi}\int_{-A}^{A}\frac{\sin\delta t}{t}\mathrm{e}^{-itx}\varphi_x(t)\,\mathrm{d}t \quad (17.10)$$

Furthermore, if $\int_{-\infty}^{\infty}|\varphi(t)|\mathrm{d}t<\infty$, the pdf $f(x)$ exists at $X = x$ and

$$f(x) = \frac{1}{2\pi} \int_{-\infty}^{\infty} e^{-itx} \varphi_x(t)\, dt. \qquad (17.11)$$

Proof For the proof of the theorem, see Wilks (1963, pp. 116, 117).

A very useful result for obtaining tail probability is given in the following theorem.

Theorem 2 (Gil-Pelaez 1951) *Let $\varphi(t)$ be the characteristic function of the random variable X. Then*

$$\begin{aligned} 1 - F(x) = P(X \geq x) &= \frac{1}{2} + \frac{1}{\pi} \int_0^{\infty} \frac{[e^{-itx}\varphi(t) - e^{itx}\varphi(-t)]}{2it} dt \\ &= \frac{1}{2} + \frac{1}{\pi} \int_0^{\infty} \mathrm{Re}\left[\frac{e^{-itx}\varphi(t)}{it}\right] dt \end{aligned} \qquad (17.12)$$

Proof See Gil-Pelaez (1951).

The above theorems will be useful in deriving the desired probabilities in Sect. 17.4.

17.4 Closed-Form Type of Option Pricing Model

Instead of the nonclosed solution of option pricing model with stochastic volatility as proposed by Scott (1987), we discuss the closed-form solution of option pricing model when the volatility of stock returns is assumed stochastic, with a special emphasis on the model proposed by Heston (1993). The results are expressed in terms of the characteristic function, which is a closed form mathematically. Of course, numerical solution is still needed in practice.

Since there are two stochastic partial differential equations as given in Eq. (17.1), we will need Lemma 17.1 with $n = 2$.

We next assume that the stock price at time t follows the diffusion

$$dS(t) = \mu(t)S(t)dt + \sqrt{v(t)}S(t)dZ_1(t). \quad (17.13)$$

where $\mu(t)$ is the instantaneous drift of stock return and $V(t)$ is the variance, and $Z_1(t)$ is a Wiener's process.

Furthermore, we assume that $V(t)$ follows

$$dV(t) = \alpha(S, v, t)dt + \eta\beta(S, v, t)dZ_2(t) \quad (17.14)$$

where Z_2 is Wiener's process with $dZ_1 dZ_2 = \rho dt$, $-1 \leq \rho \leq 1$. It is noted that η is the volatility of volatility and ρ is the correlation between dZ_1 and dZ_2.

In Black–Scholes case, there is only one source of randomness—the stock price, which can be hedged with stock. In the present case, random changes in volatility also need to be hedged in order to form a riskless portfolio. So we set up a portfolio $\prod$ containing the option being priced whose value we denote by $U(S, v, t)$, a quantity $-\Delta$ of the sock and a quantity $-\Delta_1$ of another asset with value $U_1(S, v, t)$, we have

$$\Pi = U(s, v, t) - \Delta S - \Delta_1 U_1(s, v, t)$$

By Lemma 17.1, we have

$$\begin{aligned} dU &= \frac{\partial U}{\partial S} dS + \frac{\partial U}{\partial v} dv + \left\{ \frac{\partial U}{\partial t} + \frac{1}{2} v S^2 \frac{\partial^2 U}{\partial S^2} + \rho\eta v \beta S \frac{\partial^2 U}{\partial v \partial S} + \frac{1}{2}\eta^2 v \beta^2 \frac{\partial^2 U}{\partial v^2} \right\} dt, \\ dU_1 &= \frac{\partial U_1}{\partial S} dS + \frac{\partial U_1}{\partial v} dv + \left\{ \frac{\partial U_1}{\partial t} + \frac{1}{2} v S^2 \frac{\partial^2 U_1}{\partial S^2} + \rho\eta v \beta S \frac{\partial^2 U_1}{\partial v \partial S} + \frac{1}{2}\eta^2 v \beta^2 \frac{\partial^2 U_1}{\partial v^2} \right\} dt, \end{aligned}$$

$$
\begin{aligned}
d\Pi &= dU - \Delta dS - \Delta_1 dU_1 \\
&= \left\{\frac{\partial U}{\partial t} + \frac{v}{2}S^2\frac{\partial^2 U}{\partial S^2} + \rho\eta v\beta S\frac{\partial^2 U}{\partial S\partial v} + \frac{1}{2}\eta^2 v\beta^2\frac{\partial^2 U}{\partial v^2}\right\}dt - \Delta_1\left\{\frac{\partial U_1}{\partial t} + \frac{v}{2}S^2\frac{\partial^2 U_1}{\partial S^2}\right. \\
&\quad \left. + \rho\eta v\beta S\frac{\partial^2 U_1}{\partial S\partial v} + \frac{1}{2}\eta^2 v\beta^2\frac{\partial^2 U_1}{\partial v^2}\right\}dt \\
&\quad + \left\{\frac{\partial U}{\partial S} - \Delta_1\frac{\partial U_1}{\partial S} - \Delta\right\}dS \\
&\quad + \left\{\frac{\partial U}{\partial v} - \Delta_1\frac{\partial U_1}{\partial v}\right\}dv
\end{aligned}
$$

To eliminate all randomness from the portfolio, we must choose

$$\frac{\partial U}{\partial S} - \Delta - \Delta_1\frac{\partial U_1}{\partial S} = 0$$

to eliminate dS terms, and

$$\frac{\partial U}{\partial S} - \Delta_1\frac{\partial U_1}{\partial S} = 0$$

to eliminate dv terms.

This leaves us with

$$
\begin{aligned}
d\Pi &= \left\{\frac{\partial U}{\partial t} + \frac{1}{2}vS^2\frac{\partial^2 U}{\partial S^2} + \rho\eta v\beta S\frac{\partial^2 U}{\partial v\partial S} + \frac{1}{2}\eta^2 v\beta^2\frac{\partial^2 U}{\partial v^2}\right\}dt \\
&\quad -\Delta_1\left\{\frac{\partial U_1}{\partial t} + \frac{1}{2}vS^2\frac{\partial^2 U_1}{\partial S^2} + \rho\eta v\beta S\frac{\partial^2 U_1}{\partial v\partial S}, + \frac{1}{2}\eta^2 v\beta^2\frac{\partial^2 U_1}{\partial v^2}\right\}dt \\
&= r\Pi dt \\
&= r(U - \Delta S - \Delta_1 U_1)dt
\end{aligned}
$$

where we have used the fact that the return on a risk-free portfolio must equal the risk-free rate r. Collecting all U terms on the left-hand side and all U_1 terms on the right-hand side, we find that

$$
\frac{\frac{\partial U}{\partial t} + \frac{1}{2}vS^2\frac{\partial^2 U}{\partial S^2} + \rho\eta v\beta S\frac{\partial^2 U}{\partial v\partial S} + \frac{1}{2}\eta^2 v\beta^2\frac{\partial^2 U}{\partial v^2} + rs\frac{\partial U}{\partial S} - rU}{\frac{\partial U}{\partial v}}
= \frac{\frac{\partial U_1}{\partial t} + \frac{1}{2}vS^2\frac{\partial^2 U_1}{\partial S^2} + \rho\eta v\beta S\frac{\partial^2 U_1}{\partial v\partial S} + \frac{1}{2}\eta^2 v\beta^2\frac{\partial^2 U_1}{\partial v^2} + rS\frac{U_1}{\partial S} - rU_1}{\frac{\partial U_1}{\partial v}}
$$

The left-hand side is a function of U and the right-hand side is a function of U_1 only. The only way that this can be true is for both sides to be equal to some function f of the variables S, v, t.

$$
\begin{aligned}
&\frac{\partial U}{\partial t} + \frac{1}{2}vS^2\frac{\partial^2 U}{\partial S^2} + \rho\eta v\beta S\frac{\partial^2 U}{\partial v\partial S} + \frac{1}{2}\eta^2 v\beta^2\frac{\partial^2 U}{\partial v^2} \\
&+ rs\frac{U}{\partial S} - rU = -(\alpha - \lambda\beta)\frac{\partial U}{\partial v}
\end{aligned}
\tag{17.15}
$$

where without loss of generality, we have written the arbitrary function f of S, v, t as $-(\alpha - \lambda\beta)$. Conventionally, $\lambda(S, v, t)$ is called the market price of volatility risk because it tells us how much of the expected return of U is explained by the risk of v in the capital asset pricing model framework.

The Heston Model

The Heston model (Heston 1993) corresponds to choosing $\alpha(S, v, t) = k(\theta - v(t))$, $\beta(\mathrm{S, v, t}) = 1$, $\eta = \sigma$ in Eq. (17.14). These equations become

$$
\begin{aligned}
dS(t) &= \mu\, Sdt + \sqrt{v(t)}S(t)\, dZ_1(t) \\
dv(t) &= \kappa[\theta - v(t)]dt + \sigma\sqrt{v(t)}\, dZ_2(t)
\end{aligned}
\tag{17.16}
$$

with

$$dZ_1 dZ_2 = \rho dt$$

where κ is the speed of reversion of $v(t)$ and θ represents the long-term mean of volatility process.

We substitute the above values for $\alpha(S, v, t)$ and $\beta(S, v, t)$ into the general valuation equation to obtain

$$
\begin{aligned}
&\frac{1}{2}vS^2\frac{\partial^2 U}{\partial S^2} + \rho\sigma vS\frac{\partial^2 U}{\partial S\partial v} + \frac{1}{2}\sigma^2 v\frac{\partial^2 U}{\partial v} + rS\frac{\partial U}{\partial S} \\
&+ \{\kappa[\theta - v(t)] - \lambda(S, v, t)\}\frac{\partial U}{\partial v} - rU + \frac{\partial U}{\partial t} = 0
\end{aligned}
\tag{17.17}
$$

A European call option with strike price K and maturing at time T satisfies the PDE (17.19) subject to the following boundary conditions:

$$
\begin{aligned}
&U(S,v,T)=\max(0,S-k),\\
&U(0,v,T)=0,\\
&\frac{\partial U}{\partial S}(\infty,v,T)=1,\\
&rS\frac{\partial U}{\partial S}(S,0,t)+k\theta\frac{\partial U}{\partial v}(S,0,t)-rU(S,0,t)\\
&\quad+\frac{\partial U}{\partial t}(S,0,t)=0,\\
&U(S,\infty,t)=S.
\end{aligned}
$$

By analogy with the Black–Scholes formula, we guess a solution of the form

$$U(S,v,t)=SP_1-Ke^{-r(T-t)}P_2 \tag{17.18}$$

where the first term is the present value of the spot asset upon optimal exercise, and the second term is the present value of the strike price payment.

Let $x=\ln S$, then $U(x,v,t)=e^xP_1-Ke^{-r(T-t)}P_2$.

From Eq. (17.20), we have

$$\frac{\partial U}{\partial t}=e^x\frac{\partial P_1}{\partial t}-rKe^{-r(T-t)}P_2-Ke^{-r(T-t)}\frac{\partial P_2}{\partial t}, \tag{17.18'a}$$

$$
\begin{aligned}
\frac{\partial U}{\partial S}&=\frac{\partial U}{\partial x}\frac{\partial x}{\partial S}=\frac{1}{S}\frac{\partial U}{\partial x}\\
&=\frac{1}{S}\left[e^x\frac{\partial P_1}{\partial x}+e^xP_1-ke^{-r(T-t)}\frac{\partial P_2}{\partial x}\right]
\end{aligned} \tag{17.18'b}
$$

where

$$
\begin{aligned}
\frac{\partial U}{\partial x}&=e^x\frac{\partial P_1}{\partial x}+e^xP_1-Ke^{-r(T-t)}\frac{\partial P_2}{\partial x},\\
\frac{\partial U}{\partial v}&=e^x\frac{\partial P_1}{\partial v}-ke^{-r(T-t)}\frac{\partial P_2}{\partial v}
\end{aligned} \tag{17.18'c}
$$

$$\frac{\partial^2 U}{\partial v^2}=e^x\frac{\partial^2 P_1}{\partial v^2}-ke^{-r(T-t)}\frac{\partial^2 P_2}{\partial v^2}, \tag{17.18'd}$$

$$
\begin{aligned}
\frac{\partial^2 U}{\partial S\partial v}&=\frac{\partial}{\partial v}\left(\frac{\partial U}{\partial S}\right)=\frac{1}{S}\frac{1}{\partial v}\left(\frac{\partial U}{\partial x}\right)\\
&=\frac{1}{S}\left[e^x\frac{\partial^2 P_1}{\partial x\partial v}+e^x\frac{\partial P_1}{\partial v}-ke^{-r(T-t)}\frac{\partial^2 P_2}{\partial x^2}\right],
\end{aligned} \tag{17.18'e}
$$

$$
\begin{aligned}
\frac{\partial^2 U}{\partial x^2}&=e^x\frac{\partial^2 P_1}{\partial x^2}+2e^x\frac{\partial P_1}{\partial x}+e^xP_1\\
&\quad-ke^{-r(T-t)}\frac{\partial^2 P_2}{\partial x^2},
\end{aligned} \tag{17.18'f}
$$

$$
\begin{aligned}
\frac{\partial^2 C}{\partial S^2}&=\frac{\partial}{\partial S}\left(\frac{\partial C}{\partial S}\right)=\frac{1}{\partial S}\left[\frac{\partial C}{\partial x}\frac{1}{S}\right]=\frac{-1}{S^2}\frac{\partial C}{\partial x}+\frac{1}{S^2}\frac{\partial^2 C}{\partial x^2}\\
&=\frac{-1}{S^2}\left[e^x\frac{\partial P_1}{\partial x}+e^xP_1-Ke^{-r(T-t)}\frac{\partial P_2}{\partial x}\right]\\
&\quad+\frac{1}{S^2}\left[e^x\frac{\partial^2 P_1}{\partial x^2}+e^x\frac{\partial P_1}{\partial x}+e^x\frac{\partial P_1}{\partial x}\right.\\
&\quad\left.+e^xP_1-Ke^{-r(T-t)}\frac{\partial^2 P_2}{\partial x^2}\right]\\
&=\frac{1}{S^2}\left[e^x\frac{\partial^2 P_2}{\partial x^2}+e^x\frac{\partial P_1}{\partial x}+Ke^{-r(T-t)}\frac{\partial P_2}{\partial x}\right.\\
&\quad\left.-Ke^{-r(T-t)}\frac{\partial^2 P_2}{\partial x^2}\right].
\end{aligned} \tag{17.18'g}
$$

Substituting (17.18'a)–(17.18'g) into (17.16), we obtain

$$
\begin{aligned}
&\frac{1}{2}v\frac{\partial^2 P_j}{\partial x^2}+\rho\sigma v\frac{\partial^2 P_j}{\partial x\partial v}+\frac{1}{2}\sigma^2 v\frac{\partial^2 P_j}{\partial v^2}+(r+u_jv)\frac{\partial P_j}{\partial x}\\
&\quad+(a-b_jv)\frac{\partial P_j}{\partial v}+\frac{\partial P_j}{\partial t}=0 \quad \text{for } j=1,2
\end{aligned} \tag{17.19}
$$

where

$$
\begin{aligned}
&u_1=1/2,\ u_2=-1/2,\ a=\kappa\theta,\\
&b_1=\kappa+\lambda-\rho\sigma,\ b_2=\kappa+\lambda.
\end{aligned}
$$

For the option price to satisfy the boundary condition in Eq. (17.9), these PDEs (17.12) are subject to the terminal condition.

$$P_j(x, v, T; \ln[k]) = \mathbf{1}_{(x \geq \ln[k])} = \begin{cases} 1 & \text{if } x \geq \ln[k] \\ 0 & o.w. \end{cases} \quad (17.20)$$

Let $f_j(\phi, v, t)$ denote the characteristic function of $P_j(x, v, t)$.

The inversion of $f_j(\phi, v, t)$ is given by

$$P_j(x, v, t) = \frac{1}{2\pi} \int_{-\infty}^{\infty} e^{-i\phi x} f_j(\phi, v, t) d\phi \quad (17.21)$$

$$\text{(a)}. \ \frac{\partial P_j}{\partial x} = \frac{-1}{2\pi} \int_{-\infty}^{\infty} i\varphi\, e^{i\varphi x} f_j(\varphi, v, t) d\varphi$$

$$\text{(b)}. \ \frac{\partial^2 P_j}{\partial x^2} = \frac{1}{2\pi} \int_{-\infty}^{\infty} \varphi^2 e^{i\varphi x} f_j(\varphi, v, t) d\varphi$$

$$\text{(c)}. \ \frac{\partial^2 P_j}{\partial x \partial v} = \frac{-1}{2\pi} \int_{-\infty}^{\infty} i\varphi\, e^{i\varphi x} \frac{\partial f_j(\varphi, v, t)}{\partial v} d\varphi$$

$$\text{(d)}. \ \frac{\partial^2 P_j}{\partial v^2} = \frac{1}{2\pi} \int_{-\infty}^{\infty} e^{i\varphi x} \frac{\partial^2 f_j(\varphi, v, t)}{\partial v^2} d\varphi$$

$$\text{(e)}. \ \frac{\partial P_j}{\partial t} = \frac{1}{2\pi} \int_{-\infty}^{\infty} e^{i\varphi x} \frac{\partial f_j(\varphi, v, t)}{\partial t} d\varphi$$

Substituting (17.21) into Eq. (17.19) gives

$$\frac{1}{2} v\varphi^2 f_j - i\rho\sigma v\varphi \frac{\partial f_j}{\partial v} + \frac{\sigma^2}{2} v \frac{\partial^2 f_j}{\partial v^2} - (r + u_j\sigma)\, i\varphi f_j + (a - b_j v) \frac{\partial f_j}{\partial v} + \frac{\partial f_j}{\partial t} = 0.$$

The above equation can be rewritten as

$$v\left\{\alpha f_j - \beta \frac{\partial f_j}{\partial v} + \gamma \frac{\partial^2 f_j}{\partial v^2}\right\} + a \frac{\partial f_j}{\partial v} + \frac{\partial f_j}{\partial t} - ir\phi f_j = 0 \quad (17.22)$$

where

$$\alpha = -\frac{\phi^2}{2} + u_j i\phi, \ \beta = b_j - i\rho\sigma\phi, \ \gamma = \frac{\sigma^2}{2},$$
$$a = \kappa\vartheta$$

To solve for the characteristic function explicitly, we guess the function form

$$f_j = \exp\{C(\phi, t) + D(\phi, t)v\} f_j(\phi, v, T) = \tfrac{1}{i\phi} e^{-i\phi \ln[k]} \exp\{C(\phi, t) + D(\phi, t)v\},$$

where the characteristic function of $P_j(x, v, T)$ through (17.20)

$$f_j(\phi, v, T) = \int_{-\infty}^{\infty} e^{i\phi x} P_j(x, v, T) dx = \int_{\ln[k]}^{\infty} e^{i\varphi x} dx = \tfrac{1}{i\phi} e^{i\phi \ln[k]}.$$

It follows that

$$\frac{\partial f_j}{\partial t} = \left(\frac{\partial C}{\partial t} + v\frac{\partial D}{\partial t}\right) f_j,$$
$$\frac{\partial f_j}{\partial v} = D f_j,$$
$$\frac{\partial^2 f_j}{\partial v^2} = D^2 f_j.$$

Then (*) can be repressed as

$$\left(\frac{\partial C}{\partial t} + aD - i\varphi\gamma\right) f_j + v\left(\frac{\partial D}{\partial t} + \alpha - \beta D + \gamma D^2\right) f_j = 0$$

Then Eq. (17.22) is satisfied if

$$\begin{aligned} \frac{\partial C}{\partial t} &= -aD + i\varphi\gamma, \\ \frac{\partial D}{\partial t} &= -(\alpha - \beta D + \gamma D^2) \\ &= -\gamma(D - m_+)(D - m_-), \\ m_\pm &= \frac{\beta \pm \sqrt{\beta^2 - 4\alpha\gamma}}{2\gamma} = \frac{\beta \pm d}{\sigma^2}, \end{aligned} \quad (17.23)$$

where

$$d = \sqrt{\beta^2 - 4\alpha\gamma}$$
$$= \sqrt{(b_j - \rho\sigma\phi i)^2 - 4\left(u_j i\varphi - \frac{\phi^2}{2}\right)\left(\frac{\sigma^2}{2}\right)}$$
$$= \sqrt{(\rho\sigma\phi i - b_j)^2 - \sigma^2(\phi^2 - 2u_j\phi i)}.$$

Integrating (17.23) with the terminal conditions $C(\phi, 0) = 0$ and $D(\phi, 0) = 0$ gives

$$D(\varphi, t) = m_- \frac{1 - \mathrm{e}^{-dt}}{1 - g\mathrm{e}^{-dt}},$$
$$\text{where } g = \frac{m_-}{m_+} = \frac{b_j + \rho\sigma\varphi\, i - d}{b_j + \rho\sigma\varphi\, i + d},$$
$$C(\varphi, t) = -a\left\{m_- t - \frac{2}{\sigma^2}\ln\left[\frac{1 - g\mathrm{e}^{-dt}}{1 - g}\right]\right\}$$
$$+ i\varphi\, r.$$

Using the result of Gil-Pelaez (1951), one can invert the characteristic function to get the desired probabilities. Therefore,

$$P_j(x, v, t) = \frac{1}{2} + \frac{1}{\pi}\int_0^\infty \mathrm{Re}\left[\frac{\mathrm{e}^{-i\phi x} f_j(\phi, v, t)}{i\phi}\right] \mathrm{d}\phi.$$

17.5 Conclusion

In this chapter, we assume that the volatility of underlying asset is a stochastic process. Emphasis will be placed on the nonclosed-form solution developed by Scott (1987) and the closed-form solution of Heston (1993). In both cases, we consider a model in which the variance of stock-price returns varies according to an independent diffusion process. For the closed-form option pricing model, the results are expressed in terms of the characteristic function.

Appendix: The Market Price of the Risk

The equation we used to derive a unique solution for the stock-price function is closely related to arbitrage pricing theory developed by Ross (1976). Consider some derivative depending on n state variables and time t. Suppose that there are a total of at least $n + 1$ traded securities. Some assumptions are made as:

- The short selling of securities with full use of proceeds is permitted.
- There are no transaction costs and taxes.
- All securities are perfectly divisible.
- There are no riskless arbitrage opportunities.
- Security trading is continuous.

Denote the state variables as q_i, and suppose that they follow *Itô's* processes:

$$\mathrm{d}\theta_i = m\,\theta_i \mathrm{d}t + s_i\theta_i \mathrm{d}z_i$$

and the correlation between $\mathrm{d}z_i$ and $\mathrm{d}z_j$ is r_{ij}.

Let f_j be the price of ith traded security. With *Itô's* lemma,

$$\mathrm{d}f_j = \mu_j f_j\,\mathrm{d}t + \sum_i \sigma_{ij} f_j\,\mathrm{d}z_i$$

where

$$\mu_j f_j = \sum_i \frac{\partial f_j}{\partial\theta_i} m_i\theta_i + \frac{\partial f}{\partial t} + \frac{1}{2}\sum_{i,k} \frac{\partial^2 f_j}{\partial\theta_i\partial\theta_k}\rho_{ik} s_i s_k m_i m_k$$
$$\sigma_{ij} f_j = \frac{\partial f_j}{\partial\theta_i} s_i\theta_i$$

In these equations, m_j is the instantaneous mean rate of return provided by f_j.

Since there are $n + 1$ securities and n sources of variation, an instantaneous riskless portfolio P can be obtained as

$$\Pi = \sum_j k_j f_j, \quad i = 1, \ldots, n$$

where k_j's satisfy

$$\sum_j k_j \sigma_{ij} f_j = 0. \tag{17.24}$$

The return of the portfolio is then

$$d\Pi = \sum_j k_j \mu_j f_j \, dt.$$

The cost of forming such a portfolio is $\sum_j k_j f_j$, and if there are no arbitrage opportunities, the return rate of portfolio must be equal to the risk-free interest rate, so that

$$\sum_j k_j \mu_j f_j = r \sum_j k_j f_j$$

or

$$\sum_j k_j f_j (\mu_j - r) = 0. \tag{17.25}$$

Equations (17.24) and (17.25) can be regarded as $n + 1$ homogeneous linear equations in k_j's. The k_j's are not all zero if and only if the equations are consistent. In other words, $f_j(\mu_j - r)$ is a linear combination of s_{ij} f_j's, so that

$$f_j(\mu_j - r) = \sum_i \lambda_i \sigma_{ij} f_j$$

or

$$\mu_j - r = \sum_i \lambda_i \sigma_{ij}, \tag{17.26}$$

in which l_i is the market price of the risk for q_i.

Bibliography

Cox, J. C., Ingersoll, J. E., Jr., & Ross, S. A. (1985). A theory of the term structure of interest rates. *Econometrica, 53,* 363–384.

Gil-Pelaez, J. (1951). Note on the inversion theorem. *Biometrika, 38,* 481–482.

Heston, S. (1993). A closed-form solution for options with stochastic volatility with application to bond and currency options. *The Review of Financial Studies, 6* (2), 327–343.

Karlin, S., & Taylor, H. (1981). *A second course in stochastic processes*. New York: Academic Press.

Lee, C. F., & Lee, J. C. (2010). Stochastic volatility option pricing model. In *Handbook of quantitative finance and risk management* (pp. 481–489). New York: Springer.

Lee, C. F., Lee, A. C., & Lee, J. (2010). *Handbook of quantitative finance and risk management*. New York: Springer.

Levy, P. (1925). *Calcul des probabilités*. Paris: Gauthier-Villars.

Ross, S. (1976). The arbitrage theory of capital asset pricing. *Journal of Economic Theory, 13,* 341–360.

Scott, L. (1987). Option pricing when the variance changes randomly: Theory, estimation, and an application. *Journal of Finance and Quantitative Analysis, 22*(4), 419–438.

Wilks, S. S. (1963). *Mathematical statistics* (2nd ed.). Wiley.

Alternative Methods to Estimate Implied Variance: Review and Comparison

18

Contents

Abstract

The main purpose of this chapter is to demonstrate how to estimate implied variance for the Black–Scholes option pricing model (OPM). We classify various approaches into two different estimation routines: numerical search methods and closed-form derivation approaches. Both the MATLAB approach and approximation method are used to empirically estimate implied variance for American individual stock options and Chinese ETF 50 options.

18.1 Introduction

It is well known that implied variance estimation is important for evaluating option pricing. In this chapter, we first review several alternative methods to estimate implied variance in Sect. 18.2. We classify them into two different estimation routines: numerical search methods and closed-form derivation approaches. Closed-form derivation approaches took use of either Taylor's expansion or inverse function to calculate the analytical solutions for the ISD. In Sect. 18.3, we show how the MATLAB computer program can be used to estimate implied variance. This computer program is based upon the Black–Scholes model using Newton–Raphson method. In Sect. 18.4, we discuss how the approximation method derived by Ang et al. (2013) can be used

This chapter is drawn upon the paper by Lee et al. (2018).

C.-F. Lee et al., *Financial Econometrics, Mathematics and Statistics*,
https://doi.org/10.1007/978-1-4939-9429-8_18

to estimate implied variance under the case of continuous dividends. This approximation method can also estimate implied volatility from two options with the same maturity, but different exercise prices and values. In Sect. 18.5, real data from American option markets are used to compare the performances of three typical alternative methods: regression method proposed by Lai et al. (1992), MATLAB computer program approach, and approximation method derived by Ang et al. (2013). Also, this chapter presents the empirical results of China ETF 50 options which were new in the financial markets. Section 18.6 summarizes the chapter.

18.2 Numerical Search Method and Closed-Form Derivation Method to Estimate Implied Variance

The derivation and use of the implied volatility for an option as originated by Latane and Rendleman (1976) has become a widely used methodology for variance estimation. Latane and Rendleman (1976) argued that although it is impossible to solve the BS equation directly, one can use numerical search to closely approximate the standard deviation implied by given option price. The exact form of Black and Scholes model they used is given below.

$$C = SN(d_1) - Xe^{-rT}N(d_2) \tag{18.1}$$

where

$d_1 = \frac{\ln(S/X) + \left(r + \frac{1}{2}\sigma^2\right)T}{\sigma\sqrt{T}}$;

$d_2 = d_1 - \sigma\sqrt{T}$;

S = current market price of the underlying stock;

X = exercise price;

r = continuous constant interest rate;

T = remaining life of the option.

Their procedure is to find an implied standard deviation which makes the theoretical option value, i.e., the right-hand side of Eq. (18.1), within ± 0.001 of the observed actual call price. This is a kind of trial-and-error method.

Later researchers such as Beckers (1981), Manaster and Koehler (1982), Brenner and Subrahmanyam (1988), Lai et al. (1992), Chance (1996), Corrado and Miller (1996), Hallerbach (2004), Li (2005), and Corrado and Miller (2006) have studied implied variance estimation in more detail.

Since the Black–Scholes' option pricing model is a nonlinear equation, an explicit analytic solution for the ISD is not available in the literature (except for at-the-money call) and numerical methods are generally used to approximate the ISD. Manaster and Koehler (1982) used the Newton–Raphson method to provide an iterative algorithm for the ISD.

They rewrote the Black–Scholes formula as in Eq. (18.2), for given values of S, X, r and T.

$$C = f(S, X, r, T, \sigma) = f(\sigma) \tag{18.2}$$

For given values of S, X, r and T, f is a function of σ alone, and satisfies that:

$$\lim_{\sigma \to 0^+} f(\sigma) = \max\left(0, S - Xe^{-rT}\right)$$
$$\lim_{\sigma \to \infty} f(\sigma) = S$$

Equation (18.2) will have a positive solution of implied standard deviation σ^*, if and only if the option is rationally priced that $\max(0, S - Xe^{-rT}) < C < S$. This is because function $f(\cdot)$ is strictly monotone increasing in σ

over $(0, \infty)$[1], and the monotonicity and continuity of $f(\cdot)$ guarantee there is a unique solution.

Newton–Raphson method is a common used method to solve nonlinear systems of equation. In this case, for Eq. (18.2), the method is stated as:

$$\sigma_{n+1} = \sigma_n - \frac{f(\sigma_n) - C}{f'(\sigma_n)} \tag{18.3}$$

where σ_n is the nth estimate of σ^*, and $f'(\sigma_n)$ is the first derivative of $f(\sigma)$ when $\sigma = \sigma_n$.

Mean-Value Theorem. *Let f be a continuous function on the closed interval $[a, b]$, and can be differentiable on the open interval (a, b), where $a < b$. There exists some $c \in (a, b)$ such that:*

$$f'(c) = \frac{f(b) - f(a)}{b - a} \tag{18.4}$$

Under this case, mean-value theorem is stated as:

$$\begin{aligned} f'(\lambda\sigma_n + (1-\lambda)\sigma^*) &= \frac{f(\sigma_n) - f(\sigma^*)}{\sigma_n - \sigma^*} \\ &= \frac{f(\sigma_n) - C}{\sigma_n - \sigma^*}, \text{ for some } \lambda \in (0, 1) \end{aligned} \tag{18.5}$$

[1]Here we will briefly prove that the function $f(\cdot)$is strictly monotone increasing in σ.

$$\begin{aligned} f'(\sigma) &= \frac{\partial C}{\partial \sigma} = S\frac{\partial N(d_1)}{\partial \sigma} - X\mathrm{e}^{-rT}\frac{\partial N(d_2)}{\partial \sigma} \\ &= S\frac{\partial N(d_1)}{\partial d_1}\frac{\partial d_1}{\partial \sigma} - X\mathrm{e}^{-rT}\frac{\partial N(d_2)}{\partial d_2}\frac{\partial d_2}{\partial \sigma} \\ &= S\frac{1}{\sqrt{2\pi}}\mathrm{e}^{-\frac{d_1^2}{2}}\left(\frac{\sigma^2 T^{\frac{3}{2}} - \left[\ln\left(\frac{S}{X}\right) + \left(r + \frac{1}{2}\sigma^2\right)T\right]\sqrt{T}}{\sigma^2 T}\right) \\ &\quad - X\mathrm{e}^{-rT}\left(\frac{1}{\sqrt{2\pi}}\mathrm{e}^{-\frac{d_1^2}{2}}\frac{S}{X}\mathrm{e}^{rT}\right)\left(\frac{-\left[\ln\left(\frac{S}{X}\right) + \left(r + \frac{1}{2}\sigma^2\right)T\right]\sqrt{T}}{\sigma^2 T}\right) \\ &= S\frac{1}{\sqrt{2\pi}}\mathrm{e}^{-\frac{d_1^2}{2}}\left(\frac{\sigma^2 T^{\frac{3}{2}}}{\sigma^2 T}\right) \\ &= S\sqrt{T}\frac{1}{\sqrt{2\pi}}\mathrm{e}^{-\frac{d_1^2}{2}} \\ &= S\sqrt{T}N'(d_1) \end{aligned}$$

See Chap. 20 of Lee et al. (2013) for more details if interested in derivation of partial derivatives.Since $f'(\sigma) > 0$ when S, X, r and $T > 0$, and $\sigma > 0$, we have that $f(\cdot)$ is strictly monotone increasing in σ.

Combining the above equation and Eq. (18.3) in the main text, we can easily get:

$$\begin{aligned} \frac{|\sigma_{n+1} - \sigma^*|}{|\sigma_n - \sigma^*|} &= \left|1 - \frac{f(\sigma_n) - C}{f'(\sigma_n)(\sigma_n - \sigma^*)}\right| \\ &= \left|1 - \frac{f'(\lambda\sigma^* + (1-\lambda)\sigma_n)}{f'(\sigma_n)}\right| \end{aligned} \tag{18.6}$$

This motivates to choose σ_1 as σ that maximizes $f'(\sigma)$ so that σ_2 will be closer to σ^* than σ_1. From footnote 1, we know that to maximize $f'(\sigma)$ is to maximize $N'(d_1)$, where $N'(\cdot)$ is the standard normal density function. $N'(d_1) = \frac{1}{\sqrt{2\pi}}\exp\left(-\frac{d_1^2}{2}\right)$. For simplicity of presentation, we denote $N'(d_1) = g(d_1)$. First-order conditions for maximizing $N'(d_1)$, i.e., $g(d_1)$, is: $\frac{\partial g(d_1)}{\partial \sigma} = 0$.

We then have:

$$\begin{aligned} \frac{\partial g(d_1)}{\partial \sigma} &= \frac{1}{\sqrt{2\pi}}\mathrm{e}^{-\frac{d_1^2}{2}} \cdot (-d_1) \cdot \left(\frac{\partial d_1}{\partial \sigma}\right) \\ &= \frac{1}{\sqrt{2\pi}}\mathrm{e}^{-\frac{d_1^2}{2}} \cdot (-d_1) \cdot \left(-\frac{\ln(S/X) + rT}{\sigma^2\sqrt{T}} + \frac{\sqrt{T}}{2}\right) \\ &= \frac{1}{\sqrt{2\pi}}\mathrm{e}^{-\frac{d_1^2}{2}} \cdot (d_1) \cdot \left(\frac{1}{\sigma}\right)\left(\frac{\ln(S/X) + rT}{\sigma\sqrt{T}} - \frac{\sigma\sqrt{T}}{2}\right) \\ &= \frac{1}{\sqrt{2\pi}}\mathrm{e}^{-\frac{d_1^2}{2}} d_1 d_2/\sigma \\ &= g(d_1) d_1 d_2/\sigma \\ &= 0 \end{aligned}$$

Therefore, first-order condition is simplified to: $d_1 d_2 = 0$. This happens if either $d_1 = 0$ in which case $\sigma^2 = \frac{-2(ln(S/X) + rT)}{T}$ or $d_2 = 0$ in which case $\sigma^2 = \frac{2(ln(S/X) + rT)}{T}$. Now, we are checking second-order conditions under both the two cases.

$$\begin{aligned}\frac{\partial^2 g(d_1)}{\partial\sigma^2} &= \frac{\partial g'(d_1)}{\partial\sigma}\\ &= g'(d_1)\frac{\partial d_1}{\partial\sigma}\frac{d_1 d_2}{\sigma} + g(d_1)\left(\frac{d_1 d_2}{\sigma}\right)'\\ &= g(d_1)\left(\frac{d_1 d_2}{\sigma}\right)^2 - g(d_1)\frac{d_1^2 + d_2^2 + d_1 d_2}{\sigma^2}\\ &= g(d_1)\frac{d_1^2 d_2^2 - d_1^2 - d_2^2 - d_1 d_2}{\sigma^2}\end{aligned}$$

First-order condition gives that either $d_1 = 0$ or $d_2 = 0$. When $d_1 = 0, \frac{\partial^2 g(d_1)}{\partial\sigma^2} = g(d_1)\frac{-d_2^2}{\sigma^2} < 0$. Similarly, when $d_2 = 0, \frac{\partial^2 g(d_1)}{\partial\sigma^2} = g(d_1)\frac{-d_1^2}{\sigma^2} < 0$. $\frac{\partial^2 g(d_1)}{\partial\sigma^2} < 0$ holds under both cases; therefore, $g(d_1)$ and $f'(\sigma)$ are simultaneously maximized.

From the above discussion, we know that the starting point σ_1 should be chosen by maximizing the partial derivative of call option respective to volatility $f'(\sigma)$, as given in Eq. (18.7).

$$\sigma_1^2 = \left|\ln\left(\frac{S}{X}\right) + rT\right|\frac{2}{T} \tag{18.7}$$

Manaster and Koehler (1982) claimed that by starting with the above σ_1, implied variance estimate converges monotonically quadratically.

Brenner and Subrahmanyam (1988) applied Taylor's series expansion to the cumulative normal function at zero up to the first order in the Black–Scholes option pricing model.

For at-the-money options, they set the underlying asset price S equal to the present value of exercise price Xe^{-rT}, i.e., $S = Xe^{-rT}$, then d_1 and d_2 in Eq. (18.1) are:

$$\begin{aligned} d_1 &= \frac{1}{2}\sigma\sqrt{T}\\ d_2 &= -\frac{1}{2}\sigma\sqrt{T}\end{aligned} \tag{18.8}$$

Taylor series expansion is applied to the cumulative normal function at zero, while ignoring all the remaining terms beyond d_1.

$$\begin{aligned} N(d_1) &= N(0) + N'(0)d_1 + \cdots\\ &= \frac{1}{2} + \frac{1}{\sqrt{2\pi}}d_1 + o(d_1)\end{aligned} \tag{18.9}$$

Therefore, we have:

$$\begin{aligned} N(d_1) &\approx \frac{1}{2} + \frac{1}{\sqrt{2\pi}}d_1\\ &= \frac{1}{2} + \frac{1}{2\sqrt{2\pi}}\sigma\sqrt{T}\end{aligned} \tag{18.10}$$

$$\begin{aligned} N(d_2) &\approx 1 - N(d_1)\\ &= \frac{1}{2} - \frac{1}{2\sqrt{2\pi}}\sigma\sqrt{T}\end{aligned} \tag{18.11}$$

Substituting Eqs. (18.10) and (18.11) into call option pricing equation demonstrated in Eq. (18.1), we get:

$$C = \frac{S\sigma\sqrt{T}}{\sqrt{2\pi}} \tag{18.12}$$

Implied standard deviation then can be solved from Eq. (18.13), shown below:

$$\sigma = \frac{C\sqrt{2\pi}}{S\sqrt{T}} \tag{18.13}$$

Note that Brenner and Subramanyam's method can only be used to estimate implied standard deviation from at-the-money or at least not too far in- or out-of-the-money options.

Lai et al. (1992) derived a closed-form solution for the ISD in terms of the delta $\frac{\partial C}{\partial S}$, $\frac{\partial C}{\partial X}$, and other observable variables.

From Eq. (18.1), ceteris paribus, the effects of a change in stock price S and exercise price X on the call price are determined by Smith (1976), as following equations, respectively[2,3]:

$$\frac{\partial C}{\partial S} = N(d_1) \tag{18.14}$$

$$\frac{\partial C}{\partial X} = -\mathrm{e}^{-rT} N(d_2) \tag{18.15}$$

Equations (18.14) and (18.15) can be rearranged as Eqs. (18.16) and (18.17), respectively:

$$d_1 = N^{-1}\left[\left(\frac{\partial C}{\partial S}\right)\right] \tag{18.16}$$

$$d_2 = d_1 - \sigma\sqrt{T} = N^{-1}\left[\mathrm{e}^{rT}\left(-\frac{\partial C}{\partial X}\right)\right] \tag{18.17}$$

Combining Eqs. (18.16) and (18.17) yields:

$$\sigma = \left\{N^{-1}\left(\frac{\partial C}{\partial S}\right) - N^{-1}\left[\mathrm{e}^{rT}\left(-\frac{\partial C}{\partial X}\right)\right]\right\}/\sqrt{T} \tag{18.18}$$

where $N^{-1}(\cdot)$ is the inverse cumulative normal distribution function.

Equation (18.18) shows that ISD calculation depends on two partial derivatives of the call option with respect to the stock price and exercise price, i.e., $\frac{\partial C}{\partial S}$ and $\frac{\partial C}{\partial X}$, and other two observable variables: time to maturity T and risk-free rate r.

Note that implied volatility σ should not be negative; therefore, a negative right-hand side of Eq. (18.18) is not feasible.

Lai et al. (1992) argued that although the Black–Scholes option pricing model is a function of five variables, according to Merton (1973), the BS model exhibits homogeneity of degree one in the stock price and exercise price, which is shown in Eq. (18.19).

$$C = \left(\frac{\partial C}{\partial S}\right)S + \left(\frac{\partial C}{\partial X}\right)X = \beta_S S + \beta_X X \tag{18.19}$$

where $\beta_S = \frac{\partial C}{\partial S}$ and $\beta_X = \frac{\partial C}{\partial X}$.

The two partial derivatives can be estimated by running the following linear multiple regression:

$$C_{it} = \alpha + \beta_S S_t + \beta'_X \mathrm{e}^{-rT} X_{it} + \varepsilon_{it} \tag{18.20}$$

Substituting the least square estimators $\widehat{\beta}_S$ and $\widehat{\beta}'_X$ in Eq. (18.20) into Eq. (18.18), the implied variance can be estimated as:

$$\sigma = \left[N^{-1}(\widehat{\beta}_S) - N^{-1}(-\widehat{\beta}'_X)\right]/\sqrt{T} \tag{18.21}$$

Instead of running linear regression to estimate the two partial derivatives $\frac{\partial C}{\partial S}$ and $\frac{\partial C}{\partial X}$, we can first

---

[2]The derivation of $\frac{\partial C}{\partial S}$ is as follows.

$$\begin{aligned}\frac{\partial C}{\partial S} &= N(d_1) + S\frac{\partial N(d_1)}{\partial S} - X\mathrm{e}^{-rT}\frac{\partial N(d_2)}{\partial S}\\ &= N(d_1) + S\frac{\partial N(d_1)}{\partial d_1}\frac{\partial d_1}{\partial S} - X\mathrm{e}^{-rT}\frac{\partial N(d_2)}{\partial d_2}\frac{\partial d_2}{\partial S}\\ &= N(d_1) + S\frac{1}{\sqrt{2\pi}}\mathrm{e}^{-\frac{d_1^2}{2}}\cdot\frac{1}{S\sigma\sqrt{T}}\\ &\quad - X\mathrm{e}^{-rT}\frac{1}{\sqrt{2\pi}}\mathrm{e}^{-\frac{d_1^2}{2}}\frac{S}{X}\mathrm{e}^{rT}\cdot\frac{1}{S\sigma\sqrt{T}}\\ &= N(d_1) + S\frac{1}{S\sigma\sqrt{2\pi T}}\mathrm{e}^{-\frac{d_1^2}{2}} - S\frac{1}{S\sigma\sqrt{2\pi T}}\mathrm{e}^{-\frac{d_1^2}{2}}\\ &= N(d_1)\end{aligned}$$

Please note that:

$$\begin{aligned}\frac{\partial N(d_2)}{\partial d_2} &= \frac{1}{\sqrt{2\pi}}\mathrm{e}^{-\frac{d_2^2}{2}} = \frac{1}{\sqrt{2\pi}}\mathrm{e}^{-\frac{(d_1-\sigma\sqrt{T})^2}{2}}\\ &= \frac{1}{\sqrt{2\pi}}\mathrm{e}^{-\frac{d_1^2}{2}}\cdot\mathrm{e}^{d_1\sigma\sqrt{T}}\cdot\mathrm{e}^{-\frac{\sigma^2 T}{2}}\\ &= \frac{1}{\sqrt{2\pi}}\mathrm{e}^{-\frac{d_1^2}{2}}\cdot\mathrm{e}^{\ln\left(\frac{S}{X}\right)+\left(r+\frac{\sigma^2}{2}\right)T}\cdot\mathrm{e}^{-\frac{\sigma^2 T}{2}}\\ &= \frac{1}{\sqrt{2\pi}}\mathrm{e}^{-\frac{d_1^2}{2}}\cdot\frac{S}{X}\cdot\mathrm{e}^{rT}\end{aligned}$$

See Chap. 20 of Lee et al. (2013) for details if interested.

[3]The derivation of $\frac{\partial C}{\partial X}$ is as follows.

$$\begin{aligned}\frac{\partial C}{\partial X} &= S\frac{\partial N(d_1)}{\partial X} - \mathrm{e}^{-rT}N(d_2) - X\mathrm{e}^{-rT}\frac{\partial N(d_2)}{\partial X}\\ &= S\frac{\partial N(d_1)}{\partial d_1}\frac{\partial d_1}{\partial X} - \mathrm{e}^{-rT}N(d_2) - X\mathrm{e}^{-rT}\frac{\partial N(d_2)}{\partial d_2}\frac{\partial d_2}{\partial X}\\ &= S\frac{1}{\sqrt{2\pi}}\mathrm{e}^{-\frac{d_1^2}{2}}\frac{1}{\sigma\sqrt{T}}\frac{X}{S}\left(-\frac{S}{X^2}\right) - \mathrm{e}^{-rT}N(d_2)\\ &\quad - X\mathrm{e}^{-rT}\left(\frac{1}{\sqrt{2\pi}}\mathrm{e}^{-\frac{d_1^2}{2}}\frac{S}{X}\mathrm{e}^{rT}\right)\frac{1}{\sigma\sqrt{T}}\frac{X}{S}\left(-\frac{S}{X^2}\right)\\ &= -\mathrm{e}^{-rT}N(d_2)\end{aligned}$$

See Chap. 20 of Lee et al. (2013) for details if interested.

find $\frac{\partial C}{\partial X}$ by simple or weighted averaging $\frac{\Delta C}{\Delta X}$ for various exercise prices (S is being held constant provided the call price quotes are simultaneous). Then, the other partial derivative $\frac{\partial C}{\partial S}$ is got from Eq. (18.19). Lai et al. (1992) mentioned that this alternative approach would work best for index options, where there are many simultaneous quotes.

It should also be noted that, following their method, there is an alternative way to estimate implied standard deviation only using one partial derivative $\frac{\partial C}{\partial S}$. From Eq. (18.19), we have:

$$\begin{aligned} \left(\frac{\partial C}{\partial X}\right)X &= C - \left(\frac{\partial C}{\partial S}\right)S \\ \Rightarrow \left(\frac{\partial C}{\partial X}\right) &= \frac{C}{X} - \frac{S}{X}\left(\frac{\partial C}{\partial S}\right) \end{aligned} \tag{18.22}$$

Substituting Eq. (18.22) into Eq. (18.18), we will have a new closed-form solution for the ISD only depending on delta $\frac{\partial C}{\partial S}$ and other observable variables, as shown in Eq. (18.23).

$$\sigma = \left\{N^{-1}\left(\frac{\partial C}{\partial S}\right) - N^{-1}\left[e^{rT}\left(\frac{S}{X}\left(\frac{\partial C}{\partial S}\right) - \frac{C}{X}\right)\right]\right\}/\sqrt{T} \tag{18.23}$$

Brenner and Subrahmanyam (1988)'s formula for estimating implied variance is simple, but limited only to at-the-money or at least too far in- or out-of-the-money cases. On the basis of their research, Chance (1996) developed a generalized formula so that this formula can be implemented under other cases when options are in-the-money or out-of-the-money.

Recall Brenner–Subrahmanyam formula for ISD is:

$$\sigma^* = \frac{C^*\sqrt{2\pi}}{S\sqrt{T}} \tag{18.24}$$

where C^* is the price of the at-the-money call. We assume the call has an exercise price X^*.

Chance (1996) proposed a model that starts with Eq. (18.24), and added terms to reflect both the moneyness and sensitivity of standard deviation. The option with the unknown implied standard deviation is priced at C and has an exercise price of X. By definition, the difference between the at-the-money call and the call with unknown ISD is given as:

$$\Delta C^* = C - C^* \tag{18.25}$$

He argued that the difference in the prices of the two calls comes from: (1) the difference in exercise prices, i.e., $\Delta X^* = X - X^*$; (2) the difference in standard deviation, i.e., $\Delta\sigma^* = \sigma - \sigma^*$.

He applied second-order Taylor's series expansion on ΔC^*, which yields:

$$\begin{aligned} \Delta C^* &= \frac{\partial C^*}{\partial X^*}(\Delta X^*) + \frac{1}{2}\frac{\partial^2 C^*}{\partial X^{*2}}(\Delta X^*)^2 \\ &+ \frac{\partial C^*}{\partial \sigma^*}(\Delta\sigma^*) + \frac{1}{2}\frac{\partial^2 C^*}{\partial \sigma^{*2}}(\Delta\sigma^*)^2 \\ &+ \frac{\partial^2 C^*}{\partial\sigma^*\partial X^*}(\Delta\sigma^*\Delta X^*) \end{aligned} \tag{18.26}$$

Since these partial derivatives which appear in Eq. (18.26) are for at-the-money calls, their formulas can be simplified using the following relationships.

$$S = X^* e^{-rT} \tag{18.27}$$

$$d_1^* = \frac{1}{2}\sigma^*\sqrt{T} \tag{18.28}$$

$$d_2^* = -\frac{1}{2}\sigma^*\sqrt{T} \tag{18.29}$$

Therefore, we have the following important equations for partial derivatives hold, respectively.[4]

$$\frac{\partial C^*}{\partial X^*} = -e^{-rT}N(d_2^*) \tag{18.30}$$

[4]The derivation of Eq. (18.30) has been shown in Footnote 3.

$$\frac{\partial^2 C^*}{\partial X^{*2}} = -\mathrm{e}^{-rT}\frac{\partial N(d_2^*)}{\partial d_2^*}\frac{\partial d_2^*}{\partial X^*}$$
$$= -\mathrm{e}^{-rT}\frac{\mathrm{e}^{-\frac{d_2^{*2}}{2}}}{\sqrt{2\pi}}\frac{1}{\sigma^*\sqrt{T}}\left(\frac{X^*}{S^*}\right)\left(-\frac{S^*}{X^{*2}}\right)$$
$$= \frac{\mathrm{e}^{-rT}\mathrm{e}^{-\frac{d_2^{*2}}{2}}}{X^*\sigma^*\sqrt{2\pi T}} \tag{18.31}$$

For at-the-money call, Eq. (18.31) is given as in Eq. (18.32):

$$\frac{\partial^2 C^*}{\partial X^{*2}} = \frac{\mathrm{e}^{-rT}\mathrm{e}^{-\frac{d_2^{*2}}{2}}}{X^*\sigma^*\sqrt{2\pi T}} = \frac{\mathrm{e}^{-rT}}{X^*\sigma^*\sqrt{2\pi T}}\mathrm{e}^{-\frac{\sigma^{*2}T}{8}} \tag{18.32}$$

$$\frac{\partial C^*}{\partial \sigma^*} = S\frac{\partial N(d_1^*)}{\partial \sigma^*} - X^*\mathrm{e}^{-rT}\frac{\partial N(d_2^*)}{\partial \sigma^*} = \frac{S\sqrt{T}\mathrm{e}^{-\frac{d_1^{*2}}{2}}}{\sqrt{2\pi}} \tag{18.33}$$

Given the call is at-the-money, Eq. (18.33) is given as in Eq. (18.34)[5]:

$$\frac{\partial C^*}{\partial \sigma^*} = \frac{X\mathrm{e}^{-rT}\sqrt{T}}{\sqrt{2\pi}}\mathrm{e}^{-\frac{\sigma^{*2}T}{8}} \tag{18.34}$$

$$\frac{\partial^2 C^*}{\partial \sigma^{*2}} = \frac{S\sqrt{T}\mathrm{e}^{-\frac{d_1^{*2}}{2}}}{\sqrt{2\pi}}(-d_1^*)\frac{\partial d_1^*}{\partial \sigma^*}$$
$$= -\frac{S\sqrt{T}\mathrm{e}^{-\frac{d_1^{*2}}{2}}}{\sqrt{2\pi}}d_1^*\left(-\frac{\ln(S/X^*)+rT}{\sigma^{*2}\sqrt{T}} + \frac{\sqrt{T}}{2}\right)$$
$$= \frac{S\sqrt{T}}{\sqrt{2\pi}}\mathrm{e}^{-\frac{d_1^{*2}}{2}}\frac{d_1^* d_2^*}{\sigma^*} \tag{18.35}$$

For an at-the-money call, Eq. (18.35) becomes:

$$\frac{\partial^2 C^*}{\partial \sigma^{*2}} = \frac{X^*\mathrm{e}^{-rT}T^{\frac{3}{2}}\sigma^*}{4\sqrt{2\pi}}\mathrm{e}^{-\frac{\sigma^{*2}T}{8}} \tag{18.36}$$

$$\frac{\partial^2 C^*}{\partial \sigma^*\partial X^*} = -\mathrm{e}^{-rT}\frac{\partial N(d_2^*)}{\partial d_2^*}\frac{\partial d_2}{\partial \sigma}$$
$$= -\mathrm{e}^{-rT}\left(\frac{1}{\sqrt{2\pi}}\mathrm{e}^{-\frac{d_1^{*2}}{2}}\frac{S}{X^*}\mathrm{e}^{rT}\right)$$
$$\times\left(\frac{-\left[\ln\left(\frac{S}{X^*}\right)+\left(r+\frac{1}{2}\sigma^{*2}\right)T\right]\sqrt{T}}{\sigma^{*2}T}\right)$$
$$= \frac{S\mathrm{e}^{-\frac{d_1^{*2}}{2}}d_1^*}{X^*\sigma^*\sqrt{2\pi}} \tag{18.37}$$

Given the call is at-the-money, Eq. (18.37) becomes:

$$\frac{\partial^2 C^*}{\partial \sigma^*\partial X^*} = \frac{\mathrm{e}^{-rT}\sqrt{T}}{2\sqrt{2\pi}}\mathrm{e}^{-\frac{\sigma^{*2}T}{8}} \tag{18.38}$$

Equation (18.25) can be restated as:

$$C^* - C + \Delta C^* = 0 \tag{18.39}$$

Substituting Eq. (18.26) into Eq. (18.39), Eq. (18.39) can be viewed as a quadratic equation of $\Delta\sigma^*$, written as:

$$a(\Delta\sigma^*)^2 + b(\Delta\sigma^*) + c = 0 \tag{18.40}$$

where

$$a = \frac{1}{2}\frac{\partial^2 C^*}{\partial \sigma^{*2}}$$
$$b = \frac{\partial C^*}{\partial \sigma^*} + \frac{\partial^2 C^*}{\partial \sigma^*\partial X^*}(\Delta X^*)$$
$$c = C^* - C + \frac{\partial C^*}{\partial X^*}(\Delta X^*) + \frac{1}{2}\frac{\partial^2 C^*}{\partial X^{*2}}(\Delta X^*)^2$$

Therefore, the solution of the Eq. (18.40) should be:

$$\Delta\sigma^* = \frac{-b \pm \sqrt{b^2 - 4ac}}{2a} \tag{18.41}$$

Experiments in Chance (1996) suggest that positive root for $\Delta\sigma^*$ of Eq. (18.40) gives the correct solution for implied variance when adding it to

[5]The derivation of Eq. (18.33) has been shown in Footnote 1.

the value of σ^* from Brenner–Subrahmanyam formula.

One thing that needs to be noted is that in order to apply Chance's formula to compute the ISD, the standard deviation and the option price under the at-the-money case must be given. In other words, if the underlying asset price deviates from the present value of the exercise price and the call option price is not available (or unobservable) in the market, then Chance's formula for the ISD may not apply, just as the case of Brenner–Subrahmanyam formula.

To allow for the deviation between the underlying asset price and the present value of exercise price, Corrado and Miller (1996) expanded the cumulative normal function at zero to the first-order term in the Black–Scholes OPM to derive a quadratic equation of the ISD.

Their approach followed the method employed by Brenner and Subrahmanyan, and made use of the expansion of the normal distribution function as stated in Eq. (18.42).

$$N(z) = \frac{1}{2} + \frac{1}{\sqrt{2\pi}}\left(z - \frac{z^3}{6} + \cdots\right) \quad (18.42)$$

Substituting Eq. (18.42) into the normal probabilities $N(d_1)$ and $N(d_2)$ in classic Black–Scholes model as Eq. (18.1) states, we have Eq. (18.43) hold when cubic and higher-order terms are ignored.

$$\begin{aligned} C = {} & S\left(\frac{1}{2} + \frac{d_1}{\sqrt{2\pi}}\right) \\ & - Xe^{-rT}\left(\frac{1}{2} + \frac{d_1 - \sigma\sqrt{T}}{\sqrt{2\pi}}\right) \end{aligned} \quad (18.43)$$

K is defined as the present value of strike price X, i.e., $K = Xe^{-rT}$.

Recall that the expressions for d_1 and d_2 are:

$$\begin{aligned} d_1 & = \frac{\ln(S/X) + \left(r + \frac{1}{2}\sigma^2\right)T}{\sigma\sqrt{T}} \\ & = \frac{\ln(S/K) + \frac{1}{2}\sigma^2 T}{\sigma\sqrt{T}} \end{aligned}$$

$$d_2 = d_1 - \sigma\sqrt{T}$$

Equation (18.43) can be restated as:

$$\begin{aligned} C = {} & S\left(\frac{1}{2} + \frac{\ln(S/K) + \frac{1}{2}\sigma^2 T}{\sigma\sqrt{2\pi T}}\right) \\ & - K\left(\frac{1}{2} + \frac{\ln(S/K) - \frac{1}{2}\sigma^2 T}{\sigma\sqrt{2\pi T}}\right) \end{aligned} \quad (18.44)$$

Equation (18.44) can be formulated as the quadratic equation of $\sigma\sqrt{T}$, as shown in Eq. (18.45).

$$\begin{aligned} & \sigma^2 T(S+K) - \sigma\sqrt{T}\left[(2\sqrt{2\pi}C - \sqrt{2\pi}(S-K)\right] \\ & + 2(S-K)\ln(S/X) = 0 \end{aligned} \quad (18.45)$$

Corrado and Miller (1996) proved that only the largest root for Eq. (18.45) reduced to the original Brenner–Subrahmanyan formula, which is shown in Eq. (18.46).

$$\begin{aligned} \sigma\sqrt{T} = {} & \sqrt{2\pi}\left(\frac{2C - S + K}{2(S+K)}\right) \\ & + \sqrt{\frac{\pi}{2}\left(\frac{2C - S + K}{S+K}\right)^2 - \frac{2(S-K)\ln(S/K)}{S+K}} \end{aligned} \quad (18.46)$$

After solving the quadratic equation of $\sigma\sqrt{T}$, they improve the accuracy of approximation by minimizing its concavity.[6] Therefore, their final formula to compute the implied standard deviation was given as:

[6]This approach included several steps.First, logarithmic approximation was used. $\ln(S/K) \approx 2(S-K)/(S+K)$. Secondly, they replaced the value "4" with the parameter α to restate Eq. (18.46) as:

$$\begin{aligned} \sigma\sqrt{T} = {} & \sqrt{2\pi}\left(\frac{2C - S + K}{2(S+K)}\right) \\ & + \sqrt{\frac{\pi}{2}\left(\frac{2C - S + K}{S+K}\right)^2 - \alpha\left(\frac{S-K}{S+K}\right)^2} \end{aligned}$$

They chose a value for α such that the above equation is approximately linear in the stock price when the option is near the at-the-money case. When setting the second derivative of right-hand side of the above equation with respect to stock price equal to zero, they found the realistic estimated value for α was close to 2.

$$\sigma = \sqrt{\frac{2\pi}{T}}\frac{1}{S+K}\left[C - \frac{S-K}{2} + \sqrt{\left(C - \frac{S-K}{2}\right)^2 - \frac{(S-K)^2}{\pi}}\right] \quad (18.47)$$

Li (2005) also followed Brenner and Subrahmanyan and expanded the expression to the third-order term and solved for the ISD with a cubic equation.

Taylor's expansion as Eq. (18.42) states was used in Li's paper. He retained the cubic order and substituted Eq. (18.42) into the normal probabilities in the Black–Scholes model as stated in Eq. (18.1), and this yielded:

$$C = S\left(\frac{1}{2} + \frac{d_1}{\sqrt{2\pi}} - \frac{d_1^3}{6\sqrt{2\pi}}\right) - Xe^{-rT}\left(\frac{1}{2} + \frac{d_2}{\sqrt{2\pi}} - \frac{d_2^3}{6\sqrt{2\pi}}\right) \quad (18.48)$$

For at-the-money calls, $d_1 = \frac{1}{2}\sigma\sqrt{T}, d_2 = -\frac{1}{2}\sigma\sqrt{T}, S = Xe^{-rT}$

When defining $\xi = \frac{1}{2}\sigma\sqrt{T}$, the following equation hold for at-the-money calls:

$$\frac{\sqrt{2\pi}C}{S} \approx 2\xi - \frac{1}{3}\xi^3 \quad (18.49)$$

Equation (18.49) can be solved by using the cubic formula.[7]

He went through some tedious derivations and simplifications, and finally obtained the formula to compute implied standard deviation:

$$\sigma = \frac{2\sqrt{2}}{\sqrt{T}}z - \frac{1}{\sqrt{T}}\sqrt{8z^2 - \frac{6\alpha}{\sqrt{2}z}} \quad (18.50)$$

where $\alpha = \frac{\sqrt{2\pi}C}{S}$ and $z = \cos\left[\frac{1}{3}\cos^{-1}\left(\frac{3\alpha}{\sqrt{32}}\right)\right]$.

Since Li included the third-order term in the Taylor expansion on the cumulative normal distribution in his derivation, Li claimed that his formula for ISD provided a consistently more accurate estimate of the true ISD than previous studies.

To sum up, the existing researches mainly follow two different routines to estimate implied volatility. Numerical search methods tried to find an approximate solution for implied volatility which makes the theoretical option value equal to or very close to market observed option price. These methods do not provide closed-form solution for estimated implied volatility, and need iterative algorithms to approximate the ISD. Closed-form derivation approaches took use of either Taylor's expansion or inverse function to calculate the analytical solutions for the ISD. First-order, second-order, and third-order Taylor's expansions were applied to cumulative normal distribution function, respectively, to estimate the implied volatility in previous studies. There were also studies using inverse function of normal distribution to derive closed-form solution of the ISD.

An important point to be noted is that some methods rely upon the existence of "at-the-money" options, or at least not too far in- or out-of-the-money options. These approaches include Brenner and Subrahmanyam (1988), Chance (1996), and Li (2005).

Table 18.1 classifies the existing researches of estimating implied volatility accordingly.

18.3 MATLAB Approach to Estimate Implied Variance

Usually, implied variance can be obtained from a call or put option model by an optimization technique. For each individual option, the implied variance can be obtained by first choosing an initial estimate σ_0, and then

[7]The general cubic equation has the form $ax^3 + bx^2 + cx + d = 0$, with $a \neq 0$. If the cubic equation is in the form of $t^3 + pt + q = 0$, it is called a depressed cubic equation. Please note that any general cubic equation can be reduced to the depressed cubic equation by dividing the general equation with a and substituting variable x with $t = x - \frac{b}{3a}$.For a depressed cubic equation $t^3 + pt + q = 0$, the roots are:$t_k = 2\sqrt{-\frac{p}{3}}\cos\left(\frac{1}{3}\arccos\left(\frac{3q}{2p}\sqrt{\frac{-3}{p}}\right) - k\frac{2\pi}{3}\right), k = 0, 1, 2.$

Table 18.1 Classification of the ISD estimation methods

Numerical search	Closed-form derivation
Trial and error Latane and Rendleman (1976)	*Taylor's series expansion* First-order expansion: Brenner and Subrahmanyam (1988); Corrado and Miller (1996) Second-order expansion: Chance (1996) Third-order expansion: Li (2005)
Choose an initial point, iterative algorithm Manaster and Koehler (1982)	*Inverse function* Estimate parameters by regression: Lai et al. (1992)

Eq. (18.51) is used to iterate toward the correct value.

$$C_{j,t}^{M} - C_{j,t}^{T}(\sigma_0) = \left[\frac{\partial C_{j,t}^{T}}{\partial \sigma} \bigg| \sigma_0 \right] (\sigma - \sigma_0) + e_{j,t} \tag{18.51}$$

where

$C_{j,t}^{M}$ = market price of call option j at time t;
σ = true or actual implied standard deviation;
σ_0 = initial estimate of implied standard deviation;
$C_{j,t}^{T}(\sigma_0)$ = theoretical price of call option j at time t given $\sigma = \sigma_0$;
$\frac{\partial C_{j,t}^{T}}{\partial \sigma}\Big|\sigma_0$ = partial derivative of the call option with respect to the standard deviation σ at $\sigma = \sigma_0$;
$e_{j,t}$ = error term.

The partial derivative of the call option with respective to the standard deviation $\frac{\partial C_{j,t}^{T}}{\partial \sigma}$ from Black–Scholes model is:

$$\frac{\partial C_{t,j}^{T}}{\partial \sigma} = Xe^{-r\tau}\sqrt{\tau}N'(d_1) = Xe^{-r\tau}\frac{\sqrt{\tau}}{\sqrt{2\pi}}e^{-d_1^2/2} \tag{18.52}$$

It is also called Vega of option.

The iteration proceeds by re-initializing σ_0 to equal σ_1 at each successive stage until an acceptable tolerance level is attained. The tolerance level used is:

$$\left| \frac{\sigma_1 - \sigma_0}{\sigma_0} \right| < 0.001 \tag{18.53}$$

The MATLAB finance toolbox provides a function blsimpv to search for implied volatility. The algorithm used in the blsimpv function is Newton's method, just as the procedure described in Eq. (18.51). This approach minimizes the difference between observed market option value and the theoretical value of BS model, and obtains the ISD estimate until tolerance level is attained.

The complete command of the function *blsimpv* is: Volatility = *blsimpv*(Price, Strike, Rate, Time, Value, Limit, Yield, Tolerance, Class). And the command with default setting is: Volatility = *blsimpv*(Price, Strike, Rate, Time, Value).

There are nine inputs in total, while the last four of them are optional. Detailed explanations of all the inputs are as follows:

Inputs:

Price—Current price of the underlying asset,
Strike—Strike (i.e., exercise) price of the option,
Rate—Annualized continuously compounded risk-free rate of return over the life of the option, expressed as a positive decimal number,
Time—Time to expiration of the option, expressed in years,
Value—Price (i.e., value) of a European option from which the implied volatility of the underlying asset is derived.

Optional Inputs:

Limit—Positive scalar representing the upper bound of the implied volatility search interval. If empty or missing, the default is 10, or 1000% per annum.
Yield—Annualized continuously compounded yield of the underlying asset over the life of the option, expressed as a decimal number. For example, this could represent the dividend yield and foreign risk-free interest rate for options written on stock indices and currencies, respectively. If empty or missing, the default is zero.
Tolerance—Positive scalar implied volatility termination tolerance. If empty or missing, the default is 1e − 6.
Class—Option class (i.e., whether a call or put) indicating the option type from which the implied volatility is derived. This may be either a logical indicator or a cell array of characters. To specify call options, set Class = true or Class = {'Call'}; to specify put options, set Class = false or Class = {'Put'}. If empty or missing, the default is a call option.

Output:

Volatility—Implied volatility of the underlying asset derived from European option prices, expressed as a decimal number. If no solution can be found, a NaN (i.e., Not-a-Number) is returned.

Example:

Consider a European call option trading at $5 with an exercise price of $95 and 3 months until expiration. Assume the underlying stock pays 5% annual dividends, which is trading at $90 at this moment, and the risk-free rate is 3% per annum. Under these conditions, the command used in MATLAB will be either of the following two:
Volatility = blsimpv(90, 95, 0.03, 0.25, 5, [], 0.05, [], {'Call'})
Volatility = blsimpv(90, 95, 0.03, 0.25, 5, [], 0.05, [], true)

Note that this function provided by MATLAB's toolbox can only estimate implied volatility from a single option. For more than one option, the user needs to write their own programs to estimate implied variances.

18.4 Approximation Approach to Estimate Implied Variance

In this section, we will discuss alternative method proposed by Ang et al. (2009) to use the call option model and put option model to estimate implied volatility. Our approximation approach can also estimate implied volatility from two options with the same maturity, but different exercise prices and values.

Recall the Black–Scholes call option pricing model (with continuous dividends), we have:

$$C = S'N(d_1) - KN(d_2) \tag{18.54}$$

where

$$d_1 = \frac{\ln(S/X) + (r + \frac{\sigma^2}{2} - q)T}{\sigma\sqrt{T}} = \frac{\ln(S'/K)}{\sigma\sqrt{T}} + \sigma\sqrt{T}\big/2$$

$d_2 = d_1 - \sigma\sqrt{T}$;
C = call price;
S = stock price;
q = annual dividend yield;
$S' = Se^{-qT}$;
X = exercise price;
r = risk-free interest rate;
$K = Xe^{-rT}$, present value of exercise price;
T=time to maturity of option in years;
$N(\cdot)$ = standard normal distribution;
σ = stock volatility.

We derive a formula to estimate the ISD by applying the Taylor series expansion on a single call option. We show that, following method proposed by Ang et al. (2009, 2013), the formula for ISD derived by Corrado and Miller (1996) can be improved further without any replacements.

Recall the Taylor series expansion approximating complex functions from calculus (See Appendix 5B in Lee et al. 2017), which can be mathematically written as follows:

$$F_n(x) = F(a) + F'(a)(x-a) + \frac{F''(a)}{2!}(x-a)^2 + \cdots + \frac{F^{(n)}(a)}{n!}(x-a)^2 \tag{18.55}$$

where

$F_n(x)$ is the function we are approximating,
$F'(a)$is the first derivative of the function,
$F^{(n)}(a)$ is the nth derivative of the function,
$n!$ is the factorial value of n, i.e., $n! = (n)(n-1)\cdots(2)(1)$,
a is the value near which we are making the approximation to the function $F(x)$.

Let $L' = \ln(S'/K)/\sigma\sqrt{T}$. Here, we apply the Taylor series expansion to both cumulative normal distributions in the Black–Scholes formula at L'.

Then we have

$$\begin{aligned} N\left(L' + \sigma\sqrt{T}/1\right) &= N(L') + N'(L')\left(L' + \sigma\sqrt{T}/2 - L'\right) \\ &\quad + N''(L')\frac{0(L' + \sigma\sqrt{T}/2 - L')^2}{2!} + \cdots \\ &= N(L') + N'(L')\sigma\sqrt{T}/2 \\ &\quad + N''(L')\left(\sigma\sqrt{T}/2\right)^2 \Big/ 2 + e_1 \\ &= N(L') + N'(L')\left(\sigma\sqrt{T}/2\right) \\ &\quad \times [1 - \ln(S'/K)/4] + e_1 \end{aligned} \tag{18.56}$$

and

$$\begin{aligned} N\left(L' - \sigma\sqrt{T}/2\right) &= N(L') + N'(L')\left(L' - \sigma\sqrt{T}/2 - L'\right) \\ &\quad + N''(L')\frac{(L' - \sigma\sqrt{T}/2 - L')^2}{2!} + \cdots \\ &= N(L') - N'(L')\sigma\sqrt{T}/2 \\ &\quad + N''(L')\left(\sigma\sqrt{T}/2\right)^2 \Big/ 2 + e_2 \end{aligned}$$

$$\begin{aligned} N(L' - \sigma\sqrt{T}/2) &= N(L') - N'(L')\sigma\sqrt{T}/2 \\ &\quad + N''(L')\left(\sigma\sqrt{T}/2\right)^2 \Big/ 2 + e_2 \\ &= N(L') - N'(L')\left(\sigma\sqrt{T}/2\right) \\ &\quad \times [1 + \ln(S'/K)/4] + e_2 \end{aligned} \tag{18.57}$$

where e_1 and e_2 are the remainder terms of Taylor's formulas.

The above equations can be obtained by the fact that $N''(x) = -N'(x)x$.

Given $N(0) = 1/2$, $N'(0) = 1/\sqrt{2\pi}$, $N'''(0) = -N'(0)$, and $N''(0) = N''''(0) = 0$, we expand $N(L')$ and $N'(L')$ at 0, respectively.

$$\begin{aligned} N(L') &= N(0) + N'(0)L' + N''(0)L'^2/2 + e_3 \\ &= \frac{1}{2} + L'/\sqrt{2\pi} + e_3 \end{aligned} \tag{18.58}$$

$$\begin{aligned} N'(L') &= N'(0) + N''(0)L' + N'''(0)L'^2/2 + e_4 \\ &= 1/\sqrt{2\pi} - L'^2/2\sqrt{2\pi} + e_4 \end{aligned} \tag{18.59}$$

Substituting Eqs. (18.56)–(18.59) into Eq. (18.54), dropping all of remainder terms, Eq. (18.54) becomes:

$$
\begin{aligned}
C = (S' - K)/2 + &\left(\ln(S'/K)\big/\sigma\sqrt{2\pi T}\right) \\
&\left[(S' - K)\left(1 + [\ln(S'/K)/4]^2\right)\right. \\
&-\ln(S'/K)(S' + K)/4] \\
&+\left(\sigma\sqrt{T}\big/2\sqrt{2\pi}\right)[S' + K \\
&-\ln(S'/K)(S' - K)/4]
\end{aligned} \tag{18.60}
$$

Equation (18.60) is a quadratic equation of $\sigma\sqrt{T}$ and can be rewritten as:

$$
\begin{aligned}
&\sigma^2 T[8(S' + K) - 2(S' - K)\ln(S'/K)] \\
&\quad - 8\sigma\sqrt{T}\sqrt{2\pi}(2C - S' + K) \\
&\quad + \ln(S'/K)\left[(S' - K)(16 + (\ln(S'/K))^2)\right. \\
&\quad -4(S' + K)\ln(S'/K)] = 0
\end{aligned} \tag{18.61}
$$

Solving $\sigma\sqrt{T}$ from Eq. (18.61) yields

$$
\sigma\sqrt{T} = \frac{-b \pm \sqrt{b^2 - 4ac}}{2a} \tag{18.62}
$$

where $a = 8(S' + K) - 2(S' - K)\ln(S'/K)$

$$
b = -8\sqrt{2\pi}(2C - S' + K)
$$

$$
\begin{aligned}
c = {} & [\ln(S'/K)[(S' - K)(16 + \ln(S'/K))^2] \\
& - 4(S' + K)\ln(S'/K)]
\end{aligned}
$$

A merit of Eq. (18.62) is to circumvent the ad hoc substitution present in Corrado and Miller (1996) and improve the accuracy of the ISD's estimation. Other methods to calculate the implied volatility can be found in Lai et al. (1992).

According to Lee et al. (2013), put-call parity can be defined in Eq. (18.63); we can calculate implied volatility, stock price per share, and exercise price per share in terms of put option model.

$$
P = C + Xe^{-rT} - Se^{-qT} \tag{18.63}
$$

Let $Xe^{-rT} = K$ and let $Se^{-qT} = S'$, then we have following equation.

$$
P = C + K - S' \tag{18.64}
$$

Substituting Eq. (18.60) in Eq. (18.64), we obtain following equation:

$$
\begin{aligned}
P = (K - S')/2 + &\left(\ln(S'/K)\big/\sigma\sqrt{2\pi T}\right) \\
&\left[(S' - K)\left(1 + [\ln(S'/K)/4]^2\right)\right. \\
&-\ln(S'/K)(S' + K)/4] \\
&+\left(\sigma\sqrt{T}/2\sqrt{2\pi}\right)[S' + K \\
&-\ln(S'/K)(S' - K)/4]
\end{aligned} \tag{18.65}
$$

Equation (18.65) is also a quadratic equation of $\sigma\sqrt{T}$ and can be rewritten as:

$$
\begin{aligned}
&\sigma^2 T[8(S' + K) - 2(S' - K)\ln(S'/K)] \\
&\quad - 8\sigma\sqrt{T}\sqrt{2\pi}(2P - K + S') \\
&\quad + \ln(S'/K)\left[(S' - K)(16 + (\ln(S'/K))^2)\right. \\
&\quad -4(S' + K)\ln(S'/K)] = 0
\end{aligned} \tag{18.66}
$$

Solving $\sigma\sqrt{T}$ from Eq. (18.66) yields

$$
\sigma\sqrt{T} = \frac{-b \pm \sqrt{b^2 - 4ac}}{2a} \tag{18.67}
$$

where $a = 8(S' + K) - 2(S' - K)\ln(S'/K)$, $b = -8\sqrt{2\pi}(2P - K + S')$, $c = \ln(S'/K)[(S' - K)(16 + \ln(S'/K)^2)] - 4(S' + K)\ln(S'/K)]$

We rearrange Eq. (18.61) in terms of S', then we obtain Eq. (18.68).

$$\begin{aligned}&[8\sigma^2T+8\sigma\sqrt{2\pi T}+2\sigma^2T\ln K\\&-16\ln K-4(\ln K)^2-(\ln K)^3]S'\\&+[2\sigma^2TK-16K+8K\ln K-3K(\ln K)^2]\ln S'\\&+[16-2\sigma^2T+8\ln K+3(\ln K)^2]S'\ln S'\\&+(3K\ln K-4K)(\ln S')^2-(3\ln K+4)S'(\ln S')^2\\&-K(\ln S')^3+S'(\ln S')^3\\&\quad=16C\sigma\sqrt{2\pi T}+8\sigma\sqrt{2\pi T}K-8\sigma^2TK\\&\qquad+2\sigma^2TK\ln K-16K\ln K\\&\qquad+4K(\ln K)^2-K(\ln K)^3\end{aligned}\tag{18.68}$$

Equation (18.68) can be used to estimate S' if we have the information of the other five variables. The solution of S' can only be obtained by the trial-and-error method.

Now, consider two call options, C_1 and C_2, on the same time to the maturity with exercise prices of X_1 and X_2.q is an annual dividend yield, S is the underlying asset value, and we denote $S' = Se^{-qT}$. We also denote the present values of the exercise prices $K_1 = X_1e^{-rT}$ and $K_2 = X_2e^{-rT}$, respectively.

For C_1, we apply Taylor's expansion to Eq. (18.54) at K_2. This yields the following equation:

$$\begin{aligned}C_1 &= C_2 - N(\ln(S'/K_2)\big/\sigma\sqrt{T}\\&\quad-\sigma\sqrt{T}/2)(K_1-K_2)+\varepsilon_1\end{aligned}\tag{18.69}$$

where $\varepsilon_1$is the remainder term of Taylor's formula.

Similarly, for C_2, we apply Taylor's expansion to Eq. (18.54) at K_1, which yields the following equation.

$$\begin{aligned}C_2 &= C_1 - N(\ln(S'/K_1)\big/\sigma\sqrt{T}\\&\quad-\sigma\sqrt{T}/2)(K_2-K_1)+\varepsilon_2\end{aligned}\tag{18.70}$$

where ε_2 is the remainder term of Taylor's formula.

Rearranging the equations, dividing both sides by $(K_2 - K_1)$, then applying the inverse function of cumulative normal function on both sides, we have the following two equations.

$$\begin{aligned}&N^{-1}[(C_1-C_2)/(K_2-K_1)]\\&\quad=\ln(S'/K_1)\big/\sigma\sqrt{T}-\sigma\sqrt{T}/2+\eta_1\end{aligned}\tag{18.71}$$

$$\begin{aligned}&N^{-1}[(C_1-C_2)/(K_2-K_1)]\\&\quad=\ln(S'/K_2)\big/\sigma\sqrt{T}-\sigma\sqrt{T}/2+\eta_2\end{aligned}\tag{18.72}$$

Combining the two above equations, the effect of remainder terms may be partially offset. Then, we get the quadratic equation of $\sigma\sqrt{T}$ as follows.

$$\begin{aligned}&\sigma^2T+2N^{-1}[(C_1-C_2)/(K_2-K_1)]\left(\sigma\sqrt{T}\right)\\&-\ln(S'/K_1)-\ln(S'/K_2)=0\end{aligned}\tag{18.73}$$

Then, we can solve implied volatility as:

$$\begin{aligned}\sigma\sqrt{T}&=-N^{-1}((C_1-C_2)/(K_2-K_1))\\&\quad\pm\sqrt{[N^{-1}((C_1-C_2)/(K_2-K_1))]^2+\ln(S'^2/K_1K_2)}\end{aligned}\tag{18.74}$$

Similarly, we consider two put options P_1 and P_2 on the same time to the maturity with exercise prices of X_1 and X_2.q is the annual dividend yield, S is the underlying asset value, and we denote $S' = Se^{-qT}$. We also denote the present values of the exercise prices $K_1 = X_1e^{-rT}$ and $K_2 = X_2e^{-rT}$, respectively.

According to put-call parity defined in Eq. (18.64), we have the following equations.

$$P_1 = C_1 + K_1 - S'\tag{18.75}$$

$$P_2 = C_2 + K_2 - S' \tag{18.76}$$

If we substitute the above equations into Eqs. (18.69) and (18.70), then we have the following equations:

$$P_1 = P_2 + (K_1 - K_2) - N\left(\ln(S'/K_2)\Big/\sigma\sqrt{T} - \sigma\sqrt{T}/2\right)(K_1 - K_2) + \delta_1 \tag{18.77}$$

$$P_2 = P_1 + (K_2 - K_1) - N\left(\ln(S'/K_2)\Big/\sigma\sqrt{T} - \sigma\sqrt{T}/2\right)(K_2 - K_1) + \delta_2 \tag{18.78}$$

Rearranging the equations, dividing both sides by $(K_2 - K_1)$, and then applying the inverse function of cumulative normal function on both sides, we have the following two equations.

$$N^{-1}((P_1 - P_2)/(K_2 - K_1) + 1) = \ln(S'/K_2)\Big/\sigma\sqrt{T} - \sigma\sqrt{T}/2 + \gamma_1 \tag{18.79}$$

$$N^{-1}((P_1 - P_2)/(K_2 - K_1) + 1) = \ln\left(S'/K_1\right)\Big/\sigma\sqrt{T} - \sigma\sqrt{T}/2 + \gamma_2 \tag{18.80}$$

By combining the two above equations, the effect of the remaining terms may be partially offset. Then, we get the quadratic equation of $\sigma\sqrt{T}$ as follows.

$$\sigma^2 T + 2N^{-1}[(P_1 - P_2)/(K_2 - K_1) + 1](\sigma\sqrt{T}) - \ln(S'/K_1) - \ln(S'/K_2) = 0 \tag{18.81}$$

Solving the equation for $\sigma\sqrt{T}$, we obtain:

$$\sigma\sqrt{T} = -N^{-1}((P_1 - P_2)/(K_2 - K_1) + 1) \pm \sqrt{[N^{-1}((P_1 - P_2)/(K_2 - K_1) + 1)]^2 + \ln(S'^2/K_1K_2)} \tag{18.82}$$

18.5 Some Empirical Results

18.5.1 Cases from USA—Individual Stock Options

We select 10 constituent companies of S&P 500 as American examples to compare the implied volatility estimation methods. The selected companies have relative large market values, and are from different industries. Table 18.2 shows the details of our sample.

Sample time spans from July, 2014 to August, 2014. 10-year treasury rate is used as the risk-free rate. We calculated the continuously compounded annual risk-free interest rate accordingly, as the parameter r (Table 18.3).

18.5.2 Cases from China—ETF 50 Options

In Chinese financial market, there were no stock options in the exchange until February, 2015. Now, the only traded options in China are ETF 50 options. We choose these options as our samples to compare the alternative methods of implied volatility estimation (Table 18.4).

The computer programs used to estimate the implied variance presented in this section can be found in Chap. 27 of Lee et al. (2016). Not only has this chapter explicitly discussed how to use computer programs to estimate implied variance with Black–Scholes option pricing model, but it has also discussed how to use computer programs to estimate implied variance with CEV model.

18.6 Conclusion

The main purpose of this chapter was to discuss how to use alternative methods for estimating implied variance. In this chapter, we first reviewed alternative methods to estimate implied variance. We classified them into two different

Table 18.2 Details of sample companies

Security ID	Ticker	Company name	SIC code	Industry
101594	AAPL	Apple Inc.	3571	Electronic computers
104533	XOM	Exxon Mobil Corporation	2911	Petroleum refining
121812	GOOGL	Google Inc.	7375	Information retrieval services
107525	MSFT	Microsoft Corporation	7372	Prepackaged software
106566	JNJ	Johnson & Johnson	2834	Pharmaceutical preparations
111953	WFC	Wells Fargo & Company	6022	State commercial banks
105169	GE	General Electric Company	3511	Steam, gas, and hydraulic turbines, and turbine engine
111860	WMT	Wal-Mart Stores Inc.	5331	Variety stores
102968	CVX	Chevron Corporation	2911	Petroleum refining
109224	PG	The Procter & Gamble Company	2841	Soap and other detergents
102936	JPM	JPMorgan Chase & Co.	6211	Security brokers, dealers and flotation companies
111668	VZ	Verizon Communications Inc.	4812	Radiotelephone communications
108948	PFE	Pfizer Inc.	2834	Pharmaceutical preparations
106276	IBM	International Business Machines Corporation	3571	Electronic computers
109775	T	AT&T, Inc.	4812	Radiotelephone communications

Table 18.3 Implied volatility estimation for individual stock options: comparison of alternative estimation methods

Ticker	IV-matlab	IV-approximation	IV-regression
AAPL	0.387	0.332	0.458
XOM	0.208	0.185	0.106
GOOGL	0.376	0.389	0.305
MSFT	0.327	0.341	0.298
JNJ	0.223	0.216	No positive solution
WFC	0.312	0.301	No positive solution
GE	0.176	0.142	0.224
WMT	0.127	0.124	0.164
CVX	0.306	0.285	0.367
PG	0.209	0.185	No positive solution
JPM	0.189	0.192	0.135
VZ	0.169	0.174	0.247
PFE	0.216	0.208	0.185
IBM	0.463	0.457	No positive solution
T	0.186	0.189	0.264

Table 18.4 Implied volatility estimation from ETF 50 call options

Option ticker	Exercise price	Expiration date	IV-matlab	IV-approximation
10000021.SH	2.20	2015-06-25	0.515	0.504
10000022.SH	2.25	2015-06-25	0.486	0.498
10000023.SH	2.30	2015-06-25	0.417	0.423
10000024.SH	2.35	2015-06-25	0.439	0.424
10000025.SH	2.40	2015-06-25	0.489	0.478
10000031.SH	2.20	2015-09-24	0.426	0.442
10000032.SH	2.25	2015-09-24	0.435	0.447
10000033.SH	2.30	2015-09-24	0.417	0.428
10000034.SH	2.35	2015-09-24	0.422	0.432
10000035.SH	2.40	2015-09-24	0.443	0.454
10000045.SH	2.45	2015-06-25	0.428	0.436
10000047.SH	2.45	2015-09-24	0.393	0.408
10000053.SH	2.50	2015-06-25	0.443	0.428
10000055.SH	2.50	2015-09-24	0.415	0.420
10000061.SH	2.55	2015-06-25	0.442	0.438
10000063.SH	2.55	2015-09-24	0.398	0.416
10000069.SH	2.60	2015-06-25	0.420	0.431
10000071.SH	2.60	2015-09-24	0.409	0.412
10000077.SH	2.65	2015-06-25	0.426	0.419
10000079.SH	2.65	2015-09-24	0.416	0.428
10000085.SH	2.70	2015-06-25	0.427	0.434
10000087.SH	2.70	2015-09-24	0.414	0.419
10000093.SH	2.75	2015-06-25	0.417	0.426
10000095.SH	2.75	2015-09-24	0.405	0.411
10000101.SH	2.80	2015-06-25	0.427	0.443
10000103.SH	2.80	2015-09-24	0.409	0.401
10000123.SH	2.85	2015-06-25	0.441	0.432
10000125.SH	2.85	2015-09-24	0.417	0.422

estimation routines: numerical search methods and closed-form derivation approaches, and discussed their limitations. Then, we showed how the MATLAB computer program can be used to estimate implied variance. This kind of approach used Newton–Raphson method to derive the implied variance from the standard Black–Scholes model. In addition, we also discussed how the approximation method derived by Ang et al. (2013) could be used to estimate implied variance and implied stock price per share. Not only the case of single option was presented, this approximation method also estimated implied volatility from two options with the same maturity, but different exercise prices and values. At last, we selected some large-cap stocks from S&P 500 as empirical examples. The performances of three typical alternative methods: regression method proposed by Lai et al. (1992), MATLAB computer program approach and approximation method derived by Ang et al. (2009) were compared.

Bibliography

Ang, J. S., Jou, G. D., & Lai, T. Y. (2009). Alternative formulas to compute implied standard deviation. *Review of Pacific Basin Financial Markets and Policies, 12*(02), 159–176.

Ang, J. S., Jou, G. D., & Lai, T. Y. (2013). A comparison of formulas to compute implied standard deviation. In *Encyclopedia of finance* (pp. 765–776). Berlin: Springer.

Beckers, S. (1981). Standard deviations implied in option prices as predictors of future stock price variability. *Journal of Banking & Finance, 5*(3), 363–381.

Brenner, M., & Subrahmanyam, M. G. (1988). A simple formula to compute the implied standard deviation. *Financial Analysts Journal*, 80–83.

Chance, D. M. (1996). A generalized simple formula to compute the implied volatility. *Financial Review, 31* (4), 859–867.

Corrado, C. J., & Miller, T. W. (1996). A note on a simple, accurate formula to compute implied standard deviations. *Journal of Banking & Finance, 20*(3), 595–603.

Corrado, C. J., & Miller, T. W. (2006). Estimating expected excess returns using historical and option-implied volatility. *Journal of Financial Research, 29*(1), 95–112.

Hallerbach, W. (2004). An improved estimator for Black-Scholes-Merton implied volatility. Erasmus Research Series, Erasmus University.

Lai, T.-Y., Lee, C., & Tucker, A. L. (1992). An alternative method for obtaining the implied standard deviation. *Journal of Financial Engineering, 1*, 369–375.

Latane, H. A., & Rendleman, R. J. (1976). Standard deviations of stock price ratios implied in option prices. *The Journal of Finance, 31*(2), 369–381.

Lee, A. C., Lee, J. C., & Lee, C. F. (2017). *Financial Analysis, Planning and Forecasting: Theory and Application* (3rd ed.). Singapore: World Scientific.

Lee, C. F., Lee, J. C., & Lee, A. C. (2013). *Statistics for business and financial economics*. Berlin: Springer.

Lee, C. F., Chen, Y., & Lee, J. (2018). Alternative methods to estimate implied variance: Review and comparison. *Review of Pacific Basin Financial Markets and Policies*. Forthcoming.

Lee, C. F., Lee, J., Chang, J. R., & Tai, T. (2016). *Essentials of excel, excel VBA, SAS and minitab for statistical and financial analyses*. New York: Springer.

Li, S. (2005). A new formula for computing implied volatility. *Applied Mathematics and Computation, 170* (1), 611–625.

Manaster, S., & Koehler, G. (1982). The calculation of implied variances from the Black-Scholes model: A note. *The Journal of Finance, 37*(1), 227–230.

Merton, R. C. (1973). Theory of rational option pricing. *The Bell Journal of Economics and Management Science, 4*(1), 141–183.

Smith, C. W., Jr. (1976). Option pricing: A review. *Journal of Financial Economics, 3*(1–2), 3–51.

19 Numerical Valuation of Asian Options with Higher Moments in the Underlying Distribution

Contents

Abstract

In this chapter, we develop a modified Edgeworth binomial model with higher moment consideration for pricing European or American-Asian options. If the number of the time steps increases, our numerical algorithm is as precise as that of Chalasani et al. (1999), with underlying distribution for benchmark comparison. If the underlying distribution displays a negative skewness and leptokurtosis, as often observed for stock index returns, our estimates are better and very similar to the benchmarks in Hull and White (1993). The results show that our modified Edgeworth binomial model can value European and American-Asian options with greater accuracy and speed given higher moments in their underlying distribution.

This chapter draws upon the paper by Wang and Hsu which was published as Chap. 40 of Handbook of Quantitative Finance and Risk Management edited by Lee et al. (2010).

C.-F. Lee et al., *Financial Econometrics, Mathematics and Statistics*,
https://doi.org/10.1007/978-1-4939-9429-8_19

19.1 Introduction

The payoff of an Asian option depends on the arithmetic or geometric price average of the underlying asset during the life of the option. Its value is path-dependent, normally without the closed-form solution, and therefore more difficult to calculate than that of a standard option. However, the hedging effect of an Asian option, which is specifically widely used in the foreign exchange market, is better than that of a standard option and offers convenience and lower cost.

Many researchers have applied various methods to value Asian options. There are mainly two approaches. Analytic approximations that produce closed-form solutions are proposed by Turnbull and Wakeman (1991), Lévy (1992), Zhang (2001, 2003), Curran (1994), Rogers and Shi (1995), and Thompson (2002). However, obtaining a pricing formula is still a challenging problem for many applications. In recent years, numerical methods have assumed increasing importance (Grant et al. 1997). Three popular methods, the Monte Carlo simulation, the finite difference method, and the binomial lattice, have been widely used to value Asian options. Although the Monte Carlo simulation is often used to compare with other pricing methods for its convenience and flexibility, its calculation is considered inefficient and very time-consuming. The difference equations transformed from partial differential equations can be solved quickly using numerical method. However, Barraquand and Pudet (1996) point out that for the path-dependent pricing problem, an augmentation of the state space is not viable in this approach.

Binomial tree models have been extensively applied in various option valuations (Amin 1993; Rubinstein 1994; Hilliard and Schwartz 1996; Chang and Fu 2001; Klassen 2001). In particular, Hull and White (1993), Chalasani et al. (1998, 1999), Neave and Ye (2003), and Costabile et al. (2006) all adopt extended binomial models to value European or American path-dependent options. They claim that their algorithms are considerably faster and provide more accurate results compared with the analytical approximation used by Turnbull and Wakeman (1991).

Most of the valuation methods for Asian options assume that the return distribution of the underlying asset is lognormal. However, practitioners and academics are well aware that the finite sum of the correlated lognormal random variables is not lognormal. Corrado and Su (1997), and Kim and White (2004) both find significant negative skewness and positive excess kurtosis in the implied distribution of S&P 500 index options. Ahn et al. (1999) also find non-normal distribution when using put options to reduce the cost of risk management. It is for this reason that some researchers have tried to investigate other alternatives by considering the number of moments. For example, Milevsky and Posner (1998a, b) and Posner and Milevsky (1998) apply the moment-matching analytical methods to approximate the density function of the underlying average for the European-Asian options.

The purpose of this chapter is to introduce a numerical algorithm in pricing European and American-Asian options while considering the higher moments of the underlying asset distribution. We develop a modified Edgeworth binomial model which applies the refined binomial lattice (Chalasani et al. 1998, 1999) and use the Edgeworth expansible distribution (Rubinstein 1998) to include the parameters for higher moments. The numerical results show that this approach can effectively deal with the higher moments of the underlying distribution and provide better option value estimates than those found in various studies in the literature.

This chapter introduces a numerical algorithm with higher moments in the underlying asset distribution to evaluate Asian options. Section 19.2 concisely introduces the pricing process of a binomial tree for an Asian option. Section 19.3 presents our Edgeworth binomial model for pricing Asian options with higher moment consideration. In Sect. 19.4, we discuss the lower and upper bounds of the prices of European-Asian options. In Sect. 19.5, we discuss the lower and upper bounds of the prices of American-Asian options. The numerical examples for our approach are then provided in Sect. 19.6. Especially, the algorithmic process with three periods is shown and our estimates are compared with the benchmarks discussed in the literature. Finally, Sect. 19.7 offers our conclusion.

19.2 Definitions and the Basic Binomial Model

An underlying variable, $S(t)$, of an option at time t is generally assumed to satisfy the stochastic differential equation in a risk-neutral world:

$$\mathrm{d}S(t) = \mu S(t)\mathrm{d}t + \sigma S(t)\mathrm{d}B(t),$$

where the drift μ and volatility σ are constant, and $\{B(t)\}$ denotes a Brownian motion process. Assume that the risk-free interest rate r is a constant, and that the option expires at time T.

A binomial tree can approximate the continuous-time function $S(t)$, where one divides the life of the option into n time steps of length $\Delta t = T/n$. In each time step, the underlying asset may move up by a factor u with probability p_c, or down by a factor $d = u^{-1}$ with probability $q_c = 1 - p_c$, with $0 < d < 1 < u$. Firstly, we consider the one period case, i.e., time step $k = 1$. The stock price at the end of the period will have two possible values, either up to a value $S(0)u$ with probability p_c or down to a value $S(0)u^{-1}$ with probability $1 - p_c$. These price movements can be represented in the following diagram:

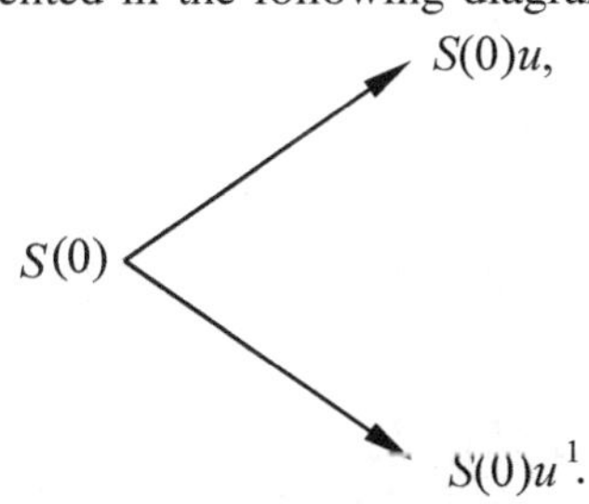

Now consider a call option with two periods ($k = 2$) before its expiry date. The price process of the stock will show three possible values after two periods,

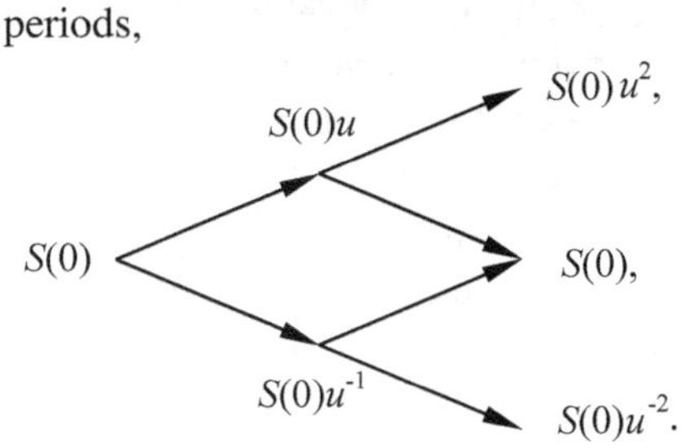

This price process of the stock can be extended to n time steps.

The stochastic differential equation describing this price process, i.e., $\mathrm{d}S(t) = \mu S(t)\mathrm{d}t + \sigma S(t)\mathrm{d}B(t)$, has the following solution,

$$S(t) = S(0)e^{(\mu - \frac{1}{2}\sigma^2)t + \sigma\Phi\sqrt{t}},$$

where Φ is a standardized normal random variable.

For a binomial random walk to have the correct drift over a time period of Δt, we need

$$\begin{aligned} p_c Su + (1 - p_c)Sd &= SE\left[e^{(\mu - \frac{1}{2}\sigma^2)\Delta t + \sigma\Phi\sqrt{\Delta t}}\right] \\ &= Se^{\mu\Delta t}, \end{aligned}$$

namely, $p_c u + (1 - p_c)d = e^{\mu\Delta t}$. Rearranging this equation, we can obtain

$$p_c = \frac{e^{\mu\Delta t} - d}{u - d}$$

with

$$u = e^{\sigma\sqrt{\Delta t}}.$$

Here, let Ω_n be a sample space of an experiment including all possible sequences of n upticks and downticks. A typical element of Ω_n is presented as $\omega = \omega_1\omega_2 \ldots \omega_n$, where ω_i denotes the ith uptick or downtick. Let $\{H_k(\omega)\}$ be an associated family of random variables, where $H_k(\omega)$ denotes the number of upticks at time k and $H_0(\omega) = 0$ for all ω. We can define a symmetric random walk X_k, such that for each $k \geq 1$, $X_k = H_k - (k - H_k) = 2H_k - k$, which represents the number of upticks minus the number of downticks up to time k. It is used to define the nodes in a binomial lattice corresponding to the possible positions of the underlying random walk at different times. Specifically, a tree path ω is displayed to pass through or reach node (k, h) if and only if $H_k(\omega) = h$ for times $k = 0, 1, \ldots, n$ and the number of possible upticks $h = 0, 1, \ldots, k$. Consequently, the underlying asset price at time k is S_k ($k = 0, 1, \ldots, n$), where

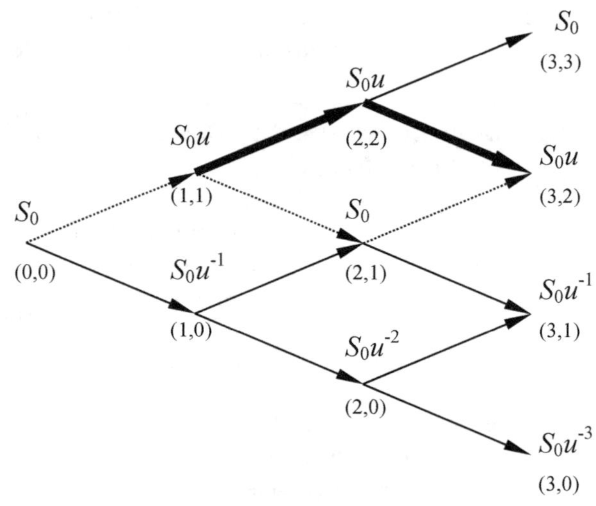

Fig. 19.1 A binomial tree. Node (3, 2) means there are 2 upticks in any path reaching this node at time 3

$$S_k = S_0 u^{X_k} = S_0 u^{2H_k - k}.$$

For example, $S_3 = S_0 u^{(2\times 2-3)} = S_0 u$ at node (3, 2) in the lattice diagram of Fig. 19.1. The underlying asset price at node (k, h) is given by $S_0 u^{2h-k}$, whose average at time k is defined as $A_k = (S_0 + S_1 + \cdots + S_k)/(k + 1)$, $k \geq 0$. Therefore, the payoff of an Asian call with strike price L at time n is $V_n^+ = (A_n - L)^+ = \max\{A_n - L, 0\}$. The price of a European option is the present value of the expected payoff discounted to time 0, i.e., $C = E[V_n^+]/(1+r)^n$. And the price of an American option at time 0 is the maximum discounted expected payoff from all possible exercise strategy τ, i.e., $C = \max_\tau E[V_n^+]/(1+r)^n$. Note that $E[V_n^+]$ is a probability-weighted average given by $\sum_k P_k (A_k - L)^+$, where P_k denotes the risk-neutral probability associated with A_k at the expiration date.

19.3 Edgeworth Binomial Model for Asian Option Valuation

To consider the higher moments, we first apply the Edgeworth binomial tree model (Rubinstein 1998). Assume that the tree has n time steps and $n + 1$ nodes ($h = 0, 1, \ldots, n$) at step n. At each node h, there is a random variable $y_h = [2h - n]/n^{1/2}$ with a standardized binomial density $b(y_h) = [n!/h!(n - h)!](1/2)^n$. Giving predetermined skewness and kurtosis, the binomial density is transformed by the Edgeworth expansion up to the fourth moment. The result is

$$\begin{aligned} F(y_h) &= f(y_h) \times b(y_h) \\ &= \left[1 + \frac{1}{6}\gamma_1(y_h^3 - 3y_h) + \frac{1}{24}(\gamma_2 - 3)(y_h^4 - 6y_h^2 + 3) \right. \\ &\quad \left. + \frac{1}{72}\gamma_1(y_h^6 - 15y_h^4 + 45y_h^2 - 15)\right] \times \left(\frac{1}{2}\right)^n \left[\frac{n!}{h!(n-h)!}\right] \end{aligned} \tag{19.1}$$

with

$$F(y_h) = \left[1 + \frac{1}{6}\gamma_1(y_h^3 - 3y_h) + \frac{1}{24}(\gamma_2 - 3)(y_h^4 - 6y_h^2 + 3) + \frac{1}{72}\gamma_1(y_h^6 - 15y_h^4 + 45y_h^2 - 15)\right]$$

where $\gamma_1 = E^Q[y_h^3]$ is the skewness and $\gamma_2 = E^Q[y_h^4]$ is the kurtosis of the underlying distribution under risk-neutral measure. While the sum of $F(y_h)$ is not one, we normalize $F(y_h)$ by $F(y_h)/\sum_j F(y_j)$ and denote it as P_h.

The variable y_h, which has probability P_h, can be standardized as $x_h = (y_h - M)/V$ with $M = \sum_h P_h y_h$ and $V^2 = \sum_h P_h (y_h - M)^2$. The variable x_h is used later in Eq. (19.2) to obtain the asset price and the corresponding risk-neutral probability, P_h, for a path to node h.

Consider a tree model of n steps. The asset price at the hth node ($h = 0, 1, \ldots, n$) during the final step, $\hat{S}_{n,h}$, is

$$\hat{S}_{n,h} = S_0 e^{\mu T + \sigma\sqrt{T}x_h} \qquad (19.2)$$

with

$$\mu = r - \frac{1}{T}\ln\sum_{h=0}^{n} P_h e^{\sigma\sqrt{T}x_h},$$

where S_0 is the initial asset price, r is the continuously compounded annual risk-free rate, T is the time for expiration of the option (in years), σ is the annualized volatility rate for the cumulative asset return, and x_h is a random variable from probability distribution P_h with mean 0 and variance 1. P_h is determined by modifying the binomial distribution using the Edgeworth expansion up to the fourth moment of $\ln(\hat{S}_{n,h}/S_0)$. Finally, μ is used to ensure that the expected risk-neutral asset return equals r. Solving backward recursively from the end of the tree, the nodal value, $S_{n-1,h}$, is

$$S_{n-1,h} = \left[p_e \hat{S}_{n,h+1} + q_e \hat{S}_{n,h}\right] \exp\left(-\frac{rT}{n}\right) \qquad (19.3)$$

with $p_e = p_{n,h+1}/(p_{n,h+1} + p_{n,h})$ and $q_e = (1 - p_e)$, where $p_{n,h}$ is $P_h/[n!/h!(n-h)!]$. To demonstrate these nodal values in the binomial lattice, a numerical example with three time steps is shown in later section.

The path dependence of Asian options is analyzed using the approach by Chalasani et al. (1999). In order to represent the refined binomial lattice, a new random variable $W_{k,h}$ denoting an area at time k is assigned. Its initial value W_0 is zero. For any node (k, h) in the tree, a lowest path reaching (k, h) is defined as the path with $k - h$ downticks followed by h upticks, and a highest path reaching (k, h) means the one with h upticks followed by $k - h$ downticks. The area $W_{k,h}$ (ω) of a path ω reaching (k, h) can be defined as the number of diamond-shaped boxes enclosed between this path ω and the lowest path reaching this node. For example, the node (5, 2) means that the paths reaching it have 2 upticks at time 5. As demonstrated in Fig. 19.2, a path passing through (5, 2) and reaching node (6, 2) is shown by the thick line segments. The area $W_{6,2}$ (ω) of this path is the number of diamond-shaped boxes, contained between this path and the lowest path reaching node (6, 2), as shown by the shaded area in the graph. The maximum area of any path reaching (k, h) is the number of boxes between the highest and the lowest paths reaching (k, h), that is, $h(k - h)$. The minimum area of any path reaching (k, h) is zero. The set of possible areas of paths reaching node (k,h) is therefore $\{0, 1, \ldots, h(k-h)\}$. Each node of the binomial lattice can be partitioned into

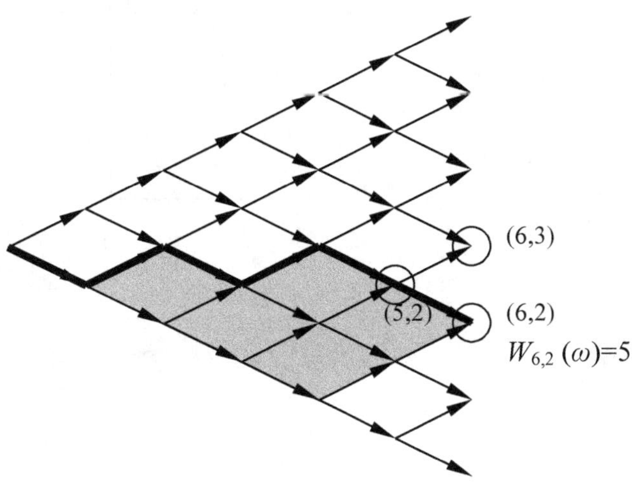

Fig. 19.2 A binomial lattice. The shaded area shows the diamond-shaped boxes for node (6, 2)

"nodelets" based on the areas of the paths reaching this node. Therefore, any path reaching a given nodelet (k, h, a) has an area $W_{k,h}(\omega) = a$ with h upticks at time k. For instance, Fig. 19.3 shows the nodelets in the nodes (5, 2), (6, 3), and (6, 2). As noted in Chalasani et al. (1999), there is a one-one correspondence between the possible areas and the possible geometric averages of underlying asset prices for paths reaching (k, h). Therefore, (k, h, a) represents all the paths in the binomial tree that reach node (k, h) and has the same geometric average asset price from time 0 to time k.

Suppose the area of a path A reaching (k, h) is $W_{k,h}(A) = a$. If A has an uptick after this point, it reaches node $(k + 1, h + 1)$ at the next time step. The path A and the lowest path B reaching $(k + 1, h + 1)$ share the same edge linking (k, h) and $(k + 1, h + 1)$ in the lattice. Hence, the number of boxes between A and B at time $k + 1$ is the same as the number at time k. In this way, the path A reaches nodelet $(k + 1, h + 1, a)$. On the other hand, if A has a downtick after time k, it will reach node $(k + 1, h)$. In this case, the number of boxes at time $k + 1$ between A and the lowest path reaching $(k + 1, h)$ will be increased by h to get $a + h$. The path A then reaches nodelet $(k + 1, h, a + h)$.

We now show how the arithmetic average of underlying asset prices over all paths reaching (k, h, a) is computed and denote it by $\overline{A}(k, h, a) = E[A_k | H_k = h, W_{k,h}(\omega) = a]$, $k = 0, 1, \ldots, n$, $h \leq k$. It is simply the average of A_k over these paths. So the arithmetic average of stock prices over all paths reaching nodelet (k, h, a) can be expressed as:

$$\overline{A}(k,h,a) = \frac{S''(k,h,a)}{(k+1)M(k,h,a)}, \qquad (19.4)$$

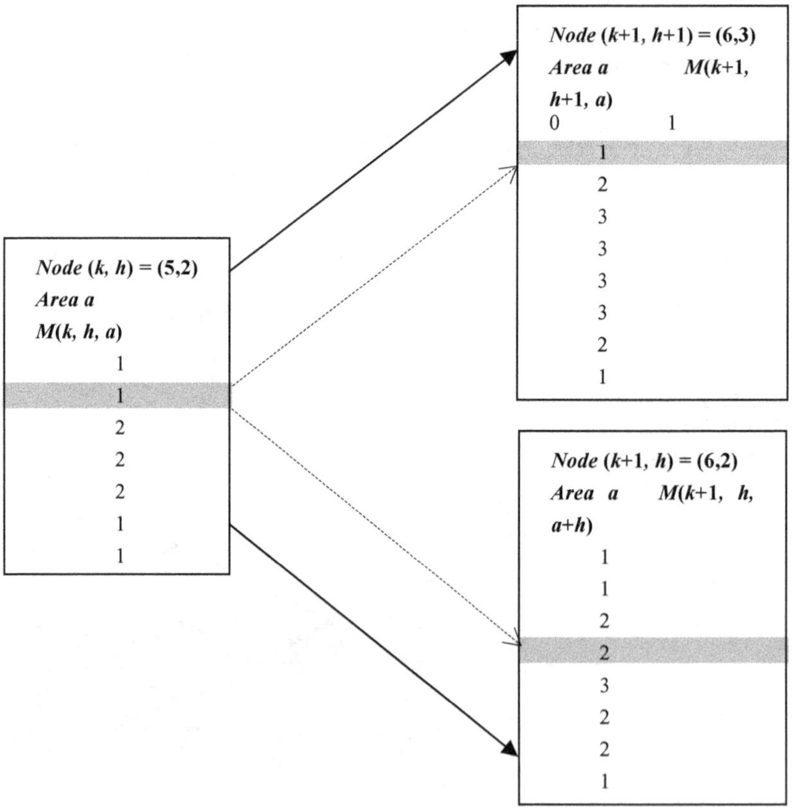

Fig. 19.3 Nodelets in the nodes (5, 2), (6, 3), and (6, 2) as circled in Fig. 19.2. This figure exhibits the number of paths $M(k, h, a)$ reaching each nodelet (k, h, a). An example is shown for nodelet (5, 2, 2) which is updated as in the nodelets (6, 3, 2) and (6, 2, 4)

where $S''(k,h,a)=\sum_{m=1}^{M}S'_m(k,h,a)$, $S'_m(k,h,a)=\sum_{i=0}^{k}S_{i,h}$, with k = 0, 1, ..., n; h = 0, 1,..., k; a = 0, 1, ..., $h(k-h)$; m = 1, 2, ..., $M(k, h, a)$; and $M(k, h, a)$ is the number of paths reaching (k, h, a) with $M(0, 0, 0) = 1$. Here, $S''(k, h, a)$ is the sum of $S'_m(k,h,a)$ over all paths passing through (k, h, a) with $S''(0, 0, 0) = S_0$, while $S'_m(k,h,a)$ is the sum of the asset prices along any possible path passing through (k, h, a) from time 0 to k.

Any path passing through nodelet (k, h, a) and having an uptick will get to nodelet $(k + 1, h + 1, a)$ at time $k + 1$. Thus, the number of paths reaching nodelet $(k + 1, h + 1, a)$, namely $M(k + 1, h + 1, a)$, should include $M(k, h, a)$ paths through (k, h, a). The sum of the prices from all the paths reaching $(k + 1, h + 1, a)$, namely $S''(k + 1, h + 1, a)$, would be $S''(k, h, a) + M(k, h, a)\, S_{k+1,h+1}$ for paths passing (k, h, a). Likewise, all paths passing through nodelet (k, h, a) with a downtick will reach nodelet $(k + 1, h, a + h)$ at time $k + 1$. Similarly, $M(k + 1, h, a + h)$ should also include $M(k, h, a)$ and $S''(k + 1, h, a + h)$ would be $S''(k, h, a) + M(k, h, a)\, S_{k+1,h}$ in the forward induction process.

19.4 Upper Bound and Lower Bound for European-Asian Options

Next, we present how the value of an Asian option after obtaining the arithmetic average of the stock prices from Eq. (19.4) is estimated. We apply the approach used by Rogers and Shi (1995), where the lower bound and the error bound are calculated for the price of an Asian option. This lower bound on the price of an Asian call option with strike L can be expressed as:

$$\begin{aligned} E\left[(A_n-L)^+\right] &= E\left[E\left[(A_n-L)^+|Z\right]\right] \\ &\geq E\left[E[A_n-L|Z]^+\right]=E\left[[E(A_n|Z)-L]^+\right] \end{aligned} \tag{19.5}$$

where $Z = (W_{n,h}, S_{n,h})$ in which the random variable $W_{n,h}$ denotes the area at node (n, h), and $S_{n,h}$ represents the stock price reaching (n, h). The composition in the lower bound, $E[A_n|W_{n,h}, S_{n,h}]$, can be expressed as $\overline{A}(n,h,a)$, as in Eq. (19.4). $\overline{A}(n,h,a)$ is the expectation of the average stock price A_n at node (n, h), where $A_n = (S_0 + S_{1,h} + ... + S_{n,h})/(n + 1)$ on a tree path passing through (n, h, a). All paths through this nodelet have the same probability $P(W_{n,h}, S_{n,h})$, which is $M(n, h, a)p_e^h q_e^{n-h}$. Thus, we can calculate the lower bound as

$$\begin{aligned} C_0^d &= E\left[\left[E\left(A_n|W_{n,h},S_{n,h}\right)-L\right]^+\right] \\ &= \sum_{h=0}^{n}\sum_{a=0}^{h(n-h)} P(W_{n,h},S_{n,h})\left[\overline{A}(n,h,a)-L\right]^+ \\ &= \sum_{h=0}^{n}\sum_{a=0}^{h(n-h)} M(n,h,a)p_e^h q_e^{n-h}\left[\overline{A}(n,h,a)-L\right]^+. \end{aligned}$$

As a result, the error bound is

$$\begin{aligned} & E\left[E\left[(A_n-L)^+|W_{n,h},S_{n,h}\right]\right]-E\left[E\left[(A_n-L)|W_{n,h},S_{n,h}\right]^+\right] \\ &= E\left[E\left[(A_n-L)^+|W_{n,h},S_{n,h}\right]-E\left[(A_n-L)|W_{n,h},S_{n,h}\right]^+\right] \\ &\leq \tfrac{1}{2}E\left[\left[\mathrm{var}\left(A_n-L|W_{n,h},S_{n,h}\right)\right]^{\frac{1}{2}}\right] \end{aligned} \tag{19.6}$$

assuming $V_n^{\min}(Z) < 0$ and $V_n^{\max}(Z) > 0$, where $V_n = A_n - L$.[1] Note that $\mathrm{var}(A_n - L) = \mathrm{var}\, A_n = E\, A_n^2 - (E\, A_n)^2$ and $\overline{A}^2(n,h,a)=E[A_n^2|W_{n,h},S_{n,h}]$. Let $A^{\min}(k, h, a)$ denote the minimal value of A_k and $A^{\max}(k, h, a)$ its maximum over all paths passing through the nodelet (k, h, a). Thus, the error bound by Eq. (19.6) equals

[1]We set the minimum and maximum values of V_n over paths ω with $Z(\omega) = z$, to be $V_n^{\max}(z)=\max_{\omega\in\Omega}\{V_n(\omega)|Z(\omega)=z\}$ and $V_n^{\min}(z)=\min_{\omega\in\Omega}\{V_n(\omega)|Z(\omega)=z\}$. If $V_n^{\max}(z_i)\leq 0$, then for all paths ω with $Z(\omega) = z_i$, we can deduce $V_n^+(\omega) = 0$, which implies $E(V_n^+|z_i)=0$, and also $E(V_n|z_i)\leq 0$, which implies $E(V_n|z_i)^+=0$. Hence, the error bound is zero. Similarly, if $V_n^{\min}(z_i) \geq 0$, then for all paths ω with $Z(\omega) = z_i$, we can deduce $V_n^+(\omega) = V_n(\omega)$, which implies $E(V_n^+|z_i)=E(V_n|z_i)$, and also $E(V_n|z_i)\geq 0$, which implies $E(V_n|z_i)^+=E(V_n|z_i)$. Therefore, the error bound is again zero.

$$\frac{1}{2}\sum_{h=0}^{n} \sum_{\substack{a=0 \\ A^{\min}(n,h,a)<L,A^{\max}(n,h,a)>L}}^{h(n-h)} P(W_{n,h}, S_{n,h})(\overline{A^2}(n,h,a) - \overline{A}(n,h,a)^2)^{\frac{1}{2}}$$
$$= \frac{1}{2}\sum_{h=0}^{n} \sum_{\substack{a=0 \\ A^{\min}(n,h,a)<L,A^{\max}(n,h,a)>L}}^{h(n-h)} M(n,h,a)p_e^h q_e^{n-h}(\overline{A^2}(n,h,a) - \overline{A}(n,h,a)^2)^{\frac{1}{2}} \quad (19.7)$$

The $\overline{A}(n,h,a)$ can be derived from Eq. (19.4). Meanwhile, $A^{\min}(k, h, a) = S^{\min}(k, h, a)/(k + 1)$ and $A^{\max}(k, h, a) = S^{\max}(k, h, a)/(k + 1)$, where $S^{\min}(k, h, a)$ and $S^{\max}(k, h, a)$ are, respectively, the minimum value and maximum value of $S_{k,h}$ over these paths reaching (k, h, a). $\overline{A}^2(n,h,a)$ can also be calculated from the following:

$$\overline{A^2}(k,h,a) = \frac{\phi(k,h,a) + 2\psi(k,h,a)}{(k+1)^2 M(k,h,a)},$$

where $\varphi(k,h,a)$ is the sum of $\sum_{i=0}^{k} S_{i,h}^2$ and $\psi(k,h,a)$ is the sum of $\sum_{0\le i\le j\le k} S_{i,h}S_{j,h}$.[2] With the lower bound and the error bound, we can obtain the upper bound.

Suppose one upward probability p_e, denoting the probability of the stock price moving up for the next step in the Edgeworth binomial tree, is lower than the other upward probability p_c, the probability of the stock price moving up in the binomial tree of Chalasani et al. (1999). The average stock price in a path with upward drift causes higher probability of $A^{\min}(k, h, a) > L$, i.e., higher probability of zero variance. So the total variance of the average stock price will be smaller. According to Eq. (19.7), the error bound of the option price with upward probability p_e will be smaller than the error bound with probability p_c. We can show in the following proposition that this can lead to tighter bounds on the error from approximating $E[V_n^+]$ if its upward probability is lower.

We first show that the error bound in approximating $E[V_n^+]$ from a modified Edgeworth binomial tree model and that from a binomial tree model employed by Chalasani et al. (1999) are proportional to their upward probabilities, respectively, in the binomial paths. For this, we use a discrete approximation method similar to the lattice approach. Let T be the time for expiration of the option. At time T, let $Y_e(T)$ denote the variance of the arithmetic average of the stock prices in our modified Edgeworth binomial tree with upward probabilities p_e, and $Y_c(T)$ denote the variance of the average price in a binomial tree from Chalasani et al. with upward probability p_c. From Eq. (19.2), the asset price in the Edgeworth model is affected by the drift with upward trend, resulting in higher average than the case in Chalasani et al. From Eq. (19.7), higher average price increases the probability of $A^{\min}(k, h, a) > L$, i.e., higher probability of zero variance based on explanations in footnote 1. We thus have $Y_c(T) \geq Y_e(T)$.

[2]To show how $\overline{A}^2(k,h,a)$ is derived, we can write $(k+1)^2\mathbf{A}_k^2 = (\sum_{i=0}^{k}\mathbf{S}_{i,h})^2 = \sum_{i=0}^{k}\mathbf{S}_{i,h}^2 + 2\sum_{0\le i\le j\le k}\mathbf{S}_{i,h}\mathbf{S}_{j,h}$. Because all paths reaching (k, h, a) have the same probability, $\overline{A}^2(k,h,a)$ is the average of $\mathbf{A}_k^2 = (\sum_{i=0}^{k}\mathbf{S}_{i,h}^2 + 2\sum_{0\le i\le j\le k}\mathbf{S}_{i,h}\mathbf{S}_{j,h}) \div (k+1)^2$ over these paths.

Assume for the moment that p_e is less than p_c. At time $t = T/3$, the conditional expectations of the variances with upward probabilities p_e and p_c are given by E_{3t} $[Y_e(T)]$ and E_{3t} $[Y_c(T)]$, respectively. We can see (ignoring the discount factors) that

$$E_{3t}[Y_c(T)]^{\frac{1}{2}} \geq E_{3t}[Y_e(T)]^{\frac{1}{2}},$$

because

$$\sum_{h=0}^{3} \sum_{a=0A^{\max} > L, A^{\min} < L}^{2} M(3,h,a)p_c^h(1-p_c)^{3-h}(\mathrm{var}(A_3|a,S_{3,h}))^{\frac{1}{2}}$$
$$\geq \sum_{h=0}^{3} \sum_{a=0A^{\max} > L, A^{\min} < L}^{2} M(3,h,a)p_e^h(1-p_e)^{3-h}(\mathrm{var}(A_3|a,S_{3,h}))^{\frac{1}{2}}.$$

Similarly, at time $t = T/4$,

$$E_{4t}[Y_c(T)]^{\frac{1}{2}} \geq E_{4t}[Y_e(T)]^{\frac{1}{2}},$$

since

$$\sum_{h=0}^{4} \sum_{a=0A^{\max} > L, A^{\min} < L}^{4} M(4,h,a)p_c^h(1-p_c)^{4-h}(\mathrm{var}(A_4|a,S_{4,h}))^{\frac{1}{2}}$$
$$\geq \sum_{h=0}^{4} \sum_{a=0A^{\max} > L, A^{\min} < L}^{4} M(4,h,a)p_e^h(1-p_e)^{4-h}(\mathrm{var}(A_4|a,S_{4,h}))^{\frac{1}{2}}.$$

Therefore, as long as the above inequality continuously holds for all time $t \leq T/5$, the error bound for a tree model with upward probability p_e will be tighter than that for a tree with upward probability p_c, given that p_e is less than p_c. The following proposition shows that this can lead to tighter bounds on the error from approximating $E[V_n^+]$ if its upward probability is lower. The detailed analytical explanation is discussed in Lo et al. (2008).

Proposition *The error bound in pricing a European-Asian option from the modified Edgeworth binomial model is tighter than the error bound from the model by Chalasani et al. (1999).*

19.5 Upper Bound and Lower Bound for American-Asian Options

Define C_U (k, h, x) as the value of an American-Asian option at time k, given that the number of upticks is h and the arithmetic asset price average A_k equals x. C_U can be expressed as follows:

$$C_U(n,h,x) = (x-L)^+, h \le n$$

$$C_U(k,h,x) = \max\{(x-L)^+, (p_e C_U(k+1,h+1, x^U(k,h)) + q_e C_U(k+1,h,x^L(k,h))) \exp(-rT/n)\}, \quad k<n, h\le[k],$$

$$\begin{aligned} C_U(n,h,x) &= (x-L)^+, h\le n \\ C_U(k,h,x) &= \max\{(x-L)^+, (p_e C_U(k+1,h+1,x^U(k,h)) + q_e C_U(k+1,h,x^L(k,h)))\exp(-rT/n)\}, \\ & k\le n \\ & h\le[k], \end{aligned} \tag{19.8}$$

where L is the exercise price, $\boldsymbol{p_e} = \boldsymbol{p}_{k+1,h+1}/(\boldsymbol{p}_{k+1,h+1} + \boldsymbol{p}_{k+1,h})$ and $\boldsymbol{q_e} = (1 - \boldsymbol{p_e})$ with $p_{k+1,h}$ being P_h $/[(k + 1)!/(h!((k + 1) - h)!)]$. Here $x^U(k, h) = [x(k + 1) + S_{k+1,h+1}]/(k + 2)$ is the arithmetic asset price average A_{k+1}, given that x is the arithmetic asset price average A_k and an uptick occurring at time $k + 1$. Similarly, $x^L(k, h) = [x(k + 1) + S_{k+1,h}]/(k + 2)$ is the arithmetic asset price average A_{k+1}, given that x equals A_k and there is a downtick at time $k + 1$. Therefore, for any path ω such that $H_k(\omega) = h$ and $A_k(\omega) = x$, the price of an American-Asian option at time k on ω is $C_U(k, h, x)$ whose value at time 0 is $C_U(0,0,S_0)$.

However, we cannot directly calculate $C_U(0,0,S_0)$. The quantity $x^U(k, h)$ may not equal any of the possible averages of asset prices over all paths reaching node $(k + 1, h + 1)$ with area a. And similarly, $x^L(k, h)$ may not equal any average at node $(k + 1, h)$. But we can use linear interpolation to derive these missing values because $\overline{A}(k,h,a)$ is a strictly increasing function in a.

Firstly, we compute the upper bound of the American-Asian option using an idea similar to Hull and White (1993). For a given $x = \overline{A}(k,h,a)$, we can find b such that for some $0 \le \lambda \le 1$,

$$\begin{aligned} x^U(k,h) &= \lambda \bar{A}(k+1,h+1,b) \\ &\quad + (1-\lambda)\bar{A}(k+1,h+1,b+1). \end{aligned}$$

Therefore, on the right-hand side of Eq. (19.8), $C_U(k + 1, h + 1, x^U(k, h))$ can be replaced by

$$W^U(k,h,a) = \lambda w_1 + (1-\lambda) w_2,$$

where w_1 is $W(k + 1, h + 1, b) = [\overline{A}(k + 1, h + 1, b) - L]^+$ and w_2 is $W(k + 1, h + 1, b + 1) = [\overline{A}(k + 1, h + 1, b + 1) - L]^+$. Similarly, $C_U(k + 1, h, x^L(k, h))$ can be substituted by $W^L(k,h,a) = \lambda w_1' + (1-\lambda) w_2'$, where w_1' is $W(k+1,h,b) = \left[\overline{A}(k+1,h,b) - L\right]^+$ and w_2' is $W(k+1,h,b+1) = \left[\overline{A}(k+1,h,b+1) - L\right]^+$. Using this procedure backward recursively, for $x = \overline{A}(k, h, a)$, $a = 0, 1,\ldots, h(k - h)$, we have

$$\begin{aligned} W(k,h,a) &= \max\{(x-L)^+, \\ & (p_e W^U(k,h,a) + q_e W^L(k,h,a)) \exp(\frac{-rT}{n})\} \end{aligned} \tag{19.9}$$

for all (k, h, a) in the tree. It follows that the estimate $W(0, 0, 0)$ is an upper bound for the value of an American-Asian option at time zero.

We now provide a lower bound of the American-Asian option which can be obtained during the process for upper bound calculation applying a proper exercise rule. Such lower bound has been proposed for European-Asian options by Rogers and Shi (1995) and generalized by Dhaene et al. (2002a, b). Let Z be a random variable with the property that all random variables $E[S_i|Z]$ are nonincreasing or nondecreasing functions of Z. The notation S^l represents a comonotonic sum of n lognormal random variables.[3] The CDF of this sum can be obtained by Theorem 5 from Dhaene et al. (2002a). If we assume that the CDFs of the random

[3]The detailed representation of comonotonic random variables is described in Sect. 4 from Dhaene et al. (2002a).

variables $E[S_i|Z]$ are strictly increasing and continuous, then the CDF of S^l is also strictly increasing and continuous. From Eq. (48) of Dhaene et al. (2002a), we obtain that for all $nL \in \left(F^{-1+}_{E[S_i|Z]}(0), F^{-1}_{E[S_i|Z]}(1)\right)$,

$$\sum_{i=0}^{n-1} F^{-1}_{E[S_i|Z]}(F_{S^l}(nL)) = nL,$$

where $F_X^{-1}(p)$ is the inverse of the cumulative distribution function of X.

Or equivalently,

$$\sum_{i=0}^{n-1} E\left[S_i|Z = F_Z^{-1+}(1 - F_{S^l}(nL))\right] = nL.$$

This determines the CDF of the convex order random variable $S^l = E\left[S'_m|Z\right]$ for $S'_m = \sum_{i=0}^{n-1} S_i$.

Under the same assumptions, the stop-loss premiums can be determined from Eq. (55) of Dhaene et al. (2002a):[4]

$$E\left[(S^l - d)^+\right] = \sum_{i=0}^{n-1} E[(E[S_i|Z] - E[S_i|Z = F_Z^{-1+}(1 - F_{S^l}(d))])^+],$$

which holds for all retentions $d \in \left(F_{S^l}^{-1+}(0), F_{S^l}^{-1}(1)\right)$. Hence, the lower bound on the price of an American-Asian call option with strike L, applying Jensen's rule, is:

$$\begin{aligned} E\left[\frac{(A_\tau - L)^+}{R^\tau}\right] &= E\left[E\left(\frac{(A_\tau - L)^+}{R^\tau}|Z_\tau\right)\right] \geq E\left[E\left(\frac{A_\tau - L}{R^\tau}|Z_\tau\right)^+\right] \\ &= E\left[\frac{(E(A_\tau|Z_\tau) - L)^+}{R^\tau}\right]E \\ &= \frac{e^{\frac{-rT}{n}}}{n}E\left[(E[S'_m(n-1,h,a)|Z_\tau] - nL)^+\right]E \\ &= \frac{e^{\frac{-rT}{n}}}{n}E\left[\left(\sum_{i=0}^{n-1} E\left[S_{i,h}|Z_\tau\right] - nL\right)^+\right] \end{aligned} \quad (19.10)$$

where τ represents a stopping time for a fixed exercise rule, Z_k is Φ_k-measurable and Φ_k represents the information set at time k, and R is the discount factor, i.e., $R = e^{-rT/n}$. Note that each possible value of the random variable Z_k corresponds to a nodelet in the refined lattice. For an exercise rule τ, the calculation of the upper bound in inequality (19.10) will generate the estimate for the lower bound.

While calculating the upper bound, whenever $W(k, h, x) > [p_e W^U(k, h, a) + q_e W^L(k, h, a)] \exp(-rT/n)$ in Eq. (19.9), the option will be exercised at nodelet (k, h, a). So $C_L(k, h, a) = [S'(k, h, a) - (k + 1)L]^+$ at each exercised nodelet (k, h, a). For a nodelet at which the option is not exercised,

$$C_L(k,h,x) = [p_e C_L(k+1,h+1,a) + q_e C_L(k+1,h,a+h)] \exp(-rT/n),$$

where $S'(k,h,a)/(k+1) = E(A_k|Z_\tau = (k,h,a)) = \sum_{i=0}^{k} E\left[S_{i,h}|Z_\tau\right]/(k+1)$, and $p_e = p_{k+1,h+1}/(p_{k+1,h+1} + p_{k+1,h})$ and $q_e = (1 - p_e)$ with $p_{k+1,h}$ being $P_h/[(k+1)!/(h!((k+1)-h)!)]$. Thus, $S'(k, h, a)$ is simply the average of $\Sigma S'_m(k, h, a)$ over all paths at nodelet (k, h, a), while $S'_m(k, h, a)$ is the sum of the asset prices along any path with exercise point in its m paths reaching (k, h, a) from time 0 to k. We can easily see that $C_L(0, 0, 0)$ equals the expected value at time 0 in Eq. (19.10), which is the lower bound.

19.6 Numerical Examples

To explain the Edgeworth binomial pricing model for American and European-Asian options under distributions with higher moments, consider the following examples of call options with three periods. The initial stock price $S_0 = 100$, the maturity $T = 1$ year, and the strike prices $L = 100$. The underlying distribution has volatility $\sigma = 0.3$ and risk-free rate $r = 0.1$.

[4]The stop-loss premium with retention d of a random variable X is shown by $E[(X - d)^+]$. See page 7 of Dhaene et al. (2002a) for detailed discussion.

19.6.1 Pricing European-Asian Options Under Lognormal Distribution

We construct the binomial lattice for the asset value process as described in the third section. Exhibit 19.1 shows the nodal values with lognormal distribution in the binomial lattice. We can write down a simple algebraic expression for underlying asset value of the nodes (3,3) and (2,2). The steps are described below:

1. We can get $F(\sqrt{3}) = b(\sqrt{3}) = 0.125$ from Eq. (19.1) because $y_3 = \sqrt{3}$ at node (3,3) when the underlying distribution is lognormal. Therefore, $P_3 = 0.125$ and $x_3 = \sqrt{3}$ can be obtained.
2. From P_3, x_3 and the assumed $\sigma = 0.3$, $r = 0.1$, and $T = 1$, the drift $\mu = 0.1 - \frac{1}{1} \ln \sum_{h=0}^{3} P_h e^{0.3\sqrt{1}x_h} \approx 0.0552$. Therefore, the asset price at node (3,3) is $S_{3,3} = 100e^{0.0552(1)+0.3\sqrt{1}(\sqrt{3})} = 177.66$. The asset prices of the other nodes at the final step in the binomial lattice can be calculated similarly.
3. Upward and downward probabilities are $p_e = q_e = 0.5$ under lognormal distribution. Solving backward recursively from the nodes (3,3) and (3,2), the nodal value, $S_{2,2}$, is $S_{2,2} = [0.5(177.66) + 0.5(125.64)]\exp(-\frac{0.1(1)}{3}) = 146.68$, according to Eq. (19.3). Using the same method, we can compute the remainder of the asset prices in the binomial lattice.

Exhibit 19.1
The nodal values in the binomial lattice under lognormal distribution.

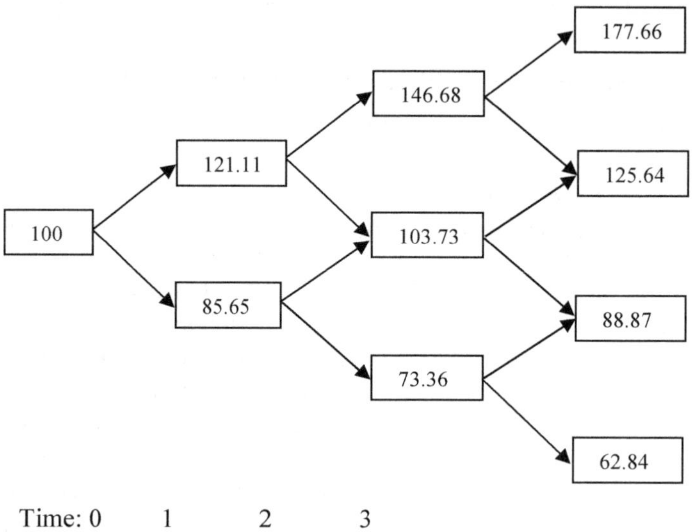

After the underlying asset value is obtained for each node in the binomial lattice, we next calculate the arithmetic average of the stock prices under lognormal distribution as shown in Exhibit 19.2. The algorithmic process for the average price at node (3,2) is described as follows:

1. We first compute the number of paths reaching node (3,2). There are three paths, i.e., (3,2,0), (3,2,1), and (3,2,2). The number

of the paths reaching these nodelets, $M(3,2,0)$, $M(3,2,1)$, and $M(3,2,2)$, is equal to one.

2. The sums of the asset prices over all paths passing through these nodelets are $S''(3, 2, 0) = 100 + 85.65 + 103.73 + 125.64 = 415.02$, $S''(3,2,1) = 450.48$ and $S''(3,2,2) = 493.43$, respectively.
3. According to Eq. (19.4), the arithmetic averages of the stock prices over all paths reaching nodelets (3,2,0), (3,2,1), and (3,2,2) can be obtained. The average price at nodelet (3,2,0), $\overline{A}(3,2,0)$, is then equal to 415.02/[(3 + 1)(1)] = 103.76. Similarly, $\overline{A}(3,2,1) = 112.62$ and $\overline{A}(3,2,2) = 123.36$.

Exhibit 19.2 The Arithmetic Averages of the Stock Prices Under Lognormal Distribution
The top number at each node denotes the arithmetic average of the stock prices, and its nodelet position in the binomial lattice is shown in parenthesis.

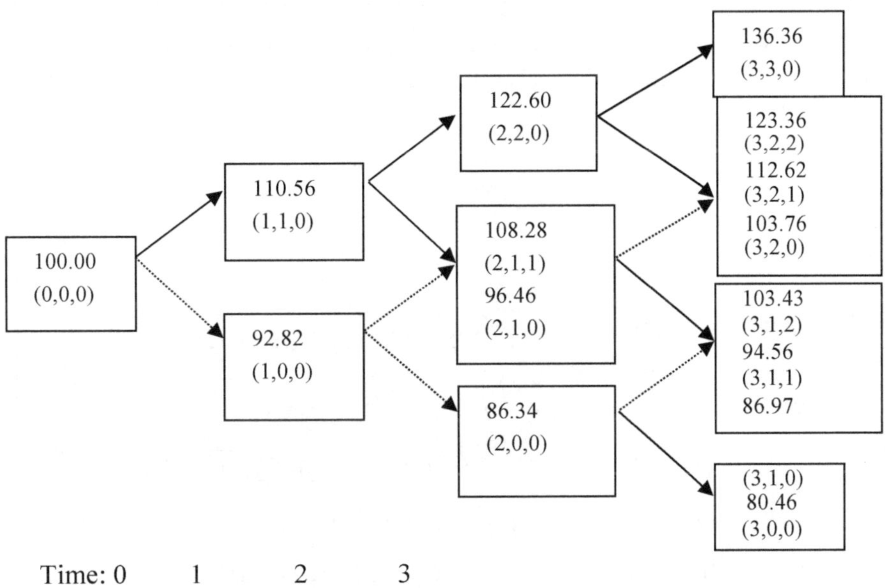

According to the average values of the underlying asset for the nodes at the end of the tree, the lower bound formula in the fourth section can be applied to calculate the price of a European-Asian option. The numerical result is:

$$
\begin{aligned}
C_0^d &= \left\{ \sum_{h=0}^{3} \sum_{a=0}^{h(3-h)} M(3,h,a) p_e^h q_e^{3-h} \left[\overline{A}(3,h,a) - L \right]^+ \right\} \times e^{-rT} \\
&= \{ M(3,0,0) p_e^0 q_e^3 \left[\overline{A}(3,0,0) - L \right] + M(3,1,0) p_e^1 q_e^2 \left[\overline{A}(3,1,0) - L \right] \\
&\quad + M(3,1,1) p_e^1 q_e^2 \left[\overline{A}(3,1,1) - L \right] + M(3,1,2) p_e^1 q_e^2 \left[\overline{A}(3,1,2) - L \right] \\
&\quad + M(3,2,0) p_e^2 q_e^1 \left[\overline{A}(3,2,0) - L \right] + M(3,2,1) p_e^2 q_e^1 \left[\overline{A}(3,2,1) - L \right] \\
&\quad + M(3,2,2) p_e^2 q_e^1 \left[\overline{A}(3,2,2) - L \right] + M(3,3,0) p_e^3 q_e^0 \left[\overline{A}(3,3,0) - L \right] \} \times e^{-rT} \\
&= \left\{ (1)(0.5)^0 (0.5)^3 (0) + (1)(0.5)^1 (0.5)^2 (0) + (1)(0.5)^1 (0.5)^2 (0) \right. \\
&\quad + (1)(0.5)^1 (0.5)^2 (3.43) + (1)(0.5)^2 (0.5)^1 (3.76) \\
&\quad + (1)(0.5)^2 (0.5)^1 (12.62) + (1)(0.5)^2 (0.5)^1 (23.36) \\
&\quad \left. + (1)(0.5)^3 (0.5)^0 (36.36) \right\} \times e^{-(0.1)(1)} \\
&= 8.99
\end{aligned}
$$

The error bound from Eq. (19.7) is then used to obtain the upper bound of this option. That is:

$$
\begin{aligned}
e_r &= \left\{ \frac{1}{2} \sum_{h=0}^{3} \sum_{\substack{a=0 \\ A^{min}(3,h,a) < L,\, A^{max}(3,h,a) > L}}^{h(3-h)} M(3,h,a) p_e^h q_e^{3-h} \left(\overline{A^2}(3,h,a) - \overline{A}(3,h,a)^2 \right)^{\frac{1}{2}} \right\} \times e^{-rT} \\
&= \left\{ \frac{1}{2} \times \left[M(3,0,0) p_e^0 q_e^3 (\overline{A^2}(3,0,0) - \overline{A}(3,0,0)^2)^{\frac{1}{2}} + M(3,1,0) p_e^1 q_e^2 (\overline{A^2}(3,1,0) \right. \right. \\
&\quad - \overline{A}(3,1,0)^2)^{\frac{1}{2}} + M(3,1,1) p_e^1 q_e^2 (\overline{A^2}(3,1,1) - \overline{A}(3,1,1)^2)^{\frac{1}{2}} \\
&\quad + M(3,1,2) p_e^1 q_e^2 (\overline{A^2}(3,1,2) - \overline{A}(3,1,2)^2)^{\frac{1}{2}} + M(3,2,0) p_e^2 q_e^1 (\overline{A^2}(3,2,0) \\
&\quad - \overline{A}(3,2,0)^2)^{\frac{1}{2}} + M(3,2,1) p_e^2 q_e^1 (\overline{A^2}(3,2,1) - \overline{A}(3,2,1)^2)^{\frac{1}{2}} \\
&\quad + M(3,2,2) p_e^2 q_e^1 (\overline{A^2}(3,2,2) - \overline{A}(3,2,2)^2)^{\frac{1}{2}} + M(3,3,0) p_e^3 q_e^0 (\overline{A^2}(3,3,0) \\
&\quad \left. \left. - \overline{A}(3,3,0)^2)^{\frac{1}{2}} \right] 1_{A^{min}(3,h,a) < L, A^{max}(3,h,a) > L} \right\} \times e^{-rT} = 0
\end{aligned}
$$

The number of all paths used in the above equation is equal to one. And the upward and downward probabilities are both 0.5 in calculating the lower bound. The error bound is zero because the conditions $A^{\min}(k,h,a)<L$ and $A^{\max}(k,h,a)>L$ are not satisfied for all the paths. We illustrate the calculating process of the $\overline{A}^2$ (3,1,0) as follows:

$$\begin{aligned}\overline{A^2}(3,1,0) &= \frac{\phi(3,1,0)+2\psi(3,2,0)}{(3+1)^2 M(3,1,0)}\\ &= \frac{30615.49+2\times 45202.5}{(3+1)^2(1)}\\ &= 7563.78\end{aligned}$$

where

$$\begin{aligned}\phi(3,1,0) &= \sum_{i=0}^{3} S_{i,1}^2\\ &= (100^2+85.65^2+73.36^2+88.87^2)\\ &= 30615.49\end{aligned}$$

and

$$\begin{aligned}\psi(3,1,0) &= \sum_{0\le i\le j\le 3} S_{i,h}S_{j,h}\\ &= 100\times 85.65+(100+85.65)\times 73.36\\ &\quad +(100+85.65+73.36)\times 88.87\\ &= 45202.5\end{aligned}$$

From the lower bound and the error bound, we can obtain the upper bound, 8.99. This result is shown in Exhibit 19.3.

Considering first the normal skewness and kurtosis, the results are tested and compared with those in the literature. The call option to be valued has the initial stock price $S_0 = 100$, the maturity $T = 1$ year, and the strike prices $L = 95$, 100, 105, and 110, respectively. The underlying distribution has volatility $\sigma = 0.1$, 0.3, and 0.5, respectively, with normal skewness $g_1 = 0$ and kurtosis $g_2 = 3$. The risk-free rate r is set to be 0.09. The time steps N equals 30, and the computing time and memory space needed in our algorithm are similar to those of Chalasani et al. (1998). We present our simulation results in Tables 19.1 and 19.2.

In Table 19.1, we compare our results with those of Rogers and Shi (1995) and of Chalasani et al. (1998). When the call is in-the-money, our valuation in general is smaller than those of Chalasani et al. (1998). However, the range of our lower and upper bounds is narrower than theirs. For at-the-money and out-of-the-money calls, our estimates are greater than theirs and closer to those of Rogers and Shi's, but the distance between our lower and upper bounds is almost the same as that of Chalasani et al. The difference between our calculations and those of Rogers and Shi's is due to our numerical approximation comparing with their continuous-time integrals.

Exhibit 19.3 Pricing European-Asian Option Under Lognormal Distribution The top number at each node denotes the value of the call option, and its nodelet position in the binomial lattice is shown in parenthesis.

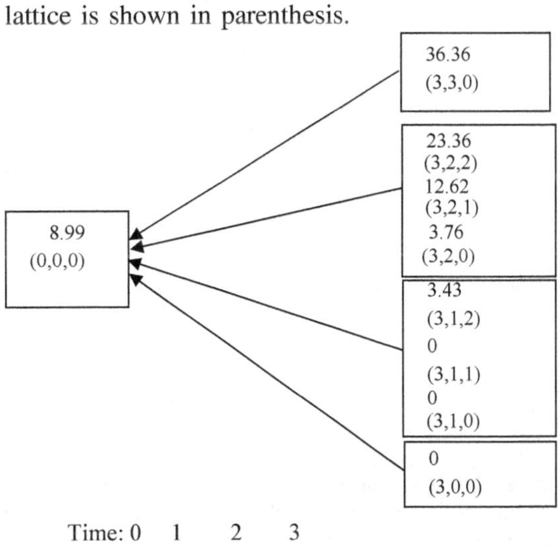

In Table 19.2, we compare our results with those of Monte Carlo simulations from Lévy and Turnbull (1992). Because Chalasani et al. (1998) claim that their results are closer to Monte Carlo estimations than those of Roger and Shi (1995), we also listed their bounds. As depicted in the table, our estimates are much closer to the results of Monte Carlo simulations, so our algorithm in pricing Asian options performs better than that of Chalasani et al. (1998).

Table 19.1 Model comparisons for European-Asian option valuations under normal skewness and kurtosis

Strike L	Vol. σ	r	E-LB	E-UB	RS-LB	RS-UB	C-LB	C-UB
95	0.05	0.05	7.177	7.177	7.178	7.183	7.178	7.178
100	0.05	0.05	2.712	2.712	2.716	2.722	2.708	2.708
105	0.05	0.05	0.332	0.332	0.337	0.343	0.309	0.309
95	0.05	0.09	8.811	8.811	8.809	8.821	8.811	8.811
100	0.05	0.09	4.306	4.306	4.308	4.318	4.301	4.301
105	0.05	0.09	0.957	0.957	0.958	0.968	0.892	0.892
95	0.05	0.15	11.100	11.100	11.094	11.114	11.100	11.100
100	0.05	0.15	6.799	6.799	6.794	6.810	6.798	6.798
105	0.05	0.15	2.745	2.745	2.744	2.761	2.667	2.667
90	0.10	0.05	11.947	11.947	11.951	11.973	11.949	11.949
100	0.10	0.05	3.635	3.635	3.641	3.663	3.632	3.632
110	0.10	0.05	0.319	0.320	0.331	0.353	0.306	0.306
90	0.10	0.09	13.385	13.385	13.385	13.410	13.386	13.386
100	0.10	0.09	4.909	4.909	4.915	4.942	4.902	4.902
110	0.10	0.09	0.621	0.621	0.630	0.657	0.582	0.583
90	0.10	0.15	15.404	15.404	15.399	15.445	15.404	15.404
100	0.10	0.15	7.024	7.024	7.028	7.066	7.015	7.015
110	0.10	0.15	1.411	1.412	1.413	1.451	1.316	1.317
90	0.30	0.05	13.928	13.936	13.952	14.161	13.929	13.938
100	0.30	0.05	7.924	7.932	7.944	8.153	7.924	7.932
110	0.30	0.05	4.041	4.051	4.070	4.279	4.040	4.049
90	0.30	0.09	14.961	14.968	14.983	15.194	14.964	14.972
100	0.30	0.09	8.811	8.818	8.827	9.039	8.807	8.815
110	0.30	0.09	4.672	4.682	4.695	4.906	4.661	4.671
90	0.30	0.15	16.494	16.500	16.512	16.732	16.499	16.506
100	0.30	0.15	10.197	10.205	10.208	10.429	10.187	10.195
110	0.30	0.15	5.715	5.725	5.728	5.948	5.685	5.696

Note The European-Asian option to be valued has initial stock price $S_0 = 100$ dollars and option life $T = 1.0$ year. Using time steps $N = 30$, the lower and upper bounds from our algorithm are indicated by E-LB and E-UB, respectively, while those from Rogers and Shi (1995) are indicated by RS-LB and RS-UB, and those from Chalasani et al.(1998) by C-LB and C-UB. We used normal skewness $g_1 = 0$ and kurtosis $g_2 = 3$ in our algorithm

19.6.2 Pricing American-Asian Options Under Lognormal Distribution

We can use the example in Exhibits 19.1 and 19.2 to explain the process for computing an American-Asian option under lognormal distribution. Equation (19.9) in the fifth section is applied to estimate an upper bound for the value of an American-Asian option at time zero. The following illustration for nodelet (2,2,0) explains this recursive backward procedure.

$$W(2,2,0) = \max\left\{ (\bar{A}(2,2,0) - L)^+, \right.$$
$$\left. (p_e W^U(2,2,0) + q_e W^L(2,2,0))\exp\left(\frac{-rT}{3}\right) \right\}$$

where $W^U(2, 2, 0) = \lambda\, w_1 + (1 - \lambda)\, w_2$ with w_1 being $W(2 + 1,\ 2 + 1,\ b) = [\bar{A}\ (2 + 1,\ 2 + 1,$

Table 19.2 Comparisons with Monte Carlo simulations under normal skewness and kurtosis

Strike L	Vol. σ	r	Monte Carlo	E-LB	E-UB	C-LB	C-UB
95	0.10	0.09	8.91	8.91	8.91	8.91	8.91
100	0.10	0.09	4.91	4.91	4.91	4.90	4.90
105	0.10	0.09	2.06	2.07	2.07	2.03	2.03
90	0.30	0.09	14.96	14.96	14.97	14.96	14.97
100	0.30	0.09	8.81	8.81	8.82	8.81	8.82
110	0.30	0.09	4.68	4.67	4.68	4.66	4.67
90	0.50	0.09	18.14	18.14	18.18	18.15	18.19
100	0.50	0.09	12.98	12.98	13.02	12.99	13.03
110	0.50	0.09	9.10	9.07	9.11	9.08	9.12

Note The European-Asian option to be valued has initial stock price $S_0 = 100$ dollars and option life $T = 1.0$ year. Using time steps $N = 30$, the lower and upper bounds from our algorithm are indicated by E-LB and E-UB, respectively, while Monte Carlo estimates from Lévy and Turnbull (1992) are indicated by Monte Carlo, and those from Chalasani et al. (1998) are indicated by C-LB and C-UB. We used normal skewness $g_1 = 0$ and kurtosis $g_2 = 3$ in our algorithm

$b) - L]^+$ and w_2 being $W(2+1,2+1,b+1) = [\bar{A}(2+1,2+1,b+1) - L]^+$, and $W^L(2,2,0) = \lambda w'_1 + (1-\lambda)w'_2$ with w'_1 being $W(2+1,2,b) = [\bar{A}(2+1,2,b) - L]^+$ and w'_2 being $W(2+1,2,b+1) = [\bar{A}(2+1,2,b+1) - L]^+$.

We need to calculate values of λ, w_1 and w_2 before $W^U(2, 2, 0)$ can be estimated.

$$\begin{aligned}
x^U(2,2) &= [\bar{A}(2,2,0)(2+1) + S_{2+1,2+1}]/(2+2) \\
&= [122.6 \times 3 + 177.66]/4 = 136.36, E \\
x^U(2,2) &= \lambda\bar{A}(2+1,2+1,0) + (1-\lambda)\bar{A}(2+1,2+1,1), \\
\lambda &= \frac{x^U(2,2) - \bar{A}(3,3,1)}{\bar{A}(3,3,0) - \bar{A}(3,3,1)} = \frac{136.36 - 0}{136.36 - 0} = 1, \\
W^U(2,2,0) &= \lambda[\bar{A}(3,3,0) - L]^+ + (1-\lambda)[\bar{A}(3,3,1) - L]^+ \\
&= (1)[136.36 - 100]^+ + (1-1)[0-100]^+ \\
&= 36.36.
\end{aligned}$$

Similarly, value of $W^L(2, 2, 0)$ is:

$$\begin{aligned}
W^L(2,2,0) &= \lambda[\bar{A}(3,2,2) - L]^+ + (1-\lambda)[\bar{A}(3,2,3) - L]^+ \\
&= (1)[123.36 - 100]^+ + (1-1)[0-100]^+ = 23.36
\end{aligned}$$

Hence, we can get $W(2, 2, 0)$ below,

$$\begin{aligned}
W(2,2,0) &= \max\{(\bar{A}(2,2,0) - L)^+, (p_e W^U(2,2,0) + q_e W^L(2,2,0))\exp(\frac{-rT}{3})\} \\
&= \max\left\{(122.6 - 100)^+, ((0.5)(36.36) + (0.5)(23.36))e(\frac{-(0.1)(1)}{3})\right\} \\
&= 28.88,
\end{aligned}$$

Values at other nodes can be calculated in the same way. As shown in Exhibit 19.4, the estimated $W(0, 0, 0) = 9.12$, which is the upper bound for the value of the American-Asian option at time zero.

The exercise rule discussed in the fifth section can be applied to find the lower bound. Whenever $W(k, h, x) > [(p_{k+1,h+1}/p)\ W^U(k, h, a) + (p_{k+1,h}/p)\ W^L(k, h, a)]\exp(-rt/n)$ in Eq. (19.9), the option

Table 19.3 Model comparisons for American-Asian option valuations under lognormal distribution

Steps n	E-LB	E-UB	C-LB	C-UB	HW (h = 0.005)	HW (h = 0.003)
20	4.811	4.813	4.812	4.815	4.815	4.814
40	4.886	4.888	4.888	4.889	4.892	4.890
60	4.916	4.917	4.917	4.918	4.924	4.920
80	4.932	4.933	4.933	4.934	4.942	4.936

Note The American-Asian options are valued with initial stock price S_0= 50 dollars, strike K = 50 dollars, option life T = 1.0 year, volatility σ = 0.3 per year, and risk-free rate r = 0.1 per year. The estimated lower and upper bounds from our Edgeworth binomial model are indicated by E-LB and E-UB, respectively, while the estimates from Hull–White (1993) with different grid-size h are denoted by HW, and those from Chalasani et al. (1999) are denoted by C-LB, C-UB. All simulations are conducted under different time steps with normal skewness $\gamma_1 = 0$ and kurtosis $\gamma_2 = 3$. Parameter values are selected so the results can be compared with those in the literature

will be exercised at nodelet (k, h, a); otherwise it will not be exercised. We again use the illustration of nodelet (2,2,0) to account for the exercise rule. At nodelet (2,2,0), $C_L(2,2,0)$ is equal to $[S(2,2,0)–(2 + 1)L]^+/(2 + 1) = [(100 + 121.11 + 146.68) - (2 + 1)(100)]/(2 + 1) = 22.6$ if the option is exercised. However, if it is not exercised, $C_L(2,2,0)$ equals $[p_e\ C_L(2 + 1,\ 2 + 1,\ 0) + q_e\ C_L(2 + 1,\ 2,\ 0 + 2)]\ \exp((-0.1)(1)/3) = [(0.5)(36.36) + (0.5)(23.36)]\ \exp\ ((-0.1)(1)/3) = 28.88$. Therefore, the option will not be exercised at nodelet (2,2,0), so its value is equal to 28.88. Because all the nodelets before the maturity time in Exhibit 19.4 are not exercised, the lower bound is equal to the upper bound.

In order to examine the possible errors from the Edgeworth approximations, we first compare our results with those of Chalasani et al. (1999) and Hull and White (1993) for lognormal underlying distribution. Table 19.3 shows the estimated value of an at-the-money American-Asian option using different time steps with yearly volatility at 0.3, the risk-free rate at 0.1, and the initial stock price at 50 dollars. Table 19.4 further provides the estimates of the values with 40 time steps but under various maturities and strike prices. Hull and White's approximation for American-Asian options is used as our benchmark.

In Table 19.3, we find that the lower and the upper price bounds of an American-Asian option using our method are lower than those from Chalasani et al. (1999) and Hull and White (1993). However, when we gradually increase the number of the time steps, our results are almost the same as theirs. Although the Edgeworth density $f(x)$ is not exactly a probability measure, the approximation errors tend to be very small if more time steps are used in the simulations.

Table 19.4 exhibits the results under different strikes and option lives using the same number of time steps. When the options are out-of-the-money, our estimates are slightly greater than those from Chalasani et al. (1999) and Hull and White (1993). However, for in-the-money options, ours are slightly lower than theirs. The differences between our results and theirs are lowered if the options are closer to the expiration. Our simulations confirm the discussions by Ju (2002) that the Edgeworth expansion method works fine for shorter maturities, but not for long maturities, while pricing Asian options.

Table 19.4 Model comparisons for American-Asian option valuations under lognormal distribution with different option lives and strikes

Option life T	Strike L	E-LB	E-UB	C-LB	C-UB	HW
0.5	40	12.105	12.105	12.111	12.112	12.115
0.5	45	7.248	7.248	7.255	7.255	7.261
0.5	50	3.268	3.269	3.269	3.270	3.275
0.5	55	1.150	1.151	1.148	1.148	1.152
0.5	60	0.323	0.323	0.320	0.320	0.322
1.0	40	13.136	13.137	13.150	13.151	13.153
1.0	45	8.535	8.537	8.546	8.547	8.551
1.0	50	4.886	4.888	4.888	4.889	4.892
1.0	55	2.537	2.539	2.532	2.534	2.536
1.0	60	1.211	1.213	1.204	1.206	1.208
1.5	40	13.967	13.969	13.984	13.985	19.988
1.5	45	9.636	9.639	9.648	9.650	9.652
1.5	50	6.193	6.195	6.195	6.197	6.199
1.5	55	3.774	3.777	3.767	3.770	3.771
1.5	60	2.201	2.204	2.190	2.193	2.194
2.0	40	14.685	14.688	14.709	14.712	14.713
2.0	45	10.605	10.609	10.620	10.623	10.623
2.0	50	7.320	7.323	7.322	7.325	7.326
2.0	55	4.889	4.893	4.881	4.885	4.886
2.0	60	3.180	3.184	3.167	3.170	3.171

Note The American-Asian options are valued with initial stock price S_0= 50 dollars, time steps n = 40, volatility σ = 0.3 per year, and risk-free rate r = 0.1 per year. The estimated lower and upper bounds from our Edgeworth binomial model are indicated by E-LB and E-UB, respectively, while the estimates from Hull–White (1993) are denoted by HW, and those from Chalasani et al. (1999) are denoted by C-LB, C-UB. All simulations are conducted under 40 time steps with normal skewness $\gamma_1 = 0$ and kurtosis $\gamma_2 = 3$, but with various option lives and strikes. Parameter values are selected so the results can be compared with those in the literature

Exhibit 19.4 Pricing American-Asian Option Under Lognormal Distribution The top number at each node denotes the value of the call option, and its nodelet position in the binomial lattice is shown in parenthesis.

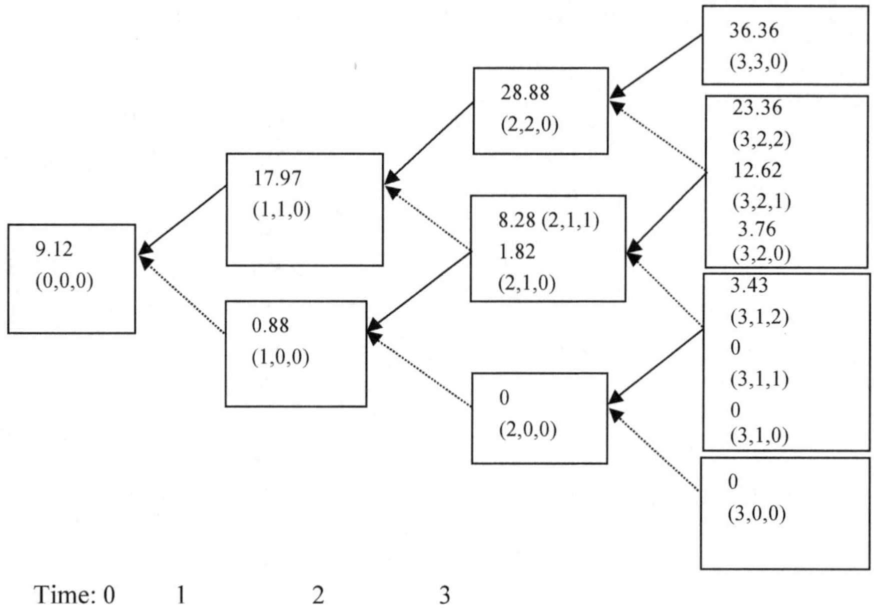

19.6.3 Pricing European-Asian Options Under Distributions with Higher Moments

We now consider the examples of call options with higher moments in the underlining distribution. Most initial presumptions are the same as those used for lognormal distribution, except that skewness $\gamma_1 = 0.05$ and kurtosis $\gamma_2 = 3.05$. We just demonstrate the process for computing the underlying values and their arithmetic averages along each path in the binomial lattice with higher moment consideration. Other pricing procedures discussed earlier for lognormal distribution can be similarly applied to the cases with higher moments in the underlining distribution.

Exhibit 19.5 provides the nodal values in the binomial lattice under distribution with higher moments. The procedure to get these values is demonstrated as below:

1. At node (3,2), the random variable $y_2 = \frac{1}{\sqrt{3}}$. After calculating the Edgeworth expansion $\left[1+(\frac{1}{6})(0.05)\left((\frac{1}{\sqrt{3}})^3-3(\frac{1}{\sqrt{3}})\right)+(\frac{1}{24})(3.05-3)\left((\frac{1}{\sqrt{3}})^4-6(\frac{1}{\sqrt{3}})^2+3\right)+(\frac{1}{72})(0.05)\left((\frac{1}{\sqrt{3}})^6-15(\frac{1}{\sqrt{3}})^4+45(\frac{1}{\sqrt{3}})^2-15\right)\right]=0.9894$, the resulting binomial density is $F(\frac{1}{\sqrt{3}})=f(\frac{1}{\sqrt{3}})\times b(\frac{1}{\sqrt{3}})=0.9894\times 0.375=0.371$. Therefore, $P_2=0.372$ and $x_2=0.585$ given $F(1/\sqrt{3})=0.371$ and $y_2=\frac{1}{\sqrt{3}}$.
2. We estimate the drift $\mu = 0.1-\frac{1}{1}\ln\sum_{h=0}^{3}P_h e^{0.3\sqrt{1}x_h}\approx 0.05509$ using P_2, x_2 and the assumed $\sigma = 0.3$, $r = 0.1$, and $T = 1$. The asset price at node (3,2) is then equal to $S_{3,2}=100e^{0.05509(1)+0.3\sqrt{1}(0.585)}=125.93$. The asset prices at other nodes in the final step of the binomial lattice can be calculated similarly.

3. The upward and downward probabilities under distribution with higher moments are $p_e = 0.4936$ and $q_e = 0.5064$. The nodal value, $S_{2,1}$, can be obtained by solving backward from the values at nodes (3,2) and (3,1):

$$S_{2,1} = [0.4936(125.93) + 0.5064(88.95)] \exp\left(-\frac{0.1(1)}{3}\right) = 103.69.$$

Exhibit 19.5 The Nodal Values in the Binomial Lattice Under Distribution with Higher Moments

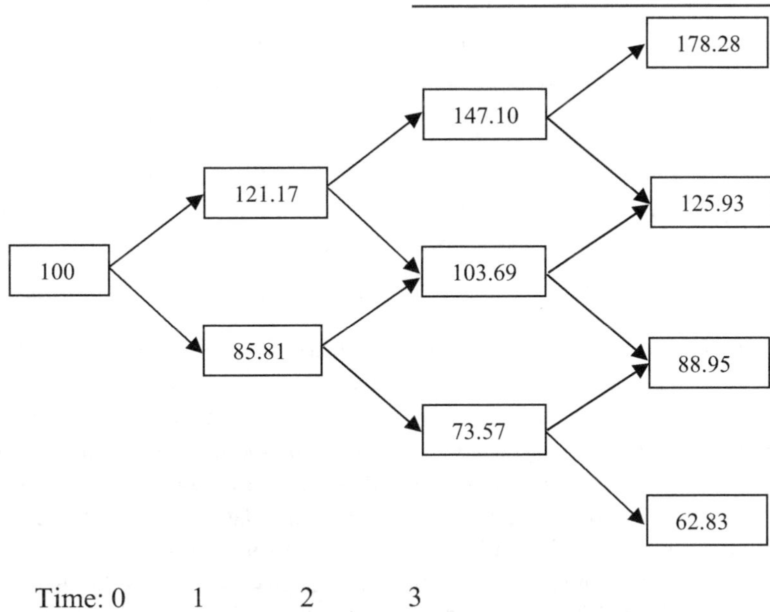

We next calculate the arithmetic average of the stock prices under distribution with higher moments. The algorithmic process for the average price at node (3,1) is explained below:

1. First, count the number of the paths reaching node (3,1). $M(3,1,0)$, $M(3,1,1)$, and $M(3,1,2)$ all equal to one.
2. The sum of the asset prices over all paths passing through the nodelet (3,1,0), $S''(3,1,0) = 100 + 85.81 + 73.57 + 88.95 = 348.33$. Similarly, $S''(3,1,1) = 378.45$ and $S''(3,1,2) = 413.8$.
3. According to Eq. (19.4), the average price at nodelet (3,1,0), $\overline{A}(3,1,0)$ is then equal to $348.33/[(3 + 1)(1)] = 87.08$. Similarly, $\overline{A}(3,1,1) = 94.61$ and $\overline{A}(3,1,2) = 103.45$.

The average of the underlying asset values at each nodelet is denoted in Exhibit 19.6.

Exhibit 19.6 The Arithmetic Averages of Stock Prices Under Distribution with Higher Moments The top number at each node denotes the arithmetic average of the stock prices, and its nodelet position in the binomial lattice is shown in parenthesis.

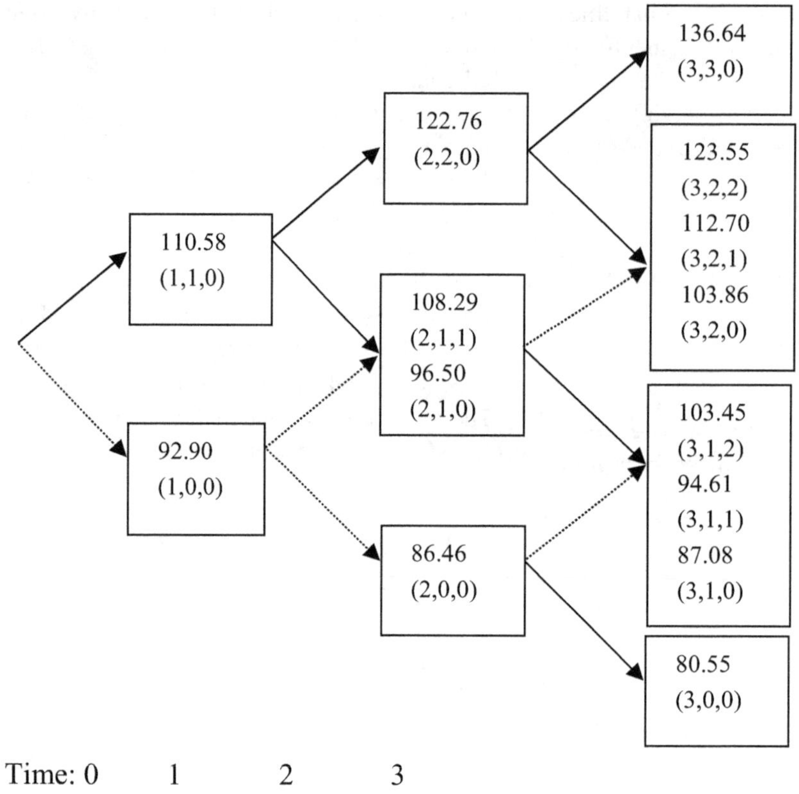

The valuation performance of the European-Asian option in Table 19.5 is based on the initial stock price $S_0 = 100$, the risk-free rate $r = 0.09$, and the maturity $T = 1$ year with varying skewness and kurtosis. The results of the numerical analysis are compared with those of the Edgeworth expansion model by Turnbull and Wakeman (1991) (TW), the modified Edgeworth expansion method by Lévy and Turnbull (1992) (LT), and the four-moment approximation model by Posner and Milevsky (1998) (PM). All these models are considered up to the fourth moments in their valuations.

For low volatility cases ($\sigma = 0.05$ and 0.1) in Table 19.5, our results from in-the-money or at-the-money calls are very close to Monte Carlo estimates, which is the benchmark used by Lévy and Turnbull (1992) under lognormal distribution. This is similar to those of LT and TW. For out-of-the-money calls, our outcomes are the same as theirs under right-skewed conditions. For high-volatility cases ($\sigma = 0.3$ and 0.5), our outcomes for at-the-money or deep-in-the-money calls approach the results from Monte Carlo simulation with positive skewness and a slight leptokurtic. Under lognormal distribution, when the call is deep-out-of-money, our lower bounds are more accurate than the estimates from all the other methods. The results from Monte Carlo method are consistent with our lower and upper bounds. Overall, our outcomes are better than those of LT and TW, and similar to PM.

Table 19.5 Model comparisons for European-Asian option valuations under various skewness and kurtosis

Strike L	Vol. σ	g_1	g_2	MC	E-LB	E-UB	LT	TW	PM
95	0.05	0	3	8.81 (0.00)	8.81	8.81	8.81	8.81	NAN
100	0.05	0	3	4.31 (0.00)	4.31	4.31	4.31	4.31	NAN
105	0.05	0.03	3	0.95 (0.00)	0.95	0.95	0.95	0.95	NAN
95	0.1	0	3	8.91 (0.00)	8.91	8.91	8.91	8.91	NAN
100	0.1	0	3	4.91 (0.00)	4.91	4.91	4.91	4.91	NAN
105	0.1	0.02	3	2.06 (0.00)	2.06	2.06	2.06	2.06	NAN
90	0.3	0	3	14.96 (0.01)	14.97	14.98	15.00	14.91	14.96
100	0.3	0.01	3	8.81 (0.01)	8.80	8.82	8.84	8.78	8.80
110	0.3	0	3	4.68 (0.01)	4.68	4.70	4.69	4.69	4.67
90	0.5	0.01	3	18.14 (0.03)	18.14	18.21	18.13	17.66	18.14
100	0.5	0	3.02	12.98 (0.03)	12.97	13.03	13.00	12.86	12.97
110	0.5	0	3	9.10 (0.03)	9.09	9.16	9.12	9.22	9.07

Note E-LB and E-UB indicate the lower and the upper bounds from our model with various skewness (g_1) and kurtosis (g_2). The approximations of Lévy and Turnbull (1992) are represented by LT, and of Turnbull and Wakeman (1991) by TW, MC represents the Monte Carlo estimates in the Lévy and Turnbull (1992), and PM represents the four-moment approximation by Posner and Milevsky (1998). The simulations assume the option life $T = 1$ year, the domestic interest rate $r = 0.09$, the time steps $N = 52$, and the initial spot price $S_0 = 100$

Table 19.6 Model comparisons for American-Asian option valuations with various time steps, skewness, and kurtosis

Steps n	γ_1	γ_2	E-LB	E-UB	C-LB	C-UB	HW ($h = 0.005$)	HW ($h = 0.003$)
20	−0.002	3.00	4.813	4.815	4.812	4.815	4.815	4.814
40	−0.046	3.06	4.890	4.891	4.888	4.889	4.892	4.890
60	−0.051	3.06	4.920	4.920	4.917	4.918	4.924	4.920
80	−0.050	3.05	4.937	4.937	4.933	4.934	4.942	4.936

Note The American-Asian options are valued with initial stock price $S_0 = 50$ dollars, strike $K = 50$ dollars, option life $T = 1.0$ year, volatility $\sigma = 0.3$ per year, and risk-free rate $r = 0.1$ per year. The estimated lower and upper bounds from our Edgeworth binomial model are indicated by E-LB and E-UB, respectively, while the estimates from Hull–White (1993) with different grid-size h are denoted by HW, and those from Chalasani et al. (1999) are denoted by C-LB, C-UB. All simulations are conducted under various time steps, skewness (γ_1), and kurtosis (γ_2). Parameter values are selected so the results can be compared with those in the literature

19.6.4 Pricing American-Asian Options Under Distributions with Higher Moments

Finally, we compare the price estimates of American-Asian options from our Edgeworth binomial model with those from Chalasani et al. (1999) and Hull and White (1993) under distributions having higher moments. As indicated in Table 19.6, if the underlying distribution has negative skewness and positive excess kurtosis while other parameters are the same as in Table 19.3, we find that the results from our model (E-LB = 4.813, 4.890, 4.920, 4.937 and E-UB = 4.815, 4.891, 4.920, 4.937) are closer to the estimates from Hull–White (with grid size $h = 0.003$ as used in Hull and White (1993)) than those from Chalasani et al. (C-LB = 4.812, 4.888, 4.917, 4.933 and C-UB = 4.815, 4.889, 4.918, 4.934). We also adopt the same parameter values used in Table 19.4 but assume an underlying distribution with left-skewness and leptokurtosis for calculating the option price. Our

Table 19.7 Model comparisons for American-Asian option valuations with various option lives, strikes, skewness, and kurtosis

Option life T	Strike L	γ_1	γ_2	E-LB	E-UB	C-LB	C-UB	HW
0.5	40	−0.030	3.06	12.115	12.115	12.111	12.112	12.115
0.5	45	−0.030	3.06	7.260	7.261	7.255	7.255	7.261
0.5	50	−0.040	3.04	3.275	3.275	3.269	3.270	3.275
0.5	55	0.000	3.00	1.150	1.151	1.148	1.148	1.152
0.5	60	−0.020	3.25	0.321	0.322	0.320	0.320	0.322
1.0	40	−0.041	3.09	13.152	13.153	13.150	13.151	13.153
1.0	45	−0.040	3.10	8.551	8.552	8.546	8.547	8.551
1.0	50	−0.040	3.05	4.891	4.892	4.888	4.889	4.892
1.0	55	−0.003	3.00	2.536	2.538	2.532	2.534	2.536
1.0	60	−0.003	3.01	1.207	1.209	1.204	1.206	1.208
1.5	40	−0.017	3.00	13.987	13.989	13.984	13.985	19.988
1.5	45	−0.017	3.02	9.651	9.653	9.648	9.650	9.652
1.5	50	−0.007	3.00	6.199	6.201	6.195	6.197	6.199
1.5	55	−0.050	3.01	3.770	3.771	3.767	3.770	3.771
1.5	60	−0.007	3.01	2.192	2.194	2.190	2.193	2.194
2.0	40	−0.036	3.05	14.712	14.715	14.709	14.712	14.713
2.0	45	−0.029	3.05	10.623	10.625	10.620	10.623	10.623
2.0	50	−0.026	3.03	7.325	7.327	7.322	7.325	7.326
2.0	55	−0.028	3.01	4.886	4.889	4.881	4.885	4.886
2.0	60	−0.001	3.02	3.168	3.172	3.167	3.170	3.171

Note The American-Asian options are valued with initial stock price S_0= 50 dollars, time steps n = 40, volatility $\sigma = 0.3$ per year, and risk-free rate $r = 0.1$ per year. The estimated lower and upper bounds from our Edgeworth binomial model are indicated by E-LB and E-UB, respectively, while the estimates from Hull–White (1993) are denoted by HW, and those from Chalasani et al. (1999) are denoted by C-LB, C-UB. All simulations are conducted under various option lives, strikes, skewness (γ_1), and kurtosis (γ_2). Parameter values are selected so the results can be compared with those in the literature

results, as shown in Table 19.7, are again closer to those from Hull–White (with $h = 0.005$ as used in Hull and White (1993)). The evidence in Tables 19.6 and 19.7 demonstrates that our modified Edgeworth binomial model performs better in simulating American-Asian options.

19.7 Conclusion

In this chapter, we have developed the modified Edgeworth binomial model to price European and American-Asian options with higher moments in the underlying distribution. Many studies in the literature have illustrated significant left-skewness and leptokurtosis for the distribution of the underlying asset. Our model combines the refined binomial lattice with the Edgeworth expansible distribution to include the high-moment parameters. As the algorithm of the Edgeworth binomial tree can greatly enhance the computing efficiency, our modified model would be able to price Asian options faster and with higher precision when the underlying distribution displays higher moments. The numerical results show that this approach can effectively deal with the higher moment issue and provide better option value estimates than those found in various studies in the literature.

Bibliography

Ahn, D.-H., Boudoukh, J., Richardson, M., & Whitelaw, R. F. (1999). Optimal risk management using options. *Journal of Finance, 54*(1), 359–375.

Amin, K. I. (1993). Jump diffusion option valuation in discrete time. *Journal of Finance, 48*(5), 1833–1863.

Barraquand, J., & Pudet, T. (1996). Pricing of American path-dependent contingent claims. *Mathematical Finance, 6,* 17–51.

Chalasani, P., Jha, S., & Varikooty, A. (1998). Accurate approximations for European Asian options. *Journal of Computational Finance, 1*(4), 11–29.

Chalasani, P., Jha, S., Egriboyun, F., & Varikooty, A. (1999). A refined binomial lattice for pricing American Asian options. *Review of Derivatives Research, 3* (1), 85–105.

Chang, C. C., & Fu, H. C. (2001). A binomial options pricing model under stochastic volatility and jump. *CJAS, 18*(3), 192–203.

Corrado, J. C., & Su, T. (1997). Implied volatility skews and stock index skewness and kurtosis implied by S&P 500 index option prices. *Journal of Derivatives, 4*(4), 8–19.

Costabile, M., Massabo, I., & Russo, E. (2006). An adjusted binomial model for pricing Asian options. *Review Quantitative Finance & Accounting, 27*(3), 285–296.

Curran, M. (1994). Valuing Asian and portfolio options by conditioning on the geometric mean. *Management Science, 40,* 1705–1711.

Dhaene, J., Denuit, M., Goovaerts, M. J., Kaas, R., & Vyncke, D. (2002a). The concept of comonotonicity in actuarial science and finance: theory. *Insurance: Mathematics & Economics, 31,* 3–33.

Dhaene, J., Denuit, M., Goovaerts, M. J., Kaas, R., & Vyncke, D. (2002b). The concept of comonotonicity in actuarial science and finance: Applications. *Insurance: Mathematics & Economics, 31,* 133–161.

Grant, D., Vora, G., & Weeks, D. (1997). Path-dependent options: Extending the Monte Carlo simulation approach. *Management Science, 43*(11), 1589–1602.

Hilliard, J. E., & Schwartz, A. L. (1996). Binomial option pricing under stochastic volatility and correlated state variable. *Journal of Derivatives*, 23–39.

Hull, J., & White, A. (1993). Efficient procedures for valuing European and American path-dependent options. *The Journal of Derivatives, 1*, 21–31.

Ju, N. (2002). Pricing Asian and basket options via Taylor expansion. *Journal of Computational Finance, 5*(3), 1–33.

Kim, T. H., & White, H. (2004). On more robust estimation of skewness and kurtosis: Simulation and application to the S&P 500 Index. *Finance Research Letters, 1,* 56–73.

Klassen, T. R. (2001). Simple, fast, and flexible pricing of Asian options. *Journal of Computational Finance, 4,* 89–124.

Lee, C.-F., Lee, A. C., & Lee, J. (2010). *Handbook of quantitative finance and risk management.* New York: Springer.

Lévy, E. (1992). Pricing European average rate currency options. *Journal of International Money & Finance, 11,* 474–491.

Lévy, E., & Turnbull, S. (1992). Average intelligence. *RISK, 5,* 53–58.

Lo, K., Wang, K., & Hsu, M. (2008). Pricing European Asian options with skewness and kurtosis in the underlying distribution. *Journal of Futures Markets, 28*(6), 598–616.

Milevsky, M. A., & Posner, S. E. (1998a). A closed-form approximation for valuing basket options. *The Journal of Derivatives, 5*(4), 54–61.

Milevsky, M. A., & Posner, S. E. (1998b). Asian options, the sum of lognormals, and the reciprocal Gamma distribution. *Journal of Financial & Quantitative Analysis, 33,* 409–422.

Neave, E. H., & Ye, G. L. (2003). Pricing Asian options in the framework of the binomial model: A quick algorithm. *Derivative Use, Trading & Regulation, 9* (3), 203–216.

Posner, S. E., & Milevsky, M. A. (1998). Valuing exotic options by approximating the SPD with higher moments. *The Journal of Financial Engineering, 7* (2), 109–125.

Rogers, L. C. G., & Shi, Z. (1995). The value of an Asian option. *Journal of Applied Probability, 32,* 1077–1088.

Rubinstein, M. (1994). Implied binomial trees. *Journal of Finance, 49*(3), 771–818.

Rubinstein, M. (1998). Edgeworth binomial trees. *The Journal of Derivatives, 5*(3), 20–27.

Thompson, G. W. P. (2002). *Fast narrow bounds on the value of Asian options.* Working Paper, Centre for Financial Research, Judge Institute of Management, University of Cambridge.

Turnbull, S., & Wakeman, L. (1991). A quick algorithm for pricing European average options. *Journal of Financial & Quantitative Analysis, 26,* 377–389.

Wang, K., & Hsu, M. F. (2010). *Handbook of Quantitative Finance and Risk Management: Numerical valuation of asian options with higher moments in the underlying distribution* (pp. 587–602). New York: Springer.

Zhang, J. E. (2001). A semi-analytical method for pricing and hedging continuously sampled arithmetic average rate options. *Journal of Computational Finance, 5,* 59–79.

Zhang, J. E. (2003). Pricing continuously sampled Asian options with perturbation method. *Journal of Futures Markets, 23*(6), 535–560.

Itô's Calculus: Derivation of the Black–Scholes Option Pricing Model

20

Contents

Abstract

The purpose of this chapter is to develop certain relatively mathematical discoveries known generally as stochastic calculus, or more specifically as Itô's calculus and to also illustrate their application in the pricing of options. The mathematical methods of stochastic calculus are illustrated in alternative derivations of the celebrated Black–Scholes–Merton model. The topic is motivated by a desire to provide an intuitive understanding of certain probabilistic methods that have found significant use in financial economics.

This chapter draws upon (1) Chap. 30 of Handbook of Quantitative Finance and Risk Management (2010) by Lee et al.; and (2) Chap. 19 of the book entitled Security Analysis, Portfolio Management, and Financial Derivatives (2013) by Lee et al.

C.-F. Lee et al., *Financial Econometrics, Mathematics and Statistics*,
https://doi.org/10.1007/978-1-4939-9429-8_20

20.1 Introduction

The purpose of this chapter is to develop certain relatively recent mathematical discoveries known generally as stochastic calculus (or mode specifically as Itô's calculus) and to illustrate their application in the pricing of options. The topics are motivated by a desire to provide an intuitive understanding of certain probabilistic methods that have found significant use in financial economics. A rigorous presentation of the same ideas is presented briefly in Malliaris and Brock (1982) and mode recently, Chalamandaris and Malliaris (2010).

Itô's calculus was prompted by purely mathematical questions originating in Wiener's work in 1923 on stochastic integrals and was developed by the Japanese probabilist Kiyosi Ito during 1944–1951. Two decades later, economists such as Merton (1973) and Black and Scholes (1973) started using Itô's stochastic differential equation to describe the behavior of asset prices. Because stochastic calculus is now used regularly by financial economists, some attention must be given to its mathematical meaning, its appropriateness in economic modeling, and its applications in economics modeling, and to finance.

We will discuss the Itô process and financial modeling in Sect. 20.2. Itô lemma will be shown in Sect. 20.3. Section 20.4 will present stochastic differential equation approach to stock-price behavior. Section 20.5 will discuss the pricing of an option. Section 20.6 will demonstrate a reexamination of option pricing. Remarks on option pricing will be presented in Sect. 20.7. Finally, Sect. 20.8 concludes. In addition, Appendix will discuss an alternative method to derive the Black–Scholes option pricing Model.

20.2 The Itô Process and Financial Modeling

Stochastic calculus is the mathematical of random change in continuous time, unlike ordinary calculus, which deals with deterministic change. A key notion in stochastic calculus is the equation:

$$\mathrm{d}S(t,w) = \mu[t,S(t,w)]\mathrm{d}t + \sigma[t,S(t,w)]\mathrm{d}Z(t,w), \tag{20.1}$$

which is analogous to the ordinary differential equation $\mathrm{d}S(t)/\mathrm{d}t = \mu[t,S(t)]$. This section defines intuitively the Itô equation in Eq. (20.1) and discusses its appropriateness to financial modeling.

A stochastic process is an Itô process if the random variable $\mathrm{d}S(t,w)$ can be represented by Eq. (20.1). The first term, $\mu[t,S(t,w)]\mathrm{d}t$, is the expected change in $S(t,w)$ at time t. The second term, $\sigma[t,S(t,w)]\mathrm{d}Z(t,w)$, reflects the uncertain term.

The Itô equation is a random equation. The domain of the equation is $[0,\infty)\times\Omega$, with the first argument t denoting time and taking values continuously in the interval $[0,\infty)$, and the second argument w denoting a random element taking values from a random set Ω. The range of the equation is the real numbers or real vectors. For simplicity, only the real numbers, denoted by R, are considered as the range of Eq. (20.1). Because time takes values continuously in $[0,\infty)$, the Itô equation is a **continuous-time random equation**.

Although at first a real random variables $S(t,w): [0,\infty)\times\Omega\to R$ is used—that is, a function having as domain $[0,\infty)\times\Omega$ and as range real numbers—Eq. (20.1) expresses not the values of $S(t,w)$, but its infinitesimal differences $\mathrm{d}S(t,w)$ as a function of two terms. For example, in finance $S(t,w)$ denotes the price of a stock at time t affected by the state of the economy described by the random element w; and (20.1) expresses the small changes in the stock price, $\mathrm{d}S(t,w)$, at time t affected by the random element w. The mathematical meaning of this small difference can be explained as follows: $\mathrm{d}S(t,w)$ may be viewed as the limit of large finite differences $\Delta S(t,w)$ as Δt approaches zero. Note that $\Delta S(t,w) = S(t+\Delta t,w) - S(t,w)$, where Δt denotes the difference in the change in time. Thus, the Itô equation expresses random changes in the values of a variable taking place continuously in time.

Moreover, these random changes are given as the sum of two terms. The first term, $\mu[t, S(t,w)]$, is called the **drift component** of the Itô equation, and in finance, it is used to compute the instantaneous expected value of the change in the random variable $S(t,w)$. Observe that $\mu[t, S(t,w)]$, as a function used in the computation of a statistical mean, is affected by both time and randomness. If at a given time t the expected change in $\mathrm{d}S(t,w)$, expressed as $E[\mathrm{d}S(t,w)]$, is desired, this can be answered by computing $E\{\mu[t, S(t,w)]\mathrm{d}t\}$.

The second term $\sigma[t, S(t,w)]\mathrm{d}Z(t,w)$ is itself the product of two factors. Each factor is important and needs special attention. The first factor, $\sigma[t, S(t,w)]$, is used in the calculation of the instantaneous standard derivation of the change in the random variable $S(t,w)$; it is a function of both time t and the range of values taken by $S(t,w)$. When $\sigma[t, S(t,w)]$ is squared to compute the instantaneous variance, it is usually called the **diffusion coefficient**; it measures the variability of $\mathrm{d}S(t,w)$ at a given instance in time.

The second factor, $\mathrm{d}Z(t,w)$, is called **white noise**; it models financial uncertainty in continuous time. Actually, $\mathrm{d}Z(t,w)$ denotes an infinitesimal change in the **Wiener process**, $Z(t,w) : [0,\infty) \times \Omega \to R$, a process with increment that are statistically independent and normally distributed with mean zero and variance equal to the increment in time. In other word, for every pair of disjoint time intervals $[t_1, t_2], [t_3, t_4]$ with, say, $t_1 < t_2 \le t_3 < t_4$, the increments $Z(t_4, w) - Z(t_3, w)$ and $Z(t_2, w) - Z(t_1, w)$ are independent and normally distributed random variables with means:

$$E[Z(t_3, w) - Z(t_4, w)] = E[Z(t_2, w) - Z(t_1, w)] = 0,$$

and respective variances:

$$\mathrm{Var}[Z(t_4, w) - Z(t_3, w)] = t_4 - t_3$$
$$\mathrm{Var}[Z(t_2, w) - Z(t_1, w)] = t_2 - t_1$$

By convention it is assumed that at time $t = 0$, the Wiener process is zero—that is, $Z(0,w) = 0$.

The two factors have been described separately; an intuitive explanation of the product of $\sigma[t, S(t,w)]$ and $\mathrm{d}Z(t,w)$ is now presented. Because $\sigma[t, S(t,w)]$ measures the instantaneous standard derivation or volatility of $\mathrm{d}S(t,w)$ and because $\mathrm{d}Z(t,w)$ is an infinitesimal increment (which is, by definition, purely random with mean zero and variance $\mathrm{d}t$), the expression of $\sigma[t, S(t,w)]\mathrm{d}Z(t,w)$ is the product of two independent random variables, with

$$E\{\sigma[t, S(t,w)]\mathrm{d}Z(t,w)\} = 0 \tag{20.2}$$

$$\begin{aligned}\mathrm{Var}\{\sigma[t, S(t,w)]\mathrm{d}Z(t,w)\} &= E\{\sigma[t, S(t,w)]\mathrm{d}Z(t,w)\}^2 \\ &= E\{\sigma^2[t, S(t,w)]\mathrm{d}t\}\end{aligned} \tag{20.3}$$

Therefore, the product $\sigma[t, S(t,w)]\mathrm{d}Z(t,w)$ is a random variable with mean and variances given by Eqs. (20.2) and (20.3), which, for a given time t and state of nature w, yields a real number. This number may be either positive or negative depending on the value of $\mathrm{d}Z(t,w)$ since $\sigma[t, S(t,w)]$ represents a measure of standard derivation and is always positive. Furthermore, the magnitude of the product $\sigma[t, S(t,w)]\mathrm{d}Z(t,w)$ depends on the magnitude of each of the two terms. Indeed, the methodological foundation of the Itô model is that the uncertainty magnitude, $\mathrm{d}Z(t,w)$, is multiplied by $\sigma[t, S(t,w)]$ to produce the **total contribution of uncertainty**. Therefore, $\sigma[t, S(t,w)]\mathrm{d}Z(t,w)$ describes the total fluctuation produced by volatility as this volatility is aggrandized or reduced by pure randomness.

This analysis explains why uncertainty given by $\mathrm{d}Z(t,w)$ is being modeled as a multiplicative factor in the product $\sigma[t, S(t,w)]\mathrm{d}Z(t,w)$: To repeat once again, the multiplicative modeling of uncertainty allows the generation of values for $\mathrm{d}S(t,w)$ that are above or below the instantaneous expected value, depending on whether uncertainty is positive or negative, respectively. In other words, the product $\sigma[t, S(t,w)]\mathrm{d}Z(t,w)$ can be viewed as the total contribution of uncertainty to $\mathrm{d}S(t,w)$, with such uncertainty being the product of two factors: the instantaneous standard derivation of changes and the purely random white noise.

The conceptual components of the Itô process can now be collected to offer a more general interpretation of its meaning in finance. If at a given time t the possible future change in the price of an asset during the next trading interval is being evaluated, this change can be decomposed into two nonoverlapping components: the *expected* change and the *unexpected* change. The expected change, $E[\mathrm{d}S(t,w)]$, is described by $E\{\sigma[t,S(t,w)]\mathrm{d}t\}$ and the unexpected change is given by $E\{\sigma[t,S(t,w)]\mathrm{d}Z(t,w)\}$. As already noted, this unexpected change depends on the asset's volatility and pure randomness, and because uncertainty cannot be anticipated, the unexpected change is zero. This holds because $\sigma[t,S(t,w)]$ and $\mathrm{d}Z(t,w)$ are independent random variables; and therefore, from Eq. (20.2):

$$E\{\sigma[t,S(t,w)]\mathrm{d}Z(t,w)\} = E\{\sigma[t,S(t,w)]\}E[\mathrm{d}Z(t,w)] = 0.$$

Because by definition $E[\mathrm{d}Z(t,w)] = 0$. Thus, Eq. (20.1), which was developed by mathematicians, captures the spirit of financial modeling admirably because it is an equation that involves three important concepts in finance: *mean, standard derivation, and randomness*.

However, Eq. (20.1) captures the spirit of finance not only because it involves means, standard derivations, and randomness but, more important, because it also expresses key methodological elements of modern financial theory. In an elegant paper, Merton (1982) identified several foundational notions of the appropriateness of the Itô equation in modern finance.

(1) Itô's equation allows uncertainty not to disappear even as trading intervals become extremely short. In the real world of financial markets, uncertainty evolves continuously, and $\mathrm{d}Z(t,w)$ captures this uncertainty because the value of $\mathrm{d}Z(\Delta t,w)$ is not zero even as Δt becomes small. On the other hand, as trading intervals increase uncertainty increase because by definition

$$\mathrm{Var}[Z(t+\Delta t,w) - Z(t,w)] = \Delta t.$$

(2) Itô's equation incorporates uncertainty for all times. This means that uncertainty is present at all trading periods.
(3) The rate of change described by $\mu[t,S(t,w)]\mathrm{d}t$ is finite, and uncertainty does not cause the term $\mu[t,S(t,w)]\mathrm{d}Z(t,w)$ to become unbounded. These notions of the Itô equation are consistent with real-world observations of finite means, finite variances, and uncertainty (which evolve nicely by obtaining continuously finite values).
(4) Finally, all that counts in an Itô equation is the present time t. In other words, the past and future are independent. This notion expresses mathematically the concept of economic markets efficiency. That is, knowledge of past price behavior does not allow for above-average returns in the future.

Sample Problems 20.1 and 20.2 provide illustrations of the concepts discussed.

Sample Problem 20.1
Consider the following example of an Itô process describing the price of given stock.

$$\mathrm{d}S(t,w) = 0.001\,S(t,w)\mathrm{d}t + 0.025\,S(t,w)\mathrm{d}Z(t,w).$$

If both sides are divided by $S(t,w)$ we get:

$$\frac{\mathrm{d}S(t,w)}{S(t,w)} = 0.001\ \mathrm{d}t + 0.025\ \mathrm{d}Z(t,w), \quad (20.4)$$

which gives the proportional change in the price of the stock. Assume that $\mathrm{d}t$ equals one trading period, such as one day. Using appropriate coefficients in Eq. (20.4), $\mathrm{d}t$ could be denoted as one second. In Eq. (20.4), the expected daily proportional change is given by

$$E\left[\frac{\mathrm{d}S(t,w)}{S(t,w)}\right] = E[0.001\ \mathrm{d}t + 0.025\ \mathrm{d}Z(t,w)]$$
$$= 0.001, \tag{20.5}$$

which means that at any given trading day, the price of the stock is expected to change by 0.1%. The standard derivation of the proportional change is $\sigma = 0.025$.

What actually occurs at a given trading period depends on the evolution of the uncertainty as modeled by the normally distributed random variable $\mathrm{d}Z(t,w)$, with mean 0 and variance 1. From the properties of the normal distribution, it can be deduced that although Eq. (20.5) says that the expected daily proportional change is 0.001 and there is a 68.26% probability that the daily proportional change will be between $0.001 \pm (1)\,(0.025)$, or a 95.46% probability that it will be between $0.001 \pm (2)\,(0.025)$, or a 99.74% probability that it will be between $0.001 \pm (3)\,(0.025)$.

Sample Problem 20.2

This second example shows how the Itô equation combines the notion of a trading interval dt, the expected change in the price of the stock μ, the volatility of the stock μ, and pure uncertainty $\mathrm{d}Z(t,w)$ to describe changes in the price of an asset. The same example will be used later to show the behavior of the price of the stock $S(t,w)$. Equation (20.5) illustrates only approximately the infinitesimal proportional change in the price of the stock and in order to obtain the solution of the stochastic differential equation in (20.5), we need the following useful lemma.

20.3 Itô Lemma

The preceding two sections dealt with the Itô process, both its intuitive mathematical meaning and its financial interpretation. This section presents briefly Itô's lemma of stochastic differentiation. By formally integrating Eq. (20.1):

$$S(t,w) = S(0,w) + \int_0^t \mu(u,w)\ \mathrm{d}u + \int_0^t \mu(u,w)\ \mathrm{d}Z(u,w), \tag{20.6}$$

This last equation describes Itô's process in terms of the original random variable $S(t,w)$ rather than infinitesimal differences (differential form), as in Eq. (20.1). Mathematically, Eqs. (20.1) and (20.6) are equivalent and (20.1) obtains clear meaning through integration, which is not developed here because of its complexity. Nevertheless, note that it is the second integral, $\int_0^t \mu(u,w)$ $\mathrm{d}Z(u,w)$, that presents the difficulties. The problem arises because uncertainty given by $\mathrm{d}Z(u,w)$ in its limit does not have a precise meaning, and therefore the second integral cannot be treated like an ordinary Riemann integral. Itô's great accomplishment was to define the second integral as a random variable, that is, the limit in probability of a certain sequence of integrals of step functions multiplied by the uncertainty $\mathrm{d}Z(u,w)$.

Suppose that a stochastic process is given by Eq. (20.6) and that a new process $Y(t,w)$ is formed by letting $Y(t,w) = u[t,S(t,w)]$. Because stochastic calculus studies random changes in continuous time, the question arises: what is $\mathrm{d}Y(t,w)$? This question is important for both mathematical analysis and finance. The answer is given in Itô's lemma.

Itô's Lemma *Consider the nonrandom continuous function $u(t,S) : [0,\infty)\times R \to R$ and suppose that it has continuous partial derivatives u_t, u_S, and u_{SS}.*

Let

$$Y(t,w) = u[t,S(t,w)] \tag{20.7}$$

with

$$\mathrm{d}S(t,w) = \mu[t,S(t,w)]\mathrm{d}t + \sigma[t,S(t,w)]\mathrm{d}Z(t,w). \tag{20.8}$$

Then, the process $Y(t,w)$ has a differential given by

$$\begin{aligned} dY(t,w) = \{&u_t[t,S(t,w)] + u_S[t,S(t,w)]\mu[t,S(t,w)] \\ &+ \tfrac{1}{2}u_{SS}[t,S(t,w)]\sigma^2[t,S(t,w)]\}\ dt \\ &+ u_S[t,S(t,w)]\sigma[t,S(t,w)]\ dZ(t,w) \end{aligned} \tag{20.9}$$

The proof is presented in Gikhman and Skorokhod (1969, pp. 387–389), and extensions of this lemma may be found in Arnold (1974, pp. 90–99). Here, the analysis is limited to three remarks.

(1) Itô's lemma is a useful result because it allows the computation of stochastic differentials of arbitrary functions having as an argument a stochastic process that itself is assumed to possess a stochastic differential. In this respect, Itô's formula is as useful as the chain rule of ordinary calculus.
(2) Given an Itô stochastic process $S(t, w)$ with respect to a given Wiener process $Z(t, w)$, and letting $Y(t,w) = u[t,S(t,w)]$ be a new process, Itô's formula gives the stochastic differential of $Y(t,w)$, where $dY(t,w)$ is given with respect to the same Wiener process—that is, both processes have the same source of uncertainty.
(3) An inspection of the proof of Itô's lemma reveals that it consists of an application of Taylor's theorem of advanced calculus and several probabilistic arguments to establish the convergence of certain quantities to appropriate integrals. Therefore, Itô's formula may be obtained by applying Taylor's theorem instead of remembering the specific result in Eq. (20.6). More specifically, the differential of $Y(t,w) = u[t,S(t,w)]$, where $S(t, w)$ is a stochastic process with differential given by Eq. (20.1), may be computed by using Taylor's theorem and the following multiplication rules:

$$dt \times dt = 0 \quad dZ \times dZ = dt \quad dt \times dZ = 0 \tag{20.10}$$

as

$$\begin{aligned} dY &= u_t dt + u_S dS + \tfrac{1}{2}(dS)^2 \\ &= u_t dt + u_S(\mu dt + \sigma dZ) + \tfrac{1}{2}u_{SS}(\mu dt + \sigma dZ)^2 \end{aligned}$$

By carrying out these multiplications and using the rules in Eq. (20.10), Eq. (20.9) is obtained.

20.4 Stochastic Differential Equation Approach to Stock-Price Behavior

This section demonstrates how a stochastic equation can be used to describe the price behavior of an asset. We begin with Sample Problem 20.3.

Sample Problem 20.3

As a special case of Eq. (20.1), consider:

$$dS(t,w) = \mu S(t,w)dt + \sigma S(t,w)dZ(t,w) \tag{20.11}$$

in which μ and σ are constants as in Sample Problem 20.1. Assume that Eq. (20.11) describes the price behavior of a certain stock with $S(0, w)$ given. The solution $S(t, w)$ of Eq. (20.11) is given by

$$S(t,\ w) = S(0,\ w)\exp\left[\left(\mu - \tfrac{1}{2}\sigma^2\right)t + \sigma Z(t,\ w)\right]. \tag{20.12}$$

To show that the behavior of Eq. (20.12) is the solution of Eq. (20.11), use Itô's lemma as follows. First, start with Eq. (20.12), which corresponds to the function $Y(t, w)$. In this case, Eq. (20.12) is a function of t and $Z(t, w)$, and instead of Eq. (20.8):

$$dZ(t,\, w) = 0 \cdot dt + 1 \cdot dZ(t,\, w).$$

Next, compute the first- and second-order partials denoted by S_t, S_Z, and S_{ZZ}:

$$S_t(t,w) = \left(\mu - \frac{\sigma^2}{2}\right)S(t,w)$$
$$S_Z(t,w) = \sigma S(t,w)$$
$$S_{ZZ}(t,w) = \sigma^2 S(t,w)$$

Collect these results and use Eq. (20.9) to conclude that Eq. (20.11) holds:

$$\begin{aligned} dS(t,w) &= \left[\left(\mu - \frac{\sigma^2}{2}\right)S(t,w) + \sigma S(t,w) \cdot 0 \right. \\ &\quad \left. + \frac{\sigma^2}{2} S(t,w) \cdot 1\right] dt \\ &\quad + \sigma S(t,w) \cdot 1 \cdot dZ(t,w) \\ &= \mu S(t,w) dt + \sigma S(t,w) dZ(t,w) \end{aligned}$$

This result is not only mathematically interesting; in finance, it means that, assuming stock price are given by an Itô process as in Eq. (20.11), then Eq. (20.12) holds as well. To see if Eq. (20.12) accurately describes stock prices in the real world, rewrite it as

$$\ln \frac{S(t,w)}{S(0,w)} = \left(\mu - \frac{\sigma^2}{2}\right)t + \sigma Z(t,w), \quad (20.13)$$

which is a random variable normally distributed with mean $\left(\mu - \sigma^2/2\right)t$ and variance $\sigma^2 t$. This means that if the stock price follows an Itô process, it has a lognormal probability distribution. Such a distribution for stock price is reasonable because it is consistent with reality, where negative stock prices are not possible; the worst that can happen is that the stock price reaches zero. For a detailed analysis of the properties of a lognormal random distribution, see Cox and Rubinstein (1985, pp. 201–204); for a list of bibliographical references on the empirical distribution of stock-price changes, see Cox and Rubinstein (1985, pp. 485–488). Sample Problem 20.4 provides further illustration.

Sample Problem 20.4

The analysis of this problem applies the results of the last two sections to the example in Sample Problem 20.1. Suppose that instead of Eq. (20.11), Eq. (20.14) is given. Then, Eq. (20.12) describes the solution of Eq. (20.14):

$$\begin{aligned} S(t,w) = S(0,w) \exp\left[\left(0.001 - \frac{0.025^2}{2}\right)t \right. \\ \left. + 0.025 Z(t,w)\right]. \end{aligned} \quad (20.14)$$

From Eq. (20.14), the exact evolution of the price of the stock as influences by a mean, a variance, time, and uncertainty is obtained. Using Eq. (20.14) for t = 1 to compute:

$$\begin{aligned} E\left[\ln \frac{S(t,w)}{S(0,w)}\right] &= \left(\mu - \frac{\sigma^2}{2}\right)t \\ &= \left[0.001 - \frac{(0.025)^2}{2}\right](0.1) \\ &= 0.0006875, \end{aligned} \quad (20.15)$$

$$\begin{aligned} \operatorname{Var}\left[\frac{S(t,w)}{S(0,w)}\right] &= \sigma^2 t = (0.025)^2(0.1) \\ &= 0.000625 \end{aligned} \quad (20.16)$$

which means that for a given trading day, the price of the stock is expected to experience a continuous change of 0.068% with a standard derivation of 0.025%. Computing Eqs. (20.15) and (20.16) for any t can be done easily; the same computations cannot be performed readily in Eq. (20.4).

Sample Problem 20.5

Here, we give an intuitive description of Eq. (20.1) with reference to Table 20.1 using Excel. In Table 20.1, we collect from Yahoo.com 40 recent daily closing Google Inc. prices. These prices are also illustrated graphically in Fig. 20.1.

For modeling purposes, the Black–Scholes equation requires a mathematical expression for prices, such as shown in Table 20.1 and

Table 20.1 Daily price data for Google Inc.

Day	Price	r_i	dZ
1	610.21		
2	630.08	0.03256	2.91992
3	625.26	−0.00765	−0.66558
4	624.22	−0.00166	−0.13180
5	624.15	−0.00011	0.00651
6	628.15	0.00641	0.58794
7	624.5	−0.00581	−0.50160
8	616.44	−0.01291	−1.13427
9	616.5	0.00010	0.02519
10	618.38	0.00305	0.28841
11	614.29	−0.00661	−0.57323
12	610.98	−0.00539	−0.46394
13	610.15	−0.00136	−0.10462
14	612	0.00303	0.28686
15	611.04	−0.00157	−0.12336
16	600.36	−0.01748	−1.54193
17	600.99	0.00105	0.11008
18	616.79	0.02629	2.36063
19	616.5	−0.00047	−0.02541
20	619.91	0.00553	0.50970
21	611.08	−0.01424	−1.25354
22	611.83	0.00123	0.12594
23	626.77	0.02442	2.19377
24	631.75	0.00795	0.72496
25	639.63	0.01247	1.12868
26	624.18	−0.02415	−2.13721
27	616.69	−0.01200	−1.05344
28	616.87	0.00029	0.04253
29	616.01	−0.00139	−0.10780
30	614.21	−0.00292	−0.24403
31	616.44	0.00363	0.34024
32	613.5	−0.00477	−0.40874
33	609.07	−0.00722	−0.62733
34	602.12	−0.01141	−1.00093
35	604.35	0.00370	0.34674
36	593.97	−0.01718	−1.51493
37	598.86	0.00823	0.75057
38	601	0.00357	0.33513
39	598.92	−0.00346	−0.29208

(continued)

Table 20.1 (continued)

Day	Price	r_i	dZ
40	602.38	0.00578	0.53162
41	604.23	0.00307	0.29035
μ_{ri}		−0.000185	
σ_{ri}		0.011215	
μ_{dz}			0
σ_{dz}			1

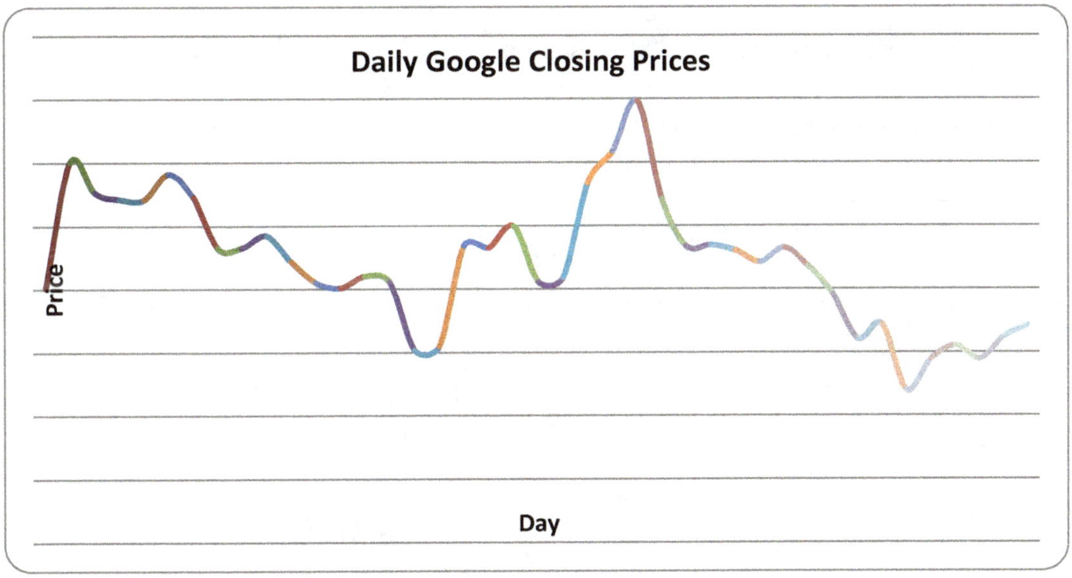

Fig. 20.1 Daily Google Inc. closing prices

Fig. 20.1. How can we check to see if the prices in these figures follow an Itô process? In column 3 of Table 20.1, we compute the daily Google returns and calculate the average historical return of the 40 daily returns. This daily average is 0.000185, which annualized becomes $0.000185 \times 250 = 0.04625$. The volatility of these 40 returns is computed as the annualized standard deviation of daily returns and is computed as $0.011215 \times \sqrt{250} = 0.177324$. Both these calculations assume 250 trading days per year. To complete checking the prices in Table 20.1 satisfy Eq. (20.8), we solve for dZ. Since these prices are daily, we use the daily constant average return and constant daily volatility to obtain daily dZs in column 4 of Table 20.1 using

$$dZ_i = \frac{r_i - \mu_{ri}}{\sigma_{ri}}.$$

We also compute the mean and variance of the dZ in the last column. Recall that earlier we postulated that E (dZ) = 0 and Var (dZ) = 1 is assumed. Indeed, the mean and variance of the dZ values in Table 20.1 are as postulated.

Note that if we were to graph the distribution of Google stock prices, it would follow a log-normal distribution while its returns would follow a normal distribution. Furthermore, the daily

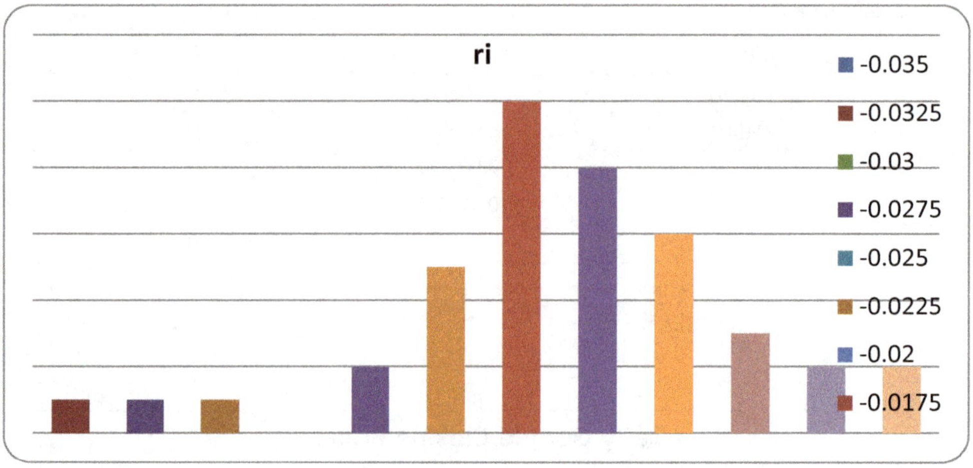

Fig. 20.2 Frequency distribution of returns

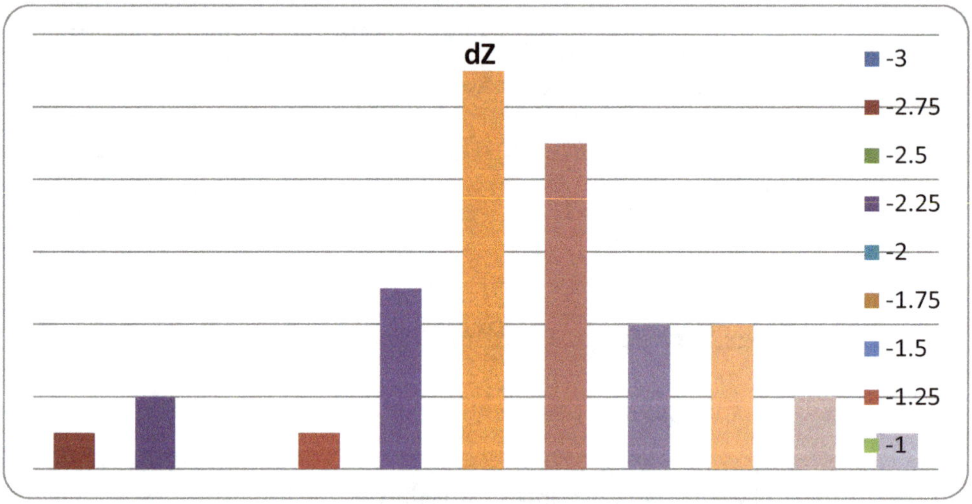

Fig. 20.3 Frequency distribution of errors

dZs, as an approximation to the continuous random walk dZ of Eq. (20.8), also are normally distributed with mean 0 and variance 1. Figures 20.2 and 20.3 are frequency approximations to a normal distribution and illustrate the distributions of returns and distributions of dZs of Table 20.1.

20.5 The Pricing of an Option

An **option** is a contract giving the right to buy or sell an asset within a specified period of time subject to certain conditions. The simplest kind of option is the **European call option**, which is a contract to buy a share of a certain stock at a given date for a specified price. The date the option expires is called the **expiration date (maturity date),** and the price that is paid for the stock when the option is exercised is called the **exercise price (striking price)**.

In terms of economic analysis, several propositions about call option pricing seem clear. The value of an option increases as the price of the stock increases. If the stock price is much greater than the exercise price, it is almost certain that the option will be exercised; and, analogously, if the price of the stock is much less than the exercise price, the value of the option will be near zero and the option will expire without being exercised. If the expiration date is very far in the future, the value of the option will be approximately equal to the price of the stock. If the expiration date is very near, the value of the option will be approximately equal to the stock price minus the exercise price, or zero if the stock price is less than the exercise price. In general, the value of the option is more volatile than the price of the stock, and the relative volatility of the option depends on both the stock price and maturity.

The first rigorous formulation and solution of the problem of option pricing was achieved by Black and Scholes (1973) and Merton (1973). Consider a stock option denoted by C whose price at time t can be written:

$$C(t,w) = C[t, S(t,w)] \tag{20.17}$$

in which C is a twice continuously differentiable function. Here, $S(t,\ w)$ is the price of some stock upon which the option is written. The price of this stock is assumed to follow Itô's stochastic differential equation:

$$\mathrm{d}S(t,w) = \mu[t, S(t,w)]\mathrm{d}t + \sigma[t, S(t,w)]\mathrm{d}Z(t,w). \tag{20.18}$$

Assume, as a simplifying case, that $\mu[t, S(t,w)] = \mu S(t,w)$ and $\sigma[t, S(t,w)] = \sigma S(t,w)$. For notational convenience, w is suppressed from the various expressions and sometimes t as well. Therefore, Eq. (20.18) becomes

$$\mathrm{d}S(t) = \mu S(t)\mathrm{d}t + \sigma S(t)\mathrm{d}Z(t). \tag{20.19}$$

Consider an investor who builds up a portfolio of stocks, options on the stocks, and a riskless asset (for example, government bonds) yielding a riskless rate r. The nominal of the portfolio, denoted by $P(t)$, is

$$P(t) = N_1(t)S(t) + N_2C(t) + Q(t), \tag{20.20}$$

where

N_1 = the number of shares of the stock;
N_2 = the number of call options; and
Q = the value of dollars invested in riskless bonds.

Assume that the stock pays no dividends or other distributions. By Itô's lemma, the differential of the call price using Eqs. (20.17) and (20.19) is

$$\begin{aligned} \mathrm{d}C &= C_t\mathrm{d}t + C_S\mathrm{d}S + \frac{1}{2}C_{SS}\mathrm{d}S^2 \\ &= \left(C_t + C_S\mu S + \frac{1}{2}C_{SS}\sigma^2S^2\right)\mathrm{d}t + C_S\sigma S\ \mathrm{d}Z \\ &= \mu_C C\mathrm{d}t + \sigma_C C\ \mathrm{d}Z \end{aligned} \tag{20.21}$$

Observe that in Eq. (20.21):

$$\mu_C C = C_t + C_S\mu S + \frac{1}{2}C_{SS}\sigma^2S^2, \tag{20.22}$$

$$\sigma_C C = C_S\sigma S. \tag{20.23}$$

In other words, $\mu_C C$ is the expected change in the call option price and $\sigma_C^2C^2$ is the variance of such a change per unit of time. Itô's lemma simply indicates that if the call option price is a function of a spot stock price that follows an Itô process, then the call option price also follows an Itô process with mean and standard derivation parameters that are more complex than those of the stock price. The Itô process for a call is given by Eqs. (20.22) and (20.23).

The change in the normal value of the portfolio dP results from the change in the prices of the assets because at a point in time the equations of option and stock are given—that is, $dN_1 = dN_2 = 0$. More precisely:

$$\begin{aligned} dP &= N_1(dS) + N_2(dS) + dQ \\ &= (\mu dt + \sigma dZ)N_1S + (\mu_C dt + \sigma_C dZ)N_2C \\ &\quad + rQ\, dt. \end{aligned} \tag{20.24}$$

Let w_1 be the fraction of the invested in stock, w_2 be the fraction invested in options, and w_3 be the fraction of the invested in the riskless asset. As before, $w_1 + w_2 + w_3 = 1$, that is, all of the funds available are invested in some type of asset. Set $w_1 = N_1S/P$, $w_2 = N_2C/P$, $w_3 = Q/P = 1 - w_1 - w_2$. Then, Eq. (20.24) becomes

$$\begin{aligned} \frac{dP}{P} &= (\mu\ dt + \sigma\ dZ)w_1 + (\mu_C\ dt + \sigma_C\ dZ)w_2 \\ &\quad + (r\ dt)w_3. \end{aligned} \tag{20.25}$$

At this point, the notion of **economic equilibrium** (also called **risk-neutral** or **preference-free pricing**) is introduced in the analysis. This notion plays an important role in modeling financing behavior, and its appropriate formulation is considered to be a major breakthrough in financial analysis.

More specifically, design the proportions w_1, w_2 so that the position is *riskless* for all $t \geq 0$—that is, let w_1 and w_2 be such that

$$\text{Var}\left(\frac{dP}{P}\right) = \text{Var}(w_1\,\sigma\ dZ + w_2\,\sigma_C\ dZ) = 0. \tag{20.26}$$

In the last equation, Var_t denotes variance conditioned on $S(t)$, $C(t)$, and $Q(t)$. In other words, choose $(w_1, w_2) = (\bar{w}_1, \bar{w}_2)$ so that

$$\bar{w}_1\sigma + \bar{w}_2\sigma_C = 0. \tag{20.27}$$

Then from Eq. (20.25), because the portfolio is riskless, it follows that the portfolio must be expected to earn the riskless rate of return, or:

$$\begin{aligned} E_t\left(\frac{dP}{P}\right) &= [\mu\bar{w}_1 + \mu_C\bar{w}_2 + r(1 - \bar{w}_1 - \bar{w}_2)]dt \\ &= r(t)dt. \end{aligned} \tag{20.28}$$

Equations (20.27) and (20.28) yield the Black–Scholes–Merton equations:

$$\frac{\bar{w}_1}{\bar{w}_2} = -\frac{\sigma_C}{\sigma}, \tag{20.29}$$

and

$$r = \mu\bar{w}_1 + \mu_C\bar{w}_2 - r\bar{w}_1 - r\bar{w}_2 + r, \tag{20.30}$$

which simplify to

$$\frac{\mu - r}{\sigma} = \frac{\mu_C - r}{\sigma_C}. \tag{20.31}$$

Because of the law of one price, Eq. (20.31) says that the net rate of return per unit of risk must be the two assets and describes an appropriate concept of economic equilibrium in this problem. If this were not the case, there would exist an arbitrage opportunity until this equality held. Using Eq. (20.31) and making the necessary substitutions from Eqs. (20.22) and (20.23), the partial differential equation of the pricing of an option is obtained:

$$\begin{aligned} &\tfrac{1}{2}\sigma^2S^2C_{SS}(t,S) + rSC_S(t,S) - rC(t,S) + C_t(t,S) \\ &= 0. \end{aligned} \tag{20.32}$$

The equation along with the boundary conditions for call options fully characterize the call price: $C(S,T) = \text{Max}(0, S - X)$, $S \geq 0$, $0 \geq E \geq T$. The solution to the differential Eq. (20.32) given these boundary conditions is the Black–Scholes formula.

20.6 A Reexamination of Option Pricing

To illustrate the notion of economic equilibrium once again, consider the nominal value of a portfolio consisting of a stock and a call option on this stock and write:

$$P(t) = N_1(t)S(t) + N_2(t)C(t) \tag{20.33}$$

using the same notation as in the previous section. Equation (20.33) differs from Eq. (20.20) because the term $Q(t)$ has been deleted. Now concentrating on the two assets of the portfolio—that is, the stock and the call option—and using Eqs. (20.21) and (20.33), the change in the value of the portfolio is given by

$$\begin{aligned} \mathrm{d}P &= N_1 \mathrm{d}S + N_2 \mathrm{d}C \\ &= N_1 \mathrm{d}S + N_2[(C_t + \frac{1}{2}C_{SS}\sigma^2 S^2)\mathrm{d}t + C_S\ \mathrm{d}S]. \end{aligned} \tag{20.34}$$

Note that $\mathrm{d}N_1 = \mathrm{d}N_2 = 0$, since at any given point in time the equations of stock and option are given. For arbitrary quantities of stock and option, Eq. (20.34) shows that the change in the nominal value of the portfolio $\mathrm{d}P$ is stochastic because $\mathrm{d}S$ is a random variable. Suppose the quantities of stock and call option are chosen as that

$$\frac{N_1}{N_2} = -C_S. \tag{20.35}$$

Note that C_S in Eq. (20.35) denotes a hedge ration and is called **delta**. Then, $N_1\mathrm{d}S + N_2C_S\mathrm{d}S = 0$, and inserting Eq. (20.35) into Eq. (20.34) yields:

$$\begin{aligned} \mathrm{d}P &= -N_2C_S\mathrm{d}S + N_2[(C_t + \frac{1}{2}C_{SS}\sigma^2 S^2)\mathrm{d}t + C_S\mathrm{d}S] \\ &= N_2(C_t + \frac{1}{2}C_{SS}\sigma^2 S^2)\mathrm{d}t \end{aligned} \tag{20.36}$$

Let $N_2 = 1$ in Eq. (20.36) and observe that in equilibrium the rate of return of the riskless portfolio must be the same as the riskless rate $r(t)$. Therefore:

$$\frac{\mathrm{d}P}{p} = r\ \mathrm{d}t. \tag{20.37}$$

Equation (20.37) can be used to derive the partial differential equation for the value of the option. Making the necessary substitutions in Eq. (20.36):

$$\frac{(C_t + \frac{1}{2}C_{SS}\sigma^2 S^2)\mathrm{d}t}{-C_S S + C} = r\ \mathrm{d}t,$$

which upon rearrangement gives Eq. (20.32). Note that the option pricing equation is a second-order linear partial differential equation of the parabolic type. The boundary conditions of Eq. (20.32) are determined by the specification of the asset. For the case of an option that can be exercised only at the expiration date t^* with an exercise price X, the boundary conditions are

$$C(t, S = 0) = 0, \tag{20.38}$$

$$C(t = t^*, S) = \mathrm{Max}(0, S - X). \tag{20.39}$$

Observe that Eq. (20.38) says that the call option price is zero if the stock price is zero at any date t; Eq. (20.39) says that the call option price at the expiration date $t = t^*$ must equal the maximum of either zero or the difference between the stock price and the exercise price.

The solution of the option pricing equation for a call and a put option subject to the boundary conditions are given in Eqs. (20.40a) and (20.40b) for $T = t^* - t$, as

$$C(T, S, \sigma^2, X, r) = SN(d_1) - X\mathrm{e}^{-rT}N(d_2), \tag{20.40a}$$

$$P(T, S, \sigma^2, X, r) = X\mathrm{e}^{-rT}(-N(d_2) - S(-N(d_1)), \tag{20.40b}$$

where N denotes the cumulative normal distribution, namely

$$N(y) = -\frac{1}{\sqrt{2\pi}} \int_{-\infty}^{y} e^{-x^2/2} dx.$$

In Eq. (20.40a), T is time to expiration (measured in years), and d_1 and d_2 are given by

$$d_1 = \frac{\ln(S/X) + (r + \frac{1}{2}\sigma^2)T}{\sigma\sqrt{T}}, \quad (20.41)$$

$$d_2 = \frac{\ln(S/X) + (r - \frac{1}{2}\sigma^2)T}{\sigma\sqrt{T}} = d_1 - \sigma\sqrt{T}. \quad (20.42)$$

It can be shown that

$$\frac{\partial C}{\partial T} > 0, \frac{\partial C}{\partial S} > 0, \frac{\partial C}{\partial \sigma^2} > 0, \frac{\partial C}{\partial X} < 0, \frac{\partial C}{\partial r} > 0. \quad (20.43)$$

These partial derivatives justify the intuitive behavior of the price of an option, as was indicated in the beginning of the previous section. More specifically, these partials show the following:

(1) As the stock price rises, so does the option price.
(2) As the variance rate of the underlying stock rises, so does the option price.
(3) With a higher exercise price, the expected payoff decreases.
(4) The value of the option increases as the interest rate rises.
(5) With a longer time to maturity, the price of the option is greater.

Before giving an example, it is appropriate to sketch the solution of Eq. (20.32) subject to the boundary conditions of Eqs. (20.38) and (20.39). Let t denote the current trading period that is prior to the expiration date t^*. At time t, two outcomes can be expected to occur at t^*. (1) $S(t^*) > X$—that is the price of the stock at the time of the expiration of the call option is greater than exercise price, or (2) $S(t^*) \leq X$. Note that the first outcome occurs with probability $P_0 \equiv P[S(t^*) > X]$ and the second occurs with probability $1 - p$. Obviously, the only interesting possibility is the first case when $S(t^*) > X$, because this is when the price of the call has positive value. If $S(t^*) \leq X$, then $C(t^*) = 0$ from Eq. (20.39). Again from Eq. (20.39), if $S(t^*) > X$, then the price of the call option at expiration $C(t^*)$ can be computed from the expiration of Eq. (20.39):

$$C(t^*) = E\{\mathrm{Max}[0, S(t^*) - X]\} = E[S(t^*) - X]. \quad (20.44)$$

What is the price of a call option if the first outcome materializes at t instead of t^*? This can be answered immediately by appropriate continuous discounting. Using Eq. (20.44):

$$C(t) = e^{-r(t^*-t)} E[S(t^*) - X]. \quad (20.45)$$

Recall, however, that $C(t)$ in Eq. (20.45) holds only with probability p. Combine both possibilities to write:

$$\begin{aligned} C(t) &= p \cdot e^{-r(t^*-t)} E[S(t^*) - X] + (1-p) \cdot 0 \\ &= p \cdot e^{-r(t^*-t)} E[S(t^*)|S(t^*) > X] \\ &\quad - p \cdot e^{-r(t^*-t)} X \end{aligned} \quad (20.46)$$

Detailed calculation in Jarrow and Rudd (1983, pp. 92–94) shows that because the price of the underlying stock is distributed lognormally, it follows that

$$\begin{aligned} p &= P[S(t^*) > X] \\ &= N\left[\frac{\ln\frac{S(t)}{X} + (r + \frac{1}{2}\sigma^2)(t^* - t)}{\sigma\sqrt{t^* - t}}\right], \end{aligned} \quad (20.47)$$

$$\begin{aligned} &p \cdot E[S(t^*)|S^* > X] \\ &= S(t)e^{r(t^*-t)} N\left[\frac{\ln\frac{S(t)}{X} + (r + \frac{1}{2}\sigma^2)(t^* - t)}{\sigma\sqrt{t^* - t}}\right]. \end{aligned} \quad (20.48)$$

Combining Eqs. (20.47) and (20.48) with $T = t^* - t$ into Eq. (20.46) yields Eq. (20.40a).

It is worth observing that two terms of Eqs. (20.40a) and (20.40b) have economic meaning. The first term, $SN(d_1)$, denotes the present value of receiving the stock provided that $S(t^*) > X$. The second term gives the present value of paying the striking price provided that $S(t^*) > X$. In the special case when there is no uncertainty and $\sigma = 0$, observe that

$$\begin{aligned} &N(d_1) = N(d_2) = N(\infty) = 1 \ ; \text{ and} \\ &C = S(t) - e^{-r(t^*-t)}X \end{aligned} \tag{20.49}$$

that is, a call is worth the difference between the current value of the stock and the discounted value of the striking price provided $S(t^*) > X$; otherwise the call price would be zero. When $\sigma \neq 0$—that is, when uncertainty exists and the stock price is volatile—the two terms in Eq. (20.49) are multiplied by $N(d_1)$ and $N(d_2)$, respectively, to adjust the call price for the prevailing uncertainties. These two probabilities can also be given an economic interpretation. As mentioned earlier, $N(d_1)$ is called delta; it is the partial derivative of the call price with respect to the stock price; $N(d_2)$ gives the probabilities that the call option will be in the money, as Eq. (20.47) shows.

Assuming that investors in the economy have risk-neutral preferences, it will be possible to derive the Black–Scholes formula without using stochastic differential equations. Garven (1986) has shown that to derive Eq. (20.40a) knowledge is required of normal and lognormal distributions and basic calculus, as presented in Appendix.

Sample Problem 20.6

Equations (20.40a) and (20.40b) indicate that the Black–Scholes option pricing model is a function of only five variables: *T*, the time to expiration, *S*, the stock price, σ^2, the instantaneous variance rate on the stock price, *X*, the exercise price, and *r*, the riskless interest rate. Of these five variables, only the variance rate must be estimated; the other four variables are directly observable. A simple example is presented to illustrate the use of Eqs. (20.40a) and (20.40b). The values of the observable variables are taken from *Yahoo! Finance*.

On Wednesday, March 16, 2011, at 3:58 p.m. EDT, IBM Corp. had a stock price of \$152.93. The July 11 call option with a strike price of 150.00 was priced at \$10.50. We estimate the riskless rate at 0.25% from the US Treasury bill rate. The only missing piece of information is the instantaneous variance of the stock price.

Several different techniques have been suggested for estimating the instantaneous variance. In this regard, the work of Latané and Rendleman (1976) must be mentioned; they derived standard derivations of continuous price-relative returns that are implied in actual call option prices on the assumption that investors behave as if they price options according to the Black–Scholes model. In the example, the implicit variance is calculated by using a numerical search to approximate the standard derivation implied by the Black–Scholes formula with these parameters: stock price $S = 152.93$, exercise price $X = 150$, time to expiration $T = 121/365 = 0.3315$, riskless rate $r = 0.0025$, and call option price $C = 10.50$. The approximated implied volatility is found to be $\sigma = 0.264$.

Sample Problem 20.7

Using the information about the implied volatility presented above and a stock price of $S = 155$, we present the following example. Given $S = 155$, $X = 150$, $T = 0.3315$, $r = 0.0025$, and $\sigma = 0.264$, use Eqs. (20.40a) and (20.40b) to compute *C*. Using Eqs. (20.41) and (20.42) calculate:

$$\begin{aligned} d_1 &= \frac{\ln\left(\frac{155}{150}\right) + \left[0.0025 + \left(\frac{0.264^2}{2}\right)\right]0.3315}{0.264\sqrt{0.3315}} \\ &= 0.29717, \end{aligned}$$

$$\begin{aligned} d_2 &= \frac{\ln\left(\frac{155}{150}\right) + \left[0.0025 - \left(\frac{0.264^2}{2}\right)\right]0.3315}{0.264\sqrt{0.3315}} \\ &= 0.14517. \end{aligned}$$

From a standard normal distribution table, giving the area of a standard normal distribution, $N(0.29717) = 0.616836$ and $N(0.14517) = 0.557923$. Finally,

$$C = 152.93 \times 0.616836 - 150e^{-.0025*0.3315} \times 0.557923 = \$10.71.$$

These calculations show that as the price of the underlying stock increases from 152.93 to 155, the call price increases as indicated in (20.43) from $10.50 to $10.71, while all other variables are the same.

Sample Problem 20.8

Using the information from the above example, we will calculate a call with a strike price of $160 using Eq. (20.40a)

$$d_1 = \frac{\ln\left(\frac{155}{160}\right) + \left[0.0025 + \left(\frac{0.264^2}{2}\right)\right]0.3315}{0.264\sqrt{0.3315}} = -0.127419,$$

$$d_2 = \frac{\ln\left(\frac{155}{160}\right) + \left[0.0025 - \left(\frac{0.264^2}{2}\right)\right]0.3315}{0.264\sqrt{0.3315}} = -0.268425.$$

From a standard normal distribution table, giving the area of a standard normal distribution, $N(-0.127419) = 0.449314$ and $N(-0.268425) = 0.394208$. Finally,

$$C = 152.93 \times 0.449314 - 160e^{-0.0025*0.3315} \times 0.394208 = \$9.63.$$

As expected, the price of this call option, C = $9.63, with X = 160 has a lower calculated price than the call option with X = 150, as indicated in (20.43).

This simple example shows how to use the Black–Scholes model to price a call option under the assumptions of the model. The example is presented for illustrative purposes only, and it relies heavily on the implicit estimate of the variance, its constancy over time, and all the remaining assumptions of the model. The appropriateness of estimating the implicit instantaneous variance is ultimately an empirical question, as is the entire Black–Scholes pricing formula. Boyle and Ananthanarayanan (1977) studied the implications of using an estimate of the variance in option valuation models and showed that this procedure produces biased option values. However, the magnitude of this bias is not large.

One additional remark must be made. The closeness of a calculated call option price to the actual call price is not necessary evidence of the validity of the Black–Scholes model. Extensive empirical work has taken place to investigate how market prices of call options compare with price predicted by Black–Scholes; see MacBeth and Merville (1979) and Bhattacharya (1980).

20.7 Remarks on Option Pricing

For a review on the literature on option pricing, see the two papers by Smith (1976, 1979). It is appropriate here to make a few remarks on the Black–Scholes option pricing model to clarify its significance and its limitation.

First, the Black–Scholes model for a European call as original derived, and as reported here, is based on several simplifying assumptions.

(1) The stock price follows an Itô equation.
(2) The market operates continuously.
(3) There are no transaction costs in buying or selling the option or the underlying stock.
(4) There are no taxes.
(5) The riskless rate is known and constant.
(6) There are no restrictions on short sales.

Several researchers have extended the original Black–Scholes model by modifying these assumptions. Merton (1973) generalized the model to include dividend payments, exercise-price changes, and the case of a stochastic interest rate. Roll (1977) had solved the problem of valuing a call option that can be exercised prior to its expiration date when the underlying stock is assumed to make known dividend payments before the option matures. Ingersoll (1976) studied the effect of differential taxes on capital gains and income, while Scholes (1976) determined the effects of the tax treatment of options on the pricing model. Furthermore, Merton

(1976) and Cox and Ross (1976) showed that if the stock-price movements are discontinuous, under certain assumptions the valuation model still holds. These and other modifications of the original Black–Scholes analysis indicate that the model is quite robust about the relaxation of its fundamental assumptions.

Second, it is worth repeating that the use of Itô's calculus and the important insight concerning the appropriate concept of an equilibrium by creating a riskless hedge portfolio have let Black and Scholes obtain a closed-form solution for option pricing. In this closed-form solution, several variables do not appear, such as (1) the expected rate of return of the stock, (2) the expected rate of return of the option, (3) a measure of investor's risk preference, (4) investor expectations, and (5) equilibrium conditions for the entire capital market.

Third, the Black–Scholes pricing model has found numerous applications. Among these are: (1) pricing the debt and equity of a firm; (2) the effects of corporate policy and, specially, the effects of mergers, acquisitions, and scale expansions on the relative values of the debt and equity of the firm; (3) the pricing of convertible bonds; (4) the pricing of underwriting contracts; (5) the pricing of leases; and (6) the pricing of insurance. Smith (1976, 1979) summarized most applications and indicates the original reference. See also Brealey and Myers (1988).

Fourth, Black (1976) showed that the original call option formula for stocks can be easily modified to be used in pricing call options on futures. The formula is

$$C\left(T,F,\sigma^2,X,r\right)=\mathrm{e}^{-rT}[FN(d_1)-XN(d_2)], \tag{20.50}$$

$$d_1=\frac{\ln(F/X)+\frac{1}{2}\sigma^2T}{\sigma\sqrt{T}}, \tag{20.51}$$

$$d_2=\frac{\ln(F/X)-\frac{1}{2}\sigma^2T}{\sigma\sqrt{T}}. \tag{20.52}$$

In Eq. (20.50), F now denotes the current futures price. The other four variables are as before—time to maturity, volatility of the underlying futures price, exercise price, and risk-free rate. Note that Eq. (20.50) differs from Eqs. (20.40a) and (20.40b) only in one respect: by substituting $\mathrm{e}^{-rT}F$ for S in the original Eqs. (20.40a) and (20.40b), Eq. (20.50) is obtained. This holds because the investment in a futures contract is zero, which causes the interest rate in Eqs. (20.51) and (20.52) to drop out.

Fifth, three important papers by Harrison and Kreps (1979) and Kreps (1981, 1982) consider some foundational issues that arise in conjunction with the arbitrage theory of option pricing. The important point to consider is this: the ability to trade securities frequently can enable a few multiperiod securities to span many states of nature. In the Black–Scholes theory, there are two securities and uncountable many state of nature, but because there are infinitely many trading opportunities and because uncertainty resolves nicely, markets are effectively complete. Thus, even though there are far fewer securities than states of nature, markets are complete and risk is allocated efficiently. An interesting result of Harrison and Kreps (1979) is that certain self-trading strategies can create something out of nothing when there are infinitely many trading opportunities. The doubling strategies are the well-known illustrations of this phenomenon. Harrison and Kreps introduce the concept of a simple strategy to eliminate free lunches and conjecture that a nonnegative wealth constraint could rule out the doubling strategies. Duffie and Huang (1985) gave an interpretation of admissible strategy as a limit of a sequence of simple strategies and use an integrability condition on the trading strategies. Dybvig and Huang (1986) showed that under certain condition and the nonnegative wealth constraint are functionally equivalent.

Finally, for an intensive survey of numerous empirical tests, see Galai (1983).

20.8 Conclusion

This chapter has discussed the basic concepts and equations of stochastic calculus (Itô's calculus), which has become a very useful tool in understanding finance theory and practice. By using these concepts and equations, the manner by which Black and Scholes derived their famous option pricing model was also illustrated. Although this chapter is not required to understand the basic ingredients of security analysis and portfolio management discussed in Chaps. 1–26, it is useful for those with trading in advanced mathematics to realize how advanced mathematics can be used in finance.

Appendix: An Alternative Method to Derive the Black–Scholes Option Pricing Model

Perhaps it is unclear why it is assumed that investors have risk-neutral preferences when the usual assumption in finance courses is that investors are risk averse. It is feasible to make this simplistic assumption because investors are able to create riskless portfolios by combining call options with their underlying securities. Since the creation of a riskless hedge places no restrictions on investor preferences other than nonsatiation, the valuation of the option and its underlying asset will be independent of investor risk preferences. Therefore, a call option will trade at the same price in risk-neutral economy as it will in a risk-averse or risk-preferent economy.

Assumptions and the Present Value of the Expected Terminal Option Price

To derive the Black–Scholes formula, it is assumed that there are no transaction costs, no margin requirements, and no taxes; that all shares are infinitely divisible, and that continuous trading can be accomplished. It is also assumed that the economy is risk neutral.

In the risk-neutral assumptions of Cox and Ross (1976) and Rubinstein (1976), today's option price can be determined by discounting the expected value of the terminal option price by the riskless rate of interest. As was seen earlier, the terminal call option price can take on only two values: $S_t - X$, if the call option expires in the money, or 0 if the call expires out of the money. So today's call option price is

$$C = \exp(-rt)\text{Max}(S_t - X, 0), \tag{20.53}$$

where

C = the market value of the call option;
r = riskless rate of interest;
t = time to expiration;
S_t = the market value of the underlying stock at time t; and
X = exercise or striking price.

Equation (20.53) says that the value of the call option today will be either $S_t - X$ or 0, whichever is greater. If the price of stock at time t is greater than the exercise price, the call option will expire in the money. This simply means that an investor who owns the call option will exercise it. The option will be exercised regardless of whether the option holder would like to take physical possession of the stock. If the investor would like to own the stock, the cheapest way to obtain the stock is by exercising the option. If the investor would not like to own the stock, he or she will still exercise the option and immediately sell the stock in the market. Since the price the investor paid (X) is lower that the price he or she can sell the stock for (S_t), the investor realizes an immediate the profit of $S_t - X$. If the price of the stock (S_t) is less than the exercise price (X), the option expires out of the money. This occurs because in purchasing shares of the stock the investor will find it cheaper to purchase the stock in the market than to exercise the option.

Assuming that the call option expires in the money, then the present value of the expected terminal option is equal to the present value of the difference between the expected terminal stock price and the exercise price, as indicated in Eq. (20.54):

$$\begin{aligned} C &= \exp(-rt)E[\mathrm{Max}(S_t - X, 0)] \\ &= \exp(-rt)\int_x^{\infty}(S_t - X)h(S_t)\mathrm{d}S_t \end{aligned} \quad (20.54)$$

where $h(S_t)$ is the lognormal density function of S_t. To evaluate the integral in (20.54), rewrite it as the difference between two integrals:

$$\begin{aligned} C &= \exp(-rt)\left[\int_x^{\infty} S_t h(S_t)\mathrm{d}S_t - X\int_x^{\infty} h(S_t)\mathrm{d}S_t\right] \\ &= E_x(S_t)\cdot\exp(-rt) - X\cdot\exp(-rt) \\ &\quad \cdot[1 - H(X)] \end{aligned} \quad (20.55)$$

where

$E_x(S_t) =$ the partial expectation of S_t, truncated from below at x; and

$H(X) =$ the probability that $S_t \le X$.

Equation (20.55) says that the value of the call option is present value of the partial expected stock price (assuming the call expires in the money) minus the present value of the exercise price (adjusted by the probability that the stock's price will be less than the exercise price at the expiration of the option). The terminal stock price S_t can be rewritten as the product of the current price (S) and the t-period lognormally distributed price ratio S_t/S, so $S_t = S(S_t/S)$. Equation (20.55) can also be rewritten as

$$\begin{aligned} C &= \exp(-rt)\left[S\int_{x/s}^{\infty}\frac{S_t}{S}g\left(\frac{S_t}{S}\right)\left(\frac{\mathrm{d}S_t}{S}\right) - X\int_{x/s}^{\infty}g\left(\frac{S_t}{S}\right)\left(\frac{\mathrm{d}S_t}{S}\right)\right] \\ &= S\exp(-rt)E_{x/s}\left(\frac{S_t}{S}\right) \\ &\quad - X\exp(-rt)\left[1 - G\left(\frac{X}{S}\right)\right] \end{aligned} \quad (20.56)$$

where

$g\left(\frac{S_t}{S}\right) =$ log normal density function of S_t/S;

$E_{x/S}\left(\frac{S_t}{S}\right) =$ the partial expextation of S_t/S, truncated from below at x/S;

$G\left(\frac{X}{S}\right) =$ the probability that $S_t/S \le X/S$.

Present Value of the Partial Expectation of the Terminal Stock Price

The right-hand side of Eq. (20.56) is evaluated by considering the two integrals separately. The first integral, $S\exp(-rt)E_{x/S}(S_t/S)$, can be solved by assuming the return on the underlying stock follows a stationary random walk. That is

$$\frac{S_t}{S} = \exp(Kt), \quad (20.57)$$

where K is the rate of return on the underlying stock per unit of time. Taking the natural logarithm of both sides of Eq. (20.57) yields:

$$\ln\left(\frac{S_t}{S}\right) = (Kt).$$

Since the ratio S_t/S is lognormally distributed, it follows that Kt is lognormally distributed with density $f(Kt)$, mean $\mu_K t$, and variance $\sigma_K^2 t$. Because $S_t/S = \exp(Kt)$, the differential can be rewritten as $dS_t/S = \exp(Kt)t\, dK$; $g(S_t/S)$ is a density function of a lognormally distributed variable S_t/S; so following Garven (1986), it can be transformed into a density function of a normally distributed variable Kt according to the relationship $S_t/S = \exp(Kt)$ as

$$g\left(\frac{S_t}{S}\right) = f(Kt)\left(\frac{S_t}{S}\right). \quad (20.58)$$

These transformations allow the first integral in Eq. (20.56) to be rewritten as

$$S\exp(-rt)\,E_{x/S}\left(\frac{S_t}{S}\right) = S\exp(-rt) \int_{\ln(x/S)}^{\infty} f(Kt)\exp(Kt)\,t\,dK.$$

Because Kt is normally distributed, the density $f(Kt)$ with mean $\mu_K t$ and variance $\sigma_K^2 t$ is

$$f(Kt) = \left(2\pi\sigma_K^2 t\right)^{-1/2}\exp\left[-\tfrac{1}{2}(Kt-\mu_K t)^2\Big/\sigma_K^2\right].$$

Substitution yields:

$$S\exp(-rt)\,E_{x/S}\left(\frac{S_t}{S}\right) = S\exp(-rt)\left(2\pi\sigma_K^2 t\right)^{-1/2} \times \int_{\ln(x/S)}^{\infty} \exp[Kt] \exp\left[-\tfrac{1}{2}(Kt-\mu_K t)^2\Big/\sigma_K^2 t\right] t\,dK \quad (20.59)$$

Equation (20.59)'s integrand can be simplified by adding the terms in the two exponents, multiplying and dividing the result by $\exp(-\frac{1}{2}\sigma_K^2 t)$. First, expand the term $(Kt-\mu_K t)^2$ and factor out t so that

$$\exp[Kt]\exp\left[-\tfrac{1}{2}(Kt-\mu_K t)^2\Big/\sigma_K^2 t\right].$$

Next, factor out t so

$$\exp(Kt)\exp\left\{-\tfrac{1}{2}t\left[\left(K^2-2\mu_K K+\mu_K^2\right)/\sigma_K^2\right]\right\}.$$

Now combine the two exponents

$$\exp\left\{-\tfrac{1}{2}t\left[\left(K^2-2\mu_K K+\mu_K^2-2\sigma_K^2 K\right)/\sigma_K^2\right]\right\}.$$

Now, multiply and divide this result by $\exp(-\frac{1}{2}\sigma_K^2 t)$ to get:

$$\exp\left\{-\tfrac{1}{2}t\left[\left(K^2-2\mu_K K+\mu_K^2-2\sigma_K^2 K+\sigma_K^4-\sigma_K^4\right)/\sigma_K^2\right]\right\}.$$

Next, rearrange and combine terms to get:

$$\begin{aligned}&\exp\left\{\left(-\tfrac{1}{2}t\right)\left[\left(K-\mu_K-\sigma_K^2\right)^2 -\sigma_K^4-2\mu_K\sigma_K^2\right]/\sigma_K^2\right\}\\ &= \exp\left[\left(\mu_K+\tfrac{1}{2}\sigma_K^2\right)t\right] \exp\left\{-\tfrac{1}{2}\left[Kt-\left(\mu_K+\sigma_K^2\right)t\right]^2\Big/\sigma_K^2 t\right\}\end{aligned} \quad (20.60)$$

In Eq. (20.60), $\exp\left[\left(\mu_K+\frac{1}{2}\sigma_K^2\right)t\right] = E(S_t/S)$, the mean of the t-period lognormally distributed price ratio S_t/S. So, Eq. (20.59) becomes:

$$S\exp(-rt)\,E_{x/S}\left(\frac{S_t}{S}\right) = SE\left(\frac{S_t}{S}\right)\exp(-rt) \left(2\pi\sigma_K^2 t\right)^{-1/2} \times \int_{\ln(x/S)}^{\infty} \exp(Kt) \exp\left\{-\tfrac{1}{2}\left[Kt-\left(\mu_K+\sigma_K^2\right)t\right]^2/\sigma_K^2 t\right\} \quad (20.61)$$

Since the equilibrium rate of return in a risk-neutral economy is the riskless rate, $E(S_t/S)$ may be rewritten as $\exp(rt)$:

$$SE\left(\frac{S_t}{S}\right)\exp(-rt) = S\exp(rt)\exp(-rt)$$
$$= S$$

So Eq. (20.61) becomes

$$S\exp(-rt)\,E_{x/S}\left(\frac{S_t}{S}\right) = S\left(2\pi\sigma_K^2 t\right)^{-1/2} \times \int_{\ln(x/S)}^{\infty} \exp\left\{-\tfrac{1}{2}[Kt - \left(\mu_K + \sigma_K^2\right)t\right]^2/\sigma_K^2 t\}\; t\; dK \qquad (20.62)$$

To complete the simplification of this part of the Black–Scholes formula, define a standard normal random variable y:

$$y = \left[Kt - \left(\mu_K + \sigma_K^2\right)t\right]/\sigma_K^2 t^{1/2}.$$

Solving for Kt yields:

$$Kt = \left(\mu_K + \sigma_K^2\right)t + \sigma_K t^{1/2} y,$$

and therefore:

$$t\; dK = \sigma_K t^{1/2}\; dy.$$

By making the transformation from Kt to y, the lower limit of integration becomes

$$\left[\ln(x/S) - \left(\mu_K + \sigma_K^2\right)t\right]/\sigma_K t^{1/2}.$$

Further simplify the integrand by noting that the assumption of a risk-neutral economy implies:

$$\exp\left[\left(\mu_K + \tfrac{1}{2}\sigma_K^2\right)t\right] = \exp(rt).$$

Taking the natural logarithm of both sides yields:

$$\left(\mu_K + \tfrac{1}{2}\sigma_K^2\right)t = (rt).$$

Hence, $\left(\mu_K + \frac{1}{2}\sigma_K^2\right)t = \left(r + \frac{1}{2}\sigma_K^2\right)t$.
The lower limit of integration is now:

$$-\left[\ln(S/x) + \left(r + \tfrac{1}{2}\sigma_K^2\right)t\right]/\sigma_K t^{1/2} = -d_1.$$

Substituting this into Eq. (20.62) and making the transformation to y yields:

$$S\exp(-rt)\,E_{x/S}\left(\frac{S_t}{S}\right) = S\int_{-d_1}^{\infty} \exp\left(-\tfrac{1}{2}y\right)^2/(2\pi)^{1/2}\; dy.$$

Since y is a standard normal random variable (distribution is symmetric around zero) the limits of integration can be exchanged:

$$S\exp(-rt)\,E_{x/S}\left(\frac{S_t}{S}\right) = S\int_{\infty}^{-d_1} \exp\left(-\tfrac{1}{2}y^2\right)/(2\pi)^{1/2}\; dy = SN(d_1) \qquad (20.63)$$

where $N(d_1)$ is the standard normal cumulative distribution function evaluated at $y = d_1$.

Present Value of the Exercise Price Under Uncertainty

To complete the derivation, the integrals that correspond to the term $X\exp(-rt)\,[1 - G(X/S)]$ must be evaluated. Start by making the logarithmic transformation:

$$\ln\left(\frac{S_t}{S}\right) = Kt.$$

This transformation allows the rewriting of $g(S_t/S)$ to $(S/S_t)f(Kt)$ as mentioned previously. The differential can be written as

$$\mathrm{d}\frac{S_t}{S} = \exp(Kt)\, t\, \mathrm{d}K.$$

Therefore,

$$\begin{aligned} X\exp(-rt)[1 - G(X/S)] &= X\exp(-rt) \int_{\ln(X/S)}^{\infty} f(Kt)\, t\, \mathrm{d}K \\ &= X\exp(-rt)\left(2\pi\sigma_K^2 t\right)^{-1/2} \int_{\ln(X/S)}^{\infty} \exp\left[-\frac{1}{2}(Kt - \mu_K t)^2/\sigma_K^2 t\right] t\, \mathrm{d}K \end{aligned} \tag{20.64}$$

The integrand is now simplified by following the same procedure used in simplifying the previous integral. Define a standard normal random variable Z:

$$Z = \left[\frac{Kt - \mu_K t}{\sigma_K t^{1/2}}\right].$$

Solving for Kt yields:

$$Kt = \mu_K t + \sigma_K t^{1/2} Z,$$

and $t\ \mathrm{d}K = \sigma_K t^{1/2}\ \mathrm{d}Z$. Making the transformation from Kt to Z means the lower limit of integration becomes

$$\frac{\ln(X/S) - \mu_K t}{\sigma_K t^{1/2}}.$$

Again, note that the assumption of a risk-neutral economy implies:

$$\exp\left(\mu_K + \frac{1}{2}\sigma_K^2 t\right) = \exp(rt).$$

Taking the natural logarithm of both sides yields:

$$\left(\mu_K + \frac{1}{2}\sigma_K^2\right)t = rt,$$

or:

$$\mu_K t = \left(r - \frac{1}{2}\sigma_K^2\right)t.$$

Therefore, the lower limit of integration becomes:

$$\begin{aligned} \frac{-\left[\ln(S/x) + \left(r - \frac{1}{2}\sigma_K^2\right)t\right]}{\sigma_K t^{1/2}} &= -\left(d_1 - \sigma_K t^{1/2}\right) \\ &= -d_2 \end{aligned}$$

Substitution yields:

$$\begin{aligned} x\exp(-rt)[1 - G(X/S)] &= x\exp(-rt) \int_{-d_2}^{\infty} \exp\left[-\frac{1}{2}Z^2 \Big/ (2\pi)^{1/2}\right] \mathrm{d}Z \\ &= x\exp(-rt) \int_{-\infty}^{d_2} \exp\left[-\frac{1}{2}Z^2 \Big/ (2\pi)^{1/2}\right] \mathrm{d}Z \\ &= x\exp(-rt)\, N(d_2) \end{aligned} \tag{20.65}$$

where $N(d_2)$ is the standard normal cumulative distribution function evaluated at $Z = d_2$.

Substituting, Eqs. (20.63) and (20.65) into Eq. (20.56) completes the derivation of the Black–Scholes formula:

$$C = SN(d_1) - X\exp(-rt)\, N(d_2). \tag{20.66}$$

This appendix provides a simple derivation of the Black–Scholes call option pricing formula. Under an assumption of risk neutrality the Black–Scholes formula was derived using only differential and integral calculus and a basic knowledge of normal and lognormal distributions.

Bibliography

Arnold, L. (1974). *Stochastic differential equation: Theory and applications*. New York: Wiley.

Bhattacharya, M. (1980). Empirical properties of the Black-Scholes formula under ideal conditions. *Journal of Financial Quantitative Analysis, 15,* 1081–1105.

Black, F. (1976). The pricing of commodity contracts. *Journal of Financial Economics, 3,* 167–178.

Black, F., & Scholes, M. (1973). The pricing of options and corporate liabilities. *Journal of Political Economy, 81,* 637–654.

Boyle, P. P., & Ananthanarayanan, A. L. (1977). The impact of variance estimation in option valuation models. *Journal of Financial Economics, 5,* 375–387.

Brealey, R., & Myers, S. (1988). *Principles of corporate finance* (3rd ed.). New York: McGraw-Hill.

Brennan, M., & Schwartz, E. (1978). Finite difference method and jump processes arising in the pricing of contingent claims: A synthesis. *Journal of Financial Quantitative Analysis, 13,* 461–474.

Brennan, M., & Schwartz, E. (1979). A continuous time approach to the pricing of bonds. *Journal of Banking & Finance, 3,* 133–155.

Chalamandaris, G., & Malliaris, A. G. (2010). Itô's calculus and the derivation of the Black-Scholes option-pricing model. In C. F. Lee, A. C. Lee, & J. Lee (Eds.), *Handbook of quantitative finance and risk management*. Singapore: Springer.

Courtadon, G. (1982). A more accurate finite difference approximation for the valuation of options. *Journal of Financial Quantitative Analysis, 17,* 697–705.

Cox, J. C., & Ross, S. A. (1976). The valuation of options for alternative stochastic processes. *Journal of Financial Economics, 3,* 145–226.

Cox, J. C., & Rubinstein, M. (1985). *Options markets*. Englewood Cliffs, NJ: Prentice-Hall.

Duffie, D., & Huang, C. (1985). Implementing Arrow-Debreu equilibria by continuous trading of few long-lived securities. *Econometrica, 53,* 1337–1356.

Dybvig, P., & Huang, C. (1986). *Optimal portfolios and positive wealth constraint* (Working paper). Yale University School of Management.

Galai, D. (1983). A survey of empirical tests of option-pricing models. In M. Brenner (Ed.), *Option pricing* (pp. 45–80). Lexington, MA: Lexington Books.

Garven, J. R. (1986). A pedagogic note on the derivation of the Black-Scholes option pricing formula. *The Financial Review, 21,* 337–344.

Geske, R., & Shostri, K. (1985). Valuation by approximation: A comparison of alternative option valuation techniques. *Journal of Financial Quantitative Analysis, 20,* 45–71.

Gikhman, I., & Skorokhod, A. V. (1969). *Investment to the theory of random processes*. Philadelphia, PA: W. B. Saunders.

Harrison, J. M., & Kreps, D. M. (1979). Martingales and arbitrage in multiperiod securities markets. *Journal of Economic Theory, 20,* 381–408.

Harrison, J., & Pliska, S. (1981). Martingales and stochastic integrals in the theory of continuous trading. *Stochastic Processes and Their Application, 2,* 261–271.

Ingersoll, J. E. (1976). A theoretical and empirical investigation of the dual purpose funds: An application of contingent claims and analysis. *Journal of Financial Economics, 3,* 83–123.

Itô, K., & McKean, H. (1964). *Diffusion processes and their sample paths*. New York: Academic Press.

Jarrow, R., & Rudd, A. (1982). Approximate option valuation for arbitrary stochastic processes. *Journal of Financial Economics, 10,* 347–370.

Jarrow, R., & Rudd, A. (1983). *Option pricing*. Homewood, IL: R. D. Irwin.

Kreps, D. M. (1981). Arbitrage and equilibrium in economics with infinitely many commodities. *Journal of Mathematical Economics, 8,* 15–35.

Kreps, D. M. (1982). Multiperiod securities and the efficient allocation of risk: A comment on the Black-Scholes option pricing model. In J. J. McCall (Ed.), *The economics of information and uncertainty* (pp. 203–232). Chicago: University of Chicago Press.

Latané, H. A., & Rendleman, R. J., Jr. (1976). Standard derivations of stock price ratios implied in option prices. *Journal of Finance, 31,* 369–381.

Lee, C.-F., Lee, A. C., & Lee, J. (2010). *Handbook of quantitative finance and risk management*. New York: Springer.

Lee, C. F., Finnerty, J., Lee, J., Lee, Alice, & Wort, D. (2013). *Security analysis, portfolio management, and financial derivatives*. Singapore: World Scientific.

MacBeth, J. D., & Merville, L. J. (1979). An empirical examination of the Black-Scholes call option pricing model. *Journal of Finance, 34,* 1173–1186.

Malliaris, A. G., & Brock, W. A. (1982). *Stochastic methods in economics and finance*. Amsterdam: North-Holland.

Merton, R. C. (1973). The theory of rational option pricing. *Bell Journal of Economics and Management Science, 4,* 141–183.

Merton, R. C. (1975). Theory of finance from the perspective of continuous time. *Journal of Financial Quantitative Analysis, 10,* 659–674.

Merton, R. C. (1976). Option pricing when underlying stock returns are discontinuous. *Journal of Financial Economics, 3,* 125–144.

Merton, R. C. (1982). On the mathematics and economics assumption of continuous-time model. In W. F. Sharpe & C. M. Cootner (Eds.), *Financial economics: Essay in honor of Paul Cootner* (pp. 19–51). Englewood Cliffs, NJ: Prentice-Hall.

Roll, R. (1977). An analytic valuation formula for unprotected American call options on stocks with known dividends. *Journal of Financial Economics, 5,* 251–281.

Rubinstein, M. (1976). The valuation of uncertain income streams and the pricing of options. *Bell Journal of Economics, 7,* 407–425.

Scholes, M. (1976). Taxes and the pricing of options. *Journal of finance, 31,* 319–332.

Shreve, S. (2010). *Stochastic calculus for finance II: Continuous-time models*. Springer.

Smith, C. W., Jr. (1976). Option pricing: A review. *Journal of Financial Economics, 3,* 3–51.

Smith, C. W., Jr. (1979). Applications of option pricing analysis. In J. L. Bicksler (Ed.), *Handbook of financial economics*. Amsterdam: North-Holland.

Alternative Methods to Derive Option Pricing Models

21

Contents

Abstract

The main purposes of this paper are: (1) to review three alternative methods for deriving option pricing models, (2) to discuss the relationship between binomial option pricing model and Black–Scholes model, (3) to compare Cox et al. method and Rendleman and Bartter method for deriving Black–Scholes model, (4) to discuss lognormal distribution method to derive Black–Scholes model, and (5) to show how the Black–Scholes model can be derived by stochastic calculus. This paper shows that the main methodologies used to derive the Black–Scholes model are: binomial distribution, lognormal distribution, and differential and integral calculus. If we assume risk neutrality, then we do not need stochastic calculus to derive the Black–Scholes model. However, the stochastic calculus approach for deriving the Black–Scholes model is still

This chapter draws upon Lee et al. (2016) which was published in Review of Quantitative Finance and Accounting.

C.-F. Lee et al., *Financial Econometrics, Mathematics and Statistics*,
https://doi.org/10.1007/978-1-4939-9429-8_21

presented in this chapter. In sum, this paper can help statisticians and mathematicians understand how alternative methods can be used to derive the Black–Scholes option pricing model.

21.1 Introduction

The use of stock options for risk reduction and return enhancement has expanded at a considerable rate over the last several decades. In 1973, the Chicago Board Option Exchange was established and brought about liquidity for successful option trading through public listing and option contract standardization. In the same year, the most famous option pricing model—Black–Scholes model was proposed and became the industry standard (Lee et al. 2013a).

Black and Scholes (1973) and Merton (1973) used stochastic calculus to derive option pricing models. Rendleman and Bartter (1979) and Cox et al. (1979) used binomial distribution to derive the Black–Scholes model. In the next several decades, a group of new models that relax some restrictive assumptions of Black–Scholes model have been proposed.

The first group of researchers developed models that allowed important parameters such as interest rate or (and) volatility, to be stochastic. Scott (1987), Wiggins (1987), Hull and White (1987), Melino and Turnbull (1990, 1995), Stein and Stein (1991), and Heston (1993) generalized the Black–Scholes model in terms of stochastic variance. While Amin and Jarrow (1992) developed the Black–Scholes model allowing stochastic interest rate. Furthermore, there are some literatures that proposed generalization method to allow interest rate and volatility to be stochastic at the same time. Examples can be found in Amin and Ng (1993), Bakshi and Chen (1997a, b), and Bailey and Stulz (1989). Similarly, Lee et al. (1991) extended the binomial option pricing model to the case where the up and down percentage changes of stock prices are stochastic. They have proved that assuming stochastic parameters in the discrete-time binomial option pricing is analogous to assuming stochastic volatility in the continuous-time option pricing.

The second group of studies introduced jump-diffusion process into the Black–Scholes model and made extensions to the original model. Several jump-diffusion models are proposed by Bates (1991), Kou (2002), Kou and Wang (2004), respectively. Psychoyios et al. (2010) well approximated the time-series behavior of VIX index by a mean reverting logarithmic diffusion process with jumps. Based on the empirical results, they derived closed-form valuation models for European options written on the spot and forward VIX, respectively. For more complicated cases, Bates (1996) introduced jump-diffusion process into the stochastic volatility model, and Scott (1997) attempted to price options in a jump-diffusion model with stochastic volatility and interest rates.

Besides the mentioned two large categories of models, an explosion of other option pricing models has also been proposed and well validated using real-world data. Examples include: (i) the constant elasticity of variance (CEV) models (Cox and Ross 1976; Beckers 1980; Davydov and Linetsky 2001; Lee et al. 2004; Chen et al. 2009); (ii) the Markovian models (Rubinstein 1994; Alt-Sahalia and Lo 1998); (iii) the GARCH models (Duan 1995; Heston and Nandi 2000; Wu 2006); and (iv) the models based on Lévy processes (Geman et al. 2001; Carr and Wu 2004).

In more recent years, much more complex cases are considered to develop new models. And these models were proved to describe the reality more precisely. Chen and Palmon (2005) proposed an empirically based, nonparametric option pricing model and used it to evaluate S&P 500 index options. As their model is derived under the real measure, an equilibrium asset pricing model rather than no-arbitrage model is assumed. Costabile et al. (2014) proposed a binomial approach for option pricing assuming the parameters governing the underlying asset process follow a regime-switching model. Lin et al. (2014) developed a currency option pricing model with regimes of high-variance or low-variance states as well as the jump nature of exchange rates. And they have proved that their model performed better than both traditional regime-switching model and the Black–Scholes model.

Overall, Black–Scholes option pricing model has been extensively studied by different researchers, and the models discussed above are by no means exhaustive. This information can be found in Hull (2014).

Black–Scholes model, the important development of option valuation theory, which relied on far fewer assumptions, shed new light on the valuation process. Subsequently, the growing popularity of the option concept is evidenced by its application to the valuation of other more abstract assets including lease contracts (Grenadier 1995) and real estate agreements (Williams 1991; Buetow and Albert 1998).

Besides assets valuation, option theory is also widely used in the field of risk management. The most important observation in Merton (1974) is that the firm's equity can be regarded as a call option on the firm's assets with exercise price equal to the liability. If the firm's assets fall below its liabilities, then the firm is in danger of bankruptcy. Under the Black–Scholes model, the probability of bankruptcy is simply the probability that the market value of assets is less than the face value of the liabilities (Hillegeist et al. 2004). Based upon the option pricing models, several commercial vendors provide default probabilities, with KMV, LLC being the best known.

Similar to the above theory, Merton (1977, 1978) first discussed the relationship between the deposit insurance and put option. If bank's assets cannot meet the amount of deposits, the bank is insolvent. Therefore, all remainders of the assets belong to depositors. And the insurer of deposit insurance should pay the difference of the bank assets and the deposits. In this case, the deposit insurance contracts can be viewed as a put option written on bank assets with the strike price equal to deposits. Marcus and Shaked (1984) used Merton's model to price fair insurance premium with constant proportional dividends and found FDIC overcharged the deposit insurance premiums in practice.

As discussed so far, it is very important to better understand the pricing theory and mechanism of option contracts, as the applications of the theory are so wide. It is well-known that binomial approach, lognormal distribution approach, and *Itô* stochastic differential approach can be used to derive option pricing model. (i) Binomial option model assumes stock price either goes up or down at each period. With no arbitrage opportunity existing, a risk-free portfolio combined with assets is constructed to produce the same return in every state over each investment period. After then, the binomial model can be generalized into n periods. (ii) As for lognormal distribution approach, the most important assumption is that the stock-price return follows a lognormal distribution. Using properties of normal distribution, lognormal distribution, and their mutual relations, Black–Scholes model can be derived without using stochastic differential. (iii) Black and Scholes have used two alternative methods to derive the well-known stochastic differential equation. By introducing boundary constraints and making variable substitutions, the stochastic differential equation evolves to the heat-transfer equation in physics. The Fourier transformation is then used to solve the heat-transfer equation under the boundary condition and finally obtain the closed-form solution, which is the famous Black–Scholes formula.

In this chapter, we are going to give a overall review and comparison of the alternative methods to derive option pricing model. This chapter will show that the main methodologies used to derive the Black–Scholes model are: binomial distribution, lognormal distribution, and differential and integral calculus. We will show that if we assume risk neutrality, then we do not need stochastic calculus to derive the Black–Scholes model. This chapter can help statisticians and mathematicians understand how alternative methods can be used to derive the Black–Scholes option model.

The rest of this chapter proceeds as follows: In Sect. 21.2, we briefly review three different approaches for deriving option model. In Sect. 21.3, we discuss the relationship between binomial option pricing model and Black–

Scholes option pricing model. In Sect. 21.4, we compare Cox et al. method and Rendleman and Bartter method for deriving Black–Scholes option pricing model. In Sect. 21.5, we discuss lognormal distribution method to derive Black–Scholes option pricing model. In Sect. 21.6, we present the stochastic calculus for deriving the Black–Scholes model. Finally, in Sect. 21.7, we summarize the chapter. In Appendix, we use the de Moivre–Laplace Theorem to prove that the best fit between the binomial and normal distributions occurs when binomial probability is 0.5.

21.2 A Brief Review of Alternative Approaches for Deriving Option Pricing Model

Binomial model, lognormal distribution approach, and the Black–Scholes model can be used to price an option. Similar results can be obtained by any of them if we assume some additional assumptions.

21.2.1 Binomial Model[1]

The binomial option pricing model derived by Rendleman and Bartter (1979) and Cox et al. (1979) is one of the most used models to price options.

In binomial model settings, stock price S either goes up with increase factor (u) to arrive uS or down with decrease factor (d) to arrive dS at each period, where $u = 1 +$ percentage of increase, $d = 1 -$ percentage of decrease.

Let $i =$ interest rate; $r = 1 + i$; $C_u = \max[uS - X, 0]$, call option price after stock price increases; $C_d = \max[dS - X, 0]$, call option price after stock price decreases.

To intuitively grasp the underlying concept of option pricing, here we set up a *risk-free portfolio*—a combination of assets that produces the same return in every state of the world over the investment horizon. The investment horizon here is assumed to be one period. We buy h shares of the stock and sell the call option at its current price of C to set up the portfolio.[2] Moreover, we choose the value of h such that our portfolio after one period will yield the same payoff whether the stock goes up or down, which is shown as follows.

$$h(uS) - C_u = h(dS) - C_d \tag{21.1}$$

By solving h, we can obtain the number of shares of stock we should buy for each call option we sell, as the following equation shows.

$$h = \frac{C_u - C_d}{(u - d)S} \tag{21.2}$$

Here, h is called the *hedge ratio*. Because our portfolio yields the same return under either of the two possible states for the stock price without risk, then it should yield the risk-free rate of return, which is equal to the risk-free borrowing and lending rate. This condition must hold; otherwise, there would be a chance to earn a risk-free profit, which is known as an arbitrage opportunity. Therefore, the ending portfolio value must be equal to r (1 + risk-free rate) times the beginning portfolio value, as defined in the following equation.

$$r(hS - C) = h(uS) - C_u = h(dS) - C_d \tag{21.3}$$

Note that S and C represent the stock price and the option price at period 0, respectively.

Substituting h as Eq. (21.2) shows, we get the expression for call option value as follows.

$$C = \left[\left(\frac{R - d}{u - d}\right)C_u + \left(\frac{u - R}{u - d}\right)C_d\right]/r \tag{21.4}$$

To simplify this equation, we set

$$p = \frac{r - d}{u - d} \tag{21.5}$$

[1]In this section, we follow the notations used by Cox et al. (1979)

[2]To sell the call option means to write the call option. If a person writes a call option on stock A, then he or she is obliged to sell at exercise price X during the contract period.

Therefore, we have

$$1 - p = \frac{u - r}{u - d} \tag{21.6}$$

Thus, we can get the option's value with one period to expiration as Eq. (21.7).

$$C = [pC_u + (1 - p)C_d]/r \tag{21.7}$$

This is the binomial call option valuation formula in its most basic form. It prices the call option with one period to expiration. In this formula, p can be viewed as the probability of stock-price increase, while $1 - p$ is the probability of stock-price decrease.

To derive the option's price with two periods to go, it is helpful as an intermediate step to derive the value of C_u and C_d with one period to expiration when the stock price is either uS or dS, respectively.

$$C_u = [pC_{uu} + (1 - p)C_{ud}]/r \tag{21.8}$$

$$C_d = [pC_{du} + (1 - p)C_{dd}]/r \tag{21.9}$$

Equation (21.8) tells us that if the value of the option after one period is C_u, the option will be worth either C_{uu} (if the stock price goes up) or C_{ud} (if stock price goes down) after one more period (at its expiration date). C_{uu} and C_{ud} are determined by: $C_{uu} = \max[u^2S - X, 0]$, and $C_{ud} = \max[udS - X, 0]$.

Similarly, Eq. (21.9) shows that if the value of the option is C_d after one period, the option will be worth either C_{du} or C_{dd} at the end of the second period. C_{du} and C_{dd} are: $C_{ud} = \max[udS - X, 0]$, and $C_{dd} = \max[d^2S - X, 0]$.

Replacing C_u and C_d in Eq. (21.4) with their expressions in Eqs. (21.8) and (21.9) respectively, we can simplify the resulting equation to yield the two-period equivalent of the one-period binomial pricing formula, which is

$$C = \left[p^2C_{uu} + 2p(1 - p)C_{ud} + (1 - p)^2C_{dd}\right]/r^2 \tag{21.10}$$

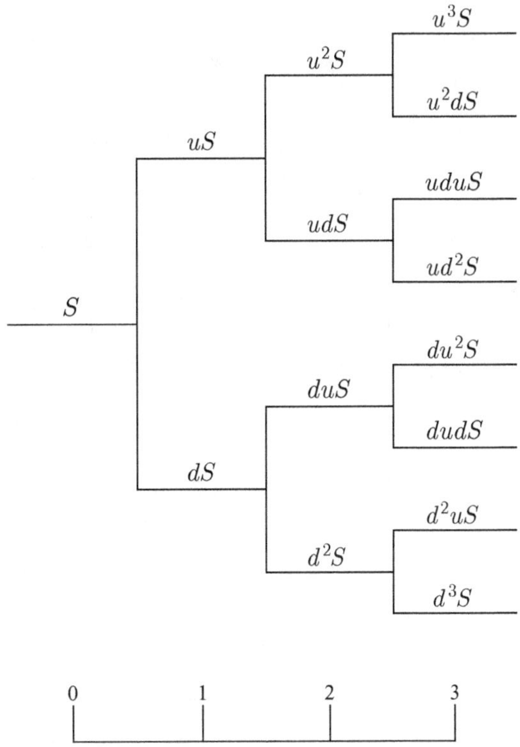

Fig. 21.1 Three-period binomial decision tree of stock price

In Eq. (21.10), we used the fact that $C_{ud} = C_{du}$ because the price will be the same in either case.

If we assume that r, u, and d will remain constant over time, deriving the option's fair value with two or more periods to maturity is a relatively simple process of working backwards from the possible maturity values. Using the same procedure, we can extend the two-period model to a three-period model as in Eq. (21.11).

$$C = \left[p^3C_{uuu} + 3p^2(1 - p)C_{uud} + 3p(1 - p)^2C_{udd} + (1 - p)^3C_{ddd}\right]/r^3 \tag{21.11}$$

A graphical interpretation of Eq. (21.11) is presented in Figs. 21.1 and 21.2. Figure 21.1 presents the stock-price binomial decision tree and Fig. 21.2 presents the call option decision tree.

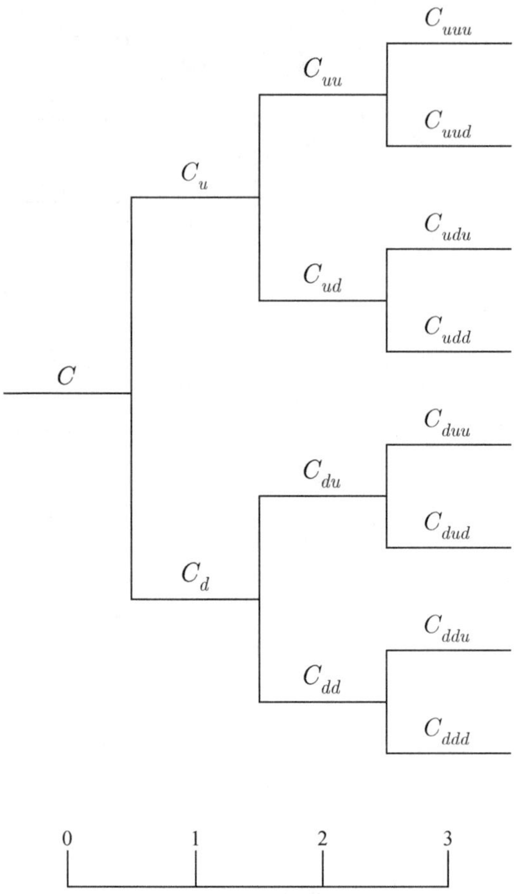

Fig. 21.2 Three-period binomial decision tree of call option

In Fig. 21.1, S represents stock price per share in period 0. In period 1, stock price can either go up (uS) or go down (dS).

Similarly, in period 2, stock prices can be u^2S, udS, duS, or d^2S.[3] In period 3, stock price has eight possible cases as presented in Fig. 21.1 in detail.

In period 3, the highest possible value for stock price based on our assumption is u^3S. We get this value first by multiplying the stock price S at period 0 by u to get the resulting value of uS of period 1. Then, we again multiply the stock price in period 1 by u to get the resulting value of u^2S of period 2. Finally, we multiply the stock price in period 2 by u to get the value of u^3S in period 3. Similarly, the lowest possible value of stock price is d^3S.

In Fig. 21.2, eight nodes in the right column represent the values of call option when the stock price is fewer than eight different possible cases. Under three-period binomial tree settings, period 3 means the maturity date. At that point, the value of the call option is determined by the relationship between stock price and exercise price X. Here, we take C_{uud}, which implies the value of the call option when the stock price is u^2dS as an example. If the stock price, u^2dS, exceeds exercise price X, then the call option value should be $C_{uud} = u^2dS - X$. Otherwise, a negative value has no value to an investor, and the call option value should be 0. All we mentioned above yields the value of option C_{uud} in period 0 as $C_{uud} = \max[u^2dS - X, 0]$. Similarly, we can determine all the option values under different stock price at expiration date.

Similarly, the binomial model can be generalized into n periods. Lee et al. (2013b) defined the pricing of a call option in a binomial OPM with n periods as Eq. (21.12).

$$C = \frac{1}{r^n}\sum_{j=0}^{n}\frac{n!}{j!(n-j)!}p^j(1-p)^{n-j} \max\left[(u)^j(d)^{n-j}S - X, 0\right] \tag{21.12}$$

We can rewrite Eq. (21.12) as:

$$C = S\left[\sum_{j=a}^{n}\frac{n!}{j!(n-j)!}p^j(1-p)^{n-j}\frac{u^jd^{n-j}}{r^n}\right] - \frac{X}{r^n}\left[\sum_{j=a}^{n}\frac{n!}{j!(n-j)!}p^j(1-p)^{n-j}\right] \tag{21.13}$$

where a denotes the minimum integer value of j for which $u^jd^{n-j} - X$ will be positive.

It is easy to observe that the second term in brackets in Eq. (21.13) is a cumulative binomial distribution with parameters n and p. If we define $p' \equiv (u/r)p$ and $1 - p' \equiv (d/r)(1-p)$, then the first term in the brackets can also become a cumulative binomial distribution with parameters n and p', as shown in Eq. (21.14).

[3]Here, we distinguish udS and duS, and count them as two possible outcomes.

$$p^j(1-p)^{n-j}\frac{u^j d^{n-j}}{r^n} = p'^j(1-p')^{n-j} \quad (21.14)$$

Therefore, Eq. (21.13) can be simplified as

$$C = SB_1(a;n,p') - \frac{X}{r^n}B_2(a;n,p) \quad (21.15)$$

where

$$B_1(a;n,p') = \sum_{j=a}^{n} {}_nC_j p'^j(1-p')^{n-j} \quad (21.15a)$$

$$B_2(a;n,p) = \sum_{j=a}^{n} {}_nC_j p^j(1-p)^{n-j} \quad (21.15b)$$

21.2.2 Black–Scholes Model

Black and Scholes (1973) and Merton (1973) have used stochastic Ito calculus to derive an option pricing model. However, if we assume risk-neutral, Lee et al. (2013b) proposed a lognormal distribution approach to derive the Black–Scholes model. In this chapter, we will discuss the lognormal distribution approach in detail in Sect. 21.5.

The most famous option pricing model is the Black–Scholes option pricing model which can be used to price European options.

The Black–Scholes model for a European call option is:

$$C = SN(d_1) - Xe^{-rT}N(d_2) \quad (21.16)$$

where

$$d_1 = \frac{\ln(S/X) + (r + \frac{\sigma^2}{2})T}{\sigma\sqrt{T}};\ d_2 = d_1 - \sigma\sqrt{T}$$

C = call price; S = stock price; X = exercise price; r = risk-free interest rate;
T = time to maturity of option in years; $N(\cdot)$ = standard normal distribution; σ = stock volatility.

This model can be used to price call option and the put option can be derived from the following put-call parity:

$$P = C + Xe^{-rT} - S \quad (21.17)$$

where P = put price, other notations are identical to those defined in Eq. (21.16).

In the following section, we will show the relationship between binomial and Black–Scholes option pricing models.

21.3 Relationship Between Binomial OPM and Black–Scholes OPM

When comparing the parameters in both models, we will find that the binomial model has an increase factor (u), a decrease factor (d), and n-period parameters that the Black–Scholes model does not have. While the Black–Scholes model has distinct parameters, σ and T do not appear in binomial model. The parameters between the two models have the links and can be translated from one to another. The derivations are as follows (Hull 2014).

As we discussed in Sect. 21.2, in binomial OPM setting, the stock price S goes up with a probability p to arrive uS, and goes down with a probability $1-p$ to arrive dS. The expected stock price is: $puS + (1-p)d$.

Assume each step is of length Δt, where $\Delta t = \frac{T}{n}$. As $n \to \infty$, $\Delta t \to 0$. The expected return on a stock (in the real world) is supposed to be r (continuously compounding). Therefore, within this small period of time, the expected price should be $e^{r\Delta t}S$.

Therefore, we have the following equation holds

$$puS + (1-p)dS = e^{r\Delta t}S \quad (21.18)$$

The volatility of a stock price is defined as σ; therefore, in the short time period Δt, the standard deviation of the stock return is $\sigma\sqrt{\Delta t}$; i.e., the variance of the return is $\sigma^2\Delta t$.

The variance of the stock-price return is[4]: $pu^2 + (1-p)d^2 - (pu + (1-p)d)^2$.

We have the following equation holds

$$pu^2 + (1-p)d^2 - (pu + (1-p)d)^2 = \sigma^2 \Delta t \qquad (21.19)$$

Combining the two equations, we get

$$e^{r\Delta t}(u+d) - ud - e^{2r\Delta t} = \sigma^2 \Delta t \qquad (21.20)$$

When ignoring terms Δt^2 and higher power of Δt, one solution of this equation is[5]:

$$\begin{aligned} u &= e^{\sigma\sqrt{\Delta t}} \\ d &= e^{-\sigma\sqrt{\Delta t}} \end{aligned} \qquad (21.21)$$

These are the values of u and d proposed by Cox et al. (1979). To summarize the main relations between the parameters of the two alternative OPMs, we have the following important equations to link the two models.

$$\begin{aligned} \Delta t &= \frac{T}{n} \\ R &= e^{r\Delta t} \\ u &= e^{\sigma\sqrt{\Delta t}} \\ d &= e^{-\sigma\sqrt{\Delta t}} \end{aligned} \qquad (21.22)$$

If n gets very large, the binomial OPM value will get close to the Black–Scholes OPM value. Benninga and Czaczkes (2000) demonstrated that the binomial value will be close to Black–Scholes when the parameter n exceeds 500.

There are two alternative methods to show how binomial OPM can be converted to the Black–Scholes OPM. These two methods are:

a. Theoretical methods proposed by Cox et al. (1979) and Rendleman and Bartter (1979). Lee and Lin (2010) have shown how these two different methods can be related. Cox et al. (1979) used the Lyapunov condition to show how the binomial OPM can be reduced to the Black–Scholes OPM. Alternatively, Rendleman and Bartter (1979) used the limited theory of the relationship between binomial and normal distribution to show how the binomial OPM can be converted to the Black–Scholes OPM. To understand this approach, we do not need to know advanced probability theory, as we will point out in the next section.
b. The Excel approach is proposed by Lee (2001). Lee has used the Excel program approach to show that the binomial model can be approached to the Black–Scholes model when n approaches 500. This approach is similar to the concept and theory used by Rendleman and Bartter (1979).

Here, we will demonstrate how to use the *binomialBS_OPM.xls* Excel file proposed by Lee (2001), to create the decision trees for call option price as an illustrative example.[6] We assume stock price is 30, strike price is 32, increase factor (u) is 1.1, and decrease factor (d) is 0.9. We are constructing a four-period binomial option pricing model with risk-free rate 3%. Decision tree of call option using binomial model is produced as shown in Fig. 21.3.

31 calculations were required to create a decision tree that has four periods. Therefore, the Excel file did $31 \times 3 = 93$ calculations to create the three decision trees for stock price, call option value, and put option value.

[4]This uses the property that the variance of a variable Q can be calculated by: $E(Q^2) - [E(Q)]^2$, where $E(\cdot)$ denotes the expected value.

[5]Here, Taylor series expansion is used: $e^x = 1 + x + \frac{x^2}{2!} + \frac{x^3}{3!} + \cdots . e^{r\Delta t} = 1 + r\Delta t$ and $e^{2r\Delta t} = 1 + 2r\Delta t$ when higher powers of Δt^2 are ignored. The solution to u and d implies that $u = 1 + \sigma\sqrt{\Delta t} + \frac{1}{2}\sigma^2\Delta t$, $d = 1 - \sigma\sqrt{\Delta t} + \frac{1}{2}\sigma^2\Delta t$. All of these expansions satisfy Eq. (21.19).

[6]The details of program presentation for stock price, call option price, and also put option price using both binomial model and Black–Scholes model are shown in Chap. 18 in Lee et al. (2013a). The Excel VBA Code for *binomialBS_OPM.xls* can be found in Appendix 18A in Lee et al. (2013a). The readers can read them if interested. Due to space limit, we only present the illustrative decision trees for call options using binomial model and Black–Scholes model in the main text.

Fig. 21.3 Call option pricing decision tree

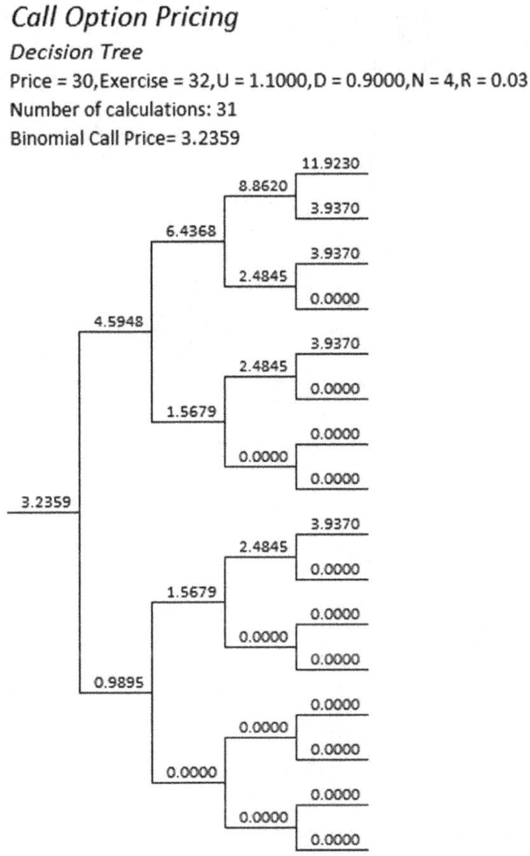

We also use the Excel program to calculate the binomial and Black–Scholes call values, which were previously illustrated. If we determine the parameter T and σ as 1 and 0.2 respectively, the increase factor (u) and decrease factor (d) will be adjusted. And we can get: $u = 1.105$ and $d = 0.905$. Figure 21.4 shows the decision tree approximation of Black–Scholes call pricing model as these parameters determined.

Notice that in Fig. 21.4, the binomial OPM value does not agree with the Black–Scholes OPM, but the values are close. The binomial OPM value will get very close to the Black–Scholes OPM value once the binomial parameter—number of periods n gets very large. Benninga and Czaczkes (2000) demonstrated that the binomial value will be close to Black–Scholes when the number of periods n is larger than 500. Here, we will use the Johnson & Johnson call option as a real example for a practical illustration. The parameters are as follows: $S = 93.45$, $X = 92.5$, $T = 0.3589$, $r = 2.75\%$. $\sigma = 13.01\%$, which is estimated from JNJ stock's daily return. The observed call price from market is $C = 3.65$.

For Black–Scholes OPM, we can get:

$$N(d_1) = 0.6175, N(d_2) = 0.5875$$

The theoretical value for the call option from Black–Scholes model should be:

$$C = SN(d_1) - e^{-rT}XN(d_2) = 3.90$$

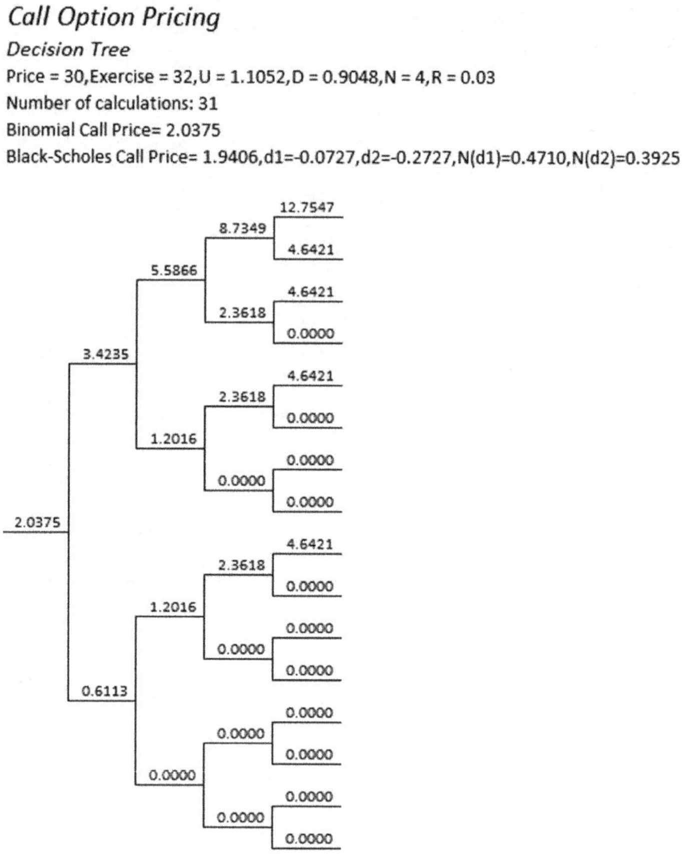

Fig. 21.4 Decision tree approximation of Black–Scholes call pricing

Table 21.1 Binomial OPM estimates of different numbers of periods

Number of period (n)	Increase factor (u)	Decrease factor (d)	Theoretical value of binomial OPM
3	1.0460	0.9560	4.1285
4	1.0397	0.9618	3.8844
5	1.0355	0.9657	4.0306
6	1.0323	0.9687	3.9122
7	1.0299	0.9710	3.9888
8	1.0279	0.9728	3.9232
9	1.0263	0.9744	3.9657
10	1.0250	0.9757	3.9282

Table 21.1 shows how the binomial OPM value converges to the Black–Scholes OPM as n gets larger, with increase factor (u) and decrease factor (d) adjusted accordingly.

Previous examples show that the Excel program can be used to demonstrate the binomial option pricing model and can converge to the

Black–Scholes model when the number of periods approaches infinity.

21.4 Compare Cox et al. and Rendleman and Bartter Methods to Derive OPM

Both Cox et al. (1979) and Rendleman and Bartter (1979) employ Eq. (21.13) to show how the binomial model can be reduced to the Black–Scholes model when the number of observation n approaches infinity. In this section, we briefly discuss the methods employed by these two papers.

21.4.1 Cox et al. Method

Cox et al. (1979) used the following binomial option pricing model to derive the Black–Scholes model.

$$\begin{aligned} C &= S\left[\sum_{j=a}^{n}\frac{n!}{j!(n-j)!}p^j(1-p)^{n-j}\frac{u^jd^{n-j}}{\hat{r}^n}\right] \\ &\quad - X\hat{r}^{-n}\left[\sum_{j=a}^{n}\frac{n!}{j!(n-j)!}p^j(1-p)^{n-j}\right] \\ &= SB_1(a;n,p') - X\hat{r}^{-n}B_2(a;n,p) \end{aligned} \tag{21.23}$$

where

$$B_1(a;n,p') = \sum_{j=a}^{n} {}_nC_j p'^j(1-p')^{n-j}$$

$$B_2(a;n,p) = \sum_{j=a}^{n} {}_nC_j p^j(1-p)^{n-j}$$

$$p' \equiv (u/\hat{r})p \text{ and } 1-p' \equiv (d/\hat{r})(1-p)$$

$$\hat{r} = 1 + \text{interest rate over one period}$$

a is the minimum number of upward stock movements necessary for the option to terminate in the money. In other words, a is the minimum value of integer j that $u^jd^{n-j}S - X > 0$ holds.

In order to show the limiting result that the binomial option pricing formula converges to the continuous version of the Black–Scholes option pricing formula, we assume that h represents the lapsed time between successive stock-price changes. Thus, if t is the fixed length of calendar time to expiration, and n is the total number of periods each with length h, then $h = \frac{t}{n}$. As the trading frequency increases, h will get closer to zero. When $h \to 0$, this is equivalent to $n \to \infty$.

$\hat{r}$ is one plus the interest rate over a trading period of length h. We not only want $\hat{r}$ to depend on n, but want it to depend on n in a particular way—so that as n changes, the total return $\hat{r}^n$ remains the same. We denote r as one plus the rate over a fixed unit of calendar time, then over time t, the total return should be r^t. Then, we will have following equation:

$$\hat{r}^n = r^t \tag{21.24}$$

for any choice of n. Therefore, $\hat{r} = r^{\frac{t}{n}}$. Let S^* be the stock price at the end of the nth period with the initial price S. If there are j upwards move, then the generalized expression should be:

$$\begin{aligned} \log(S^*/S) &= j\log u + (n-j)\log d \\ &= j\log(u/d) + n\log d \end{aligned} \tag{21.25}$$

Therefore, j is the realization of a binomial random variable with probability of a success being p. We have the expectation of $\log(S^*/S)$ as

$$E(\log(S^*/S)) = [p\log(u/d) + \log d]n \equiv \tilde{\mu}n \tag{21.26}$$

and its variance

$$\operatorname{var}(\log S^*/S) = [\log(u/d)]^2p(1-p)n \equiv \tilde{\sigma}^2n \tag{21.27}$$

We are considering dividing up the original time period t into many shorter subperiods of length h so that $t = nh$. Our procedure calls for making n larger while keeping the original time period t fixed. As $n \to \infty$, we would at least like the mean and the variance if the continuously compounded return rate of the assumed stock-price movement coincided with that of actual stock price. Label the actual empirical values of $\tilde{\mu}n$ and $\tilde{\sigma}^2 n$ as μt and $\sigma^2 t$, respectively. Then, we want to choose u, d, and p so that $\tilde{\mu}n \to \mu t$ and $\tilde{\sigma}^2 n \to \sigma^2 t$ as $n \to \infty$.

A little algebra shows that we can accomplish this by letting

$$u = e^{\sigma\sqrt{\frac{t}{n}}}, \quad d = e^{-\sigma\sqrt{\frac{t}{n}}}$$
$$p = \frac{1}{2} + \frac{1}{2}\left(\frac{\mu}{\sigma}\right)\sqrt{\frac{t}{n}} \tag{21.28}$$

At this point, in order to proceed further, we need the Lyapunov condition of central limit theorem as follows (Ash and Doleans-Dade 1999; Billingsley 2008).

Lyapunov's Condition *Suppose* $X_1, X_2, \ldots$ *are independent and uniformly bounded with* $E(X_i) = 0$, $Y_n = X_1 + \cdots + X_n$, *and* $s^2 = E(Y_n^2) = \text{Var}(Y_n)$.

If $\lim_{n\to\infty} \sum_{k=1}^{n} \frac{1}{s_n^{2+\delta}} E|X_k|^{2+\delta} = 0$ *for some* $\delta > 0$, *then the distribution of* $\frac{Y_n}{s_n}$ *converges to the standard normal distribution as* $n \to \infty$.

Theorem *If*

$$\frac{p|\log u - \tilde{\mu}|^3 + (1-p)|\log d - \tilde{\mu}|^3}{\tilde{\sigma}^3\sqrt{n}} \to 0 \text{ as } n \to \infty \tag{21.29}$$

then

$$\Pr\left[\frac{\log\left(\frac{S^*}{S}\right) - \tilde{\mu}n}{\tilde{\sigma}\sqrt{n}} \le z\right] \to N(z) \tag{21.30}$$

where N(z) is the cumulative standard normal distribution function.

Proof Since

$$p|\log u - \tilde{\mu}|^3 = p\left|\log u - p\log\frac{u}{d} - \log d\right|^3 = p(1-p)^3\left|\log\frac{u}{d}\right|^3$$

and

$$(1-p)|\log d - \tilde{\mu}|^3 = (1-p)\left|\log d - p\log\frac{u}{d} - \log d\right|^3 = p^3(1-p)\left|\log\frac{u}{d}\right|^3,$$

we have

$$p|\log u - \tilde{\mu}|^3 + (1-p)|\log d - \tilde{\mu}|^3 = p(1-p)[(1-p)^2 - p^2]\left|\log\frac{u}{d}\right|^3.$$

Thus

$$\frac{p|\log u - \tilde{\mu}|^3 + (1-p)|\log d - \tilde{\mu}|^3}{\tilde{\sigma}^3\sqrt{n}} = \frac{p(1-p)\left[(1-p)^2 - p^2\right]\left|\log\frac{u}{d}\right|^3}{\left(\sqrt{p(1-p)}\log\left(\frac{u}{d}\right)\right)^3\sqrt{n}} = \frac{(1-p)^2 + p^2}{\sqrt{np(1-p)}}$$

Recall that $p = \frac{\hat{r}-d}{u-d}$ with $\hat{r} = r^{\frac{t}{n}}$, $u = e^{\sigma\sqrt{\frac{t}{n}}}$, $d = e^{-\sigma\sqrt{\frac{t}{n}}}$, we have:

$$p = \frac{e^{\frac{t}{n}\log r} - e^{-\sigma\sqrt{\frac{t}{n}}}}{e^{\sigma\sqrt{\frac{t}{n}}} - e^{-\sigma\sqrt{\frac{t}{n}}}} = \frac{1 + \frac{t}{n}\log r - [1 - \sigma\sqrt{\frac{t}{n}} + \frac{1}{2}\sigma^2\frac{t}{n}] + O(n^{-\frac{3}{2}})}{1 + \sigma\sqrt{\frac{t}{n}} - [1 - \sigma\sqrt{\frac{t}{n}}] + O(n^{-\frac{3}{2}})} = \frac{1}{2} + \frac{1}{2}\left[\frac{\log r - \frac{1}{2}\sigma^2}{\sigma}\right]\sqrt{\frac{t}{n}} + O(n^{-1})$$

Therefore, $\frac{(1-p)^2 + p^2}{\sqrt{np(1-p)}} \to 0 \quad \text{as } n \to \infty$.

Hence, the condition for the theorem to hold as stated in Eq. (21.29) is satisfied. It is noted that the condition (21.29) is a special case of Lyapunov's condition where $\delta = 1$. Next, we will show that the binomial option pricing model as given in Eq. (21.23) will indeed coincide with the Black–Scholes option pricing formula. We can see that there are apparent similarities in Eq. (21.23). In order to show the limiting result, we need to show that:

As $n \to \infty$, $B_1(a;n,p') \to N(x)$ and $B_2(a;n,p) \to N(x-\sigma\sqrt{t})$.

In this section, we will only show the second convergence result, as the same argument will hold true for the first convergence. From the definition of $B_2(a;n,p)$, it is clear that

$$\begin{aligned} 1 - B_2(a;n,p) &= \Pr(j \le a-1) \\ &= \Pr\left(\frac{j-np}{\sqrt{np(1-p)}} \le \frac{a-1-np}{\sqrt{np(1-p)}}\right) \end{aligned} \tag{21.31}$$

Recall that we consider a stock to move from S to uS with probability p and dS with probability $1-p$. The mean and variance of the continuously compounded rate of return for this stock are $\tilde{\mu}_p$ and $\tilde{\sigma}_p^2$ where

$$\begin{aligned} \tilde{\mu}_p &= p\log\left(\frac{u}{d}\right) + \log d \quad \text{and} \\ \tilde{\sigma}_p^2 &= \left[\log\left(\frac{u}{d}\right)\right]^2 p(1-p) \end{aligned} \tag{21.32}$$

From Eq. (21.25) and the definitions for $\tilde{\mu}_p$ and $\tilde{\sigma}_p^2$, we have

$$\frac{j-np}{\sqrt{np(1-p)}} = \frac{\log\left(\frac{S^*}{S}\right) - \tilde{\mu}_p n}{\tilde{\sigma}_p\sqrt{n}} \tag{21.33}$$

Also, from the binomial option pricing formula, we have

$$\begin{aligned} a-1 &= \frac{\log\left(\frac{X}{Sd^n}\right)}{\log\left(\frac{u}{d}\right)} - \varepsilon \\ &= \left[\log\frac{X}{S} - n\log d\right] \Big/ \log\left(\frac{u}{d}\right) - \varepsilon \end{aligned} \tag{21.34}$$

where ε is a real number between 0 and 1.

From the definitions of $\tilde{\mu}_p$ and $\tilde{\sigma}_p^2$, it is easy to show that

$$\frac{a-1-np}{\sqrt{np(1-p)}} = \frac{\log\left(\frac{X}{S}\right) - \tilde{\mu}_p n - \varepsilon\log\left(\frac{u}{d}\right)}{\tilde{\sigma}_p\sqrt{n}} \tag{21.35}$$

Thus, from Eq. (21.31), we have

$$\begin{aligned} 1 - B_2(a;n,p) = \Pr\left(\frac{\log\frac{S^*}{S} - \tilde{\mu}_p n}{\tilde{\sigma}_p\sqrt{n}} \le \frac{\log\frac{X}{S} - \tilde{\mu}_p n - \varepsilon\log\left(\frac{u}{d}\right)}{\tilde{\sigma}_p\sqrt{n}}\right) \end{aligned} \tag{21.36}$$

We have checked the condition given by Eq. (21.29) in order to apply the central limit theorem. In addition, we have to evaluate $\tilde{\mu}_p n$, $\tilde{\sigma}_p^2 n$, and $\log\left(\frac{u}{d}\right)$ as $n \to \infty$. $\tilde{\mu}_p n \to \left(\log r - \frac{1}{2}\sigma^2\right)t$, which can be derived from the property of the lognormal distribution that $\log E(S^*/S) = \mu_p t + \frac{1}{2}\sigma^2 t$, and $E(S^*/S) = [pu + (1-p)d]^n = \hat{r}^n = r^t$. It is also clear that $n\tilde{\sigma}_p^2 \to \sigma^2 t$ and $\log\left(\frac{u}{d}\right) \to 0$.

Hence, in order to evaluate the asymptotic probability in Eq. (21.30), we have

$$\begin{aligned} &\frac{\log\left(\frac{X}{S}\right) - \tilde{\mu}_p n - \varepsilon\log\left(\frac{u}{d}\right)}{\tilde{\sigma}_p\sqrt{n}} \\ &\to z = \frac{\log\left(\frac{X}{S}\right) - \left(\log r - \frac{1}{2}\sigma^2\right)t}{\sigma\sqrt{t}} \end{aligned} \tag{21.37}$$

Using the fact that $1 - N(z) = N(-z)$, we have, as $n \to \infty$, $B_2(a;n,p) \to N(-z) = N\left(x - \sigma\sqrt{t}\right)$, where $x = \frac{\log\left(\frac{S}{Xr^{-t}}\right)}{\sigma\sqrt{t}} + \frac{1}{2}\sigma\sqrt{t}$. A similar argument holds for $B_1(a;n,p')$, and hence, we completed the proof that the binomial option pricing formula as given in Eq. (21.23) includes the Black–Scholes option pricing formula as a limiting case.

Lyapunov's condition requires that $X_1, X_2, \ldots$ are independent and uniformly bounded with $E(X_i) = 0$, $Y_n = X_1 + \cdots + X_n$, and $s^2 = E(Y_n^2) = \operatorname{Var}(Y_n)$. However, rates of return are generally not independent over time and not necessarily uniformly bounded by the condition required. This is the potential limitation of proof by Cox et al. (1979). We found that the derivation methods proposed by Rendleman and Bartter (1979), which will be discussed in next section, are not so restrictive as the proof discussed in this section.

21.4.2 Rendleman and Bartter Method

In Rendleman and Bartter (1979), a stock price can either advance or decline during the next period. Let H_T^+ and H_T^- represent the returns per dollar invested in the stock if the price rises (the + state) or falls (the − state), respectively, from time $T-1$ to time T(maturity of the option). V_T^+ and V_T^- the corresponding end-of-period values of the option.

Let R be the riskless interest rate, they showed that the price of the option can be represented as a recursive form as:

$$W_{T-1} = \frac{W_T^+(1+R-H_T^-)+W_T^-(H_T^+-1-R)}{(H_T^+-H_T^-)(1+R)} \tag{21.38}$$

Equation (21.37) can be applied at any time $T-1$ to determine the price of the option as a function of its value at time T.[7] By using recursive substitution as discussed in Sect. 21.2.1, they derived the binomial option pricing model as defined in Eq. (21.38).[8]

$$W_0 = S_0 B_1(a;T,\varphi) - \frac{X}{(1+R)^T} B_2(a;T,\phi) \tag{21.39}$$

where pseudo probabilities φ and ϕ are defined as:

$$\varphi = \frac{(1+R-H^-)H^+}{(1+R)(H^+-H^-)} \tag{21.40}$$

$$\phi = \frac{(1+R-H^-)}{(H^+-H^-)} \tag{21.41}$$

Please note that ϕ and φ are identical to p and p', which are defined as $p = \frac{r-d}{u-d}$ and $p' \equiv (u/r)p$ in Sect. 21.2.1.

a denotes the minimum integer value of i for which $S_0 H^{+^i} H^{-^{T-i}} > X$ will be satisfied. This value is given by[9]:

$$a = 1 + \mathrm{INT}\left[\frac{\ln(X/S_0) - T\ln(H^-)}{\ln H^+ - \ln H^-}\right] \tag{21.42}$$

where INT[·] is the integer operator.

$B_1(a;T,\varphi)$ and $B_2(a;T,\varphi)$ are the cumulative binomial probability. The number of successes will fall between a and T after T trials, φ and ϕ represent the probability associated with a success after one trial.

In each period, the stock price rises with the probability θ. We assume the distribution of returns, which is generated after T periods, will follow a log-binomial distribution. Then, the mean of the stock-price return is:

$$\begin{aligned}\mu &= T[h^+\theta + h^-(1-\theta)] \\ &= T[(h^+-h^-)\theta + h^-]\end{aligned} \tag{21.43}$$

And the variance of stock-price return is:

$$\sigma^2 = T(h^+-h^-)^2\theta(1-\theta) \tag{21.44}$$

where

θ = probability that the price of the stock will rise.

$$h^+ = \ln(H^+) \tag{21.45}$$

$$h^- = \ln(H^-) \tag{21.46}$$

Please note that in Cox et al. (1979), they assume log-binomial distribution with mean μt, and variance $\sigma^2 t$. Apparently, Rendleman and Bartter (1979) assumed that $t = 1$. Therefore, the Black–Scholes model derived by them is not exactly identical to the original Black–Scholes

[7]Please note that notation T used here is the number of periods rather than calendar time.

[8]Please note that some of the variables used in this section are different from those used in Sect. 21.2.1 and Sect. 21.4.1.

[9]We first solve equality $S_0 H^{+^i} H^{-^{T-i}} = X$. This yields $i = \frac{\ln(X/S_0) - T\ln(H^-)}{\ln H^+ - \ln H^-}$. To get a, the minimum integer value of i for which $S_0 H^{+^i} H^{-^{T-i}} > X$ will be satisfied, we should note a as: $a = 1 + \mathrm{INT}\left[\frac{\ln(X/S_0) - T\ln(H^-)}{\ln H^+ - \ln H^-}\right]$.

model. The implied values of H^+ and H^- are then determined by solving Eqs. (21.42)–(21.45), shown as Eqs. (21.46) and (21.47), respectively.

$$H^+ = \exp\left(\mu/T + \left(\sigma/\sqrt{T}\right)\sqrt{\frac{(1-\theta)}{\theta}}\right) \tag{21.47}$$

$$H^- = \exp\left(\mu/T - \left(\sigma/\sqrt{T}\right)\sqrt{\frac{\theta}{(1-\theta)}}\right) \tag{21.48}$$

As T becomes larger, the cumulative binomial density function can be approximated by the cumulative normal density function. When $T \to \infty$, the approximation will be exact, and Eq. (21.38) evolves to Eq. (21.48).[10]

$$W_0 \sim S_0 N(Z_1, Z_1') - \frac{X}{(1+R)^T} N(Z_2, Z_2') \tag{21.49}$$

In this equation, $N(Z, Z')$ is the probability that a random variable from a standard normal distribution will take on values between a lower limit Z and an upper limit Z'. According to the property of binomial probability distribution function, we have:

$$Z_1 = \frac{a - T\varphi}{\sqrt{T\varphi(1-\varphi)}}, Z_1' = \frac{T - T\varphi}{\sqrt{T\varphi(1-\varphi)}}$$
$$Z_2 = \frac{a - T\phi}{\sqrt{T\phi(1-\phi)}}, Z_2' = \frac{T - T\phi}{\sqrt{T\phi(1-\phi)}}$$

Thus, the price of option when the two-state process evolves continuously is presented as:

$$W_0 = S_0 N\left(\lim_{T\to\infty} Z_1, \lim_{T\to\infty} Z_1'\right) - \frac{X}{\lim_{T\to\infty}(1+R)^T} N\left(\lim_{T\to\infty} Z_2, \lim_{T\to\infty} Z_2'\right) \tag{21.50}$$

Let $1+R = e^{r/T}$ reflect the continuous compounding of interest, then $\lim_{T\to\infty}(1+R)^T = e^r$. It is obvious that $\lim_{T\to\infty} Z_1' = \lim_{T\to\infty} Z_2' = \infty$; therefore, all that needs to be determined is $\lim_{T\to\infty} Z_1$ and $\lim_{T\to\infty} Z_2$ in the derivation of the two-state model under a continuous-time case. Substituting H^+ and H^- in Eqs. (21.46) and (21.47) into Eq. (21.41), we have: $a = 1 + \text{INT}\left[\frac{\ln(X/S_0) - \mu + \sigma\sqrt{T}\sqrt{\frac{\theta}{(1-\theta)}}}{\sigma/\sqrt{T\theta(1-\theta)}}\right]$.

Then, we have Eq. (21.50) holds

$$Z_1 = \frac{a - T\varphi}{\sqrt{T\varphi(1-\varphi)}} = \frac{1 + INT\left[\frac{\ln\left(\frac{X}{S_0}\right) - \mu + \sigma\sqrt{T}\sqrt{\frac{\theta}{1-\theta}}}{\frac{\sigma}{\sqrt{T\theta(1-\theta)}}}\right] - T\varphi}{\sqrt{T\varphi(1-\varphi)}} \tag{21.51}$$

In the limit, the term $1 + INT[\cdot]$ will be simplified to $[\cdot]$. Therefore, Z_1 can be restated as:

$$Z_1 \sim \frac{\ln\left(\frac{X}{S_0}\right) - \mu}{\sigma\sqrt{\frac{\varphi(1-\varphi)}{\theta(1-\theta)}}} + \frac{\sqrt{T}(\theta - \varphi)}{\sqrt{\varphi(1-\varphi)}} \tag{21.52}$$

Substituting H^+ and H^- in Eqs. (21.46) and (21.47) and $1+R = e^{r/T}$ into Eq. (21.39), we have:

$$\varphi = \frac{(1+R-H^-)H^+}{(1+R)(H^+ - H^-)} = \frac{\left(e^{\frac{r}{T}} - e^{\frac{\mu}{T} - (\sigma/\sqrt{T})\sqrt{\frac{\theta}{1-\theta}}}\right)e^{\frac{\mu}{T} + (\sigma/\sqrt{T})\sqrt{\frac{1-\theta}{\theta}}}}{e^{\frac{r}{T}}\left(e^{\frac{\mu}{T} + (\sigma/\sqrt{T})\sqrt{\frac{1-\theta}{\theta}}} - e^{\frac{\mu}{T} - (\sigma/\sqrt{T})\sqrt{\frac{\theta}{1-\theta}}}\right)} = \frac{e^{(\sigma/\sqrt{T})\sqrt{\frac{1-\theta}{\theta}}} - e^{\frac{\mu}{T} - \frac{r}{T} + (\sigma/\sqrt{T})\left(\sqrt{\frac{1-\theta}{\theta}} - \sqrt{\frac{\theta}{1-\theta}}\right)}}{e^{(\sigma/\sqrt{T})\sqrt{\frac{1-\theta}{\theta}}} - e^{-(\sigma/\sqrt{T})\sqrt{\frac{\theta}{1-\theta}}}} \tag{21.53}$$

[10]In Appendix, we will use de Moivre–Laplace Theorem to show that the best fit between the binomial and normal distributions occurs when the binomial probability (or pseudo probability in this case) is $\frac{1}{2}$. In addition, we also present the Excel program.

Now, we expand Taylor's series[11] in $\frac{1}{\sqrt{T}}$ and obtain:

$$\varphi = \frac{(\sigma/\sqrt{T})\sqrt{\frac{1-\theta}{\theta}} - (\mu - r)/T - (\sigma/\sqrt{T})\left(\sqrt{\frac{1-\theta}{\theta}} - \sqrt{\frac{\theta}{1-\theta}}\right) + O\left(\frac{1}{\sqrt{T}}\right)}{(\sigma/\sqrt{T})\left(\sqrt{\frac{1-\theta}{\theta}} + \sqrt{\frac{\theta}{1-\theta}}\right) + O\left(\frac{1}{\sqrt{T}}\right)} = \frac{(\sigma/\sqrt{T})\sqrt{\frac{\theta}{1-\theta}} + O\left(\frac{1}{\sqrt{T}}\right)}{(\sigma/\sqrt{T})\left(\sqrt{\frac{1-\theta}{\theta}} + \sqrt{\frac{\theta}{1-\theta}}\right) + O\left(\frac{1}{\sqrt{T}}\right)} \quad (21.54)$$

where $O\left(\frac{1}{\sqrt{T}}\right)$ denotes a function tending to zero more rapidly than $\frac{1}{\sqrt{T}}$.

It can be shown that

$$\lim_{T\to\infty} \varphi = \frac{\sqrt{\frac{\theta}{1-\theta}}}{\sqrt{\frac{1-\theta}{\theta}} + \sqrt{\frac{\theta}{1-\theta}}} = \frac{\sqrt{\frac{\theta}{1-\theta}}}{\frac{1-\theta+\theta}{\sqrt{\theta(1-\theta)}}} = \theta \quad (21.55)$$

Similarly, we have:

We also expand Taylor's series in $\frac{1}{\sqrt{T}}$, and we can obtain:

$$\sqrt{T}(\theta - \varphi) = \sqrt{T}\left(\theta - \frac{e^{(\sigma/\sqrt{T})\sqrt{\frac{1-\theta}{\theta}}} - e^{\frac{\mu}{T} - \frac{r}{T} + (\sigma/\sqrt{T})\left(\sqrt{\frac{1-\theta}{\theta}} - \sqrt{\frac{\theta}{1-\theta}}\right)}}{e^{(\sigma/\sqrt{T})\sqrt{\frac{1-\theta}{\theta}}} - e^{-(\sigma/\sqrt{T})\sqrt{\frac{\theta}{1-\theta}}}}\right)$$
$$= \frac{\theta\sqrt{T}\left(e^{(\sigma/\sqrt{T})\sqrt{\frac{1-\theta}{\theta}}} - e^{-(\sigma/\sqrt{T})\sqrt{\frac{\theta}{1-\theta}}}\right) - \sqrt{T}\left(e^{(\sigma/\sqrt{T})\sqrt{\frac{1-\theta}{\theta}}} - e^{\frac{\mu}{T} - \frac{r}{T} + (\sigma/\sqrt{T})\left(\sqrt{\frac{1-\theta}{\theta}} - \sqrt{\frac{\theta}{1-\theta}}\right)}\right)}{e^{(\sigma/\sqrt{T})\sqrt{\frac{1-\theta}{\theta}}} - e^{-(\sigma/\sqrt{T})\sqrt{\frac{\theta}{1-\theta}}}} \quad (21.56)$$

[11]Using Taylor expansion, we have $e^x = 1 + x + \frac{x^2}{2!} + O(x^2)$.

$$
\begin{aligned}
\sqrt{T}(\theta-\varphi) &= \frac{\theta\sqrt{T}\left[(\sigma/\sqrt{T})\left(\sqrt{\frac{1-\theta}{\theta}}+\sqrt{\frac{\theta}{1-\theta}}\right)+\frac{1}{2}(\sigma/\sqrt{T})^2\left(\frac{1-\theta}{\theta}-\frac{\theta}{1-\theta}\right)+O\left(\frac{1}{T}\right)\right]}{(\sigma/\sqrt{T})\left(\sqrt{\frac{1-\theta}{\theta}}+\sqrt{\frac{\theta}{1-\theta}}\right)+O\left(\frac{1}{\sqrt{T}}\right)} \\
&- \frac{\sqrt{T}\left[(\sigma/\sqrt{T})\left(\sqrt{\frac{1-\theta}{\theta}}\right)+\frac{1}{2}(\sigma/\sqrt{T})^2\left(\frac{1-\theta}{\theta}\right)-\left(\frac{\mu-r}{T}+(\sigma/\sqrt{T})\left(\sqrt{\frac{1-\theta}{\theta}}-\sqrt{\frac{\theta}{1-\theta}}\right)\right)\right]}{(\sigma/\sqrt{T})\left(\sqrt{\frac{1-\theta}{\theta}}+\sqrt{\frac{\theta}{1-\theta}}\right)+O\left(\frac{1}{\sqrt{T}}\right)} \\
&+ \frac{\frac{1}{2}(\sigma/\sqrt{T})^2\left(\sqrt{\frac{1-\theta}{\theta}}-\sqrt{\frac{\theta}{1-\theta}}\right)^2+O\left(\frac{1}{T}\right)}{(\sigma/\sqrt{T})\left(\sqrt{\frac{1-\theta}{\theta}}+\sqrt{\frac{\theta}{1-\theta}}\right)+O\left(\frac{1}{\sqrt{T}}\right)} \\
&= \frac{\frac{1}{2}\theta\frac{\sigma^2}{\sqrt{T}}\left(\frac{1-\theta}{\theta}-\frac{\theta}{1-\theta}\right)+\frac{\mu-r}{\sqrt{T}}+\frac{1}{2}\frac{\sigma^2}{\sqrt{T}}\left(\frac{\theta}{1-\theta}-2\right)+O\left(\frac{1}{\sqrt{T}}\right)}{(\sigma/\sqrt{T})\left(\sqrt{\frac{1-\theta}{\theta}}+\sqrt{\frac{\theta}{1-\theta}}\right)+O\left(\frac{1}{\sqrt{T}}\right)}
\end{aligned}
\tag{21.57}
$$

Therefore, we have:

$$
\begin{aligned}
&\lim_{T\to\infty}\sqrt{T}(\theta-\varphi) \\
&= \lim_{T\to\infty}\frac{\frac{1}{2}\theta\frac{\sigma^2}{\sqrt{T}}\left(\frac{1-\theta}{\theta}-\frac{\theta}{1-\theta}\right)+\frac{\mu-r}{\sqrt{T}}+\frac{1}{2}\frac{\sigma^2}{\sqrt{T}}\left(\frac{\theta}{1-\theta}-2\right)+O\left(\frac{1}{\sqrt{T}}\right)}{(\sigma/\sqrt{T})\left(\sqrt{\frac{1-\theta}{\theta}}+\sqrt{\frac{\theta}{1-\theta}}\right)+O\left(\frac{1}{\sqrt{T}}\right)} \\
&= \frac{\frac{1}{2}\theta\sigma^2\left(\frac{1-\theta}{\theta}-\frac{\theta}{1-\theta}\right)+\mu-r+\frac{1}{2}\sigma^2\left(\frac{\theta}{1-\theta}-2\right)}{\sigma\left(\sqrt{\frac{1-\theta}{\theta}}+\sqrt{\frac{\theta}{1-\theta}}\right)} \\
&= \frac{\mu-r-\frac{1}{2}\sigma^2}{\sigma\left(\sqrt{\frac{1-\theta}{\theta}}+\sqrt{\frac{\theta}{1-\theta}}\right)} = \frac{\left(\mu-r-\frac{1}{2}\sigma^2\right)\sqrt{\theta(1-\theta)}}{\sigma}
\end{aligned}
\tag{21.58}
$$

Now substituting $\lim_{T\to\infty}\varphi$ for φ and $\lim_{T\to\infty}\sqrt{T}(\theta-\varphi)$ for $\sqrt{T}(\theta-\varphi)$ into Eq. (21.51). Then, we have Eq. (21.58) holds.

$$
\begin{aligned}
\lim_{T\to\infty} Z_1 &= \frac{\ln\left(\frac{X}{S_0}\right)-\mu}{\sigma\sqrt{\frac{\theta(1-\theta)}{\theta(1-\theta)}}} - \frac{\sqrt{\theta(1-\theta)}\left(r-\mu+\frac{1}{2}\sigma^2\right)}{\sigma\sqrt{\theta(1-\theta)}} \\
&= \frac{\ln\left(\frac{X}{S_0}\right)-r-\frac{1}{2}\sigma^2}{\sigma}
\end{aligned}
\tag{21.59}
$$

Similarly, we can also prove that

$$
\lim_{T\to\infty} Z_2 = \frac{\ln\left(\frac{X}{S_0}\right)-r+\frac{1}{2}\sigma^2}{\sigma}
\tag{21.60}
$$

According to the property of normal distribution, $N(Z,\infty)=N(-\infty,-Z)$. Let $d_1 = -\lim_{T\to\infty} Z_1$, $d_2 = -\lim_{T\to\infty} Z_2$, the continuous-time version of the two-state model is obtained:

$$
\begin{aligned}
w_0 &= S_0N(-\infty,d_1) - Xe^{-r}N(-\infty,d_2) \\
&= S_0N(d_1) - Xe^{-r}N(d_2) \\
d_1 &= \frac{\ln\left(\frac{S_0}{X}\right)+r+\frac{1}{2}\sigma^2}{\sigma} \\
d_2 &= d_1 - \sigma
\end{aligned}
\tag{21.61}
$$

Equation (21.60) is not exactly identical to the original Black–Scholes model because of the assumed log-binomial distribution with mean μ and variance σ^2. If they assume a log-binomial distribution with mean μt and variance $\sigma^2 t$, then d_1 and d_2 should be rewritten as:

Table 21.2 Comparison between Rendleman and Bartter's and Cox et al.'s approaches

Model	Rendleman and Bartter (1979)	Cox et al. (1979)
Mathematical and probability theory knowledge	Basic algebra Taylor expansion Binomial theorem Central limit theorem Properties of binomial distribution	Basic algebra Taylor expansion Binomial theorem Central limit theorem Properties of binomial distribution Lyapunov's condition
Assumption	The mean and variance of logarithmic returns of the stock are held constant over the life of the option	The stock follows a binomial process from one period to the next. It can only go up by a factor of u with probability p or go down by a factor of d with probability $1-p$ In order to apply the central limit theorem, u, d, and p are needed to be chosen
Advantage and disadvantage	1. Readers who have undergraduate-level training in mathematics and probability theory can follow this approach 2. The approach is intuitive, but the derivation is more complicated and tedious	1. Readers who have advanced-level knowledge in probability theory can follow this approach, but for those who do not, it may be difficult to follow 2. The assumption on the parameters u, d, and p makes this approach more restricted

$$d_1 = \frac{\ln\left(\frac{S_0}{X}\right) + \left(r + \frac{1}{2}\sigma^2\right)t}{\sigma\sqrt{t}} \qquad (21.57)$$
$$d_2 = d_1 - \sigma\sqrt{t}$$

Lee and Lin (2010) have theoretically compared these two derivation methods. Based upon (i) mathematical and probability theory knowledge, (ii) assumption and (iii) advantage and disadvantage, the comparison results are listed in Table 21.2. The main differences of assumptions between two approaches are: Under Cox et al. (1979) method, the stock price's increase factor and decrease factor are expressed as: $u = e^{\sigma\sqrt{\frac{t}{n}}}$ and $d = e^{-\sigma\sqrt{\frac{t}{n}}}$, respectively, which implies the restraints equality $ud = 1$ holds. While under the Rendleman and Bartter (1979) method, the increase factor and decrease factor are: $H^+ = \exp\left(\mu/T + (\sigma/\sqrt{T})\sqrt{\frac{(1-\theta)}{\theta}}\right)$ and $H^+ = \exp\left(\mu/T - (\sigma/\sqrt{T})\sqrt{\frac{\theta}{(1-\theta)}}\right)$, respectively. In the Rendleman and Bartter (1979) method's settings, time to maturity is settled as "1." With the number of periods $T \to \infty$, we can find that the expressions are similar to the Cox et al. (1979) method. They still have the "adjusted factor" $\sqrt{\frac{(1-\theta)}{\theta}}$ and $\sqrt{\frac{\theta}{(1-\theta)}}$ before $\frac{\sigma}{\sqrt{T}}$ in the exponential expression for increase factor and decrease factor. Under the Rendleman and Bartter (1979) method, $H^+H^- \neq 1$.

Hence, like we indicate in Table 21.2, the Cox et al. method is easy to follow if one has the advanced-level knowledge in probability theory, but the assumptions on the model parameters make its applications limited. On the other hand, the Rendleman and Bartter model is intuitive and does not require higher-level knowledge in probability theory. However, the derivation is more complicated and tedious. In Appendix, we show that the best fit between binomial distribution and normal distribution will occur when binomial probability is 0.5.

21.5 Lognormal Distribution Approach to Derive Black–Scholes Model[12]

To derive the option pricing model in terms of lognormal distribution, we begin by assuming that the stock price follows a lognormal

[12]The presentation and derivation of this section follow Garven (1986), Lee et al. (2013a, b).

distribution (Lee et al. 2013b). Denote the current stock price by S and the stock price at the end of t-th period by S_t. Then, $\frac{S_t}{S_{t-1}} = \exp(K_t)$ is a random variable with a lognormal distribution, where K_t is the rate of return in t-th period and is assumed as a random variable with normal distribution. Assume K_t has the same expected value μ_k and variance σ_k^2 for each. Then $K_1 + K_2 + \cdots + K_T$ is a normal random variable with expected value $T\mu_k$ and variance $T\sigma_k^2$.

Property of lognormal distribution *If a continuous random variable y is normally distributed, then the continuous variable x defined in Eq. (21.61) is lognormally distributed.*

$$x = e^y \tag{21.62}$$

If the variable y has mean μ and variance σ^2, then the mean μ_x and variance σ_x^2 of variable x are defined as follows, respectively.

$$\mu_x = e^{\mu + 1/2\sigma^2} \tag{21.63}$$

$$\sigma_x^2 = e^{2\mu+\sigma^2}\left(e^{\sigma^2} - 1\right) \tag{21.64}$$

Following the property, we then can define the expected value of $\frac{S_T}{S} = \exp(K_1 + K_2 + \cdots + K_T)$ as:

$$E\left(\frac{S_T}{S}\right) = \exp\left(T\mu_k + \frac{T\sigma_k^2}{2}\right) \tag{21.65}$$

Under the assumption of a risk-neutral investor, the expected return $E\left(\frac{S_T}{S}\right)$ is assumed to be e^{rT} (where r is the riskless rate of interest). In other words, we have the following equality holds.

$$\mu_k = r - \sigma_k^2/2 \tag{21.66}$$

The call option price C can be determined by discounting the expected value of the terminal option price by the risk-free rate.

$$C = e^{-rT}E[\text{Max}(S_T - X, 0)] \tag{21.67}$$

Note that in Eq. (21.66):

$$\begin{aligned}\text{Max}(S_T - X, 0) &= S_T - X \quad \text{for } S_T > X \\ &= 0 \qquad\qquad \text{otherwise}\end{aligned}$$

where T is the time of expiration and X is the exercise price.

Let $x = \frac{S_T}{S}$ be a lognormal distribution. Then, we have:

$$\begin{aligned}C &= e^{-rT}E[\text{Max}(S_T - X)] \\ &= e^{-rT}\int_{\frac{X}{S}}^{\infty} S\left[x - \frac{X}{S}\right] g(x)\mathrm{d}x \\ &= e^{-rT}S\int_{\frac{X}{S}}^{\infty} xg(x)\mathrm{d}x - e^{-rT}S\frac{X}{S}\int_{\frac{X}{S}}^{\infty} g(x)\mathrm{d}x\end{aligned} \tag{21.68}$$

where $g(x)$ is the probability density function $x = \frac{S_T}{S}$.

Here, we will use properties of normal distribution, lognormal distribution, and their mutual relations to derive the Black–Scholes model. We continue with variable settings in Eq. (21.61), where y is normally distributed and x is lognormally distributed.

The PDF of x is:

$$f(x) = \frac{1}{x\sigma\sqrt{2\pi}}\exp\left[-\frac{1}{2\sigma^2}(x - \mu)^2\right], \quad x > 0 \tag{21.69}$$

The PDF of y can be defined as:

$$f(y) = \frac{1}{\sigma\sqrt{2\pi}}\exp\left[-\frac{1}{2\sigma^2}(y - \mu)^2\right], \quad -\infty < y < \infty \tag{21.70}$$

By comparing the PDF of normal distribution and the PDF of lognormal distribution, we know that

$$f(x) = \frac{f(y)}{x} \tag{21.71}$$

In addition, it can be shown that[13]

$$\mathrm{d}x = x\mathrm{d}y \tag{21.72}$$

The CDF of lognormal distribution can be defined as

$$\int_a^\infty f(x)\mathrm{d}x \tag{21.73}$$

If we transform variable x in Eq. (21.72) into variable y, then the upper and lower limits of integration for a new variable are ∞ and $\ln a$, respectively. Then, the CDF for lognormal distribution can be written in terms of the CDF for normal distribution as

$$\int_a^\infty f(x)\mathrm{d}x = \int_{\ln a}^\infty \left(\frac{f(y)}{x}\right) x\,\mathrm{d}y = \int_{\ln(a)}^\infty f(y)\mathrm{d}y \tag{21.74}$$

We can rewrite Eq. (21.73) in a standard normal distribution form, by substituting the variable.

$$\int_a^\infty f(x)\mathrm{d}x = \int_{\ln(a)}^\infty f(y)\mathrm{d}y = N(d) \tag{21.75}$$

where $d = \frac{\mu - \ln(a)}{\sigma}$.

Similarly, the mean of a lognormal variable can be defined as:

$$\int_0^\infty xf(x)\mathrm{d}x = e^{\mu + 1/2\sigma^2} \tag{21.76}$$

If the lower bound a is greater than 0; then the partial mean of x can be shown as[14]:

$$\int_0^\infty xf(x)\mathrm{d}x = \int_{\ln(a)}^\infty f(y)e^y\mathrm{d}y = e^{\mu + \sigma^2/2}N(d) \tag{21.77}$$

where $d = \frac{\mu - \ln(a)}{\sigma} + \sigma$.

Substituting $\mu = r - \sigma^2/2$ and $a = \frac{X}{S}$ into Eq. (21.74), and we obtain:

$$\int_{\frac{X}{S}}^\infty g(x)\mathrm{d}x = N(d_2) \tag{21.78}$$

where $d_2 = \frac{r - (1/2)\sigma^2 - \ln\frac{X}{S}}{\sigma}$.

Similarly, we substitute $\mu = r - \sigma^2/2$ and $a = \frac{X}{S}$ into Eq. (21.76), we obtain:

$$\int_{\frac{X}{S}}^\infty xg(x)\mathrm{d}x = e^r N(d_1) \tag{21.79}$$

where $d_1 = \frac{r - (1/2)\sigma^2 - \ln\frac{X}{S}}{\sigma} + \sigma$.

Substituting Eqs. (21.77) and (21.78) into Eq. (21.67), we obtain Eq. (21.79), which is identical to the Black–Scholes formula.

$$\begin{aligned} C &= SN(d_1) - Xe^{-rT}N(d_2) \\ d_1 &= \frac{\ln\left(\frac{S}{X}\right) + \left(r + \frac{1}{2}\sigma^2\right)T}{\sigma\sqrt{T}} \\ d_2 &= \frac{\ln\left(\frac{S}{X}\right) + \left(r - \frac{1}{2}\sigma^2\right)T}{\sigma\sqrt{T}} = d_1 - \sigma\sqrt{T} \end{aligned} \tag{21.80}$$

In this section, we show that the Black–Scholes model can be derived by differential and integral calculus without using stochastic

[13]Now that $x = e^y$, then $\mathrm{d}x = d(e^y) = e^y\mathrm{d}y = x\mathrm{d}y$.

[14]The second equality is obtained by substituting the PDF of normal distribution into $\int_{\ln(a)}^{\infty} f(y)e^y\mathrm{d}y$ and does the appropriate manipulation.

calculus. However, it should be noted that we assume risk neutrality instead of risk averse in the derivation of this section.

21.6 Using Stochastic Calculus to Derive Black–Scholes Model

Black and Scholes (1973) have used two alternative approaches to derive the well-known stochastic differential equation defined in Eq. (21.80)[15]:

$$\begin{aligned}\frac{1}{2}\sigma^2 S^2 C_{SS}(t,S) + rSC_S(t,S)\\ -rC(t,S) + C_t(t,S) = 0\end{aligned} \tag{21.81}$$

where

t = passage of time;
S = stock price, which is a function of time t;
$C(t,S)$ = call price, which is a function of time t and stock price S;
$C_t(t,S)$ is the first-order partial derivative of $C(t,S)$ with respect to t;
$C_S(t,S)$ is the first-order partial derivative of $C(t,S)$ with respect to S;
$C_{SS}(t,S)$ is the second-order partial derivative of $C(t,S)$ with respect to S;
r = risk-free interest rate;
σ = stock volatility.

We rewrite it in a simpler way, as shown in Eq. (21.81).

$$\frac{\partial C}{\partial t} + rS\frac{\partial C}{\partial S} + \frac{1}{2}\sigma^2 S^2 \frac{\partial^2 C}{\partial S^2} = rC \tag{21.82}$$

where
$\frac{\partial C}{\partial t} = C_t(t,S)$; $\frac{\partial C}{\partial S} = C_S(t,S)$; $\frac{\partial^2 C}{\partial S^2} = C_{SS}(t,S)$ in Eq. (21.80).

To derive the Black–Scholes model, we need to solve this differential equation under the boundary condition:

$$\begin{aligned} C(S,T) &= S - X \quad \text{if } S \geq X \\ &= 0 \qquad\quad\ \ \text{otherwise} \end{aligned} \tag{21.83}$$

where T is the maturity date of the option, and X is the exercise price.

By introducing boundary constraints and making variable substitutions, they obtained a differential equation, which is the heat-transfer equation in physics (Joshi 2003). They used the Fourier transformation to solve the heat-transfer equation under the boundary condition, and finally obtain the solution. Here, we will demonstrate the main procedures to obtain the heat-transfer equation, and then get the closed-form solution under the boundary condition.

Let $Z = \ln S$, using the chain rule of partial derivatives, then we have the following equations hold.

$$\frac{\partial C}{\partial S} = \frac{\partial C}{\partial Z}\frac{\partial Z}{\partial S} = \frac{\partial C}{\partial Z}\frac{1}{S} \tag{21.84}$$

$$\begin{aligned}\frac{\partial^2 C}{\partial S^2} &= \frac{\partial\left(\frac{\partial C}{\partial Z}\right)}{\partial S}\frac{1}{S} - \frac{\partial C}{\partial Z}\frac{1}{S^2}\\ &= \frac{\partial^2 C}{\partial Z^2}\frac{1}{S^2} - \frac{\partial C}{\partial Z}\frac{1}{S^2}\end{aligned} \tag{21.85}$$

Then, we changed Eq. (21.81) into Eq. (21.85).

$$\frac{\partial C}{\partial t} + \left(r - \frac{1}{2}\sigma^2\right)\frac{\partial C}{\partial Z} + \frac{1}{2}\sigma^2\frac{\partial^2 C}{\partial Z^2} = rC \tag{21.86}$$

Let $\tau = T - t$.

[15]Black and Scholes have used two alternative methods to derive this equation. In addition, the careful derivation of this equation can be found in Chap. 27 of Lee et al. (2013a), which was written by Professor A. G. Malliaris, Loyola University of Chicago. Beck (1993) has proposed an alternative way to derive this equation, and raised questions about the methods used by Black and Scholes. In the summary of his paper, he mentioned that the traditional derivation of the Black–Scholes formula is mathematically unsatisfactory. The hedge portfolio is not a hedge portfolio since it is neither self-financing nor riskless. Due to compensating inconsistencies, the final result obtained is nevertheless correct. In his paper, these inconsistencies, which abound in the literature, were pointed out and an alternative, more rigorous derivation avoiding these problems is presented.

$$\frac{\partial C}{\partial \tau} - \left(r - \frac{1}{2}\sigma^2\right)\frac{\partial C}{\partial Z} - \frac{1}{2}\sigma^2\frac{\partial^2 C}{\partial Z^2} = -rC \tag{21.87}$$

Let $D = e^{r\tau}C$, i.e., $C = e^{-r\tau}D$, and redefine three partial derivatives in Eq. (21.86). We have:

$$\begin{aligned}\frac{\partial C}{\partial \tau} &= -re^{-r\tau}D + e^{-r\tau}\frac{\partial D}{\partial \tau}\\ &= -rC + e^{-r\tau}\frac{\partial D}{\partial \tau}\end{aligned} \tag{21.88}$$

$$\frac{\partial C}{\partial Z} = e^{-r\tau}\frac{\partial D}{\partial Z} \tag{21.89}$$

$$\frac{\partial^2 C}{\partial Z^2} = e^{-r\tau}\frac{\partial^2 D}{\partial Z^2} \tag{21.90}$$

If we substitute Eqs. (21.87), (21.88), and (21.89) into Eq. (21.86), we obtain:

$$\frac{\partial D}{\partial \tau} - \left(r - \frac{1}{2}\sigma^2\right)\frac{\partial D}{\partial Z} - \frac{1}{2}\sigma^2\frac{\partial^2 D}{\partial Z^2} = 0 \tag{21.91}$$

We introduce a new variable Y to replace Z, and then we have:

$$\begin{aligned}Y &= \ln(S/X) + \left(r - \frac{1}{2}\sigma^2\right)\tau\\ &= Z + (r - \frac{1}{2}\sigma^2)\tau - \ln X\end{aligned} \tag{21.92}$$

Since $D = e^{r\tau}C$, and it is a function of Z and τ, we explicitly rewrite D as $D^Z(Z,\tau)$. Equation (21.91) implies that D is also a function of Y and τ. We define $D^Z(Z,\tau)$ and $D^Y(Y,\tau)$ as follows.

$$D^Z(Z,\tau) = D^Z\left(Y - \left(r - \frac{1}{2}\sigma^2\right)\tau + \ln X, \tau\right) \tag{21.93}$$

$$D^Y(Y,\tau) = D^Y\left(Z + \left(r - \frac{1}{2}\sigma^2\right)\tau - \ln X, \tau\right) \tag{21.94}$$

Taking partial derivatives of $D^Z(Z,\tau)$ with respect to Z, we obtain:

$$\frac{\partial D^Z(Z,\tau)}{\partial Z} = \frac{\partial D^Y(Y,\tau)}{\partial Y}\frac{\partial Y}{\partial Z} = \frac{\partial D^Y(Y,\tau)}{\partial Y} \tag{21.95}$$

Similarly, we have:

$$\frac{\partial^2 D^Z(Z,\tau)}{\partial Z^2} = \frac{\partial^2 D^Y(Y,\tau)}{\partial Y^2} \tag{21.96}$$

$$\begin{aligned}\frac{\partial D^Z(Z,\tau)}{\partial \tau} &= \frac{\partial D^Y(Y,\tau)}{\partial Y}\frac{\partial Y}{\partial \tau} + \frac{\partial D^Y(Y,\tau)}{\partial \tau}\\ &= \frac{\partial D^Y(Y,\tau)}{\partial Y}\frac{\partial(Z + (r - \frac{1}{2}\sigma^2)\tau - \ln X)}{\partial \tau}\\ &\quad + \frac{\partial D^Y(Y,\tau)}{\partial \tau}\\ &= \frac{\partial D^Y(Y,\tau)}{\partial Y}\left(r - \frac{1}{2}\sigma^2\right) + \frac{\partial D^Y(Y,\tau)}{\partial \tau}\end{aligned} \tag{21.97}$$

Substituting Eqs. (21.94), (21.95), and (21.96) into Eq. (21.90), we can get:

$$\frac{\partial D^Y}{\partial \tau} - \frac{1}{2}\frac{\partial^2 D^Y}{\partial Y^2}\sigma^2 = 0 \tag{21.98}$$

Equation (21.97) is almost close to the heat-transfer equation used by Black and Scholes.

Let $u = \frac{2}{\sigma^2}\left(r - \frac{1}{2}\sigma^2\right)Y$, $v = \frac{2}{\sigma^2}\left(r - \frac{1}{2}\sigma^2\right)^2\tau$, and redenote $D(u,v)$ as the function of u and v.

$$\begin{aligned}\frac{\partial D^Y(Y,\tau)}{\partial Y} &= \frac{\partial D(u,v)}{\partial u}\frac{\partial u}{\partial Y}\\ &= \frac{\partial D(u,v)}{\partial u}\frac{2}{\sigma^2}\left(r - \frac{1}{2}\sigma^2\right)\end{aligned} \tag{21.99}$$

$$\frac{\partial^2 D^Y(Y,\tau)}{\partial Y^2} = \frac{\partial^2 D(u,v)}{\partial u^2}\left(\frac{2}{\sigma^2}\left(r - \frac{1}{2}\sigma^2\right)\right)^2 \tag{21.100}$$

$$\begin{aligned}\frac{\partial D^Y(Y,\tau)}{\partial \tau} &= \frac{\partial D(u,v)}{\partial u}\frac{\partial u}{\partial \tau} + \frac{\partial D(u,v)}{\partial v}\frac{\partial v}{\partial \tau}\\ &= \frac{\partial D(u,v)}{\partial v}\frac{\partial v}{\partial \tau}\\ &= \frac{\partial D(u,v)}{\partial v}\frac{2}{\sigma^2}\left(r - \frac{1}{2}\sigma^2\right)^2\end{aligned} \tag{21.101}$$

We finally reach the heat-transfer equation derived by Black and Scholes, when we substitute Eqs. (21.99) and (21.100) into Eq. (21.97).

$$\frac{\partial D(u,v)}{\partial v} = \frac{\partial^2 D(u,v)}{\partial u^2} \tag{21.102}$$

Equation (21.101) is identical to Eq. (21.10) of Black–Scholes (1973). In terms of the Black–Scholes notation, Eq. (21.101) can be written as

$$y_2 = y_{11} \tag{21.101$'$}$$

Now, we need to find a function $D(u,v)$ that satisfies both of the boundary conditions and the partial differential equation as shown in Eq. (21.101).[16] The general solution given in Churchill (1963) is as follows.

If:

$$v_t(x,t) = kv_{xx}(x,t) \quad (-\infty < x < \infty, t > 0) \tag{21.103}$$

$$v(x,0) = f(x) \quad (-\infty < x < \infty) \tag{21.104}$$

Then, the general solution for $v_t(x,t)$ is[17]:

$$v(x,t) = 1/\sqrt{\pi} \int_{-\infty}^{\infty} f(x + 2\eta\sqrt{kt}) e^{-\eta^2} d\eta \tag{21.105}$$

In our notation, $D(u,v) = v(x,t)$, and $k = 1$, which makes Eq. (21.102) equivalent to the partial differential equation as shown in Eq. (21.101). Moreover, we have the boundary condition:

$$\begin{aligned} C(S,T) &= S - X \quad \text{if } S \geq X \\ &= 0 \quad \text{otherwise} \end{aligned}$$

At maturity date, $t = T$, $v = 0$. Then we have $D(u,0) = C(S,T)$. $f(u)$ must be determined to make Eq. (21.103) satisfied. Black and Scholes choose:

$$\begin{aligned} f(u) &= X\left(e^{u\left(\frac{1}{2}\sigma^2\right)/\left(r-\frac{1}{2}\sigma^2\right)} - 1\right) \quad \text{if } u \geq 0 \\ &= 0 \quad \text{if } u < 0 \end{aligned} \tag{21.106}$$

Note that, when $t = T$, $u = \frac{2}{\sigma^2}\left(r - \frac{1}{2}\sigma^2\right) \ln(S/X)$, then we have:

$$\begin{aligned} D(u,0) &= X\left(e^{\ln(S/X)} - 1\right) = S - X \quad \text{if } u \geq 0 \\ &= 0 \quad \text{if } u < 0 \end{aligned} \tag{21.107}$$

which is identical to the boundary condition. Therefore, the determined $f(u)$ makes Eq. (21.103) hold.

Now that Eqs. (21.102) and (21.103) are satisfied, the solution to the differential equation is given by:[18]

$$D(u,v) = 1/\sqrt{\pi} \int_{-u/2\sqrt{v}}^{\infty} X\left(e^{(u+2\eta\sqrt{v})\left(\frac{1}{2}\sigma^2\right)/\left(r-\frac{1}{2}\sigma^2\right)} - 1\right) \times e^{-\eta^2} \mathrm{d}\eta \tag{21.108}$$

Let $\eta = q/\sqrt{2}$, and substitute it and $C = e^{-r\tau}D$ into Eq. (21.107), and then we have:

$$C(S,t) = e^{-r\tau}\frac{1}{\sqrt{2\pi}} \int_{-u/\sqrt{2v}}^{\infty} X\left(e^{\left(u+q\sqrt{2v}\right)\left(\frac{1}{2}\sigma^2\right)\left(r-\frac{1}{2}\sigma^2\right)} - 1\right) \times e^{-q^2/2}\mathrm{d}q \tag{21.109}$$

Note that $-u/\sqrt{2v} = -\frac{\ln(S/X) + \left(r-\frac{1}{2}\sigma^2\right)\tau}{\sigma\sqrt{\tau}} = -d_2$. Therefore, Eq. (21.108) can evolve into:

$$\begin{aligned} C(S,t) = {} & Xe^{-r\tau}\frac{1}{\sqrt{2\pi}} \int_{-d_2}^{\infty} e^{\left[\left(u+q\sqrt{2v}\right)\left(\frac{1}{2}\sigma^2\right)/\left(r-\frac{1}{2}\sigma^2\right)\right] - q^2/2}\mathrm{d}q \\ & - Xe^{-r\tau}\frac{1}{\sqrt{2\pi}} \int_{-d_2}^{\infty} e^{-q^2/2}\mathrm{d}q \end{aligned} \tag{21.110}$$

---

[16]The following procedure has closely related to Kutner (1988). Therefore, we strongly suggest the readers read his paper.

[17]The solution is obtained as an application of the general Fourier integral. See Churchill (1963, pp. 154–155) for more details.

[18]The lower limit exists since if $u < 0, f(u) = 0$. Therefore, we require $u + 2\eta\sqrt{v}$, i.e., $\eta \geq -u/2\sqrt{v}$.

We observe the second term in Eq. (21.109). Recall that the cumulative standard normal density function is defined as:

$$N(x) = \int_{-\infty}^{x} \frac{1}{\sqrt{2\pi}} e^{-\frac{t^2}{2}} dt \tag{21.111}$$

Therefore, the second term of Eq. (21.109) is:

$$\begin{aligned} Xe^{-r\tau} \int_{-d_2}^{\infty} \frac{1}{\sqrt{2\pi}} e^{-q^2/2} \mathrm{d}q &= Xe^{-r\tau}(1 - N(-d_2)) \\ &= Xe^{-r\tau} N(d_2) \end{aligned} \tag{21.112}$$

Deriving the first term in Eq. (21.109) is much more tedious and difficult. Recall the expressions for u and v, $u = \frac{2}{\sigma^2}\left(r - \frac{1}{2}\sigma^2\right)\left(\ln(S/X) + \left(r - \frac{1}{2}\sigma^2\right)\tau\right)$, $v = \frac{2}{\sigma^2}\left(r - \frac{1}{2}\sigma^2\right)^2 \tau$. Therefore, we have Eq. (21.112) and (21.113) hold.

$$u\left(\frac{1}{2}\sigma^2\right) / \left(r - \frac{1}{2}\sigma^2\right) = \left(\ln(S/X) + \left(r - \frac{1}{2}\sigma^2\right)\tau\right) \tag{21.113}$$

$$q\sqrt{2v}\left(\frac{1}{2}\sigma^2\right) / \left(r - \frac{1}{2}\sigma^2\right) = q\sigma\sqrt{\tau} \tag{21.114}$$

Therefore, the first term in Eq. (21.109) is:

$$\begin{aligned} & Xe^{-r\tau} \frac{1}{\sqrt{2\pi}} \int_{-d_2}^{\infty} e^{\left[\left(u + q\sqrt{2v}\right)\left(\frac{1}{2}\sigma^2\right)/\left(r - \frac{1}{2}\sigma^2\right)\right] - q^2/2} \mathrm{d}q \\ & = Xe^{-r\tau} e^{\ln(S/X)} \frac{1}{\sqrt{2\pi}} \int_{-d_2}^{\infty} e^{r\tau} e^{-\frac{1}{2}\left(q^2 - 2q\sigma\sqrt{\tau} + \sigma^2\tau\right)} \mathrm{d}q \\ & = S \frac{1}{\sqrt{2\pi}} \int_{-d_2}^{\infty} e^{-\frac{1}{2}(q - \sigma\sqrt{\tau})^2} \mathrm{d}q \end{aligned} \tag{21.115}$$

Here, again we apply variable substitution. Let $q' = q - \sigma\sqrt{\tau}$, then $\mathrm{d}q' = \mathrm{d}q$. Therefore, Eq. (21.114) evolves to:

$$S \frac{1}{\sqrt{2\pi}} \int_{-d_2 - \sigma\sqrt{\tau}}^{\infty} e^{-\frac{1}{2}q'^2} \mathrm{d}q' \tag{21.116}$$

Let $d_1 = d_2 + \sigma\sqrt{\tau}$, then we obtain:

$$\begin{aligned} S \frac{1}{\sqrt{2\pi}} \int_{-d_1}^{\infty} e^{-\frac{1}{2}q'^2} \mathrm{d}q' &= S \int_{-d_1}^{\infty} \frac{1}{\sqrt{2\pi}} e^{-\frac{1}{2}q'^2} \mathrm{d}q' \\ &= S(1 - N(-d_1)) = SN(d_1) \end{aligned} \tag{21.117}$$

Finally, when combining the first and second terms in Eq. (21.109), simplified by Eqs. (21.116) and (21.111) respectively, we reach the Black–Scholes formula.

$$C(S,t) = SN(d_1) - Xe^{-r\tau}N(d_2) \tag{21.118}$$

where

$$\begin{aligned} d_1 &= \frac{\ln(S/X) + \left(r + \frac{1}{2}\sigma^2\right)\tau}{\sigma\sqrt{\tau}} \\ d_2 &= \frac{\ln(S/X) + \left(r - \frac{1}{2}\sigma^2\right)\tau}{\sigma\sqrt{\tau}} = d_1 - \sigma\sqrt{\tau} \\ \tau &= T - t \end{aligned} \tag{21.119}$$

21.7 Conclusion

In this chapter, we have reviewed three alternative approaches to derive option pricing models. We have discussed how the binomial model can be used to derive the Black–Scholes model in detail. In addition, we also show how the Excel program in terms of decision tree can be used to empirically show how binomial model can be converted to Black–Scholes model when observations approach infinity. Under an assumption of risk neutrality, we show that the Black–Scholes formula can be derived using only differential and integral calculus and a basic knowledge of normal and lognormal distributions. In Appendix, we use the de Moivre–Laplace Theorem to prove that the best fit between the binomial and normal distributions occurs when binomial probability is $\frac{1}{2}$. Overall, this chapter can help statisticians and mathematicians better understand how alternative methods can be used to derive the Black–Scholes option model.

Appendix: The Relationship Between Binomial Distribution and Normal Distribution

In this appendix, we will use the de Moivre–Laplace Theorem to prove that the best fit between the binomial and normal distributions occurs when the binomial probability is $\frac{1}{2}$.

de Moivre–Laplace Theorem *As n grows larger and approaches infinity, for k in the neighborhood of np we can approximate*

$$\binom{n}{k} p^k q^{n-k} \simeq \frac{1}{\sqrt{2\pi npq}} e^{-\frac{(k-np)^2}{2npq}}, \quad p+q=1,\ p,q>0. \tag{21.120}$$

Proof According to Stirling's approximation (or Stirling's formula) for factorials approximation, we can replace the factorial of large number n with the following:

$$n! \simeq n^n e^{-n}\sqrt{2\pi n} \text{ as } n \to \infty \tag{21.121}$$

Then, $\binom{n}{k} p^k q^{n-k}$ can be approximated as shown in the following procedures.

$$\begin{aligned}\binom{n}{k} p^k q^{n-k} &= \frac{n!}{k!(n-k)!} p^k q^{n-k} \\ &\simeq \frac{n^n e^{-n}\sqrt{2\pi n}}{k^k e^{-k}\sqrt{2\pi k}(n-k)^{n-k} e^{-k}\sqrt{2\pi k}} p^k q^{n-k} \\ &= \sqrt{\frac{n}{2\pi k(n-k)}}\left(\frac{k}{np}\right)^{-k}\left(\frac{n-k}{nq}\right)^{-(n-k)}\end{aligned} \tag{21.122}$$

Let $x = \frac{k-np}{\sqrt{npq}}$, we obtain:

$$\begin{aligned}&\sqrt{\frac{n}{2\pi k(n-k)}}\left(\frac{k}{np}\right)^{-k}\left(\frac{n-k}{nq}\right)^{-(n-k)} \\ &= \sqrt{\frac{n}{2\pi k(n-k)}}\left(1+x\sqrt{\frac{q}{np}}\right)^{-k}\left(1-x\sqrt{\frac{p}{nq}}\right)^{-(n-k)} \\ &= \sqrt{\frac{n^{-1}}{2\pi\frac{k}{n}\frac{(n-k)}{n}}}\left(1+x\sqrt{\frac{q}{np}}\right)^{-k}\left(1-x\sqrt{\frac{p}{nq}}\right)^{-(n-k)} \\ &= \sqrt{\frac{n^{-1}}{2\pi\frac{k}{n}(1-\frac{k}{n})}}\left(1+x\sqrt{\frac{q}{np}}\right)^{-k}\left(1-x\sqrt{\frac{p}{nq}}\right)^{-(n-k)}\end{aligned} \tag{21.123}$$

As $k \to np$, we get $\frac{k}{n} \to p$. Then, Eq. (21.123) can be approximated as:

$$\begin{aligned}&\sqrt{\frac{n^{-1}}{2\pi\frac{k}{n}(1-\frac{k}{n})}}\left(1+x\sqrt{\frac{q}{np}}\right)^{-k}\left(1-x\sqrt{\frac{p}{nq}}\right)^{-(n-k)} \\ &\simeq \sqrt{\frac{n^{-1}}{2\pi pq}}\left(1+x\sqrt{\frac{q}{np}}\right)^{-k}\left(1-x\sqrt{\frac{p}{nq}}\right)^{-(n-k)} \\ &= \sqrt{\frac{1}{2\pi npq}}\exp\left\{\ln\left[\left(1+x\sqrt{\frac{q}{np}}\right)^{-k}\right] + \ln\left[\left(1-x\sqrt{\frac{p}{nq}}\right)^{-(n-k)}\right]\right\} \\ &= \sqrt{\frac{1}{2\pi npq}}\exp\left\{-k\ln\left(1+x\sqrt{\frac{q}{np}}\right) - (n-k)\ln\left(1-x\sqrt{\frac{p}{nq}}\right)\right\} \\ &= \sqrt{\frac{1}{2\pi npq}}\exp\left\{-(np+x\sqrt{npq})\ln\left(1+x\sqrt{\frac{q}{np}}\right) - (nq-x\sqrt{npq})\ln\left(1-x\sqrt{\frac{p}{nq}}\right)\right\}\end{aligned} \tag{21.124}$$

We are considering the term in exponential function, i.e.,

$$-(np+x\sqrt{npq})\ln\left(1+x\sqrt{\frac{q}{np}}\right) - (nq-x\sqrt{npq})\ln\left(1-x\sqrt{\frac{p}{nq}}\right) \tag{21.125}$$

Here, we are using the Taylor series expansions of functions $\ln(1 \pm x)$:

$$\ln(1+x) = x - \frac{x^2}{2} + \frac{x^3}{3} + o(x^3)$$
$$\ln(1-x) = -x - \frac{x^2}{2} - \frac{x^3}{3} + o(x^3)$$

Then, we expand Eq. (21.125) with respect to x and obtain:

$$\begin{aligned} &-(np + x\sqrt{npq})\left(x\sqrt{\frac{q}{np}} - \frac{x^2 q}{2np} + \frac{x^3 q^{\frac{3}{2}}}{3n^{\frac{3}{2}}p^{\frac{3}{2}}} + o(x^3)\right) \\ &-(nq - x\sqrt{npq})\left(-x\sqrt{\frac{p}{nq}} - \frac{x^2 p}{2nq} - \frac{x^3 p^{\frac{3}{2}}}{3n^{\frac{3}{2}}q^{\frac{3}{2}}} + o(x^3)\right) \\ &= -\left(x\sqrt{npq} + x^2 q - \frac{1}{2}x^2 q - \frac{x^3 p^{-\frac{1}{2}}q^{\frac{3}{2}}}{2\sqrt{n}} + \frac{x^3 p^{-\frac{1}{2}}q^{\frac{3}{2}}}{3\sqrt{n}}\right) \\ &-\left(-x\sqrt{npq} + x^2 p - \frac{1}{2}x^2 q + \frac{x^3 p^{\frac{3}{2}}q^{-\frac{1}{2}}}{2\sqrt{n}} - \frac{x^3 p^{\frac{3}{2}}q^{-\frac{1}{2}}}{3\sqrt{n}}\right) \\ &+ o(x^3) \\ &= -\frac{1}{2}x^2(p+q) - \frac{1}{6\sqrt{npq}}x^3(p^2 - q^2) + o(x^3) \end{aligned} \tag{21.126}$$

Since we have $p + q = 1$ when we ignore the higher order of x, Eq. (21.126) can be simply approximated to:

$$-\frac{1}{2}x^2 - \frac{1}{6\sqrt{npq}}x^3(p - q) \tag{21.127}$$

Then, we replace Eq. (21.127) in the exponential function in Eq. (21.124), and we obtain:

$$\binom{n}{k}p^k q^{n-k} \simeq \frac{1}{\sqrt{2\pi pq}}\exp\left[-\frac{1}{2}x^2 - \frac{1}{6\sqrt{npq}}x^3(p-q)\right] \tag{21.128}$$

Although $-\frac{1}{6\sqrt{npq}}x^3(p-q) \to 0$ as $n \to \infty$, the term $-\frac{1}{6\sqrt{npq}}x^3(p-q)$ will be exactly zero if and only if $p = q$. Under this condition,

$$\binom{n}{k}p^k q^{n-k} \simeq \frac{1}{\sqrt{2\pi pq}}\exp\left(-\frac{1}{2}x^2\right) = \frac{1}{\sqrt{2\pi pq}}\exp\left[-\frac{(k-np)^2}{2npq}\right].$$

Thus, it is shown that the best fit between the binomial and normal distribution occurs when $p = q = \frac{1}{2}$.

If $p \neq q$, then there exists an additional term $-\frac{1}{6\sqrt{npq}}x^3(p-q)$. It is obvious that $\sqrt{pq} = \sqrt{p(p-q)}$ will reach maximum if and only if $p = q = \frac{1}{2}$. Therefore, when n is fixed, if the difference between p and q becomes larger, the absolute value of an additional term $-\frac{1}{6\sqrt{npq}}x^3(p-q)$ will be larger. This implies that the magnitude of absolute value of the difference between p and q is an important factor to make the approximation to normal distribution less precise. We use the following figures to demonstrate how the absolute number of differences between p and q affect the precision of using binomial distribution to approximate normal distribution.

From Figs. 21.5 and 21.6, we find that when $p \neq q$, the absolute magnitude does affect the estimated continuous distribution as indicated by red solid curves. For example, when $n = 30$, the red solid curve when $p = 0.9$ is very much different from $p = 0.5$. In other words, when $p = 0.9$, the red solid curve is not as similar to the normal curve as $p = 0.5$. If we increase n from 30 to 100, the solid red curve from $p = 0.9$ is less different from the solid red curve when $p = 0.5$. In sum, both the magnitude of n and p will affect the shape of using normal distribution to approximate binomial distribution.

From Eqs. (21.15) and (21.16) in the text, we can define the binomial OPM and the Black–Scholes OPM as follows:

$$C = SB_1(a; n, p') - \frac{X}{r^n}B_2(a; n, p) \tag{21.15}$$

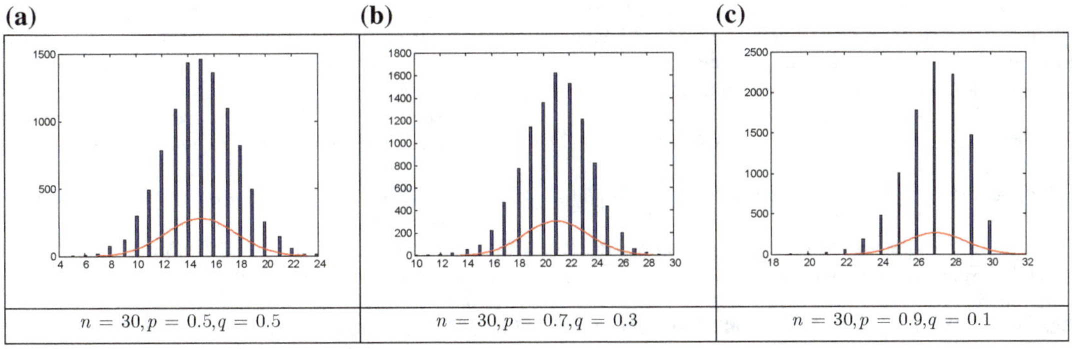

Fig. 21.5 Binomial distributions to approximate normal distributions ($n = 30$)

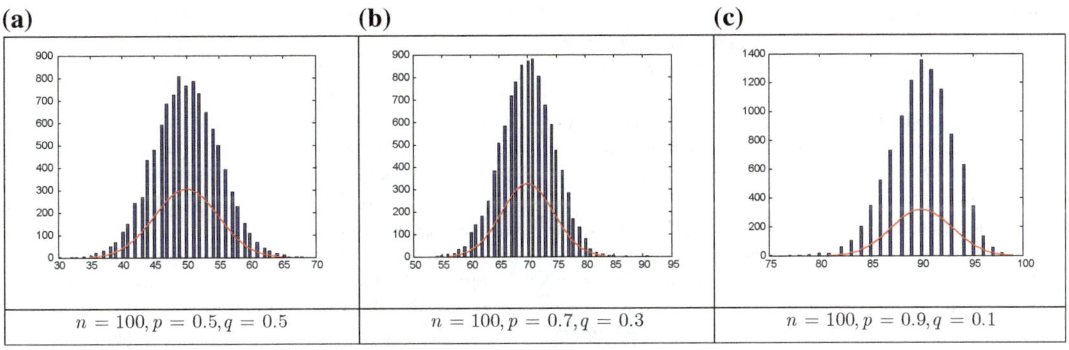

Fig. 21.6 Binomial distributions to approximate normal distributions ($n = 100$)

$$C = SN(d_1) - Xe^{-rT}N(d_2) \tag{21.16}$$

Both Cox et al. and Rendleman and Bartter tried to show the binomial cumulative functions of Eq. (21.15) will converge to the normal cumulative function of Eq. (21.16) when n approaches infinity. In this appendix, we have mathematically and graphically showed that the relative magnitude between p and q is the important factor to determine this approximation when n is constant. In addition, we also demonstrate the size of n which also affects the precision of this approximation process.

Bibliography

Alt-Sahalia., & Lo, Y. (1998), Nonparametric estimation of state-price density implicit asset prices. *The Journal of Finance, 53*(2), 499–547.

Amin, K. I., & Jarrow, R. A. (1992). Pricing options on risky assets in a stochastic interest rate economy. *Mathematical Finance, 2*(4), 217–237.

Amin, K. I., & Ng, V. K. (1993). Option valuation with systematic stochastic volatility. *The Journal of Finance, 48*(3), 881–910.

Ash, R. B., & Doleans-Dade, C. (1999). *Probability and measure theory* (2nd ed.). Academic Press.

Bailey, W., & Stulz, R. M. (1989). The pricing of stock index options in a general equilibrium model. *Journal of Financial and Quantitative Analysis, 24*(01), 1–12.

Bakshi, G. S., & Chen, Z. (1997a). An alternative valuation model for contingent claims. *Journal of Financial Economics, 44*(1), 123–165.

Bakshi, G. S., & Chen, Z. (1997b). Equilibrium valuation of foreign exchange claims. *The Journal of Finance, 52*(2), 799–826.

Bakshi, G., Cao, C., & Chen, Z. (1997). Empirical performance of alternative option pricing models. *The Journal of Finance, 52*(5), 2003–2049.

Bates, D. S. (1991). The crash of '87: Was it expected? The evidence from options markets. *The Journal of Finance, 46*(3), 1009–1044.

Bates, D. S. (1996). Jumps and stochastic volatility: Exchange rate processes implicit in deutsche mark options. *Review of Financial Studies, 9*(1), 69–107.

Beck, T. M. (1993). Black–Scholes revisited: Some important details. *Financial Review, 28*(1), 77–90.

Beckers, S. (1980). The constant elasticity of variance model and its implications for option pricing. *The Journal of Finance, 35*(3), 661–673.

Benninga, S., & Czaczkes, B. (2000). *Financial modeling*. MIT press.

Billingsley, P. (2008). *Probability and measure* (3rd ed.). Wiley.

Black, F., & Scholes, M. (1973). The pricing of options and corporate liabilities. *The Journal of Political Economy*, 637–654.

Buetow, G. W., & Albert, J. D. (1998). The pricing of embedded options in real estate lease contracts. *Journal of Real Estate Research, 15*(3), 253–266.

Cakici, N., & Topyan, K. (2000). The GARCH option pricing model: A lattice approach. *Journal of Computational Finance, 3*(4), 71–85.

Carr, P., & Wu, L. (2004). Time-changed Lévy processes and option pricing. *Journal of Financial Economics, 71*(1), 113–141.

Chen, R. R., & Palmon, O. (2005). A non-parametric option pricing model: Theory and empirical evidence. *Review of Quantitative Finance and Accounting, 24* (2), 115–134.

Chen, R. R., Lee, C. F., & Lee, H. H. (2009). Empirical performance of the constant elasticity variance option pricing model. *Review of Pacific Basin Financial Markets and Policies, 12*(2), 177–217.

Churchill, R. V. (1963). *Fourier series and boundary value problems* (2nd ed.). McGraw-Hill Companies.

Costabile, M., Leccadito, A., Massabó, I., & Russo, E. (2014). A reduced lattice model for option pricing under regime-switching. *Review of Quantitative Finance and Accounting, 42*(4), 667–690.

Cox, J. C., & Ross, S. A. (1976). The valuation of options for alternative stochastic processes. *Journal of Financial Economics, 3*(1), 145–166.

Cox, J. C., Ross, S. A., & Rubinstein, M. (1979). Option pricing: A simplified approach. *Journal of Financial Economics, 7*(3), 229–263.

Davydov, D., & Linetsky, V. (2001). Pricing and hedging path-dependent options under the CEV process. *Management Science, 47*(7), 949–965.

Duan, J. C. (1995). The GARCH option pricing model. *Mathematical Finance, 5*(1), 13–32.

Garven, J. R. (1986). A pedagogic note on the derivation of the Black–Scholes option pricing formula. *Financial Review, 21*(2), 337–348.

Geman, H., Madan, D. B., & Yor, M. (2001). Time changes for Lévy processes. *Mathematical Finance, 11*(1), 79–96.

Grenadier, S. R. (1995). Valuing lease contracts: A real-options approach. *Journal of Financial Economics, 38*(3), 297–331.

Heston, S. L. (1993). A closed-form solution for options with stochastic volatility with applications to bond and currency options. *Review of Financial Studies, 6*(2), 327–343.

Heston, S. L., & Nandi, S. (2000). A closed-form GARCH option valuation model. *Review of Financial Studies, 13*(3), 585–625.

Hillegeist, S. A., Keating, E. K., Cram, D. P., & Lundstedt, K. G. (2004). Assessing the probability of bankruptcy. *Review of Accounting Studies, 9*(1), 5–34.

Hull, J. C. (2014). *Options, futures, and other derivatives* (9th ed.). Prentice Hall.

Hull, J., & White, A. (1987). The pricing of options on assets with stochastic volatilities. *The Journal of Finance, 42*(2), 281–300.

Joshi, M. S. (2003). *The concepts and practice of mathematical finance* (Vol. 1). Cambridge University Press.

Kou, S. G. (2002). A jump-diffusion model for option pricing. *Management Science, 48*(8), 1086–1101.

Kou, S. G., & Wang, H. (2004). Option pricing under a double exponential jump diffusion model. *Management Science, 50*(9), 1178–1192.

Kutner, G. W. (1988). Black–Scholes revisited: Some important details. *Financial Review, 23*(1), 95–104.

Lee, J. C. (2001). Using microsoft excel and decision trees to demonstrate the binomial option pricing model. *Advances in Investment Analysis and Portfolio Management, 8,* 303–329.

Lee, C.-F., & Lin, C. S.-M. (2010). Two alternative binomial option pricing model approaches to derive Black–Scholes option pricing model. In *Handbook of quantitative finance and risk management* (pp. 409–419). Springer.

Lee, J. C., Lee, C. F., & Wei, K. J. (1991). Binomial option pricing with stochastic parameters: A beta distribution approach. *Review of Quantitative Finance and Accounting, 1*(4), 435–448.

Lee, C. F., Wu, T. P., & Chen, R. R. (2004). The constant elasticity of variance models: New evidence from S&P 500 index options. *Review of Pacific Basin Financial Markets and Policies, 7*(2), 173–190.

Lee, C.-F., Finnerty, J., Lee, J., Lee, A. C., & Wort, D. (2013a). *Security analysis, portfolio management, and financial derivatives*. World Scientific.

Lee, C.-F., Lee, J., & Lee, A. C. (2013b). *Statistics for business and financial economics* (3rd ed.). Springer.

Lee, C. F., Chen, Y., & Lee, J. (2016). Alternative methods to derive option pricing models: Review and comparison. *Review of Quantitative Finance and Accounting, 47*(2), 417–451.

Lin, C. H., Lin, S. K., & Wu, A. C. (2014). Foreign exchange option pricing in the currency cycle with jump risks. *Review of Quantitative Finance and Accounting*, 1–35.

Madan, D. B., Carr, P. P., & Chang, E. C. (1998). The variance gamma process and option pricing. *European Finance Review, 2*(1), 79–105.

Marcus, A. J., & Shaked, I. (1984). The valuation of FDIC deposit insurance using option-pricing estimates. *Journal of Money, Credit and Banking*, 446–460.

Melino, A., & Turnbull, S. M. (1990). Pricing foreign currency options with stochastic volatility. *Journal of Econometrics, 45*(1), 239–265.

Melino, A., & Turnbull, S. M. (1995). Misspecification and the pricing and hedging of long-term foreign currency options. *Journal of International Money and Finance, 14*(3), 373–393.

Merton, R. C. (1973). Theory of rational option pricing. *The Bell Journal of Economics and Management Science, 4*(1), 141–183.

Merton, R. C. (1974). On the pricing of corporate debt: The risk structure of interest rates. *The Journal of Finance, 29*(2), 449–470.

Merton, R. C. (1977). An analytic derivation of the cost of deposit insurance and loan guarantees: An application of modern option pricing theory. *Journal of Banking & Finance, 1*(1), 3–11.

Merton, R. C. (1978). On the cost of deposit insurance when there are surveillance costs. *Journal of Business*, 439–452.

Psychoyios, D., Dotsis, G., & Markellos, R. N. (2010). A jump diffusion model for VIX volatility options and futures. *Review of Quantitative Finance and Accounting, 35*(3), 245–269.

Rendleman, R. J., & Bartter, B. J. (1979). Two-state option pricing. *The Journal of Finance, 34*(5), 1093–1110.

Rubinstein, M. (1994). Implied binomial trees. *The Journal of Finance, 49*(3), 771–818.

Scott, L. O. (1987). Option pricing when the variance changes randomly: Theory, estimation, and an application. *Journal of Financial and Quantitative Analysis, 22*(4), 419–438.

Scott, L. O. (1997). Pricing stock options in a jump-diffusion model with stochastic volatility and interest rates: Applications of Fourier inversion methods. *Mathematical Finance, 7*(4), 413–426.

Smith, C. W., Jr. (1976). Option pricing: A review. *Journal of Financial Economics, 3*(1), 3–51.

Stein, E. M., & Stein, J. C. (1991). Stock price distributions with stochastic volatility: An analytic approach. *Review of Financial Studies, 4*(4), 727–752.

Wiggins, J. B. (1987). Option values under stochastic volatility: Theory and empirical estimates. *Journal of Financial Economics, 19*(2), 351–372.

Williams, J. T. (1991). Real estate development as an option. *The Journal of Real Estate Finance and Economics, 4*(2), 191–208.

Wu, C. C. (2006). The GARCH option pricing model: A modification of lattice approach. *Review of Quantitative Finance and Accounting, 26*(1), 55–66.

Constant Elasticity of Variance Option Pricing Model: Integration and Detailed Derivation

22

Contents

Abstract

In this chapter, we review the renowned constant elasticity of variance (CEV) option pricing model and the detailed derivations. We first show the details of the formulae needed in deriving the option pricing and bridge the gaps in deriving the necessary formulae for the model. Second, we use a result by Feller to obtain the transition probability density function of the stock price at time T given its price at time t with $t < T$. In addition, some computational considerations are given which will facilitate the computation of the CEV option pricing formula.

22.1 Introduction

Cox (1975) has derived the renowned constant elasticity of variance (CEV) option pricing model, and Schroder (1989) has subsequently extended the model by expressing the CEV option pricing formula in terms of the noncentral

This chapter draws upon the paper by Hsu, Lin, and Lee which was published as Chap. 31 of Handbook of Quantitative Finance and Risk Management Edited by Lee et al. (2010).

C.-F. Lee et al., *Financial Econometrics, Mathematics and Statistics*,
https://doi.org/10.1007/978-1-4939-9429-8_22

chi-square distribution. However, neither of them has given details of their derivations as well as the mathematical and statistical tools in deriving the formulae.

In Sect. 22.2, we will discuss the CEV diffusion and its transition probability density function. Section 22.3 will discuss noncentral chi-square distribution. Section 22.4 will demonstrate the noncentral chi-square approach to option pricing model. Finally, Sect. 22.5 concludes. In addition, Appendix will provide proof of Feller's lemma.

22.2 The CEV Diffusion and Its Transition Probability Density Function

The CEV option pricing model assumes that the stock price is governed by the diffusion process

$$dS = \mu S dt + \sigma S^{\frac{\beta}{2}} dZ, \quad \beta < 2, \tag{22.1}$$

where dZ is a Wiener process and σ is a positive constant. The elasticity is $\beta - 2$ since the return variance. $\upsilon(S,t) = \sigma^2 S^{\beta-2}$ with respect to price S has the following relationship

$$\frac{d\upsilon(S,t)/dS}{\upsilon(S,t)/S} = \beta - 2$$

which implies that $d\upsilon(S,t)/\upsilon(S,t) = \beta - 2dS/S$. Upon integration on both sides, we have $\log \upsilon(S,t) = (\beta-2)\log S + \log \sigma^2$, or $\upsilon(S,t) = \upsilon(S,t) = \sigma^2 S^{\beta-2}$.

If $\beta = 2$, then the elasticity is zero and the stock prices are lognormally distributed as in the Black and Scholes model. If $\beta = 1$, then Eq. (22.1) is the model proposed by Cox and Ross (1976).

In this article, we will focus on the case of $\beta < 2$ since many empirical evidences (see Campbell 1987; Glosten et al. 1993; Brandt and Kang 2004) have shown that the relationship between the stock price and its return volatility is negative. The transition density for $\beta > 2$ is given by Emanuel and Macbeth (1982) and the corresponding CEV option pricing formula can be derived through a similar strategy. For more details, see Chen and Lee (1993).

In order to derive the CEV option pricing model, we need the transition probability density function $f(S_T|S_t, T > t)$ of the stock price at time T, given the current stock price S_t. For the transition probability density function $f(S_T|S_t)$, we will start with the Kolmogorov forward and backward equations.

Assume X_t follows the diffusion process

$$dX = \mu(X,t)dt + \sigma(X,t)dZ \tag{22.2}$$

and $P = P(X_t, t)$ is the function of X_t and t, then P satisfies the partial differential equations of motion. From Eq. (22.2), we have the Kolmogorov backward equation,

$$\frac{1}{2}\sigma^2(X_0,t_0)\frac{\partial^2 P}{\partial X_0^2} + \mu(X_0,t_0)\frac{\partial P}{\partial X_0} + \frac{\partial P}{\partial t_0} = 0 \tag{22.3}$$

and the Kolmogorov forward (or Fokker–Planck) equation

$$\frac{1}{2}\frac{\partial^2}{\partial X_t^2}\left[\sigma^2(X,t)P\right] - \frac{\partial P}{\partial X_t}\left[\mu(X_t,t)P\right] - \frac{\partial P}{\partial t} = 0. \tag{22.4}$$

Consider the following parabolic equation

$$(P)_t = (axP)_{xx} - ((bx+h)P)_x, \quad 0 < x < \infty, \tag{22.5}$$

where $P = P(x,t)$ and are constants with $a > 0$, $(P)_t$ is the partial derivative of P with respect to t, $()_x$ and $()_{xx}$ are the first and second partial derivatives of () with respect to x. This can be interpreted as the Fokker–Planck equation of a diffusion problem in which $bx + h$ represents the drift, and a x represents the diffusion coefficient.

Lemma 22.1. (Feller (1951)) *Let $xf(x,t|x_0)$ be the probability density function for x and t conditional on x_0. The explicit form of the fundamental solution to the above parabolic equation is given by*

$$f(t,x|x_0)=\frac{b}{a(\mathrm{e}^{-bt}-1)}\left(\frac{\mathrm{e}^{-bt}x}{x_0}\right)^{(h-a)/2a}\exp\\ \times\left\{\frac{-b\left(x+x_0\mathrm{e}^{bt}\right)}{a(\mathrm{e}^{bt}-1)}\right\}\\ \times I_{1-h/a}\left(\frac{2b}{a(1-\mathrm{e}^{-bt})}\left(\mathrm{e}^{-bt}xx_0\right)^{\frac{1}{2}}\right),\tag{22.6}$$

where $I_x(x)$ is the modified Bessel function of the first kind of order k and is defined as

$$I_k(x)=\sum_{r=0}^{\infty}\frac{(x/2)^{2r+k}}{r!\Gamma(r+1+k)}.\tag{22.7}$$

Proof See Appendix.

Before pursuing further, we will first consider the special case in which $\beta=1$ which is the model considered by Cox and Ross (1976). In this situation, we have

$$\mathrm{d}S=\mu(S,t)\mathrm{d}t+\sigma(S,t)\mathrm{d}Z,\tag{22.8}$$

where $\sigma(S,t)=\sigma\sqrt{S}$.

Now suppose also that each unit of the stock pays out in dividends in the continuous stream $b(S,t)$ so that the required mean becomes $\mu(S,t)=rS-b(S,t)=rS-(aS+h)$, where $b(S,t)=aS+h$ and r is the risk-free interest rate. Then, $\mathrm{d}S=[(r-a)S-h]\mathrm{d}t+\sigma\sqrt{S}\mathrm{d}z$ and the differential option price equation becomes

$$\frac{1}{2}\sigma^2S\frac{\partial^2P}{\partial S^2}+[(r-a)S-h]\frac{\partial P}{\partial S}+\frac{\partial P}{\partial t}=rP,\tag{22.9}$$

and the corresponding Kolmogorov forward equation for the diffusion process (Eq. 22.8) is

$$\frac{1}{2}\frac{\partial^2}{\partial S_t^2}(\sigma^2S_TP)+\frac{\partial}{\partial S_T}[((r-a)S-h)P]-\frac{\partial P}{\partial t}=0,\tag{22.10}$$

which is obtained by using (22.4) with $\mu(x_t,t)=(r-a)S-h$.

Comparing with Eq. (22.6), we set $a=\sigma^2/2$, $x=S_T, x_0=S_t$, $b=r-\sigma^2/2, h=-h$, and $t=\tau=(T-t)$. Thus, we have the following transition probability density function for the Cox–Ross model:

$$f(S_T|S_t,T>t)\\ =\frac{2(r-\sigma^2/2)}{\sigma^2[\mathrm{e}^{(r-\sigma^2/2)\tau}-1]}\times\left(\frac{S_t\mathrm{e}^{(r-\sigma^2/2)\tau}}{S_T}\right)^{\left(1+2h/\sigma^2\right)/2}\\ \times\left\{\frac{-2(r-\sigma^2/2)\left[S_T+S_t\mathrm{e}^{(r-\sigma^2/2)\tau}\right]}{\sigma^2[\mathrm{e}^{(r-\sigma^2/2)\tau}-1]}\right\}\\ \times I_{1+2h/\sigma^2}\\ \times\left(\frac{4(r-\sigma^2/2)\left(S_tS_T\mathrm{e}^{(r-\sigma^2/2)\tau}\right)^{\frac{1}{2}}}{\sigma^2[\mathrm{e}^{(r-\sigma^2/2)\tau}-1]\sigma^2}\right).\tag{22.11}$$

We next consider the constant elasticity of variance diffusion,

$$\mathrm{d}S=\mu(S,t)+\sigma(S,t)\mathrm{d}Z,\tag{22.12}$$

where

$$\mu(S,t)=rS-aS,\tag{22.13}$$

and

$$\sigma(S,t)=\sigma S^{\beta/2}\mathrm{d}Z,\quad 0\le\beta<2.\tag{22.14}$$

Then,

$$\mathrm{d}S=(r-a)S\mathrm{d}t+\sigma S^{\frac{\beta}{2}}\mathrm{d}Z.\tag{22.15}$$

Let $Y=Y(S,t)=S^{2-\beta}$. By Ito's lemma with

$$\frac{\partial Y}{\partial S}=(2-\beta)S^{1-\beta},\ \frac{\partial Y}{\partial t}=0,\ \frac{\partial^2Y}{\partial S^2}\\ =(2-\beta)(1-\beta)S^{-\beta},$$

we have

$$\mathrm{d}Y = \left[(r-a)(2-\beta)Y - \frac{1}{2}\sigma^2(\beta-1)(\beta-2)\right]\mathrm{d}t + \sigma^2(2-\beta)^2 Y\mathrm{d}Z. \tag{22.16}$$

The Kolmogorov forward equation for Y becomes

$$\frac{\partial P}{\partial t} = \frac{1}{2}\frac{\partial^2}{\partial Y^2}\left[\sigma^2(2-\beta)YP\right] - \frac{\partial}{\partial Y}\left\{ \begin{array}{c} [(r-a)(2-\beta)Y \\ + \frac{1}{2}\sigma^2(\beta-1)(\beta-2)\Big]P \end{array} \right\}. \tag{22.17}$$

Then, $f(S_T|S_t, T>t) = f(Y_T|y_t, T>t)|J|$ where $J = (2-\beta)S^{1-\beta}$. By Feller's lemma with $a = \frac{1}{2}\sigma^2(2-\beta)^2, b = (r-a)(2-\beta),\ h = \frac{1}{2}\sigma^2(\beta-2)(1-\beta), x = \frac{1}{T}, x_0 = 1$ and $t = \tau = (T-t)$, we have

$$f(S_T|S_t, T>t) = (2-\beta)k^{*1/(2-\beta)}\left(xz^{1-\beta}\right)^{1/(2(2-\beta))} \times \mathrm{e}^{-x-z} I_{1/(2-\beta)}\left(2(xz)^{1/2}\right), \tag{22.18}$$

where

$$k^* = \frac{2(r-a)}{\sigma^2(2-\beta)[\mathrm{e}^{(r-a)(2-\beta)\tau}-1]},$$
$$x = k^* S_t^{2-\beta}\mathrm{e}^{(r-a)(2-\beta)\tau},\quad z = k^* S_T^{2-\beta}$$

Cox (1975) obtained the following option pricing formula:

$$C = S_t\mathrm{e}^{-r\tau}\sum_{n=0}^{\infty}\frac{\mathrm{e}^{-x}x^n G(n+1+1/(2-\beta), k^*K^{2-\beta})}{\Gamma(n+1)} - K\mathrm{e}^{-r\tau}\sum_{n=0}^{\infty}\frac{\mathrm{e}^{-x}x^{n+1/(2-\beta)}G(n+1, k^*K^{2-\beta})}{\Gamma(n+1+1/(2-\beta))}, \tag{22.19}$$

where $G(m,v) = [\Gamma(m)]^{-1}\int_v^\infty \mathrm{e}^{-u}u^{m-1}\mathrm{d}u$ is the standard complementary gamma distribution function.

For a proof of the above formula, see Chen and Lee (1993). We next present the detailed derivations of the option pricing formula as presented by Schroder (1989). Since the option pricing formula is expressed in terms of the noncentral chi-square complementary distribution function, a brief review of the noncentral chi-square distribution is presented in the next section.

22.3 Review of Noncentral Chi-Square Distribution

If $Z_1\ldots, Z_v$ are standard normal random variables, and $\delta_1\ldots, \delta_v$ are constants, then

$$Y = \sum_{i=1}^{v}(Z_i+\delta_i)^2 \tag{22.20}$$

is the noncentral chi-square distribution with degrees of freedom and noncentrality parameter $\lambda = \sum_{j=1}^{v}\delta_j^2$ and is denoted as $\chi_v'^2(\lambda)$. When $\delta_j = 0$ for all j, then Y is distributed as the central chi-square distribution with v degrees of freedom and is denoted as χ_v^2. The cumulative distribution function of $\chi_v'^2(\lambda)$ is

$$F(x;v,\lambda) = P\left(\chi_v'^2(\lambda)\le x\right) = \mathrm{e}^{-\lambda/2}\sum_{j=0}^{\infty}\frac{(\lambda/2)j}{j!2^{v/2}\Gamma(v/2+j)} \times \int_0^x y^{v/2+j-1}\mathrm{e}^{-\frac{y}{2}}\mathrm{d}y,\ x>0. \tag{22.21}$$

An alternative expression for $F(x;v,\lambda)$ is

$$F(x;v,\lambda) = \sum_{j=0}^{\infty}\left(\frac{(\lambda/2)^j\mathrm{e}^{-\lambda/2}}{j!}\right)P\left(\chi_{v+2j}^2\le x\right). \tag{22.22}$$

The complementary distribution function of $\chi_v'^2(\lambda)$ is

$$Q(x;v,\lambda) = 1 - F(x;v,\lambda), \qquad (22.23)$$

where $F(x;v,\lambda)$ is given in either Eqs. (22.21) or (22.22).

The probability density function of $\chi_v'^2(\lambda)$ can be expressed as a mixture of central chi-square probability density functions.

$$P_{\chi_v'^2(\lambda)}(x) = \mathrm{e}^{-\lambda/2}\sum_{j=0}^{\infty}\frac{\left(\frac{1}{2}\lambda\right)^j}{j!}p_{\chi^2_{v+2j}}(x) = \frac{\mathrm{e}^{-(x+\lambda)/2}}{2^{v/2}}\sum_{j=0}^{\infty}\frac{x^{v/2+j-1}\lambda^j}{\Gamma(v/2+j)2^{2j}j!}. \qquad (22.24)$$

An alternative expression for the probability density function of $\chi_v'^2(\lambda)$ is

$$P_{\chi_v'^2(\lambda)}(x) = \frac{1}{2}\left(\frac{x}{\lambda}\right)^{(v-2)/4}\exp\left\{-\frac{1}{2}(\lambda+x)\right\} \times I_{(v-2)/2}\left(\sqrt{\lambda x}\right),\ x > 0, \qquad (22.25)$$

where I_k is the modified Bessel function of the first kind of order k and is defined as

$$I_k(z) = \left(\frac{1}{2}z\right)^k\sum_{j=0}^{\infty}\frac{(z^2/4)^j}{j!\Gamma(k+j+1)}. \qquad (22.26)$$

It is noted that for integer k,

$$I_k(z) = \frac{1}{\pi}\int_0^{\pi}\mathrm{e}^{z\cos(k\theta)}\cos(k\theta)\,\mathrm{d}\theta = I_{-k}(z). \qquad (22.27)$$

The noncentral chi-square distribution satisfies the reproductivity property with respect to n and λ. If $X_1,\ldots,X_k$ are independent random variables with X_t distributed as $\chi_{n_i}'^2(\lambda_i)$ then

$$Y = \sum_{i=1}^{k}X_i \sim \chi'^2_{\sum_{i=1}^{k}n_i}\left(\sum_{i=1}^{k}\lambda_i\right). \qquad (22.28)$$

22.4 The Noncentral Chi-Square Approach to Option Pricing Model

Following Schroder (1989), with the transition probability density function given in (22.18), the option pricing formula under the CEV model is

$$\begin{aligned}C &= E(\max(0, S_T - K)) \quad \tau = T - t\\ &= \mathrm{e}^{-r\tau}\int_K^{\infty} f(S_T|S_t, T > t)(S_T - K)\mathrm{d}S_T\\ &= \mathrm{e}^{-r\tau}\int_K^{\infty} S_T f(S_T|S_t, T > t)\mathrm{d}S_T \qquad (22.29)\\ &\quad - \mathrm{e}^{-r\tau}K\int_K^{\infty} f(S_T|S_t, T > t)\mathrm{d}S_T\\ &= C_1 - C_2.\end{aligned}$$

22.4.1 Detailed Derivations of C_1 and C_2

Making the change of variable $w = k^* S_T^{2-\beta}$, we have

$$\mathrm{d}S_T = (2-\beta)^{-1}k^{*-1/(2-\beta)}w^{(\beta-1)/(2-\beta)}\mathrm{d}w.$$

Thus, with $y = k^* K^{2-\beta}$, we have,

$$
\begin{aligned}
C_1 &= \mathrm{e}^{-r\tau} \int_y^{\infty} \mathrm{e}^{-x-w} (x/w)^{1/(4-2\beta)} \\
&\quad \times I_{\frac{1}{2-\beta}}(2\sqrt{xw})(w/k^*)^{1/(2-\beta)} \mathrm{d}w \\
&= \mathrm{e}^{-r\tau} \int_y^{\infty} \mathrm{e}^{-x-w} (x/w)^{1/(4-2\beta)} \\
&\quad \times I_{\frac{1}{2-\beta}}(2\sqrt{xw})(w/x)^{1/(2-\beta)}(x/k^*)^{1/(2-\beta)} \mathrm{d}w \\
&= \mathrm{e}^{-r\tau}(x/k^*)^{1/(2-\beta)} \int_y^{\infty} \mathrm{e}^{-x-w} (x/w)^{-1/(4-2\beta)} \\
&\quad \times I_{\frac{1}{2-\beta}}(2\sqrt{xw}) \mathrm{d}w \\
&= \mathrm{e}^{-r\tau} S_t \mathrm{e}^{(r-a)\tau} \int_y^{\infty} \mathrm{e}^{-x-w} (w/x)^{1/(4-2\beta)} \\
&\quad \times I_{\frac{1}{2-\beta}}(2\sqrt{xw}) \mathrm{d}w, \\
&= \mathrm{e}^{-a\tau} S_t \int_y^{\infty} \mathrm{e}^{-x-w} (w/x)^{1/(4-2\beta)} \\
&\quad \times I_{\frac{1}{2-\beta}}(2\sqrt{xw}) \mathrm{d}w,
\end{aligned}
\tag{22.30}
$$

and

$$
\begin{aligned}
C_2 &= K\mathrm{e}^{-r\tau} \int_y^{\infty} (2-\beta) k^{*1/(2-\beta)} \left(xw^{1-2\beta}\right)^{\frac{1}{4-2\beta}} \mathrm{e}^{-x-w} \\
&\quad \times I_{\frac{1}{2-\beta}}(2\sqrt{xw}) \frac{k^{*-1/(2-\beta)}}{2-\beta} w^{\frac{\beta-1}{2-\beta}} \mathrm{d}w \\
&= K\mathrm{e}^{-r\tau} \int_y^{\infty} x^{\frac{1}{4-2\beta}} w^{(1-2\beta+2\beta-2)/(4-2\beta)} \mathrm{e}^{-x-w} \\
&\quad \times I_{\frac{1}{2-\beta}}(2\sqrt{xw}) \mathrm{d}w \\
&= K\mathrm{e}^{-r\tau} \int_y^{\infty} \mathrm{e}^{-x-w} (x/w)^{1/(4-\beta)} I_{\frac{1}{2-\beta}}(2\sqrt{xw}) \mathrm{d}w.
\end{aligned}
\tag{22.31}
$$

Recall that the probability density function of the noncentral chi-square distribution with noncentrality λ and degree of freedom υ is

$$
\begin{aligned}
P_{\chi_\upsilon'^2(\lambda)}(x) &= \frac{1}{2}(x/\lambda)^{(\upsilon-2)/4} I_{(\upsilon-2)/2}\left(\sqrt{\lambda x}\right) \mathrm{e}^{-(\lambda+x)/2} \\
&= P(x;\upsilon,\lambda).
\end{aligned}
$$

Let $Q(x;\upsilon,\lambda) = \int_x^{\infty} P_{\chi_\upsilon'^2(\lambda)}(y)\,\mathrm{d}y$. Then, letting $w' = 2w$ and $x' = 2x$, we have

$$
\begin{aligned}
C_1 &= S_t \mathrm{e}^{-a\tau} \int_y^{\infty} \mathrm{e}^{-(x+w)/2} \left(\frac{w}{x}\right)^{1/(4-2\beta)} \\
&\quad \times I_{\frac{1}{2-\beta}}(2\sqrt{xw}) \mathrm{d}w \\
&= S_t \mathrm{e}^{-a\tau} \int_{2y}^{\infty} \mathrm{e}^{-(x'+w')/2} \left(\frac{w'}{x'}\right)^{1/(4-2\beta)} \\
&\quad \times I_{\frac{1}{2-\beta}}\left(2\sqrt{x'w'}\right) \mathrm{d}w' \\
&= S_t \mathrm{e}^{-a\tau} Q(2y;\upsilon,x'), \\
&= S_t \mathrm{e}^{-a\tau} Q\left(2y; 2+\frac{2}{2-\beta}, 2x\right),
\end{aligned}
\tag{22.32}
$$

obtained by noting that $(\upsilon-2)/2, 1/(2-\beta)$, implying $\upsilon = 2+2/(2-\beta)$. Analogously, with $w' = 2w$, $x' = 2x$, and $I_n(z) = I_{-n}(z)$, we have

$$
\begin{aligned}
C_2 &= K\mathrm{e}^{-r\tau} \int_y^{\infty} \mathrm{e}^{-x-w} \left(\frac{x}{w}\right)^{1/(4-2\beta)} I_{\frac{1}{2-\beta}}(2\sqrt{xw}) \mathrm{d}w \\
&= K\mathrm{e}^{-r\tau} \int_{2y}^{\infty} \mathrm{e}^{-(x'+w')/2} \left(\frac{x'}{w'}\right)^{1/(4-2\beta)} \\
&\quad \times I_{\frac{1}{2-\beta}}\left(2\sqrt{x'w'}\right) \mathrm{d}w' \\
&= Q\left(2y; 2-\frac{2}{2-\beta}, 2x\right),
\end{aligned}
\tag{22.33}
$$

obtained by noting that $(\upsilon^*-2)/2 = -1/(2-\beta)$, implying $\upsilon^* = 2-2/(2-\beta)$. Thus,

$$
\begin{aligned}
C &= S_t \mathrm{e}^{-a\tau} Q\left(2y; 2+\frac{2}{2-\beta}, 2x\right) \\
&\quad - K\mathrm{e}^{-r\tau} Q\left(2y; 2+\frac{2}{2-\beta}, 2x\right).
\end{aligned}
\tag{22.34}
$$

It is noted that $2 - 2/(2 - \beta)$ can be negative for $\beta < 2$. Thus, further work is needed. Using the monotone convergence theorem and the integration by parts, we have

$$\begin{aligned}
\int_y^\infty P(2y, 2\upsilon, 2k)\,\mathrm{d}k &= \int_y^\infty \mathrm{e}^{-z-k}(z/k)^{\upsilon-1}\sqrt{kz}^{\upsilon-1} \\
&\quad \times \sum_{n=0}^\infty \frac{(zk)^n}{n!\Gamma(n+\upsilon-1+1)}\mathrm{d}k \\
&= \int_y^\infty \frac{\mathrm{e}^{-z}z^{n+\upsilon-1}}{\Gamma(n+\upsilon)} \int_y^\infty \frac{\mathrm{e}^{-k}k^n}{\Gamma(n+1)}\mathrm{d}k \\
&= \sum_{n=0}^\infty g(n+\upsilon, z)G(n+1, y) \\
&= \sum_{n=0}^\infty g(n+\upsilon-1, z)\sum_{i=1}^\infty g(i, y).
\end{aligned} \tag{22.35}$$

Now we also have the result $G(n, y) = \sum_{i=1}^\infty g(i, y)$, which can be shown by observing that

$$\begin{aligned}
G(n, y) &= \int_y^\infty \frac{e^{-k}k^{n-1}}{\Gamma(n)}\mathrm{d}k = -\int_y^\infty \frac{k^{n-1}}{\Gamma(n)}\mathrm{d}\mathrm{e}^{-k} \\
&= \frac{y^{n-1}\mathrm{e}^{-y}}{\Gamma(n)} + \int_y^\infty \frac{k^{n-2}\mathrm{e}^{-k}}{\Gamma(n-1)}\mathrm{d}k \\
&= \sum_{i=1}^n \frac{y^{i-1}\mathrm{e}^{-y}}{\Gamma(i)} = \sum_{i=1}^n g(i, y).
\end{aligned}$$

The above result can also be expressed as

$$G(m+1, t) = G(m+1, t) + G(m, t). \tag{22.36}$$

Next, applying the monotone convergence theorem, we have

$$\begin{aligned}
Q(z; \upsilon, k) &= \int_z^\infty \frac{1}{2}\left(\frac{y}{k}\right)^{(\upsilon-2)/4} I_{\frac{\upsilon-2}{2}}\left(\sqrt{ky}\right)\mathrm{e}^{-(k+y)/2}\mathrm{d}y \\
&= \int_z^\infty \frac{1}{2}\left(\frac{y}{k}\right)^{(\upsilon-2)/4}\left(\frac{1}{2}\sqrt{ky}\right)^{(\upsilon-2)/2} \\
&\quad \times \sum_{n=0}^\infty \frac{(ky/4)^n}{n!\Gamma\left(\frac{\upsilon+2n}{2}\right)}\mathrm{e}^{-(k+y)/2}\mathrm{d}y \\
&= \sum_{n=0}^\infty \mathrm{e}^{-k/2}\frac{k^n\left(\frac{1}{2}\right)^n}{\Gamma(n+1)} \\
&\quad \times \int_z^\infty \frac{(1/2)^{(\upsilon+2n)/2}}{\left(\frac{1}{2}\right)^{-(\upsilon+2n)/2}\Gamma\left(\frac{\upsilon+2n}{2}\right)}\mathrm{d}y \\
&= \sum_{n=0}^\infty \mathrm{e}^{-k/2}\frac{(k/2)^n}{\Gamma(n+1)} \\
&\quad \times \int_z^\infty \frac{(1/2)^{(\upsilon+2n)/2}}{\Gamma\left(\frac{\upsilon-2n}{2}\right)}\mathrm{e}^{-y/2}y^{\frac{\upsilon+2n}{2}-1}\mathrm{d}y \\
&= \sum_{n=0}^\infty \mathrm{e}^{-k/2}\frac{(k/2)^n}{\Gamma(n+1)}Q(z; \upsilon+2n, 0),
\end{aligned} \tag{22.37}$$

where

$$\begin{aligned}
Q(x; \upsilon+2n, 0) &= \int_z^\infty \frac{(1/2)^{(\upsilon+2n)/2}}{\Gamma\left(\frac{\upsilon+2n}{2}\right)}\mathrm{e}^{-y/2}y^{\frac{\upsilon+2n}{2}-1}\mathrm{d}y \\
&= \int_{z/2}^\infty \frac{1}{\Gamma\left(\frac{\upsilon+2n}{2}\right)}\mathrm{e}^{-y}y^{\frac{\upsilon+2n}{2}-1}\mathrm{d}y \\
&= G(n+\upsilon/2, z/2).
\end{aligned}$$

Furthermore, from the property of $G(\cdot,\cdot)$ as shown in Eq. (22.36), we have

$$Q(z;\upsilon,k)=\sum_{n=0}^{\infty}g(n+1,k/2)G(n+\upsilon/2,z/2)$$
$$=\sum_{n=0}^{\infty}g(n,k/2)G\left(n+\frac{\upsilon-2}{2},z/2\right). \tag{22.38}$$

Hence,

$$Q(2z;2\upsilon,2k)=\sum_{n=0}^{\infty}g(n,k)G(n+\upsilon-1,z).$$

Again from the property of $G(\cdot,\cdot)$ as given by (22.36), we have

$$\begin{aligned}
&Q(2z;2\upsilon,2k)\\
&\quad=g(1,k)G(\upsilon,z)+g(2,k)G(\upsilon+1,z)\\
&\qquad+g(3,k)G(\upsilon+2,z)+\cdots\\
&\quad=g(1,k)[G(\upsilon-1,z)+g(\upsilon,z)]\\
&\qquad+g(2,k)[G(\upsilon-1,z)+g(\upsilon+1,z)]\\
&\qquad+g(3,k)[G(\upsilon-1,z)+g(\upsilon,z)\\
&\qquad+g(\upsilon+1,z)+g(\upsilon+2,z)]+\cdots\\
&\quad=[G(\upsilon-1,z)+g(\upsilon,z)]\sum_{n=1}^{\infty}g(n,k)\\
&\qquad+g(\upsilon+1,z)\sum_{n=2}^{\infty}g(n,k)\\
&\qquad+g(\upsilon+2,z)\sum_{n=3}^{\infty}g(n,k)+\cdots\\
&\quad=G(\upsilon-1,z)+g(\upsilon,z)\\
&\qquad+g(\upsilon+1,z)[1-g(1,k)]\\
&\qquad+g(\upsilon+2,z)[1-g(1,k)-g(2,k)]+\cdots\\
&\quad=G(\upsilon-1,z)+\sum_{n=0}^{\infty}g(\upsilon+n,z)\\
&\qquad-g(\upsilon+1,z)[g(1,k)]\\
&\qquad-g(\upsilon+2,z)[g(1,k)+g(2,k)]+\cdots\\
&\quad=1-g(\upsilon+1,z)[g(1,k)]\\
&\qquad-g(\upsilon+2,z)[g(1,k)+g(2,k)]-\cdots
\end{aligned}$$

We conclude that

$$Q(2z;2\upsilon,2k)=1-\sum_{n=1}^{\infty}g(n+\upsilon,z)\sum_{i=1}^{n}g(i,k). \tag{22.39}$$

From (22.35) and (22.39) we observe that

$$\int_{y}^{\infty}P(2z;2\upsilon,2k)\mathrm{d}k=1-Q(2z;2(\upsilon-1),2y). \tag{22.40}$$

Thus, we can write C_2 as

$$\begin{aligned}
C_2&=K\mathrm{e}^{-r\tau}\int_{y}^{\infty}P\left(2x;2+\frac{2}{2-\beta},2w\right)\mathrm{d}w\\
&=K\mathrm{e}^{-r\tau}Q\left(2y;2-\frac{2}{2-\beta},2x\right)\\
&=K\mathrm{e}^{-r\tau}\left(1-Q\left(2x;\frac{2}{2-\beta},2y\right)\right).
\end{aligned} \tag{22.41}$$

From (22.41), we immediately obtain

$$Q\left(2y;2-\frac{2}{2-\beta},2x\right)+Q\left(2x;\frac{2}{2-\beta},2y\right)=1 \tag{22.42}$$

Implying

$$Q(z;2n,k)+Q(k;2-2n,z)=1 \tag{22.43}$$

with degrees of freedom $2-2n$ of $Q(k;2-2n,z)$ can be a noninteger.

From Eq. (22.42), we can obtain that the noncentral chi-square $Q(2y;2-2/(2-\beta),2x)$ with $2-2/(2-\beta)$ degrees of freedom and the noncentrality parameter $2x$ can be represented by another noncentral chi-square distribution $1-Q(2x;2-2/(2-\beta),2y)$ with degrees of freedom $2/(2-\beta)$ and the noncentrality parameter $2y$. The standard definition of noncentral chi-square distribution in Sect. 22.3 has integer degrees of freedom. If the degree of freedom is not an integer, we can use Eq. (22.43) to transfer the original noncentral chi-square distribution into another noncentral chi-square distribution. Thus, we obtain an option pricing formula for the CEV model in terms of the complementary noncentral chi-square distribution function $Q(z;\upsilon,k)$ which is valid for any value of β less than 2, as required by the model.

Substituting Eq. (22.41) into Eq. (22.34), we obtain

$$C = S_t e^{-a\tau} Q\left(2y; 2 + \frac{2}{2-\beta}, 2x\right) - Ke^{-r\tau}\left(1 - Q\left(2x; \frac{2}{2-\beta}, 2y\right)\right). \tag{22.44}$$

where $y = k^* K^{2-\beta}, x = k^* S_t^{2-\beta} e^{(r-a)(2-\beta)\tau}$, $k^* = 2(r-a)/\left(\sigma^2(2-\beta)\left(e^{(r-a)(2-\beta)\tau} - 1\right)\right)$, and a is the continuous proportional dividend rate. The corresponding CEV option pricing formula for $\beta > 2$ can be derived through a similar manner. When $\beta > 2$ (see, Emanuel and Macheth 1982; Chen and Lee 1993), the call option formula is as follows:

$$C = S_t e^{-a\tau} Q\left(2x; \frac{2}{\beta-2}, 2y\right) - Ke^{-r\tau}\left(1 - Q\left(2y; 2 + \frac{2}{\beta-2}, 2x\right)\right). \tag{22.45}$$

We note that from the evaluation of the option pricing formula C, especially C_2, as given in (22.34), we have

$$2k^* S_T^{2-\beta} \sim \chi_\upsilon^{\prime 2}(\lambda), \tag{22.46}$$

where

$$\upsilon = 2 - \frac{2}{2-\beta},$$
$$\lambda = 2k^* S_t^{2-\beta} e^{(r-a)(2-\beta)\tau},$$

Thus, the option pricing formula for the CEV model as given in (22.44) that can be obtained directly from the payoff function

$$\max(S_T - K, 0) = \begin{cases} S_T - K, & \text{if } S_T > K \\ 0, & \text{otherwise} \end{cases} \tag{22.47}$$

by taking the expectation of (22.47), with S_T having the distribution given by (22.46).

Before concluding this subsection, we consider that the noncentral chi-square distribution will approach lognormal as β tends to 2. Since when either λ or υ approaches to infinity, the standardized variable

$$\frac{\chi_\upsilon^{\prime 2}(\lambda) - (\upsilon + \lambda)}{\sqrt{2(\upsilon + 2\lambda)}}$$

tends to $N(0,1)$ as either $\upsilon \to \infty$ or $\lambda \to \infty$. Using the fact that $(x^a - 1)/a$ will approach to $\ln x$ as $a \to \infty$, it can be verified that

$$\begin{aligned}\frac{\chi_\upsilon^{\prime 2}(\lambda) - (\upsilon + \lambda)}{\sqrt{2(\upsilon + 2\lambda)}} &= \frac{2k^* S_T^{2-\beta} - (\upsilon + \lambda)}{\sqrt{2(\upsilon + 2\lambda)}} \\ &= \frac{2r^\bullet S_T^{2-\beta} - (1-\beta)\sigma^2\left(e^{r^\bullet\tau(2-\beta)} - 1\right) - 2r^\bullet S_t^{2-\beta} e^{r^\bullet\tau(2-\beta)}}{\sigma^2\left(e^{r^\bullet\tau(2-\beta)} - 1\right)} \\ &\quad \bullet \sqrt{\frac{\sigma^2\left(e^{r^\bullet\tau(2-\beta)} - 1\right)/(2-\beta)}{(1-\beta)\sigma^2\left(e^{r^\bullet\tau(2-\beta)} - 1\right) + 4r^\bullet S_t^{2-\beta} e^{r^\bullet\tau(2-\beta)}}} \\ &= \frac{\ln S_T - \left[\ln S_t + (r^\bullet - \sigma^2/2)\tau\right]}{\sigma\sqrt{\tau}},\end{aligned}$$

where $r^\bullet = r - a$. Thus,

$$\ln S_T | \ln S_t \sim N\left(\ln S_t + \left(r - a - \sigma^2/2\right)\tau, \sigma^2\tau\right) \tag{22.48}$$

as $\beta \to 2^-$. Similarly, Eq. (22.48) also holds when $\beta \to 2^+$. From Eq. (22.45), we have $2k^* S_T^{2-\beta} \sim \chi_\upsilon^{\prime 2}(\lambda)$, where $\upsilon = 2 + 2/(\beta - 2)$ if $\beta > 2$. Thus, we clarify the result of Eq. (22.48).

22.4.2 Some Computational Considerations

As noted by Schroder (1989), (22.39) allows the following iterative algorithm to be used in computing the infinite sum when z and k are not large. First, initialize the following four variables (with $n = 1$)

$$
\begin{aligned}
gA &= \frac{e^{-z} z^{\upsilon}}{\Gamma(1+\upsilon)} = g(1+\upsilon, z),\\
gB &= e^{-k} = g(1,k),\\
Sg &= gB,\\
R &= 1 - (gA)(Sg).
\end{aligned}
$$

Then, repeat the following loop beginning with $n = 2$ and increase increment n by one after each iteration. The loop is terminated when the contribution to the sum, R, is declining and is very small.

$$
\begin{aligned}
gA &= gA\left(\frac{z}{n+\upsilon-1}\right) = g(n+\upsilon, z),\\
gB &= gB\left(\frac{k}{n-1}\right) = g(n,k),\\
Sg &= Sg + gB = g(1,k) + g(n,k),\\
R &= R - (gA)(Sg) = \text{the } n\text{th partial sum.}
\end{aligned}
$$

As each iteration, gA equals $g(n+\upsilon, z), gB$ equals $g(n,k)$, and Sg equals $g(1,k) + \cdots + g(n,k)$. The computation is easily done.

As for an approximation, Sankaran (1963) showed that the distribution of $\left(\chi_{\upsilon}^{\prime 2}/(\upsilon+k)\right)^h$ is approximately normal with the expected value $\mu = 1 + h(h-1)P - h(2-h)mP^2/2$ and variance $\sigma^2 = h^2 P(1+mP)$, where $h = 1 - \frac{2}{3}(\upsilon+k)(\upsilon+3k)(\upsilon+2k)^{-2}$, $P = (\upsilon+2k)/(\upsilon+k)^2$, and $m = (h-1)(1-3h)$. Using the approximation, we have approximately

$$
\begin{aligned}
Q(z,\upsilon,k) &= Pr(\chi'^2 > z)\\
&= Pr\left(\frac{\chi'^2}{\upsilon+k} > \frac{z}{\upsilon+k}\right)\\
&= Pr\left(\left(\frac{\chi'^2}{\upsilon+k}\right)^h > \left(\frac{z}{\upsilon+k}\right)^h\right)\\
&= \Phi\left(\frac{1 - hP[1-h+0.5(2-h)mP] - \left(\frac{z}{\upsilon+k}\right)^h}{h\sqrt{2P(1+mP)}}\right).
\end{aligned}
$$

22.5 Conclusion

The option pricing formula under the CEV model is quite complex because it involves the cumulative distribution function of the noncentral chi-square distribution $Q(z,\upsilon,k)$. Some computational considerations are given in the article which will facilitate the computation of the CEV option pricing formula. Hence, the computation will not be a difficult problem in practice.

Appendix: Proof of Feller's Lemma

We need some preliminary results in order to prove Eq. (22.6).

Proposition 22.1. $f(z) = e^{A\upsilon/z} z^{-1}$ *is the Laplace transformation of* $I_0\left(2(A\upsilon x)^{1/2}\right)$, *where* $I_k(x)$ *is the Bessel function*

$$I_k(x) = \sum_{r=0}^{\infty} \frac{(x/2)^{2r+k}}{r!(r+1+k)}.$$

Proof. By the definition of the Laplace transformation and the monotone convergence theorem, we have

$$
\begin{aligned}
f(z) &= \int_0^{\infty} e^{-zx} I_0\left(2(A\upsilon x)^{1/2}\right) dx\\
&= \int_0^{\infty} e^{-zx} \sum_0^{\infty} \frac{(A\upsilon x)^{1/2}}{r!\Gamma(r+1)}\\
&= \int_0^{\infty} e^{-zx} \left\{1 + \frac{(A\upsilon x)}{\Gamma(2)} + \frac{(A\upsilon x)^2}{2!\Gamma(3)} + \cdots \right.\\
&\quad \left. + \frac{(A\upsilon x)^n}{n!\Gamma(n+1)} + \cdots \right.\\
&= \frac{1}{z} + \frac{A\upsilon}{z^2} + \frac{(A\upsilon)^2}{2!z^3} + \cdots + \frac{(A\upsilon)^n}{n!z^{n+1}} + \cdots\\
&= \frac{1}{z}\left\{1 + \frac{A\upsilon}{z} + \frac{(A\upsilon)^2}{2!z^2} + \cdots + \frac{(A\upsilon)^n}{n!z^n}\right\}\\
&= e^{A\upsilon/z} z^{-1}.
\end{aligned}
$$

Proposition 22.2. *Consider the parabolic differential equation*

$$P_t = (axP)_{xx} - ((bx+h)P)_x, \quad 0<x<\infty \tag{22.49}$$

where a, b, h are constants, $0<h<a$, then the Laplace transformation of $f(t,x,x_0)$ with respect to x takes the form

$$\begin{aligned} w(t,s;x_0) &= \int_0^\infty e^{-sx} f(t,x;x_0)\mathrm{d}x \\ &= \left(\frac{b}{sa(e^{bt}-1)+b}\right)^{h/a} \exp\left\{\frac{-sbx_0e^{bt}}{sa(e^{bt}-1)+b}\right\} \\ &\quad \times \Gamma\left(1-\frac{h}{a};\frac{b^2x_0e^{bt}}{a(e^{bt}-1)(sa(e^{bt}-1)+b)}\right), \end{aligned} \tag{22.50}$$

where $\Gamma(n;z) = \Gamma^{-1}(n)\int_0^z e^{-x}x^{n-1}\mathrm{d}x$.

Proof. The proof of the lemma is too tedious and hence is omitted. For more details, please see Lemma 7 of Feller (1951).

We now turn to prove Eq. (22.6). From Eq. (22.50), let $A = \frac{bx_0}{a(1-e^{-bt})}$ and $z = \frac{1}{b}\left(sa\left(e^{bt}-1\right)+b\right)$. The $w(t,s;x_0)$ in Eq. (22.50) can be rewritten as

$$\begin{aligned} w(t,s;x_0) &= \frac{z^{-h/a}e^{-(1-1/z)A}}{\Gamma(1-h/a)}\int_0^{A/z} e^{-x}x^{-h/a}\mathrm{d}x \\ &= \frac{z^{-h/a}e^{-(1-1/z)A}}{\Gamma(1-h/a)} \times \int_0^1 e^{-Ax'/z}(Ax'/z)^{-h/a} \\ &\quad \times (A/z)^{1-h/a}\mathrm{d}x' \\ &= \frac{A^{1-h/a}e^{-A}}{\Gamma(1-h/a)}\int_0^1 e^{A(1-x)'/z}(x')^{-h/a}z^{-1}\mathrm{d}x' \\ &= \frac{A^{1-h/a}e^{-A}}{\Gamma(1-h/a)}\int_0^1 e^{A\upsilon/z}(1-\upsilon)^{-h/a}z^{-1}\mathrm{d}\upsilon \end{aligned}$$

By Proposition 22.1, we know that $f(z) = e^{A\upsilon/z}z^{-1}$ is the Laplace transformation of $I_0\left(2(A\upsilon X)^{1/2}\right)$ and by the Fubini theorem, we have

$$\begin{aligned} &w(t,s;x_0) \\ &= \frac{A^{1-h/a}e^{-A}}{\Gamma(1-h/a)}\int_0^1 (1-\upsilon)^{-h/a} \\ &\quad \times \left[\int_0^\infty e^{-zx}I_0\left(2(A\upsilon x)^{1/2}\right)\mathrm{d}x\right]\mathrm{d}\upsilon \\ &= \frac{A^{1-h/a}e^{-A}}{\Gamma(1-h/a)} \\ &\quad \times \int_0^\infty\int_0^1 (1-\upsilon)^{-h/a}e^{-zx}I_0\left(2(A\upsilon x)^{1/2}\right)\mathrm{d}x\mathrm{d}\upsilon \\ &= \frac{A^{2-h/a}e^{-A}}{\Gamma(1-h/a)x_0e^{bt}} \\ &\quad \times \int_0^\infty\int_0^1 (1-\upsilon)^{-h/a}e^{-sx'}e^{-Ax'/\left(x_0e^{bt}\right)} \\ &\quad \times I_0\left(2A\left(e^{-bt}\upsilon x'/x_0\right)^{1/2}\right)\mathrm{d}\upsilon\mathrm{d}x'. \end{aligned}$$

Hence, upon comparing the two formulae for $w(t,s;x_0)$ and by the monotone convergence theorem, we have

$$\begin{aligned} &f(t,x;x_0) \\ &= \frac{b}{\Gamma(1-h/a)a(e^{bt}-1)}\left[\frac{bx_0}{a(e^{bt}-1)}\right]^{1-h/a} \\ &\quad \times \exp\left\{\frac{-b\left(x+x_0e^{bt}\right)}{a(e^{bt}-1)}\right\}\int_0^1 (1-\upsilon)^{-h/a}I_0 \\ &\quad \times \left(\frac{2b\left(e^{-bt}\upsilon xx_0\right)^{1/2}}{a(1-e^{-bt})}\right)\mathrm{d}\upsilon \\ &= \frac{b}{\Gamma(1-h/a)a(e^{bt}-1)}\left[\frac{bx_0}{a(e^{bt}-1)}\right]^{1-h/a} \\ &\quad \times \exp\left\{\frac{-b\left(x+x_0e^{bt}\right)}{a(e^{bt}-1)}\right\}\int_0^1 (1-\upsilon)^{-h/a} \end{aligned}$$

$$\times \sum_{r=0}^{\infty} \frac{\left[b\left(\mathrm{e}^{-bt}\upsilon x x_0\right)^{1/2}/\left(a\left(1-\mathrm{e}^{-bt}\right)\right)\right]^{2r}}{r!\Gamma(r+1)}$$

$$= \frac{b}{\Gamma(1-h/a)a(\mathrm{e}^{bt}-1)}\left[\frac{bx_0}{a(\mathrm{e}^{bt}-1)}\right]^{1-h/a} \times \exp\left\{\frac{-b\left(x+x_0\mathrm{e}^{bt}\right)}{a\left(\mathrm{e}^{bt}-1\right)}\right\} \times \sum_{r=0}^{\infty} \frac{\left[b\left(\mathrm{e}^{-bt}x x_0\right)^{1/2}/\left(a\left(1-\mathrm{e}^{-bt}\right)\right)\right]^{2r}}{r!\Gamma(r+1)} \times \int_0^1 (1-\upsilon)^{-h/a}\upsilon^r \mathrm{d}\upsilon$$

$$= \frac{b}{\Gamma(1-h/a)a(\mathrm{e}^{bt}-1)}\left[\frac{bx_0}{a(\mathrm{e}^{bt}-1)}\right]^{1-h/a} \times \exp\left\{\frac{-b\left(x+x_0\mathrm{e}^{bt}\right)}{a\left(\mathrm{e}^{bt}-1\right)}\right\} \times \sum_{r=0}^{\infty} \frac{\left[b\left(\mathrm{e}^{-bt}x x_0\right)^{1/2}/\left(a\left(1-\mathrm{e}^{-bt}\right)\right)\right]^{2r}}{r!\Gamma(r+1)} \times \frac{\Gamma(r+1)\Gamma(1-h/a)}{\Gamma(r+1+1-h/a)}.$$

This completes the proof.

Bibliography

Brandt, M. W., & Kang, Q. (2004). On the relationship between the conditional mean and volatility of stock returns: A latent VAR approach. *Journal of Financial Economics, 72,* 217–257.

Campbell, J. (1987). Stock returns and the term structure. *Journal of Financial Economics, 18,* 373–399.

Chen, R. R., & Lee, C. F. (1993). A constant elasticity of variance (CEV) family of stock price distributions in option pricing: Review and integration. *Journal of Financial Studies, 1,* 29–51.

Cox, J. (1975). *Notes on option pricing I: Constant elasticity of variance diffusion*. Standford University, Graduate School of Business (Unpublished note). Also, *Journal of Portfolio Management* (1996) 23, 5–17.

Cox, J., & Ross, S. A. (1976). The valuation of options for alternative stochastic processes. *Journal of Financial Economics, 3,* 145–166.

Emanuel, D., & MacBeth, J. (1982). Further results on the constant elasticity of variance call option pricing formula. *Journal of Financial and Quantitative Analysis, 17,* 533–554.

Feller, W. (1951). Two singular diffusion problems. *Annals of Mathematics, 54,* 173–182.

Glostern, L., Jagannathan, R., & Runkle, D. (1993). On the relation between the expected value and the volatility of the nominal excess returns on stocks. *Journal of Finance, 48,* 1779–1802.

Lee, C. F., Hsu, Y. L., & Lin, T. I. (2010). Constant elasticity of variance (CEV) option pricing model: Integration and detailed derivations. In *Handbook of quantitative finance and risk management* (pp. 471–478).

Lee, C.-F., Lee, A. C., & Lee, J. (2010b). *Handbook of quantitative finance and risk management*. New York: Springer.

Sankaran, M. (1963). Approximations to the non-central Chi-square distribution. *Biometrika, 50,* 199–204.

Schroder, M. (1989). Computing the constant elasticity of variance option pricing formula. *Journal of Finance, 44,* 211–219.

Option Pricing and Hedging Performance Under Stochastic Volatility and Stochastic Interest Rates

23

Contents

Abstract

Recent studies have extended the Black–Scholes model to incorporate either stochastic interest rates or stochastic volatility. However, there is not yet any comprehensive empirical study demonstrating whether and by how much each generalized feature will improve option pricing and hedging performance. This chapter fills this gap by first developing an implementable option model in closed form that admits both stochastic volatility and stochastic interest rates and that is parsimonious in the number of parameters. The model includes many known ones as special cases. Both delta-neutral and single-instrument minimum-variance hedging strategies are derived analytically. Using S&P 500 option prices, we then compare the pricing and hedging performance of this model with that of three existing ones that respectively allow for (i) constant volatility and constant interest rates (the Black–Scholes), (ii) constant

This chapter draws upon the paper by Bakshi et al. which was published as Chap. 37 of Handbook of Quantitative Finance and Risk Management edited by Lee et al. (2010).

C.-F. Lee et al., *Financial Econometrics, Mathematics and Statistics*,
https://doi.org/10.1007/978-1-4939-9429-8_23

volatility but stochastic interest rates, and (iii) stochastic volatility but constant interest rates. Overall, incorporating stochastic volatility and stochastic interest rates produces the best performance in pricing and hedging, with the remaining pricing and hedging errors no longer systematically related to contract features. The second performer in the horse race is the stochastic volatility model, followed by the stochastic interest rates model and then by the Black–Scholes.

23.1 Introduction

Option pricing has, in the last two decades, witnessed an explosion of new models that each relaxes some of the restrictive assumptions underlying the seminal Black–Scholes (1973) model. In doing so, most of the focus has been on the counterfactual constant volatility and constant interest rates assumptions. For example, Merton's (1973) option pricing model allows interest rates to be stochastic but keeps a constant volatility for the underlying asset, while Amin and Jarrow (1992) develop a similar model where, unlike in Merton's, interest rate risk is also priced. A second class of option models admits stochastic conditional volatility for the underlying asset, but maintains the constant interest rates assumption. These include the Cox and Ross (1976) constant elasticity of variance model and the stochastic volatility models of Bailey and Stulz (1989), Bates (1996b, 2000), Heston (1993), Hull and White (1987a), Scott (1987), Stein and Stein (1991), and Wiggins (1987). Recently, Bakshi and Chen (1997) and Scott (1997) have developed closed-form equity option formulas that admit both stochastic volatility and stochastic interest rates.[1] Their efforts have, in some sense, helped reach the ultimate possibility of completely relaxing the Black–Scholes assumptions of constant volatility and constant interest rates. As a practical matter, these sufficiently general pricing formulas should in principle result in significant improvement in pricing and hedging performance over the Black–Scholes model. While option pricing theory has made such impressive progress, the empirical front is nonetheless far behind.[2] Will incorporating these general features improve both pricing and hedging effectiveness? If so, by how much? Can these relaxed assumptions help resolve the well-known empirical biases associated with the Black–Scholes formula, such as the volatility smiles [e.g., Rubinstein (1985, 1994)]? —These empirical questions must be answered before the potential of the general models can be fully realized in practical applications.

In this chapter, we first develop a practically implementable version of the general equity option pricing models in Bakshi and Chen (1997) and Scott (1997), that admits stochastic interest rates and stochastic volatility, yet resembles to the extent possible the Black–Scholes model in its implementability. We present procedures for applying the resulting model to price and hedge option-like derivative products. Next, we conduct a complete analysis of the relative empirical performance, in both pricing and hedging, of the four classes of models that respectively allow for (i) constant volatility and constant interest rates (*the BS model*), (ii) constant volatility but stochastic interest rates (*the SI model*),

[1]Amin and Ng (1993), Bailey and Stulz (1989), and Heston (1993) also incorporate both stochastic volatility and stochastic interest rates, but their option pricing formulas are not given in closed form, which makes applications difficult. Consequently, comparative statics and hedge ratios are difficult to obtain in their cases.

[2]There have been a few empirical studies that investigate the pricing, but not the hedging, performance of versions of the stochastic volatility model, relative to the Black–Scholes model. These include Bates (1996b, 2000), Dumas et al (1998), Madan et al. (1998), Nandi (1996), and Rubinstein (1985). In Bates' work, currency and equity index options data are respectively used to test a stochastic volatility model with Poisson jumps included. Nandi does investigate the pricing and hedging performance of Heston's stochastic volatility model, but he focuses exclusively on a single-instrument minimum-variance hedge that involves only the S&P 500 futures. As will be clear shortly, we address in this chapter both the pricing and the hedging effectiveness issues from different perspectives and for four distinct classes of option models.

(iii) stochastic volatility but constant interest rates (*the SV model*), and (iv) stochastic volatility and stochastic interest rates (*the SVSI model*). As the SVSI model has all the other three models nested, one should expect its static pricing and dynamic hedging performance to surpass that of the other classes. But, this performance improvement must come at the cost of potentially more complex implementation steps. In this sense, conducting such a horse-race study can at least offer a clear picture of possible tradeoffs between costs and benefits that each model may present.

Specifically, the SVSI option pricing formula is expressed in terms of the underlying stock price, the stock's volatility, and the short-term interest rate. The spot volatility and the short interest rate are each assumed to follow a Markov mean-reverting square root process. Consequently, seven structural parameters need to be estimated as input to the model. These parameters can be estimated using the generalized method of moments (GMM) of Hansen (1982), as is done in, for instance, Andersen and Lund (1997), Chan et al. (1992), and Day and Lewis (1995). Or, they can be backed out from the pricing model itself by using observed option prices, as is similarly done for the BS model both in the existing literature and in Wall Street practice.

In our empirical investigation, we will adopt this implied parameter approach to implement the four models. In this regard, it is important to realize that the BS model is implemented as if the spot volatility and the spot interest rates were assumed to be time-varying within the model, that is, the spot volatility is backed out from option prices each day and used, together with the current yield curve, to price the following day's options. The SI and the SV models are implemented with a similarly internally inconsistent treatment, though to a lesser degree. Since this implementation is how one would expect each model to be applied, we chose to follow this convention in order to give the alternatives to the standard BS model the "toughest hurdle." Clearly, such a treatment works in the strongest favor of the BS model and is especially biased against the SVSI model.

Based on 38,749 S&P 500 call (and put) option prices for the sample period from June 1988 to May 1991, our empirical investigation leads to the following conclusions. First, on the basis of two out-of-sample pricing error measures, the SVSI model is found to perform slightly better than the SV model, while they both perform substantially better than the SI (the third-place performer) and the BS model. That is, when volatility is kept constant, allowing interest rates to vary stochastically can produce respectable pricing improvement over the BS model. However, in the presence of stochastic volatility, doing so no longer seems to improve pricing performance much further. Thus, modeling stochastic volatility is far more important than stochastic interest rates, at least for the purpose of pricing options. It is nonetheless encouraging to know that based on our sample both the SVSI and the SV models typically reduce the BS model's pricing errors by more than half, whereas the SI model helps reduce the BS pricing errors by 20% or more. While all four models inherit moneyness- and maturity-related pricing biases, the severity of these types of bias is increasingly reduced by the SI, the SV, and the SVSI models. In other words, the SVSI model produces pricing errors that are the least moneyness or maturity-related. This conclusion is also confirmed when the Rubinstein (1985) implied volatility smile diagnostic is adopted to examine each model.

Two types of hedging strategy are employed in this study to gauge the relative hedging effectiveness. The first type is the conventional delta-neutral hedge, in which as many distinct hedging instruments as the number of risk sources affecting the hedging target's value are used so as to make the net position completely risk-immunized (locally). Take the SVSI model as an example. The call option value is driven by three risk sources: the underlying price shocks, volatility shocks, and shocks to interest rates. Accordingly, we employ the underlying stock, a different call option, and a position in a discount bond to create a delta-neutral hedge for a target call option. That closed-form expressions are derived for each hedge ratio which is of great value for devising hedging strategies analytically. Similarly, for the SV model, we only need to rely on the underlying stock and an option contract to design a delta-neutral hedge. Based on the delta-neutral hedging errors, the same performance ranking of the four models obtains as that determined by their static pricing performance, except that now the SVSI and the SV models, and the SI and the BS models, are respectively pairwise virtually indistinguishable. This reinforces the view that adding stochastic interest rates may not affect performance much. However, it is found that the average hedging errors by the SVSI and the SV models are typically less than one-third of the corresponding BS model's hedging errors. Furthermore, reducing the frequency of hedge rebalancing does not tend to reduce the SV and the SVSI models' hedging effectiveness, whereas the BS and the SI models' hedging errors are often doubled when rebalancing frequency changes from daily to once every 5 days. Therefore, after stochastic volatility is controlled for, the frequency of hedge rebalancing will have relatively little impact on hedging effectiveness. This finding is in accord with Galai's (1983a) results that in any hedging scheme it is probably more important to control for stochastic volatility than for discrete hedging [see Hull and White (1987b) for a similar, simulation-based result for currency options].

To see how the models perform under different hedging schemes, we also look at minimum-variance hedges involving only a position in the underlying asset. As argued by Ross (1995), the need for this type of hedges may arise in contexts where a perfect delta-neutral hedge may not be feasible, either because some of the underlying risks are not traded or even reflected in any traded financial instruments, or because model misspecifications and transaction costs render it undesirable to use as many instruments to create a perfect hedge. In the present context, both volatility risk and interest rate risk are, of course, traded and hence can, as indicated above, be controlled for by employing an option and a bond. But, a point can be made that it is sometimes more preferable to adopt a single-instrument minimum-variance hedge. To study this type of hedges, we again calculate the absolute and the dollar-value hedging errors for each model. Results from this exercise indicate that the SV model performs the best among all four, while the BS and the SV models outperform their respective stochastic interest rates counterparts, the SI and the SVSI models. Therefore, under the single-instrument hedges, incorporating stochastic interest rates actually worsens hedging performance. It is also true that hedging errors under this type of hedges are always significantly higher than those under the conventional delta-neutral hedges, for each given moneyness and maturity option category. Thus, whenever possible, including more instruments in a hedge will in general produce better hedging effectiveness.

While our discussion is mainly focused on results obtained using the entire sample period and under specific model implementation designs, robustness of these empirical results is also checked by examining alternative implementation designs, different subperiods as well as option transaction price data. Especially, given the popularity of the "implied volatility matrix" method among practitioners, we will also implement each of the four models, and compare their pricing and hedging performance, by using only option contracts from a given moneyness-maturity category. It turns out that this alternative implementation scheme does not change the rankings of the four models.

In this chapter, we discuss the option pricing and hedging performance under stochastic volatility and stochastic interest rate. Section 23.2 develops the SVSI option pricing formula. It discusses issues pertaining to the implementation of the formula and derives the hedge ratios analytically. Section 23.3 provides a description of the S&P 500 option data. In Sect. 23.4, we evaluate the static pricing and the dynamic hedging performance of the four models. Concluding remarks are offered in Sect. 23.5. In addition, Appendix 1 presents derivation of stochastic interest model and stochastic volatility model.

23.2 The Option Pricing Model

Consider an economy in which the instantaneous interest rate at time *t*, denoted *R*(*t*), follows a Markov diffusion process:

$$dR(t) = [\theta_R - \bar{\kappa}_R R(t)]dt + \sigma_R\sqrt{R(t)}d\omega_R(t) \quad t \in [0, T], \tag{23.1}$$

where $\bar{\kappa}_R$ regulates the speed at which the interest rate adjusts to its long-run stationary value $\frac{\theta_R}{\bar{\kappa}_R}$, and $\omega_R = \{\omega_R(t) : t \in [0, T]\}$ is a standard Brownian motion.[3] This single-factor interest rate structure of Cox et al. (1985) is adopted as it requires the estimation of only three structural parameters. Adding more factors to the term structure model will of course lead to more plausible formulas for bond prices, but it can make the resulting option formula harder to implement.

Take a generic nondividend-paying stock whose price dynamics are described by

$$\frac{dS(t)}{S(t)} = \mu(S,t)dt + \sqrt{V(t)}d\omega(t) \quad t \in [0, T], \tag{23.2}$$

where $\mu(S,t)$, which is left unspecified, is the instantaneous expected return, and !*S* a standard Brownian motion. The instantaneous stock return variance, *V*(*t*), is assumed to follow a Markov process:

$$dV(t) = [\theta_\upsilon - \bar{\kappa}_\upsilon V(t)]dt + \sigma_\upsilon\sqrt{V(t)}d\omega_\upsilon(t) \quad t \in [0, T], \tag{23.3}$$

where again ω_υ is a standard Brownian motion and the structural parameters have the usual interpretation. We refer to $V(t)$ as the spot volatility or, simply, volatility. This process is also frugal in the number of parameters to be estimated and is similar to the one in Heston (1993). Letting ρ denote the correlation coefficient between ω_S and ω_υ, the covariance between changes in $S(t)$ and in $V(t)$ is $\mathrm{Co}\upsilon_t[dS(t), dV(t)] = \rho\sigma_S\sigma_\upsilon S(t)V(t)dt$, which can take either sign and is time-varying. According to Bakshi et al. (1997, 2000a, b), Bakshi and Chen (1997), Bates (1996a), Cao and Huang (2008), and Rubinstein (1985), this additional feature is important for explaining the skewness and kurtosis-related biases associated with the BS formula. Finally, for ease of presentation, assume that the equity-related shocks and the interest rate shocks are uncorrelated:[4] $\mathrm{Co}\upsilon_t(d\omega_S, d\omega_R) = \mathrm{Co}\upsilon_t(d\omega_\upsilon, d\omega_R) = 0$.

By a result from Harrison and Kreps (1979), there are no free-lunches in the economy if and only if there exists an equivalent martingale measure with which one can value claims as if the economy were risk neutral. For instance, the

[3]Here we follow a common practice to assume from the outset a structure for the underlying price and rate processes, rather than derive them from a full-blown general equilibrium. See Bates (1996a), Heston (1993), Melino and Turnbull (1990, 1995), and Scott (1987, 1997). The simple structure assumed in this section can, however, be derived from the general equilibrium model of Bakshi and Chen (1997).

[4]This assumption on the correlation between stock returns and interest rates is somewhat severe and likely counterfactual. To gauge the potential impact of this assumption on the resulting option model's performance, we initially adopted the following stock-price dynamics:

$$\frac{dS(t)}{S(t)} = \mu(S,t)dt + \sqrt{V(t)}d\omega_S(t) + \sigma_{S,R}\sqrt{R(t)}d\omega_R(t) \quad t \in [0, T],$$

with the rest of the stochastic structure remaining the same as given above. Under this more realistic structure, the covariance between stock-price changes and interest rate shocks is $\mathrm{Co}\upsilon_t[dS(t), dR(t)] = \sigma_{S,R}\sigma_R R(t)S(t)dt$, so bond market innovations can be transmitted to the stock market and vice versa. The obtained closed-form option pricing formula under this scenario would have one more parameter $\sigma_{S,R}$ than the one presented shortly, but when we implemented this slightly more general model, we found its pricing and hedging performance to be indistinguishable from that of the SVSI model studied in this chapter. For this reason, we chose to present the more parsimonious SVSI model derived under the stock-price process in (23.2). We could also make both the drift and the diffusion terms of $V(t)$ a linear function of $R(t)$ and $\omega_R(t)$. In such cases, the stock returns, volatility and interest rates would all be correlated with each other (at least globally), and we could still derive the desired equity option valuation formula. But, that would again make the resulting formula more complex while not improving its performance.

time-t price $B(t,\tau)$ of a zero-coupon bond that pays \$1 in periods can be determined via

$$B(t,\tau) = E_Q\left\{\exp\left(-\int_t^{t+\tau} R(s)\mathrm{d}s\right)\right\}, \quad (23.4)$$

where E_Q denotes the expectation with respect to an equivalent martingale measure and conditional on the information generated by $R(t)$ and $V(t)$. Assume that the factor risk premiums for $R(t)$ and $V(t)$ are, respectively, given by $\lambda_R R(t)$ and $\lambda_\upsilon V(t)$, for two constants λ_R and λ_υ. Bakshi and Chen (1997) provide a general equilibrium model in which risk premiums have precisely this form and in which the interest rate and stock-price processes are as assumed here. Under this assumption, we obtain the risk-neutralized processes for $R(t)$ and $V(t)$ below:

$$\mathrm{d}R(t) = [\theta_R - \bar{\kappa}_R R(t)]\mathrm{d}t + \sigma_R\sqrt{R(t)}\mathrm{d}\omega_R(t) \quad (23.5)$$

$$\mathrm{d}V(t) = [\theta_\upsilon - \bar{\kappa}_\upsilon V(t)]\mathrm{d}t + \sigma_\upsilon\sqrt{V(t)}\mathrm{d}\omega_\upsilon(t), \quad (23.6)$$

where $\kappa_R \equiv \bar{\kappa}_R + \lambda_R$ and $\kappa_\upsilon \equiv \bar{\kappa}_\upsilon + \lambda_\upsilon$. The risk-neutralized stock-price process becomes

$$\frac{\mathrm{d}S(t)}{S(t)} = R(t)\mathrm{d}t + \sqrt{V(t)}\mathrm{d}\omega_S(t), \quad (23.7)$$

That is, under the martingale measure, the stock should earn no more and no less than the risk-free rate. With these adjustments, we solve the conditional expectation in (23.4) and obtain the familiar bond price equation below:

$$B(t,\tau) = exp[-\varphi(\tau) - \vartheta(\tau)R(\tau)], \quad (23.8)$$

where
$\varphi(\tau) = \frac{\theta_R}{\sigma_R^2}\left\{(\varsigma - \kappa_R)\tau + 2\ln\left[1 - \frac{(1-e^{-\varsigma\tau})(\varsigma-\kappa_R)}{2\varsigma}\right]\right\}$,
$\vartheta(\tau) = \frac{2(1-e^{-\varsigma\tau})}{2\varsigma-[\varsigma-\kappa_R](1-e^{-\varsigma\tau})}$ and $\varsigma \equiv \sqrt{\kappa_R^2 + 2\sigma_R^2}$. Cox et al. (1985) for an analysis of this class of term structure models.

23.2.1 Pricing Formula for European Options

Now, consider a European call option written on the stock, with strike price K and term-to-expiration. Let its time-t price be denoted by $C(t,\tau)$. As (S,R,V) form a joint Markov process, the price $C(t,\tau)$ must be a function of $S(t)$, $R(t)$ and $V(t)$ (in addition to). By a standard argument, the option price must solve

$$\begin{aligned}&\frac{1}{2}VS^2\frac{\partial^2 C}{\partial S^2} + RS\frac{\partial C}{\partial S} + \rho\sigma_\upsilon VS\frac{\partial^2 C}{\partial S\partial V}\\ &\quad + \frac{1}{2}\sigma_\upsilon^2 V\frac{\partial^2 C}{\partial V^2} + [\theta_\upsilon - \kappa_\upsilon V]\frac{\partial C}{\partial V}\\ &\quad + \frac{1}{2}\sigma_R^2 R\frac{\partial^2 C}{\partial R^2} + [\theta_R - \kappa_R R]\frac{\partial C}{\partial R} - \frac{\partial C}{\partial \tau} - RC = 0,\end{aligned} \quad (23.9)$$

subject to $C(t+\tau,0) = \max\{S(t+\tau) - K, 0\}$. In the Appendix, it is shown that

$$\begin{aligned}C(t,\tau) &= S(t)\prod_1(t,\tau;S,R,V)\\ &\quad - KB(t,\tau)\prod_2(t,\tau;S,R,V),\end{aligned} \quad (23.10)$$

where the risk-neutral probabilities, 1 and 2, are recovered from inverting the respective characteristic functions [see Heston (1993), and Scott (1997) for similar treatments]:

$$\begin{aligned}&\prod_j(t,\tau,S(t),R(t),V(t)) = \frac{1}{2}\\ &\quad + \frac{1}{\pi}\int_0^\infty Re\left[\frac{e^{-i\phi ln[K]}f_j(t,\tau,S(t),R(t),V(t);\phi)}{i\phi}\right]\mathrm{d}\phi\end{aligned} \quad (23.11)$$

for $j = 1, 2$. The characteristic functions f_j are respectively given by

$$f_1(t,\tau)=\exp\left\{\begin{array}{l}-\frac{\theta_R}{\sigma_R^2}\left[2\ln\left(1-\frac{[\xi_R-\kappa_R](1-e^{-\xi_R\tau})}{2\xi_\upsilon}\right)+[\xi_R-\kappa_R]\tau\right]\\ -\frac{\theta_\upsilon}{\sigma_\upsilon^2}\left[2\ln\left(1-\frac{[\xi_\upsilon-\kappa_\upsilon+(1+i\phi)\rho\sigma_\upsilon](1-e^{-\xi_\upsilon\tau})}{2\xi_\upsilon}\right)\right]\\ -\frac{\theta_\upsilon}{\sigma_\upsilon^2}[\xi_\upsilon-\kappa_\upsilon+(1+i\phi)\rho\sigma_\upsilon]\tau+i\phi\ln[S(t)]+\frac{2i\phi(1-e^{\xi_R\tau})}{2\xi_R-[\xi_R-\kappa_R](1-e^{\xi_R\tau})}R(t)\\ +\frac{i\phi(i\phi+1)(1-e^{-\xi_\upsilon\tau})}{2\xi_\upsilon-[\xi_\upsilon-\kappa_\upsilon+(1+i\phi)\rho\sigma_\upsilon](1-e^{-\xi_\upsilon\tau})}V(t)\end{array}\right\}, \tag{23.12}$$

and,

$$f_2(t,\tau)=\exp\left\{\begin{array}{l}-\frac{\theta_R}{\sigma_R^2}\left[2\ln\left(1-\frac{[\xi_R^*-\kappa_R](1-e^{-\xi_R^*\tau})}{2\xi_R^*}\right)+[\xi_R^*-\kappa_R]\tau\right]\\ -\frac{\theta_\upsilon}{\sigma_\upsilon^2}\left[2\ln\left(1-\frac{[\xi_\upsilon^*+i\phi\rho\sigma_\upsilon](1-e^{-\xi_\upsilon^*\tau})}{2\xi_\upsilon^*}\right)+\left[\xi_\upsilon^*-\kappa_\upsilon+i\phi\rho\sigma_\upsilon\right]\tau\right]\\ +i\phi\ln[S(t)]-\ln[B(t,\tau)]+\frac{2(i\phi-1)(1-e^{\xi_R^*\tau})}{2\xi_R^*-[\xi_R^*-\kappa_R](1-e^{\xi_R^*\tau})}R(t)\\ +\frac{i\phi(i\phi-1)(1-e^{-\xi_\upsilon^*\tau})}{2\xi_\upsilon^*-[\xi_\upsilon^*-\kappa_\upsilon+i\phi\rho\sigma_\upsilon](1-e^{-\xi_\upsilon^*\tau})}V(t)\end{array}\right\}, \tag{23.13}$$

where $\xi_R=\sqrt{\kappa_R^2-2\sigma_R^2 i\phi}$, $\xi_\upsilon=\sqrt{[\kappa_\upsilon-(1+i\phi)\rho\sigma_\upsilon]^2-i\phi(i\phi+1)\sigma_\upsilon^2}$, $\xi_R^*=\sqrt{\kappa_R^2-2\sigma_R^2(i\phi-1)}$, and $\xi_\upsilon^*=\sqrt{\kappa_\upsilon^2-i\phi(i\phi-1)\sigma_\upsilon^2}$. The price of a European put on the same stock can be determined from the put-call parity.

The option valuation model in (23.10) has several distinctive features. First, it applies to cases with stochastically varying interest rates and volatility. It contains as special cases most existing models, such as the SV models, the SI models, and clearly the BS model. Second, as mentioned earlier, it allows for a flexible correlation structure between the stock return and its volatility, as opposed to the perfect correlation assumed in, for instance, Heston's (1993) model. Furthermore, the volatility risk premium is time-varying and state-dependent. This is a departure from Hull and White (1987a), Scott (1987), Stein and Stein (1991), and Wiggins (1987) where the volatility risk premium is either a constant or zero. Third, when compared to the general models in Bakshi and Chen (1997) and Scott (1997), the formula in (23.10) is parsimonious in the number of parameters. Especially, it is given only as a function of identifiable variables such that all parameters can be estimated based on available financial market data.

The pricing formula in (23.10) applies to European equity options. But, in reality most of the traded option contracts are American in nature. While it is beyond the scope of this chapter to derive a model for American options, it is nevertheless possible to capture the first-order effect of early exercise in the following manner. For options with early exercise potential, compute the Barone-Adesi and Whaley (1987) or Kim (1990) early exercise premium, treating it as if the stock volatility and the yield curve were time-invariant. Adding this early exercise adjustment component to the European option price in (23.10) should deliver a reasonable approximation of the corresponding American option price [e.g., Bates (1996b)].

23.2.2 Hedging and Hedge Ratios

One appealing feature of a closed-form option pricing formula, such as the one in (23.10), is the possibility of deriving comparative statics and hedge ratios analytically. In the present context, there are three sources of stochastic variations over time, price risk *S*(*t*), volatility risk *V*(*t*), and interest rate risk *R*(*t*). Consequently, there are three deltas:

$$\Delta_s(t,\tau;K) \equiv \frac{\partial C(t,\tau)}{\partial S} = \prod\nolimits_1 > 0 \quad (23.14)$$

$$\Delta_V(t,\tau;K) \equiv \frac{\partial C(t,\tau)}{\partial V} = S(t)\frac{\partial \prod_1}{\partial V} - KB(t,\tau)\frac{\partial \prod_2}{\partial V} > 0 \quad (23.15)$$

$$\Delta_R(t,\tau;K) \equiv \frac{\partial C(t,\tau)}{\partial R} = S(t)\frac{\partial \prod_1}{\partial R} - KB(t,\tau)\left\{\frac{\partial \prod_2}{\partial R} - \vartheta(\tau)\prod\nolimits_2\right\} > 0, \quad (23.16)$$

where, for $\vartheta = V, R$ and $j = 1, 2$,

$$\frac{\partial \prod_j}{\partial_g} = \frac{1}{\pi}\int_0^{\infty} Re\left[(i\phi)^{-1}e^{-i\phi;\ln[K]}\frac{\partial f_j}{\partial g}\right]\mathrm{d}\phi. \quad (23.17)$$

The second-order partial derivatives with respect to these variables are provided in the Appendix.

As $V(t)$ and $R(t)$ are both stochastic in our model, these deltas will in general differ from their Black–Scholes counterpart. To see how they may differ, let us resort to an example in which we set $R(t) = 6.27\%, S(t) = 270, \sqrt{V(t)} = 22.12\%, \kappa_R = 0.481, \theta_R = 0.037, \sigma_R = 0.043, \kappa_\upsilon = 1.072, \theta_\upsilon = 0.041, \sigma_\upsilon = 0.284$, and $\rho = -0.60$. These values are backed out from the S&P 500 option prices as of July 5, 1988. Fix $K = \$270$ and $\tau = 45$ days. Let Δ_S be as given in (23.14) for the SVSI model and Δ_S^{bs} its BS counterpart, with Δ_S^{bs} calculated using the same implied volatility. Figure 23.1 plots the difference between Δ_S and Δ_S^{bs}, across different spot price levels and different correlation values. The correlation coefficient ρ is chosen to be the focus as it is known to play a crucial role in determining the skewness of the stock return distribution. When ρ is respectively at −0.50 and −1.0 (see the 2{curve and the {curve), the difference between the deltas is W-shaped, and it reaches the highest value when the option is at-the-money. The reverse is true when is positive. Thus, Δ_S is generally different from Δ_S^{bs}. Analogous difference patterns emerge when the other option deltas are compared with their respective BS counterpart. From Figs. 23.2 and 23.3, one can observe the following. (i) The volatility hedge ratio Δ_V from the SVSI model is, at each spot price, lower than its BS counterpart (except for deep in-the-money options when $\rho<0$, and for deep out-of-the-money options when $\rho > 0$).[5] (ii) The interest rate delta, Δ_R, and its BS counterpart, Δ_R^{bs}, are almost not different from each other for slightly out-of-the-money options, but can be dramatically different for at-the-money options as well as for sufficiently deep in-the-money or deep out-of-the-money calls. For example, pick $\rho = -1.0$. When $S = \$315$, we have $\Delta_R = 30.94$ and $\Delta_R^{bs} = 32.35$; when $S = \$226$, we have $\Delta_R = 0.003$ and $\Delta_R^{bs} = 0.430$. (iii) As expected,

[5] In making such a comparison, one should apply sufficient caution. In the BS model, the volatility delta is only a comparative static, not a hedge ratio, as volatility is assumed to be constant. In the context of the SVSI model, however, Δ_V is time-varying hedge ratio as volatility is stochastic. This distinction also applies to the case of the interest rate delta Δ_R.

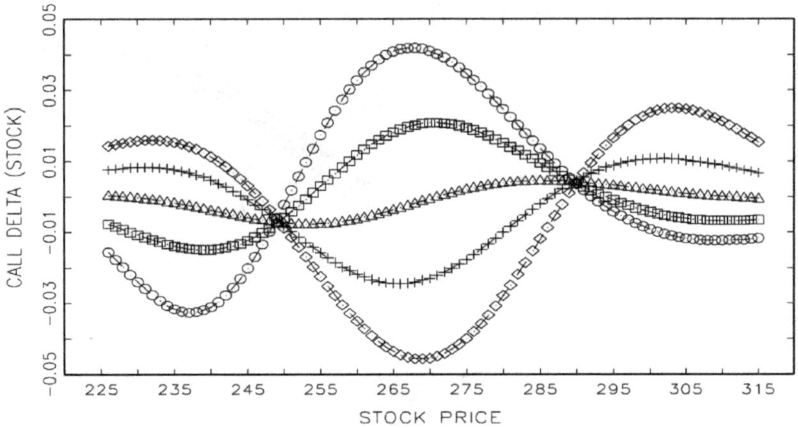

Fig. 23.1 Call delta with respect to stock. The white circle-curve, the white square-curve, the white up-pointing triangle-curve, the plus sign-curve, and the white diamond-curve, respectively, plot the difference between the SVSI call option delta (with respect to stock) and its Black–Scholes counterpart, as ρ varies from −1.0, −0.50, 0, 0.50, to 1.0. The structural parameter values used in the computation of the delta in (23.14) are backed out using Procedure B described in Sect. 2.3 and correspond to the calendar date July 5, 1988. The values of the structural parameters are: $\kappa R = 0.4811$, $\theta_R = 0.0370$, $\sigma_R = 0.0429$, $\kappa_v = 1.072$, $\theta_v = 0.0409$, $\sigma_v = 0.284$, $\rho = -0.60$. The initial (time-t) $R = 0.062733$, $\sqrt{V} = 22.12\%$, $B(t,\ 0.1232) = 0.99163$. The strike price is fixed at \$270 and the term-to-expiration of the option is 45 days

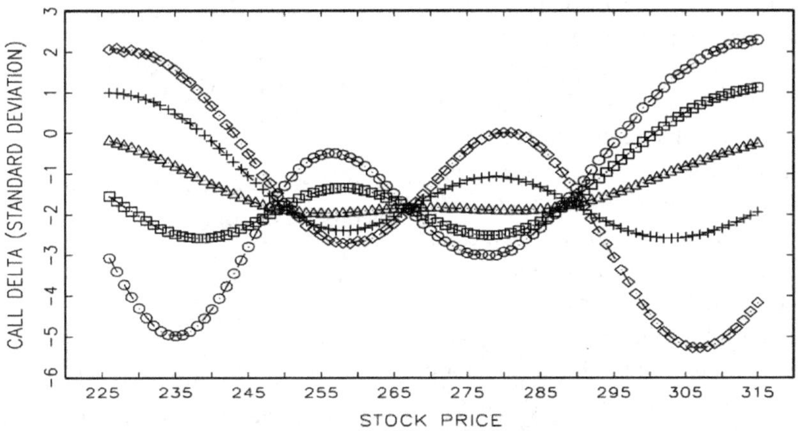

Fig. 23.2 Call delta with respect to standard deviation. The open circle-curve, the open square-curve, the white up-pointing triangle-curve, the plus sign-curve, and the white diamond-curve, respectively, plot the difference between the SVSI call option delta (with respect to the standard deviation) and the Black–Scholes counterpart, as ρ varies from −1.0, −0.50, 0, 0.50, to 1.0. The strike price is fixed at \$270, and the term-to-expiration of the option is 45 days. All computations are based on the parameter values given in the note to Fig. 23.1

out-of-the-money options are overall less sensitive to changes in the spot interest rate, regardless of the model used. In summary, if a portfolio manager/trader relies, in an environment with stochastic interest rates and stochastic volatility, on the BS model to design a hedge for option positions, the manager/trader will likely fail. Analytical expressions for the deltas are useful for constructing hedges based on an option formula. Below, we present two types of hedges by using the SVSI model as an example.

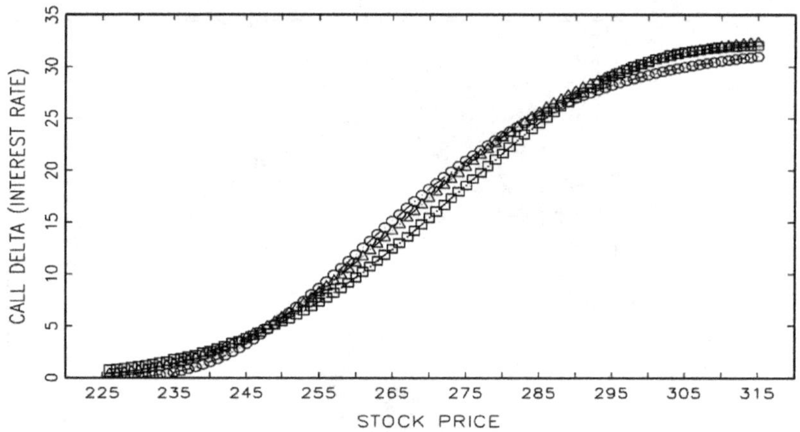

Fig. 23.3 Call delta with respect to interest rate. The white circle-curve, the white square-curve, the white up-pointing triangle-curve, the plus sign-curve, and the white diamond-curve, respectively, plot the difference between the SVSI call option delta (with respect to the spot interest rate) and the Black–Scholes counterpart, as ρ varies from −1.0, −0.50, 0, 0.50, to 1.0. The strike price is fixed at \$270, and the term-to-expiration of the option is 45 days. All computations are based upon the parameter values reported in the note to Fig. 23.1

23.2.2.1 Delta-Neutral Hedges

To demonstrate how the deltas may be used to construct a delta-neutral hedge, consider an example in which a financial institution intends to hedge a short position in a call option with τ periods to expiration and strike price K. In the stochastic interest rate-stochastic volatility environment, a perfectly delta-neutral hedge can be achieved by taking a long position in the replicating portfolio of the call. As three traded assets are needed to control the three sources of uncertainty, the replicating portfolio will involve a position in (i) some $X_S(t)$ shares of the underlying stock (to control for the $S(t)$ risk), (ii) some $X_B(t)$ units of a -period discount bond (to control for the $R(t)$ risk), and (iii) some $X_C(t)$ units of another call option with strike price $\overline{K}$ (or any option on the stock with a different maturity) in order to control for the volatility risk $V(t)$. Denote the time-t price of the replicating portfolio by $G(t) : G(t) = X_0(t) + X_S(t)S(t) + X_B(t)B(t,\tau) + X_C(t)C(t,\tau;\overline{K})$, where $X_0(t)$ denotes the amount put into the instantaneously maturing risk-free bond and it serves as a residual "cash position." Deriving the dynamics for $G(t)$ and comparing them with those of $C(t,\tau;K)$, we find the following solution for the delta-neutral hedge:

$$X_C(t) = \frac{\Delta_V(t,\tau;K)}{\Delta_V(t,\tau;\overline{K})} \tag{23.18}$$

$$X_S(t) = \Delta_S(t,\tau;K) - \Delta_S(t,\tau;\overline{K})X_C(t) \tag{23.19}$$

$$X_B(t) = \frac{1}{B(t,\tau)\vartheta(\tau)}\{\Delta_R(t,\tau;\overline{K})X_C(t) - \Delta_R(t,\tau;K)\} \tag{23.20}$$

and the residual amount put into the instantaneously maturing bond is

$$X_0(t) = C(t,\tau;K) - X_S(t)S(t) - X_C(t)C(t,\tau;\overline{K}) - X_B(t)B(t,\tau), \tag{23.21}$$

where all the primitive deltas, Δ_S, Δ_R and Δ_V, are as determined in Eqs. (23.14–23.16). Like the option prices, these hedge ratios all depend on the values taken by $S(t)$, $V(t)$, and $R(t)$ and those by the structural parameters. Such a hedge created using the general option pricing model should in principle perform better than using the BS model. In the latter case, only the underlying price uncertainty is controlled for, but not the uncertainties associated with volatility and interest rate fluctuations.

In theory, this delta-neutral hedge requires continuous rebalancing to reflect the changing market conditions. In practice, of course, only discrete rebalancing is possible. To derive a hedging effectiveness measure, suppose that portfolio rebalancing takes place at intervals of length Δ_t Then, precisely as described above, at time t short the call option, go long in (i) $X_S(t)$ shares of the underlying asset, (ii) $X_B(t)$ units of the $\tau-$ period bond, and (iii) $X_C(t)$ contracts of a call option with the same term-to-expiration but a different strike price $\overline{K}$, and invest the residual, X_0, in an instantaneously maturing risk-free bond. After the next interval, compute the hedging error according to

$$\begin{aligned} H(t+\Delta t) &= X_0 e^{R(t)\Delta t} + X_S(t)S(t+\Delta t) \\ &+ X_B(t)B(t+\Delta t, \tau-\Delta t) + X_C(t)C(t+\Delta t, \tau-\Delta t; \overline{K}) \\ &- C(t+\Delta t, \tau-\Delta t; K) \end{aligned} \tag{23.22}$$

Then, at time $t+\Delta t$, reconstruct the self-financed portfolio, repeat the hedging error calculation at time $t+2\Delta t$, and so on. Record the hedging errors $H(t+j\Delta t)$ for $j=1,\cdots,J\equiv\frac{\tau-t}{\Delta t}$. Finally, compute the average absolute hedging error as a function of rebalancing frequency $\Delta t : H(\Delta t) = \frac{1}{J}\sum_{j=1}^{J}|H(t+j\Delta t)|$, and the average dollar-value hedging error: $\overline{H}(\Delta t) = \frac{1}{J}\sum_{j=1}^{J}H(t+j\Delta t)$.

In comparison, if one relies on the BS model to construct a delta-neutral hedge, the hedging error measures can be similarly defined as in (23.22), except that $X_B(t)$ and $X_C(t)$ must be restricted to zero and $X_S(t)$ must be the BS delta. Likewise, if the SI model is applied, the only change is to set $X_C(t)$ to zero with Δ_S and Δ_R determined by the SI model; in the case of the SV model, set $X_B(t)=0$ and let Δ_S and Δ_V be as determined in the SV model. The Appendix provides in closed form a SI option pricing formula and a SV option formula.

23.2.2.2 Single-Instrument Minimum-Variance Hedges

As discussed before, consideration of such factors as model misspecification and transaction costs may render it more practical to use only the underlying asset of the target option as the hedging instrument. Under this single-instrument constraint, a standard design is to choose a position in the underlying stock so as to minimize the variance of instantaneous changes in the value of the hedge. Letting $X_S(t)$ again be the number of shares of the stock to be purchased, solving the standard minimum-variance hedging problem under the SVSI model gives

$$\begin{aligned} X_S(t) &= \frac{\mathrm{Cov}_t[\mathrm{d}S(t), \mathrm{d}C(t,\tau)]}{\mathrm{Var}[\mathrm{d}S(t)]} \\ &= \Delta_s + \rho\sigma_v\frac{\Delta_V(t,\tau)}{S(t)}, \end{aligned} \tag{23.23}$$

and the resulting residual cash position for the replicating portfolio is

$$X_0(t) = C(t,\tau) - X_S(t)S(t). \tag{23.24}$$

This minimum-variance hedge solution is quite intuitive, as it says that if stock volatility is deterministic (i.e., $\sigma_v = 0$), or if stock returns are not correlated with volatility changes (i.e., $\rho = 0$), one only needs to long $\Delta_s(t)$ shares of the stock and no other adjustment is necessary. However, if volatility is stochastic and correlated with stock returns, the position to be taken in the stock must control not only for the direct impact of underlying stock-price changes on the target option value, but also for the indirect impact of that part of volatility changes which is correlated with stock-price fluctuations. This effect is reflected in the last term in (23.23), which shows that the additional number of shares needed besides Δ_s is increasing in ρ (assuming $\sigma_v > 0$).

As for the previous case, suppose that the target call is shorted and that $X_S(t)$ shares are bought and $X_0(t)$ dollars are put into the instantaneous risk-free bond, at time t. The combined position is a self-financed portfolio. At time $t+\Delta t$, the hedging error of this minimum-variance hedge is calculated as

$$\begin{aligned} H(t+\Delta t) = X_S(t)S(t+\Delta t) + X_0(t)e^{R(t)\Delta t} \\ - C(t+\Delta t, \tau-\Delta t). \end{aligned} \tag{23.25}$$

Unlike in Nandi (1996) where he uses the remaining variance of the hedge as a hedging effectiveness gauge, we compute, based on the entire sample period, the average absolute and the average dollar hedging errors to measure the effectiveness of the hedge.

Minimum-variance hedging errors under the SV model as well as under the SI model can be similarly determined accounting for their modeling differences. In the case of the SV model, there is still an adjustment term for the single stock position as in (23.23). But, for the SI model, the corresponding $X_S(t)$ is the same as its Δ_s. For the BS model, this single-instrument minimum-variance hedge is the same as the delta-neutral hedge. Both types of hedging strategy will be examined under each of the four alternative models.

23.2.3 Implementation

In addition to the strike price and the term-to-expiration (which are specified in the contract), the SVSI pricing formula in (23.10) requires the following values as input:

- The spot stock price. If the stock pays dividends, the stock price must be adjusted by the present value of future dividends;
- The spot volatility;
- The spot interest rate;
- The matching-period yield-to-maturity (or the bond price);
- The seven structural parameters $\{\kappa_R, \theta_R, \sigma_R, \kappa_v, \theta_v, \sigma_v, \rho\}$.

For computing the price of a European option, we offer two alternative two-step procedures below. One can implement these steps on any personal computer:

Procedure A:

Step 1. Obtain a time-series each for the short rate, the stock return, and the stock volatility. Jointly estimate the structural parameters, $\{\kappa_R, \theta_R, \sigma_R, \kappa_v, \theta_v, \sigma_v, \rho\}$., using Hansen's (1982) GMM.

Step 2. Determine the risk-neutral probabilities, $\prod_1$ and $\prod_2$, from the characteristic functions in (23.12) and (23.13). Substitute (i) the two probabilities, (ii) the stock price, and (iii) the yield-to-maturity, into (23.10) to compute the option price.

While offering an econometrically rigorous method to estimate the structural parameters, Step 1 in Procedure A may not be as practical or convenient, because of its requirement on historical data. A further difficulty with this approach is its dependence on the measurement of stock volatility. In implementing the BS model, practitioners predominantly use the implied volatility from the model itself, rather than relying on historical data. This practice has not only reduced data requirement dramatically but also resulted in significant performance improvement [e.g., Bates (2000), and Melino and Turnbull (1990, 1995)]. Clearly, one can also follow this practice to implement the SVSI model:

Procedure B:

Step 1. Collect N option prices on the same stock and taken from the same point in time (or same day), for any $N \geq 8$. Let $\hat{C}_n(t, \tau_n, K_n)$ be the observed price, and $C_n(t, \tau_n, K_n)$ the model price as determined by (23.10) with $S(t)$ and $R(t)$ taken from the market, for the nth option with τ_n periods to expiration and strike price K_n and for each $n = 1, \cdots, N$. Clearly, the difference between $\hat{C}_n$ and C_n is a function of the values taken by $V(t)$ and by $\Phi = \{\kappa_R, \theta_R, \sigma_R, \kappa_\upsilon, \theta_\upsilon, \sigma_\upsilon, \rho\}$. Define

$$\varepsilon_n[V(t), \Phi] \equiv \hat{C}_n(t, \tau_n, K_n) - C_n(t, \tau_n, K_n), \tag{23.26}$$

for each n. Then, find $V(t)$ and parameter vector Φ (a total of eight), so as to minimize the sum of squared errors:

$$\sum_{n=1}^{N} |\varepsilon_n[V(t), \Phi]|^2 \qquad (23.27)$$

The result from this step is an estimate of the implied spot variance and seven structural parameter values, for date t. See Bates (1996b, 2000), Day and Lewis (1995), Dumas et al. (1998), Longsta (1995), Madan et al. (1998), and Nandi (1996) where they adopt this technique for similar purposes.

Step 2. Based on the estimate from the first step, follow Step 2 of Procedure A to compute date-$t+1$'s option prices on the same stock.

In the existing literature, the performance of a new option pricing model is often judged relative to that of the BS model when the latter is implemented using the model's own implied volatility and the time-varying interest rates. Since volatility and interest rates in the BS are assumed to be constant over time, this internally inconsistent practice will clearly and significantly bias the application results in favor of the BS model. But, as this is the current standard in judging performance, we will follow Procedure B to implement the SVSI model and similar procedures to implement the BS, the SV, and the SI models. Then, the models will be ranked relative to each other according to their performance so determined.

23.3 Data Description

For all the tests to follow, we use, based on the following considerations, S&P 500 call option prices as the basis. First, options written on this index are the most actively traded European-style contracts. Recall that like the BS model, Formula (23.10) applies to European options. Second, the daily dividend distributions are available for the index (from the S&P 500 Information Bulletin). Harvey and Whaley (1992a, b), for instance, emphasize that critical pricing errors can result when dividends are omitted from empirical tests of any option valuation model. Furthermore, S&P 500 options and options on S&P 500 futures have been the focus of many existing empirical investigations including, among others, Bates (2000), Dumas et al. (1998), Madan et al. (1998), Nandi (1996), and Rubinstein (1994). Finally, we also used S&P 500 put option prices to estimate the pricing and hedging errors of all four models and found the results to be similar, both qualitatively and quantitatively. To save space, we chose to focus on the results based on the call option prices.

The sample period extends from June 1, 1988 through May 31, 1991. The intradaily transaction prices and bid–ask quotes for S&P 500 options are obtained from the Berkeley Option Database. Note that the recorded S&P 500 index values are not the daily closing index levels. Rather, they were the corresponding index levels at the moment when the recorded option transaction took place or when an option price quote was recorded. Thus, there is no nonsynchronous price issue here, except that the S&P 500 index level itself may contain stale component stock prices at each point in time.

The data on the daily Treasury-bill bid and ask discounts with maturities up to one year are hand-collected from the Wall Street Journal and provided to us by Hyuk Choe and Steve Freund. By convention, the average of the bid and ask Treasury-bill discounts is used and converted to an annualized interest rate. Careful attention is given to this construction since Treasury bills mature on Thursdays while index options expire on the third Friday of the month. In such cases, we utilize the two Treasury-bill rates straddling the option's expiration date to obtain the interest rate of that maturity, which is done for each contract and each day in the sample. The Treasury-bill rate with 30-days to maturity is the surrogate used for the short rate in (23.1) [and in the determination of the probabilities in (23.10)].

For European options, the spot stock price must be adjusted for discrete dividends. For each option contract with τ periods to expiration from time t, we first obtain the present value of the daily dividends $D(t)$ by computing

$$\overline{D}(t,\tau) \equiv \sum_{s=1}^{\tau-t} e^{-R(t,s)s} D(t+s), \qquad (23.28)$$

where $R(t, s)$ is the s-period yield-to-maturity. This procedure is repeated for all option maturities and for each day in our sample. In the next step, we subtract the present value of future dividends from the time-t index level, in order to obtain the dividend-exclusive S&P 500 spot index series that is later used as input into the option models.

Several exclusion filters are applied to construct the option price data set. First, option prices that are time-stamped later than 3:00 PM Central Daytime are eliminated. This ensures that the spot price is recorded synchronously with its option counterpart. Second, as options with less than six days to expiration may induce liquidity-related biases, they are excluded from the sample. Third, to mitigate the impact of price discreteness on option valuation, option prices lower than $\$ = \frac{3}{8}$ are not included. Finally, quote prices that are less than the intrinsic value of the option are taken out of the sample.

We divide the option data into several categories according to either moneyness or term-to-expiration. A call option is said to be at-the-money (ATM) if its $\frac{S}{K} \in (0.97; 1.03)$, where S is the spot price and K the strike; out-of-the-money (OTM) if $\frac{S}{K} \leq 0.97$; and in-the-money (ITM) if $\frac{S}{K} \geq 1.03$. A finer partition resulted in nine moneyness categories. By the term-to-expiration, each option can be classified as [e.g., Rubinstein (1985)] (i) extremely

Table 23.1 Sample properties of S&P 500 index options

Moneyness $\frac{S}{K}$	Term-to-expiration (days)						Subtotal
	<30	30–60	60–90	90–120	120–180	≥ 180	
<0.93	0.78	1.33	1.99	2.84	4.88	7.82	
	{23}	{246}	{266}	{431}	{1080}	{1538}	{3584}
0.93–0.95	1.02	1.91	3.30	5.08	8.14	12.86	
	{121}	{595}	{267}	{319}	{596}	{646}	{2544}
0.95–0.97	1.35	3.05	5.35	7.45	10.87	15.91	
	{488}	{1012}	{316}	{351}	{670}	{628}	{3465}
0.97–0.99	2.47	5.53	8.23	10.83	14.19	19.33	
	{838}	{1020}	{312}	{336}	{676}	{706}	{3888}
0.99–1.01	5.27	8.99	11.96	14.55	17.95	23.20	
	{776}	{954}	{285}	{308}	{629}	{631}	{3583}
1.01–1.03	9.65	13.17	15.99	18.84	22.06	27.74	
	{752}	{906}	{276}	{283}	{607}	{597}	{3421}
1.03–1.05	14.79	17.80	20.80	23.36	26.39	31.91	
	{675}	{844}	{241}	{264}	{542}	{501}	{3067}
1.05–1.07	20.20	22.63	25.83	27.83	30.69	35.70	
	{620}	{760}	{224}	{242}	{449}	{473}	{2818}
>1.07	41.23	42.28	47.50	49.27	51.34	59.82	
	{2143}	{2350}	{1284}	{1355}	{2184}	{3063}	{12379}
Subtotal	{6436}	{8687}	{3471}	{3889}	{7483}	{8783}	{38749}

The reported numbers are, respectively, the average quoted bid–ask mid-point price and the number of observations. Each option contract is consolidated across moneyness and term-to-expiration categories. The sample period extends from June 1, 1988 through May 31, 1991 for a total of 38,749 calls. Daily information from the last quote of each option contract is used to obtain the summary statistics. S denotes the spot S&P 500 index level, and K is the exercise price

short-term (<30 days); (ii) short-term (30–60 days); (iii) near-term (60–120 days); (iv) middle-maturity (120–180 days); and (v) long-term (>180 days). The proposed moneyness and term-to-expiration classifications resulted in 54 categories for which the empirical results will be reported.

Table 23.1 describes sample properties of the S&P 500 call option prices used in the tests. Summary statistics are reported for the average bid–ask mid-point price and the total number of observations, for each moneyness-maturity category. Note that there are a total of 38,749 call price observations, with deep in-the-money and at-the-money options, respectively, taking up 32 and 28% of the total sample, and that the average call price ranges from \$0.78 for extremely short-term, deep out-of-the-money options to \$59.82 for long-term, deep in-the-money options.

23.4 Empirical Tests

This section examines the relative empirical performance of the four models. The analysis is intended to present a complete picture of what each generalization of the benchmark BS model can really buy in terms of performance improvement and whether each generalization produces a worthy tradeoff between benefits and costs. We will pursue this analysis by using three yardsticks: (i) the size of the out-of-sample cross-sectional pricing errors (static performance); (ii) the size of model-based hedging errors (dynamic performance); and (iii) the existence of systematic biases across strike prices or across maturities (i.e., Does the implied volatility still smile?).

Based on Procedure B of Sect. 2.3, Table 23.2 reports the summary statistics for the daily estimated structural parameters and the implied spot standard deviation, respectively, for the SVSI, the SV, the SI, and the BS models. Take the SVSI model as an example. Over the entire sample period 06:1988–05:1991, $\kappa_\upsilon = 0.906, \theta_\upsilon = 0.042$ and $\sigma_\upsilon = 0.414$. These estimates imply a long-run mean of 21:53% for the volatility process. The implicit (average) half-life for variance mean-reversion is 9.18 months. These estimates are similar in magnitude to those reported in Bates (1996b, 2000) for S&P 500 futures options. The estimated parameters for the (risk-neutralized) short rate process are also reasonable and comparable to those in Chan et al. (1992). The presented correlation estimate for ρ is −0.763. The average implied standard deviation is 19.27%. As seen from the reported standard errors in Table 23.2, for each given model the daily parameter and spot volatility estimates are quite stable from subperiod to subperiod. Histogram-based inferences (not reported) indicate that the majority of the estimated values are centered around the mean.

In estimating the structural parameters and the implied volatility for a given day, we used all S&P 500 options collected in the sample for that day (regardless of maturity and moneyness). This is the treatment applied to the SI, the SV, and the SVSI models. For the BS model, however, Whaley (1982) makes the point that ATM options may give an implied volatility estimate which produces the best pricing and hedging results. Based on his justification, we used, for each given day, one ATM option that had at least 15 days to expiration to back out the BS model's implied volatility value. This estimate was then used to determine the next day's pricing and hedging errors of the BS model. See Bates (1996a) for a review of alternative approaches to estimating the BS model's implied volatility.

Observe in Table 23.2 that for the overall sample period, the average implied standard deviation is 19.27% by the SVSI model, 19.02% by the SV, 18.14% by the SI, and 18.47% by the BS model, where the difference between the highest and the lowest is only 1.13%. For each subperiod, the implied volatility estimates are similarly close across the four models. This is somewhat surprising. It should, however, be recognized that this comparison is based only on the average estimates over a given period. When we examined the day-to-day time-series paths of the four models' implied volatility estimates, we found the difference between two models' implied standard deviations to be sometimes as

Table 23.2 Estimates of the structural parameters for stochastic interest rates (SI), stochastic volatility (SV), and stochastic volatility and stochastic interest rates (SVSI) models

Parameters	SI	SV	SVSI	SI	SV	SVSI	SI	SV	SVSI	SI	SV	SVSI
	06:1988–05:1991			06:1988–05:1989			06:1989–05:1990			06:1990–05:1991		
κ_R	2.35		0.61	2.51		0.49	2.29		0.61	2.03		0.76
	(0.03)		(0.02)	(0.06)		(0.02)	(0.05)		(0.03)	(0.05)		(0.04)
θ_R	0.35		0.02	0.33		0.02	0.34		0.02	0.35		0.02
	(0.00)		(0.00)	(0.01)		(0.00)	(0.01)		(0.00)	(0.01)		(0.00)
σ_R	0.04		0.03	0.04		0.04	0.04		0.03	0.04		0.03
	(0.00)		(0.00)	(0.00)		(0.00)	(0.00)		(0.00)	(0.01)		(0.00)
κ_υ		1.10	0.91		1.29	1.16		1.05	0.78		0.94	0.76
		(0.02)	(0.03)		(0.05)	(0.07)		(0.03)	(0.05)		(0.04)	(0.04)
θ_υ		0.04	0.04		0.04	0.04		0.04	0.04		0.04	0.04
		(0.00)	(0.01)		(0.00)	(0.00)		(0.00)	(0.00)		(0.00)	(0.00)
σ_υ		0.38	0.41		0.30	0.35		0.42	0.44		0.43	0.46
		(0.00)	(0.00)		(0.01)	(0.01)		(0.01)	(0.01)		(0.00)	(0.00)
ρ		−0.64	−0.76		−0.54	−0.73		−0.62	−0.74		−0.76	−0.82
		(0.01)	(0.01)		(0.01)	(0.01)		(0.01)	(0.01)		(0.01)	(0.01)
$\sqrt{V(t)}(\%)$	18.14	19.02	19.27	17.22	18.41	18.50	17.43	17.61	18.10	20.14	21.19	21.47
	(0.13)	(0.15)	(0.15)	(0.15)	(0.17)	(0.18)	(0.13)	(0.18)	(0.21)	(0.25)	(0.35)	(0.35)

Each day in the sample, the structural parameters of a given model are estimated by minimizing the sum of squared pricing errors between the market price and the model-determined price for each option. The daily average of the estimated parameters is reported first, followed by its standard error in parentheses. The average implied volatility obtained from inverting the Black–Scholes model (using a short-term at-the-money option) is, respectively, 18.47, 17.72, 17.41, and 20.52% over the sample periods: 06:1988–05:1991, 06:1988–05:1989, 06:1989–05:1990, and 06:1990–05:1991

high as 6%. Economically, option prices and hedge ratios are generally quite sensitive to the volatility input [see Figlewski (1989)]. Even small differences in the implied volatility estimate can lead to significantly different pricing and hedging results.

23.4.1 Static Performance

To examine out-of-sample cross-sectional pricing performance for each model, we use *previous day's* option prices to back out the required parameter values and then use them as input to compute current day's model-based option prices. Next, subtract the model-determined price from the observed market price, to compute both the absolute pricing error and the percentage pricing error. This procedure is repeated for each call and each day in the sample, to obtain the average absolute and the average percentage pricing errors and their associated standard errors. These steps are separately followed for each of the BS, the SI, the SV, and the SVSI models. The results from this exercise are reported in Table 23.3.

Let us first examine the relative performance in pricing OTM options. Overpricing of OTM options is often considered a critical problem for the BS model [e.g., McBeth and Merville (1979) and Rubinstein (1985)]. Panel A of Table 23.3 reports the absolute and the percentage pricing error estimates for OTM options. According to both error measures, the overall ranking of the four models is consistent with our priors: the SVSI model outperforms all others, followed by the SV, the SI, and finally the BS model. For extremely short-term (<30 days) and extremely out-of-the-money ($\frac{S}{K}<0.93$) options, for example, the average absolute pricing error by the SVSI model is \$0.23 versus \$0.53 by the BS, \$0.28 by the SI, and \$0.25 by the SV model. For this category, the

Table 23.3 Out-of-sample pricing errors

Moneyness $\frac{S}{K}$	Model	Percentage pricing error						Absolute pricing error					
		Term-to-expiration (days)						Term-to-expiration (days)					
		<30	30–60	60–90	90–120	120–180	≥ 180	<30	30–60	60–90	90–120	120–180	≥ 180
Panel A: Out-of-the-money options													
<0:93	BS	−65.99 (12.02)	−86.80 (4.51)	−62.45 (2.96)	−57.63 (2.92)	−47.71 (1.37)	−33.72 (1.05)	0.53 (0.10)	1.00 (0.04)	1.14 (0.05)	1.50 (0.06)	1.96 (0.05)	2.36 (0.06)
	SI	−24.53 (6.59)	−58.13 (3.81)	−40.04 (2.60)	−28.43 (1.67)	−16.70 (0.95)	−3.92 (0.63)	0.24 (0.04)	0.66 (0.03)	0.72 (0.04)	0.80 (0.03)	0.91 (0.05)	0.96 (0.05)
	SV	−22.08 (6.90)	−30.38 (3.07)	−12.43 (1.54)	−4.02 (0.90)	0.89 (0.47)	6.08 (0.39)	0.25 (0.04)	0.44 (0.03)	0.34 (0.02)	0.33 (0.02)	0.43 (0.04)	0.62 (0.05)
	SVSI	−16.29 (7.79)	−21.96 (2.64)	−5.68 (1.40)	−1.68 (0.93)	0.92 (0.51)	0.18 (0.64)	0.23 (0.05)	0.38 (0.02)	0.29 (0.02)	0.33 (0.02)	0.46 (0.04)	0.66 (0.04)
0.93–0.95	BS	−53.68 (5.31)	−54.50 (2.08)	−33.82 (1.79)	−21.88 (1.25)	−16.43 (0.61)	−11.25 (0.56)	0.56 (0.04)	0.90 (0.03)	1.05 (0.04)	1.24 (0.06)	1.38 (0.04)	1.80 (0.06)
	SI	−42.06 (5.32)	−49.30 (2.18)	−32.22 (2.07)	−15.78 (1.07)	−10.18 (0.55)	−5.91 (0.43)	0.42 (0.03)	0.77 (0.02)	0.92 (0.05)	0.83 (0.05)	0.85 (0.03)	0.98 (0.05)
	SV	−25.68 (4.61)	−26.16 (1.43)	−8.83 (0.81)	−3.39 (0.61)	−0.55 (0.30)	1.23 (0.24)	0.40 (0.03)	0.48 (0.02)	0.35 (0.02)	0.39 (0.02)	0.39 (0.02)	0.52 (0.02)
	SVSI	−22.50 (4.53)	−18.85 (1.43)	−4.84 (0.85)	−2.39 (0.74)	0.66 (0.32)	0.71 (0.26)	0.38 (0.03)	0.44 (0.02)	0.31 (0.02)	0.42 (0.03)	0.44 (0.02)	0.58 (0.03)
0.95–0.97	BS	−36.61 (2.33)	−28.83 (0.93)	−16.21 (0.95)	−9.91 (0.84)	−7.75 (0.41)	−5.77 (0.45)	0.55 (0.03)	0.81 (0.02)	0.87 (0.04)	1.03 (0.05)	1.05 (0.04)	1.44 (0.06)
	SI	−35.83 (2.45)	−30.09 (1.09)	−18.97 (1.30)	−7.44 (0.68)	−5.70 (0.41)	−3.62 (0.32)	0.51 (0.04)	0.81 (0.02)	0.92 (0.05)	0.69 (0.04)	0.79 (0.03)	0.86 (0.04)
	SV	−23.68 (2.06)	−16.94 (0.68)	−5.63 (0.58)	−1.63 (0.42)	−0.26 (0.22)	0.56 (0.20)	0.45 (0.03)	0.51 (0.02)	0.40 (0.02)	0.40 (0.02)	0.41 (0.02)	0.49 (0.02)
	SVSI	−16.90 (2.01)	−13.53 (0.72)	−3.59 (0.60)	−1.80 (0.51)	0.05 (0.23)	0.30 (0.21)	0.42 (0.03)	0.49 (0.02)	0.38 (0.02)	0.47 (0.03)	0.45 (0.02)	0.56 (0.02)

(continued)

Panel B: At-the-money options													
0.97–0.99	BS	−19.94 (1.03)	−10.16 (0.49)	−4.83 (0.58)	−2.66 (0.56)	−2.21 (0.30)	−1.42 (0.35)	0.51 (0.02)	0.66 (0.02)	0.63 (0.03)	0.77 (0.05)	0.82 (0.03)	1.18 (0.05)
	SI	−22.85 (1.16)	−11.38 (0.59)	−7.76 (0.70)	−1.69 (0.46)	−2.29 (0.30)	−1.76 (0.23)	0.53 (0.02)	0.68 (0.02)	0.71 (0.04)	0.58 (0.03)	0.67 (0.03)	0.74 (0.03)
	SV	−18.93 (1.01)	−8.37 (0.40)	−2.76 (0.37)	−0.44 (0.28)	−0.16 (0.17)	−0.04 (0.15)	0.50 (0.02)	0.52 (0.02)	0.37 (0.02)	0.37 (0.02)	0.40 (0.02)	0.48 (0.02)
	SVSI	−17.10 (1.05)	−7.74 (0.45)	−2.36 (0.39)	−1.04 (0.37)	−0.42 (0.17)	−0.29 (0.15)	0.49 (0.02)	0.53 (0.02)	0.38 (0.02)	0.46 (0.03)	0.44 (0.02)	0.52 (0.02)
0.99–1.01	BS	−4.42 (0.47)	−1.29 (0.30)	1.14 (0.39)	1.44 (0.45)	1.06 (0.24)	0.86 (0.26)	0.44 (0.01)	0.57 (0.02)	0.54 (0.03)	0.74 (0.04)	0.77 (0.03)	0.97 (0.04)
	SI	−5.97 (0.57)	−2.28 (0.38)	−1.08 (0.48)	1.81 (0.39)	0.29 (0.25)	−0.63 (0.22)	0.50 (0.02)	0.63 (0.02)	0.59 (0.04)	0.68 (0.05)	0.70 (0.03)	0.75 (0.04)
	SV	−7.93 (0.47)	−3.72 (0.26)	−0.76 (0.27)	0.51 (0.25)	0.13 (0.14)	0.05 (0.13)	0.50 (0.02)	0.50 (0.02)	0.38 (0.02)	0.43 (0.02)	0.42 (0.02)	0.51 (0.02)
	SVSI	−8.22 (0.49)	−3.93 (0.29)	−0.74 (0.30)	−0.40 (0.31)	−0.31 (0.15)	−0.25 (0.15)	0.51 (0.02)	0.55 (0.02)	0.42 (0.03)	0.51 (0.03)	0.46 (0.02)	0.59 (0.03)
1.01–1.03	BS	2.43 (0.23)	3.01 (0.19)	4.15 (0.29)	3.76 (0.38)	3.36 (0.19)	2.47 (0.28)	0.46 (0.01)	0.63 (0.02)	0.77 (0.04)	0.96 (0.05)	1.01 (0.03)	1.42 (0.05)
	SI	1.65 (0.27)	2.29 (0.23)	2.08 (0.38)	3.52 (0.30)	1.76 (0.19)	−0.34 (0.20)	0.47 (0.02)	0.65 (0.02)	0.70 (0.04)	0.88 (0.05)	0.76 (0.03)	0.81 (0.04)
	SV	−0.99 (0.23)	−0.69 (0.17)	0.46 (0.22)	1.04 (0.20)	0.38 (0.11)	−0.21 (0.11)	0.39 (0.01)	0.42 (0.02)	0.37 (0.02)	0.44 (0.03)	0.42 (0.02)	0.49 (0.02)
	SVSI	−1.49 (0.23)	−1.16 (0.18)	0.12 (0.24)	0.27 (0.24)	−0.19 (0.12)	−0.55 (0.11)	0.41 (0.02)	0.46 (0.02)	0.39 (0.03)	0.48 (0.03)	0.46 (0.02)	0.54 (0.02)

(continued)

Panel C: In-the-money options

1.03–1.05	BS	3.69 (0.13)	4.45 (0.14)	5.31 (0.24)	4.76 (0.29)	4.38 (0.17)	2.98 (0.24)	0.59 (0.02)	0.85 (0.03)	1.14 (0.05)	1.20 (0.07)	1.29 (0.04)	1.40 (0.06)
	SI	3.37 (0.15)	3.83 (0.18)	3.64 (0.28)	4.08 (0.25)	2.51 (0.17)	0.17 (0.19)	0.57 (0.02)	0.82 (0.03)	0.92 (0.05)	1.04 (0.06)	0.90 (0.04)	0.83 (0.04)
	SV	1.27 (0.13)	0.79 (0.13)	1.09 (0.17)	1.11 (0.15)	0.43 (0.10)	−0.20 (0.11)	0.38 (0.01)	0.42 (0.02)	0.42 (0.03)	0.43 (0.03)	0.42 (0.02)	0.50 (0.02)
	SVSI	0.84 (0.13)	0.32 (0.14)	0.83 (0.19)	0.29 (0.20)	−0.03 (0.10)	−0.41 (0.11)	0.37 (0.01)	0.45 (0.02)	0.42 (0.03)	0.49 (0.03)	0.45 (0.02)	0.54 (0.03)
1.05–1.07	BS	3.37 (0.10)	4.54 (0.11)	5.57 (0.22)	5.08 (0.27)	4.82 (0.14)	4.27 (0.22)	0.70 (0.02)	1.06 (0.02)	1.46 (0.05)	1.47 (0.06)	1.56 (0.04)	1.73 (0.06)
	SI	3.28 (0.12)	4.02 (0.14)	4.08 (0.24)	4.47 (0.22)	2.65 (0.15)	0.59 (0.18)	0.69 (0.02)	0.97 (0.03)	1.13 (0.06)	1.29 (0.06)	0.99 (0.04)	0.88 (0.05)
	SV	1.82 (0.09)	1.41 (0.09)	1.47 (0.16)	1.44 (0.14)	0.54 (0.09)	−0.40 (0.11)	0.45 (0.02)	0.46 (0.02)	0.53 (0.03)	0.50 (0.03)	0.45 (0.02)	0.57 (0.03)
	SVSI	1.59 (0.09)	1.12 (0.10)	1.35 (0.17)	0.83 (0.17)	0.17 (0.09)	−0.52 (0.11)	0.42 (0.01)	0.46 (0.02)	0.52 (0.03)	0.51 (0.04)	0.44 (0.02)	0.58 (0.03)
> 1.07	BS	1.79 (0.04)	2.65 (0.05)	2.96 (0.07)	3.10 (0.08)	3.36 (0.05)	2.61 (0.05)	0.60 (0.01)	0.95 (0.01)	1.22 (0.02)	1.35 (0.02)	1.56 (0.02)	1.58 (0.02)
	SI	1.86 (0.05)	2.50 (0.06)	2.14 (0.07)	2.45 (0.08)	1.63 (0.05)	−0.76 (0.05)	0.59 (0.01)	0.89 (0.02)	0.88 (0.02)	1.04 (0.03)	0.86 (0.02)	1.09 (0.02)
	SV	1.36 (0.13)	1.33 (0.13)	1.06 (0.17)	0.92 (0.15)	0.45 (0.10)	−0.64 (0.11)	0.50 (0.01)	0.55 (0.01)	0.52 (0.01)	0.49 (0.02)	0.42 (0.01)	0.64 (0.01)
	SVSI	1.29 (0.03)	1.26 (0.03)	1.18 (0.04)	0.81 (0.04)	0.40 (0.03)	−0.37 (0.03)	0.46 (0.01)	0.55 (0.01)	0.58 (0.01)	0.52 (0.02)	0.44 (0.01)	0.57 (0.01)

For a given model, compute the price of each option using previous day's implied parameters and implied stock volatility. The reported percentage pricing error is the sample average of the market price minus the model price divided by the market price. The reported absolute pricing error is the sample average of the absolute difference between the market price and the model price for each call. The corresponding standard errors are recorded in parentheses. The sample period is 06:1988–05:1991, with a total of 38,749 call option prices

BS model's absolute pricing error is cut by more than a half by each of the other three models. Fix the moneyness category at $\frac{S}{K} \in (0.93;\ 0.95)$. Then, for medium-term (120–180 days) options, the SVSI model produces an average absolute pricing error of \$0.44 versus \$1.38 by the BS, \$0.72 by the SI, and \$0.39 by the SV model. For short-term (30–60 day) calls, the absolute pricing errors are \$0.44 by the SVSI, \$0.48 by the SV, \$0.73 by the SI, and \$0.90 by the BS model. Clearly, the performance improvement is significant for each moneyness and maturity category in Panel A, from the BS to the SI, to the SV, and to the SVSI model. This pricing performance ranking of the four models can also be seen using the average percentage pricing errors, as given in the same table. Here, the SVSI model produces percentage pricing errors that are the lowest in magnitude. As an example, take OTM options with term-to-expiration of 30–60 days and with $\frac{S}{K} \in (0.93;\ 0.95)$. In this category, the BS, the SI, the SV, and the SVSI models, respectively, have average percentage pricing errors of −54.50%, −46.20%, −26.16%, and −18.85%. For long-term options with $\frac{S}{K} \in (0.93;\ 0.95)$ and with $\frac{S}{K} \in (0.95;\ 0.97)$, the SVSI model results in a percentage pricing error that is as low as 0.71 and 0.30%, respectively.

For ATM calls, recall that the BS model's implied volatility input is backed out from the (previous day's) short-term ATM options, which should give the BS model a relative advantage in pricing ATM options. In contrast, the implied spot variance for the other models is obtained by minimizing the sum of squared errors for all options of the previous day. Thus, for ATM options, one would expect the BS model to perform relatively better. As seen from Panel B of Table 23.3, except for the shortest-term ATM calls, the SVSI model typically generates the lowest absolute and percentage pricing errors (especially for longer-term options), followed by the SV, by the SI and finally by the BS model. For the shortest-term options with $\frac{S}{K} \in (0.97;\ 0.99)$ and $\frac{S}{K} \in (0.99;\ 1.01)$, the BS and the SI models perform somewhat better than the other two.

Panel C of Table 23.3 reports the average absolute and percentage pricing errors of ITM calls by all four models. While the previous ranking of the models based on OTM and ATM options is preserved by Panel C, it can be noted that the average percentage pricing error is below 1.0% for 12 out of the 18 categories in the case of the SVSI model, for 8 out of the 18 categories in the case of the SV model, for 3 categories out of 18 for the SI model, and for none of the 18 categories in the case of the BS model. The pricing improvement by the SV and the SVSI models over the BS and the SI is quite substantial for ITM options, especially for long-term options.

Some patterns of mispricing can, however, be noted across all moneyness-maturity categories. First, all four models produce negative percentage pricing errors for options with moneyness $\frac{S}{K} \leq 0.99$, and positive percentage pricing errors for options with $\frac{S}{K} \geq 1.03$, subject to their time-to-expiration not exceeding 120 days. This means that the models systematically overprice OTM call options while underprice ITM calls. But the magnitude of such mispricing varies dramatically across the models, with the BS producing the strongest and the SVSI model the weakest systematic biases. Next, according to the absolute pricing error measure, the SV model seems to perform slightly better than the SVSI in pricing calls with more than 90 days to expiration. This pattern is, however, not supported by the percentage pricing errors reported in Table 23.3, possibly because for these relatively long-term calls the two models produce pricing errors that have mixed signs, in which case taking the average absolute value of the pricing errors can sometimes distort the picture. According to the percentage pricing errors, the SVSI model does slightly better than the SV in pricing those longer-term options. Finally, for the BS model, its absolute pricing error has a U-shaped relationship (i.e., "smile") with moneyness, and the magnitude of its percentage pricing error increases as the call goes from deep in-the-money to deep out-of-the-money, regardless of time to expiration. These patterns are reduced by each relaxation of the BS model assumptions.

23.4.2 Dynamic Hedging Performance

Recall that in implementing a hedge using any of the four models, we follow three basic steps. First, based on Procedure B of Sect. 2.3, estimate the structural parameters and spot variance by using day 1's option prices. Next, on day 2, use previous day's parameter and spot volatility estimates and current day's spot price and interest rates, to construct the desired hedge as given in Sect. 2.2. Finally, rely on either Eq. (23.22) or Eq. (23.25) to calculate the hedging error as of day 3. We then compute both the average absolute and the average dollar hedging errors of all call options in a given moneyness-maturity category, to gauge the relative hedging performance of each model.

It should be recognized that in both the delta-neutral and the minimum-variance hedging exercises conducted in the two subsections below, the spot S&P 500 index, rather than an S&P 500 futures contract, is used in place of the "spot asset" for the hedges devised in Sect. 2.2. This is done out of two considerations. First, the spot S&P 500 and the immediate-expiration month S&P 500 futures price generally have a correlation coefficient close to one. This means that whether the spot index or the futures price is used in the hedging exercises, the qualitative as well as the quantitative conclusions are most likely the same. In other words, if it is demonstrated using the spot index that one model results in better hedging performance than another, the same hedging performance ranking of the two models will likely be achieved by using an S&P 500 futures contract. After all, our main interest here lies in the relative performance of the models. Second, when a futures contract is used in constructing a hedge, a futures pricing formula has to be adopted. That will introduce another dimension of model misspecification (due to stochastic interest rates), which can in turn produce a compounded effect on the hedging results. For these reasons, using the spot index may lead to a cleaner comparison among the four option models.

23.4.2.1 Effectiveness of Delta-Neutral Hedges

Observe that the construction and the execution of the hedging strategy in Eq. (23.22) requires, in the cases of the SV and the SVSI models, (i) the availability of prices for four time-matched target and hedging-instrumental options: $C(t,\tau,\overline{K})$, $C(t,\tau;K)$, $C(t,\Delta t,\tau-\Delta t;K)$, $C(t,\Delta t,\tau-\Delta t;\overline{K})$ and (ii) the computation of S, V, and R for the target and the instrumental option. Due to this requirement, it is important to match as closely as possible the time points at which the target and the instrumental option prices were respectively taken, in order to ensure that the hedge ratios are properly determined. For this reason, we use as hedging instruments only options whose prices on both the hedge construction day and the following liquidation day were quoted no more than 15 s apart from the times when the respective prices for the target option were quoted. This requirement makes the overall sample for the hedging exercise smaller than that used for the preceding pricing exercise, but it nonetheless guarantees that the deltas for the target and instrumental options on the same day are computed based on the same spot price. The remaining sample contains 15,041 matched pairs when hedging revision occurs at one-day intervals, and 11,704 matched pairs when rebalancing takes place at five-day intervals. In addition, we partition the target options into three maturity classes: less than 60 days, 60–180 days, and greater than 180 days, and report hedging results accordingly.

In theory, a call option with any expiration date and any strike price can be chosen as a hedging instrument for any given target option. In practice, however, different choices can mean different hedging effectiveness, even for the same option pricing model. Out of this consideration, we employ as a hedging instrument the call option which has the same expiration date as the target option and whose strike price is the closest, but not identical, to the target options.

Table 23.4 presents delta-neutral hedging results for the four models. Several patterns emerge from Table 23.4. First, the BS model

Table 23.4 Delta-neutral hedging errors

Moneyness $\frac{S}{K}$	Model	Dollar hedging error						Absolute hedging error					
		1-day revision			5-day revision			1-day revision			5-day revision		
		Term-to-expiration (days)						Term-to-expiration (days)					
		<60	60–180	180	<60	60–180	$\geq$ 180	<60	60–180	180	<60	60–180	$\geq$ 180
Panel A: Out-of-the-money options													
<0.93	BS	NA	−0.06 (0.03)	−0.04 (0.02)	NA	−0.33 (0.09)	−0.21 (0.06)	NA	0.37 (0.01)	0.45 (0.01)	NA	0.91 (0.06)	0.87 (0.04)
	SI		−0.08 (0.02)	−0.05 (0.02)		−0.40 (0.05)	−0.36 (0.06)		0.35 (0.01)	0.40 (0.01)		0.69 (0.03)	0.81 (0.04)
	SV		0.02 (0.02)	0.01 (0.02)		0.03 (0.02)	0.03 (0.02)		0.15 (0.01)	0.14 (0.01)		0.17 (0.02)	0.21 (0.02)
	SVSI		0.02 (0.02)	0.00 (0.02)		0.03 (0.02)	0.00 (0.03)		0.16 (0.01)	0.14 (0.01)		0.17 (0.01)	0.22 (0.02)
0.93–0.95	BS	NA	−0.06 (0.02)	−0.01 (0.03)	NA	−0.24 (0.07)	−0.02 (0.07)	NA	0.32 (0.01)	0.46 (0.02)	NA	0.79 (0.04)	0.82 (0.05)
	SI		−0.08 (0.02)	0.00 (0.04)		−0.29 (0.05)	−0.22 (0.07)		0.33 (0.01)	0.43 (0.02)		0.67 (0.03)	0.66 (0.05)
	SV		−0.01 (0.01)	−0.01 (0.02)		−0.00 (0.01)	0.00 (0.03)		0.12 (0.01)	0.23 (0.01)		0.13 (0.01)	0.18 (0.02)
	SVSI		−0.01 (0.01)	−0.01 (0.02)		−0.00 (0.01)	0.00 (0.03)		0.12 (0.01)	0.24 (0.02)		0.13 (0.01)	0.18 (0.02)
0.95–0.97	BS	−0.08 (0.06)	−0.06 (0.02)	−0.01 (0.03)	−0.55 (0.16)	−0.21 (0.06)	−0.12 (0.08)	0.23 (0.04)	0.33 (0.02)	0.45 (0.03)	0.66 (0.13)	0.77 (0.04)	0.85 (0.05)
	SI	−0.06 (0.02)	−0.06 (0.02)	−0.06 (0.04)	−0.22 (0.05)	−0.31 (0.05)	−0.34 (0.09)	0.27 (0.01)	0.34 (0.01)	0.41 (0.02)	0.61 (0.03)	0.71 (0.03)	0.89 (0.06)
	SV	0.03 (0.03)	−0.01 (0.01)	−0.01 (0.02)	−0.03 (0.03)	−0.00 (0.01)	−0.02 (0.03)	0.10 (0.02)	0.13 (0.01)	0.19 (0.01)	0.10 (0.02)	0.14 (0.01)	0.26 (0.02)
	SVSI	(0.03) 0.02	−0.01 (0.01)	−0.01 (0.02)	−0.02 (0.02)	−0.01 (0.01)	−0.01 (0.03)	0.09 (0.02)	0.13 (0.01)	0.20 (0.01)	0.08 (0.02)	0.14 (0.01)	0.27 (0.02)

(continued)

Table 23.4 (continued)

Panel B: At-the-money options													
Panel B: At-the-money options													
0.97–0.99	BS	−0.01 (0.05)	−0.04 (0.02)	−0.04 (0.03)	−0.36 (0.08)	−0.11 (0.05)	−0.21 (0.07)	0.34 (0.03)	0.34 (0.01)	0.46 (0.02)	0.54 (0.06)	0.75 (0.03)	0.89 (0.05)
	SI	0.08 (0.01)	−0.05 (0.02)	−0.07 (0.02)	−0.30 (0.05)	−0.25 (0.05)	−0.44 (0.08)	0.29 (0.01)	0.36 (0.01)	0.41 (0.02)	0.73 (0.03)	0.70 (0.03)	0.79 (0.06)
	SV	−0.03 (0.02)	0.01 (0.01)	0.00 (0.01)	−0.04 (0.03)	−0.00 (0.01)	−0.01 (0.02)	0.12 (0.01)	0.13 (0.01)	0.17 (0.01)	0.14 (0.02)	0.14 (0.01)	0.23 (0.01)
	SVSI	−0.02 (0.02)	0.01 (0.01)	0.00 (0.01)	0.02 (0.03)	0.00 (0.01)	−0.01 (0.02)	0.12 (0.01)	0.13 (0.01)	0.17 (0.01)	0.13 (0.02)	0.14 (0.01)	0.24 (0.01)
0.99–1.01	BS	−0.10 (0.01)	−0.02 (0.02)	−0.01 (0.03)	−0.43 (0.09)	−0.08 (0.05)	−0.10 (0.07)	0.37 (0.01)	0.37 (0.01)	0.47 (0.02)	0.80 (0.06)	0.77 (0.03)	0.77 (0.05)
	SI	−0.08 (0.02)	−0.05 (0.02)	−0.03 (0.03)	−0.29 (0.05)	−0.24 (0.05)	−0.18 (0.08)	0.36 (0.01)	0.37 (0.01)	0.42 (0.02)	0.81 (0.03)	0.67 (0.03)	0.61 (0.05)
	SV	0.01 (0.02)	−0.00 (0.01)	−0.01 (0.01)	0.02 (0.02)	0.01 (0.01)	0.03 (0.02)	0.14 (0.01)	0.13 (0.01)	0.17 (0.01)	0.15 (0.02)	0.15 (0.01)	0.25 (0.02)
	SVSI	0.01 (0.02)	0.00 (0.01)	−0.01 (0.01)	0.02 (0.02)	−0.00 (0.01)	0.04 (0.02)	0.14 (0.03)	0.13 (0.01)	0.17 (0.01)	0.16 (0.02)	0.15 (0.01)	0.25 (0.01)
1.01–1.03	BS	−0.09 (0.03)	−0.02 (0.02)	−0.01 (0.03)	−0.40 (0.08)	−0.11 (0.05)	−0.09 (0.07)	0.40 (0.02)	0.39 (0.01)	0.46 (0.02)	0.82 (0.05)	0.75 (0.03)	0.82 (0.05)
	SI	−0.09 (0.02)	−0.05 (0.02)	−0.07 (0.04)	−0.30 (0.05)	−0.25 (0.05)	−0.27 (0.09)	0.38 (0.01)	0.36 (0.01)	0.43 (0.01)	0.75 (0.03)	0.65 (0.03)	0.71 (0.06)
	SV	0.00 (0.03)	−0.00 (0.01)	0.03 (0.01)	0.01 (0.02)	−0.01 (0.01)	0.05 (0.02)	0.13 (0.01)	0.14 (0.01)	0.17 (0.01)	0.13 (0.01)	0.17 (0.01)	0.24 (0.02)
	SVSI	0.00 (0.01)	−0.00 (0.01)	0.03 (0.01)	−0.00 (0.01)	−0.01 (0.01)	0.05 (0.02)	0.14 (0.01)	0.14 (0.01)	0.17 (0.01)	0.13 (0.01)	0.17 (0.01)	0.24 (0.02)

(continued)

Table 23.4 (continued)

Panel C: In-the-money options

Panel C: In-the-money options

1.03–1.05	BS	−0.06 (0.02)	−0.03 (0.02)	−0.05 (0.03)	−0.36 (0.05)	−0.09 (0.04)	−0.23 (0.08)	0.40 (0.02)	0.38 (0.01)	0.47 (0.02)	0.70 (0.03)	0.69 (0.03)	0.90 (0.05)
	SI	−0.09 (0.02)	−0.05 (0.02)	−0.07 (0.04)	−0.31 (0.06)	−0.19 (0.05)	−0.35 (0.09)	0.39 (0.01)	0.36 (0.01)	0.41 (0.02)	0.65 (0.03)	0.61 (0.03)	0.81 (0.06)
	SV	0.01 (0.03)	0.01 (0.01)	−0.01 (0.02)	0.00 (0.01)	0.00 (0.01)	−0.03 (0.03)	0.15 (0.01)	0.13 (0.01)	0.15 (0.01)	0.17 (0.01)	0.15 (0.01)	0.24 (0.02)
	SVSI	0.01 (0.01)	0.00 (0.01)	−0.00 (0.02)	0.00 (0.01)	0.00 (0.01)	−0.03 (0.02)	0.15 (0.01)	0.12 (0.01)	0.16 (0.01)	0.16 (0.01)	0.14 (0.01)	0.25 (0.01)
1.05–1.07	BS	−0.05 (0.02)	−0.02 (0.02)	−0.06 (0.04)	−0.35 (0.04)	−0.06 (0.04)	−0.22 (0.07)	0.41 (0.01)	0.40 (0.01)	0.47 (0.02)	0.68 (0.02)	0.64 (0.03)	0.77 (0.04)
	SI	−0.07 (0.02)	−0.04 (0.02)	−0.11 (0.04)	−0.26 (0.04)	−0.12 (0.04)	−0.56 (0.08)	0.40 (0.01)	0.37 (0.02)	0.44 (0.02)	0.57 (0.03)	0.51 (0.03)	0.59 (0.01)
	SV	−0.00 (0.01)	−0.00 (0.01)	−0.01 (0.02)	−0.05 (0.01)	−0.02 (0.01)	0.00 (0.02)	0.16 (0.01)	0.13 (0.01)	0.18 (0.01)	0.18 (0.01)	0.15 (0.01)	0.22 (0.02)
	SVSI	−0.00 (0.01)	0.00 (0.01)	−0.00 (0.02)	−0.03 (0.01)	−0.02 (0.00)	0.00 (0.02)	0.15 (0.01)	0.12 (0.01)	0.17 (0.01)	0.17 (0.01)	0.15 (0.01)	0.22 (0.01)
> 1.07	BS	−0.04 (0.01)	−0.03 (0.00)	−0.02 (0.01)	−0.15 (0.02)	−0.07 (0.02)	−0.10 (0.03)	0.36 (0.01)	0.39 (0.01)	0.48 (0.01)	0.51 (0.01)	0.58 (0.01)	0.72 (0.02)
	SI	−0.05 (0.03)	−0.04 (0.01)	−0.03 (0.02)	−0.18 (0.02)	0.08 (0.02)	−0.21 (0.08)	0.35 (0.01)	0.37 (0.01)	0.43 (0.01)	0.45 (0.01)	0.51 (0.01)	0.66 (0.02)
	SV	−0.01 (0.01)	−0.01 (0.00)	−0.00 (0.01)	−0.03 (0.01)	−0.01 (0.00)	0.01 (0.01)	0.15 (0.01)	0.14 (0.01)	0.20 (0.01)	0.17 (0.00)	0.18 (0.00)	0.27 (0.00)
	SVSI	−0.00 (0.00)	−0.00 (0.00)	0.00 (0.01)	−0.01 (0.01)	0.00 (0.00)	0.01 (0.01)	0.15 (0.00)	0.14 (0.00)	0.20 (0.00)	0.16 (0.00)	0.17 (0.00)	0.27 (0.01)

For each call option, calculate the hedging error, which is the difference between the market price of the call and the replicating portfolio. The average dollar hedging error and the average absolute hedging error are reported for each model. The standard errors are given in parentheses. The sample period is 06:1988–05:1991. In calculating the hedging errors generated with daily (once every five days) hedge rebalancing, 15,041 (11,704) observations are used

produces the worst hedging performance by most measures, the SI shows noticeable improvement according to the average dollar hedging errors (especially in the five-day hedging revision categories) but not so according to the average absolute hedging errors, while the SV and the SVSI models have average absolute and average dollar hedging errors that are typically one-third of the corresponding BS hedging errors, or lower. The improvement by the SV and the SVSI is thus remarkable. Second, as portfolio adjustment frequency decreases from daily to once every 5 days, hedging effectiveness deteriorates, regardless of the model used. The deterioration is especially apparent for OTM and ATM options with $\frac{S}{K} \leq 1.05$. It should, however, be noted with emphasis that for both the SV and the SVSI models, their hedging effectiveness is relatively stable, whether the hedges are rebalanced each day or once every 5 days. For the BS and the SI models, such a change in revision frequency can mean doubling their hedging errors. This finding is strong evidence in support of the SV and the SVSI models for hedging.

Third, the BS model-based delta-neutral hedging strategy always overhedges a target call option, as its average dollar hedging error is negative for each moneyness-maturity category and at either frequency of portfolio rebalancing. In contrast, the dollar hedging errors based on the SV and the SVSI models are more random and can take either sign. Therefore, the BS formula has a systematic hedging bias pattern, whereas the SV and the SVSI do not.

Fourth, the SVSI model is indistinguishable from the SV according to their absolute hedging errors, but is slightly better than the latter when judged using their average dollar hedging errors. Similarly, the SI model has worse hedging performance than the BS according to their absolute hedging error values, but the reverse is true according to their dollar hedging errors. This phenomenon exists possibly because with stochastic interest rates there are larger hedging errors of opposite signs, so that when added together, these errors cancel out, but the sum of their absolute values is nonetheless large.

Finally, no matter which model is used, there do not appear to be moneyness- or maturity-related bias patterns in the hedging errors. In other words, hedging errors do not seem to "smile" across exercise prices or times to expiration as pricing errors do. This is a striking disparity between pricing and hedging results.

23.4.2.2 Effectiveness of Single-Instrument Minimum-Variance Hedges

If one is, for reasons given before, constrained to using only the underlying stock to hedge a target call option, dimensions of uncertainty that move the target option value but are uncorrelated with the underlying stock price cannot be hedged by any position in the stock and will necessarily be uncontrolled for in such a single-instrument minimum-variance hedge. Based on the sample option data, the average absolute and the average dollar hedging errors, with either a daily or a five-day rebalancing frequency, are given in Table 23.5 for each of the four models and each of the moneyness-maturity categories. With this type of hedges, the relative performance of the models is no longer clear-cut. For OTM options with $\frac{S}{K} \leq 0.97$, the SV model has, regardless of the hedging error measure used and the hedge revision frequency adopted, the lowest hedging errors, followed by the SVSI, then by the BS, and lastly by the SI model. For ATM options, the hedging performance by the BS and the SV models is almost indistinguishable, but still better, by a small margin, than that by both the SI and the SVSI models, whereas the latter two models' performance is also indistinguishable. Finally, for ITM options, the BS model has the best hedging performance, followed by the SV, the SVSI, and then by the SI model. Having said the above, it should nonetheless be noted that for virtually all cases in Table 23.5 the hedging error differences among the BS, the SV, and the SVSI models are economically insignificant because of their low magnitude. Only the SI model's performance appears to be significantly poorer than the others.

Table 23.5 Single-instrument hedging errors

Moneyness $\frac{S}{K}$	Model	Dollar hedging error						Absolute hedging error					
		1-day revision			5-day revision			1-day revision			5-day revision		
		Term-to-expiration (days)						Term-to-expiration (days)					
		<60	60–180	180	<60	60–180	≥ 180	<60	60–180	180	<60	60–180	≥ 180
Panel A: Out-of-the-money options													
<0.93	BS	NA	−0.06 (0.03)	−0.04 (0.02)	NA	−0.33 (0.09)	−0.21 (0.06)	NA	0.37 (0.01)	0.45 (0.01)	NA	0.91 (0.06)	0.87 (0.04)
	SI		−0.09 (0.05)	−0.06 (0.04)		−0.51 (0.17)	−0.45 (0.08)		0.49 (0.03)	0.52 (0.03)		1.33 (0.10)	0.91 (0.05)
	SV		−0.02 (0.03)	−0.05 (0.02)		−0.03 (0.08)	−0.09 (0.05)		0.30 (0.02)	0.39 (0.02)		0.75 (0.05)	0.75 (0.04)
	SVSI		0.02 (0.03)	−0.04 (0.03)		0.11 (0.08)	−0.13 (0.05)		0.35 (0.03)	0.43 (0.02)		0.76 (0.05)	0.82 (0.04)
0.93–0.95	BS	NA	−0.06 (0.02)	−0.01 (0.03)	NA	−0.24 (0.07)	−0.02 (0.07)	NA	0.32 (0.01)	0.46 (0.02)	NA	0.79 (0.04)	0.82 (0.05)
	SI		−0.10 (0.03)	−0.00 (0.04)		−0.36 (0.10)	−0.38 (0.08)		0.35 (0.02)	0.52 (0.03)		0.97 (0.06)	0.72 (0.06)
	SV		−0.06 (0.02)	0.00 (0.03)		−0.14 (0.07)	−0.00 (0.06)		0.33 (0.02)	0.43 (0.02)		0.79 (0.04)	0.73 (0.04)
	SVSI		−0.04 (0.03)	−0.00 (0.03)		−0.17 (0.07)	−0.17 (0.07)		0.36 (0.02)	0.47 (0.02)		0.82 (0.04)	0.78 (0.04)

(continued)

Table 23.5 (continued)

Moneyness $\frac{S}{K}$	Model	Dollar hedging error						Absolute hedging error					
		1-day revision			5-day revision			1-day revision			5-day revision		
		Term-to-expiration (days)						Term-to-expiration (days)					
		<60	60–180	180	<60	60–180	≥ 180	<60	60–180	180	<60	60–180	≥ 180
Panel A: Out-of-the-money options													
0.95–0.97	BS	−0.08 (0.06)	−0.06 (0.02)	−0.01 (0.03)	−0.55 (0.16)	−0.21 (0.06)	−0.12 (0.08)	0.23 (0.04)	0.33 (0.02)	0.45 (0.03)	0.66 (0.13)	0.77 (0.04)	0.85 (0.05)
	SI	−0.04 (0.11)	−0.09 (0.03)	0.04 (0.05)	−0.44 (0.05)	−0.38 (0.09)	−0.42 (0.10)	0.33 (0.01)	0.39 (0.01)	0.47 (0.03)	0.74 (0.09)	0.94 (0.06)	0.92 (0.07)
	SV	−0.03 (0.06)	−0.05 (0.02)	−0.01 (0.03)	−0.06 (0.16)	−0.14 (0.06)	−0.16 (0.07)	0.21 (0.04)	0.32 (0.02)	0.42 (0.02)	0.48 (0.09)	0.77 (0.04)	0.74 (0.04)
	SVSI	−0.04 (0.01)	−0.02 (0.02)	−0.00 (0.03)	−0.21 (0.03)	−0.14 (0.04)	−0.23 (0.06)	0.22 (0.03)	0.35 (0.01)	0.46 (0.02)	0.65 (0.11)	0.83 (0.04)	0.79 (0.04)
Panel B: At-the-money options													
0.97–0.99	BS	−0.01 (0.05)	−0.04 (0.02)	−0.04 (0.03)	−0.36 (0.08)	−0.11 (0.05)	−0.21 (0.07)	0.34 (0.03)	0.34 (0.01)	0.46 (0.02)	0.54 (0.06)	0.75 (0.03)	0.89 (0.05)
	SI	0.00 (0.06)	−0.06 (0.03)	−0.06 (0.04)	−0.34 (0.05)	−0.11 (0.09)	−0.42 (0.08)	0.34 (0.05)	0.40 (0.02)	0.51 (0.03)	0.82 (0.09)	0.90 (0.06)	0.82 (0.06)
	SV	−0.01 (0.06)	−0.05 (0.02)	−0.06 (0.03)	−0.20 (0.12)	−0.11 (0.06)	−0.19 (0.07)	0.35 (0.04)	0.35 (0.01)	0.44 (0.02)	0.56 (0.07)	0.81 (0.03)	0.84 (0.04)
	SVSI	−0.03 (0.06)	−0.05 (0.02)	−0.03 (0.03)	−0.15 (0.11)	−0.16 (0.06)	−0.27 (0.07)	0.36 (0.04)	0.37 (0.01)	0.45 (0.02)	0.59 (0.07)	0.86 (0.03)	0.89 (0.05)
0.99–1.01	BS	−0.10 (0.01)	−0.02 (0.02)	−0.01 (0.03)	−0.43 (0.09)	−0.08 (0.05)	−0.10 (0.07)	0.37 (0.01)	0.37 (0.01)	0.47 (0.02)	0.80 (0.06)	0.77 (0.03)	0.77 (0.05)
	SI	−0.15 (0.05)	−0.04 (0.03)	0.03 (0.05)	−0.34 (0.15)	−0.18 (0.07)	−0.31 (0.09)	0.39 (0.03)	0.41 (0.02)	0.55 (0.03)	0.77 (0.09)	0.82 (0.05)	0.78 (0.06)
	SV	−0.12 (0.04)	−0.02 (0.02)	−0.00 (0.03)	−0.22 (0.11)	−0.12 (0.05)	−0.06 (0.06)	0.38 (0.03)	0.37 (0.01)	0.45 (0.02)	0.78 (0.06)	0.79 (0.03)	0.69 (0.04)
	SVSI	−0.04	−0.01	−0.00	−0.27	−0.13	−0.15	0.38	0.40	0.47	0.88	0.89	0.80

(continued)

Table 23.5 (continued)

Panel B: At-the-money options													
		(0.01)	(0.02)	(0.03)	(0.11)	(0.05)	(0.06)	(0.03)	(0.01)	(0.02)	(0.07)	(0.04)	(0.04)
1.01–1.03	BS	−0.09 (0.03)	−0.03 (0.02)	−0.01 (0.03)	−0.40 (0.08)	−0.11 (0.05)	−0.09 (0.07)	0.40 (0.02)	0.39 (0.01)	0.46 (0.02)	0.82 (0.05)	0.75 (0.03)	0.82 (0.05)
	SI	−0.10 (0.04)	−0.05 (0.03)	−0.04 (0.05)	−0.43 (0.11)	−0.16 (0.07)	−0.34 (0.10)	0.41 (0.02)	0.41 (0.01)	0.51 (0.03)	0.86 (0.07)	0.83 (0.05)	0.89 (0.07)
	SV	−0.06 (0.03)	−0.04 (0.02)	0.00 (0.03)	−0.21 (0.09)	−0.15 (0.05)	−0.03 (0.06)	0.38 (0.02)	0.39 (0.02)	0.43 (0.02)	0.84 (0.05)	0.81 (0.03)	0.72 (0.04)
	SVSI	0.06 (0.03)	−0.04 (0.02)	−0.01 (0.03)	−0.29 (0.08)	−0.18 (0.05)	−0.17 (0.06)	0.41 (0.02)	0.42 (0.01)	0.45 (0.02)	0.91 (0.05)	0.85 (0.03)	0.76 (0.04)
Panel C: In-the-money options													
1.03–1.05	BS	−0.06 (0.02)	−0.03 (0.02)	−0.05 (0.03)	−0.36 (0.05)	−0.09 (0.04)	−0.23 (0.08)	0.40 (0.02)	0.38 (0.01)	0.47 (0.02)	0.70 (0.03)	0.69 (0.03)	0.90 (0.05)
	SI	−0.08 (0.03)	−0.03 (0.03)	−0.04 (0.05)	−0.47 (0.08)	−0.11 (0.06)	−0.52 (0.09)	0.43 (0.02)	0.40 (0.02)	0.48 (0.03)	0.84 (0.05)	0.77 (0.04)	0.97 (0.07)
	SV	−0.04 (0.03)	−0.02 (0.02)	−0.07 (0.04)	−0.23 (0.06)	−0.09 (0.05)	−0.23 (0.08)	0.41 (0.02)	0.39 (0.01)	0.46 (0.02)	0.70 (0.03)	0.76 (0.03)	0.88 (0.05)
	SVSI	−0.05 (0.02)	−0.02 (0.02)	−0.04 (0.03)	−0.27 (0.05)	−0.14 (0.05)	−0.31 (0.08)	0.41 (0.01)	0.40 (0.01)	0.47 (0.02)	0.77 (0.03)	0.80 (0.03)	0.90 (0.05)
1.05–1.07	BS	−0.05 (0.02)	−0.02 (0.02)	−0.06 (0.04)	−0.35 (0.04)	−0.06 (0.04)	−0.22 (0.07)	0.41 (0.01)	0.40 (0.01)	0.47 (0.02)	0.68 (0.02)	0.64 (0.03)	0.77 (0.04)
	SI	−0.07 (0.03)	−0.04 (0.03)	−0.07 (0.05)	−0.37 (0.05)	−0.09 (0.06)	−0.55 (0.08)	0.45 (0.02)	0.42 (0.02)	0.50 (0.03)	0.74 (0.04)	0.66 (0.04)	0.81 (0.06)
	SV	−0.06 (0.02)	−0.03 (0.02)	−0.05 (0.04)	−0.32 (0.04)	−0.09 (0.04)	−0.10 (0.07)	0.43 (0.02)	0.40 (0.01)	0.44 (0.03)	0.71 (0.02)	0.68 (0.03)	0.69 (0.04)
	SVSI	−0.05 (0.02)	−0.02 (0.02)	−0.02 (0.04)	−0.31 (0.04)	−0.09 (0.04)	−0.09 (0.07)	0.42 (0.01)	0.43 (0.01)	0.46 (0.02)	0.74 (0.03)	0.79 (0.03)	0.75 (0.05)

(continued)

Table 23.5 (continued)

Panel C: In-the-money options

>1.07	BS	−0.04 (0.01)	−0.03 (0.00)	−0.02 (0.01)	−0.15 (0.02)	−0.07 (0.02)	−0.10 (0.03)	0.36 (0.01)	0.39 (0.01)	0.48 (0.01)	0.51 (0.01)	0.58 (0.01)	0.72 (0.02)
	SI	−0.05 (0.01)	−0.04 (0.01)	−0.04 (0.02)	−0.17 (0.03)	−0.07 (0.03)	−0.26 (0.03)	0.40 (0.01)	0.41 (0.01)	0.47 (0.01)	0.61 (0.02)	0.66 (0.02)	0.79 (0.02)
	SV	−0.04 (0.04)	−0.04 (0.02)	−0.02 (0.02)	−0.18 (0.02)	−0.14 (0.02)	−0.09 (0.02)	0.35 (0.01)	0.39 (0.01)	0.44 (0.01)	0.50 (0.01)	0.58 (0.01)	0.64 (0.02)
	SVSI	−0.04 (0.01)	−0.03 (0.01)	−0.01 (0.01)	−0.18 (0.01)	−0.14 (0.02)	−0.10 (0.02)	0.36 (0.01)	0.41 (0.01)	0.46 (0.01)	0.50 (0.01)	0.63 (0.01)	0.70 (0.02)

For each call option, calculate the hedging error, which is the difference between the market price of the call and the replicating portfolio. The average dollar hedging error and the average absolute hedging error are reported for each model. The standard errors are shown in parentheses. The sample period is 06:1988–05:1991. In calculating the hedging errors generated with daily (once every five days) hedge rebalancing, 15,041 (11,704) observations are used

The fact that the SI model performs worse than the BS and that the SVSI model performs slightly worse than the SV suggests that adding stochastic interest rates to the option pricing framework actually make the single-instrument hedge's performance worse. This can be explained as follows. In the setup of the present chapter, interest rate shocks are assumed to be independent of shocks to the stock price and/or to the stochastic volatility. Therefore, in the single-instrument minimum-variance hedges, there is no adjustment in the optimal position in the underlying stock to be taken. The hedging results in Table 23.5 have shown that if interest rate risk is not to be controlled by any position in the hedging instrument, then it is perhaps better to design a single-instrument hedge based on an option model that assumes no interest rate risk. Assuming interest rate risk in an option pricing model and yet not controlling for this risk in a hedge can make the hedging effectiveness worse.

In the case of the SV versus the BS model, the situation is somewhat different from the above. As volatility shocks are assumed to be correlated with stock-price shocks, the position to be taken in the underlying stock (i.e., the hedging instrument) needs to be adjusted relative to the BS model-determined hedge, so that this single position not only helps contain the underlying stock's price risk but also neutralize that part of volatility risk which is related to stock-price fluctuations [see Eq. (23.23)]. Thus, by rendering it possible to use the single hedging position to control for both stock-price risk and volatility risk, introducing stochastic volatility into the BS framework helps improve the single-instrument hedging performance, albeit by a small amount. Nandi (1996) uses the remaining variance of a hedged position as a hedging effectiveness measure, according to which he finds the SV model performs better than the BS model. Our single-instrument hedging results are hence consistent with his, regarding the SV versus the BS model.

It is useful to recall that all four models are implemented allowing both the spot volatility and the spot interest rates to vary from day to day, which is, except in the sole case of the SVSI model, not consistent with the models' assumptions. Given this practical ad hoc treatment, it may not come as a surprise that when only the underlying asset is used as the hedging instrument, the four models performed virtually indifferently, with the magnitude of their hedging error differences being generally small. As easily seen, if all four models were implemented in a way consistent with the respective model setups, the single-instrument hedges based on the SVSI model would for sure perform the best.

Comparing Tables 23.4 and 23.5, one can conclude that based on a given option model, the conventional delta-neutral hedges perform far better than their single-instrument counterparts, for every moneyness-maturity category. This is not surprising as the former type of hedges involves more hedging instruments (except under the BS model).

23.4.3 Regression Analysis of Option Pricing and Hedging Errors

So far we have examined pricing and hedging performance according to option moneyness-maturity categories. The purpose was to see whether the errors have clear moneyness- and maturity-related biases. By appealing to a regression analysis, we can more rigorously study the association of the errors with factors that are either contract-specific or market condition-dependent. Fix an option pricing model, and let $\varepsilon_n(t)$ denote the $n-$th call option's percentage pricing error on day t. Then, run the regression below for the entire sample:

$$\varepsilon_n(t) = \beta_0 \frac{S(t)}{K_n} + \beta_2 \tau_n + \beta_3 \text{SPREAD}_n(t) + \beta_5 \text{LAGVOL}(t-1) + \beta_4 \text{SLOPE}(t) + \eta_n t \tag{23.29}$$

where K_n is the strike price of the call, n the remaining time to expiration, and $\text{SPREAD}_n(t)$ the percentage bid–ask spread at date t of the call (constructed by computing $\frac{\text{Ask}-\text{Bid}}{0.5(\text{Ask}-\text{Bid})}$, all of which are contract-specific variables. The variable, $\text{LAGVOL}(t-1)$, is the (annualized) standard deviation of the previous day's intraday S&P 500 returns computed over five-min intervals, and it is included in the regression to see whether the previous day's volatility of the underlying may cause systematic pricing biases. The variable, $\text{SLOPE}(t)$ represents the yield differential between one-year and 30-day Treasury bills. This variable can provide information on whether the single-factor Cox–Ingersoll–Ross (1985) term structure model assumed in this chapter is sufficient to make the resulting option formula capture all term structure-related effects on the S&P 500 index options. In some sense, the contract-specific variables help detect the existence of cross-sectional pricing biases, whereas $\text{LAGVOL}(t-1)$ and $\text{SLOPE}(t)$ serve to indicate whether the pricing errors over time are related to the dynamically changing market conditions. Similar regression analyses have been done for the BS pricing errors in, for example, Galai (1983b), George and Longsta (1993), and Madan et al. (1998). For each given option model, the same regression as in (23.29) is also run for the conventional delta-neutral hedging errors, with $\varepsilon_n(t)$ in (23.29) replaced by the dollar hedging error for the nth option on day t. Table 23.6 reports the regression results based on the entire sample period, where the standard error for each coefficient estimate is adjusted according to the White (1980) heteroscedasticity-consistent estimator and is given in the parentheses. Let us first examine the pricing error regressions. For every option model, each independent variable has statistically significant explanatory power of the remaining pricing errors. That is, the pricing errors from each model have some moneyness, maturity, intraday volatility, bid–ask spread, and term structure-related biases. The magnitude of each such bias, however, decreases from the BS to the SI, to the SV, and to the SVSI model. For instance, the BS percentage pricing errors will on the average be 2.29 points higher when the yield spread $\text{SLOPE}(t)$ increases by one point, whereas the SV and the SVSI percentage errors will only be, respectively, 0.32 and 0.34 points

Table 23.6 Regression analysis of pricing and hedging errors

Coefficient	BS	SI	SV	SVSI	BS	SI	SV	SVSI
	Percentage pricing errors				Hedging errors			
Constant	−0.05 (0.03)	0.28 (0.03)	0.24 (0.02)	0.11 (0.02)	−0.41 (0.11)	−0.30 (0.10)	0.00 (0.05)	−0.03 (0.05)
$\frac{S}{K}$	0.22 (0.03)	−0.18 (0.02)	−0.20 (0.01)	−0.09 (0.02)	0.34 (0.09)	0.29 (0.08)	0.00 (0.04)	0.03 (0.04)
τ	−0.04 (0.01)	0.04 (0.01)	0.08 (0.01)	0.05 (0.01)	0.03 (0.02)	0.08 (0.02)	0.00 (0.01)	0.00 (0.01)
SPREAD	−5.24 (0.12)	−4.48 (0.11)	−2.13 (0.07)	−1.57 (0.08)	2.26 (0.48)	1.04 (0.32)	0.32 (0.23)	0.34 (0.23)
SLOPE	2.29 (0.16)	1.33 (0.13)	0.32 (0.08)	0.34 (0.11)	−2.09 (0.58)	−2.01 (0.65)	−0.39 (0.26)	−0.39 (0.25)
LAGVOL	−0.16 (0.02)	0.12 (0.02)	0.06 (0.01)	0.04 (0.01)	−0.31 (0.07)	−0.51 (0.05)	−0.06 (0.02)	−0.05 (0.02)
Adj. R^2	0.29	0.22	0.12	0.07	0.01	0.01	0.00	0.00

The regression results below are based on the equation
$\epsilon_n(t) = \beta_1 \frac{S(t)}{K_n} + \beta_2 \tau_n + \beta_3 \mathrm{SPREAD}_n(t) + \beta_4 \mathrm{SLOPE}(t) + \beta_4 \mathrm{LAGVOL}(\mathrm{t}-1) + \eta_n(t)$
where $\epsilon_n(t)$ denotes either the percentage pricing error or the dollar hedging error of the nth call on date-t; $\frac{S}{K_n}$ and τ_n, respectively, represent the moneyness and the term-to-expiration of the option contract; the variable $\mathrm{SPREAD}_n(t)$ is the percentage bid–ask spread; $\mathrm{SLOPE}(t)$ the yield differential between the 1-year and the 30-day Treasury-bill rates; and $\mathrm{LAGVOL}(t-1)$ the previous day's (annualized) standard deviation of S&P 500 index returns computed from 5-min intradaily returns. The standard errors, reported in parenthesis, are White's (1980) heteroscedastically consistent estimator. The sample period is 06:1988–05:1991 for a total of 38,749 observations

higher in response. Thus, a higher yield spread on the term structure means higher pricing errors, regardless of the option model used. This points out that *a possible direction to further improve pricing performance is to include the yield spread as a second factor in the term structure model of interest rates*. Other noticeable patterns include the following. The BS pricing errors are decreasing, while the SI, the SV, and the SVSI pricing errors are increasing, in both the option's time-to-expiration and the underlying stock's volatility on the previous day. The deeper in-the-money the call or the wider its bid–ask spread, the lower the SI's, the SV's, and the SVSI model's mispricing. But, for the BS model, its mispricing increases with moneyness and decreases with bid–ask spread.

Even though all four models' pricing errors are significantly related to each independent variable, the collective explanatory power of these variables is not so impressive. The adjusted R^2 is 29% for the BS formula's pricing errors, 22% for the SI's, 12% for the SV's, and 7% for the SVSI model's. Therefore, while both the BS and the SI models have significant overall biases related to contract terms and market conditions (indicating systematic model misspecifications), the remaining pricing errors under the SV and the SVSI are not as significantly associated with these variables. About 93% of the SVSI model's pricing errors cannot be explained by these variables!

As reported in Table 23.6, delta-neutral hedging errors by the BS and the SI model tend to increase with the moneyness and the bid–ask spread of the target call, but decrease with the noncontract-specific yield spread and lagged stock volatility variables. Therefore, the two models are misspecified for hedging purposes and they lead to systematic hedging biases. But, overall, these variables can explain only 1% of the hedging errors by the two models. And, even more impressively, none of the included independent variables can explain any of the remaining hedging errors by the SV and the SVSI model, as their R^2 values are both zero. Finally, when the dollar pricing errors are used to replace the percentage pricing errors or when the percentage hedging errors are employed to replace the dollar hedging errors in the above regressions, the sign of each resulting coefficient estimate and the magnitude of each R2 value in Table 23.6 remain unchanged. Thus, the conclusions drawn from Table 23.6 are independent of the choice of the pricing or hedging error measure. Results from these exercises are not reported here but available upon request.

23.4.4 Robustness of Empirical Results

Using the entire sample period data, we have concluded that the evidence, based on both static performance and dynamic performance measures, is in favor of both the SVSI and the SV models. However, it is important to demonstrate that this conclusion still holds when alternative test designs and different sample periods are used. Below we briefly report results from two controlled experiments.

According to Rubinstein (1985), the volatility smile pattern and the nature of pricing biases are time period-dependent. To see whether our conclusion may be reversed, we separately examined the pricing and hedging performance of the models in three subperiods: 06:1988–05:1989, 06:1989–05:1990, and 06:1990–05:1991. Each subperiod contains about 10,000 call option observations. As the results are similar for each subperiod, we provide the percentage pricing errors in Panel A and the absolute delta-neutral hedging errors in Panel B of Table 23.7, for the subperiod 06:1990–05:1991. It is seen that these results are qualitatively the same as those in Tables 23.3 and 23.4.

We examined the pricing and hedging error measures of each model when the structural parameters were not updated daily. Rather, retain the structural parameter values estimated from the options of the first day of each month and then, for the remainder of the month, use them as input to compute the corresponding model-based price for each traded option, except that the implied spot volatility is updated each day based on the previous day's option prices. The obtained absolute pricing errors for the subperiod 06:1990–05:1991 indicate that the performance ranking of the four models remains the same as before.

In addition, when we used only ATM (or only ITM or only OTM) option prices to back out each model's parameter values, the resulting pricing and hedging errors did not change the performance ranking of the models either. This means that even if one would estimate and use a matrix of implied volatilities (across moneynesses and maturities) to accordingly price and hedge options in different moneyness-maturity categories, it would still not change the fact that the SV and the SVSI models are better specified than the other two for pricing and hedging. Given that the implied volatility matrix method has gained some popularity among practitioners, our results should be appealing. On the one hand, they suggest that with the SV and the SVSI models there is far less a need to engage in moneyness- and maturity-related fitting. On the other hand, if one is still interested in the matrix method, the SV and the SVSI models should be better model choices.

Early in the project, we used only option transaction price data for the pricing and hedging estimations. But, that meant a far smaller data set, especially for the hedging estimations. Nonetheless, the results obtained from the transaction prices were similar to these presented and discussed in this chapter.

Table 23.7 Robustness analysis

Moneyness $\frac{S}{K}$	Model	Term-to-expiration (days)					
		<30	30–60	60–90	90–120	120–180	≥ 180
Panel A: Percentage pricing errors, 06:1990–05:1991							
<0.93	BS	−76.51	−96.75	−74.73	−78.71	−61.36	−46.83
	SI	−26.88	−62.89	−38.89	−34.40	−18.68	−5.14
	SV	−25.05	−35.80	−12.59	−1.25	1.64	8.51
	SVSI	−20.56	−31.82	−8.15	−3.00	0.58	3.57
0.93–0.95	BS	−54.99	−59.06	−31.97	−24.29	−19.28	−12.36
	SI	−46.25	−46.77	−28.99	−10.62	−9.55	−5.37
	SV	−26.57	−25.32	−8.57	−1.60	−0.33	1.50
	SVSI	−28.25	−21.71	−6.26	−2.18	−0.04	0.79
0.95–0.97	BS	−34.72	−29.97	−16.71	−12.74	−10.27	−7.33
	SI	−31.85	−24.18	−16.79	−4.92	−5.35	−4.49
	SV	−20.09	−13.89	−5.54	−1.68	−0.56	0.61
	SVSI	−15.83	−13.08	−4.20	−3.29	−0.91	0.18
0.97–0.99	BS	−15.93	−10.36	−5.84	−3.22	−3.46	−2.14
	SI	−15.27	−7.45	−7.22	2.36	−2.05	−1.24
	SV	−13.09	−7.04	−3.64	0.77	−0.47	0.15
	SVSI	−12.38	−7.49	−3.37	−0.90	−0.83	−0.33
0.99–1.01	BS	−3.92	−1.23	0.62	1.54	0.65	1.99
	SI	−3.24	−0.09	−0.87	5.22	0.38	−0.09
	SV	−6.69	−3.43	−1.23	1.22	−0.10	0.14
	SVSI	−7.46	−4.17	−1.49	−0.33	−0.48	−0.27
1.01–1.03	BS	2.48	3.36	4.17	4.05	3.28	2.93
	SI	2.73	3.75	2.78	5.57	1.82	−0.60
	SV	−0.92	−0.56	0.24	1.44	0.19	−0.24
	SVSI	−1.41	−1.02	−0.09	−0.59	−0.16	−0.53
1.03–1.05	BS	3.93	4.86	5.35	5.57	4.62	3.41
	SI	3.95	4.92	4.08	6.62	2.61	−0.08
	SV	1.21	0.74	0.82	1.87	0.39	−0.17
	SVSI	0.85	0.19	0.48	0.79	0.08	−0.40
1.05–1.07	BS	3.69	5.07	5.84	6.36	5.12	4.93
	SI	3.82	5.08	4.67	6.55	2.87	1.29
	SV	1.83	1.52	1.49	2.07	0.51	−0.54
	SVSI	1.68	1.18	1.17	1.38	0.32	−0.57
> 1.07	BS	1.99	2.98	3.58	4.35	4.00	3.59
	SI	2.44	2.90	2.77	4.08	1.93	−1.01
	SV	1.43	1.40	1.21	1.47	0.45	−0.69
	SVSI	1.38	1.30	1.18	1.18	0.46	−0.40

(continued)

Table 23.7 (continued)

Panel B: Absolute hedging errors (1 and 5 day), 06:1990–05:1991

Panel B: Absolute hedging errors (1 and 5 day), 06:1990–05:1991

<0.93	BS	NA	0.42	0.48	NA	1.13	0.99
	SI		0.46	0.45		0.77	0.82
	SV		0.17	0.22		0.18	0.30
	SVSI		0.17	0.22		0.18	0.30
0.93–0.95	BS	NA	0.40	0.50	NA	0.97	0.95
	SI		0.45	0.47		0.73	0.75
	SV		0.13	0.25		0.15	0.30
	SVSI		0.13	0.25		0.15	0.30
0.95–0.97	BS	NA	0.37	0.44	NA	0.96	0.85
	SI		0.45	0.44		0.77	0.86
	SV		0.16	0.22		0.16	0.29
	SVSI		0.16	0.22		0.16	0.29
0.97–0.99	BS	0.39	0.42	0.47	0.72	0.95	0.97
	SI	0.33	0.45	0.41	0.66	0.74	0.76
	SV	0.14	0.17	0.17	0.16	0.17	0.24
	SVSI	0.14	0.17	0.16	0.15	0.17	0.23
0.99–1.01	BS	0.41	0.43	0.50	0.99	0.91	0.89
	SI	0.40	0.48	0.50	0.79	0.71	0.78
	SV	0.16	0.16	0.17	0.20	0.17	0.28
	SVSI	0.16	0.16	0.17	0.19	0.17	0.26
1.01–1.03	BS	0.40	0.46	0.47	0.99	0.89	0.83
	SI	0.45	0.44	0.45	0.74	0.71	0.73
	SV	0.17	0.17	0.18	0.19	0.20	0.25
	SVSI	0.17	0.17	0.17	0.19	0.20	0.25
1.03–1.05	BS	0.45	0.43	0.50	0.88	0.85	0.97
	SI	0.46	0.44	0.48	0.71	0.72	0.68
	SV	0.17	0.14	0.17	0.18	0.16	0.27
	SVSI	0.17	0.14	0.17	0.18	0.16	0.27
1.05–1.07	BS	0.46	0.47	0.51	0.73	0.78	0.77
	SI	0.47	0.45	0.50	0.61	0.67	0.68
	SV	0.18	0.14	0.22	0.19	0.16	0.24
	SVSI	0.17	0.14	0.21	0.19	0.16	0.22
>1.07	BS	0.41	0.45	0.53	0.62	0.70	0.81
	SI	0.38	0.46	0.50	0.48	0.64	0.75
	SV	0.17	0.15	0.22	0.18	0.19	0.32
	SVSI	0.16	0.15	0.21	0.18	0.18	0.31

Panel A: The reported percentage pricing error is the sample average of the market price minus the model price divided by the market price. The sample period is 06:1990–05:1991 for a total of 11,979 call options

Panel B: The average absolute hedging error for each model is reported based on the subsample period 06:1990–05:1991 (with a total of 6440 observations)

23.5 Conclusion

We have developed and analyzed a simple option pricing model that admits both stochastic volatility and stochastic interest rates. It is shown that this closed-form pricing formula is practically implementable, leads to useful analytical hedge ratios, and contains many known option formulas as special cases. This last feature has made it relatively straightforward to conduct a comparative empirical study of the four classes of option pricing models.

According to the pricing and hedging performance measures, the SVSI and the SV models both perform much better than the SI and the BS models, as the former typically reduce the pricing and hedging errors of the latter by more than a half. These error reductions are also economically significant. Furthermore, the hedging errors by the SV and the SVSI models are relatively insensitive to the frequency of portfolio revision, whereas those of the SI and the BS models are sensitive. Given that both the SV and the SVSI models can be easily implemented on a personal computer, they should thus be better alternatives to the widely applied BS formula. A regression-based analysis of the pricing and hedging errors indicates that while the BS and the SI models show significant pricing biases related to moneyness, time-to-expiration, bid–ask spread, lagged stock volatility and interest rate term spread, pricing errors by the SV and the SVSI models are not as systematically related to either contract-specific or market-dependent variables. Overall, the results lend empirical support to the claim that incorporating stochastic interest rates and, especially, stochastic volatility can both improve option pricing and hedging performance substantially and resolve some known empirical biases associated with the BS model.

The empirical issues and questions addressed in this chapter can also be re-examined using data from individual stock options, American-style index options, options on futures, currency and commodity options, and so on. Eventually, the acceptability of option pricing models with the added features will be judged not only by its easy implementability or even its impressive pricing and hedging performance as demonstrated in this chapter using European-style index calls, but also by its success or failure in pricing and hedging other types of options. These extensions are left for future research.

Acknowledgments Bakshi is at Department of Finance, College of Business, University of Maryland; Cao at Department of Finance, Smeal College of Business, Pennsylvania State University; and Chen at School of Management, Yale University. We would like to thank Sanjiv Das, Ranjan D'Mello, Helyette Geman, Eric Ghysels, Frank Hatheway, Steward Hodges, Ravi Jagannathan, Andrew Karolyi, Bill Kracaw, C. F. Lee, Dilip Madan, Louis Scott, Rene Stulz, Stephen Taylor, Siegfried Trautmann, Alex Triantis, and Alan White for their helpful suggestions. Any remaining errors are our responsibility alone. Address correspondence to Charles Cao, Smeal College of Business, Pennsylvania State University, University Park, PA 16802, or email: qxc2@psu.edu.

Appendix 1: Derivation of Stochastic Interest Model and Stochastic Volatility Model

Proof of the option pricing formula in (23.10). The valuation PDE in (23.9) can be rewritten as:

$$\begin{aligned}&\frac{1}{2}\frac{\partial^2 C}{\partial L^2}+\left(R-\frac{1}{2}V\right)\frac{\partial C}{\partial L}+\rho\sigma_\upsilon V\frac{\partial^2 C}{\partial L\partial V}\\&+\frac{1}{2}\sigma_\upsilon^2 V\frac{\partial^2 C}{\partial V^2}+[\theta_\upsilon-\kappa_\upsilon V]\frac{\partial C}{\partial V}\\&+\frac{1}{2}\sigma_R^2 R\frac{\partial^2 C}{\partial R^2}+[\theta_R-\kappa_R R]\frac{\partial C}{\partial R}-\frac{\partial C}{\partial \tau}-RC=0,\end{aligned} \tag{23.30}$$

where we have applied the transformation $L(t)\equiv\ln[S(t)]$ Inserting the conjectured solution in (23.10) into (23.30) produces the PDEs for the risk-neutralized probabilities, $\prod_j j=1,2$:

$$\frac{1}{2}\frac{\partial^2\prod_1}{\partial L^2}+\left(R-\frac{1}{2}V\right)\frac{\partial\prod_1}{\partial L}+\rho\sigma_\upsilon V\frac{\partial^2\prod_1}{\partial L\partial V}$$
$$+\frac{1}{2}\sigma_\upsilon^2 V\frac{\partial^2\prod_1}{\partial V^2}+[\theta_\upsilon-(\kappa_\upsilon-\rho\sigma_\upsilon)V]\frac{\partial\prod_1}{\partial V}$$
$$+\frac{1}{2}\sigma_R^2 R\frac{\partial^2\prod_1}{\partial R^2}+[\theta_R-\kappa_R R]\frac{\partial\prod_1}{\partial R}-\frac{\partial\prod_1}{\partial\tau}=0, \tag{23.31}$$

and

$$\frac{1}{2}\frac{\partial^2\prod_2}{\partial L^2}+\left(R-\frac{1}{2}V\right)\frac{\partial\prod_2}{\partial L}+\rho\sigma_\upsilon V\frac{\partial^2\prod_2}{\partial L\partial V}+\frac{1}{2}\sigma_\upsilon^2 V\frac{\partial^2\prod_2}{\partial V^2}$$
$$+[\theta_\upsilon-\kappa_\upsilon V]\frac{\partial\prod_2}{\partial V}+\frac{1}{2}\sigma_R^2 R\frac{\partial^2\prod_2}{\partial R^2}$$
$$+\left[\theta_R-\left(\kappa_R-\frac{\sigma_R^2}{B(t,\tau)}\frac{\partial B(t,\tau)}{\partial R}\right)R\right]\frac{\partial\prod_2}{\partial R}-\frac{\partial\prod_2}{\partial\tau}=0, \tag{23.32}$$

Observe that (23.31) and (23.32) are the Fokker–Planck forward equations for probability functions. This implies that $\prod_1$ and $\prod_2$ must indeed be valid probability functions, with values bounded between 0 and 1. These PDEs must be separately solved subject to the terminal condition:

$$\prod\nolimits_j(t+\tau,0)=1_{L(t+\tau)\geq K}\quad j=1,2. \tag{23.33}$$

The corresponding characteristic functions for $\prod_1$ and $\prod_2$ will also satisfy similar PDEs:

$$\frac{1}{2}\frac{\partial^2 f_1}{\partial L^2}+\left(R-\frac{1}{2}V\right)\frac{\partial f_1}{\partial L}+\rho\sigma_\upsilon V\frac{\partial^2 f_1}{\partial L\partial V}+\frac{1}{2}\sigma_\upsilon^2 V\frac{\partial^2 f_1}{\partial V^2}$$
$$+[\theta_\upsilon-(\kappa_\upsilon-\rho\sigma_\upsilon)V]\frac{\partial f_1}{\partial V}+\frac{1}{2}\sigma_R^2 R\frac{\partial^2 f_1}{\partial R^2}$$
$$+[\theta_R-\kappa_R R]\frac{\partial f_1}{\partial R}-\frac{\partial f_1}{\partial\tau}=0, \tag{34.34}$$

and

$$\frac{1}{2}\frac{\partial^2 f_2}{\partial L^2}+\left(R-\frac{1}{2}V\right)\frac{\partial f_2}{\partial L}+\rho\sigma_\upsilon V\frac{\partial^2 f_2}{\partial L\partial V}+\frac{1}{2}\sigma_\upsilon^2 V\frac{\partial^2 f_2}{\partial V^2}$$
$$+[\theta_\upsilon-\kappa_\upsilon V]\frac{\partial f_2}{\partial V}+\frac{1}{2}\sigma_R^2 R\frac{\partial^2 f_2}{\partial R^2}$$
$$+\left[\theta_R-\left(\kappa_R-\frac{\sigma_R^2}{B(t,\tau)}\frac{\partial B(t,\tau)}{\partial R}\right)R\right]\frac{\partial f_2}{\partial R}-\frac{\partial f_2}{\partial\tau}=0, \tag{23.35}$$

with the boundary condition:

$$f_j(t+\tau.0;\phi)=e^{i\phi L(t+\tau)}\quad j=1,2. \tag{23.36}$$

Conjecture that the solution to the PDEs (23.34) and (23.35) is respectively given by

$$f_1(t,\tau,S(t),V(t),R(t);\phi)=\exp\{u_r(\tau)+u_\upsilon(\tau)$$
$$+x_r(\tau)R(t)+x_\upsilon(\tau)V(t)+i\phi\ln[S(t)]\} \tag{23.37}$$

$$f_2(t,\tau,S(t),V(t),R(t);\phi)$$
$$=\exp\left\{\begin{array}{l} z_r(\tau)+z_\upsilon(\tau)+y_r(\tau)R(t)+y_\upsilon(\tau)V(t)\\ +i\phi\ln[S(t)]-\ln[B(t,\tau)]\end{array}\right\} \tag{23.38}$$

with $u_r(0)=u_\upsilon(0)=x_r(0)=x_\upsilon(0)=0$ and $z_r(0)=z_\upsilon(0)=y_r(0)=y_\upsilon(0)=0$. Solving the resulting system of differential equations and noting that $B(t+\tau,0)=1$ will, respectively, produce the desired characteristic functions in (23.12) and (23.13).

Both the constant interest rate-stochastic volatility and constant volatility-stochastic interest rate option pricing models are nested in (34.10). In the constant interest rate-stochastic volatility model, for instance, the partial derivatives with respect to R vanish in (23.30). The general solution in (23.37–23.38) will still apply except that now $R(t)=R$ (a constant), $B(t,\tau)=e^{-R\tau}, x_r(\tau)=i\phi\tau, y_r(\tau)=(i\phi-1)\tau$, and $u_r(\tau)=z_r(\tau)=0$. The final characteristic functions $\hat{f}_j$ for the constant interest rate-stochastic volatility option model are respectively given by

$$\hat{f}_1=\exp\left\{\begin{array}{l} -i\phi\ln[B(t,\tau)]\\ -\frac{\theta_\upsilon}{\sigma_\upsilon^2}\left[2\ln\left(1-\frac{[\xi_\upsilon-\kappa_\upsilon+(1+i\phi)\rho\sigma_\upsilon]\left(1-e^{-\xi_\upsilon\tau}\right)}{2\xi_\upsilon}\right)\right]\\ -\frac{\theta_\upsilon}{\sigma_\upsilon^2}[\xi_\upsilon-\kappa_\upsilon+(1+i\phi)\rho\sigma_\upsilon]\tau+i\phi\ln[S(t)]\\ +\frac{i\phi(i\phi+1)(1-e^{-\xi_\upsilon\tau})}{2\xi_\upsilon-[\xi_\upsilon-\kappa_\upsilon+(1+i\phi)\rho\sigma_\upsilon](1-e^{-\xi_\upsilon\tau})}V(t)\end{array}\right\}, \tag{23.39}$$

and,

$$\hat{f}_2 = \exp\left\{\begin{array}{l} -i\phi \ln[B(t,\tau)] \\ -\frac{\theta_v}{\sigma_v^2}\left[2\ln\left(1-\frac{[\xi_v^*-\kappa_v+i\phi\rho\sigma_v](1-e^{-\xi_v^*\tau})}{2\xi_v^*}\right)\right] \\ -\frac{\theta_v}{\sigma_v^2}\left[\xi_v^*-\kappa_v+(1+i\phi)\rho\sigma_v\right]\tau+i\phi\ln[S(t)] \\ +\frac{i\phi(i\phi+1)(1-e^{-\xi_v^*\tau})}{2\xi_v^*-[\xi_v^*-\kappa_v+(1+i\phi)\rho\sigma_v](1-e^{-\xi_v^*\tau})}V(t) \end{array}\right\}, \tag{23.40}$$

Similarly, the constant volatility-stochastic interest rate option model obtains with $V(t) = V$ (a constant), $x_v(\tau) = \frac{1}{2}i\phi(1+i\phi)\tau, y_v(\tau) = \frac{1}{2}i\phi(i\phi-1)\tau$, and $u_v(\tau) = z_v(\tau) = 0$. The final characteristic functions $\tilde{f}_j$ for the stochastic interest rate-constant volatility model are:

$$\tilde{f}_1 = \exp\left\{\begin{array}{l} \frac{1}{2}i\phi(1+i\phi)V\tau+i\phi\ln[S(t)] \\ -\frac{\theta_R}{\sigma_R^2}\left[2\ln\left(1-\frac{[\xi_R-\kappa_R](1-e^{-\xi_R\tau})}{2\xi_R}\right)\right] \\ +[\xi_R-\kappa_R]\tau+\frac{2i\phi(1-e^{-\xi_R\tau})}{2\xi_R-[\xi_R-\kappa_R](1-e^{-\xi_R\tau})}R(t) \end{array}\right\}, \tag{23.41}$$

and,

$$\tilde{f}_2 = \exp\left\{\begin{array}{l} \frac{1}{2}i\phi(i\phi-1)V\tau+i\phi\ln[S(t)]-\ln[B(t,\tau)] \\ -\frac{\theta_R}{\sigma_R^2}\left[2\ln\left(1-\frac{[\xi_R^*-\kappa_R](1-e^{-\xi_R^*\tau})}{2\xi_R^*}\right)\right] \\ +[\xi_R^*-\kappa_R]\tau+\frac{2i\phi(1-e^{-\xi_R\tau})}{2\xi_R^*-[\xi_R^*-\kappa_R](1-e^{-\xi_R^*\tau})}R(t) \end{array}\right\}, \tag{23.42}$$

Expressions for the Gamma Measures. The various second-order partial derivatives of the call price in (10), which are commonly referred to as Gamma measures, are given below:

$$\Gamma_S = \frac{\partial^2 C(t,\tau)}{\partial S^2} = \frac{\partial \prod_1}{\partial S} = \frac{1}{\pi}\int_0^\infty Re\left[(i\phi)^{-1}e^{-i\phi\ln[K]}f_1\frac{i\phi}{S}\right]d\phi. > 0. \tag{23.43}$$

$$\Gamma_V = \frac{\partial^2 C(t,\tau)}{\partial V^2} = S(t)\frac{\partial^2 \prod_1}{\partial V^2} - KB(t,\tau)\frac{\partial^2 \prod_2}{\partial V^2}. \tag{23.44}$$

$$\begin{aligned}\Gamma_R &= \frac{\partial^2 C(t,\tau)}{\partial R^2} \\ &= S(t)\frac{\partial^2 \prod_1}{\partial R^2} \\ &\quad - KB(t,\tau)\left\{\frac{\partial^2 \prod_2}{\partial R^2} - 2\vartheta(\tau)\frac{\partial^2 \prod_2}{\partial R} + \vartheta^2(\tau)\prod_2\right\}.\end{aligned} \tag{23.45}$$

$$\begin{aligned}\Gamma_{S,V} &= \frac{\partial^2 C(t,\tau)}{\partial S\partial V} = \frac{\partial \prod_1}{\partial V} \\ &= \frac{1}{\pi}\int_0^\infty Re\left[(i\phi)^{-1}e^{-i\phi\ln[K]}\frac{\partial f_1}{\partial V}\right]t]d\phi.\end{aligned} \tag{23.46}$$

where for $g = V, R$ and $j = 1, 2$

$$\frac{\partial^2 \prod_j}{\partial g^2} = \frac{1}{\pi}\int_0^\infty Re\left[(i\phi)^{-1}e^{-i\phi\ln[K]}\frac{\partial^2 f_j}{\partial g^2}\right]d\phi. \tag{23.47}$$

Bibliography

Amin, K., & Jarrow, R. (1992). Pricing options on risky assets in a stochastic interest rate economy. *Mathematical Finance, 2,* 217–237.

Amin, K., & Ng, V. (1993). Option valuation with systematic stochastic volatility. *Journal of Finance, 48,* 881–910.

Andersen, T., & Lund, J. (1997). Estimating continuous time stochastic volatility models of the short term interest rate. *Journal of Econometrics, 77,* 343–377.

Bailey, W., & Stulz, R. (1989). The pricing of stock index options in a general equilibrium model. *Journal of Financial and Quantitative Analysis, 24,* 1–12.

Bakshi, G., Cao, C., & Chen, Z. (1997). Empirical performance of alternative option pricing models. *Journal of Finance, 52,* 2003–2049.

Bakshi, G., Cao, C., & Chen, Z. (2000a). Do call prices and the underlying stock always move in the same direction? *Review of Financial Studies, 13,* 549–584.

Bakshi, G., Cao, C., & Chen, Z. (2000b). Pricing and hedging long-term options. *Journal of Econometrics, 94,* 277–318.

Bakshi G., Cao, C., & Chen, Z. (2010). Option pricing and hedging performance under stochastic volatility and stochastic interest rate. In C. F. Lee, A. C. Lee, & J. Lee (Ed.), *Handbook of quantitative finance and risk management*. Singapore: Springer.

Bakshi, G., & Chen, Z. (1997). An alternative valuation model for contingent claims. *Journal of Financial Economics, 44,* 123–165.

Barone-Adesi, G., & Whaley, R. (1987). Efficient analytic approximation of American option values. *Journal of Finance, 42,* 301–320.

Bates, D. (1996a). Testing option pricing models. In G. S. Maddala & C. R. Rao (Eds.), *Handbook of Statistics, Vol. 14: Statistical methods in finance* (pp. 567–611). Amsterdam: Elsevier.

Bates, D. (1996b). Jumps and stochastic volatility: Exchange rate processes implicit in deutschemark options. *Review of Financial Studies, 9,* 69–108.

Bates, D. (2000). Post-87 crash fears in S&P 500 futures options. *Journal of Econometrics, 94,* 181–238.

Black, F. (1975). Fact and fantasy in the use of options. *Financial Analyst Journal, 31,* 899–908.

Black, F., & Scholes, M. (1973). The pricing of options and corporate liabilities. *Journal of Political Economy, 81,* 637–659.

Cao, C., & Huang, J. (2008). Determinants of S&P 500 index option returns. *Review of Derivatives Research* (forthcoming).

Chan, K., Karolyi, A., Longsta, F., & Sanders, A. (1992). An empirical comparison of alternative models of the short-term interest rate. *Journal of Finance, 47,* 1209–1227.

Cox, J., Ingersoll, J., & Ross, S. (1985). A theory of the term structure of interest rates. *Econometrica, 53,* 385–408.

Cox, J., & Ross, S. (1976). The valuation of options for alternative stochastic processes. *Journal of Financial Economics, 3,* 145–166.

Day, T., & Lewis, C. (1995). Initial margin policy and stochastic volatility in the crude oil futures market. *Review of Financial Studies* (Forthcoming).

Dumas, B., Fleming, J., & Whaley, R. (1998). Implied volatility functions: Empirical tests. *Journal of Finance, 53,* 2059–2106.

Figlewski, S. (1989). Option arbitrage in imperfect markets. *Journal of Finance, 44,* 1289–1311.

Galai, D. (1983a). The components of the return from hedging options against stocks. *Journal of Business, 56,* 45–54.

Galai, D. (1983b). A survey of empirical tests of option pricing models. In M. Brenner (Ed.), *Option pricing* (pp. 45–80). Lexington, MA: Heath.

George, T., & Longstaff, F. (1993). Bid-ask spreads and trading activity in the S&P 100 Index options market. *Journal of Financial and Quantitative Analysis, 28,* 381–397.

Hansen, L. (1982). Large sample properties of generalized method of moments estimators. *Econometrica,* 1029–1054.

Harrison, M., & Kreps, D. (1979). Martingales and arbitrage in multiperiod securities markets. *Journal of Economic Theory, 20,* 381–408.

Harvey, C., & Whaley, R. (1992a). Market volatility and the efficiency of the S&P 100 Index option market. *Journal of Financial Economics, 31,* 43–73.

Harvey, C., & Whaley, R. (1992b). Dividends and S&P 100 Index option valuation. *Journal of Futures Markets, 12,* 123–137.

Heston, S. (1993). A closed-form solution for options with stochastic volatility with applications to bond and currency options. *Review of Financial Studies, 6,* 327–343.

Hull, J., & White, A. (1987a). The pricing of options with stochastic volatilities. *Journal of Finance, 42,* 281–300.

Hull, J., & White, A. (1987b). Hedging the risks from writing foreign currency options. *Journal of International Money and Finance, 6,* 131–152.

Kim, I.-J. (1990). The analytical valuation of American options. *Review of Financial Studies, 3*(4), 547–572.

Lee, C.-F., Lee, A. C., & Lee, J. (2010). *Handbook of quantitative finance and risk management*. New York: Springer.

Longstaff, F. (1995). Option pricing and the martingale restriction. *Review of Financial Studies, 8*(4), 1091–1124.

Madan, D., Carr, P., & Chang, E. (1998). The variance gamma process and option pricing. *European Finance Review, 2,* 79–105.

McBeth, J., & Merville, L. (1979). An empirical examination of the Black-Scholes call option pricing model. *Journal of Finance, 34,* 1173–1186.

Melino, A., & Turnbull, S. (1990). Pricing foreign currency options with stochastic volatility. *Journal of Econometrics, 45,* 239–265.

Melino, A., & Turnbull, S. (1995). Misspecication and the pricing and hedging of long-term foreign currency options. *Journal of International Money and Finance, 45,* 239–265.

Merton, R. (1973). Theory of rational option pricing. *Bell Journal of Economics, 4,* 141–183.

Nandi, S. (1996). *Pricing and hedging index options under stochastic volatility.* Working Paper, Federal Reserve Bank of Atlanta.

Ross, S. (1995). *Hedging long-run commitments: Exercises in incomplete market pricing*. Working Paper, Yale School of Management.

Rubinstein, M. (1985). Nonparametric tests of alternative option pricing models using all reported trades and quotes on the 30 most active CBOE options classes from August 23,1976 through August 31, 1978. *Journal of Finance*, 455–480.

Rubinstein, M. (1994). Implied binomial trees. *Journal of Finance, 49,* 771–818.

Scott, L. (1987). Option pricing when the variance changes randomly: Theory, estimators, and applications. *Journal of Financial and Quantitative Analysis, 22,* 419–438.

Scott, L. (1997). Pricing stock options in a jump-diffusion model with stochastic volatility and interest rates: Application of fourier inversion methods. *Mathematical Finance, 7,* 413–426.

Stein, E., & Stein, J. (1991). Stock price distributions with stochastic volatility. *Review of Financial Studies, 4,* 727–752.

Whaley, R. (1982). Valuation of American call options on dividend paying stocks. *Journal of Financial Economics, 10,* 29–58.

White, H. (1980). A heteroskedasticity-consistent covariance matrix estimator and a direct test for heteroskedasticity. *Econometrica, 48*(4), 817–838.

Wiggins, J. (1987). Option values under stochastic volatilities. *Journal of Financial Economics, 19,* 351–372.

24 Nonparametric Method for European Option Bounds

Contents

Abstract

There is much research whose efforts have been devoted to discovering the distributional defects in the Black–Scholes model, which are known to cause severe biases. However, with a free specification for the distribution, one can only find upper and lower bounds for option prices. In this chapter, we derive a new nonparametric lower bound and provide an alternative interpretation of Ritchken's (J Finance 40:1219–1233, 1985) upper bound to the price of European option. In a series of numerical examples, our new lower bound is substantially tighter than previous lower bounds.

This chapter is an update and expansion of the paper by Lin et al. which was published as Chap. 7 of Handbook of Financial Econometrics and Statistics edited by Lee and Lee (2015).

24.1 Introduction

Based upon the information of statistical distributions and option pricing models, in this chapter, we will discuss how nonparametric statistic can be used to do option pricing model analysis. In Sect. 24.2, we will discuss related literature and the model of bounds. Section 24.3 will

C.-F. Lee et al., *Financial Econometrics, Mathematics and Statistics*,
https://doi.org/10.1007/978-1-4939-9429-8_24

compare our lower bound to several lower bounds from previous studies. Section 24.4 will extend previous studies in option bounds. We perform some empirical study in Sect. 24.5. Finally, in Sect. 24.6, we summarize the chapter. Appendix 1 reviews related option studies adopting nonparametric method, and Appendix 2 discusses asset pricing model with a stochastic kernel.

24.2 The Bounds

In a seminal paper, Merton (1973) presents for the first time the no arbitrage bounds of European call and put options. These bounds are nonparametric and do not rely on any assumption.[1] Exact pricing formulas such as the Black–Scholes (1973) model and its variants, on the other hand, rely on strong assumptions on the asset price process and continuous trading. Due to the discreteness of actual trading opportunities, Perrakis and Ryan (1984) point out that option analyses in continuous time limit the accuracy and applicability of the Black–Scholes and related formulas. Relying on Rubinstein's (1976) approach, the single-price law, and arbitrage arguments, they derive upper and lower bounds for option prices with both a general price distribution and discrete trading opportunities. Their lower bound is tighter than that of Merton.

Levy (1985) applies stochastic dominance rules with borrowing and lending at the risk-free interest rate to derive upper and lower option bounds for all unconstrained utility functions and alternatively for concave utility functions. The derivation of these bounds can be applied to any kinds of stock-price distribution as long as the stock is "non-negative beta," which is identical to the assumption of Perrakis and Ryan (1984). Moreover, Levy claims that Perrakis and Ryan's bounds can be obtained by applying the second-degree stochastic dominance rule. However, Perrakis and Ryan do not cover all possible combinations of the risky asset with the riskless asset, and their bounds are therefore wider than those of Levy. Levy also applies the first-degree stochastic dominance rule (FSDR) with riskless assets to prove that Merton's bounds are in fact FSDR bounds and applies the second-degree stochastic dominance rule to strengthen Merton's bounds on the option value. At the same time, Ritchken (1985) uses a linear programming methodology to derive option bonds based on primitive prices in incomplete markets and claims that his bounds are tighter than those of Perrakis and Ryan (1984).

With an additional restriction that the range of the distribution of the one-period returns per dollar invested in the optioned stock is finite and has a strictly positive lower limit, Perrakis (1986) extends Perrakis and Ryan (1984) to provide bounds for American options. Instead of assuming that no opportunities exist to revise positions prior to expiration in Levy (1985) and Ritchken (1985), Ritchken and Kuo (1988) obtain tighter bounds on option prices under an incomplete market by allowing for a finite number of opportunities to revise position before expiration and making more restrictive assumptions on probabilities and preferences. The single-period linear programming option model is extended to handle multiple periods and the stock price is assumed to follow a multiplicative multinomial process. Their results show that the upper bounds are identical to those of Perrakis while the lower bounds are tighter. Later, Ritchken and Kuo (1989) also add suitable constraints to a linear programming problem to derive option bounds under higher orders of stochastic dominance preferences. Their results show that while the upper bounds remain unchanged beyond the second-degree stochastic dominance, the lower bounds become sharper as the order of stochastic dominance increases.[2]

[1]The only assumption is that both option and its underlying stock are traded securities.

[2]To further explore the research work of Ritchken and Kuo (1989) under the decreasing absolute risk aversion dominance rule, Basso and Pianca (1997) obtain efficient lower and upper option pricing bounds by solving nonlinear optimization problem. Unfortunately, neither model provides enough information of their numerical examples for us to compare our model with. The

Claiming that Perrakis and Ryan (1984), Levy (1985), Ritchken (1985), and Perrakis (1986) are all parametric models, Lo (1987) derives semiparametric upper bounds for the expected payoff of call and put options. These upper bounds are semiparametric because they depend on the mean and variance of the stock price at maturity but not on its entire distribution. In addition, the derivation of corresponding semiparametric upper bounds for option prices is shown by adopting the risk-neutral pricing approach of Cox and Ross (1976).[3] To continue the work of Lo (1987), Zhang (1994) and De La Pena et al. (2004), both of which assume that the underlying asset price must be continuously distributed, sharpen the upper option bounds of Lo (1987). Boyle and Lin (1997) extend the results of Lo (1987) to contingent claims on multiple underlying assets. Under an intertemporal setting, Constantinides and Zariphopoulou (2001) derive bounds for derivative prices with proportional transaction costs and multiple securities. Frey and Sin (1999) examine the sufficient conditions of Merton's bounds on European option prices under random volatility. More recently, Gotoh and Konno (2002) use the semidefinite programming and a cutting plane algorithm to study upper and lower bounds of European call option prices. Rodriguez (2003) uses a nonparametric method to derive lower and upper bounds and contributes a new tighter lower bound than previous work. Huang (2004) puts restrictions on the representative investor's relative risk aversion and produces a tighter call option bound than that of Perrakis and Ryan (1984). Hobson et al. (2005) derive arbitrage-free upper bounds for the prices of basket options. Peña et al. (2010) conduct static-arbitrage lower bounds on the prices of basket options via linear programming. Broadie and Cao (2008) introduce new and improved methods based on simulation to obtain tighter lower and upper bounds for pricing American options. Lately, Chung et al. (2010) also use an exponential function to approximate the early exercise boundary to obtain tighter bounds on American option prices. Lee et al. (2019) provide a more complete review and comparison of theoretical and empirical development on option bounds.[4] Some recent studies also adopt nonparametric methods in the related option investigation which are introduced in Appendix 1.

In this chapter, we derive a new and tighter lower bound for European option prices under a nonparametric framework. We show that Ritchken's (1985) upper bound is consistent with our nonparametric framework. Both bounds are nonparametric because the price distribution of underlying asset is totally flexible, can be arbitrarily chosen, and is consistent with any utility preference.[5] We compare our lower bound with those in previous studies and show that ours dominate those models by a wide margin. We also present the lower bound result on the model with random volatility and random interest rates (Bakshi et al. 1997; Scott 1997) to demonstrate how easily our model can be made consistent with any parametric structure.[6] Finally, we present how our bounds can be derived from histograms which are nonparametric in an empirical study. We discover violations of our lower bound and show that those violations present arbitrage

Ritchken–Kuo model provides no Black–Scholes comparison and the Basso–Pianca model provides only some partial information on the Black–Scholes model (we find the Black–Scholes model under 0.2 volatility to be 13.2670 and under the 0.4 volatility to be 20.3185, which are different from what are reported in their paper (12.993 and 20.098, respectively)) which is insufficient for us to provide any comparison.

[3]Inspired by Lo (1987), Grundy (1991) derives semiparametric upper bounds on the moments of the true, other than risk-neutral, distribution of underlying assets and obtain lower bounds by using observed option prices.

[4]This chapter only provides a nonparametric method on examining European option bounds. For a more complete review and comparison on prior studies of option bounds, please see Chuang et al. (2011).

[5]Christoffersen et al. (2010) provide results for the valuation of European-style contingent claims for a large class of specifications of the underlying asset returns.

[6]Given that our upper bound turns out to be identical to Ritchken's (1985), we do not compare with those upper bound models that dominate Ritchken (e.g., Huang 2004; Zhang 1994; De La Pena et al. 2004). Also, we do not compare our model with those models that require further assumptions to carry out exact results (e.g., Huang 2004; Frey and Sin 1999), since it is technically difficult to do.

profits.[7] In particular, our empirical results show that out-of-the money calls are substantially overpriced (violate the lower bound).

A generic and classical asset pricing model with a stochastic kernel is:

$$S_t = E_t[M_{t,T}S_T] \tag{24.1}$$

where $M_{t,T}$ is the marginal rate of substitution, also known as the pricing kernel that discounts the future cash flow at time T, $E_t[\cdot]$ is the conditional expectation under the physical measure $\mathbb{P}$ taken at time t, and S_t is the value of an arbitrary asset at time t. The standard kernel pricing theory (e.g., Ingersoll 1989) demonstrates that:

$$D_{t,T} = E_t[M_{t,T}], \tag{24.2}$$

where $D_{t,T}$ is the risk-free discount factor that gives the present value of \$1 over the period (t, T). The usual separation theorem gives rise to the well-known risk-neutral pricing result:

$$\begin{aligned} S_t &= E_t[M_{t,T}S_T] \\ &= E_t[M_{t,T}]\hat{E}_t^{(T)}[S_T], \\ &= D_{t,T}\hat{E}_t^{(T)}[S_T] \end{aligned} \tag{24.3}$$

If the risk-free interest rate is stochastic, then $\hat{E}_t^{(T)}[\cdot]$ is the conditional expectation under the T-forward measure $\hat{\mathbb{P}}^{(T)}$. When the risk free rate is nonstochastic, then the forward measure reduces to the risk-neutral measure $\hat{\mathbb{P}}$ and will not depend upon maturity time, i.e., $E_t^{(T)}[\cdot] \to \hat{E}_t[\cdot]$.[8]

Note that Eqs. (24.1) and (24.3) can be applied to both the stock and the option prices. This leads to the following theorem which is the main result of this chapter.

Theorem 24.1 *The following formula provides a lower bound for the European call option $\underline{C}_t$,*

$$\underline{C}_t = D_{t,T}E_t[C_T] + \beta_C\{S_t - D_{t,T}E_t[S_T]\} \tag{24.4}$$

where $\beta_C = \frac{\text{cov}[C_T, S_T]}{\text{var}[S_T]}$.

Proof By Eq. (24.1), the option price must follow $C_t = E_t[M_{t,T}C_T]$, and hence:

$$\begin{aligned} C_t &= E_t[M_{t,T}C_T] \\ &= E_t[M_{t,T}]E_t[C_T] + \text{cov}[M_{t,T}, C_T] \\ &= D_{t,T}E_t[C_T] + \text{cov}[M_{t,T}, C_T] \end{aligned} \tag{24.5}$$

or

$$C_t - D_{t,T}E_t[C_T] = \text{cov}[M_{t,T}, C_T] \tag{24.6}$$

Similarly,

$$S_t - D_{t,T}E_t[S_T] = \text{cov}[M_{t,T}, S_T] \tag{24.7}$$

Hence, to prove Eq. (24.4), we only need to prove:

$$\text{cov}[M_{t,T}, C_T] \geq \beta_C \text{cov}[M_{t,T}, S_T] \tag{24.8}$$

Note that $\text{cov}[M_{t,T}, C_T] = \text{cov}[M_{t,T}, \max\{S_T - K, 0\}]$ is monotonic in strike price K and has a minimum value when $K = 0$ in which case $\text{cov}[M_{t,T}, C_T] = \text{cov}[M_{t,T}, S_T]$ and a maximum value when $K \to \infty$ in which case $\text{cov}[M_{t,T}, C_T] = 0$. Hence, $\text{cov}[M_{t,T}, S_T] \leq \text{cov}[M_{t,T}, C_T]$ which is less than 0. Note that $0 < \beta_C < 1$ (see the proof in the following Corollary). In the Appendix 2, it is proved that $\text{cov}[M_{t,T}, C_T] \geq \beta_C \text{cov}[M_{t,T}, S_T]$.

The put lower bound takes the same form and is provided in the following corollary.

Corollary 24.1 *The lower bound of the European put option $\underline{P}_t$ can be obtained by the put-call parity and satisfies the same functional form in Theorem* 24.1:

$$\underline{P}_t = D_{t,T}E_t[P_T] + \beta_P\{S_t - D_{t,T}E_t[S_T]\} \tag{24.9}$$

where $\beta_P = \frac{\text{cov}[P_T, S_T]}{\text{var}[S_T]}$.

Proof By the put-call parity:

[7]For the related empirical studies of S&P 500 index options, see Constantinides et al. (2009, 2011).

[8]Without loss of generality and for the ease of exposition, we take nonstochastic interest rates and proceed with the risk-neutral measure $\hat{\mathbb{P}}$ for the rest of the paper.

$$\begin{aligned} P_t &= C_t + D_{t,T}K - S_t \\ &\geq \underline{C}_t + D_{t,T}K - S_t = \underline{P}_t \end{aligned} \quad (24.10)$$

We then substitute in the result of Theorem 24.1 to get:

$$\begin{aligned} \underline{P}_t &= D_{t,T}E_t[C_T] + \beta_C\{S_t - D_{t,T}E_t[S_T]\} \\ &\quad + D_{t,T}K - S_t \\ &= D_{t,T}E_t[S_T + P_T - K] + \beta_C\{S_t - D_{t,T}E_t[S_T]\} \\ &\quad + D_{t,T}K - S_t \\ &= D_{t,T}E_t[P_T] + \beta_P\{S_t - D_{t,T}E_t[S_T]\} \end{aligned} \quad (24.11)$$

where $\beta_P = \beta_C - 1$. Note that $\beta_P < 0 < \beta_C$. This also implies that $\beta_C < 1$. Finally, it is straightforward to show that $\beta_P = \frac{\mathrm{cov}[P_T, S_T]}{\mathrm{var}[S_T]}$ using the put-call parity.

Therefore, for both call and put options, since the relationship between the pricing kernel M and stock price S is convex, the theorem provides a lower bound.[9] Merton (1973) shows that the stock price will be the upper bound for the call option and the strike price should be the upper bound for the put option, otherwise arbitrage should occur. Ritchken (1985) provides a tighter upper bound[10] than that of Merton, which is stated in the following theorem, although we provide an alternative proof to Ritchken that is consistent with our derivation of the lower bound.

Theorem 24.2 Ritchken (1985)[11]

The following formulas provide upper bounds for the European call and put options,

$$\begin{aligned} \overline{C}_t &= \frac{S_t}{E_t[S_T]}E_t[C_T] \\ \overline{P}_t &= \frac{S_t}{E_t[S_T]}E_t[P_T] + \left(D_{t,T} - \frac{S_t}{E_t[S_T]}\right)K \end{aligned} \quad (24.12)$$

Proof Similar to the proof of the lower bound, the upper bound of the call option is provided as follows:

$$\begin{aligned} \frac{S_t}{E_t[S_T]}E_t[C_T] &= \frac{E_t[M_{t,T}S_T]}{E_t[S_T]}E_t[C_T] \\ &= \frac{E_t[M_{t,T}S_TC_T] - \mathrm{cov}[M_{t,T}S_T, C_T]}{E_t[S_T]} \\ &> \frac{E_t[M_{t,T}S_TC_T]}{E_t[S_T]} \\ &= C_t\frac{E_t^{(C)}[S_T]}{E_t[S_T]} \\ &> C_t \end{aligned} \quad (24.13)$$

The third line of the above equation is a result from the fact that $\mathrm{cov}[M_{t,T}S_T, C_T] < 0$. The fourth line of the above equation is a change of measure with the call option being the numeraire. The last line of the above equation is a result based upon $E_t^{(C)}[S_T]/E_t[S_T] > 1$.[12]

By the put-call parity, we can show that the upper bound of the put option requires an additional term:

$$\begin{aligned} \overline{P}_t &= \overline{C}_t + D_{t,T}K - S_t \\ &= \frac{S_t}{E_t[S_T]}E_t[P_T + S_T - K] + D_{t,T}K - S_t \\ &= \frac{S_t}{E_t[S_T]}E_t[P_T] + \left(D_{t,T} - \frac{S_t}{E_t[S_T]}\right)K \end{aligned} \quad (24.14)$$

The lower and upper bounds we show in this chapter have two important advantages over the existing bounds. The bounds will converge to the true value of the option if:

- the expected stock return, $\frac{E_t[S_T]}{S_t}$, approaches the risk free rate, or
- the correlation between the stock and the call or put option (ρ_{SC} or ρ_{SP}) approaches 1 or -1.

[9] In the Appendix 2, $\varepsilon > 0$.

[10] Perrakis and Ryan (1984) and Ritchken (1985) obtain the identical upper bound.

[11] This is same as Proposition 3-i (Eq. 7.26) in Ritchken (1985).

[12] By the definition of measure change, we have $E_t[C_TS_T] = E_t[C_T]E_t^{(C)}[S_T]$ which implies $E_t^{(C)}[S_T]/E[S_T] = E_t[C_TS_T]/\{E_t[C_T]E_t[S_T]\} > 1$.

These advantages help us identify when the bounds are tight and when they are not. The first advantage indicates that the bounds are tight for low-risk stocks and not tight for high-risk stocks. The second advantage indicates that the bounds are tighter for in-the-money options than out-of-the-money options.

24.3 Comparisons

The main purpose of this section is to compare our lower bound to several lower bounds in previous studies, namely Merton (1973), Perrakis and Ryan (1984), Ritchken (1985), Ritchken and Kuo (1988), Gotoh and Konno (2002), and Rodriguez (2003) using the Black–Scholes model as the benchmark for its true option value. We also compare Ritchken's upper bound (which is also our upper bound) with more recent works by Gotoh and Konno (2002) and Rodriguez (2003).

The Black–Scholes model has five variables: stock price, strike price, volatility (standard deviation), risk-free rate (constant), and time to maturity. In addition to the five variables, the lower bound models need the physical expected stock return. The following is the base case for the comparison:

Current stock S_0	50
Strike K	50
Volatility σ	0.2
Risk-free rate r	0.1
Time to maturity T	1
Stock expected return μ	0.2

In the Black–Scholes model, stock-price (S) evolution follows a lognormal process:

$$dS = \mu S dt + \sigma S dW \tag{24.15}$$

where instantaneous expected rate of stock return μ and volatility of stock price σ are assumed to be constants and where dW is a wiener process. The call option price is computed as:

$$C = S_0 N(d_+) - e^{-rT} K N(d_-) \tag{24.16}$$

where

$$d_\pm = \frac{\ln S_0 - \ln K + (r \pm 0.5\sigma^2)T}{\sigma\sqrt{T}}.$$

The lower bound is computed by simulating the stock price using Eq. (24.15) via a binomial distribution approximation:

$$S_j = S_0 u^j d^{n-j} \tag{24.17}$$

As n approaches infinity, S_j approaches the lognormal distribution and the binomial model converges to the Black–Scholes model. Under the risk-neutral measure, the probability associated with the jth state is set as:

$$\widehat{\Pr}[j] = \binom{n}{j} \hat{p}^j (1 - \hat{p})^{n-j} \tag{24.18}$$

where

$$\hat{p} = \frac{e^{r\Delta t} - d}{u - d},$$

$$u = e^{\sigma\sqrt{\Delta t}},$$

$$d = e^{-\sigma\sqrt{\Delta t}},$$

and $\Delta t = T/n$ represents the length of the partition. Under the actual measure, the formula will change to:

$$\Pr[j] = \binom{n}{j} p^j (1 - p)^{n-j} \tag{24.19}$$

where

$$p = \frac{e^{\mu\Delta t} - d}{u - d}.$$

Finally, the pricing kernel in our model is set as:

$$M_{0T}[j] = \frac{\widehat{\Pr}[j]}{\Pr[j]} e^{-rT} \tag{24.20}$$

In our results, we let n be great enough so that the binomial model price is 4-digits accurate to the Black–Scholes model. We hence set n to be 1000. The results are reported in Table 24.1. The first panel presents the results for various moneyness levels, the second panel presents the results for various volatility levels, the third panel presents the results for various interest rates, the fourth panel presents the results for various maturity times, and the last presents the results for various stock expected returns. In general, the lower bounds are tighter (for all models) when the moneyness is high (in-the-money), volatility is high, risk-free rate is high, time to maturity is short, and the expected return of stock is low.

This table presents comparisons between our lower bound and existing lower bounds by Merton (1973), Perrakis and Ryan (1984), and Ritchken (1985). The base-case parameter values are:

- stock price = 50,
- strike price = 50,
- volatility = 0.2,
- risk-free rate = 10%,
- time to maturity = 1 year,
- stock expected return (μ) = 20%.

The italicised represent the base case.

As we can easily see, universally our model for the lower bound is tighter than any of the comparative models. One result particularly worth mentioning is that our lower bound performs better than the other lower bound models in out-of-the-money options. For example, our lower bound is much better than Ritchken's (1985) lower bound when the option is 20% out-of-the-money and continues to show value

Table 24.1 Lower bound comparison

S	Blk-Sch	Merton	PR	Ritch.	Our	$error	%error
20	0.0000	0	0	0	0	0.0000	
25	0.0028	0	0	0	0	0.0028	
30	0.0533	0	0	0	0	0.0533	
35	0.3725	0	0	0	0.0008	0.3717	99.78
40	1.3932	0	0	0.272	0.8224	0.5708	40.97
45	3.4746	0	1.441	2.278	2.8957	0.5789	16.66
50	*6.6322*	*4.758*	*5.449*	*5.791*	*6.1924*	*0.4398*	*6.63*
55	10.6248	9.758	10.017	10.143	10.3535	0.2714	2.55
60	15.1288	14.758	14.849	14.888	14.9851	0.1437	0.95
65	19.9075	19.758	19.788	19.797	19.8395	0.0680	0.34
70	24.8156	24.758	24.768	24.767	24.7860	0.0296	0.12
75	29.7794	29.758	29.761	29.76	29.7673	0.0121	0.04
80	34.7658	34.758	34.759	34.754	34.7611	0.0047	0.01
Vol	Blk-Sch	Merton	PR	Ritch.	Our	$error	%error
0.1	5.1526	4.7580	4.7950	4.8630	4.8930	0.2596	5.04
0.15	5.8325	4.7580	5.0290	5.2400	5.4367	0.3958	6.79
0.2	*6.6322*	*4.7580*	*5.4490*	*5.7890*	*6.1924*	*0.4398*	*6.63*
0.25	7.4847	4.7580	5.9700	6.4360	7.0247	0.4600	6.15
0.3	8.3633	4.7580	6.5370	7.1010	7.8864	0.4769	5.70

(continued)

Table 24.1 (continued)

Vol	Blk-Sch	Merton	PR	Ritch.	Our	$error	%error
0.35	9.2555	4.7580	7.1180	7.7760	8.7601	0.4954	5.35
0.4	10.1544	4.7580	7.6990	8.4570	9.6382	0.5162	5.08
0.45	11.0559	4.7580	8.2680	9.1250	10.5170	0.5389	4.87
0.5	11.9574	4.7580	8.8220	9.7780	11.3945	0.5629	4.71
0.55	12.8569	4.7580	9.3570	10.4240	12.2692	0.5877	4.57
Rate	Blk-Sch	Merton	PR	Ritch.	Our	$error	%error
0.02	4.4555	0.9901	1.7380	2.7680	3.0243	1.4313	32.12
0.04	4.9600	1.9605	2.6940	3.5010	3.8402	1.1198	22.58
0.06	5.4923	2.9118	3.6310	4.2490	4.6400	0.8523	15.52
0.08	6.0504	3.8442	4.5490	5.0200	5.4240	0.6264	10.35
0.1	*6.6322*	*4.7581*	*5.4490*	*5.7970*	*6.1924*	*0.4398*	*6.63*
0.12	7.2355	5.6540	6.3310	6.5810	6.9456	0.2900	4.01
0.14	7.8578	6.5321	7.1960	7.3670	7.6839	0.1739	2.21
0.16	8.4965	7.3928	8.0430	8.1470	8.4076	0.0889	1.05
0.18	9.1488	8.2365	8.8740	8.9350	9.1169	0.0319	0.35
0.2	9.8122	9.0635	9.6880	9.7170	9.8122	0.0000	0.00
0.22	10.4841	9.8741	10.4870	10.4830	10.4841	0.0000	0.00
T	Blk-Sch	Merton	PR	Ritch.	Our	$error	%error
0.1	1.5187	0.4975	1.2850	1.3600	1.4976	0.0211	1.39
0.2	2.3037	0.9901	1.8870	2.0240	2.2469	0.0568	2.47
0.3	2.9693	1.4777	2.4000	2.5940	2.8696	0.0997	3.36
0.4	3.5731	1.9605	2.8730	3.0990	3.4266	0.1465	4.10
0.5	4.1371	2.4385	3.3250	3.5830	3.9416	0.1955	4.72
0.6	4.6726	2.9118	3.7640	4.0510	4.4272	0.2454	5.25
0.7	5.1862	3.3803	4.1940	4.5020	4.8909	0.2953	5.69
0.8	5.6822	3.8442	4.6170	4.9460	5.3375	0.3447	6.07
0.9	6.1635	4.3034	5.0350	5.3710	5.7705	0.3930	6.38
1	*6.6322*	*4.7581*	*5.4490*	*5.7890*	*6.1924*	*0.4398*	*6.63*
Mu (μ)	Blk-Sch	Merton	PR	Ritch.	Our	$error	%error
0.1	6.6322	4.7580	6.3740	6.4310	6.6322	0.0000	0.00
0.15	6.6322	4.7580	5.8380	6.0610	6.4703	0.1620	2.44
0.2	*6.6322*	*4.7580*	*5.4490*	*5.7890*	*6.1924*	*0.4398*	*6.63*
0.25	6.6322	4.7580	5.1800	5.5820	5.8689	0.7633	11.51
0.3	6.6322	4.7580	5.0040	5.4360	5.5572	1.0750	16.21
0.35	6.6322	4.7580	4.8940	5.3090	5.2936	1.3386	20.18
0.4	6.6322	4.7580	4.8300	5.2130	5.0929	1.5393	23.21
0.45	6.6322	4.7580	4.7940	5.1330	4.9536	1.6786	25.31

Note S is the stock price; *vol* is the volatility; *rate* is the risk-free rate; *Mu* (μ) is the expected rate of return of stock; *Blk-Sch* is the Black–Scholes (1973) solution; *Merton* is the Merton (1973) model; *PR* is Perrakis and Ryan (1984) model; *Ritch.* is the Ritchken (1985) model; *Our* is our model; *$error* is error in dollar; *%error* is error in percentage

Table 24.2 Comparison of upper and lower bounds with the Gotoh and Konno (2002) model

Stk	Lower bound		Blk-Sch	Upper bound	
	Our	GK		Our	GK
S = 40; rate = 6%; vol = 0.2; t = 1 week					
30	10.0346	10.0346	10.0346	10.1152	10.0349
35	5.0404	5.0404	5.0404	5.1344	5.0428
40	0.4628	0.3425	0.4658	0.5225	0.5771
45	0.0000	0.0000	0.0000	0.0000	0.0027
50	0.0000	0.0000	0.0000	0.0000	0.0003
S = 40; rate = 6%; vol = 0.8; t = 1 week					
30	10.0400	10.0346	10.0401	10.1202	10.1028
35	5.2644	5.0404	5.2663	5.3483	5.4127
40	1.7876	1.2810	1.7916	1.8428	2.2268
45	0.3533	0.0015	0.3548	0.3717	0.5566
50	0.0412	0.0000	0.0419	0.0444	0.1021
S = 40; rate = 6%; vol = 0.8; t = 12 week					
30	11.9661	10.4125	12.0278	12.7229	12.8578
35	8.7345	6.2980	8.8246	9.4774	9.7658
40	6.2141	3.8290	6.3321	6.8984	7.5165
45	4.3432	2.5271	4.4689	4.9421	6.8726
50	2.9948	1.5722	3.1168	3.4990	4.5786

Note S is the stock price; *Stk* is the strike price; *vol* is the volatility; *rate* is the risk-free rate; *Blk-Sch* is the Black–Scholes (1973) solution; *GK* is the Gotoh and Konno (2002) model; *Our* is our model; *$error* is error in dollar; *%error* is error in percentage

when Ritchken's lower bound returns 0 (see the first panel of Table 24.1).

While Ritchken and Kuo (1988) claim to obtain tighter lower bounds than Perrakis (1986) and Perrakis and Ryan (1984), they do not show direct comparisons in their paper. Rather, they present through a convergence plot (Fig. 3 on page 308 in Ritchken and Kuo (1988)) of a Black–Scholes example with the true value being $5.4532 and the lower bound approaching roughly $5.2. The same parameter values with our lower bound show a lower bound of $5.4427, which demonstrates a substantial improvement over the Ritchken and Kuo model.

The comparisons with more recent studies of Gotoh and Konno (2002) and Rodriguez (2003) are given in Tables 24.2 and 24.3.[13] Gotoh and Konno use semidefinite programming and a cutting plane algorithm to study upper and lower bounds of European call option prices. Rodriguez uses a nonparametric method to derive lower and upper bounds. As we can see in Tables 24.2 and 24.3, except for very few upper bound cases, none of the bounds under the Gotoh and Konno's model and Rodriguez's model is very tight, compared to our model. Furthermore, note that our model requires no moments of the underlying distribution.[14]

[13]We also compare with the upper bound by Zhang (1994), which is an improved upper bound by Lo (1987), and show overwhelming dominance of our upper bound. The results (comparison to Tables 24.1, 24.2, and 24.3 in Zhang) are available upon request.

[14]The upper bounds by the Gotoh and Konno model perform well in only in-the-money, short maturity, and low volatility scenarios and these scenarios are where the option prices are close to their intrinsic values and hence the percentage errors are small.

Table 24.3 Comparison of upper and lower bounds with the Rodriguez (2003) model

S	Lower bound		Blk-Sch	Upper bound	
	Our	Rodriguez		Our	Rodriguez
30	0.0221	0	0.0538	0.1001	0.1806
32	0.0725	0.0000	0.1284	0.2244	0.3793
34	0.1828	0.0171	0.2692	0.4451	0.7090
36	0.3878	0.1158	0.5072	0.7973	1.2044
38	0.7224	0.3598	0.8735	1.3100	1.8900
40	1.2177	0.7965	1.3950	2.0044	2.7767
42	1.8982	1.4521	2.0902	2.8927	3.8619
44	2.7711	2.3329	2.9676	3.9697	5.1315
46	3.8319	3.4286	4.0255	5.2211	6.5640
48	5.0709	4.7177	5.2535	6.6302	8.1337
50	6.4703	6.1724	6.6348	8.1753	9.8149
52	8.0072	7.7635	8.1494	9.8327	11.5835
54	9.6574	9.4631	9.7758	11.5794	13.4187
56	11.3974	11.2462	11.4933	13.3950	15.3032
58	13.2067	13.0922	13.2832	15.2621	17.2235
60	15.0683	14.9845	15.1292	17.1674	19.1693
62	16.9716	16.9101	17.0179	19.1024	21.1328
64	18.9035	18.8593	18.9384	21.0573	23.1086
66	20.8559	20.8250	20.8822	23.0262	25.0926
68	22.8239	22.8020	22.8429	25.0056	27.0822
70	24.8018	24.7867	24.8157	26.9915	29.0754

Note S is the stock price; *Blk-Sch* is the Black–Scholes (1973) solution; *Rodriguez* is the Rodriguez (2003) model; *Our* is our model; *$error* is error in dollar; *%error* is error in percentage

The base-case parameter values are:

- stock price = 40,
- risk-free rate = 6%,

stock expected return (μ) = 20%.
The base-case parameter values are:

- strike price = 50,
- volatility = 0.2,
- risk-free rate = 10%,
- time to maturity = 1 year,

stock expected return (μ) = 20%.

24.4 Extensions

In addition to a tight lower bound, another major contribution of our model is that it makes no assumption on the distribution of the underlying stock (unlike Lo 1987 and Gotoh and Konno 2002 who require moments of the underlying distribution), or any assumption on interest rates and volatility (unlike Rodriguez 2003 who requires constant interest rates). As a result, our lower bound can be used with models that assume random volatility and random interest rates, or any arbitrary specification of the underlying stock. Note that our model needs only the dollar beta of the option and expected payoffs of the stock and the option.[15] In this section, we extend our numerical experiment to a model with random volatility and random interest rates.

Option models with random volatility and random interest rates can be derived with closed-form solutions under the Scott (1997) and Bakshi et al. (1997) specifications. However, here, given that there is no closed-form solution to the covariance our model requires, we shall use Monte Carlo to simulate the lower bound. In order to be consistent with the lower bound, we must use the same Monte Carlo paths for the valuation of the option. For the ease of exposition and simplicity, we assume the following joint stochastic processes of stock price S, interest rate r, and volatility V under the actual measure, respectively:

$$\begin{aligned} dS &= \mu S dt + \sqrt{V} S dW_1 \\ dr &= \alpha(\theta - r)dt + v dW_2 \\ dV &= \eta V dW_3 \end{aligned} \tag{24.21}$$

where dW is a wiener process, $dW_i dW_j = 0$, μ, α, θ, v, and η are constants. The processes under the actual measure are used for simulating the lower and upper bounds. The Monte Carlo simulations are performed under the risk-neutral measure in order to compute the option price:

$$\begin{aligned} dS &= rS dt + \sqrt{V} S d\hat{W}_1 \\ dr &= \alpha(\theta - r)dt + v d\hat{W}_2 \\ dV &= \eta V d\hat{W}_3 \end{aligned} \tag{24.22}$$

To simplify the problem without loss of generosity, we assume that investors charge no risk premiums on interest rate risk and volatility risk, i.e., $dW_2 = d\hat{W}_2$ and $dW_3 = d\hat{W}_3$.

The simulations are done by the following integrals under the actual (top equation) and risk-neutral (bottom equation) measures:

$$\begin{aligned} S_t &= S_0 \exp\left(\int_0^t \mu_u du - \int_0^t 1/2 V_u du + \int_0^t \sqrt{V_u} dW_u \right) \\ \hat{S}_t &= S_0 \exp\left(\int_0^t r_u du - \int_0^t 1/2 V_u du + \int_0^t \sqrt{V_u} d\hat{W}_u \right) \end{aligned} \tag{24.23}$$

The no-arbitrage price of the call option is computed under the risk-neutral measure as:

$$C = E\left[\exp\left(- \int_0^t r_u du \right) \max\{\hat{S}_t - K, 0\} \right] \tag{24.24}$$

The bounds are computed under the actual measure. For example,

$$E[C_t] = E[\max\{S_t - K, 0\}] \tag{24.25}$$

Given that $dW_i dW_j = 0$, we can simulate the interest rates and volatility separately and then simulate the stock price. That is, conditional on known interest rates and volatility, under independence, the stock price is lognormally distributed.

We perform our simulations using 10,000 paths over 52 weekly periods. The parameters are given as follows:

Strike K	50
Time to maturity T	1
Stock expected return μ	0.2
Reverting speed α	0.5
Reverting level θ	0.1
Interest rate volatility v	0.03
Initial interest rate r_0	0.1
Initial variance V_0	0.04[a]
Volatility on variance η	0.2

[a]This is so because the initial volatility is 0.2

[15]The term "dollar beta" is originally from page 173 of Black (1976). Here, we mean β_C and β_P.

Table 24.4 Lower bound under the random volatility and random interest rate model

S	BCC/Scott	Our	$error	%error
25	1.9073	1.1360	0.7713	40.44
30	3.2384	2.4131	0.8253	25.49
35	5.2058	4.1891	1.0167	19.53
40	7.5902	6.5864	1.0039	13.23
45	10.2962	9.5332	0.7630	7.41
50	*13.6180*	*12.8670*	*0.7510*	*5.52*
55	17.1579	16.5908	0.5671	3.30
60	21.1082	20.5539	0.5543	2.63
65	25.0292	24.7251	0.3040	1.21
70	29.3226	29.0742	0.2484	0.85

Note S is the stock price; *BCC/Scott* are Bakshi et al. (1997) and Scott (1997) models; *Our* is our model; *$error* is error in dollar; *%error* is error in percentage

Note that implicitly we assume the price of risk for both interest rate process and volatility process to be 0, for simplicity and without loss of generality. The results are shown in Table 24.4. Compared to the model of the Black–Scholes (i.e., the first panel of Table 24.4), the lower bound performs similarly in the random volatility and random interest rate model. Take the base case as an example where the Black–Scholes price is 6.6322, the Bakshi–Cao–Chen/Scott price is 13.6180 as a result of extra uncertainty in the stock price due to random volatility and interest rates. The error of the lower bound of our model is 0.7510 in the Bakshi–Cao–Chen/Scott case as opposed to 0.4398 in the Black–Scholes case.[16] The percentage error is 5.52% in the Bakshi–Cao–Chen/Scott case versus 6.63% in the Black–Scholes case. The in-the-money options have larger percentage errors than those of the out-of-the-money options.

The parameters are given as follows:

Strike price	50
Time to maturity	1
Stock expected return	0.2
Reverting speed	0.5
Reverting level	0.1
Interest rate volatility	0.03
Price of risk	0
Initial interest rate	0.1
Initial volatility	0.2
Volatility on volatility	0.2

The Monte Carlo paths are 10,000. The stock price, volatility, and interest rate processes are assumed to be independent.

24.5 Empirical Study

In this section, we test the lower and upper bounds against data. Charles Cao has generously provided us with the approximated prices of S&P 500 index call option contracts, matched levels of S&P 500 index, and approximated risk-free 90-day T-Bill rates for the period of June 2, 1988 through December 31, 1991.[17] For each day, the approximated option prices are calculated as the average of the last bid and ask quotes. Index returns are computed using daily closing levels for the S&P 500 index that are collected and confirmed using data obtained from

[16]This Black–Scholes case is from the italicised in the first panel of Table 24.1.

[17]The data are used in Bakshi et al. (1997).

Standard and Poor's, CBOE, Yahoo, and Bloomberg.[18]

The data set contains 46,540 observations over 901 days (from June 2, 1988 to December 31, 1991). Hence, on average, there are over 50 options for various maturities and strikes. The shortest maturity of the data set is 7 days and the longest is 367 days. 15% of the data are less than 30 days to maturity, 32% are between 30 and 60 days, 30% are between 60 and 180 days, and 24% are more than 180 days to maturity. Hence, these data do not have maturity bias.

The deepest out of money option is −18.33% and the deepest in-the-money is 47.30%. Roughly half of the data are at the money options (46% of the data are within 5% in and out-of-the-money). 10% are deep-in-the-money (more than 15% in-the-money) but less than 1% of the data are deep-out–of-the-money (more than 15% out-of-the-money). Hence, the data have disproportional fraction of in-the-money options. This is clearly a reflection of the bull market in the sample period.

The best way to test the lower and upper bounds derived in this chapter is to use a non-parametric, distribution-free model. Note that the lower bound in Theorem 24.1 requires only the expected return of the underlying stock and the covariance between the stock and the option. There is no further requirement for the lower and the upper bounds. In other words, our lower bound model can permit any arbitrary distribution of the underlying stock and any parametric specification of the underlying stock such as random volatility, random interest rates, and jumps. Hence, to best test the bounds with a parsimonious empirical design, we adopt the histogram method introduced by Lin et al. (2016) where the underlying asset is modeled by past realizations, i.e., histogram.

We construct histograms from realizations of S&P 500 (SPX) returns. We calculate the price on day t of an option that settles on day T using a histogram of S&P 500 index returns for a holding period of $T - t$, taken from a five-year window immediately preceding time t.[19] For example, an x-calendar-day option price on any date is evaluated using a histogram of $\text{round}\left[\frac{252}{365}x\right]$-trading-day holding period returns where $\text{round}[\cdot]$ is rounding the nearest integer.[20] The index levels used to calculate these returns are taken from a window that starts on the 1260th ($\approx 5 \times 252$) trading day before the option trading date and ends one day before the trading date. Thus, this histogram contains 1260-$\text{round}\left[\frac{252}{365}x\right]$-trading-day return realizations. Formally, we compute histogram of the (unannualized) returns by the following equation:

$$R_{t,t+x,i} = \ln S_{t-i-x} - \ln S_{t-i}, \qquad (24.26)$$

where each i is an observation in time and t is the last i. For example, if t is 1988/06/02 and x is 15 calendar days (or 10 business days). We further choose our histogram horizon to be 5 years, or 1260 business days. Fifteen business days after 1988/06/02 is 1988/06/17. To estimate a distribution of the stock return for 1988/06/17, we look back a series of 10-business-day returns. Since we choose a 5-year historical window, or 1260-business-day window, the histogram will contain 1260 observations. The first return in the histogram, $R_{1988/06/02,1988/06/17,1}$ is the difference between the log of the stock price on 1988/06/01, $\ln S_{t-1}$, and the log of the stock price 15 calendar days (10 business days) earlier on 1988/05/17, $\ln S_{t-1-x}$. The second observation in the his-

[18]The (Ex-dividend) S&P 500 index we use is the index that serves as an underlying asset for the option. For option evaluation, realized returns of this index need not be adjusted for dividends unless the timing of the evaluated option contract is correlated with lumpy dividends. Because we use monthly observations, we think that such correlation is not a problem. Furthermore, in any case, this should not affect the comparison of the volatility smile between our model and the Black–Scholes model.

[19]We use three alternative time windows, 2-year, 10-year, and 30-year, to check the robustness of our procedure and results.

[20]The conversion is needed because we use trading day intervals to identify the appropriate return histograms and calendar-day intervals to calculate the appropriate discount factor.

togram, $R_{1988/06/02,1988/06/17,2}$ is computed as $\ln S_{t-2} - \ln S_{t-2-x}$.

After we complete the histogram of returns, we then convert it to the histogram of prices by multiplying every observation in the histogram by the current stock price:

$$S_{t+x,i} = S_t R_{t,t+x,i}. \quad (24.27)$$

The expected option payoff is calculated as the average payoff where all the realizations in the histogram are given equal weights. Thus, $E_t[C_{T,T,K}]$ and $E_t[S_T]$ are calculated as:

$$\begin{cases} E_t[C_{T,T,K}] = \frac{1}{N}\sum_{i=1}^{N} \max\{S_{T,i} - K, 0\} \\ E_t[S_T] = \frac{1}{N}\sum_{i=1}^{N} S_{T,i} \end{cases}, \quad (24.28)$$

where N is the total number of realized returns and $C_{t,T,K}$ is the price observed at time t, of an option that expires at time T with strike price K. Substituting the results in Eq. (24.28) in the approximation pricing formula of Eq. (24.4), we obtain our empirical model:

$$\begin{aligned} \underline{C}_t &= P_{t,T}E_t[C_{T,T,K}] + \beta_C\{S_t - P_{t,T}E_t[S_T]\} \\ &= P_{t,T}\frac{1}{N}\sum_{i=1}^{N}\max\{S_{T,i} - K, 0\} \\ &\quad + \beta_C\left\{S_t - P_{t,T}\frac{1}{N}\sum_{i=1}^{N}S_{T,i}\right\} \end{aligned} \quad (24.29)$$

where the dollar beta is defined as, $\beta_C = \frac{\mathrm{cov}[C_T,S_T]}{\mathrm{var}[S_T]}$ as defined in Eq. (24.4).

Note that option prices should be based upon projected future volatility levels rather than historical estimates. We assume that investors believe that the distribution of index returns over the time to maturity follows the histogram of a particular horizon with a projected volatility. In practice, traders obtain this projected volatility by calibrating the model to the market price. We incorporate the projected volatility, $v^*_{t,T,K}$, into the histogram by adjusting it returns:

$$\begin{aligned} R^*_{t,T,K,i} &= \frac{v^*_{t,T,K}}{v_{t,T}}(R_{t,T,i} - \overline{R}_{t,T}) + \overline{R}_{t,T}, \\ i &= 1, \cdots, N, \end{aligned} \quad (24.30)$$

where the historical volatility $v_{t,T}$ is calculated as the standard deviation of the historical returns as follows,

$$v^2_{t,T} = \frac{1}{N-1}\sum_{i=1}^{N}(R_{t,T,i} - \overline{R}_{t,T})^2 \quad (24.31)$$

where $R_{t,T,i} = S_{T,i}/S_t$ and $\overline{R}_{t,T} = \frac{1}{N}\Sigma_{i=1}^{N} R_{t,T,i}$ is the mean return.

Note that the transformation from R to R^* changes the standard deviation from $v_{t,T}$ to $v^*_{t,T,K}$, but does not change the mean, skewness, or kurtosis. In our empirical study, we approximate the true volatility by the Black–Scholes implied volatility. For the upper bound calculations, we also need an expected mean return of the stock. In our empirical study, we simply use the histogram mean for it.

The selection of the time horizon is somewhat arbitrary. Empiricists know well that too long horizons reduce the impact of recent events and yet too short understate the impact of long-time effects. Given that there is no consensus on a most proper horizon, we perform our test over a variety of choices, namely 5-year, 10-year, and 30-year, and the results are similar. To conserve space, we provide the results on the 10-year and leave the others available on request.

The results are shown in Table 24.5. Columns (1) and (2) define the maturity and moneyness buckets. Short maturity is less than 30 days to maturity, medium is between 31 and 90 days, long is between 91 and 180 days, and real long is over 180 days to maturity. At the money is between 5% in-the-money and 5% out-of-the-money (or −5%), near-in-the-money/near-out-of-the-money is between 5 and 15% and deep-in-the-money/deep-out-of-the-money is over 15%. Moneyness

Table 24.5 Empirical results on the lower bound

(1)	(2)	(3)	(4)	(5)
Maturity	Money	Total	% lbdd	Violation
Short	Deep out	0		0
Medium	Deep out	0		0
Long	Deep out	84	57.26	0
Real long	Deep out	295	61.18	0
Short	Near out	145	65.33	0
Medium	Near out	1925	69.27	5
Long	Near out	3554	82.29	30
Real long	Near out	3317	86.88	271
Short	At	4106	89.73	492
Medium	At	7732	92.45	962
Long	At	5650	92.82	380
Real long	At	3857	92.61	584
Short	Near in	2220	*97.17*	218
Medium	Near in	3924	97.08	785
Long	Near in	2951	94.50	234
Real long	Near in	2018	90.31	130
Short	Deep in	660	92.48	0
Medium	Deep in	1296	94.52	29
Long	Deep in	1352	93.14	46
Real long	Deep in	1454	89.42	67
Total		46,540	90.43	4233

Notes

1. This is based upon 2520 business-day (10 years) horizon. Results of other horizons are similar and are available on request
2. Columns (1) and (2) define the maturity and moneyness buckets. Short maturity is less than 30 days to maturity, medium is between 31 and 90 days, long is between 91 and 180 days, and real long is over 180 days to maturity. At the money is between 5% in-the-money and 5% out-of-the-money (or −5%), near-in-the-money/near-out-of-the-money is between 5 and 15%, and deep-in-the-money/deep-out-of-the-money is over 15%. Moneyness is defined as $S/K - 1$. Column (3) is a frequency count of the number of observations in each bucket. Column (4) represents the average value of the ratios of the lower bound over the market price of the option. Column (5) shows the number of violations when the lower bound is higher than the market price
3. The best lower bound performance is when the option is near in-the-money and short-term maturity (2.83% below market value)

is defined as $S/K - 1$. Column (3) is a frequency count of the number of observations in each bucket. Column (4) represents the average value of the ratios of the lower bound over the market price of the option. Column (5) shows the number of violations when the lower bound is higher than the market price.

The underlying stock return distribution is 10-year historical return histogram with the volatility replaced by the Black–Scholes implied volatility.

Out of the entire sample (46,540 observations), on average, the lower bound is 9.57% below the market value and the upper bound is 9.28% above the market value. When we look into subsamples, the performances vary. In general, the lower bound performs better in-the-money than out-of-the-money and

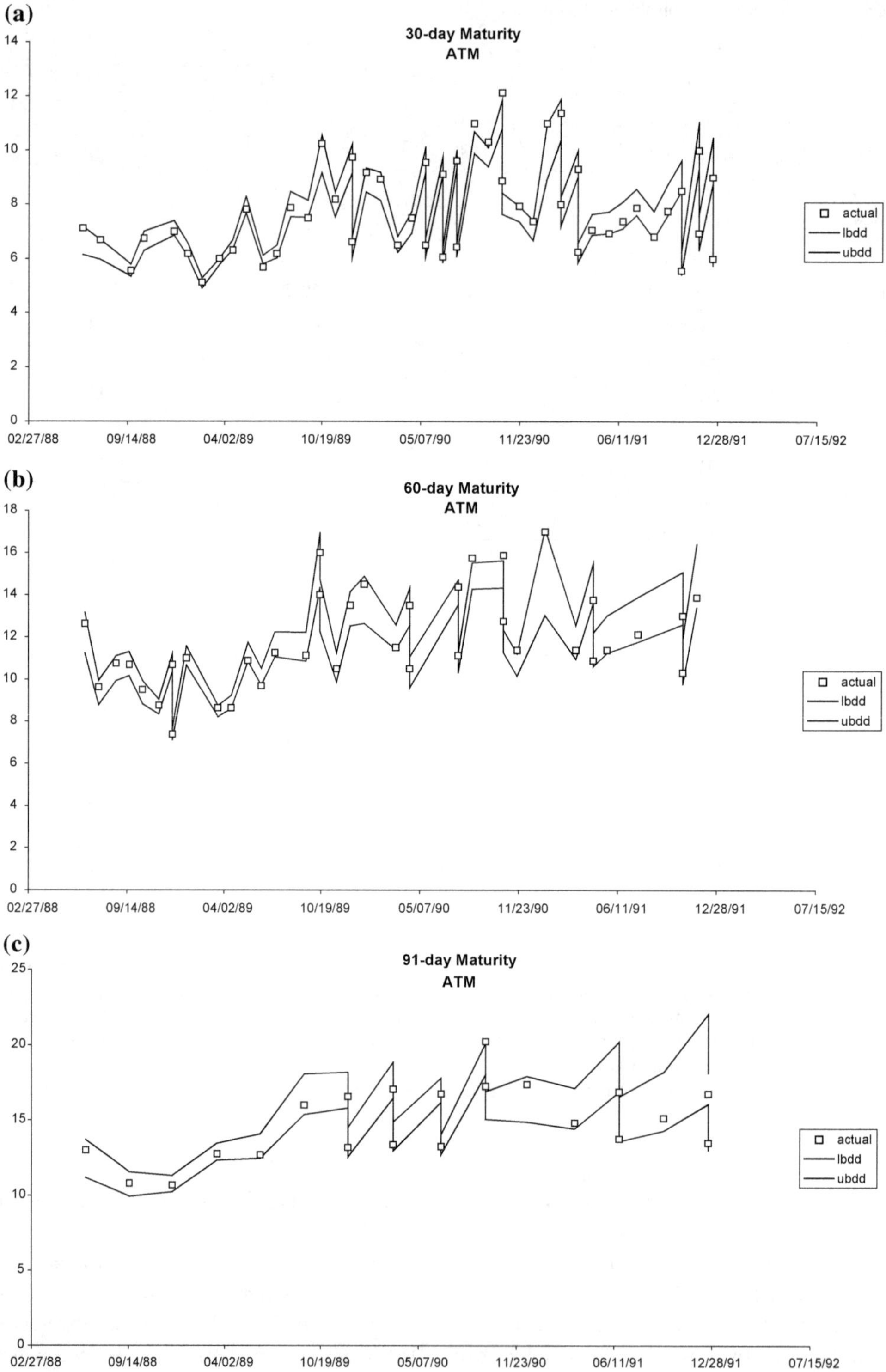

Fig. 24.1 **a** Plot of 30-day maturity actual ATM option prices and its upper and lower bound values. **b** Plot of 60-day maturity actual ATM option prices and its upper and lower bound values. **c** Plot of 91-day maturity actual ATM option prices and its upper and lower bound values. **d** Plot of 182-day maturity actual ATM option prices and its upper and lower bound values

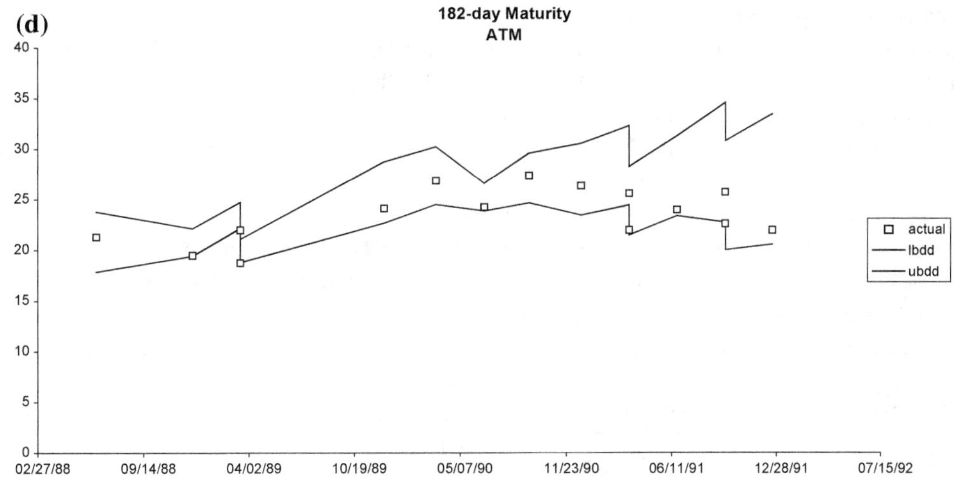

Fig. 24.1 (continued)

medium maturity than other maturities. The best lower bound performance is when the option is near in-the-money and short-term maturity (2.83% below market value).

To visualize the upper and lower bounds, we plot selected contracts in Fig. 24.1. In Fig. 24.1, we plot at the money (ATM) options from four maturities, 1 month (Fig. 24.1a), 2 months (Fig. 24.1b), 3 months (Fig. 24.1c), and 6 months (Fig. 24.1d). As we move from short maturity to long maturity, the number of observations drops (51, 38, 21, and 15 observations, respectively). The bounds are wider as we move from short maturity to long maturity, consistent with the analysis in the previous sections.

As we can see, in general, the lower bound using histograms is best for in-the-money and short-dated options and worst for out-of-the-money options. On average the lower bound is 9.57% below the market value (the ratio is 90.43%). However, there are violations. For example, medium term, near out-of-the-money options have five violations of the lower bound and there are a total of 4233 violations. Theoretically, there should be arbitrage opportunities when the bounds are violated. To test the lower bound does imply arbitrage opportunities, we perform a simple buy and hold trading strategy. If the lower bound is violated, we will buy the option and hold it till maturity. For the 4233 (out of 46,540) violations, the buy and hold strategy generated $22,809, or an average of $5.39 per contract. Given that the buy and hold strategy can be profitable simply due to the bull market, we compute those that had no violation of the lower bond. For the 42,307 cases that violated no lower bound, the average profit is $1.83. Hence, the options that violated the lower bound imply a trading profit 200% above the average.

24.6 Conclusion

In this chapter, we derive a new and tighter lower bound for European option prices comparing with those of previous studies. We further reinterpret Ritchken's (1985) upper bound under a nonparametric framework. Our model contributes to the literature in two different ways. First, our bounds require no parametric assumption of the underlying stock or the moments of the distribution. Furthermore, our bounds require no assumptions on interest rates or volatility. The only requirements of our model are the dollar beta of the option and expected payoffs of the stock and the option. Hence, our bounds can be applied to any model such as the random volatility and random interest rate model by Bakshi et al. (1997) and Scott (1997). Second, despite of much looser and flexible assumptions, our bounds are significantly tighter than the existing upper and lower bound models. Most importantly, our bounds are tighter for the out-of-the-money options that cannot be bounded efficiently by previous models. Finally, we apply our model to real data using histograms of the realized stock returns. The results show that nearly 10% of the observations violate the lower bound. These violations are shown to generate significant arbitrage profits, after correction of the bull market in the sample period.

Acknowledgements The financial support of National Science Council, Taiwan, Republic of China (NSC 96-2416-H-006-039-) is gratefully acknowledged.

Appendix 1: Related Option Studies Adopting Nonparametric Method

The nonparametric method is adopted in many important prior option-related studies. Besides the studies which examine option bounds by using the semiparametric or nonparametric method reviewed in introduction, several studies also adopt the nonparametric method in option-related research more recently. Kuang and Lai (2015) develop a novel method to price basket options by using an application-driven approach to estimate the state-price density of the basket or the joint state-price density of the asset prices in the basket. Fengler and Hin (2015) suggest a semi-nonparametric estimator for the call option price surface, families of first-order strike derivatives, and state-price densities for S&P 500 option data. Lin et al. (2016) employ a "nonparametric" pricing approach of European options to explain the volatility smile. Using historical put and call prices on the S&P 500 during the year 2012, Marinelli and d'Addona (2017) analyze the empirical performance of several nonparametric estimators of the pricing functional for European options. Zu and Boswijk (2017) develop nonparametric specification tests for stochastic volatility models by comparing the nonparametrically estimated return density and distribution functions with their parametric counterparts. Studying short-maturity ("weekly") S&P 500 index options and adopting a novel semi-nonparametric approach, Anderson et al. (2017) uncover variation in the negative jump tail risk, which is not spanned by market volatility and helps predict future equity returns.

Appendix 2: Asset Pricing Model with a Stochastic Kernel

In this appendix, we prove Theorem 24.1. Without loss of generality, we prove Theorem 24.1 by a three-point convex function. The extension of the proof to multiple points is straightforward but tedious. Let the distribution be trinomial and the relationship between the pricing kernel M and the stock price S be convex. In the following table, $x \leq 0$; $y > \varepsilon \geq 0$.

Probability	M	S	$C = \max\{S - K, 0\}$
p^2	$M + x$	$S + y > K$	$S + y - K$
$2p(1-p)$	M	$S - \varepsilon < K$	0
$(1-p)^2$	$M - x$	$S - y < K$	0

When $\varepsilon = 0$, the relationship between the pricing kernel M and stock price S is linear and we obtain equality. When $\varepsilon > 0$, the relationship is convex. We first calculate the mean values:

$$E[M] = M - x + 2px$$
$$E[S] = S - y + 2p(y - \varepsilon) + 2p^2\varepsilon$$
$$E[C] = p^2(S + y - K)$$

The three covariances are computed as follows:

$$\text{cov}[M, S] = 2p(1-p)x(y + \varepsilon(2p-1))$$
$$\text{cov}[M, C] = 2p^2(1-p)x(S + y - K)$$
$$\text{cov}[S, C] = 2p^2(1-p)(S + y - K)(y + p\varepsilon)$$

The variance of the stock price is more complex:

$$\text{var}[S] = 2p(1-p)z$$

where

$$z = (y^2 + \varepsilon^2(1 - 2p + 2p^2) + 2\varepsilon y(2p - 1)) > 0$$

As a result, it is straightforward to show that:

$$\begin{aligned} \frac{\text{cov}[S, C]}{\text{var}[S]}\text{cov}[M, S] &= \frac{2p^2(1-p)(S+y-K)(y+p\varepsilon)}{2p(1-p)z} \\ &\quad 2p(1-p)x(y+\varepsilon(2p-1)) \\ &= 2p^2(1-p)x(S+y-K) \\ &\quad \frac{(y+p\varepsilon)(y+\varepsilon(2p-1))}{z} \\ &= 2p^2(1-p)x(S+y-K) \\ &\quad \left[1 + \frac{\varepsilon(1-p)(y-\varepsilon)}{z}\right] \\ &\leq 2p^2(1-p)x(S+y-K) \\ &=: \text{cov}[M, C] \end{aligned}$$

The fourth line is obtained because $\text{cov}[M, C] < 0$ and $1 + \frac{\varepsilon(1-p)(y-\varepsilon)}{z} > 1$. Note that the result is independent of p since all it needs is $0 < p < 1$ for $1 + \frac{\varepsilon(1-p)(y-\varepsilon)}{z}$ to be greater than 1. Also note that when $\varepsilon = 0$ the equality holds.

Bibliography

Anderson, T. G., Fusari, N., & Todorov, V. (2017). Short-term market risks implied by weekly options. *Journal of Finance, 72,* 1335–1386.

Bakshi, G., Cao, C., & Chen, Z. (1997). Empirical performance of alternative option pricing models. *Journal of Finance, 52,* 2003–2049.

Basso, A., & Pianca, P. (1997). Decreasing absolute risk aversion and option pricing bounds. *Management Science, 43,* 206–216.

Black, F. (1976). The pricing of commodity contacts. *Journal of Financial Economics, 3,* 167–179.

Black, F., & Scholes, M. (1973). The pricing of options and corporate liabilities. *Journal of Political Economy, 81,* 637–654.

Boyle, P., & Lin, X. (1997). Bounds on contingent claims based on several assets. *Journal of Financial Economics, 46,* 383–400.

Broadie, M., & Cao, M. (2008). Improved lower and upper bound algorithms for pricing American options by simulation. *Quantitative Finance, 8,* 845–861.

Christoffersen, P., Elkamhi, R., Feunou, B., & Jacobs, K. (2010). Option valuation with conditional heteroskedasticity and nonnormality. *Review of Financial Studies, 23,* 2139–2189.

Chung, S. L., Hung, M. W., & Wang, J. Y. (2010). Tight bounds on American option prices. *Journal of Banking & Finance, 34,* 77–89.

Chuang, H., Lee, C. F., & Zhong, Z. K. (2011). Option bounds: A review and comparison. Working paper.

Constantinides, G., & Zariphopoulou, T. (2001). Bounds on derivative prices in an intertemporal setting with proportional transaction costs and multiple securities. *Mathematical Finance, 11,* 331–346.

Constantinides, G. M., Jackwerth, J. C., & Perrakis, S. (2009). Mispricing of S&P 500 index options. *Review of Financial Studies, 22,* 1247–1277.

Constantinides, G. M., Czerwonko, M., Jackwerth, J. C., & Perrakis, S. (2011). Are options on index futures profitable for risk-averse investors? Empirical evidence. *Journal of Finance, 66,* 1407–1437.

Cox, J., & Ross, S. (1976). The valuation of options for alternative stochastic process. *Journal of Finance Economics, 3,* 145–166.

De La Pena, V., Ibragimov, R., & Jordan, S. (2004). Option bounds. *Journal of Applied Probability, 41,* 145–156.

Fengler, M. R., & Hin, L. Y. (2015). Semi-nonparametric estimation of the call-option price surface under strike and time-to-expiry no-arbitrage constraints. *Journal of Econometrics, 184,* 242–261.

Frey, R., & Sin, C. (1999). Bounds on European option prices under stochastic volatility. *Mathematical Finance, 9,* 97–116.

Gotoh, J. Y., & Konno, H. (2002). Bounding option prices by semidefinite programming: A cutting plane algorithm. *Management Science, 48,* 665–678.

Grundy, B. (1991). Option prices and the underlying assets return distribution. *Journal of Finance, 46,* 1045–1069.

Hobson, D., Laurence, P., & Wang, T. H. (2005). Static-arbitrage upper bounds for the prices of basket options. *Quantitative Finance, 5,* 329–342.

Huang, J. (2004). Option pricing bounds and the elasticity of the pricing kernel. *Review of Derivatives Research, 7,* 25–51.

Ingersoll, J. (1989). *Theory of financial decision making*. Totowa, New Jersey: Rowman & Littlefield.

Kuang, Y., & Lai, T. L. (2015). State price density estimation and nonparametric pricing of basket options. *Journal of Mathematical Finance, 5,* 448–456.

Lee, C. F., & Lee, J. C. (2015). *Handbook of financial econometrics and statistics* (Vol. 1). New York: Springer Reference.

Lee, C. F., Tai, T., Zhong, Z. K., & Chuang, H. W. (2019). Option bounds: A review and comparison. In *Handbook of financial econometrics mathematics, statistics, and technology* (Chap. 24). Forthcoming.

Levy, H. (1985). Upper and lower bounds of put and call option value: Stochastic dominance approach. *Journal of Finance, 40,* 1197–1217.

Lin, H. C., Chen, R. R., & Palmon, O. (2015). Nonparametric bounds for European option prices. In *Handbook of financial econometrics and statistics* (Vol. 1, Chap. 7, pp. 207–232). New York: Springer Reference.

Lin, H. C., Chen, R. R., & Palmon, O. (2016). Explaining the volatility smile: Nonparametric versus parametric option models. *Review of Quantitative Finance and Accounting, 46,* 907–935.

Lo, A. (1987). Semi-parametric upper bounds for option prices and expected payoff. *Journal of Financial Economics, 19,* 373–388.

Marinelli, C., & d'Addona, S. (2017). Nonparametric estimates of pricing functionals. *Journal of Empirical Finance, 44,* 19–35.

Merton, R. (1973). The theory of rational option pricing. *The Bell Journal of Economics and Management Science, 4,* 141–183.

Peña, J., Vera, J. C., & Zuluaga, L. F. (2010). Static-arbitrage lower bounds on the prices of basket options via linear programming. *Quantitative Finance, 10,* 819–827.

Perrakis, S. (1986). Option pricing bounds in discrete time: Extensions and the pricing of the American put. *Journal of Business, 59,* 119–141.

Perrakis, S., & Ryan, J. (1984). Option pricing in discrete time. *Journal of Finance, 39,* 519–525.

Ritchken, P. (1985). On option pricing bounds. *Journal of Finance, 40,* 1219–1233.

Ritchken, P., & Kuo, S. (1988). Option bounds with finite revision opportunities. *Journal of Finance, 43,* 301–308.

Ritchken, P., & Kuo, S. (1989). On stochastic dominance and decreasing absolute risk averse option pricing bounds. *Management Science, 35,* 51–59.

Rodriguez, R. (2003). Option pricing bounds: Synthesis and extension. *Journal of Financial Research, 26,* 149–164.

Rubinstein, M. (1976). The valuation of uncertain income stream and the pricing of options. *The Bell Journal of Economics, 7,* 407–425.

Scott, L. (1997). Pricing stock options in a jump diffusion model with stochastic volatility and interest rates: Applications of Fourier inversion methods. *Mathematical Finance, 7,* 413–424.

Zhang, P. (1994). Bounds for option prices and expected payoffs. *Review of Quantitative Finance and Accounting, 4,* 179–197.

Zu, Y., & Boswijk, H. P. (2017). Consistent nonparametric specification tests for stochastic volatility models based on the return distribution. *Journal of Empirical Finance, 41,* 53–75.

Author Index

© Springer Science+Business Media, LLC, part of Springer Nature 2019
C.-F. Lee et al., *Financial Econometrics, Mathematics and Statistics*,
https://doi.org/10.1007/978-1-4939-9429-8

Subject Index

C.-F. Lee et al., *Financial Econometrics, Mathematics and Statistics*,
https://doi.org/10.1007/978-1-4939-9429-8

CPSIA information can be obtained
at www.ICGtesting.com
Printed in the USA
BVHW010226060619
550340BV00002B/6/P

9 781493 994274